Introduction to Rational Elasticity

Monographs and textbooks on mechanics of solids and fluids

Editor-in-chief G. Æ. Oravas

Mechanics of continua

Editor W. O. Williams

Introduction to Rational Elasticity

by

C.-C. WANG & C. TRUESDELL

Rice University, Houston, USA
&
The Johns Hopkins University, Baltimore, USA

NOORDHOFF INTERNATIONAL PUBLISHING
LEYDEN

A division of A. W. Sijthoff International Publishing Company B.V.,
Leyden, The Netherlands

ISBN 90 01 93710 1

Library of Congress Catalog Card Number 72-97240

Printed in The Netherlands by
Drukkerij Dijkstra Niemeyer bv – Groningen

Contents

Dedicated to

Walter Noll

Our Teacher

Acknowledgment

Our debt to many authors is made plain by citations, not only directly in this book but also indirectly through passages of NFTM where their work is presented. In addition, we thank our editor, W. O. WILLIAMS, for his painstaking and helpful criticism of the manuscript and galley proofs. Finally we thank the U.S. National Science Foundation for its support of our own work for many years, as well as of many of the researches by others whose discoveries we incorporate in this treatise.

Preface

The theory of elasticity was created by the mathematicians JAMES BERNOULLI, EULER, and CAUCHY. For a long time it was a favorite subject of mathematicians and was regularly taught in mathematics departments. In this century both HADAMARD and HILBERT lectured upon it, as had POINCARÉ and many others in the last. Of the mathematicians of that time who are best known for their work in what is now called "pure" mathematics, we may collect a long list naming those who made at least one important addition to elasticity: BELTRAMI, BETTI, BIRKHOFF, CESÀRO, CHRISTOFFEL, CLEBSCH, FREDHOLM, HADAMARD, KORN, LAMÉ, LEVI-CIVITA, LIPSCHITZ, MORERA, VOLTERRA, WEINGARTEN, WEYL. To these we may add some distinguished Italian mathematicians who specialized in elasticity: ALMANSI, CERRUTI, LAURICELLA, PIOLA, SIGNORINI, SOMIGLIANA, TEDONE, as well as those great British mathematicians whose main interest lay in physics: GREEN, KELVIN, STOKES, and MAXWELL.

Now, unfortunately, this has changed. Although some papers by FRIEDRICHS, JOHN, and FICHERA prove that the rule is not without exceptions, generally mathematicians of our time have abandoned elasticity to the tillers of application. In virtue of GRESHAM'S law, the quality of research on the subject has worsened, whereupon knowledge of it has shrivelled and withered.

This book is written in an attempt to show that now elasticity is again a living branch of mathematics.

In the middle 1940's began a revival and renascence in rational continuum mechanics as a whole. Although this springtide was brought in by a few men who were provoked in the main by a desire to understand and represent the inelastic phenomena of everyday life, which had been foolishly neglected or botched by the professionalized custodians of "application" for several decades, conceptual analysis of those phenomena prerequired reascent of the then almost forgotten rational structure of elasticity itself. While the backgrounds and titles of the men – lonely scouts and barons they were, not "leaders" – were various, respect for mathematics united them. Among the products of the new work is a refounding and recasting of classical elasticity, enriched by a flood of splendid theorems theretofore undreamt of. This rebirth began with the thesis of NOLL in 1954 and owes much to his later work as well as to his strong and benign influence on all who are able to understand his thought.

This is the subject, the new rational elasticity, which is presented in this book. The task has been lightened by the previous composition of two other works. First, NOLL & I published as Chapter D of *The Non-Linear Field Theories of Mechanics*, in FLÜGGE'S *Encyclopedia of Physics*, Vol. III/3, 1965, a general and for that time complete treatise of non-linear elasticity. Since that treatise, as its place of appearance suggests, emphasizes the physical aspects of the subject, in this book the mathematical structure can justly take the center of the stage. Some special cases valuable for their enlightenment of general ideas have indeed been inserted into the mathematical text which follows, but for most applications the reader is referred to the earlier work, cited as "NFTM", which we do not here attempt to subsume. Second, a masterly exposition of the principles of the linearized theory of elasticity as a branch of mathematics has been composed by GURTIN and has appeared in Volume VIa/2 of the *Encyclopedia of Physics*. Thus our omission of linear elasticity from this book is neither more nor less than an acknowledgment that it lies in better hands.

The text below was projected and draughted by C.–C. WANG. It emphasizes his own work on uniform but inhomogeneous elastic bodies, the theory of which, entirely new, was created by NOLL scarcely more than six years ago and was thereupon extended and developed by WANG. It provides a rational basis for some of the word-pictures limned by solid-state physicists in terms of small, round, hard balls. To the reader of this book, however, its interest should be intrinsic.

The passages concerning manifolds and Lie groups, in particular, Chapters 1 and 5, are wholly the work of WANG. My part lies mainly in the earlier treatise NFTM, which is drawn upon so frequently as to be the primary source of this book, but also I have revised and augmented a few sections and made some suggestions here and there.

The term "Introduction" in the title is meant as truth, not paedagogy. WANG & I hope that this book will be so good an introduction to the real problems of elasticity that mathematicians will put it quite out of date in a few years.

TRUESDELL

CHAPTER I

Mathematical Preliminaries

In this first chapter we review the essential mathematical concepts which we shall use frequently in the theory of elasticity. Our purpose is to provide an outline, not a textbook, so we leave out the proofs of most theorems. Thus he who has some knowledge of modern mathematics, especially tensor analysis, differential geometry, and group theory, may pass over this chapter entirely, using it only for reference later.

We shall denote the field of real numbers by $\mathscr{R}$.

1. Finite-Dimensional Vector Spaces

DEFINITION. A *vector space* $\mathscr{V}$ is a set equipped with two operations:

$$\mathbf{A} : \mathscr{V} \times \mathscr{V} \to \mathscr{V}, \tag{1.1}$$

called *addition*, and

$$\mathbf{S} : \mathscr{R} \times \mathscr{V} \to \mathscr{V}, \tag{1.2}$$

called *scalar multiplication*, which satisfy the following rules:

$$\mathbf{A}(\boldsymbol{u}, \boldsymbol{v}) = \mathbf{A}(\boldsymbol{v}, \boldsymbol{u}) \quad \forall \boldsymbol{u}, \boldsymbol{v} \in \mathscr{V}. \tag{1.3}$$

$$\mathbf{A}(\boldsymbol{u}, \mathbf{A}(\boldsymbol{v}, \boldsymbol{w})) = \mathbf{A}(\mathbf{A}(\boldsymbol{u}, \boldsymbol{v}), \boldsymbol{w}) \quad \forall \boldsymbol{u}, \boldsymbol{v}, \boldsymbol{w} \in \mathscr{V}. \tag{1.4}$$

$$\exists \mathbf{0} \in \mathscr{V} : \mathbf{A}(\boldsymbol{u}, \mathbf{0}) = \boldsymbol{u} \quad \forall \boldsymbol{u} \in \mathscr{V}. \tag{1.5}$$

$$\boldsymbol{u} \in \mathscr{V} \Rightarrow \exists -\boldsymbol{u} \in \mathscr{V} : \mathbf{A}(\boldsymbol{u}, -\boldsymbol{u}) = \mathbf{0}. \tag{1.6}$$

$$\mathbf{S}(a, \mathbf{S}(b, \boldsymbol{u})) = \mathbf{S}(ab, \boldsymbol{u}) \quad \forall a, b \in \mathscr{R}, \ \forall \boldsymbol{u} \in \mathscr{V}. \tag{1.7}$$

$$\mathbf{S}(a+b, \boldsymbol{u}) = \mathbf{A}(\mathbf{S}(a, \boldsymbol{u}), \ \mathbf{S}(b, \boldsymbol{u})) \quad \forall a, b \in \mathscr{R}, \ \forall \boldsymbol{u} \in \mathscr{V}. \tag{1.8}$$

$$\mathbf{S}(a, \mathbf{A}(\boldsymbol{u}, \boldsymbol{v})) = \mathbf{A}(\mathbf{S}(a, \boldsymbol{u}), \mathbf{S}(a, \boldsymbol{v})) \quad \forall a \in \mathscr{R}, \ \forall \boldsymbol{u}, \ \boldsymbol{v} \in \mathscr{V}. \tag{1.9}$$

$$\mathbf{S}(1, \boldsymbol{u}) = \boldsymbol{u} \quad \forall \boldsymbol{u} \in \mathscr{V}. \tag{1.10}$$

The rule (1.3) is called the *commutative law* of addition; (1.4), the *associative law* of addition; (1.7), the *associative law* of scalar multiplication; while (1.8) and (1.9) are called the *distributive laws* of scalar multiplication over scalar addition and vector addition, respectively. The elements of $\mathscr{V}$ are called *vectors*; those of $\mathscr{R}$, *scalars*. The traditional notations for the values of the operations $\mathbf{A}$ and $\mathbf{S}$, which we shall use henceforth, are a plus sign and juxtaposition:

$$\boldsymbol{u}+\boldsymbol{v} \equiv \mathbf{A}(\boldsymbol{u}, \boldsymbol{v}) \tag{1.11}$$

and

$$a\boldsymbol{u} \equiv \mathbf{S}(a, \boldsymbol{u}) \tag{1.12}$$

The vector $\mathbf{0}$ satisfying rule (1.5) is called the *null vector* or *zero* of $\mathscr{V}$, and the vector $-\boldsymbol{u}$ in rule (1.6) is called the *opposite* or *negative* of $\boldsymbol{u}$. We define *subtraction* for any pair of vectors $\boldsymbol{u}, \boldsymbol{v} \in \mathscr{V}$ by

$$\boldsymbol{u}-\boldsymbol{v} \equiv \boldsymbol{u}+(-\boldsymbol{v}). \tag{1.13}$$

Then any composition of the three basic operations – addition, subtraction, and scalar multiplication – is called a *linear operation*.

From the basic rules (1.3)–(1.10), the linear operations satisfy many rules familiar from the special cases which occur in common arithmetic. For example, if $\boldsymbol{v}_1, \ldots, \boldsymbol{v}_k \in \mathscr{V}$ and $a_1, \ldots, a_k \in \mathscr{R}$, then the vector

$$a_1\boldsymbol{v}_1 + \ldots + a_k\boldsymbol{v}_k, \tag{1.14}$$

which is defined by the above-stated rules, is called a *linear combination* of $\boldsymbol{v}_1, \ldots, \boldsymbol{v}_k$. A scalar multiple of a linear combination is again a linear combination, as is a sum of two linear combinations, *etc.*

A particular and important example of a vector space is furnished by the set $\mathscr{R}^n$ of lists of n elements of $\mathscr{R}$, with addition and scalar multiplication in $\mathscr{R}^n$ defined by addition and multiplication of the components in $\mathscr{R}$:

$$\begin{aligned}(a^1, \ldots, a^n)+(b^1, \ldots, b^n) &\equiv (a^1+b^1, \ldots, a^n+b^n),\\ b(a^1, \ldots, a^n) &\equiv (ba^1, \ldots, ba^n).\end{aligned} \tag{1.15}$$

$\mathscr{R}^n$ is generally called the *Cartesian n-space*. In this notation, $\mathscr{R} = \mathscr{R}^1$.

We call a subset $\mathscr{U}$ of $\mathscr{V}$ a *subspace* if it is closed with respect to linear combination. In particular, $\mathscr{V}$ is a subspace of itself.

Given a subspace $\mathscr{U}$ of $\mathscr{V}$, we define an equivalence relation $\sim$ on $\mathscr{V}$ by

$$\boldsymbol{u} \sim \boldsymbol{v} \Leftrightarrow \boldsymbol{u}-\boldsymbol{v} \in \mathscr{U}. \tag{1.16}$$

On the collection $\mathscr{V}/\mathscr{U}$ of equivalence classes with respect to $\sim$, we introduce the operations

$$\{\boldsymbol{u} \sim \bar{\boldsymbol{u}} \sim \ldots\}+\{\boldsymbol{v} \sim \bar{\boldsymbol{v}} \sim \ldots\} \equiv \{\boldsymbol{u}+\boldsymbol{v} \sim \bar{\boldsymbol{u}}+\bar{\boldsymbol{v}} \sim \ldots\}, \tag{1.17}$$

and

$$a\{\boldsymbol{u} \sim \bar{\boldsymbol{u}} \sim \ldots\} \equiv \{a\boldsymbol{u} \sim a\bar{\boldsymbol{u}} \sim \ldots\}. \tag{1.18}$$

Then $\mathscr{V}/\mathscr{U}$ becomes a vector space, called the *quotient space* of $\mathscr{V}$ over $\mathscr{U}$. The notation $\mathscr{V}/\mathscr{U}$ is defined only for subspaces $\mathscr{U}$ of $\mathscr{V}$.

Since the intersection of all members of a given set of subspaces is itself a subspace, for any subset $\mathscr{A} \subset \mathscr{V}$ there exists a unique minimal subspace Sp $(\mathscr{A})$ containing $\mathscr{A}$. Indeed, Sp $(\mathscr{A})$ is the subspace consisting in all linear combinations of the elements of $\mathscr{A}$. Thus Sp $(\mathscr{A})$ is called the *spread* of $\mathscr{A}$, or, often, the subspace *spanned* by $\mathscr{A}$, and $\mathscr{A}$ is called a *generating set* of Sp $(\mathscr{A})$. A minimal generating set of a subspace $\mathscr{U}$ is called a *basis* of $\mathscr{U}$.

PROPOSITION. *Every subspace has a basis.*

We call a subspace *finite-dimensional* if it has a finite basis. A vector-space that is not finite-dimensional is *infinite-dimensional.*

PROPOSITION. *If m vectors constitute a basis of a subspace $\mathscr{U}$, every other basis of $\mathscr{U}$ consists in precisely m vectors.*

We denote the positive integer delivered by this proposition by dim $\mathscr{U}$ and call it the *dimension* of $\mathscr{U}$. For the mathematical formulation of the theory of elasticity, finite-dimensional vector spaces suffice. Thus, henceforth, we assume that

$$\dim \mathscr{V} = n, \quad 0 < n < \infty. \tag{1.19}$$

For the Cartesian n-space $\mathscr{R}^n$, a basis is $\{(1, 0, \ldots, 0), (0, 1, \ldots, 0), \ldots, (0, \ldots, 0, 1)\}$. Hence dim $\mathscr{R}^n = n$. The particular basis just exhibited is called the *standard basis* of $\mathscr{R}^n$.

PROPOSITION. *The subspaces of $\mathscr{V}$ are all finite-dimensional; further, if $\mathscr{U}$ is a subspace of $\mathscr{V}$, then*

$$\dim \mathscr{U} \leqq \dim \mathscr{V}, \tag{1.20}$$

where equality holds if and only if $\mathscr{U}$ and $\mathscr{V}$ coincide.

We call a set of vectors $\{\boldsymbol{v}_1, \ldots, \boldsymbol{v}_k\}$ *linearly independent* if the equation

$$a_1\boldsymbol{v}_1 + \ldots + a_k\boldsymbol{v}_k = \boldsymbol{0} \tag{1.21}$$

is satisfied by no other a_i than

$$a_1 = \ldots = a_k = 0. \tag{1.22}$$

A set that is not linearly independent is called *linearly dependent.*

PROPOSITION. *Let* $\{\boldsymbol{v}_1, \ldots, \boldsymbol{v}_k\}$ *be a set of vectors. Then*

$$\dim \mathrm{Sp}\,(\boldsymbol{v}_1, \ldots, \boldsymbol{v}_k) \leqq k, \tag{1.23}$$

where equality holds if and only if $\{\boldsymbol{v}_1, \ldots, \boldsymbol{v}_k\}$ *is linearly independent.*

PROPOSITION. *A generating set* $\{\boldsymbol{v}_1, \ldots, \boldsymbol{v}_k\}$ *of* $\mathscr{U}$ *is a basis if and only if it is linearly independent.*

PROPOSITION. *Let* $\{\boldsymbol{v}_1, \ldots, \boldsymbol{v}_k\}$ *be a basis of* $\mathscr{U}$, *and let* $\{\boldsymbol{u}_1, \ldots, \boldsymbol{u}_j\}$ *be a linearly independent set in* $\mathscr{U}$. *Then* $j \leqq k$; *further, we can choose vectors* $\boldsymbol{v}_{i_1}, \ldots, \boldsymbol{v}_{i_{k-j}}$ *in* $\{\boldsymbol{v}_1, \ldots, \boldsymbol{v}_k\}$ *such that* $\{\boldsymbol{u}_1, \ldots, \boldsymbol{u}_j, \boldsymbol{v}_{i_1}, \ldots, \boldsymbol{v}_{i_{k-j}}\}$ *is a basis of* $\mathscr{U}$.

Now suppose that $\{\boldsymbol{e}_1, \ldots, \boldsymbol{e}_n\}$ is a basis of $\mathscr{V}$. Then

$$\mathscr{V} = \mathrm{Sp}\,(\boldsymbol{e}_1, \ldots, \boldsymbol{e}_n), \tag{1.24}$$

whence every vector $\boldsymbol{v} \in \mathscr{V}$ is a linear combination of the vectors constituting the basis, say

$$\boldsymbol{v} = v^1\boldsymbol{e}_1 + \ldots + v^n\boldsymbol{e}_n, \quad v^i \in \mathscr{R}. \tag{1.25}$$

The coefficients v^i are called the *components* of $\boldsymbol{v}$ with respect to $\{\boldsymbol{e}_i\}$; they are unique, since the basis $\{\boldsymbol{e}_i\}$ is linearly independent.

We shall encounter frequently sums like the right-hand side of (1.25) and more general ones in which several indices occur. To shorten the writing, we adopt henceforth the usual *summation convention.* Namely, any term containing the same literal index once as a subscript and once as a superscript is summed with respect to that index from 1 to n, where n is the range of the repeated literal index. Under this convention, (1.25) is written simply as

$$\boldsymbol{v} = v^i\boldsymbol{e}_i. \tag{1.26}$$

In particular,

$$\boldsymbol{e}_i = \delta_i^j\boldsymbol{e}_j, \text{ where } \delta_i^j \equiv \begin{cases} 1 \text{ if } i = j, \\ 0 \text{ if } i \neq j. \end{cases} \tag{1.27}$$

Let $\{e_i\}$ and $\{\bar{e}_j\}$ be two bases of $\mathscr{V}$. We define the scalars e^j_k and $\bar{e}^j_k$ as, respectively, the jth components of the kth members of these bases with respect to each other:

$$e_k = e^j_k \bar{e}_j, \quad \bar{e}_k = \bar{e}^j_k e_j. \tag{1.28}$$

Then

$$e_i = e^j_i \bar{e}^k_j e_k, \quad \bar{e}_i = \bar{e}^j_i e^k_j \bar{e}_k. \tag{1.29}$$

Comparing these formulae with (1.27), we see that

$$e^i_j \bar{e}^j_k = \bar{e}^i_j e^j_k = \delta^i_k. \tag{1.30}$$

In the terminology of matrices, these equations state that the matrix $[e^i_j]$ is the inverse of the matrix $[\bar{e}^i_j]$, and *vice versa*.

PROPOSITION. *Let v^i and $\bar{v}^i$ be the components of a vector $\boldsymbol{v}$ relative to the bases $\{e_i\}$ and $\{\bar{e}_i\}$, respectively. Then*

$$v^i = \bar{e}^i_j \bar{v}^j, \tag{1.31}$$

where $\bar{e}^i_j$ is defined by $(1.28)_2$.

Now we consider mappings of vector spaces.

DEFINITION. Let $\mathscr{U}$ and $\mathscr{V}$ be vector spaces, and let $\boldsymbol{L}$ be a mapping of $\mathscr{U}$ into $\mathscr{V}$,

$$\boldsymbol{L} : \mathscr{U} \to \mathscr{V}. \tag{1.32}$$

If $\boldsymbol{u} \in \mathscr{U}$, the value $\boldsymbol{L}(\boldsymbol{u}) \in \mathscr{V}$ is called the *image* of $\boldsymbol{u}$ under $\boldsymbol{L}$; if $\boldsymbol{v} \in \mathscr{V}$ and $\boldsymbol{v} = \boldsymbol{L}(\boldsymbol{u})$, the argument $\boldsymbol{u}$ is called a *pre-image* of $\boldsymbol{v}$ under $\boldsymbol{L}$.

The *kernel* ker $\boldsymbol{L}$ of $\boldsymbol{L}$ is the set of all pre-images of the null vector under $\boldsymbol{L}$:

$$\ker \boldsymbol{L} \equiv \{\boldsymbol{u} \in \mathscr{U} : \boldsymbol{L}(\boldsymbol{u}) = \boldsymbol{0}\}, \tag{1.33}$$

while the *image* im $\boldsymbol{L}$ of $\boldsymbol{L}$ is the set of all images under $\boldsymbol{L}$:

$$\operatorname{im} \boldsymbol{L} \equiv \{\boldsymbol{v} \in \mathscr{V} : \exists \boldsymbol{u} \in \mathscr{U},\ \boldsymbol{L}(\boldsymbol{u}) = \boldsymbol{v}\}. \tag{1.34}$$

$\boldsymbol{L}$ is called a *linear map* if

$$\boldsymbol{L}(a\boldsymbol{u}+b\boldsymbol{v}) = a\boldsymbol{L}(\boldsymbol{u})+b\boldsymbol{L}(\boldsymbol{v}) \quad \forall \boldsymbol{u}, \boldsymbol{v} \in \mathscr{U},\ \forall a, b \in \mathscr{R}. \tag{1.35}$$

PROPOSITION. *Let* $\boldsymbol{L}$ *be a linear map from* $\mathscr{U}$ *to* $\mathscr{V}$. *Then* ker $\boldsymbol{L}$ *is a subspace of* $\mathscr{U}$, im $\boldsymbol{L}$ *is a subspace of* $\mathscr{V}$, *and*

$$\dim \ker \boldsymbol{L} + \dim \operatorname{im} \boldsymbol{L} = \dim \mathscr{U}. \tag{1.36}$$

A linear map $\boldsymbol{L} : \mathscr{U} \to \mathscr{V}$ is called an *isomorphism* if it is one-to-one and surjective [1]. If such a map exists, the two vector spaces $\mathscr{U}$ and $\mathscr{V}$ are said to be *isomorphic.*

PROPOSITION. *Two vector spaces $\mathscr{U}$ and $\mathscr{V}$ are isomorphic if and only if they have the same dimension.*

PROPOSITION. *If $\mathscr{U}$ and $\mathscr{V}$ are isomorphic, then a linear map $\mathbf{L} : \mathscr{U} \to \mathscr{V}$ is an isomorphism if and only if any one, and hence all three, of the following conditions are satisfied:*

i) $$\ker \mathbf{L} = \{\mathbf{0}\}, \tag{1.37}$$

ii) $$\operatorname{im} \mathbf{L} = \mathscr{V}, \tag{1.38}$$

iii) *$\mathbf{L}$ maps a basis of $\mathscr{U}$ onto a basis of $\mathscr{V}$.*

In particular, if $\dim \mathscr{V} = n$, then $\mathscr{V}$ is isomorphic to the Cartesian n-space $\mathscr{R}^n$. Specifically, if we set

$$\boldsymbol{K}(\boldsymbol{v}) \equiv (v^1, \ldots, v^n), \tag{1.39}$$

where v^i is the ith component of $\boldsymbol{v}$ relative to the basis $\{\boldsymbol{e}_j\}$, then $\boldsymbol{K}$ is linear, and it maps $\{\boldsymbol{e}_j\}$ onto the standard basis of $\mathscr{R}^n$, so by condition (iii) of the foregoing proposition, $\boldsymbol{K}$ is an isomorphism.

For two given spaces $\mathscr{U}$ and $\mathscr{V}$, we denote the collection of all linear maps from $\mathscr{U}$ to $\mathscr{V}$ by $\mathscr{L}(\mathscr{U};\mathscr{V})$. We can introduce the operations of addition and scalar multiplication on $\mathscr{L}(\mathscr{U};\mathscr{V})$ in a natural way. Namely,

$$(\boldsymbol{L}_1+\boldsymbol{L}_2)(\boldsymbol{u}) \equiv \boldsymbol{L}_1(\boldsymbol{u})+\boldsymbol{L}_2(\boldsymbol{u}) \quad \forall \boldsymbol{u} \in \mathscr{U}, \tag{1.40}$$

and

$$(a\boldsymbol{L})(\boldsymbol{u}) \equiv a\boldsymbol{L}(\boldsymbol{u}) \quad \forall \boldsymbol{u} \in \mathscr{U},\ \forall a \in \mathscr{R}. \tag{1.41}$$

Relative to these operations, the set $\mathscr{L}(\mathscr{U};\mathscr{V})$ is a vector space. Without fear of confusion, to denote the zero of $\mathscr{L}(\mathscr{U};\mathscr{V})$ we shall use $\mathbf{0}$, the same symbol as for the zeros of $\mathscr{U}$ and $\mathscr{V}$.

PROPOSITION.

$$\dim \mathscr{L}(\mathscr{U};\mathscr{V}) = (\dim \mathscr{U})(\dim \mathscr{V}). \tag{1.42}$$

Indeed, let $\{\boldsymbol{f}_1, \ldots, \boldsymbol{f}_m\}$ be a basis of $\mathscr{U}$, and $\{\boldsymbol{e}_1, \ldots, \boldsymbol{e}_n\}$ a basis of $\mathscr{V}$. Then we define linear maps

$$\boldsymbol{e}_i \otimes \boldsymbol{f}^\alpha : \mathscr{U} \to \mathscr{V} \tag{1.43}$$

---

[1] French: "surjectif"; jargon: "onto".

by the condition

$$(\boldsymbol{e}_i \otimes \boldsymbol{f}^\alpha)(\boldsymbol{f}_\beta) = \delta^\alpha_\beta \boldsymbol{e}_i. \tag{1.44}$$

Here the Latin indices range from 1 to n, while the Greek indices range from 1 to m. (At this stage the symbol $\boldsymbol{e}_i \otimes \boldsymbol{f}^\alpha$ denotes a linear map from $\mathscr{U}$ to $\mathscr{V}$ defined by (1.44), not an operation $\otimes$ on $\boldsymbol{e}_i$ and $\boldsymbol{f}^\alpha$. We shall define $\boldsymbol{f}^\alpha$ and $\otimes$ below, after (1.59) and (1.106), respectively. Then the symbol $\boldsymbol{e}_i \otimes \boldsymbol{f}^\alpha$ will acquire the meaning of the *tensor product* of $\boldsymbol{e}_i$ and $\boldsymbol{f}^\alpha$.) We claim that the set $\{\boldsymbol{e}_i \otimes \boldsymbol{f}^\alpha,\ \alpha = 1, \ldots, m,\ i = 1, \ldots, n\}$ forms a basis of $\mathscr{L}(\mathscr{U}; \mathscr{V})$. To see this, we prove first that that set is linearly independent. Indeed, if

$$a^i_\alpha \boldsymbol{e}_i \otimes \boldsymbol{f}^\alpha = \boldsymbol{0}, \tag{1.45}$$

by (1.40) and (1.41) it follows that

$$(a^i_\alpha \boldsymbol{e}_i \otimes \boldsymbol{f}^\alpha)(\boldsymbol{u}) = \boldsymbol{0} \quad \forall \boldsymbol{u} \in \mathscr{U}. \tag{1.46}$$

In particular

$$(a^i_\alpha \boldsymbol{e}_i \otimes \boldsymbol{f}^\alpha)(\boldsymbol{f}_\beta) = \boldsymbol{0}, \quad \beta = 1, \ldots, m. \tag{1.47}$$

But by (1.44) this condition implies that

$$a^i_\alpha \delta^\alpha_\beta \boldsymbol{e}_i = a^i_\beta \boldsymbol{e}_i = \boldsymbol{0}, \quad \beta = 1, \ldots, m; \tag{1.48}$$

equivalently, since $\{\boldsymbol{e}_i\}$ is a basis of $\mathscr{V}$,

$$a^i_\beta = 0, \quad i = 1, \ldots, n,\ \beta = 1, \ldots, m. \tag{1.49}$$

Next we prove that $\{\boldsymbol{e}_i \otimes \boldsymbol{f}^\alpha\}$ is a generating set of $\mathscr{L}(\mathscr{U}; \mathscr{V})$. Let $\mathbf{L}$ be an arbitrary element of $\mathscr{L}(\mathscr{U}; \mathscr{V})$. Then $\mathbf{L}(\boldsymbol{f}_\beta)$ is a vector in $\mathscr{V}$, say

$$\mathbf{L}(\boldsymbol{f}_\beta) = L^i_\beta \boldsymbol{e}_i. \tag{1.50}$$

But by (1.44) this equation means that

$$\mathbf{L}(\boldsymbol{f}_\beta) = (L^i_\alpha \boldsymbol{e}_i \otimes \boldsymbol{f}^\alpha)(\boldsymbol{f}_\beta), \quad \beta = 1, \ldots, m, \tag{1.51}$$

so that

$$\mathbf{L} = L^i_\alpha \boldsymbol{e}_i \otimes \boldsymbol{f}^\alpha. \tag{1.52}$$

Since, then, $\{\boldsymbol{e}_i \otimes \boldsymbol{f}^\alpha\}$ is a linearly independent generating set, it is a basis of $\mathscr{L}(\mathscr{U}; \mathscr{V})$, and (1.42) follows.

The basis $\{\boldsymbol{e}_i \otimes \boldsymbol{f}^\alpha\}$ of $\mathscr{L}(\mathscr{U}; \mathscr{V})$ just described is called the *product basis* with respect to the bases $\{\boldsymbol{e}_i\}$ and $\{\boldsymbol{f}_\alpha\}$, and the components L^i_α of $\mathbf{L}$ relative to $\{\boldsymbol{e}_i \otimes \boldsymbol{f}^\alpha\}$ are called also the *components of* $\mathbf{L}$ *relative to* $\{\boldsymbol{e}_i\}$ *and* $\{\boldsymbol{f}_\alpha\}$. Equation (1.50) characterizes these components. For an arbitrary vector $\boldsymbol{u} \in \mathscr{U}$ with components u^α relative to $\{\boldsymbol{f}_\alpha\}$, the image $\boldsymbol{L}(\boldsymbol{u})$ of $\boldsymbol{u}$ under $\boldsymbol{L}$ has the component form

$$\boldsymbol{L}(\boldsymbol{u}) = L^i_\alpha u^\alpha \boldsymbol{e}_i \tag{1.53}$$

relative to $\{\boldsymbol{e}_i\}$. As a consequence of (1.43) and (1.44), the basis $\{\boldsymbol{e}_i \otimes \boldsymbol{f}^\alpha\}$ of $\mathscr{L}(\mathscr{U};\mathscr{V})$ gives rise to an isomorphism

$$\boldsymbol{K} : \mathscr{L}(\mathscr{U};\mathscr{V}) \to \mathscr{R}^{m\times n} \tag{1.54}$$

such that $\boldsymbol{K}(\boldsymbol{L})$ is the matrix $[L^i_\alpha]$. In particular, $\boldsymbol{K}(\boldsymbol{e}_i \otimes \boldsymbol{f}^\alpha)$ is the matrix with entries 1 at the $\binom{i}{\alpha}^{\text{th}}$ position and 0 elsewhere.

Proposition. *Relative to change of bases from* $\{\boldsymbol{e}_i\}$ *to* $\{\bar{\boldsymbol{e}}_i\}$ *and from* $\{\boldsymbol{f}_\alpha\}$ *to* $\{\bar{\boldsymbol{f}}_\alpha\}$, *the components of* **L** *satisfy the following transformation rule*:

$$L^i_\alpha = \bar{e}^i_j \bar{L}^j_\beta f^\beta_\alpha . \tag{1.55}$$

Here $\bar{e}^i_j$ is defined by $(1.28)_2$, and f^β_α is defined by analogy to $(1.28)_1$.

We now consider some examples of the space $\mathscr{L}(\mathscr{U};\mathscr{V})$ arising from various choices of $\mathscr{U}$ and $\mathscr{V}$.

Example 1. $\mathscr{L}(\mathscr{R};\mathscr{V})$, illustrating the concept of canonical isomorphism.

When $\mathscr{U} = \mathscr{R}$, the vector space $\mathscr{L}(\mathscr{R};\mathscr{V})$ is just a copy of $\mathscr{V}$. Indeed, if $L \in \mathscr{L}(\mathscr{R};\mathscr{V})$, we put

$$\boldsymbol{I}(\boldsymbol{L}) \equiv \boldsymbol{L}(1). \tag{1.56}$$

Then the mapping

$$\boldsymbol{I} : \mathscr{L}(\mathscr{R};\mathscr{V}) \to \mathscr{V} \tag{1.57}$$

is an isomorphism. Since it is established without use of a basis, it may be regarded as "natural". This kind of isomorphism is called *canonical* and is denoted by the abridged symbol $\cong$. Thus (1.57) may be written as

$$\mathscr{L}(\mathscr{R};\mathscr{V}) \cong \mathscr{V}. \tag{1.58}$$

Two vector spaces between which there is a canonical isomorphism are said to be *canonically isomorphic*. If, on the other hand, a basis is used in setting up a particular isomorphism, we regard that isomorphism as not being canonical until such time as the contrary can be proved. No effective criterion is known for determining whether or not a particular isomorphism originally established by use of a basis be in fact canonical or not.

EXAMPLE 2. Dual space, contraction.

Reversing the order of $\mathscr{R}$ and $\mathscr{V}$, we obtain the vector space $\mathscr{L}(\mathscr{V}; \mathscr{R})$, which is called the *dual space* of $\mathscr{V}$ and is denoted by $\mathscr{V}^*$. The elements of $\mathscr{V}^*$ are real-valued linear functions on $\mathscr{V}$. As before, let $\{\boldsymbol{e}_i\}$ be a basis of $\mathscr{V}$. Then $\{1 \otimes \boldsymbol{e}^i\}$ is a product basis of $\mathscr{V}^*$, where 1 is the standard basis of $\mathscr{R}$. For simplicity, we suppress the standard basis 1 and the symbol $\otimes$ from the notation of the product basis, writing it simply as $\{\boldsymbol{e}^i\}$, which we shall call the *dual basis* of $\{\boldsymbol{e}_i\}$. From (1.44), $\{\boldsymbol{e}_i\}$ and $\{\boldsymbol{e}^i\}$ are related by

$$\boldsymbol{e}^i(\boldsymbol{e}_j) = \delta^i_j. \tag{1.59}$$

By linearity, this relation characterizes the dual basis completely. The symbol $\boldsymbol{e}^i$, with a superscript i, is now defined in general relative to any basis $\{\boldsymbol{e}_i\}$ for any vector space $\mathscr{V}$. In particular, relative to a basis $\{\boldsymbol{f}_\alpha\}$ for a vector space $\mathscr{U}$, $\boldsymbol{f}^\alpha$ is defined by a condition similar to (1.59) with $\boldsymbol{e}$ replaced by $\boldsymbol{f}$ and i, j replaced by α, β; further, $\{\boldsymbol{f}^\alpha\}$ is a basis of $\mathscr{U}^*$.

So as to distinguish the vector spaces $\mathscr{V}$ and $\mathscr{V}^*$, only the elements of $\mathscr{V}$ are called *vectors*, while the elements of $\mathscr{V}^*$ are called *covectors*. Further, if the components of a vector are taken relative to a basis $\{\boldsymbol{e}_i\}$, then those of a covector are taken relative to the dual basis $\{\boldsymbol{e}^i\}$.

PROPOSITION. *Let $\{\boldsymbol{e}_i\}$ and $\{\bar{\boldsymbol{e}}_i\}$ be two bases of $\mathscr{V}$ related by* (1.28). *Then the corresponding dual bases $\{\boldsymbol{e}^i\}$ and $\{\bar{\boldsymbol{e}}^i\}$ are related by*

$$\boldsymbol{e}^i = \bar{e}^i_j \bar{\boldsymbol{e}}^j, \quad \bar{\boldsymbol{e}}^i = e^i_j \boldsymbol{e}^j. \tag{1.60}$$

Hence the transformation rule for the components of a covector $\boldsymbol{u}^*$ is

$$u_i = e^j_i \bar{u}_j, \tag{1.61}$$

where

$$\boldsymbol{u}^* = u_i \boldsymbol{e}^i = \bar{u}_i \bar{\boldsymbol{e}}^i. \tag{1.62}$$

The dual space of the dual space $\mathscr{V}^*$ is canonically isomorphic to $\mathscr{V}$ in the following way: For any fixed vector $\boldsymbol{u} \in \mathscr{V}$, we define a real-valued linear function $\boldsymbol{u}^{**}$ on $\mathscr{V}^*$ by

$$\boldsymbol{u}^{**}(\boldsymbol{w}) \equiv \boldsymbol{w}(\boldsymbol{u}) \quad \forall \boldsymbol{w} \in \mathscr{V}^*. \tag{1.63}$$

Then the $**$-operation:

$$** : \mathscr{V} \to (\mathscr{V}^*)^*, \tag{1.64}$$

is the canonical isomorphism. Because of this result, we say that the relation between $\mathscr{V}$ and $\mathscr{V}^*$ is *reflexive*, as indeed is suggested also by the name "dual".

The condition (1.63) can be expressed by a more symmetric notation. Namely, if $\boldsymbol{u} \in \mathscr{V}$ and $\boldsymbol{w} \in \mathscr{V}^*$, we put

$$\langle \boldsymbol{u}, \boldsymbol{w} \rangle \equiv \boldsymbol{w}(\boldsymbol{u}) = \boldsymbol{u}^{**}(\boldsymbol{w}) \equiv \langle \boldsymbol{w}, \boldsymbol{u} \rangle. \tag{1.65}$$

The mapping

$$\langle , \rangle : \mathscr{V} \times \mathscr{V}^* \to \mathscr{R}, \tag{1.66}$$

which is called the *contraction* on $\mathscr{V}$ and $\mathscr{V}^*$, satisfies the following two conditions:

i) $\langle , \rangle$ is *bilinear*, *i.e.*, it is linear with respect to each of its two variables.
ii) $\langle , \rangle$ is *definite*, *i.e.*, if $\boldsymbol{u}$ is fixed and $\langle \boldsymbol{u}, \boldsymbol{w} \rangle = 0\ \forall \boldsymbol{w}$, then $\boldsymbol{u} = \boldsymbol{0}$; conversely, if $\boldsymbol{w}$ is fixed and $\langle \boldsymbol{u}, \boldsymbol{w} \rangle = 0\ \forall \boldsymbol{u}$, then $\boldsymbol{w} = \boldsymbol{0}$.

If $\langle \boldsymbol{u}, \boldsymbol{w} \rangle = 0$, the vector $\boldsymbol{u}$ is said to be *orthogonal* to the covector $\boldsymbol{w}$, and *vice versa*. Motivation for this usage will be given in § 2.

EXAMPLE 3. The algebra of $\mathscr{V}$.

The vector space $\mathscr{L}(\mathscr{V}; \mathscr{V})$ consists of linear transformations of $\mathscr{V}$ itself. Sometimes these transformations are the only ones called *tensors*, but in other discussions that term is applied more generally, as we shall see later on in this section. As before, let $\{\boldsymbol{e}_i\}$ be a basis of $\mathscr{V}$. Then a product basis of $\mathscr{L}(\mathscr{V}; \mathscr{V})$ is $\{\boldsymbol{e}_i \otimes \boldsymbol{e}^j, i, j = 1, \ldots, n\}$ as defined by (1.44) with $\boldsymbol{f}_\beta$ replaced by $\boldsymbol{e}_k$, so that the component form of $\boldsymbol{L}$ is

$$\boldsymbol{L} = L^i_j \boldsymbol{e}_i \otimes \boldsymbol{e}^j, \quad \boldsymbol{L} \in \mathscr{L}(\mathscr{V}; \mathscr{V}). \tag{1.67}$$

For transformations in $\mathscr{L}(\mathscr{V}; \mathscr{V})$ it is customary to omit the parentheses around the argument vector, writing $\boldsymbol{Lv} \equiv \boldsymbol{L}(\boldsymbol{v})$. In this notation the components L^i_j are characterized by the conditions

$$\boldsymbol{L}\boldsymbol{e}_i = L^j_i \boldsymbol{e}_j, \quad i = 1, \ldots, n. \tag{1.68}$$

Therefore, if $\boldsymbol{v} \in \mathscr{V}$ and has components v^i relative to $\{\boldsymbol{e}_i\}$,

$$\boldsymbol{L}\boldsymbol{v} = L^i_j v^j \boldsymbol{e}_i. \tag{1.69}$$

Thus the components of the image of $\boldsymbol{v}$ under $\boldsymbol{L}$ are $L^i_j v^j$, $i = 1, \ldots, n$. In particular, the identity transformation $\boldsymbol{1}$ has the components δ^i_j relative to any basis of $\mathscr{V}$.

The vector space $\mathscr{L}(\mathscr{V}; \mathscr{V})$ is equipped with a natural multiplication. Namely, for any $\boldsymbol{L}$, $\bar{\boldsymbol{L}} \in \mathscr{L}(\mathscr{V}; \mathscr{V})$ we define $\boldsymbol{L}\bar{\boldsymbol{L}}$ as the composition of

$\boldsymbol{L}$ and $\bar{\boldsymbol{L}}$:

$$\boldsymbol{L}\bar{\boldsymbol{L}}\boldsymbol{v} \equiv \boldsymbol{L}(\bar{\boldsymbol{L}}\boldsymbol{v}) \quad \forall \boldsymbol{v} \in \mathscr{V}. \tag{1.70}$$

The reader will verify easily that this multiplication is associative and is also distributive with respect to the linear operations but generally fails to be commutative. A vector space equipped with such a multiplication is called an *algebra.*

As usual, two algebras are said to be *isomorphic* if there exists an isomorphism that is also a homomorphism with respect to the multiplication operations of the algebras. Here the algebra $\mathscr{L}(\mathscr{V}; \mathscr{V})$ is isomorphic to the matrix algebra $\mathscr{R}^{n^2}$. Indeed, relative to any basis $\{\boldsymbol{e}_i\}$ of $\mathscr{V}$, a linear transformation $\boldsymbol{L}$ corresponds to its component matrix $[L^i_j]$. Further, from (1.70), the multiplication of linear transformations corresponds to the matrix product. That is to say, if with respect to the product basis $\{\boldsymbol{e}_i \otimes \boldsymbol{e}^j\}$ of $\mathscr{L}(\mathscr{V}; \mathscr{V})$ the linear transformations $\boldsymbol{L}$ and $\boldsymbol{N}$ have the components L^k_m and N^p_q, respectively, then with respect to the same basis the components of $\boldsymbol{LN}$ are given by

$$(LN)^i_j = L^i_k N^k_j, \tag{1.71}$$

which means that the matrix $[(LN)^i_j]$ is the matrix product of the matrices $[L^i_j]$ and $[N^i_j]$.

PROPOSITION. *Under a change of basis from* $\{\boldsymbol{e}_i\}$ *to* $\{\bar{\boldsymbol{e}}_i\}$ *the components of a linear transformation* $\boldsymbol{L} \in \mathscr{L}(\mathscr{V}; \mathscr{V})$ *satisfy the following transformation rule*:

$$L^i_j = \bar{e}^i_l \bar{L}^l_k e^k_j, \tag{1.72}$$

where $\bar{e}^i_l$ *and* e^k_j *are defined by* (1.28).

In view of this transformation rule, the formulae

$$\operatorname{tr} \boldsymbol{L} \equiv L^i_i = \bar{L}^i_i, \tag{1.73}$$

and

$$\det \boldsymbol{L} \equiv \det [L^i_j] = \det [\bar{L}^i_j] \tag{1.74}$$

define real-valued scalar functions of $\boldsymbol{L}$. They are called the *trace* and the *determinant* of $\boldsymbol{L}$, respectively. The reader will verify the following statements easily:

i) Trace is a linear operation but determinant is not (unless $\dim \mathscr{V} = 1$).

ii) $$\operatorname{tr}(\boldsymbol{LN}) = \operatorname{tr}(\boldsymbol{NL}), \tag{1.75}$$

iii) $$\det(\boldsymbol{LN}) = (\det \boldsymbol{L})(\det \boldsymbol{N}) = \det(\boldsymbol{NL}). \tag{1.76}$$

We call $\boldsymbol{L}$ *non-singular* if it is an isomorphism of $\mathscr{V}$; otherwise, we call $\boldsymbol{L}$ *singular*. From (1.69) and Cramer's rule, $\boldsymbol{L}$ is singular if and only if its determinant vanishes. The collection of all non-singular linear transformations of $\mathscr{V}$ forms a group with respect to multiplication in $\mathscr{L}(\mathscr{V};\mathscr{V})$. We denote this group by $\mathfrak{GL}(\mathscr{V})$ and call it the *general linear group* of $\mathscr{V}$. The sets $\mathfrak{U}(\mathscr{V})$ and $\mathfrak{SL}(\mathscr{V})$ consisting in all transformations the determinants of which equal ± 1 and 1, respectively, are subgroups of $\mathfrak{GL}(\mathscr{V})$; they are called the *unimodular* and the *special linear groups* of $\mathscr{V}$, respectively. We can express the relation among $\mathscr{L}(\mathscr{V};\mathscr{V})$, $\mathfrak{GL}(\mathscr{V})$, and $\mathfrak{SL}(\mathscr{V})$ by the following diagram:

$$\begin{array}{c} \underbrace{\mathscr{L}(\mathscr{V};\mathscr{V}) \supset \mathfrak{GL}(\mathscr{V}) \supset \mathfrak{U}(\mathscr{V}) \supset \mathfrak{SL}(\mathscr{V})} \\ \downarrow \det \\ \overbrace{\mathscr{R} \supset \mathscr{R}-\{0\} \supset \{+1,\ -1\} \supset \{1\}.} \end{array} \qquad (1.77)$$

A non-zero vector $\boldsymbol{v} \in \mathscr{V}$ is called a *proper vector* of $\boldsymbol{L}$ if the image of $\boldsymbol{v}$ under $\boldsymbol{L}$ is a scalar multiple of $\boldsymbol{v}$, say

$$\boldsymbol{L}\boldsymbol{v} = l\boldsymbol{v}. \qquad (1.78)$$

A scalar l such that (1.78) holds for some $\boldsymbol{v}$ other than $\boldsymbol{0}$ is called a *proper number* of $\boldsymbol{L}$. Since $\boldsymbol{v} \neq \boldsymbol{0}$, (1.78) shows the linear transformation $\boldsymbol{L}-l\boldsymbol{1}$ to be singular, so that

$$\det(\boldsymbol{L}-l\boldsymbol{1}) = 0. \qquad (1.79)$$

Conversely, if l satisfies the equation (1.79), then $\ker(\boldsymbol{L}-l\boldsymbol{1}) \neq \boldsymbol{0}$, so that there is a non-zero vector $\boldsymbol{v}$ which satisfies (1.78). Thus (1.79) is a necessary and sufficient condition that l be a proper number of $\boldsymbol{L}$. From (1.74), the left-hand side of (1.79) is a polynomial of degree n in l, so we may write it in the form

$$(-l)^n + I_1(-l)^{n-1} + \ldots + I_{n-1}(-l) + I_n = 0, \qquad (1.80)$$

which is called the *characteristic equation* of $\boldsymbol{L}$. Its real roots are the proper numbers of $\boldsymbol{L}$. The coefficients $I_1, I_2, \ldots, I_n$ of the characteristic equation are real-valued scalar functions of $\boldsymbol{L}$ and are called the *principal invariants* of $\boldsymbol{L}$.

PROPOSITION. *In terms of the components* L^i_j *and the roots* $l_1, l_2, \ldots, l_n$ *of* (1.80), *the principal invariants are given by*

$$
\begin{aligned}
I_1 &= \frac{1}{1!}\,\delta^{i_1}_{j_1} L^{j_1}_{i_1} = l_1 + l_2 + \ldots + l_n,\\
I_2 &= \frac{1}{2!}\,\delta^{i_1 i_2}_{j_1 j_2} L^{j_1}_{i_1} L^{j_2}_{i_2} = l_1 l_2 + l_1 l_3 + \ldots + l_{n-1} l_n,\\
&\ldots\ldots\ldots\ldots\ldots\ldots \\
I_k &= \frac{1}{k!}\,\delta^{i_1 \ldots i_k}_{j_1 \ldots j_k} L^{j_1}_{i_1} \ldots L^{j_k}_{i_k} = l_1 l_2 \ldots l_k + \ldots + l_{n-k} l_{n-k+1} \ldots l_n,\\
&\ldots\ldots\ldots\ldots\ldots\ldots \\
I_n &= \frac{1}{n!}\,\delta^{i_1 \ldots i_n}_{j_1 \ldots j_n} L^{j_1}_{i_1} \ldots L^{j_n}_{i_n} = l_1 l_2 \ldots l_n,
\end{aligned}
\tag{1.81}
$$

where

$$
\delta^{i_1 \ldots i_k}_{j_1 \ldots j_k} \equiv \begin{cases} 1 \text{ if } (i_1 \ldots i_k) \text{ is an even permutation of } (j_1 \ldots j_k),\\ -1 \text{ if } (i_1 \ldots i_k) \text{ is an odd permutation of } (j_1 \ldots j_k),\\ 0 \text{ otherwise,} \end{cases} \tag{1.82}
$$

and where a k-fold root l of (1.80) *is considered to be k equal roots* $l_{i_1}, l_{i_2}, \ldots, l_{i_k}$. *In particular,*

$$I_1 = \operatorname{tr} \mathbf{L}, \tag{1.83}$$

and

$$I_n = \det \mathbf{L}, \tag{1.84}$$

so that 0 *is a proper number of* **L** *if and only if* det **L** *vanishes.*

PROPOSITION (HAMILTON, CAYLEY). **L** *satisfies its own characteristic equation*:

$$(-\mathbf{L})^n + I_1(-\mathbf{L})^{n-1} + \ldots + I_{n-1}(-\mathbf{L}) + I_n \mathbf{1} = \mathbf{0}. \tag{1.85}$$

Here $\mathbf{L}^k$ denotes the kth power of **L**, *i.e.*,

$$\mathbf{L}^k \equiv \underbrace{\mathbf{L} \ldots \mathbf{L}}_{k \text{ times}}. \tag{1.86}$$

Let l be a particular proper number of **L**. Then the set of all vectors (including **0**) which satisfy the condition (1.78) forms a subspace $\mathscr{V}_l(\mathbf{L})$ called the *characteristic subspace* of **L** associated with the proper number l. The characteristic subspaces of **L** corresponding to different proper numbers are independent subspaces of $\mathscr{V}$:

$$\mathscr{V}_l(\mathbf{L}) \cap \mathscr{V}_m(\mathbf{L}) = \{\mathbf{0}\} \tag{1.87}$$

if $l \neq m$.

Two linear transformations $\mathbf{L}$ and $\bar{\mathbf{L}}$ are called *similar* if there exists an isomorphism $\mathbf{E}$ of $\mathscr{V}$ such that

$$\bar{\mathbf{L}} = \mathbf{E}\mathbf{L}\mathbf{E}^{-1}. \tag{1.88}$$

Comparing this condition with the transformation rule (1.72), we see that $\mathbf{L}$ and $\bar{\mathbf{L}}$ are similar if and only if there exist bases $\{\mathbf{e}_i\}$ and $\{\bar{\mathbf{e}}_i\}$ such that the components of $\mathbf{L}$ relative to $\{\mathbf{e}_i\}$ are the same as those of $\bar{\mathbf{L}}$ relative to $\{\bar{\mathbf{e}}_i\}$. The isomorphism $\mathbf{E}$ carrying $\{\mathbf{e}_i\}$ onto $\{\bar{\mathbf{e}}_i\}$ then satisfies the condition (1.88).

PROPOSITION. *If* $\mathbf{L}$ *and* $\bar{\mathbf{L}}$ *are similar, then they have the same principal invariants and hence the same characteristic equation. Further, if* $\mathscr{V}_l(\mathbf{L})$ *is a characteristic subspace of* $\mathbf{L}$, *then*

$$\mathscr{V}_l(\bar{\mathbf{L}}) = \mathbf{E}(\mathscr{V}_l(\mathbf{L})) \tag{1.89}$$

is the corresponding characteristic subspace of $\bar{\mathbf{L}}$, *where* $\mathbf{E}$ *is any isomorphism satisfying the condition* (1.88).

A linear transformation $\mathbf{L}$ is called a *dilatation*[1] if there exists a basis $\{\mathbf{e}_i\}$ consisting entirely of proper vectors of $\mathbf{L}$. Such a basis is called a *principal basis* of $\mathbf{L}$. Relative to a principal basis $\{\mathbf{e}_i\}$ the components of $\mathbf{L}$ form a diagonal matrix, say

$$[L^i_j] = \operatorname{diag}[l_1, \ldots, l_n], \tag{1.90}$$

where $l_1, \ldots, l_n$ are the roots of the characteristic equation of $\mathbf{L}$, which must all be real numbers in this case but need not be distinct.

PROPOSITION. $\mathbf{L}$ *is a dilatation if and only if the following two conditions are satisfied*:

i) *The characteristic equation of* $\mathbf{L}$ *has real roots only.*

ii) *The dimension of each characteristic space of* $\mathbf{L}$ *is equal to the multiplicity of the corresponding proper number as a root of the characteristic equation.*

---

[1] In continuum mechanics the term "dilatation" usually refers to the special case when $\mathbf{L} = l\mathbf{1}$, $l \neq 0$.

In particular, if the characteristic equation of $\boldsymbol{L}$ has distinct real roots only, then $\boldsymbol{L}$ is a dilatation.

EXAMPLE 4. Adjoint Operators.

$\mathscr{L}(\mathscr{V}^*; \mathscr{V}^*)$, the space of linear transformations of $\mathscr{V}^*$, is isomorphic to $\mathscr{L}(\mathscr{V}; \mathscr{V})$, the isomorphism being the operation

$$A : \mathscr{L}(\mathscr{V}; \mathscr{V}) \to \mathscr{L}(\mathscr{V}^*; \mathscr{V}^*) \tag{1.91}$$

characterized by the condition

$$\langle \boldsymbol{w}, \boldsymbol{L}\boldsymbol{v} \rangle = \langle \boldsymbol{L}^A \boldsymbol{w}, \boldsymbol{v} \rangle \tag{1.92}$$

for all $\boldsymbol{v} \in \mathscr{V}$ and all $\boldsymbol{w} \in \mathscr{V}^*$. We do not suppress the notation A here, since $\mathscr{V}^*$ is not canonically isomorphic to $\mathscr{V}$. We call $\boldsymbol{L}^A$ the *adjoint* of $\boldsymbol{L}$.

EXAMPLE 5. Transposition, Bilinear Functions.

Consider the space $\mathscr{L}(\mathscr{V}; \mathscr{V}^*)$. Relative to a basis $\{\boldsymbol{e}^i\}$ of $\mathscr{V}^*$, the component form of an element $\boldsymbol{A} \in \mathscr{L}(\mathscr{V}; \mathscr{V}^*)$ is

$$\boldsymbol{A} = A_{ij} \boldsymbol{e}^i \otimes \boldsymbol{e}^j, \tag{1.93}$$

where the components A_{ij} are characterized by the condition

$$\boldsymbol{A}\boldsymbol{e}_i = A_{ji} \boldsymbol{e}^j, \quad i = 1, \ldots, n, \tag{1.94}$$

$\{\boldsymbol{e}_i\}$ being, as usual, the dual basis of $\{\boldsymbol{e}^i\}$. The notation $\boldsymbol{e}^i \otimes \boldsymbol{e}^j$ is consistent with the notation $\boldsymbol{e}_i \otimes \boldsymbol{f}^\alpha$ defined in general by (1.44), since, here, $\{\boldsymbol{e}^i\}$ is a basis for the image $\mathscr{V}^*$ and $\{\boldsymbol{e}_i\}$ is a basis for the domain $\mathscr{V}$ of any element $\boldsymbol{A} \in \mathscr{L}(\mathscr{V}; \mathscr{V}^*)$. As in the case of $\mathscr{L}(\mathscr{V}; \mathscr{V})$, we omit parentheses around the argument of $\boldsymbol{A}$, writing $\boldsymbol{A}\boldsymbol{v} \equiv \boldsymbol{A}(\boldsymbol{v})$. If $\boldsymbol{v} \in \mathscr{V}$, and if v^i is the ith component of $\boldsymbol{v}$ with respect to $\{\boldsymbol{e}_i\}$, then

$$\boldsymbol{A}\boldsymbol{v} = A_{ij} v^j \boldsymbol{e}^i. \tag{1.95}$$

The transformation rule for the components of $\boldsymbol{A}$ relative to the change of basis (1.28) from $\{\boldsymbol{e}_i\}$ to $\{\bar{\boldsymbol{e}}_i\}$ is

$$A_{ij} = e_i^k \bar{A}_{kl} e_j^l. \tag{1.96}$$

The space $\mathscr{L}(\mathscr{V}; \mathscr{V}^*)$ is non-trivially isomorphic to itself with respect to the operation called *transposition* and defined as follows:

$$T : \mathscr{L}(\mathscr{V}; \mathscr{V}^*) \to \mathscr{L}(\mathscr{V}; \mathscr{V}^*) \tag{1.97}$$

such that

$$\langle \boldsymbol{u}, \boldsymbol{A}\boldsymbol{v}\rangle = \langle \boldsymbol{A}^T\boldsymbol{u}, \boldsymbol{v}\rangle \quad \forall \boldsymbol{u}, \ \boldsymbol{v} \in \mathscr{V}. \tag{1.98}$$

Relative to any basis $\{\boldsymbol{e}_i\}$, the component matrix of $\boldsymbol{A}^T$ is the transpose of that of $\boldsymbol{A}$:

$$A_{ij}^T = A_{ji}, \quad i, j = 1, \ldots, n. \tag{1.99}$$

We call $\boldsymbol{A}$ *symmetric* if it is equal to its transpose, *skew* if it is equal to the negative of its transpose.

As before, a function

$$f: \mathscr{V} \times \mathscr{V} \to \mathscr{R} \tag{1.100}$$

is called *bilinear* if it is linear with respect to each of its two arguments. The collection $\mathscr{L}(\mathscr{V}, \mathscr{V}; \mathscr{R})$ consisting in all bilinear functions on $\mathscr{V} \times \mathscr{V}$ forms a vector space in an obvious way; further, $\mathscr{L}(\mathscr{V}, \mathscr{V}; \mathscr{R})$ is canonically isomorphic to $\mathscr{L}(\mathscr{V}; \mathscr{V}^*)$, and the canonical isomorphism

$$\wedge: \mathscr{L}(\mathscr{V}; \mathscr{V}^*) \to \mathscr{L}(\mathscr{V}, \mathscr{V}; \mathscr{R}) \tag{1.101}$$

is given by

$$\hat{A}(\boldsymbol{u}, \boldsymbol{v}) \equiv \boldsymbol{A}\boldsymbol{v}(\boldsymbol{u}) = \langle \boldsymbol{u}, \boldsymbol{A}\boldsymbol{v}\rangle \quad \forall \boldsymbol{A} \in \mathscr{L}(\mathscr{V}; \mathscr{V}^*), \ \forall \boldsymbol{u}, \ \boldsymbol{v} \in \mathscr{V}. \tag{1.102}$$

The component form of the third member of this equation relative to $\{\boldsymbol{e}_i\}$ is $A_{ij}u^iv^j$. The notation $\wedge$ will be suppressed from now on, and, as is customary for real functions, the light-faced letter A will stand for the bilinear function corresponding to $\boldsymbol{A}$.

A bilinear function $A(\boldsymbol{u}, \boldsymbol{v})$ is called *positive-definite* if

$$A(\boldsymbol{u}, \boldsymbol{u}) > 0 \qquad \forall \boldsymbol{u} \neq \boldsymbol{0}, \tag{1.103}$$

positive-semidefinite if $A(\boldsymbol{u}, \boldsymbol{u}) \geqq 0 \ \forall \boldsymbol{u} \in \mathscr{V}$. These same terms are applied also the the mapping $\boldsymbol{A}$ itself. Similarly, A is said to be *symmetric* if $\boldsymbol{A}$ is. The function $A(\boldsymbol{u}, \boldsymbol{u})$ is called the *quadratic form* of $\boldsymbol{A}$.

A symmetric, positive-definite, bilinear function on $\mathscr{V} \times \mathscr{V}$ is called an *inner product* on $\mathscr{V}$. The concept of inner product is an important one, which we shall develop in the next section.

The spaces of linear maps considered so far are canonically isomorphic to certain *tensor spaces* over the vector space $\mathscr{V}$. In general, let $\mathscr{U}$ and $\mathscr{V}$ be any two vector spaces. Then we define the *tensor product* $\mathscr{U} \otimes \mathscr{V}$ as the dual space of the space of all bilinear functions on $\mathscr{U} \times \mathscr{V}$, *i.e.*,

$$\mathscr{U} \otimes \mathscr{V} \equiv \mathscr{L}(\mathscr{U}, \mathscr{V}; \mathscr{R})^*. \tag{1.104}$$

The terms *tensor product* and *simple tensor* are applied also to those elements $\boldsymbol{u} \otimes \boldsymbol{v}$ of $\mathscr{U} \otimes \mathscr{V}$ which are defined by ordered pairs $\boldsymbol{u} \in \mathscr{U}$ and $\boldsymbol{v} \in \mathscr{V}$ as the linear functions

$$\boldsymbol{u} \otimes \boldsymbol{v} : \mathscr{L}(\mathscr{U}, \mathscr{V}; \mathscr{R}) \rightarrow \mathscr{R} \tag{1.105}$$

such that

$$[\boldsymbol{u} \otimes \boldsymbol{v}](A) = A(\boldsymbol{u}, \boldsymbol{v}) \quad \forall A \in \mathscr{L}(\mathscr{U}, \mathscr{V}; \mathscr{R}). \tag{1.106}$$

In particular, the tensor product $\boldsymbol{f}_\alpha \otimes \boldsymbol{e}_i$ is defined relative to any bases $\{\boldsymbol{f}_\alpha\}$ and $\{\boldsymbol{e}_i\}$ in $\mathscr{U}$ and $\mathscr{V}$, respectively. Using an argument similar to that of the proof of (1.42), we can show that $\{\boldsymbol{f}_\alpha \otimes \boldsymbol{e}_i, \alpha = 1, \ldots, m, i = 1, \ldots, n\}$ form a basis for $\mathscr{U} \otimes \mathscr{V}$.

Having defined the tensor product $\otimes$ in general, we can now reinterpret the notation $\boldsymbol{e}_i \otimes \boldsymbol{f}^\alpha$ in (1.43) as denoting the *tensor product* of $\boldsymbol{e}_i$ with $\boldsymbol{f}^\alpha$. To do so, we use the fact that $\mathscr{L}(\mathscr{U}; \mathscr{V})$ is canonically isomorphic to $\mathscr{V} \otimes \mathscr{U}^*$, the canonical isomorphism being characterized by the condition that the simple tensor $\boldsymbol{v} \otimes \boldsymbol{w}$ in $\mathscr{V} \otimes \mathscr{U}^*$ is identified with the linear map from $\mathscr{U}$ to $\mathscr{V}$ such that $(\boldsymbol{v} \otimes \boldsymbol{w})(\boldsymbol{u}) = \langle \boldsymbol{w}, \boldsymbol{u} \rangle \boldsymbol{v}\ \forall \boldsymbol{u} \in \mathscr{U}$.

Using the general definition (1.104), we can verify also that the vector spaces $\mathscr{L}(\mathscr{V}; \mathscr{V})$, $\mathscr{L}(\mathscr{V}; \mathscr{V}^*)$, and $\mathscr{L}(\mathscr{V}^*; \mathscr{V})$ are canonically isomorphic to the tensor spaces $\mathscr{V} \otimes \mathscr{V}^*$, $\mathscr{V}^* \otimes \mathscr{V}^*$, and $\mathscr{V} \otimes \mathscr{V}$, respectively.

We now generalize the definition (1.104) to tensor products of more than two vector spaces $\mathscr{U}, \mathscr{V}, \ldots, \mathscr{W}$. We denote the space of all multilinear functions on $\mathscr{U} \times \mathscr{V} \times \ldots \times \mathscr{W}$ by $\mathscr{L}(\mathscr{U}, \mathscr{V}, \ldots, \mathscr{W}; \mathscr{R})$, and we put

$$\mathscr{U} \otimes \mathscr{V} \otimes \ldots \otimes \mathscr{W} \equiv \mathscr{L}(\mathscr{U}, \mathscr{V}, \ldots, \mathscr{W}; \mathscr{R})^*. \tag{1.07}$$

As before, if $\boldsymbol{u} \in \mathscr{U}, \boldsymbol{v} \in \mathscr{V}, \ldots, \boldsymbol{w} \in \mathscr{W}$, we define the simple tensor $\boldsymbol{u} \otimes \boldsymbol{v} \otimes \ldots \otimes \boldsymbol{w}$ by the condition

$$[\boldsymbol{u} \otimes \boldsymbol{v} \otimes \ldots \otimes \boldsymbol{w}](\Psi) \equiv \Psi(\boldsymbol{u}, \boldsymbol{v}, \ldots, \boldsymbol{w}) \quad \forall \Psi \in \mathscr{L}(\mathscr{U}, \mathscr{V}, \ldots, \mathscr{W}; \mathscr{R}). \tag{1.08}$$

These simple tensors form a generating set of $\mathscr{U} \otimes \mathscr{V} \otimes \ldots \otimes \mathscr{W}$. Symbols of the planets, Ψ being one, are used in this book to denote a vector or tensor of unspecified order.

We define the *direct sum* $\mathscr{U} \oplus \mathscr{V} \oplus \ldots \oplus \mathscr{W}$ of $\mathscr{U}, \mathscr{V}, \ldots, \mathscr{W}$ as the Cartesian product $\mathscr{U} \times \mathscr{V} \times \ldots \times \mathscr{W}$ together with the linear operation

$$a(\boldsymbol{u}, \boldsymbol{v}, \ldots, \boldsymbol{w}) + \bar{a}(\bar{\boldsymbol{u}}, \bar{\boldsymbol{v}}, \ldots, \bar{\boldsymbol{w}}) \equiv (a\boldsymbol{u} + \overline{a\boldsymbol{u}}, a\boldsymbol{v} + \overline{a\boldsymbol{v}}, \ldots, a\boldsymbol{w} + \overline{a\boldsymbol{w}}). \tag{1.109}$$

Then to within a canonical isomorphism the operations $\otimes$ and $\oplus$ are commutative, associative, and distributive with respect to each other:

$$\mathscr{U} \otimes \mathscr{V} \cong \mathscr{V} \otimes \mathscr{U}, \quad \mathscr{U} \oplus \mathscr{V} \cong \mathscr{V} \oplus \mathscr{U}, \tag{1.110}$$

$$\mathscr{U} \otimes (\mathscr{V} \otimes \mathscr{W}) \cong (\mathscr{U} \otimes \mathscr{V}) \otimes \mathscr{W} \cong \mathscr{U} \otimes \mathscr{V} \otimes \mathscr{W},$$

$$\mathscr{U} \oplus (\mathscr{V} \oplus \mathscr{W}) \cong (\mathscr{U} \oplus \mathscr{V}) \oplus \mathscr{W} \cong \mathscr{U} \oplus \mathscr{V} \oplus \mathscr{W}, \tag{1.111}$$

and

$$\mathscr{U} \otimes (\mathscr{V} \oplus \mathscr{W}) \cong \mathscr{U} \otimes \mathscr{V} \oplus \mathscr{U} \otimes \mathscr{W}, \tag{1.112}$$

where $\cong$ denotes a canonical isomorphism. Also, $\otimes$ and $\oplus$ commute with the operation * of taking the dual space:

$$(\mathscr{U} \otimes \mathscr{V})^* \cong \mathscr{U}^* \otimes \mathscr{V}^*, \quad (\mathscr{U} \oplus \mathscr{V})^* \cong \mathscr{U}^* \oplus \mathscr{V}^*. \tag{1.113}$$

Having defined the tensor product in general, we can now introduce the *tensor space* $\mathscr{T}^r_s$ of *order* (r, s) over $\mathscr{V}$ by

$$\mathscr{T}^r_s \equiv \underbrace{\mathscr{V} \otimes \dots \otimes \mathscr{V}}_{r} \otimes \underbrace{\mathscr{V}^* \otimes \dots \otimes \mathscr{V}^*}_{s}. \tag{1.114}$$

Relative to a basis $\{\boldsymbol{e}_i\}$ of $\mathscr{V}$, a tensor in $\mathscr{T}^r_s$ has the component form

$$\Psi = \Psi^{i_1 \dots i_r}_{j_1 \dots j_s} \boldsymbol{e}_{i_1} \otimes \dots \otimes \boldsymbol{e}_{i_r} \otimes \boldsymbol{e}^{j_1} \otimes \dots \otimes \boldsymbol{e}^{j_s}, \tag{1.115}$$

where under a change of basis from $\{\boldsymbol{e}_i\}$ to $\{\bar{\boldsymbol{e}}_i\}$ the components $\Psi^{i_1 \dots i_r}_{j_1 \dots j_s}$ satisfy the transformation rule

$$\Psi^{i_1 \dots i_r}_{j_1 \dots j_s} = \bar{e}^{i_1}_{k_1} \dots \bar{e}^{i_r}_{k_r} e^{l_1}_{j_1} \dots e^{l_s}_{j_s} \overline{\Psi}^{k_1 \dots k_r}_{l_1 \dots l_s}. \tag{1.116}$$

A simple tensor $\boldsymbol{v}_1 \otimes \dots \otimes \boldsymbol{v}_r \otimes \boldsymbol{u}^1 \otimes \dots \otimes \boldsymbol{u}^s$ of order (r, s) has the components

$$(v_1 \otimes \dots \otimes v_r \otimes u^1 \otimes \dots \otimes u^s)^{i_1 \dots i_r}_{j_1 \dots j_s} = v^{i_1}_1 \dots v^{i_r}_r u^1_{j_1} \dots u^s_{j_s}. \tag{1.117}$$

In (1.115)–(1.117) the numbers r and s are the *contravariant* and *covariant orders*, respectively, of the tensors in question. If one of these orders is 0, it is usually left unmentioned. Thus, for example, a vector $\boldsymbol{v}$ is a contravariant tensor of order 1, while a covector $\boldsymbol{w}$ is a covariant tensor of order 1.

We can define various *contractions* on the tensor spaces over $\mathscr{V}$. A typical one is

$$\mathbf{C}^1_1 : \mathscr{T}^r_s \to \mathscr{T}^{r-1}_{s-1}, \tag{1.118}$$

which is characterized by the condition

$$\mathbf{C}_1^1(\boldsymbol{v}_1 \otimes \ldots \otimes \boldsymbol{v}_r \otimes \boldsymbol{u}^1 \otimes \ldots \otimes \boldsymbol{u}^s) = \langle \boldsymbol{v}_1, \boldsymbol{u}^1 \rangle \boldsymbol{v}_2 \otimes \ldots \otimes \boldsymbol{v}_r \otimes \boldsymbol{u}^2 \otimes \ldots \otimes \boldsymbol{u}^s \quad (1.119)$$

for simple tensors and then extended by linearity. Equivalently, this contraction has the component form

$$[C_1^1 \Psi]^{i_2 \ldots i_r}_{j_2 \ldots j_s} = \Psi^{k i_2 \ldots i_r}_{k j_2 \ldots j_s}, \quad (1.120)$$

for all $\Psi \in \mathscr{T}^r_s$. The contraction (1.66) is a special case.

In an obvious way, we can define tensor products of tensors also. For example, for simple tensors $\boldsymbol{u} \otimes \boldsymbol{w} \in \mathscr{T}^1_1$ and $\boldsymbol{a} \otimes \boldsymbol{b} \otimes \boldsymbol{c} \in \mathscr{T}^2_1$ we define

$$(\boldsymbol{u} \otimes \boldsymbol{w}) \otimes (\boldsymbol{a} \otimes \boldsymbol{b} \otimes \boldsymbol{c}) \equiv \boldsymbol{u} \otimes \boldsymbol{a} \otimes \boldsymbol{b} \otimes \boldsymbol{w} \otimes \boldsymbol{c}. \quad (1.121)$$

Then this operation can be extended to all tensors in $\mathscr{T}^1_1$ and $\mathscr{T}^2_1$ by linearity.

2. Inner-Product Spaces

In the previous section we remarked that any two vector spaces of the same dimension are isomorphic, so that, in particular, a vector space $\mathscr{V}$ is isomorphic to its dual space $\mathscr{V}^*$, but in general the isomorphism is not a canonical one. However, we can choose to contemplate a particular isomorphism of $\mathscr{V}$ with $\mathscr{V}^*$, say

$$\boldsymbol{g} : \mathscr{V} \to \mathscr{V}^*. \quad (2.1)$$

When we suppose $\mathscr{V}$ to be identified with $\mathscr{V}^*$ in this way by a particular $\boldsymbol{g}$, we need no longer distinguish between a vector and a covector. As also explained in Example 5 in the previous section, $\boldsymbol{g}$ corresponds to a unique bilinear function g on $\mathscr{V} \times \mathscr{V}$:

$$g : \mathscr{V} \times \mathscr{V} \to \mathscr{R}. \quad (2.2)$$

We call such a function g a *bilinear product* on $\mathscr{V}$. (To conform with classical precedents, we use minuscule $\boldsymbol{g}$ and g rather than the majuscules or bold Greek letters used in general in this book for mappings of vector spaces and bilinear forms.) Clearly a bilinear function is a bilinear product if and only if it is definite: For given $\boldsymbol{u}$, if $g(\boldsymbol{u}, \boldsymbol{v}) = 0\ \forall \boldsymbol{v}$ or $g(\boldsymbol{v}, \boldsymbol{u}) = 0\ \forall \boldsymbol{v}$, then $\boldsymbol{u} = \mathbf{0}$. A bilinear product need not be symmetric.

The inverse of $\boldsymbol{g}$ is an isomorphism of $\mathscr{V}^*$ with $\mathscr{V}$:

$$\boldsymbol{g}^{-1} : \mathscr{V}^* \to \mathscr{V}. \tag{2.3}$$

It corresponds to a unique bilinear function, the *dual* g^* of g on $\mathscr{V}^* \times \mathscr{V}^*$, namely

$$g^* : \mathscr{V}^* \times \mathscr{V}^* \to \mathscr{R}. \tag{2.4}$$

The operations $\boldsymbol{g}$ and $\boldsymbol{g}^{-1}$ in (2.1) and (2.3) are called the operations of *lowering* and *raising of indices* associated with the bilinear product, since they give rise to the component relations

$$v_i = g_{ij} v^j \tag{2.5}$$

and

$$v^j = g^{ij} v_j, \tag{2.6}$$

where v^i and v_i are the ith components of $\boldsymbol{v}$ and $\boldsymbol{gv}$ relative to some basis $\{\boldsymbol{e}_i\}$ and its dual basis $\{\boldsymbol{e}^i\}$, respectively. The covector $\boldsymbol{gv}$ is said to be *associated* with $\boldsymbol{v}$ by $\boldsymbol{g}$, and when some particular $\boldsymbol{g}$ is chosen and held fixed, we generally regard $\boldsymbol{v}$ and $\boldsymbol{gv}$ as being the same object, since each determines the other uniquely, and since any property of one is readily expressed as a formally analogous property of the other. More than this, we can systematically replace all concepts and operations that in general involve the dual space $\mathscr{V}^*$ by counterparts in the original space $\mathscr{V}$: The contraction becomes a function of elements of $\mathscr{V} \times \mathscr{V}$, tensors over $\mathscr{V}^*$ are replaced by counterpart tensors over $\mathscr{V}$, *etc.*

For example, if $\boldsymbol{A}$ has the components $A_i{}^{kl}$, we can calculate several associated tensors from it:

$$A_{ij}{}^{l} = g_{jk} A_i{}^{kl}, \quad A^j{}_i{}^l{}_q = g^{ji} g_{qk} A_i{}^{kl}, \tag{2.7}$$

etc. Because of these operations, the positions of the superscripts relative to those of the subscripts is important in the notation.

A bilinear product is called an *inner product* if it is symmetric and positive-definite. A vector space equipped with a particular inner product is called an *inner-product space.* Of course, no matter what be the vector space, an inner product g may be defined for it. To see this, we choose a basis $\{\boldsymbol{e}_i\}$ and define the isomorphism $\boldsymbol{g}$ by the component form

$$g = \delta_{ij} e^i \otimes e^j = e^1 \otimes e^1 + \ldots + e^n \otimes e^n. \tag{2.8}$$

Then g is symmetric and positive-definite. In fact, every inner product on $\mathscr{V}$ can be obtained in this manner. Let g be a particular inner product on $\mathscr{V}$. Then a basis $\{e_i\}$ satisfying the condition (2.8) is called an *orthonormal basis* of the inner product space $\mathscr{V}$. We have the following results:

PROPOSITION. *Given any basis $\{f_i\}$ in $\mathscr{V}$, there exists an orthonormal basis $\{e_i\}$ such that*

$$\mathrm{Sp}\,(f_1, \ldots, f_k) = \mathrm{Sp}\,(e_1, \ldots, e_k), \quad k = 1, 2, \ldots, n. \tag{2.9}$$

Here, as in § 1, Sp ($\mathscr{A}$) is the spread of $\mathscr{A}$.

PROPOSITION. *A basis $\{e_i\}$ is orthonormal if and only if it is the same as its dual basis $\{e^i\}$ under the operation g:*

$$e^i = g e_i, \quad i = 1, \ldots, n. \tag{2.10}$$

Since the components of g relative to an orthonormal basis $\{e_i\}$ are δ_{ij}, for components relative to an orthonormal basis there is no need to distinguish between subscripts and superscripts. For example,

$$\mathsf{A}_{ij}{}^{k} = \delta_{jl}\mathsf{A}_i{}^{lk} = \mathsf{A}_i{}^{jk}. \tag{2.11}$$

When the inner product defined by a particular g is laid down once and for all, we do not need to recall g in the notation, so we use the classical symbol of a dot:

$$u \cdot v \equiv g(u, v) = \langle u, gv \rangle. \tag{2.12}$$

Relative to some basis $\{e_i\}$, we then have

$$u \cdot v = g_{ij} u^i v^j = u_i v^i = u^i v_i. \tag{2.13}$$

In particular, if $\{e_i\}$ is orthonormal, then

$$u \cdot v = u^1 v^1 + \ldots + u^n v^n. \tag{2.14}$$

A linear transformation Q of $\mathscr{V}$ itself is called *orthogonal* if it preserves the inner product, *i.e.*,

$$u \cdot v = Qu \cdot Qv \qquad \forall u, v \in \mathscr{V}. \tag{2.15}$$

Relative to an orthonormal basis this condition means that

$$\delta_{ij} = \boldsymbol{e}_i \cdot \boldsymbol{e}_j = \boldsymbol{Q}\boldsymbol{e}_i \cdot \boldsymbol{Q}\boldsymbol{e}_j = \sum_{k=1}^{n} Q_{ki} Q_{kj}. \tag{2.16}$$

Thus $\boldsymbol{Q}$ is orthogonal if and only if the matrix of its components relative to an orthonormal basis is what is called in matrix algebra an orthogonal matrix.

The set of all orthogonal transformations of $\mathscr{V}$ forms a group $\mathfrak{O}(\mathscr{V})$ called the *orthogonal group*, a subgroup of the general linear group $\mathfrak{GL}(\mathscr{V})$. From (2.16), the determinant of an orthogonal transformation must equal $+1$ or -1. Thus $\mathfrak{O}(\mathscr{V})$ is in fact a subgroup of the unimodular group $\mathfrak{U}(\mathscr{V})$. An orthogonal transformation is called *proper* if its determinant is positive. The proper orthogonal transformations form a subgroup $\mathfrak{SO}(\mathscr{V})$ of $\mathfrak{SL}(\mathscr{V})$.

As explained in Example 4 of the previous section, for any linear transformation $\boldsymbol{L}$ on $\mathscr{V}$ there is an adjoint transformation $\boldsymbol{L}^A$ on $\mathscr{V}^*$. When an inner product g has been laid down, we can regard $\boldsymbol{L}^A$ equivalently as a transformation on $\mathscr{V}$. Indeed, we can define a linear transformation $\boldsymbol{L}^T$ on $\mathscr{V}$ by

$$\boldsymbol{L}^T = \boldsymbol{g}^{-1} \circ \boldsymbol{L}^A \circ \boldsymbol{g}, \tag{2.17}$$

as indicated in the following diagram:

$$\begin{array}{ccc} \mathscr{V}^* & \xrightarrow{\ \boldsymbol{L}^A\ } & \mathscr{V}^* \\ {\scriptstyle \boldsymbol{g}^{-1}}\big\downarrow & & \big\downarrow{\scriptstyle \boldsymbol{g}^{-1}} \\ \mathscr{V} & \xrightarrow[\ \boldsymbol{L}^T\ ]{} & \mathscr{V} \end{array} \tag{2.18}$$

In component form,

$$(L^T)^i{}_j = g^{il}(L^A)_l^{\ k}\, g_{kj}, \tag{2.19}$$

but since $\boldsymbol{L}^A$ is the adjoint of $\boldsymbol{L}$, we have also

$$(L^A)_i^{\ j} = L^j{}_i. \tag{2.20}$$

Consequently, the components of $\boldsymbol{L}$ and $\boldsymbol{L}^T$ are related by

$$(L^T)^i{}_j = g^{il} L^k{}_l\, g_{kj}. \tag{2.21}$$

In particular, if the basis is orthonormal, then the component matrix of $\boldsymbol{L}^T$ is the same as that of $\boldsymbol{L}^A$ and hence is the transpose of that of $\boldsymbol{L}$. Further, the bilinear function corresponding to $\boldsymbol{L}^T$ is the transpose, as defined by (1.98), of that corresponding to $\boldsymbol{L}$.

As this last remark suggests, we can characterize the transpose of $\boldsymbol{L}$ also by the following condition:

$$\boldsymbol{u} \cdot \boldsymbol{L}\boldsymbol{v} = \boldsymbol{v} \cdot \boldsymbol{L}^T \boldsymbol{u} \quad \forall \boldsymbol{u}, \boldsymbol{v} \in \mathscr{V}. \tag{2.22}$$

Comparing this condition (2.22) with (2.15), we see that the tensor $\boldsymbol{Q}$ is orthogonal if and only if

$$\boldsymbol{Q}\boldsymbol{Q}^T = \boldsymbol{Q}^T\boldsymbol{Q} = \boldsymbol{1}, \tag{2.23}$$

or equivalently,

$$\boldsymbol{Q}^T = \boldsymbol{Q}^{-1}. \tag{2.24}$$

The reader will verify easily the following properties of the transpose:

$$(\boldsymbol{L}^T)^T = \boldsymbol{L}, \ (\boldsymbol{L}\boldsymbol{N})^T = \boldsymbol{N}^T\boldsymbol{L}^T, \tag{2.25}$$

$$(\ker \boldsymbol{L}^T) \cdot (\operatorname{im} \boldsymbol{L}) = 0, \tag{2.26}$$

$$\dim \ker \boldsymbol{L}^T = \dim \ker \boldsymbol{L}, \tag{2.27}$$

$$\operatorname{tr} \boldsymbol{L}^T = \operatorname{tr} \boldsymbol{L}, \tag{2.28}$$

and

$$\det \boldsymbol{L}^T = \det \boldsymbol{L}. \tag{2.29}$$

As the foregoing definitions show, in an inner-product space we may obviate the adjoint altogether, and the dual space need never be used or even mentioned. The inner product g on $\mathscr{V}$ gives rise not only to an inner product g^* on $\mathscr{V}^*$ but also to an inner product on each tensor space over $\mathscr{V}$. For example, we can define an inner product on $\mathscr{T}^1_2$ by the condition

$$(\boldsymbol{v} \otimes \boldsymbol{u} \otimes \boldsymbol{w}) \cdot (\boldsymbol{a} \otimes \boldsymbol{b} \otimes \boldsymbol{c}) \equiv (\boldsymbol{v} \cdot \boldsymbol{a})(\boldsymbol{u} \cdot \boldsymbol{b})(\boldsymbol{w} \cdot \boldsymbol{c}) \tag{2.30}$$

for all simple tensors $\boldsymbol{v} \otimes \boldsymbol{u} \otimes \boldsymbol{w}$ and $\boldsymbol{a} \otimes \boldsymbol{b} \otimes \boldsymbol{c}$ in $\mathscr{T}^1_2$. For the space of linear transformations on $\mathscr{V}$, the inner product is given by the formula

$$\boldsymbol{L} \cdot \boldsymbol{N} = \operatorname{tr}(\boldsymbol{L}\boldsymbol{N}^T), \tag{2.31}$$

which is consistent with the condition that $(\boldsymbol{u} \otimes \boldsymbol{w}) \cdot (\boldsymbol{v} \otimes \boldsymbol{z}) = (\boldsymbol{u} \cdot \boldsymbol{v})(\boldsymbol{w} \cdot \boldsymbol{z})$ for all simple tensors $\boldsymbol{u} \otimes \boldsymbol{w}$ and $\boldsymbol{v} \otimes \boldsymbol{z}$ in $\mathscr{T}^1_1(\cong \mathscr{L}(\mathscr{V}; \mathscr{V}))$.

Symmetric and skew elements of $\mathscr{L}(\mathscr{V}; \mathscr{V}^*)$ were defined following (1.99). Similarly, in an inner-product space a linear transformation in $\mathscr{L}(\mathscr{V}; \mathscr{V})$ is called *symmetric* if it is equal to its transpose, and *skew* if it is equal to the negative of its transpose. The set $\mathscr{L}_+(\mathscr{V}; \mathscr{V})$ of all symmetric transformations and the set $\mathscr{L}_-(\mathscr{V}; \mathscr{V})$ of all skew transformations are subspaces of $\mathscr{L}(\mathscr{V};\mathscr{V})$. Further, $\mathscr{L}_+(\mathscr{V};\mathscr{V})$ is the orthogonal complement of $\mathscr{L}_-(\mathscr{V};\mathscr{V})$ in $\mathscr{L}(\mathscr{V};\mathscr{V})$:

$$\mathscr{L}(\mathscr{V};\mathscr{V}) = \mathscr{L}_+(\mathscr{V};\mathscr{V}) \oplus \mathscr{L}_-(\mathscr{V};\mathscr{V}), \tag{2.32}$$

and

$$\mathscr{L}_+(\mathscr{V};\mathscr{V}) \cdot \mathscr{L}_-(\mathscr{V};\mathscr{V}) = 0. \tag{2.33}$$

A more compact notation for these two conditions is

$$\mathscr{L}(\mathscr{V};\mathscr{V}) = \mathscr{L}_+(\mathscr{V};\mathscr{V}) \oplus\!\!\!\!\perp \mathscr{L}_-(\mathscr{V};\mathscr{V}), \tag{2.34}$$

where the symbol $\oplus\!\!\!\!\perp$ is called the *orthogonal sum.*

PROPOSITION. *Let $\boldsymbol{A}$ be a symmetric transformation of $\mathscr{V}$. Then $\boldsymbol{A}$ is a dilatation and possesses an orthonormal principal basis $\{\boldsymbol{e}_i\}$. In particular, the characteristic subspaces of $\boldsymbol{A}$ are mutually orthogonal, and their orthogonal sum is $\mathscr{V}$:*

$$\mathscr{V} = \mathscr{V}_a(\boldsymbol{A}) \oplus\!\!\!\!\perp \mathscr{V}_b(\boldsymbol{A}) \oplus\!\!\!\!\perp \ldots \oplus\!\!\!\!\perp \mathscr{V}_c(\boldsymbol{A}), \tag{2.35}$$

where $\{a, b, \ldots, c\}$ is the set of distinct proper numbers of $\boldsymbol{A}$.

This proposition implies that every dilatation on $\mathscr{V}$ is similar to some symmetric transformation, since every set of n real numbers, distinct or not, can be regarded as the set of proper numbers (counting each proper number as many times as its multiplicity) of some symmetric transformation of $\mathscr{V}$.

PROPOSITION. *An orthogonal transformation $\boldsymbol{Q}$ commutes with a symmetric transformation $\boldsymbol{A}$ if and only if $\boldsymbol{Q}$ preserves all characteristic subspaces of $\boldsymbol{A}$.*

In general, a linear transformation $\boldsymbol{L}$ is called *positive-definite* if its corresponding bilinear function is positive-definite:

$$\boldsymbol{u} \cdot \boldsymbol{L}\boldsymbol{u} > 0 \quad \forall \boldsymbol{u} \neq \boldsymbol{0}. \tag{2.36}$$

Relative to the decomposition (2.34), the *symmetric part* $\boldsymbol{L}_+$ and the *skew art* $\boldsymbol{L}_-$ of $\boldsymbol{L}$ are given by

$$\begin{aligned} \boldsymbol{L}_+ &= \tfrac{1}{2}(\boldsymbol{L}+\boldsymbol{L}^T), \\ \boldsymbol{L}_- &= \tfrac{1}{2}(\boldsymbol{L}-\boldsymbol{L}^T). \end{aligned} \tag{2.37}$$

so $\boldsymbol{L} = \boldsymbol{L}_+ + \boldsymbol{L}_-$. Clearly the quadratic form of $\boldsymbol{L}$ is the same as that of $\boldsymbol{L}_+$, while the quadratic form of $\boldsymbol{L}_-$ vanishes identically. Consequently $\boldsymbol{L}$ is positive-definite if and only if $\boldsymbol{L}_+$ is.

Proposition. *A symmetric transformation is positive-definite if and only if its proper numbers are all positive. Any positive-definite and symmetric transformation* $\boldsymbol{A}$ *has a unique positive-definite and symmetric square root* $\boldsymbol{A}^{\frac{1}{2}}$ *such that*

$$(\boldsymbol{A}^{\frac{1}{2}})^2 = \boldsymbol{A}. \tag{2.38}$$

Further, the proper numbers of $\boldsymbol{A}^{\frac{1}{2}}$ *are the positive square roots of those of* $\boldsymbol{A}$.

For any non-singular transformation $\boldsymbol{F}$, the transformations $\boldsymbol{B}$ and $\boldsymbol{C}$ defined by

$$\boldsymbol{B} = \boldsymbol{F}\boldsymbol{F}^T,\ \boldsymbol{C} = \boldsymbol{F}^T\boldsymbol{F}, \tag{2.39}$$

are positive-definite and symmetric. We call their square roots, namely

$$\boldsymbol{V} \equiv \boldsymbol{B}^{\frac{1}{2}},\ \boldsymbol{U} \equiv \boldsymbol{C}^{\frac{1}{2}}, \tag{2.40}$$

the *left* and the *right stretch tensors* of $\boldsymbol{F}$. We have also the following important result, due in principle to Cauchy:

Proposition. *There exist unique positive-definite and symmetric transformations* $\boldsymbol{V}$ *and* $\boldsymbol{U}$ *and a unique orthogonal transformation* $\boldsymbol{R}$ *such that*

$$\boldsymbol{F} = \boldsymbol{V}\boldsymbol{R} = \boldsymbol{R}\boldsymbol{U}. \tag{2.41}$$

Further, $\boldsymbol{U}$ *and* $\boldsymbol{V}$ *are given by* (2.40).

The factorizations of $\boldsymbol{F}$ given by (2.41) are called the *polar decompositions* of $\boldsymbol{F}$. From (2.41), we see that

$$\boldsymbol{V} = \boldsymbol{R}\boldsymbol{U}\boldsymbol{R}^T, \tag{2.42}$$

and hence

$$\boldsymbol{B} = \boldsymbol{R}\boldsymbol{C}\boldsymbol{R}^T, \tag{2.43}$$

so that the principal invariants of $\boldsymbol{V}$ and $\boldsymbol{B}$ are the same as those of $\boldsymbol{U}$ and $\boldsymbol{C}$, respectively.

3. Euclidean Spaces

An inner product induces in a natural way the concepts of a *norm* and a *distance function* on a vector space. Let $\mathscr{V}$ be an inner-product space, and let $\boldsymbol{u}, \boldsymbol{v}, \boldsymbol{w}, \ldots$ be vectors in $\mathscr{V}$. Then we define the *norm* $\|\boldsymbol{u}\|$ of $\boldsymbol{u}$ by

$$\|\boldsymbol{u}\| \equiv \sqrt{\boldsymbol{u} \cdot \boldsymbol{u}}, \tag{3.1}$$

and the *distance* $d(\boldsymbol{u}, \boldsymbol{v})$ from $\boldsymbol{u}$ to $\boldsymbol{v}$ by

$$d(\boldsymbol{u}, \boldsymbol{v}) \equiv \|\boldsymbol{u}-\boldsymbol{v}\|. \tag{3.2}$$

Clearly, the norm and the distance function defined in this way satisfy the axioms standard for such functions, namely,

$$\|\boldsymbol{u}\| > 0 \text{ unless } \boldsymbol{u} = \boldsymbol{0}, \tag{3.3}$$

$$\|a\boldsymbol{u}\| = |a|\|\boldsymbol{u}\| \quad \forall a \in \mathscr{R},\ \forall \boldsymbol{u} \in \mathscr{V}, \tag{3.4}$$

$$\|\boldsymbol{u}+\boldsymbol{v}\| \leqq \|\boldsymbol{u}\|+\|\boldsymbol{v}\| \quad \forall \boldsymbol{u},\ \boldsymbol{v} \in \mathscr{V}, \tag{3.5}$$

$$d(\boldsymbol{u}, \boldsymbol{v}) > 0 \text{ unless } \boldsymbol{u} = \boldsymbol{v}, \tag{3.6}$$

$$d(\boldsymbol{u}, \boldsymbol{v}) = d(\boldsymbol{v}, \boldsymbol{u}) \quad \forall \boldsymbol{u}, \boldsymbol{v} \in \mathscr{V}, \tag{3.7}$$

and

$$d(\boldsymbol{u}, \boldsymbol{v})+d(\boldsymbol{v}, \boldsymbol{w}) \geqq d(\boldsymbol{u}, \boldsymbol{w}) \quad \forall \boldsymbol{u}, \boldsymbol{v}, \boldsymbol{w} \in \mathscr{V}. \tag{3.8}$$

The inequalities (3.5) and (3.8) are satisfied by the norm and the distance function defined by (3.1) and (3.2) because the inner product obeys the *Cauchy inequality*:

$$|\boldsymbol{u} \cdot \boldsymbol{v}| \leqq \|\boldsymbol{u}\|\|\boldsymbol{v}\| \quad \forall \boldsymbol{u},\ \boldsymbol{v} \in \mathscr{V}. \tag{3.9}$$

An inner product induces also the concept of the *angle* between two vectors. Let $\boldsymbol{u}$ and $\boldsymbol{v}$ be non-zero vectors. Then we define the angle $\theta(\boldsymbol{u}, \boldsymbol{v})$ between them by the formula

$$\cos \theta(\boldsymbol{u}, \boldsymbol{v}) \equiv \frac{\boldsymbol{u} \cdot \boldsymbol{v}}{\|\boldsymbol{u}\|\|\boldsymbol{v}\|}, \quad 0 \leqq \theta \leqq \pi. \tag{3.10}$$

From the Cauchy inequality (3.9) and from the properties of the function $\cos x$, we see that $\theta(\boldsymbol{u}, \boldsymbol{v})$ is well defined. As before, we say that $\boldsymbol{u}$ and $\boldsymbol{v}$ are *orthogonal* if

$$\boldsymbol{u} \cdot \boldsymbol{v} = 0. \tag{3.11}$$

Thus the angle between two orthogonal vectors is $\frac{1}{2}\pi$. Also, $\boldsymbol{u}$ is called a *unit* vector or a *normalized* vector if

$$\|\boldsymbol{u}\| = 1. \tag{3.12}$$

In terms of these conditions, an orthonormal basis introduced in the preceding section (*cf.* $(2.16)_1$) is a basis consisting of mutually orthogonal normalized vectors.

A distance function gives rise to the concept of a *topology* in a natural way. If $\boldsymbol{v} \in \mathscr{V}$, we define a *neighborhood* of $\boldsymbol{v}$ as being any subset $\mathscr{N}$ of $\mathscr{V}$ that contains an open ball $\mathscr{B}$ centered at $\boldsymbol{v}$, say

$$\mathscr{B}(\boldsymbol{v}, R) \equiv \{\boldsymbol{u} : d(\boldsymbol{u}, \boldsymbol{v}) < R\}, \tag{3.13}$$

for some positive radius R. A subset $\mathscr{D}$ of $\mathscr{V}$ is called an *open set* if it is a neighborhood of each of its points. For example, the open ball $\mathscr{B}(\boldsymbol{v}, R)$ is an open set. We call the collection $\mathfrak{J}$ consisting of all open sets of $\mathscr{V}$ the *topology* of $\mathscr{V}$ induced by the distance function. Clearly, $\mathfrak{J}$ satisfies the standard conditions of a topology, namely,

$$\mathscr{V} \in \mathfrak{J}, \quad \emptyset \in \mathfrak{J}, \tag{3.14}$$

$$\mathscr{D}_1, \ldots, \mathscr{D}_\Gamma \in \mathfrak{J} \Rightarrow \bigcap_{\Lambda=1}^{\Gamma} \mathscr{D}_\Lambda \in \mathfrak{J}, \tag{3.15}$$

and

$$\{\mathscr{D}_\Lambda, \Lambda \in I\} \subset \mathfrak{J} \Rightarrow \bigcup_{\Lambda \in I} \mathscr{D}_\Lambda \in \mathfrak{J}, \tag{3.16}$$

where I is an arbitrary (finite or infinite) index set, and where $\emptyset$ denotes the empty set.

Among the three structures induced by an inner product – the norm, the distance function, and the topology – the first two clearly depend on the choice of the inner product. The topology $\mathfrak{J}$, however, turns out to be independent of that choice. This fact is a direct consequence of the following

PROPOSITION. *Let d_1 and d_2 be the distance functions induced by the inner products g_1 and g_2, respectively, on $\mathscr{V}$. Then there exist positive constants A and B, determined by the inner products g_1 and g_2, such that*

$$A d_1(\boldsymbol{u}, \boldsymbol{v}) \leqq d_2(\boldsymbol{u}, \boldsymbol{v}) \leqq B d_1(\boldsymbol{u}, \boldsymbol{v}) \tag{3.17}$$

for all $\boldsymbol{u}, \boldsymbol{v} \in \mathscr{V}$. In particular, $\mathscr{D}$ is an open set with respect to d_1 if and only if it is an open set with respect to d_2.

We call the topology $\mathfrak{J}$ induced by any inner product on $\mathscr{V}$ the *Euclidean topology* of $\mathscr{V}$.

As usual, we can define the concepts of *convergence* and *limit* with respect to a topology. A sequence $\boldsymbol{v}_k$, $k = 1, 2, \ldots$, is called *convergent* if there exists a *limit* $\boldsymbol{v} \in \mathscr{V}$ such that every neighborhood of $\boldsymbol{v}$ contains $\boldsymbol{v}_k$ for all but a finite number of indices $k = 1, 2, \ldots$. We denote the convergence of $\boldsymbol{v}_k$ to $\boldsymbol{v}$ by the standard notations

$$\boldsymbol{v} = \lim_{k \to \infty} \boldsymbol{v}_k, \tag{3.18}$$

or

$$\boldsymbol{v}_k \to \boldsymbol{v} \text{ as } k \to \infty. \tag{3.19}$$

In terms of some particular inner product g on $\mathscr{V}$, a condition of convergence is

$$d(\boldsymbol{v}_k, \boldsymbol{v}) \to 0 \text{ as } k \to \infty. \tag{3.20}$$

While $d(\boldsymbol{v}_k, \boldsymbol{v})$ depends on the choice of g, the condition (3.20) of convergence is independent of g in the sense that if it holds for one g, it holds for all, as follows from (3.17).

A subset $\mathscr{C} \subset \mathscr{V}$ is called a *closed set* if it contains the limits of all convergent sequences in $\mathscr{C}$. The reader will verify easily that a subset $\mathscr{C}$ is closed if and only if its complement $\mathscr{C}^*$ (with $\mathscr{C} \cup \mathscr{C}^* = \mathscr{V}$) is an open set. Thus the collection $\mathfrak{J}^*$ consisting of all closed subsets of $\mathscr{V}$ satisfies the following three conditions:

$$\mathscr{V} \in \mathfrak{J}^*, \quad \emptyset \in \mathfrak{J}^*, \tag{3.21}$$

$$\mathscr{C}_1, \ldots, \mathscr{C}_\Gamma \in \mathfrak{J}^* \Rightarrow \bigcup_{\Lambda=1}^{\Gamma} \mathscr{C}_\Lambda \in \mathfrak{J}^*, \tag{3.22}$$

and

$$\{\mathscr{C}_\Lambda, \Lambda \in I\} \subset \mathfrak{J}^* \Rightarrow \bigcap_{\Lambda \in I} \mathscr{C}_\Lambda \in \mathfrak{J}^*, \tag{3.23}$$

where I is again an arbitrary, finite or infinite index set.

As usual, a set equipped with a topology is called a *topological space*. A mapping

$$\boldsymbol{f}: \mathscr{X} \to \mathscr{Y} \tag{3.24}$$

from a topological space $\mathscr{X}$ to a topological space $\mathscr{Y}$ is called *continuous* if for every open set $\mathscr{N} \subset \mathscr{Y}$ the pre-image

$$\boldsymbol{f}^{-1}(\mathscr{N}) \equiv \{\boldsymbol{x} : \boldsymbol{f}(\boldsymbol{x}) \in \mathscr{N}, \boldsymbol{x} \in \mathscr{X}\} \tag{3.25}$$

is an open set in $\mathscr{X}$. A continuous map is called a *homeomorphism* if it is invertible and if its inverse map is also continuous. Hence from (3.25), a homeomorphism and its inverse map carry open sets onto open sets. In this sense, a homeomorphism is a mapping that preserves the topology. Consequently, we call a property of a topological space a *topological property* if it is preserved by any homeomorphism of the space, and we call two topological spaces *homeomorphic* if there exists a homeomorphism mapping one onto the other.

The Euclidean topological space $\mathscr{V}$ has many important topological properties. Here we mention only two:

1. $\mathscr{V}$ is a *Hausdorff space*. That is, any two distinct points $\boldsymbol{u}, \boldsymbol{v} \in \mathscr{V}$ can be separated by open sets: $\exists \mathscr{A} \ni \boldsymbol{u}, \mathscr{B} \ni \boldsymbol{v}$, such that $\mathscr{A} \cap \mathscr{B} = \emptyset$.

2. $\mathscr{V}$ has a *countable base*. That is, there exists a countable collection of open sets in $\mathscr{V}$ such that every open set is the union of elements of that collection.

Within the class of Euclidean topological spaces, the dimension is also a topological property. More specifically, we have the following

PROPOSITION. *A linear map is a continuous map, and a linear isomorphism is a homeomorphism. Further, two Euclidean topological spaces are homeomorphic if and only if they have the same dimension; if they are homeomorphic, they are in fact linearly isomorphic.*

So as to indicate that the Euclidean topology is now defined on the vector space, henceforth we shall denote an n-dimensional Euclidean topological space by $\mathscr{E}^n$.

So far, we have considered the topological structure of a vector space. Next we consider its differentiable structure. Our starting point is the concept of a *co-ordinate system*. Let $\mathscr{D}$ be an open set in $\mathscr{E}^n$. Then a *co-ordinate system* on $\mathscr{D}$ is a homeomorphism from $\mathscr{D}$ into $\mathscr{R}^n$, say

$$\boldsymbol{\kappa} : \mathscr{D} \to \mathscr{R}^n. \tag{3.26}$$

For any $\boldsymbol{v} \in \mathscr{D}$, the list of real numbers $(v^1, \ldots, v^n)$ given by

$$\boldsymbol{\kappa}(\boldsymbol{v}) = (v^1, \ldots, v^n) \in \mathscr{R}^n \tag{3.27}$$

are then called the *co-ordinates* of $\boldsymbol{v}$ relative to $\boldsymbol{\kappa}$. We can regard the co-ordinates as the values of the *co-ordinate functions* v^i at the point $\boldsymbol{v}$. Since $\boldsymbol{\kappa}$ is a homeomorphism, the co-ordinate functions are necessarily continuous on $\mathscr{D}$. We call $\mathscr{D}$ the *co-ordinate neighborhood* of $\boldsymbol{\kappa}$.

A special co-ordinate system on the whole space $\mathscr{E}^n$ is provided by the components of the vectors relative to a basis $\{\boldsymbol{e}_i\}$. As usual, we can express any vector $\boldsymbol{v} \in \mathscr{E}^n$ in component form relative to $\{\boldsymbol{e}_i\}$, *viz*,

$$\boldsymbol{v} = v^i \boldsymbol{e}_i. \tag{3.28}$$

Then the mapping

$$\boldsymbol{\varphi} : \mathscr{E}^n \to \mathscr{R}^n \tag{3.29}$$

defined by

$$\boldsymbol{\varphi}(\boldsymbol{v}) = (v^1, \ldots, v^n) \tag{3.30}$$

is a linear homeomorphism of $\mathscr{E}^n$ with $\mathscr{R}^n$, and hence it is a co-ordinate system on $\mathscr{E}^n$. We call $\boldsymbol{\varphi}$ the *Cartesian co-ordinate system* induced by the basis $\{\boldsymbol{e}_i\}$.

In the terminology of differentiable manifolds, a pair $(\mathscr{D}, \boldsymbol{\kappa})$ consisting in an open set $\mathscr{D} \subset \mathscr{E}^n$ and a co-ordinate system $\boldsymbol{\kappa}$ on $\mathscr{D}$ is called a (*co-ordinate*) *chart*. Two charts $(\mathscr{D}, \boldsymbol{\kappa})$ and $(\overline{\mathscr{D}}, \overline{\boldsymbol{\kappa}})$ of $\mathscr{E}^n$ are C^∞*-related* if the co-ordinate transformation

$$\overline{\boldsymbol{\kappa}} \circ \boldsymbol{\kappa}^{-1} : \boldsymbol{\kappa}(\mathscr{D} \cap \overline{\mathscr{D}}) \to \overline{\boldsymbol{\kappa}}(\mathscr{D} \cap \overline{\mathscr{D}}) \tag{3.31}$$

on the overlap $\mathscr{D} \cap \overline{\mathscr{D}}$ of their co-ordinate neighborhoods is given by C^∞ functions

$$\bar{v}^i = \bar{v}^i(v^1, \ldots, v^n), \quad i = 1, \ldots, n, \tag{3.32}$$

where (v^i) and $(\bar{v}^i)$ denote the co-ordinates of a typical point $\boldsymbol{v} \in \mathscr{D} \cap \overline{\mathscr{D}}$ relative to $\boldsymbol{\kappa}$ and $\overline{\boldsymbol{\kappa}}$, respectively. From the transformation rule (1.31), we see that any two Cartesian charts are C^∞-related.

If a collection of pairwise C^∞-related co-ordinate charts cannot be enlarged to a bigger collection satisfying the same smoothness condition, then it is said to be *maximal* or *complete*. Such a collection is called an *atlas*. The atlas $\mathfrak{D}$ that contains all Cartesian charts is called the *Euclidean atlas* or the *Euclidean differentiable structure*. Unless the contrary is stated, henceforth we use only charts that belong to $\mathfrak{D}$.

In a later section we shall define precisely the concept of a *differentiable manifold*, which is, roughly speaking, a topological space equipped with an atlas. A vector space equipped with the Euclidean topology and the Euclidean atlas, of course, turns out to be a particular differentiable manifold, and for definiteness it is called a *Euclidean space*. An open set $\mathscr{O}$ of a Euclidean space clearly has a natural topology and a natural atlas consisting of open sets in $\mathscr{O}$ and charts with domain in $\mathscr{O}$, respectively. We call such an open set an *open submanifold* of the Euclidean space.

We now give definitions of some basic concepts induced by the differentiable structure of an arbitrary open submanifold of a Euclidean space. All these definitions are easily extended to a general differentiable manifold, as we shall see in the next section.

First we define the concepts of a *differentiable map* and a *diffeomorphism*.

Let $\mathscr{A}$ and $\mathscr{B}$ be open submanifolds of Euclidean spaces, let $\boldsymbol{f}$ be a mapping from $\mathscr{A}$ to $\mathscr{B}$,

$$\boldsymbol{f} : \mathscr{A} \rightarrow \mathscr{B}, \tag{3.33}$$

and let $\boldsymbol{f}$ be expressed by the co-ordinate relations

$$x^i = f^i(y^1, \ldots, y^m), \; i = 1, \ldots, n, \tag{3.34}$$

relative to some charts in $\mathscr{A}$ and $\mathscr{B}$ with co-ordinates $(x^1, \ldots, x^m)$ and $(y^1, \ldots, y^n)$, respectively, where m and n are the dimensions of the two spaces. If the functions f^i are C^∞, the mapping $\boldsymbol{f}$ is called a *differentiable map*. Since the co-ordinate transformation from one chart to another is of class C^∞, this definition does not depend upon the choice of charts.

PROPOSITION. *The mapping* $\boldsymbol{f}$ *in* (3.33) *is continuous if and only if the functions* f^i, $i = 1, \ldots, n$, *in* (3.34) *are continuous. In particular, a differentiable map is necessarily a continuous map.*

A differentiable map is called a *diffeomorphism* if it is invertible and if its inverse map is also differentiable. From the preceding proposition, we see that a diffeomorphism is necessarily a homeomorphism. Like a homeomorphism, a diffeomorphism preserves the differentiable structure in the sense that it and its inverse carry a chart into a chart. A property preserved by all diffeomorphisms is called a *differentiable property*. We now consider some differentiable properties of a Euclidean space.

First, we define the concept of the *differential*[1] of a differentiable function. Let $\boldsymbol{v}$ be a given point in $\mathscr{E}^n$. Then a function f is said to be *defined near* $\boldsymbol{v}$ if the domain of f is a neighborhood of $\boldsymbol{v}$. Two functions defined near $\boldsymbol{v}$ are regarded as equivalent at $\boldsymbol{v}$ if they are identical on a neighborhood of $\boldsymbol{v}$. We denote the collection of equivalence classes of differentiable functions defined near $\boldsymbol{v}$ as $\mathfrak{F}(\boldsymbol{v})$, which forms an infinite-dimensional vector space in an obvious way. Let f be an element of $\mathfrak{F}(\boldsymbol{v})$. Then f is said to have *differential zero* at $\boldsymbol{v}$ if the first partial derivatives of its representation in some co-ordinate system defined near $\boldsymbol{v}$ are zero

---

[1] Readers whose background is in classical analysis and its applications should note that the differential corresponds to what is commonly called the gradient. (See the remark following (3.66), below.) When $n = 1$, it reduces to the derivative df/dx, not the classical differential df. The classical differential has outlived its usefulness, and the new usage of the word is closer to the spirit of the language in that a differentiable function is now one which has a (unique) differential, just as an integrable function is one which has a (unique) integral.

at $\boldsymbol{v}$. By the chain rule, this condition does not depend on the choice of the co-ordinate system. Further, the collection of all elements of $\mathfrak{F}(\boldsymbol{v})$ having differential zero at $\boldsymbol{v}$ forms an infinite-dimensional subspace $\mathfrak{Z}(\boldsymbol{v})$ in $\mathfrak{F}(\boldsymbol{v})$. We call the projection map from $\mathfrak{F}(\boldsymbol{v})$ to the quotient space $\mathfrak{F}(\boldsymbol{v})/\mathfrak{Z}(\boldsymbol{v})$ the *differential* at $\boldsymbol{v}$, *viz*[1]

$$\partial_{\boldsymbol{v}} : \mathfrak{F}(\boldsymbol{v}) \to \mathfrak{F}(\boldsymbol{v})/\mathfrak{Z}(\boldsymbol{v}). \tag{3.35}$$

Since a linear function is a differentiable function, we can regard the dual space $\mathscr{V}^*$ as a subspace of $\mathfrak{F}(\boldsymbol{v})$, and the restriction of the map $\partial_{\boldsymbol{v}}$ to $\mathscr{V}^*$ is a linear map. In fact, it is an isomorphism; *the quotient space* $\mathfrak{F}(\boldsymbol{v})/\mathfrak{Z}(\boldsymbol{v})$ *is canonically isomorphic to* $\mathscr{E}^{n*}$. As usual, we suppress the notation for the canonical isomorphism of $\mathscr{E}^{n*}$ with $\mathfrak{F}(\boldsymbol{v})/\mathfrak{Z}(\boldsymbol{v})$ and regard the differential $\partial_{\boldsymbol{v}}$ as a mapping into $\mathscr{E}^{n*}$:

$$\partial_{\boldsymbol{v}} : \mathfrak{F}(\boldsymbol{v}) \to \mathscr{E}^{n*}. \tag{3.36}$$

For any $f \in \mathfrak{F}(\boldsymbol{v})$, we call the covector $\partial_{\boldsymbol{v}} f$ the *differential of f at $\boldsymbol{v}$*. The differential of a linear function is that function itself.

PROPOSITION. *The differential of f at $\boldsymbol{v}$ can be characterized by the condition*

$$\langle \partial_{\boldsymbol{v}} f, \boldsymbol{u} \rangle = \frac{d}{dt} f(\boldsymbol{v}+t\boldsymbol{u})\bigg|_{t=0} \qquad \forall \boldsymbol{u} \in \mathscr{E}^n. \tag{3.37}$$

This equation means that $\langle \partial_{\boldsymbol{v}} f, \boldsymbol{u} \rangle$ is the *directional derivative* of f in the direction of $\boldsymbol{u}$ at the point $\boldsymbol{v}$.

Now let (x^1) be a co-ordinate system defined near $\boldsymbol{v}$. Then the co-ordinate functions x^i are elements of $\mathfrak{F}(\boldsymbol{v})$. In particular, each x^i has a differential at $\boldsymbol{v}$.

PROPOSITION. *The set of differentials $\{\partial_{\boldsymbol{v}} x^i\}$ is a basis of $\mathscr{E}^{n*}$. Further, the component form of any differential $\partial_{\boldsymbol{v}} f$ relative to $\{\partial_{\boldsymbol{v}} x^i\}$ is*

$$\partial_{\boldsymbol{v}} f = \frac{\partial f}{\partial x^i}\bigg|_{\boldsymbol{v}} \partial_{\boldsymbol{v}} x^i, \tag{3.38}$$

where $\partial f/\partial x^i$ is the partial derivative of the representation of f in terms of (x^i).

[1] The common notation for the differential in the modern theory of manifolds is $d_{\boldsymbol{v}}$. We have chosen $\partial_{\boldsymbol{v}}$ as a compromise with the classical ones; *e.g.*, $\nabla_{\boldsymbol{x}} f(\boldsymbol{x})|_{\boldsymbol{v}}$ is one classical notation for what we here denote by $\partial_{\boldsymbol{v}} f$. *Cf.* (3.38), below.

We call $\{\partial_{\boldsymbol{v}} x^i\}$ the *natural cobasis* of the co-ordinate system (x^i) at $\boldsymbol{v}$. For a Cartesian co-ordinate system induced by a basis $\{\boldsymbol{e}_i\}$, the natural cobasis at any point is simply the dual basis $\{\boldsymbol{e}^j\}$ of $\{\boldsymbol{e}_i\}$. In this case, the natural cobasis is the same at all points. For a co-ordinate system in general, however, the natural cobasis at some point $\boldsymbol{v}$ need not be the same as that at some other point $\boldsymbol{u}$. Considering the natural cobases at all points, we then obtain a *field* of cobases on the co-ordinate neighborhood.

In general, let $\boldsymbol{w}$ be a covector field on an open submanifold $\mathscr{D}$:

$$\boldsymbol{w} : \mathscr{D} \to \mathscr{E}^{n*}. \tag{3.39}$$

Then we define the components of $\boldsymbol{w}$ relative to any co-ordinate system (x^i) on $\mathscr{D}$ as the fields whose values at any $\boldsymbol{v} \in \mathscr{D}$ are the components of $\boldsymbol{w}$ at $\boldsymbol{v}$ relative to the natural cobasis $\{\partial_{\boldsymbol{v}} x^j\}$. Thus we have the field equation

$$\boldsymbol{w} = w_i \partial x^i, \tag{3.40}$$

where the components w_i are scalar fields. Now suppose that $(\bar{x}^i)$ is another co-ordinate system on $\mathscr{D}$. Then by (3.38) the fields of natural cobases $\{\partial x^i\}$ and $\{\partial \bar{x}^j\}$ are related by

$$\partial \bar{x}^i = \frac{\partial \bar{x}^i}{\partial x^j} \partial x^j. \tag{3.41}$$

Hence the transformation rule yielding the components $\bar{w}_j$ of a covector field $\boldsymbol{w}$ relative to $(\bar{x}^i)$ is

$$w_i = \frac{\partial \bar{x}^j}{\partial x^i} \bar{w}_j, \tag{3.42}$$

familiar in classical vector analysis as the definition of a "covariant vector field".

So far, we have considered the differential of a differentiable function. The counterpart of a differentiable function is a differentiable curve, which is a differentiable map from $\mathscr{R}$, or an interval thereof, into $\mathscr{E}^n$, say

$$\boldsymbol{\lambda} : (a, b) \to \mathscr{E}^n. \tag{3.43}$$

We say that $\boldsymbol{\lambda}$ *passes through* a point $\boldsymbol{v} \in \mathscr{E}^n$ if $\boldsymbol{v}$ is a point in the image of the curve, say

$$\boldsymbol{\lambda}(c) = \boldsymbol{v} \tag{3.44}$$

for some $c \in (a, b)$.

The counterpart of the differential is the tangent. If $\boldsymbol{\lambda}$ is a differentiable curve passing through $\boldsymbol{v}$, we define the *tangent*[1] $\partial_v \boldsymbol{\lambda}$ of $\boldsymbol{\lambda}$ at $\boldsymbol{v}$ by the condition

$$\langle \partial_v \boldsymbol{\lambda}, \partial_v f \rangle = \frac{d}{dt} f(\boldsymbol{\lambda}(t)) \bigg|_{t=c} \quad \forall f \in \mathfrak{F}(\boldsymbol{v}). \tag{3.45}$$

The reader will verify easily that the right-hand side of (3.45) depends linearly on the differential of f. The condition (3.37) can now be interpreted as saying that the tangent of the straight line $\boldsymbol{v}+t\boldsymbol{u}$ at $t = 0$ is $\boldsymbol{u}$.

Let (x^i) be a co-ordinate system defined near $\boldsymbol{v}$. Then the co-ordinate curves are differentiable. We write $\partial_v x_i$ for the tangent[2] of the ith co-ordinate curve at $\boldsymbol{v}$.

PROPOSITION. *The set of tangents $\{\partial_v x_i\}$ is the dual basis of the natural cobasis $\{\partial_v x^i\}$. Further, the component form of any tangent $\partial_v \boldsymbol{\lambda}$ relative to $\{\partial_v x_i\}$ is*

$$\partial_v \boldsymbol{\lambda} = \frac{d\lambda^i}{dt} \bigg|_v \partial_v x_i \equiv \dot{\boldsymbol{\lambda}}|_v, \tag{3.46}$$

where the derivatives on the right-hand side are those of the co-ordinate functions of $\boldsymbol{\lambda}(t)$ in (x^i).

Of course, we call $\{\partial_v x_i\}$ the *natural basis* of the co-ordinate system (x^i). For a Cartesian co-ordinate system induced by a basis $\{\boldsymbol{e}_i\}$, the natural basis is simply $\{\boldsymbol{e}_i\}$ itself. In this case the natural basis is a constant field, while for a general co-ordinate system it need not be.

Now let $\boldsymbol{u}$ be an arbitrary vector field. Then, as before, we define the components of $\boldsymbol{u}$ relative to (x^i) to be the fields whose values at any

---

[1] The tangent as defined here is a particular vector. Different parametrizations of the curve $\boldsymbol{\lambda}$ give rise to different tangents. In elementary geometry the term "tangent" refers to the tangent line, which in the present notation is the one-dimensional subspace Sp $(\partial_v \boldsymbol{\lambda})$.

[2] So as to avoid confusing the reader with a classical background, and also to avoid spreading lines of text, we depart here from the notation

$$\frac{\partial}{\partial x^i} \bigg|_v$$

which is used in the theory of manifolds, and which suggests an operator rather than a vector.

point $\boldsymbol{v}$ are the components of $\boldsymbol{u}$ at $\boldsymbol{v}$ relative to $\{\partial_{\boldsymbol{v}} x_i\}$. Thus we have the field equation

$$\boldsymbol{u} = u^i \partial x_i. \tag{3.47}$$

Under a change of co-ordinates from (x^i) to $(\bar{x}^i)$ the components u^i satisfy the transformation rule familiar in classical vector analysis as the definition of a "contravariant vector field", namely

$$u^i = \frac{\partial x^i}{\partial \bar{x}^j} \bar{u}^j, \tag{3.48}$$

since from (3.46) the natural basis satisfies the transformation rule

$$\partial \bar{x}_i = \frac{\partial x^j}{\partial \bar{x}^i} \partial x_j. \tag{3.49}$$

Naturally, we define the (r, s) *product basis* of a co-ordinate system (x^i) as the field

$$\{\partial x_{i_1} \otimes \ldots \otimes \partial x_{i_r} \otimes \partial x^{j_1} \otimes \ldots \otimes \partial x^{j_s}\}, \tag{3.50}$$

where i and j take integer values from 1 to n. This field furnishes a basis for tensors of type (r, s):

$$\Psi = \Psi^{i_1 \ldots i_r}_{j_1 \ldots j_s} \partial x_{i_1} \otimes \ldots \otimes \partial x_{i_r} \otimes \partial x^{j_1} \otimes \ldots \otimes \partial x^{j_s}, \tag{3.51}$$

where the components $\Psi^{i_1 \ldots i_r}_{j_1 \ldots j_r}$ are scalar fields. The transformation rule for these components is

$$\Psi^{i_1 \ldots i_r}_{j_1 \ldots j_s} = \overline{\Psi}^{k_1 \ldots k_r}_{l_1 \ldots l_s} \frac{\partial x^{i_1}}{\partial \bar{x}^{k_1}} \cdots \frac{\partial x^{i_r}}{\partial \bar{x}^{k_r}} \frac{\partial \bar{x}^{l_1}}{\partial x^{j_1}} \cdots \frac{\partial \bar{x}^{l_s}}{\partial x^{j_s}}, \tag{3.52}$$

familiar in classical tensor analysis.

Differentiable functions and differentiable curves are special cases of differentiable maps from an open set of some Euclidean space into some other Euclidean space. Let $\boldsymbol{\kappa}$ be a differentiable map in general, say from an open set $\mathscr{D}$ in $\mathscr{E}^n$ into a Euclidean space $\mathscr{E}^m$,

$$\boldsymbol{\kappa} : \mathscr{D} \to \mathscr{E}^m. \tag{3.53}$$

Then $\boldsymbol{\kappa}$ carries any differentiable curve $\boldsymbol{\lambda}$ in $\mathscr{D}$ into a differentiable curve $\boldsymbol{\kappa} \circ \boldsymbol{\lambda}$ in $\mathscr{E}^m$. Likewise, if f is any differentiable function on $\mathscr{E}^m$, then $f \circ \boldsymbol{\kappa}$ is a differentiable function on $\mathscr{D}$. We now consider the relation between the tangents vectors of $\boldsymbol{\lambda}$ and $\boldsymbol{\kappa} \circ \boldsymbol{\lambda}$ and the relation between the differentials of f and $f \circ \boldsymbol{\kappa}$.

PROPOSITION. *Suppose that $\boldsymbol{\lambda}_1$ and $\boldsymbol{\lambda}_2$ be differentiable curves passing through a point $\boldsymbol{v} \in \mathscr{D}$ and having the same tangent there:*

$$\partial_{\boldsymbol{v}} \boldsymbol{\lambda}_1 = \partial_{\boldsymbol{v}} \boldsymbol{\lambda}_2 . \tag{3.54}$$

Then

$$\partial_{\boldsymbol{u}}(\boldsymbol{\kappa} \circ \boldsymbol{\lambda}_1) = \partial_{\boldsymbol{u}}(\boldsymbol{\kappa} \circ \boldsymbol{\lambda}_2), \tag{3.55}$$

where

$$\boldsymbol{u} = \boldsymbol{\kappa}(\boldsymbol{v}). \tag{3.56}$$

Thus there exists a mapping

$$\boldsymbol{\kappa}_{*\boldsymbol{v}} : \mathscr{E}^n \to \mathscr{E}^m \tag{3.57}$$

such that

$$\boldsymbol{\kappa}_{*\boldsymbol{v}} \partial_{\boldsymbol{v}} \boldsymbol{\lambda} = \partial_{\boldsymbol{u}}(\boldsymbol{\kappa} \circ \boldsymbol{\lambda}) \tag{3.58}$$

for any differentiable curve $\boldsymbol{\lambda}$ passing through $\boldsymbol{v}$. The mapping $\boldsymbol{\kappa}_{*\boldsymbol{v}}$ defined in this way is a linear map.

Similarly, suppose that f_1 and f_2 be differentiable functions defined near $\boldsymbol{u}$ and having the same differential there:

$$\partial_{\boldsymbol{u}} f_1 = \partial_{\boldsymbol{u}} f_2 . \tag{3.59}$$

Then

$$\partial_{\boldsymbol{v}}(f_1 \circ \boldsymbol{\kappa}) = \partial_{\boldsymbol{v}}(f_2 \circ \boldsymbol{\kappa}). \tag{3.60}$$

The mapping

$$\boldsymbol{\kappa}_{\boldsymbol{v}}^* : \mathscr{E}^{m*} \to \mathscr{E}^{n*} \tag{3.61}$$

defined by

$$\boldsymbol{\kappa}_{\boldsymbol{v}}^*(\partial_{\boldsymbol{u}} f) = \partial_{\boldsymbol{v}}(f \circ \boldsymbol{\kappa}) \tag{3.62}$$

is also a linear map.

Moreover, $\boldsymbol{\kappa}_{\boldsymbol{v}}^*$ is the adjoint of $\boldsymbol{\kappa}_{*\boldsymbol{v}}$:

$$\boldsymbol{\kappa}_{\boldsymbol{v}}^* = (\boldsymbol{\kappa}_{*\boldsymbol{v}})^A, \tag{3.63}$$

which means that

$$\langle \boldsymbol{\kappa}_{\boldsymbol{v}}^*(\boldsymbol{z}), \boldsymbol{w} \rangle = \langle \boldsymbol{z}, \boldsymbol{\kappa}_{*\boldsymbol{v}}(\boldsymbol{w}) \rangle \quad \forall \boldsymbol{z} \in \mathscr{E}^{m*}, \ \forall \boldsymbol{w} \in \mathscr{E}^{n*}. \tag{3.64}$$

The mappings $\boldsymbol{\kappa}_{*\boldsymbol{v}}$ and $\boldsymbol{\kappa}_{\boldsymbol{v}}^*$ are called the *induced linear maps* of $\boldsymbol{\kappa}$ at $\boldsymbol{v}$. The tensor corresponding to them is called the *gradient* of $\boldsymbol{\kappa}$ at $\boldsymbol{v}$ and is denoted by $\operatorname{grad}_{\boldsymbol{v}} \boldsymbol{\kappa}$. This tensor is an element of $\mathscr{E}^m \otimes \mathscr{E}^{n*}$.

PROPOSITION. *Relative to any co-ordinate systems* (x^i) *in* $\mathscr{D}$ *and* (y^α) *in* $\mathscr{E}^m$,

$$\operatorname{grad} \boldsymbol{\kappa} = \frac{\partial \kappa^\alpha}{\partial x^i} \partial y_\alpha \otimes \partial x^i, \tag{3.65}$$

where the partial derivative on the right-hand side is that of the representation

$$y^\alpha = \kappa^\alpha(x^1, \ldots, x^n) \tag{3.66}$$

of $\boldsymbol{\kappa}$ *relative to* (x^i) *and* (y^α).

The reader should verify that for the special cases in which $\boldsymbol{\kappa} = \boldsymbol{\lambda}$ or $\kappa = f$ the gradient of $\boldsymbol{\kappa}$ reduces to the tangent $\partial\boldsymbol{\lambda}$ and to the differential ∂f, respectively. Also, the classical inverse function theorem is equivalent to the following

PROPOSITION. *If* $\boldsymbol{\kappa}_{*\mathbf{v}}$ *is an isomorphism, then there exists a neighborhood of* $\mathbf{v}$ *on which the restriction of* $\boldsymbol{\kappa}$ *is a diffeomorphism. Conversely, if* $\boldsymbol{\kappa}$ *is a diffeomorphism near* $\mathbf{v}$, *then* $\boldsymbol{\kappa}_{*\mathbf{v}}$ *is an isomorphism.*

It should be noted that the component matrix of the gradient of $\boldsymbol{\kappa}$ is the Jacobian matrix of the transformation (3.66). Thus the statement that $\boldsymbol{\kappa}_{*\mathbf{v}}$ is an isomorphism means that

$$\dim \mathscr{E}^n = \dim \mathscr{E}^m, \quad \text{that is,} \quad m = n, \tag{3.67}$$

and that the Jacobian matrix of (3.66) has a non-zero determinant at the point $\mathbf{v}$.

The classical chain rule of partial differentiation is equivalent to the following

PROPOSITION. *Let* $\boldsymbol{\kappa}$ *and* ξ *be differentiable maps,*

$$\boldsymbol{\kappa} : \mathscr{D} \to \mathscr{E}, \tag{3.68}$$

and

$$\xi : \mathscr{E} \to \mathscr{F}, \tag{3.69}$$

where $\mathscr{D}$, $\mathscr{E}$, *and* $\mathscr{F}$ *are open submanifolds of Euclidean spaces. Then*

$$(\xi \circ \boldsymbol{\kappa})_* = \xi_* \circ \boldsymbol{\kappa}_*, \tag{3.70}$$

and

$$(\xi \circ \boldsymbol{\kappa})^* = \boldsymbol{\kappa}^* \circ \xi^*. \tag{3.71}$$

Next we consider the concept of the *Euclidean parallelism* of a Euclidean space $\mathscr{E}^n$. We said above that the quotient space $\mathfrak{F}(\boldsymbol{v})/\mathfrak{Z}(\boldsymbol{v})$ is canonically isomorphic to the dual space $\mathscr{E}^{n*}$. In the terminology of differentiable manifolds, the quotient space $\mathfrak{F}(\boldsymbol{v})/\mathfrak{Z}(\boldsymbol{v})$ is called the *cotangent space* at $\boldsymbol{v}$, and its dual space is called the *tangent space*. The tangent space at $\boldsymbol{v}$ is then a copy of $\mathscr{E}^n$. But since $\boldsymbol{v}$ is arbitrary, the isomorphism of the tangent space with $\mathscr{E}^n$ induces isomorphisms among the tangent spaces of different points of $\mathscr{E}^n$. Such isomorphisms give the *Euclidean parallellism* on $\mathscr{E}^n$.

Let (x^i) be a co-ordinate system on $\mathscr{D}$ as before. We have defined the vector field

$$\partial x_i : \mathscr{D} \to \mathscr{E}^n, \tag{3.72}$$

and the covector field

$$\partial x^i : \mathscr{D} \to \mathscr{E}^{n*}. \tag{3.73}$$

Here the Euclidean parallelism is used in an implicit way. Since the fields ∂x_i and ∂x^i are differentiable, their gradients exist. We put

$$\operatorname{grad} \partial x_i \equiv \boldsymbol{\Gamma}_i, \tag{3.74}$$

and

$$\operatorname{grad} \partial x^i \equiv \boldsymbol{\Gamma}^i. \tag{3.75}$$

Expressing $\boldsymbol{\Gamma}_i$ and $\boldsymbol{\Gamma}^i$ in terms of the natural basis and cobasis of (x^i), we write

$$\boldsymbol{\Gamma}_i = \Gamma_i{}^j{}_k \partial x_j \otimes \partial x^k \tag{3.76}$$

and

$$\boldsymbol{\Gamma}^i = \Gamma^i{}_{jk} \partial x^j \otimes \partial x^k, \tag{3.77}$$

and we call $\Gamma_i{}^j{}_k$ and $\Gamma^i{}_{jk}$ the *connection symbols* of the Euclidean parallelism in (x^i). As we remarked before, ∂x_i and ∂x^i are constant fields if (x^i) is a Cartesian co-ordinate system, so that $\boldsymbol{\Gamma}_i$ and $\boldsymbol{\Gamma}^i$ vanish then. Thus in a Cartesian co-ordinate system the connection symbols of the Euclidean parallelism are all zero.

PROPOSITION. *The connection symbols of the Euclidean parallelism satisfy the following transformation rules*:

$$\bar{\Gamma}^j{}_{kl} = \Gamma^r{}_{st} \frac{\partial \bar{x}^j}{\partial x^r} \frac{\partial x^s}{\partial \bar{x}^k} \frac{\partial x^t}{\partial \bar{x}^l} - \frac{\partial^2 x^r}{\partial \bar{x}^k \partial \bar{x}^l} \frac{\partial \bar{x}^j}{\partial x^r}, \tag{3.78}$$

and

$$\bar{\Gamma}_{i\ k}^{\ j} = \Gamma_{r\ t}^{\ s} \frac{\partial x^r}{\partial \bar{x}^i} \frac{\partial \bar{x}^j}{\partial x^s} \frac{\partial x^t}{\partial \bar{x}^k} + \frac{\partial^2 x^r}{\partial \bar{x}^i \partial \bar{x}^k} \frac{\partial \bar{x}^j}{\partial x^r}, \tag{3.79}$$

with respect to a change of co-ordinate system from (x^i) *to* $(\bar{x}^i)$.

From (3.78) and (3.79), we see that if (x^i) is a Cartesian co-ordinate system, then

$$\bar{\Gamma}^j_{\ kl} = - \frac{\partial^2 x^r}{\partial \bar{x}^k \partial \bar{x}^l} \frac{\partial \bar{x}^j}{\partial x^r}, \tag{3.80}$$

and

$$\bar{\Gamma}_{k\ l}^{\ j} = + \frac{\partial^2 x^r}{\partial \bar{x}^k \partial \bar{x}^l} \frac{\partial \bar{x}^j}{\partial x^r}. \tag{3.81}$$

From these formulae we see that the connection symbols of the Euclidean parallelism satisfy the identities

$$\Gamma^j_{\ kl} = -\Gamma_{k\ l}^{\ j}, \tag{3.82}$$

and

$$\Gamma^j_{\ kl} = \Gamma^j_{\ lk}. \tag{3.83}$$

Relative to the Euclidean parallelism, the gradient of a differentiable tensor field is again a tensor field. In the terminology of differentiable manifolds, the gradient of a tensor field Ψ is called the *covariant derivative* of Ψ since the covariant order of grad Ψ is one greater than that of Ψ. Using (3.74) and (3.75), we can express the covariant derivative of a tensor field Ψ of type (r, s) by the component form

$$\text{grad } \Psi = \Psi^{i_1 \ldots i_r}_{j_1 \ldots j_s | k} \partial x_{i_1} \otimes \ldots \otimes \partial x_{i_r} \otimes \partial x^{j_1} \otimes \ldots \otimes \partial x^{j_s} \otimes \partial x^k, \tag{3.84}$$

where

$$\Psi^{i_1 \ldots i_r}_{j_1 \ldots j_s | k} = \frac{\partial \Psi^{i_1 \ldots i_r}_{j_1 \ldots j_s}}{\partial x^k} + \Psi^{l i_2 \ldots i_r}_{j_1 \ldots j_s} \Gamma_{l\ k}^{\ i_1} + \ldots + \Psi^{i_1 \ldots i_{r-1} l}_{j_1 \ldots j_s} \Gamma_{l\ k}^{\ i_r}$$
$$- \Psi^{i_1 \ldots i_r}_{l j_2 \ldots j_s} \Gamma_{j_1\ k}^{\ l} - \ldots - \Psi^{i_1 \ldots i_r}_{j_1 \ldots j_{s-1} l} \Gamma_{j_s\ k}^{\ l}. \tag{3.85}$$

In a Cartesian co-ordinate system (x^i), since the connection symbols vanish,

$$\Psi^{i_1 \ldots i_r}_{j_1 \ldots j_s | k} = \frac{\partial \Psi^{i_1 \ldots i_r}_{j_1 \ldots j_s}}{\partial x^k}. \tag{3.86}$$

As the covariant derivative of any constant tensor field vanishes, we have

$$g_{ij|k} = \frac{\partial g_{ij}}{\partial x^k} - g_{lj}\Gamma^{\,l}_{i\,k} - g_{il}\Gamma^{\,l}_{j\,k} = 0 \tag{3.87}$$

for any inner product g. This equation can be solved for the connection symbols, yielding

$$\Gamma^{\,i}_{j\,k} = \tfrac{1}{2} g^{iq}\left(\frac{\partial g_{qj}}{\partial x^k} + \frac{\partial g_{qk}}{\partial x^j} - \frac{\partial g_{jk}}{\partial x^q}\right). \tag{3.88}$$

The quantity on the right-hand side of this formula is called a *Christoffel symbol* of g and is denoted by $\left\{{}^{\,i}_{j\,k}\right\}$. Hence for a Euclidean space the connection symbols of the Euclidean parallelism are equal to the Christoffel symbols of any inner product (regarded as a constant tensor field on the Euclidean space).

4. Differentiable Manifolds

As we have seen in the preceding section, a Euclidean space has four basic structures:

i) Vector-space structure
ii) Topological structure
iii) Differentiable structure
iv) Euclidean parallelism

The vector-space structure is characterized by the linear operations presented in § 1. The topological structure and the differentiable structure are characterized by the Euclidean topology and the Euclidean atlas, respectively. Finally, the Euclidean parallelism is characterized by the canonical isomorphism of the underlying vector space with the tangent space at any point. In this section, we focus our attention to the topological structure and the differentiable structure only. A set having these two structures is called a *differentiable manifold.* In general, a manifold lacks the structure of a vector space, and there is no intrinsic parallelism among its tangent spaces at two different points.

More specifically, a set $\mathscr{M}$ is a *differentiable manifold* if it is equipped with a Hausdorff topology having a countable base and with an atlas

$$\mathfrak{D} = \{(\mathscr{D}_\Lambda, \kappa_\Lambda),\ \Lambda \in I\} \tag{4.1}$$

such that the following four conditions are satisfied:

i) $\{\mathscr{D}_\Lambda, \Lambda \in I\}$ is an open covering of $\mathscr{M}$.

ii) $\boldsymbol{\kappa}_\Lambda$ is a homeomorphism from $\mathscr{D}_\Lambda$ into $\mathscr{R}^n$, where the positive integer n, which is the same for all Λ, is called the *dimension* of $\mathscr{M}$.

iii) For any two indices Λ and Γ in I the co-ordinate transformation

$$\boldsymbol{\kappa}_\Gamma \circ \boldsymbol{\kappa}_\Lambda^{-1} : \boldsymbol{\kappa}_\Lambda(\mathscr{D}_\Lambda \cap \mathscr{D}_\Gamma) \to \boldsymbol{\kappa}_\Gamma(\mathscr{D}_\Lambda \cap \mathscr{D}_\Gamma) \tag{4.2}$$

is a diffeomorphism of $\mathscr{R}^n$ in the sense defined in the preceding section.

iv) The collection $\mathfrak{D}$ is maximal with respect to the preceding three conditions.

As before, we call the pair $(\mathscr{D}_\Gamma, \boldsymbol{\kappa}_\Gamma)$ a *(co-ordinate) chart* for $\mathscr{M}$ and the mapping $\boldsymbol{\kappa}_\Gamma$ a *co-ordinate system* on $\mathscr{D}_\Gamma$. However, since the underlying set of a differentiable manifold need not be a Euclidean space, in general no Cartesian co-ordinate system can be defined for all of it. Other than this difference, the differentiable structure of a differentiable manifold is essentially the same as that of a Euclidean space. Hence the definitions of many concepts introduced in the preceding section can be extended in a straightforward way to differentiable manifolds in general. For example, a *differentiable map*

$$\boldsymbol{\kappa} : \mathscr{M} \to \mathscr{N} \tag{4.3}$$

from a differentiable manifold $\mathscr{M}$ of dimension m into a differentiable manifold $\mathscr{N}$ of dimension n is again characterized by the condition that the representation of $\boldsymbol{\kappa}$,

$$y^\alpha = \kappa^\alpha(x^1, \ldots, x^m),\ \alpha = 1, \ldots, n, \tag{4.4}$$

relative to any co-ordinate systems (x^i) and (y^α) in $\mathscr{M}$ and $\mathscr{N}$, respectively, be of class C^∞.

Let $\boldsymbol{x}$ be a point in $\mathscr{M}$. Then, as before, the collection of all differentiable functions defined near $\boldsymbol{x}$ is denoted by $\mathfrak{F}(\boldsymbol{x})$. The condition for a function to have differential zero at $\boldsymbol{x}$ is also the same as in the preceding section, and we denote the collection of functions having differential zero at $\boldsymbol{x}$ by $\mathfrak{Z}(\boldsymbol{x})$. The quotient space of $\mathfrak{F}(\boldsymbol{x})$ over $\mathfrak{Z}(\boldsymbol{x})$ is called the *cotangent space* of $\mathscr{M}$ at $\boldsymbol{x}$ and is denoted by

$$\mathscr{M}_{\boldsymbol{x}}^* \equiv \mathfrak{F}(\boldsymbol{x})/\mathfrak{Z}(\boldsymbol{x}). \tag{4.5}$$

The dual space of $\mathscr{M}_{\boldsymbol{x}}^*$ is called the *tangent space* of $\mathscr{M}$ at $\boldsymbol{x}$ and is denoted by $\mathscr{M}_{\boldsymbol{x}}$. Unlike a Euclidean space, a differentiable manifold in general is not equipped with a parallelism, so that the tangent spaces or the cotangent spaces at different points of a manifold are not canonically isomorphic to one another.

We define the differential

$$\partial_{\boldsymbol{x}} : \mathfrak{F}(\boldsymbol{x}) \to \mathscr{M}^*_{\boldsymbol{x}} \tag{4.6}$$

in exactly the same way as before. For any co-ordinate system (x^i) near $\boldsymbol{x}$, the set of differentials $\{\partial_{\boldsymbol{x}} x^i\}$ is a basis of $\mathscr{M}^*_{\boldsymbol{x}}$ and is called the *natural cobasis* of (x^i) at $\boldsymbol{x}$. Relative to the natural cobasis, the component form of the differential $\partial_{\boldsymbol{x}} f$ is

$$\partial_{\boldsymbol{x}} f = \left.\frac{\partial f}{\partial x^i}\right|_{\boldsymbol{x}} \partial_{\boldsymbol{x}} x^i, \tag{4.7}$$

analogous to (3.38). We define a cotangent vector field $\boldsymbol{w}$ by assigning a cotangent vector $\boldsymbol{w}(\boldsymbol{x}) \in \mathscr{M}^*_{\boldsymbol{x}}$ for all $\boldsymbol{x}$ in the domain of $\boldsymbol{w}$. The component form of $\boldsymbol{w}$ is again (3.40), and the rule of transformation for the components is again (3.42).

For a differentiable curve $\boldsymbol{\lambda}$ passing through $\boldsymbol{x}$, we define the tangent $\partial_{\boldsymbol{x}} \boldsymbol{\lambda}$ by a condition generalizing (3.45), namely,

$$\langle \partial_{\boldsymbol{x}} \boldsymbol{\lambda}, \partial_{\boldsymbol{x}} f \rangle = \left.\frac{d}{dt} f(\boldsymbol{\lambda}(t))\right|_{\boldsymbol{x}} \quad \forall f \in \mathfrak{F}(\boldsymbol{x}). \tag{4.8}$$

Of course, the tangent $\partial_{\boldsymbol{x}} \boldsymbol{\lambda}$ belongs to the tangent space $\mathscr{M}_{\boldsymbol{x}}$, and the set of tangent vectors $\{\partial_{\boldsymbol{x}} x_i\}$ of the co-ordinate curves is a basis of $\mathscr{M}_{\boldsymbol{x}}$. In component form, $\partial_{\boldsymbol{x}} \boldsymbol{\lambda}$ is given by

$$\partial_{\boldsymbol{x}} \boldsymbol{\lambda} = \left.\frac{d\lambda^i}{dt}\right|_{\boldsymbol{x}} \partial_{\boldsymbol{x}} x_i, \tag{4.9}$$

where $\lambda^i(t)$ is the ith co-ordinate of $\boldsymbol{\lambda}(t)$ with respect to (x^i). We define a tangent vector field $\boldsymbol{u}$ by assigning a tangent vector $\boldsymbol{u}(\boldsymbol{x}) \in \mathscr{M}_{\boldsymbol{x}}$ for all $\boldsymbol{x}$ in the domain of $\boldsymbol{u}$. Then the component form of $\boldsymbol{u}$ relative to (x^i) is again (3.47), and the transformation rule for the components is again (3.48).

Alternatively we can define a tangent vector $\partial_{\boldsymbol{x}} \boldsymbol{\lambda}$ at $\boldsymbol{x}$ as an equivalence class of differentiable curves $\boldsymbol{\lambda}$ passing through $\boldsymbol{x}$ such that $\partial_{\boldsymbol{x}} \boldsymbol{\lambda} = \partial_{\boldsymbol{x}} \bar{\boldsymbol{\lambda}}$ if and only if $d\lambda^i/dt = d\bar{\lambda}^i/dt$ at $\boldsymbol{x}$, for all $i = 1, \ldots, n$, relative to any co-ordinate system defined near $\boldsymbol{x}$. Then we define the tangent space $\mathscr{M}_{\boldsymbol{x}}$ as the space of all tangent vectors at $\boldsymbol{x}$, and we define the cotangent space $\mathscr{M}_x^*$ as the dual space of the tangent space $\mathscr{M}_x$. In this sense, (4.8) becomes a definition for the differential $\partial_x f$ if it is required to hold for all $\boldsymbol{\lambda}$ passing through $\boldsymbol{x}$.

In general, we define the *tensor space* of type (r, s) at $\boldsymbol{x}$ by

$$\mathscr{M}_x^{r,s} \equiv \underbrace{\mathscr{M}_x \otimes \ldots \otimes \mathscr{M}_x}_{r} \otimes \underbrace{\mathscr{M}_x^* \otimes \ldots \otimes \mathscr{M}_x^*}_{s}. \tag{4.10}$$

Then a tensor field of type (r, s) is defined by assigning a tensor $\Psi(\boldsymbol{x}) \in \mathscr{M}_x^{r,s}$ for all $\boldsymbol{x}$ in the domain of Ψ. Any co-ordinate system (x^i) induces a natural product basis of the form (3.50) for $\mathscr{M}_x^{r,s}$, and any tensor field Ψ has the component form (3.51), where the components satisfy the transformation rule (3.52). Naturally, we call Ψ a *differentiable field* if its components are differentiable functions. The *orders* of a tensor are defined as before.

The concept of the induced linear maps, introduced in the preceding section, can be extended to differentiable maps of manifolds. For the differentiable map $\boldsymbol{\kappa}$ in (4.3), the induced linear maps are

$$\boldsymbol{\kappa}_{*x} : \mathscr{M}_x \to \mathscr{N}_y, \tag{4.11}$$

and

$$\boldsymbol{\kappa}_x^* : \mathscr{N}_y^* \to \mathscr{M}_x^*, \tag{4.12}$$

where $\boldsymbol{y}$ is the image of $\boldsymbol{x}$ under $\boldsymbol{\kappa}$:

$$\boldsymbol{y} = \boldsymbol{\kappa}(\boldsymbol{x}). \tag{4.13}$$

As before, $\boldsymbol{\kappa}_{*x}$ and $\boldsymbol{\kappa}_y^*$ satisfy the following adjoint relation:

$$\langle \boldsymbol{\kappa}_{*x}(\boldsymbol{v}), \boldsymbol{w} \rangle = \langle \boldsymbol{v}, \boldsymbol{\kappa}_x^*(\boldsymbol{w}) \rangle \quad \forall \boldsymbol{v} \in \mathscr{M}_x, \ \forall \boldsymbol{w} \in \mathscr{N}_y^*. \tag{4.14}$$

The tensor (in $\mathscr{N}_y \otimes \mathscr{M}_x^*$) corresponding to $\boldsymbol{\kappa}_{*x}$ and $\boldsymbol{\kappa}_x^*$ is denoted by $\text{grad}_x \boldsymbol{\kappa}$ and is called the *gradient* of $\boldsymbol{\kappa}$ at $\boldsymbol{x}$. In component form relative to the co-ordinate systems (x^i) and (y^α), the tensor field grad $\boldsymbol{\kappa}$ is given by

$$\text{grad}\, \boldsymbol{\kappa} = \frac{\partial \kappa^\alpha}{\partial x^i} \partial y_\alpha \otimes \partial x^i. \tag{4.15}$$

As before, the classical implicit function theorem implies that $\boldsymbol{\kappa}$ is a local diffeomorphism near $\boldsymbol{x}$ if and only if $\boldsymbol{\kappa}_{*x}$ is an isomorphism.

Since we have not introduced a parallelism, the concept of the covariant derivative cannot yet be extended to a manifold. We shall treat the general theory of parallelism on a manifold in the next section. Here we consider only a special type of parallel transport that is induced by a vector field. Specifically, let $\boldsymbol{u}$ be a vector field on $\mathscr{M}$. Then a curve $\boldsymbol{\lambda}$ whose tangent vector at each point is the value of $\boldsymbol{u}$ there is called an *integral curve* or *orbit* of $\boldsymbol{u}$:

$$\partial_{\boldsymbol{x}} \boldsymbol{\lambda} = \boldsymbol{u}(\boldsymbol{x}) \quad \forall \boldsymbol{x} \in \operatorname{im} \boldsymbol{\lambda}. \tag{4.16}$$

From (4.9) and (3.47), we can express the condition (4.16) in component form by

$$\frac{d\lambda^i}{dt} = u^i(\lambda^1(t), \ldots, \lambda^n(t)), \; i = 1, 2, \ldots, n. \tag{4.17}$$

For simplicity, we assume that the vector field $\boldsymbol{u}$ is a differentiable field. Then from a standard result in differential equations, there exists a unique solution $\lambda^i(x^1 \ldots, x^n, t)$ of (4.17) subject to the initial condition

$$\lambda^i(x^1, \ldots, x^n, 0) = x^i, \; i = 1, 2, \ldots, n. \tag{4.18}$$

The solution $\lambda^i(x^1, \ldots, x^n, t)$ is defined uniformly on an interval $t \in (-a, a)$ for all $\boldsymbol{x}$ belonging to some neighborhood of a given point $\boldsymbol{x}_0$ and is of class C^∞ with respect to all x^j and to t. Further, the solution gives a local 1-parameter group of transformations of a neighborhood of $\boldsymbol{x}_0$ in the sense that

$$\lambda^i(\lambda^j(x^k, s), t) = \lambda^i(x^k, t+s) \tag{4.19}$$

for all $s, t \in (-a, a)$ with $t+s \in (-a, a)$ and for all $\boldsymbol{x}$ in a neighborhood of $\boldsymbol{x}_0$. Hence the mapping $\boldsymbol{\rho}_t$ carrying the initial point $\boldsymbol{x}$ into the solution $\boldsymbol{\lambda}(\boldsymbol{x}, t)$ is a local 1-parameter group of diffeomorphisms. We call the local 1-parameter group of induced isomorphisms of the tangent spaces, *viz*

$$\rho_{t*x_0} : \mathscr{M}_{x_0} \to \mathscr{M}_{\rho_t(x_0)}, \tag{4.20}$$

the *parallel transport induced by* $\boldsymbol{u}$ *along the orbit at* $\boldsymbol{x}_0$. As usual, ρ_{t*x_0} then gives rise also to parallel transports of the various types of tensors along the orbit at $\boldsymbol{x}_0$. We denote such a parallel transport by ζ_t.

Next we define the *Lie derivative* $\mathscr{L}_{\boldsymbol{u}}\Psi$ of a tensor field Ψ with respect to $\boldsymbol{u}$:

$$\mathscr{L}_{\boldsymbol{u}}\Psi \equiv \lim_{t \to 0} \frac{1}{t}(\zeta_t\Psi - \Psi). \tag{4.21}$$

If both $\boldsymbol{u}$ and Ψ are differentiable fields, the Lie derivative $\mathscr{L}_{\boldsymbol{u}}\Psi$ exists and is also a differentiable field. In component form relative to any coordinate system (x^i), $\mathscr{L}_{\boldsymbol{u}}\Psi$ is given by

$$[\mathscr{L}_{\boldsymbol{u}}\Psi]^{i_1\ldots i_r}_{j_1\ldots j_s} = \frac{\partial\Psi^{i_1\ldots i_r}_{j_1\ldots j_s}}{\partial x^k}u^k + \Psi^{i_1\ldots i_r}_{kj_2\ldots j_s}\frac{\partial u^k}{\partial x^{j_1}} + \ldots + \Psi^{i_1\ldots i_r}_{j_1\ldots j_{s-1}k}\frac{\partial u^k}{\partial x^{j_s}} -$$

$$-\Psi^{ki_2\ldots i_r}_{j_1\ldots j_s}\frac{\partial u^{i_1}}{\partial x^k} - \ldots - \Psi^{i_1\ldots i_{r-1}k}_{j_1\ldots j_s}\frac{\partial u^{i_r}}{\partial x^k}. \tag{4.22}$$

In particular, when Ψ is a vector field $\boldsymbol{v}$,

$$\mathscr{L}_{\boldsymbol{u}}\boldsymbol{v} = \left(\frac{\partial v^i}{\partial x^j}u^j - \frac{\partial u^i}{\partial x^j}v^j\right)\partial x_i. \tag{4.23}$$

While we have introduced the Lie derivative $\mathscr{L}_{\boldsymbol{u}}$ as an operation defined in terms of an assigned vector field $\boldsymbol{u}$, the particular case $\mathscr{L}_{\boldsymbol{u}}\boldsymbol{v}$ may be regarded as the result of an operation $[\cdot,\cdot]$ on a pair of differentiable vector fields $\boldsymbol{u}$ and $\boldsymbol{v}$:

$$[\boldsymbol{u}, \boldsymbol{v}] \equiv \mathscr{L}_{\boldsymbol{u}}\boldsymbol{v}. \tag{4.24}$$

By (4.23),

$$\begin{gathered}[\boldsymbol{u}, \boldsymbol{v}] = -[\boldsymbol{v}, \boldsymbol{u}],\\ [\boldsymbol{u}, [\boldsymbol{v}, \boldsymbol{w}]] + [\boldsymbol{v}, [\boldsymbol{w}, \boldsymbol{u}]] + [\boldsymbol{w}, [\boldsymbol{u}, \boldsymbol{v}]] = \boldsymbol{0}.\end{gathered} \tag{4.25}$$

Generally, any vector-valued bilinear operation that satisfies the identities (4.25) is called a *bracket product*, and any vector space equipped with a bracket product is called a *Lie algebra*. Thus we have shown that the space of all differentiable vector fields on a manifold is a particular Lie algebra, its bracket product being defined in terms of the Lie derivative by (4.24).

Another differential operation that we shall use later is the *exterior derivative*. While the Lie derivative $\mathscr{L}_{\boldsymbol{u}}$ may be applied to any differentiable field, its definition rests upon use of a particular, presumably given vector field $\boldsymbol{u}$. The value of the exterior derivative, on the contrary, does not depend upon any preferred field, but the operation itself may be applied only to certain particular tensor fields called *differential forms*. A *differential r-form* is a differentiable, completely skew, covariant tensor field of order r. A covariant tensor is *completely skew* if its components are completely skew with respect to their indices. Clearly this condition is an intrinsic property of the tensor, and it is equivalent to the condition that the multilinear function corresponding to the covariant tensor be a completely skew function. Of course a differentiable covector field is a 1-form, and a differentiable scalar function is a 0-form. Since a completely skew covariant tensor of order greater than the dimension of the vector space must vanish, so must an r-form if $r > \dim \mathcal{M}$.

We denote the space of all r-forms defined near a point $\boldsymbol{x}$ by $\mathfrak{F}^r(\boldsymbol{x})$, and the space of all completely skew covariant tensors of order r at $\boldsymbol{x}$ by $\hat{\mathcal{M}}^r_{\boldsymbol{x}}$. Then the differential $\partial_{\boldsymbol{x}}$, defined above, can be regarded as a linear map defined for all 0-forms:

$$\partial_{\boldsymbol{x}} : \mathfrak{F}^0(\boldsymbol{x}) \to \hat{\mathcal{M}}^1_{\boldsymbol{x}}. \tag{4.26}$$

We seek to generalize this operation by a linear map defined for all r-forms:

$$\partial_{\boldsymbol{x}} : \mathfrak{F}^r(\boldsymbol{x}) \to \hat{\mathcal{M}}^{r+1}_{\boldsymbol{x}}. \tag{4.27}$$

The definition is motivated by the "curl" in classical vector analysis. If $\Psi \in \mathfrak{F}^r(\boldsymbol{x})$, say

$$\Psi = \Psi_{i_1 \ldots i_r} \partial x^{i_1} \otimes \ldots \otimes \partial x^{i_r}, \tag{4.28}$$

where $\Psi_{i_1 \ldots i_r}$ is completely skew in the indices $i_1, \ldots, i_r$, we define

$$\partial_{\boldsymbol{x}} \Psi \equiv \frac{1}{(r+1)!} \delta^{ki_1 \ldots i_r}_{j_1 \ldots j_{r+1}} \left. \frac{\partial \Psi_{i_1 \ldots i_r}}{\partial x^k} \right|_{\boldsymbol{x}} \partial_{\boldsymbol{x}} x^{j_1} \otimes \ldots \otimes \partial_{\boldsymbol{x}} x^{j_{r+1}}, \tag{4.29}$$

where δ denotes the generalized Kronecker's symbol. When $r = 0$, this definition reduces to the previous one, (4.7). Further, for any r, the right-hand side of (4.29) does not depend on the choice of the co-ordinate system, and the components satisfy the transformation rule for a covariant tensor of order $r+1$. We call $\partial_{\boldsymbol{x}} \Psi$ the *exterior derivative of* Ψ *at* $\boldsymbol{x}$, and we call $\partial \Psi$ the field of exterior derivatives of Ψ. Also we shall write $\mathbf{0}$ for the zero differential r-form.

The exterior derivative operation has many important properties. We mention some of them here.

PROPOSITION. *For any differential form* Ψ

$$\partial(\partial \Psi) = \mathbf{0}. \tag{4.30}$$

Conversely, let Φ *be a differential form such that*

$$\partial \Phi = \mathbf{0}. \tag{4.31}$$

Then, locally, there exists a differential form ♆ *such that*

$$\text{☿} = \partial \text{♆}. \tag{4.32}$$

Now suppose that f is a differentiable map from $\mathscr{M}$ to $\mathscr{N}$. We have shown that the induced linear map f^* carries a 1-form on $\mathscr{N}$ into a 1-form on $\mathscr{M}$. This induced linear map can be extended to any covariant tensor field on $\mathscr{N}$ in an obvious way. Let ♂ be a covariant tensor field of order r, say

$$\text{♂} = \text{♂}_{\alpha_1 \ldots \alpha_r} \partial y^{\alpha_1} \otimes \cdots \otimes \partial y^{\alpha_r}. \tag{4.33}$$

Then we simply put

$$\otimes f^*(\text{♂}) \equiv \text{♂}_{\alpha_1 \ldots \alpha_r} f^*(\partial y^{\alpha_1}) \otimes \ldots \otimes f^*(\partial y^{\alpha_r}). \tag{4.34}$$

Clearly, this extension does not depend on the choice of the co-ordinate system (y^α).

PROPOSITION. *Let* ♂ *be a differential form on* $\mathscr{N}$, *and let* f *be a differentiable map from* $\mathscr{M}$ *to* $\mathscr{N}$. *Then*

$$\partial(\otimes f^*(\text{♂})) = \otimes f^*(\partial \text{♂}). \tag{4.35}$$

In particular, ∂ *commutes with the Lie derivative with respect to any vector field*:

$$\mathscr{L}_u(\partial \text{♆}) = \partial(\mathscr{L}_u \text{♆}). \tag{4.36}$$

Now let ♆ and ♂ be an r-form and an s-form, respectively, on $\mathscr{M}$. Then we define their *exterior product* $\text{♆} \wedge \text{♂}$ by

$$\text{♆} \wedge \text{♂} = \frac{1}{(r+s)!} \delta^{i_1 \ldots i_r j_1 \ldots j_s}_{k_1 \ldots k_{r+s}} \text{♆}_{i_1 \ldots i_r} \text{♂}_{j_1 \ldots j_s} \partial x^{k_1} \otimes \ldots \otimes \partial x^{k_{r+s}}. \tag{4.37}$$

Clearly, this operation does not depend on the choice of the co-ordinate system, and $\text{♆} \wedge \text{♂}$ is an $(r+s)$-form. Like the tensor product, the exterior product is a bilinear operation and satisfies the associative rule, so that we can define the multiple products by

$$\text{♆} \wedge \text{♂} \wedge \text{☿} \equiv (\text{♆} \wedge \text{♂}) \wedge \text{☿} = \text{♆} \wedge (\text{♂} \wedge \text{☿}), \tag{4.38}$$

etc. In this sense, we can write the component form (4.28) as

$$\Psi = \Psi_{i_1 \ldots i_r} \partial x^{i_1} \wedge \ldots \wedge \partial x^{i_r}, \tag{4.39}$$

and the exterior derivative (4.29) as

$$\partial \Psi = \frac{\partial \Psi_{i_1 \ldots i_r}}{\partial x^k} \partial x^k \wedge \partial x^{i_1} \wedge \ldots \wedge \partial x^{i_r}, \tag{4.40}$$

since we have

$$\partial x^{i_1} \wedge \ldots \wedge \partial x^{i_r} = \frac{1}{r!} \delta^{i_1 \ldots i_r}_{j_1 \ldots j_r} \partial x^{j_1} \otimes \ldots \otimes \partial x^{j_r}. \tag{4.41}$$

PROPOSITION. *The exterior derivative satisfies the following product rule with respect to the exterior product:*

$$\partial(\Psi \wedge \mathfrak{F}) = (\partial \Psi) \wedge \mathfrak{F} + (-1)^r \Psi \wedge (\partial \mathfrak{F}), \tag{4.42}$$

where r is the order of Ψ.

The exterior derivative is related to the bracket product (4.31) by the following

PROPOSITON. *Let $\boldsymbol{u}$ and $\boldsymbol{v}$ be differentiable vector fields, and let $\boldsymbol{z}$ be a 1-form. Then*

$$\langle \boldsymbol{u}, \partial\langle \boldsymbol{v}, \boldsymbol{z}\rangle\rangle - \langle \boldsymbol{v}, \partial\langle \boldsymbol{u}, \boldsymbol{z}\rangle\rangle = \langle [\boldsymbol{u}, \boldsymbol{v}], \boldsymbol{z}\rangle + 2\partial \boldsymbol{z}(\boldsymbol{u}, \boldsymbol{v}), \tag{4.43}$$

where $\partial \boldsymbol{z}(\boldsymbol{u}, \boldsymbol{v})$ denotes the bilinear function of the 2-form $\partial \boldsymbol{z}$.

Next, we give versions of the classical Frobenius theorem in terms of the Lie derivative and the exterior derivative.

First, we define the concept of a *distribution* on a differentiable manifold. If an r-dimensional subspace $\mathscr{W}_{\boldsymbol{x}}$ of $\mathscr{M}_{\boldsymbol{x}}$ is assigned for all $\boldsymbol{x} \in \mathscr{M}$, we call such a field $\mathscr{W}$ an *r-dimensional distribution* on $\mathscr{M}$ if the following smoothness condition is satisfied: *Locally*, there exist r differentiable vector fields, say $\boldsymbol{v}_1, \ldots, \boldsymbol{v}_r$, such that

$$\mathscr{W}_{\boldsymbol{x}} = \mathrm{Sp}\,(\boldsymbol{v}_1(\boldsymbol{x}), \ldots, \boldsymbol{v}_r(\boldsymbol{x})) \tag{4.44}$$

for all $\boldsymbol{x}$ in the domain of $\boldsymbol{v}_1, \ldots, \boldsymbol{v}_r$. This smoothness condition is strictly a *local* condition, since a distribution, in general, need not be generated by r global differentiable vector fields.

We call a distribution $\mathscr{W}$ *integrable* if locally we can choose a coordinate system (x^i) such that the basis vectors $\partial x_1, \ldots, \partial x_r$ form a generating set of $\mathscr{W}$. In this case, the r-dimensional surfaces characterized by the conditions

$$x^{r+1} = \text{constant}, \ldots, x^n = \text{constant} \tag{4.45}$$

are called the *integral surfaces* of $\mathscr{W}$.

The classical Frobenius theorem gives a necessary and sufficient condition of integrability for any distribution. In terms of the Lie derivative, this condition is given by the following.

PROPOSITION. *A distribution $\mathscr{W}$ is integrable if and only if its local generating set $\{\boldsymbol{v}_1, \ldots, \boldsymbol{v}_r\}$ satisfies the condition*

$$[\boldsymbol{v}_\rho, \boldsymbol{v}_\xi] = C^\eta_{\rho\xi} \boldsymbol{v}_\eta, \ \rho, \xi = 1, \ldots, r. \tag{4.46}$$

Here the bracket product is defined by (4.24), η is summed from 1 to r, and the coefficients $C^\eta_{\rho\xi}$ are certain differentiable functions on the domain of $\boldsymbol{v}_1, \ldots, \boldsymbol{v}_r$.

Next, we state the condition of integrability in terms of the exterior derivative. For this purpose, we consider the orthogonal complement $\mathscr{W}^\perp$ of $\mathscr{W}$ in the cotangent space. Locally, $\mathscr{W}^\perp$ is generated by $(n-r)$ 1-forms, say $\boldsymbol{z}_1, \ldots, \boldsymbol{z}_{n-r}$, such that

$$\mathscr{W}^\perp_{\boldsymbol{x}} = \mathrm{Sp}\,(\boldsymbol{z}_1(\boldsymbol{x}), \ldots, \boldsymbol{z}_{n-r}(\boldsymbol{x})) \tag{4.47}$$

for all $\boldsymbol{x}$ in the domain of $\boldsymbol{z}_1, \ldots, \boldsymbol{z}_{n-r}$. We call $\{\boldsymbol{z}_1, \ldots, \boldsymbol{z}_{n-r}\}$ a local *cogenerating set* of $\mathscr{W}$.

PROPOSITION. *A distribution $\mathscr{W}$ is integrable if and only if its local cogenerating set $\{\boldsymbol{z}_1, \ldots, \boldsymbol{z}_{n-r}\}$ satisfies the following condition*:

$$\partial \boldsymbol{z}_\nu \wedge \boldsymbol{z}_1 \ldots \wedge \boldsymbol{z}_{n-r} = \boldsymbol{0}, \ \nu = 1, \ldots, n-r. \tag{4.48}$$

Next, we review the theory of integration on a differentiable manifold. We call an n-dimensional manifold *orientable* if it can be equipped with a global n-form that is non-zero at every point. Then we *orient* such a manifold by assigning a particular n-form having the aforesaid property, and we call that n-form the *volume tensor* of the oriented manifold.

Let $\boldsymbol{V}$ be the volume tensor of $\mathscr{M}$. Then a co-ordinate system (x^i) is called a *positive co-ordinate system* if

$$\boldsymbol{V} = V_{i_1 \ldots i_n} \partial x^{i_1} \otimes \ldots \otimes \partial x^{i_n} \tag{4.49}$$

with

$$V_{1\ldots n} > 0. \tag{4.50}$$

Since $V_{i_1\ldots i_n}$ is completely skew with respect to $i_1 \ldots i_n$,

$$V_{i_1\ldots i_n} = \delta^{1\ldots n}_{i_1\ldots i_n} V_{1\ldots n}. \tag{4.51}$$

Thus $\mathbf{V} \neq \mathbf{0}$ if and only if $V_{1\ldots n} \neq 0$. The scalar v, defined by

$$v \equiv n! V_{1\ldots n}, \tag{4.52}$$

is called the *specific volume* of $\mathbf{V}$ relative to (x^i). Thus

$$\mathbf{V} = v\,\partial x^1 \wedge \ldots \wedge \partial x^n. \tag{4.53}$$

PROPOSITION. *The transformation rule for the specific volume is*

$$\bar{v} = \det\left[\frac{\partial x^i}{\partial \bar{x}^j}\right] v. \tag{4.54}$$

Now for an open subset $\mathscr{D}$ that can be covered by a positive co-ordinate system $\boldsymbol{\kappa} = (x^i)$ we define the *volume* of $\mathscr{D}$ by

$$V(\mathscr{D}) \equiv \int_{\mathscr{D}} \mathbf{V} \equiv \int_{\boldsymbol{\kappa}(\mathscr{D})} v\,dx^1 \ldots dx^n, \tag{4.55}$$

where the right-hand side is the usual Riemann integral of the differentiable function v over the open set $\boldsymbol{\kappa}(\mathscr{D})$ in $\mathscr{R}^n$. The integral of v over $\boldsymbol{\kappa}(\mathscr{D})$ need not be finite, of course. However, if that integral is finite, then its value is independent of the choice of the positive co-ordinate system, since the transformation rule (4.54) is equivalent to the condition

$$\int_{\boldsymbol{\kappa}(\mathscr{D})} v\,dx^1 \ldots dx^n = \int_{\bar{\boldsymbol{\kappa}}(\mathscr{D})} \bar{v}\,d\bar{x}^1 \ldots d\bar{x}^n, \tag{4.56}$$

provided that

$$\det\left[\frac{\partial x^i}{\partial \bar{x}^j}\right] > 0. \tag{4.57}$$

Next, suppose that $\mathscr{D}$ cannot be covered by one co-ordinate system. In order to define the volume of $\mathscr{D}$, we introduce the concept of a *partition of unity* on $\mathscr{D}$. First, we choose a covering of $\mathscr{D}$ by positive co-ordinate charts $(\mathscr{D}_1, \boldsymbol{\kappa}_1), (\mathscr{D}_2, \boldsymbol{\kappa}_2), \ldots$, such that each point $\boldsymbol{x} \in \mathscr{D}$ has a neighborhood which meets only a finite number of charts in the covering.

In topology it is known that a covering of this kind, which is called *locally finite*, exists. A *partition of unity*[1] associated with the covering $\{(\mathscr{D}_a, \boldsymbol{\kappa}_a)\}$ is a collection of differentiable functions $\{f_a\}$ such that

i) Each f_a vanishes outside $\mathscr{D}_a$, and

$$0 \leqq f_a \leqq 1. \tag{4.58}$$

ii) For each $\boldsymbol{x} \in \mathscr{D}$,

$$\sum_{a=1}^{\infty} f_a(\boldsymbol{x}) = 1. \tag{4.59}$$

Since the covering is locally finite, only finitely many summands in the infinite sum differ from zero.

Now we define the volume of $\mathscr{D}$ simply by

$$V(\mathscr{D}) \equiv \int_{\mathscr{D}} \mathbf{V} \equiv \sum_{a=1}^{\infty} \int_{\boldsymbol{\kappa}_a(\mathscr{D} \cap \mathscr{D}_a)} f_a v_a \, dx_a^1 \ldots dx_a^n, \tag{4.60}$$

where v_a denotes the specific volume relative to the co-ordinate system $\boldsymbol{\kappa}_a = (x_a^i)$, $a = 1, 2, \ldots$. Of course, the series on the right-hand side of (4.60) need not converge. However, if it does converge, then its value is independent of the choice of the partition of unity and of the locally finite covering.

Having defined the volume $V(\mathscr{D})$ of any open set $\mathscr{D}$ of $\mathscr{M}$, we can proceed to define the integral of a continuous function on $\mathscr{D}$ in the standard way. Namely, for the simple case when $\mathscr{D}$ can be covered by as positive co-ordinate system $\boldsymbol{\kappa} = (x^i)$, we put

$$\int_{\mathscr{D}} f\mathbf{V} \equiv \int_{\boldsymbol{\kappa}(\mathscr{D})} f v \, dx^1 \ldots dx^n; \tag{4.61}$$

for the general case $\mathscr{D}$ can be equipped with a covering $\{(\mathscr{D}_a, \boldsymbol{\kappa}_a)\}$ and a partition of unity $\{f_a\}$, and we put

$$\int_{\mathscr{D}} f\mathbf{V} \equiv \sum_{a=1}^{\infty} \int_{\boldsymbol{\kappa}_a(\mathscr{D} \cap \mathscr{D}_a)} f f_a v_a \, dx_a^1 \ldots dx_a^n. \tag{4.62}$$

In both cases, the integral is well defined and is independent of the choice of the co-ordinate system.

---

[1] For a proof of the existence of a partition of unity, we refer the reader to Appendix 3 of Volume 1 of *Foundations of Differential Geometry* by S. Kobayashi & K. Nomizu, Interscience Publishers, John Wiley & Sons, 1963.

5. Parallelisms on Manifolds

As remarked in the preceding section, a Euclidean space is equipped with a Euclidean parallelism which assigns a linear isomorphism between the tangent spaces of any two points. Such a parallelism induces the operation of covariant differentiation of vector or tensor fields. In this section we examine first the structure of the Euclidean parallelism within the context of a differentiable manifold. Then we extend the concept of parallelism to manifolds in general.

We define first the *bundle of tangent spaces* $\mathscr{T}(\mathscr{V})$, or more briefly the *tangent bundle* of a Euclidean space $\mathscr{V}$ as being the union of all tangent spaces of points of $\mathscr{V}$, *viz*,

$$\mathscr{T}(\mathscr{V}) = \bigcup_{x \in \mathscr{V}} \mathscr{V}_x. \tag{5.1}$$

We call $\mathscr{V}_x$ a *fibre* of $\mathscr{T}(\mathscr{V})$ and $\mathscr{V}$ itself the *base space* of $\mathscr{T}(\mathscr{V})$. The *projection map*

$$\pi : \mathscr{T}(\mathscr{V}) \to \mathscr{V} \tag{5.2}$$

is defined naturally by

$$\pi(\mathscr{V}_x) = x \;\; \forall x \in \mathscr{V}. \tag{5.3}$$

Thus the fibre $\mathscr{V}_x$ is given by

$$\mathscr{V}_x = \pi^{-1}(x). \tag{5.4}$$

These terms suggest that we can visualize $\mathscr{T}(\mathscr{V})$ literally as a bundle of fibres with the base space as the set of indices for them.

A (*cross*) *section* of $\mathscr{T}(\mathscr{V})$ is a mapping

$$\sigma : \mathscr{V} \to \mathscr{T}(\mathscr{V}) \tag{5.5}$$

such that

$$\pi \circ \sigma = \mathrm{id}_{\mathscr{V}}. \tag{5.6}$$

The condition (5.6) means simply that the image of any point x under σ belongs to the fibre $\mathscr{V}_x$ at x for all $x \in \mathscr{V}$. Since the fibre $\mathscr{V}_x$ is the tangent space at x, a section σ assigns a tangent vector at every point. Thus a section of $\mathscr{T}(\mathscr{V})$ is nothing but a vector field on $\mathscr{V}$.

From the Euclidean parallelism, we can identify the fibres of $\mathscr{T}(\mathscr{V})$ with $\mathscr{V}$ and hence the tangent bundle $\mathscr{T}(\mathscr{V})$ with the product space $\mathscr{V} \times \mathscr{V}$. Equivalently, this condition means that there exists a family of *constant sections* of $\mathscr{T}(\mathscr{V})$ such that each point $(x, v) \in \mathscr{T}(\mathscr{V})$, where $v \in \mathscr{V}_x$, is contained in one and only one such section. Of course there are sections of $\mathscr{T}(\mathscr{V})$ which are not constant; these sections correspond to vector fields in general. As we shall see, the covariant derivative of vector field $\boldsymbol{v}$ in general can be identified as the *slope* of the section $\boldsymbol{\sigma}$ corresponding to $\boldsymbol{v}$ relative to the constant sections. In other words, if we regard the constant sections to be sections having slope zero, then we can use them as the datum sections for taking the slope of other sections. For this reason the constant sections are also called the *horizontal sections* of $\mathscr{T}(\mathscr{V})$. The fibre $\mathscr{V}_{\boldsymbol{x}}$ then becomes the *vertical space*. It should be noted, however, that both the horizontal and the vertical spaces here are n-dimensional.

To illustrate the concept of the *slope* of a section relative to a given datum, we consider first the simple case of the *slope* of a function. We define the *bundle of scalars*, or more briefly the *scalar bundle*, $\mathscr{R}(\mathscr{V})$ of $\mathscr{V}$ by

$$\mathscr{R}(\mathscr{V}) \equiv \bigcup_{\boldsymbol{x}\in\mathscr{V}} \mathscr{R}_{\boldsymbol{x}} = \mathscr{R} \times \mathscr{V}, \tag{5.7}$$

where

$$\mathscr{R}_{\boldsymbol{x}} = \mathscr{R} \quad \forall \boldsymbol{x} \in \mathscr{V}. \tag{5.8}$$

Then a function on $\mathscr{V}$ corresponds to a section of $\mathscr{R}(\mathscr{V})$. Of course, we require the sections corresponding to functions belonging to $\mathfrak{Z}(\boldsymbol{x})$ to be sections with *slope zero*. In particular, constant sections corresponding to constant functions are sections with slope zero. In general, we say that two sections have the same *slope* at $\boldsymbol{x}$ if their corresponding functions belong to the same equivalence class in $\mathfrak{F}(\boldsymbol{x})/\mathfrak{Z}(\boldsymbol{x})$. Thus the slope of a section corresponds to the differential of the function of the section, and *vice versa*.

Now we go back to the discussion of the covariant derivative of a vector field. In order to define the concept of the *slope* of a section precisely, we must introduce first a differentiable structure on $\mathscr{T}(\mathscr{V})$. Such a structure is easily found from the isomorphism of $\mathscr{T}(\mathscr{V})$ with $\mathscr{V} \times \mathscr{V}$. In fact, let (x^i) be a Cartesian co-ordinate system on $\mathscr{V}$ induced by the basis $\{\boldsymbol{e}_i\}$. Then (x^i, v^j), $i = 1, \ldots, n$, $j = 1, \ldots, n$, is a global co-ordinate system on $\mathscr{T}(\mathscr{V})$ with

$$x^i(\boldsymbol{x}, \boldsymbol{v}) = x^i(\boldsymbol{x}) = \langle \boldsymbol{x}, \boldsymbol{e}^i \rangle, \tag{5.9}$$

and

$$v^i(\boldsymbol{x}, \boldsymbol{v}) = \langle \boldsymbol{v}, \boldsymbol{e}^i \rangle, \tag{5.10}$$

for all $(\boldsymbol{x}, \boldsymbol{v}) \in \mathscr{T}(\mathscr{V})$. We call (x^i, v^j) the *lifted co-ordinate system* of the co-ordinate system (x^i).

Relative to the lifted co-ordinate system (x^i, v^j), the horizontal section is spanned by the vectors $\{\partial_{\boldsymbol{x},\boldsymbol{v}} x_i\}$, and the vertical space is spanned by the vectors $\{\partial_{\boldsymbol{x},\boldsymbol{v}} v_i\}$ for all $(\boldsymbol{x}, \boldsymbol{v}) \in \mathscr{T}(\mathscr{V})$. The subscripts $(\boldsymbol{x}, \boldsymbol{v})$ in these notations are important, since they indicate that the vectors themselves are tangents of $\mathscr{T}(\mathscr{V})$, not tangents of $\mathscr{V}$. Of course, the complete set

$$\{\partial_{\boldsymbol{x},\boldsymbol{v}} x_1, \ldots, \partial_{\boldsymbol{x},\boldsymbol{v}} x_n, \partial_{\boldsymbol{x},\boldsymbol{v}} v_1, \ldots, \partial_{\boldsymbol{x},\boldsymbol{v}} v_n\} \tag{5.11}$$

is the natural basis of the co-ordinate system (x^i, v^j) at $(\boldsymbol{x}, \boldsymbol{v})$.

Now let $\boldsymbol{u}$ be a differentiable vector field. Then the section of $\mathscr{T}(\mathscr{V})$ corresponding to $\boldsymbol{u}$ carries any differentiable curve $\boldsymbol{\lambda}$ in $\mathscr{V}$ into a differentiable curve $\boldsymbol{u} \circ \boldsymbol{\lambda}$ in $\mathscr{T}(\mathscr{V})$. We call $\boldsymbol{u} \circ \boldsymbol{\lambda}$ the ***u**-lift* of $\boldsymbol{\lambda}$. The relation between $\boldsymbol{\lambda}$ and $\boldsymbol{u} \circ \boldsymbol{\lambda}$ has been explained in general in connection with an arbitrary differentiable map $\boldsymbol{\kappa}$ (*cf.* (3.53) and (4.3)). Here $\boldsymbol{u}$ maps $\mathscr{V}$ to $\mathscr{T}(\mathscr{V})$, and $\boldsymbol{\lambda}$ is a curve in $\mathscr{V}$, so that the tangent $\partial\boldsymbol{\lambda}$ belongs to $\mathscr{V}$ also. On the other hand, $\boldsymbol{u} \circ \boldsymbol{\lambda}$ is a curve in $\mathscr{T}(\mathscr{V})$, so that its tangent $\partial(\boldsymbol{u} \circ \boldsymbol{\lambda})$ belongs to the tangent space of $\mathscr{T}(\mathscr{V})$. Since we have already set up a vertical space and a horizontal space at any point of $\mathscr{F}(\mathscr{V})$, this tangent $\partial(\boldsymbol{u} \circ \boldsymbol{\lambda})$ can be decomposed uniquely into a horizontal component and a vertical component. The *slope of **u** with respect to the parameter of* $\boldsymbol{\lambda}$ is then defined simply as the vertical component of $\partial(\boldsymbol{u} \circ \boldsymbol{\lambda})$ and is denoted by $d\boldsymbol{u}/dt$.

Relative to the lifted co-ordinate system (x^i, v^j) the co-ordinates of $\boldsymbol{u} \circ \boldsymbol{\lambda}$ are

$$x^i(\boldsymbol{u} \circ \boldsymbol{\lambda}(t)) = x^i(\boldsymbol{\lambda}(t)) = \lambda^i(t), \tag{5.12}$$

and

$$v^i(\boldsymbol{u} \circ \boldsymbol{\lambda}(t)) = u^i(\lambda^1(t), \ldots, \lambda^n(t)), \tag{5.13}$$

where $\lambda^i(t)$ and $u^i(x^1, \ldots, x^n)$ are the co-ordinates of $\boldsymbol{\lambda}(t)$ and the components of $\boldsymbol{u}$, respectively, in the co-ordinate system (x^i). As usual, we can compute the tangent vector of $\boldsymbol{u} \circ \boldsymbol{\lambda}$ by the formula

$$\partial(\boldsymbol{u} \circ \boldsymbol{\lambda}) = \frac{d\lambda^i}{dt} \partial x_i + \frac{\partial u^i}{\partial x^j} \frac{d\lambda^j}{dt} \partial v_i. \tag{5.14}$$

But since $\{\partial_{\boldsymbol{u}\circ\boldsymbol{\lambda}} x_i\}$ spans the horizontal space at $\boldsymbol{u} \circ \boldsymbol{\lambda}$, while $\{\partial_{\boldsymbol{u}\circ\boldsymbol{\lambda}} v_j\}$

spans the vertical space at $\boldsymbol{u} \circ \boldsymbol{\lambda}$, the component form (5.14) shows clearly that

$$\frac{d\boldsymbol{u}}{dt} = \frac{\partial u^i}{\partial x^j} \frac{d\lambda^j}{dt} \partial v_i. \tag{5.15}$$

Thus the directional slope $d\boldsymbol{u}/dt$ depends linearly on the direction vector $\partial\boldsymbol{\lambda}$.

The vertical vector $d\boldsymbol{u}/dt$ can be regarded as a vector in $\mathscr{V}$ also, since the vertical space is canonically isomorphic to $\mathscr{V}$. Under this canonical isomorphism the natural basis vector ∂v_j of the Cartesian co-ordinate system (v^j) on the vertical space is identified with the original basis vector $\boldsymbol{e}_j$. Then the component form (5.15) becomes

$$\frac{d\boldsymbol{u}}{dt} = \frac{\partial u^i}{\partial x^j} \frac{d\lambda^j}{dt} \boldsymbol{e}_i. \tag{5.16}$$

Now we define the *slope* of $\boldsymbol{u}$ simply as the linear map $D\boldsymbol{u}$ that carries the direction vector $\partial\boldsymbol{\lambda}$ into the directional slope $d\boldsymbol{u}/dt$ for all directions. From (5.16), the component form of $D\boldsymbol{u}$ relative to $\{\boldsymbol{e}_i\}$ is

$$D\boldsymbol{u} = \frac{\partial u^i}{\partial x^j} \boldsymbol{e}_i \otimes \boldsymbol{e}^j. \tag{5.17}$$

Thus the slope is nothing but the covariant derivative as defined in § 3.

In the special case of a vector field $\boldsymbol{u}$ on a Euclidean space, the gradient and the slope coincide : grad $\boldsymbol{u} = D\boldsymbol{u}$, as explained in § 4. For a vector field on a general manifold, the gradient is not defined, since a manifold is not equipped with an intrinsic parallelism.

So far, we have used only a Cartesian co-ordinate system (x^i) and its lifted co-ordinate system (x^i, v^j). The advantage of such a choice is evident, since the natural basis of (x^i, v^j) splits into a generating set $\{\partial_{\boldsymbol{x},\boldsymbol{v}} x_i\}$ for the horizontal space and a generating set $\{\partial_{\boldsymbol{x},\boldsymbol{v}} v_i\}$ for the vertical space at any point $(\boldsymbol{x}, \boldsymbol{v})$ in $\mathscr{T}(\mathscr{V})$. In a general co-ordinate system (x^i), this simple decomposition does not hold. We define the *lifted co-ordinate system* (x^i, v^j) of (x^i) by

$$x^i(\boldsymbol{x}, \boldsymbol{v}) = x^i(\boldsymbol{x}), \tag{5.18}$$

and

$$v^i(\boldsymbol{x}, \boldsymbol{v}) \equiv \langle \boldsymbol{v}, \partial x^i \rangle. \tag{5.19}$$

Thus (v^i) remains a Cartesian co-ordinate system on the fibre. However, since the natural basis of (x^i) is not a constant basis in general, the section with

$$v^i = \text{constant} \tag{5.20}$$

for all $i = 1, \ldots, n$ need not be horizontal. Consequently the set $\{\partial_{\boldsymbol{x},\boldsymbol{v}} x_i\}$ no longer forms a generating set for the horizontal space at $(\boldsymbol{x}, \boldsymbol{v})$. On the other hand, the set $\{\partial_{\boldsymbol{x},\boldsymbol{v}} v_i\}$ still spans the vertical space, since the fibre is still characterized by

$$x^i = \text{constant}, \quad i = 1, \ldots, n. \tag{5.21}$$

PROPOSITION. *Relative to the lifted co-ordinate system* (x^i, v^j) *of a co-ordinate system* (x^i), *the horizontal space at any point* $(\boldsymbol{x}, \boldsymbol{v})$ *is spanned by the set*

$$\{\partial_{\boldsymbol{x},\boldsymbol{v}} x_i - \Gamma_k{}^j{}_i(\boldsymbol{x}) v^k \partial_{\boldsymbol{x},\boldsymbol{v}} v_j\}, \tag{5.22}$$

where the coefficients $\Gamma_i{}^j{}_k$ *are the connection symbols of the Euclidean parallelism relative to* (x^i), *defined by* (3.76).

Using this set of generators, we can recover the formula for the covariant derivative in the following way. First, let $\boldsymbol{u}$ be an arbitrary differentiable vector field, and let $\boldsymbol{\lambda}$ be a differentiable curve. From (5.18) and (5.19), the co-ordinates of the $\boldsymbol{u}$-lift $\boldsymbol{u} \circ \boldsymbol{\lambda}$ in (x^i, v^j) are again given by (5.12) and (5.13), and hence the tangent $\partial \boldsymbol{u} \circ \boldsymbol{\lambda}$ is again given by (5.14). But now the leading term of the component form (5.14) is not a horizontal vector any more. Indeed, from (5.22), the horizontal part of $\partial(\boldsymbol{u} \circ \boldsymbol{\lambda})$ is

$$\frac{d\lambda^i}{dt}(\partial x_i - \Gamma_k{}^j{}_i u^k \partial v_j), \tag{5.23}$$

while the vertical part is

$$\frac{d\boldsymbol{u}}{dt} = \left(\frac{\partial u^i}{\partial x^j} + \Gamma_k{}^i{}_j u^k\right) \frac{d\lambda^j}{dt} \partial v_i. \tag{5.24}$$

Since $\boldsymbol{\lambda}$ is arbitrary, we recover the formula

$$u^i{}_{|j} = \frac{\partial u^i}{\partial x^j} + \Gamma_k{}^i{}_j u^k, \tag{5.25}$$

which is the special case of (3.85) when $r = 1$ and $s = 0$.

Having considered the Euclidean parallelism as a structure on the tangent bundle of a Euclidean space, we can now generalize the concept to an arbitrary differentiable manifold. Let $\mathscr{M}$ be a differentiable manifold. Then we define the *bundle of tangent spaces* or briefly the *tangent bundle* $\mathscr{T}(\mathscr{M})$ of $\mathscr{M}$ by

$$\mathscr{T}(\mathscr{M}) = \bigcup_{\boldsymbol{x} \in \mathscr{M}} \mathscr{M}_{\boldsymbol{x}}. \tag{5.26}$$

The *fibres* of the tangent bundle are again the tangent spaces of the base space, and we have the projection map

$$\pi : \mathscr{T}(\mathscr{M}) \to \mathscr{M} \tag{5.27}$$

with

$$\pi(\mathscr{M}_{\boldsymbol{x}}) = \boldsymbol{x} \tag{5.28}$$

for all $\boldsymbol{x} \in \mathscr{M}$.

We can introduce a differentiable structure on $\mathscr{T}(\mathscr{M})$ in the same way as before. Let (x^i) be a co-ordinate system defined on the co-ordinate neighborhood $\mathscr{D} \subset \mathscr{M}$. Then we define a lifted co-ordinate system (x^i, v^j) on $\mathscr{T}(\mathscr{D})$ again by the condition (5.18) and (5.19). The reader will verify easily that the lifted co-ordinate systems are C^∞-related. Consequently, $\mathscr{T}(\mathscr{M})$ becomes a differentiable manifold of dimension $2n$ with respect to the differentiable structure generated by the lifted co-ordinate systems. As before, a vector field on $\mathscr{M}$ corresponds to a section of $\mathscr{T}(\mathscr{M})$, and a vector field is differentiable if and only if its corresponding section is a differentiable map.

Unlike the tangent bundle of a Euclidean space, the tangent bundle of a differentiable manifold need not be diffeomorphic to a product space of the manifold with a vector space. The condition of smoothness is crucial here. Individually, the fibres of $\mathscr{T}(\mathscr{M})$ are all *isomorphic* to the vector space $\mathscr{R}^n$, but the bundle $\mathscr{T}(\mathscr{M})$ need not be *diffeomorphic* (or even *homeomorphic*) to $\mathscr{M} \times \mathscr{R}^n$. The simplest counterexample for such a differentiable manifold is the ordinary 2-dimensional sphere $\mathscr{S}^2$. It can be shown[1] that the tangent bundle $\mathscr{T}(\mathscr{S}^2)$ is not homeomorphic to the product space $\mathscr{S}^2 \times \mathscr{R}^2$.

In view of this counterexample, we see that a manifold in general cannot be equipped with a Euclidean parallelism. However, from the meaning of the covariant derivative, explained earlier in this section, we can define a more general concept of parallelism that is applicable to all manifolds.

---

[1] *Cf.* § 27 of N. Steenrod, *The Topology of Fibre Bundles*. Princeton University Press, 1951.

Instead of the horizontal sections of the Euclidean parallelism, we introduce a distribution of horizontal spaces on the tangent bundle of a differentiable manifold. While the Euclidean parallelism corresponds to an integrable distribution of horizontal spaces on the tangent bundle of a Euclidean space, a distribution of horizontal spaces on the tangent bundle of a manifold need not be integrable in general. We call any distribution of horizontal spaces on the tangent bundle a *connection*. As we shall see later, the tangent bundle of any manifold can be equipped with many connections. We shall prove a general existence theorem for a class of *affine connections* that satisfy some conditions in addition to that of smoothness of a distribution (*cf.* § 4).

Let us now consider the tangent bundle $\mathscr{T}(\mathscr{M})$ of a manifold $\mathscr{M}$ in general. For any point $\boldsymbol{x} \in \mathscr{M}$ and any tangent vector $\boldsymbol{v} \in \mathscr{M}_{\boldsymbol{x}}$ the pair $(\boldsymbol{x}, \boldsymbol{v})$ constitutes a typical point of $\mathscr{T}(\mathscr{M})$. As usual, we denote the tangent space of $\mathscr{T}(\mathscr{M})$ at $(\boldsymbol{x}, \boldsymbol{v})$ by $\mathscr{T}(\mathscr{M})_{(\boldsymbol{x},\boldsymbol{v})}$. Let n be the dimension of $\mathscr{M}$. Then

$$\dim \mathscr{T}(\mathscr{M})_{(\boldsymbol{x},\boldsymbol{v})} = 2n. \tag{5.29}$$

Relative to the lifted co-ordinate system (x^i, v^j) of a co-ordinate system (x^i), the projection map $\boldsymbol{\pi}$ in (5.27) has the representation

$$\pi^i(x^1, \ldots, x^n, v^1, \ldots, v^n) = x^i, \quad i = 1, 2, \ldots, n, \tag{5.30}$$

so that $\boldsymbol{\pi}$ is a differentiable map and grad $\boldsymbol{\pi}$ has the component matrix

$$[\operatorname{grad} \pi^i_\alpha] = \begin{bmatrix} 1 & 0 \ldots 0 & 0 & 0 \ldots 0 \\ 0 & 1 \quad \vdots & \vdots & \vdots \quad\; \vdots \\ \vdots & \quad\;\; \vdots & \vdots & \vdots \quad\; \vdots \\ \vdots & \quad\;\; \vdots & \vdots & \vdots \quad\; \vdots \\ 0 & \ldots 1 & 0 & \ldots\ldots \\ 0 & \ldots 0 & 1 & 0 \ldots 0 \end{bmatrix}. \tag{5.31}$$

In other words, $\boldsymbol{\pi}_*$ satisfies the conditions

$$\boldsymbol{\pi}_{*(\boldsymbol{x},\boldsymbol{v})} \partial_{\boldsymbol{x},\boldsymbol{v}} x_i = \partial_{\boldsymbol{x}} x_i \tag{5.32}$$

and

$$\boldsymbol{\pi}_{*(\boldsymbol{x},\boldsymbol{v})} \partial_{\boldsymbol{x},\boldsymbol{v}} v_i = \mathbf{0}, \quad i = 1, \ldots, n. \tag{5.33}$$

As before, we define the *vertical space* $\mathscr{V}_{(\boldsymbol{x},\boldsymbol{v})}$ at $(\boldsymbol{x}, \boldsymbol{v})$ by

$$\mathscr{V}_{(\boldsymbol{x},\boldsymbol{v})} = \ker \boldsymbol{\pi}_{*(\boldsymbol{x},\boldsymbol{v})}. \tag{5.34}$$

Of course, $\mathscr{V}_{(\boldsymbol{x},\boldsymbol{v})}$ is nothing but the tangent space of the fibre $\mathscr{M}_{\boldsymbol{x}}$ at the position $\boldsymbol{v} \in \mathcal{M}_x$. Relative to the Euclidean parallelism of the Euclidean space $\mathcal{M}_x$, $\mathscr{V}_{(x,v)}$ is canonically isomorphic to $\mathcal{M}_x$.

An n-dimensional subspace $\mathscr{H}_{(x,v)}$ of $\mathscr{T}(\mathcal{M})_{(x,v)}$ is called a *horizontal space* at (x, v) if $\mathscr{V}_{(x,v)} \cup \mathscr{H}_{(x,v)}$ generates $\mathscr{T}(\mathcal{M})_{(x,v)}$. In other words, $\mathscr{T}(\mathcal{M})_{(x,v)}$ may be decomposed as follows into a direct sum:

$$\mathscr{T}(\mathcal{M})_{(x,v)} = \mathscr{V}_{(x,v)} \oplus \mathscr{H}_{(x,v)}. \tag{5.35}$$

This decomposition is similar to the isomorphism $\mathscr{T}(\mathscr{V}) \cong \mathscr{V} \oplus \mathscr{V}$, which pertains to the Euclidean case. In general, of course, there is no intrinsic horizontal space on the tangent bundle of an arbitrary differentiable manifold. At any point $(\boldsymbol{x}, \boldsymbol{v}) \in \mathscr{T}(\mathcal{M})$, we can find infinitely many subspaces that satisfy the condition (5.35) for a horizontal space. As we mentioned before, a distribution of horizontal spaces on the tangent bundle is called a *connection*.

Before showing the existence of a connection in general, we explain first the concept of a *path-dependent parallelism* induced by a connection. Let $\mathscr{H}$ be a connection on $\mathscr{T}(\mathcal{M})$, and let $\boldsymbol{\lambda}$ be a differentiable curve in $\mathcal{M}$. Then a differentiable curve $\tilde{\boldsymbol{\lambda}}$ in $\mathscr{T}(\mathcal{M})$ is called a *horizontal lift* (or an *$\mathscr{H}$-lift*) of $\boldsymbol{\lambda}$ if the following two conditions are satisfied:

$$\partial\tilde{\boldsymbol{\lambda}}(t) \in \mathscr{H}_{\tilde{\boldsymbol{\lambda}}(t)}, \tag{5.36}$$

and

$$\pi(\tilde{\boldsymbol{\lambda}}(t)) = \boldsymbol{\lambda}(t), \tag{5.37}$$

for all values t of the parameter of the curve $\boldsymbol{\lambda}$. Plainly, the first condition means that $\tilde{\boldsymbol{\lambda}}$ is a horizontal curve with respect to the particular connection $\mathscr{H}$ on $\mathscr{T}(\mathcal{M})$, while the second condition means that $\tilde{\boldsymbol{\lambda}}$ lies directly above $\boldsymbol{\lambda}$ with respect to the projection map π.

The conditions of smoothness for $\mathscr{H}$ and $\boldsymbol{\lambda}$ ensure the truth of the following

PROPOSITION. *Let $\boldsymbol{\lambda}(0)$ be any fixed reference point on $\boldsymbol{\lambda}$. Then for any $\boldsymbol{v}(0) \in \mathcal{M}_{\boldsymbol{\lambda}(0)}$ there exists a unique $\mathscr{H}$-lift $\tilde{\boldsymbol{\lambda}}(t, \boldsymbol{v}(0))$ such that*

$$\tilde{\boldsymbol{\lambda}}(0, \boldsymbol{v}(0)) = (\boldsymbol{\lambda}(0), \boldsymbol{v}(0)). \tag{5.38}$$

In other words, there exists a family of $\mathscr{H}$-lifts of $\boldsymbol{\lambda}$ such that each point $(\boldsymbol{\lambda}(t), \boldsymbol{v}(t)) \in \pi^{-1}(\boldsymbol{\lambda}(t))$ is contained in one and only one of the $\mathscr{H}$-lifts. We can visualize the $\mathscr{H}$-lifts of $\boldsymbol{\lambda}$ as the graph of a 1-parameter family of mappings

$$\rho_{0,t} : \mathscr{M}_{\lambda(0)} \to \mathscr{M}_{\lambda(t)} \tag{5.39}$$

such that

$$\tilde{\lambda}(t, \mathbf{v}(0)) = (\lambda(t), \rho_{0,t}(\mathbf{v}(0))) \tag{5.40}$$

for all $\mathbf{v}(0) \in \mathscr{M}_{\lambda(0)}$. The mappings $\rho_{0,t}$ defined in this manner are diffeomorphisms of $\mathscr{M}_{\lambda(0)}$ with the fibres along λ and are called the *parallel transport induced by $\mathscr{H}$ along λ.*

The proof of this proposition follows as a direct consequence of the standard existence and uniqueness theorem for ordinary differential equations. From the condition (5.35) we know that the restriction of $\pi_{*(\mathbf{x},\mathbf{v})}$ to $\mathscr{H}_{(\mathbf{x},\mathbf{v})}$ is a linear isomorphism from $\mathscr{H}_{(\mathbf{x},\mathbf{v})}$ to $\mathscr{M}_{\mathbf{x}}$. Hence relative to any lifted co-ordinate system (x^i, v^j), a basis of $\mathscr{H}_{(\mathbf{x},\mathbf{v})}$ has the component form

$$\{\partial_{\mathbf{x},\mathbf{v}} x_i - \Gamma_i^j(\mathbf{x}, \mathbf{v})\partial_{\mathbf{x},\mathbf{v}} v_j\}, \tag{5.41}$$

where Γ_i^j denotes a C^∞ function of (x^k, v^l). As before, let the co-ordinates of λ relative to (x^i) be $\lambda^i(t)$, and let the co-ordinates of $\tilde{\lambda}$ relative to (x^i, v^j) be $\lambda^i(t)$, $v^j(t)$. Then the condition (5.37) is automatically satisfied, while the condition (5.36) is equivalent to the system of differential equations

$$\dot{v}^j(t) + \Gamma_i^j(\lambda^1(t), \ldots, \lambda^n(t), v^1(t), \ldots, v^n(t))\dot{\lambda}^i(t) = 0, \quad j = 1, \ldots, n. \tag{5.42}$$

We call this system the *equation of parallel transport* along λ in the co-ordinate system (x^i). The solution of this system with the initial condition $v^i(0)$ gives the parallel transport $\rho_{0,t}(\mathbf{v}(0))$ of $\mathbf{v}(0)$.

If the connection $\mathscr{H}$ is integrable, then *locally* there exists a horizontal section passing through each point $(\mathbf{x}, \mathbf{v}) \in \mathscr{T}(\mathscr{M})$. In this case, the $\mathscr{H}$-lift of any curve λ is simply the intersection of the vertical cylinder $\pi^{-1}(\lambda)$ with the horizontal sections. Consequently, the parallel transport is locally path-independent. Conversely, if the parallel transport is locally path-independent, then the local sections of parallel vectors are horizontal, and hence the connection is integrable. Thus the path-independence of the parallel transport induced by $\mathscr{H}$ is characterized by the integrability condition of $\mathscr{H}$. As we remarked before, a manifold in general need not have any integrable connections. Thus we cannot require the parallel transport to be path-independent in general.

Next we observe that the parallel transport induced by a connection generally need not be linear. This remark is easily seen from the equation of parallel transport (5.42), which need not be a linear system for $\boldsymbol{v}$. In order that the parallel transports along all curves be linear isomorphisms, it is necessary and sufficient that the functions $\Gamma^i_j(x^1, \ldots, x^n, v^1, \ldots, v^n)$ be linear in v^k, *viz*

$$\Gamma^j_i(x^1, \ldots, x^n, v^1, \ldots, v^n) = \Gamma_k{}^j{}_i(x^1, \ldots, x^n)v^k. \tag{5.43}$$

A connection satisfying this condition is then called an *affine connection.*

While an integrable connection exists only for certain manifolds, affine connections exist for all, as we shall prove shortly. Also, in the applications in this book we shall use affine connections only.

As before, the functions $\Gamma_k{}^j{}_i(x^1, \ldots, x^n)$ that characterize an affine connection $\mathscr{H}$ are called the *connection symbols* of $\mathscr{H}$ relative to the co-ordinate system (x^i). We proceed to derive the transformation rule of the connection symbols with respect to a change of co-ordinate system. From (5.41) and (5.43), the horizontal space $\mathscr{H}_{(x,v)}$ is spanned by the set

$$\{\partial_{x,v}x_i - \Gamma_j{}^k{}_i(x^1, \ldots, x^n)v^j\,\partial_{x,v}v_k\}, \tag{5.44}$$

Let $\bar{\Gamma}_j{}^k{}_i$ be the connection symbols of $\mathscr{H}$ relative to the co-ordinate system $(\bar{x}^i)$. Then $\mathscr{H}_{(x,v)}$ is spanned also by the set

$$\{\partial_{x,v}\bar{x}_i - \bar{\Gamma}_l{}^m{}_i(\bar{x}^1, \ldots, \bar{x}^n)\bar{v}^l\partial_{x,v}\bar{v}_m\}. \tag{5.45}$$

Since the lifted co-ordinate systems (x^i, v^j) and $(\bar{x}^i, \bar{v}^j)$ are related by

$$\bar{x}^i = \bar{x}^i(x^1, \ldots, x^n), \tag{5.46}$$

and

$$\bar{v}^i = \frac{\partial \bar{x}^i}{\partial x^j}\, v^j, \tag{5.47}$$

their natural bases obey the usual transformation rules:

$$\partial_{x,v}x_i = \left.\frac{\partial \bar{x}^j}{\partial x^i}\right|_x \partial_{x,v}\bar{x}_j + \left.\frac{\partial^2 \bar{x}^j}{\partial x^i \partial x^k}\right|_x v^k\partial_{x,v}\bar{v}_j, \tag{5.48}$$

and

$$\partial_{x,v}v_i = \left.\frac{\partial \bar{x}^j}{\partial x^i}\right|_x \partial_{x,v}\bar{v}_j. \tag{5.49}$$

Substituting (5.48) and (5.49) into (5.44) and using the fact that the Jacobian matrix $[\partial\bar{x}^i/\partial x^j]$ is non-singular, we see that $\mathscr{H}_{(x,v)}$ is spanned also by the set

$$\left\{\partial_{x,v}\bar{x}_i - \left(\Gamma_{p\ q}^{\ k}\frac{\partial x^q}{\partial \bar{x}^i}\frac{\partial x^p}{\partial \bar{x}^l}\frac{\partial \bar{x}^m}{\partial x^k} - \frac{\partial^2 \bar{x}^m}{\partial x^p \partial x^k}\frac{\partial x^k}{\partial \bar{x}^l}\frac{\partial x^p}{\partial \bar{x}^i}\right)\bar{v}^l\partial_{x,v}\bar{v}_m\right\}. \tag{5.50}$$

Comparing this set with the set (5.45) yields directly the transformation rule

$$\bar{\Gamma}_{l\ i}^{\ m} = \Gamma_{p\ q}^{\ k}\frac{\partial x^q}{\partial \bar{x}^i}\frac{\partial x^p}{\partial \bar{x}^l}\frac{\partial \bar{x}^m}{\partial x^k} - \frac{\partial^2 \bar{x}^m}{\partial x^p \partial x^k}\frac{\partial x^k}{\partial \bar{x}^l}\frac{\partial x^p}{\partial \bar{x}^i}, \tag{5.51}$$

which is equivalent to (3.78) and (3.79), the rules of transformation for the connection symbols of the Euclidean connection.

Like the transformation rule for the components of a tensor field, the transformation rule (5.51) can be used to define an affine connection in terms of the connection symbols. Suppose that the functions $\Gamma_{j\ k}^{\ i}$ be assigned in each co-ordinate system (x^i), and suppose that (5.51) be satisfied. Then we define $\mathscr{H}_{(x,v)}$ simply as the subspace spanned by the set (5.44). The transformation rule ensures that the subspace $\mathscr{H}_{(x,v)}$ thus defined be independent of the choice of the co-ordinate system. We now follow this approach to prove a general existence theorem for affine connections.

On any co-ordinate neighborhood $\mathscr{D}$ with a co-ordinate system $\kappa = (x^i)$, an affine connection $\mathscr{H}_\kappa$ clearly exists. Indeed, we can choose $\mathscr{H}_\kappa$ as the Euclidean connection whose horizontal sections on $\mathscr{D}$ are the sections with constant components with respect to (x^i). In other words the connection symbols of $\mathscr{H}_\kappa$ relative to (x^i) vanish identically, so that

$$\{\partial_{x,v}x_i\} \tag{5.52}$$

generates $\mathscr{H}_{\kappa(x,v)}$ for all $x \in \mathscr{D}$. Such a choice of $\mathscr{H}_\kappa$, however, does not give rise to a connection $\mathscr{H}$ globally, since the connection symbols of the $\mathscr{H}_\kappa$ corresponding to the various co-ordinate systems do not satisfy the transformation rules (5.51). In order to construct a global affine connection, we use the technique of a partition of unity, introduced in the preceding section.

Let $\{(\mathscr{D}_a, \kappa_a)\}$ be a sequence of charts that cover $\mathscr{M}$, and let $\{f_a\}$ be a partition of unity associated with the covering. We introduce a local affine connection $\mathscr{H}_a$ for each chart $(\mathscr{D}_a, \kappa_a)$ of the covering as before. Let x be an arbitrary point in $\mathscr{M}$, and let (x^i) be an arbitrary co-ordinate system defined near x. We denote the connection symbols of $\mathscr{H}_a$ relative to (x^i) by $\Gamma_j{}^i{}_k(\boldsymbol{x}, a)$. Of course, $\boldsymbol{x}$ belongs to only a finite number of $\mathscr{H}_a$, since the covering $\{(\mathscr{D}_a, \boldsymbol{\kappa}_a)\}$ is locally finite. Then we put

$$\Gamma_j{}^i{}_k(\boldsymbol{x}) \equiv \sum_{a=1}^{\infty} f_a(\boldsymbol{x}) \Gamma_j{}^i{}_k(\boldsymbol{x}, a), \tag{5.53}$$

where the number of non-vanishing summands is finite, and the sum is a differentiable function of $(x^1, \ldots, x^n)$. We claim that the functions $\Gamma_j{}^i{}_k(\boldsymbol{x})$ defined in this way relative to any co-ordinate system satisfy the transformation rule (5.51), so that they give rise to a global affine connection as explained before.

To prove that the functions $\Gamma_j{}^i{}_k(\boldsymbol{x})$ defined by (5.53) obey the transformation rule (5.51), we write down first the transformation rule of $\Gamma_j{}^i{}_k(\boldsymbol{x}, a)$ for each a, *viz*,

$$\bar{\Gamma}_l{}^m{}_i(\boldsymbol{x}, a) = \Gamma_p{}^k{}_q(\boldsymbol{x}, a) \frac{\partial x^q}{\partial \bar{x}^i} \frac{\partial x^p}{\partial \bar{x}^l} \frac{\partial \bar{x}^m}{\partial x^k} + \frac{\partial^2 x^k}{\partial \bar{x}^l \partial \bar{x}^i} \frac{\partial \bar{x}^m}{\partial x^k}. \tag{5.54}$$

Multiplying this equation by $f_a(\boldsymbol{x})$ and then summing with respect to a yields

$$\bar{\Gamma}_l{}^m{}_i(\boldsymbol{x}) = \Gamma_p{}^k{}_q(\boldsymbol{x}) \frac{\partial x^q}{\partial \bar{x}^i} \frac{\partial x^p}{\partial \bar{x}^l} \frac{\partial \bar{x}^m}{\partial x^k} + \sum_{a=1}^{\infty} f_a(\boldsymbol{x}) \frac{\partial^2 x^k}{\partial \bar{x}^l \partial \bar{x}^i} \frac{\partial \bar{x}^m}{\partial x^k}. \tag{5.55}$$

But since $\sum_{a=1}^{\infty} f_a = 1$, the last term of (5.55) is the same as that of (5.51). Thus the proof is complete.

It should be noted that Condition i) on the partition of unity $\{f_a\}$ ensures that the right-hand side of (5.53) is smooth with respect to $\boldsymbol{x}$. Also, although the local connections $\mathscr{H}_a$ used in the preceding construction of $\mathscr{H}$ are all integrable, the global connection $\mathscr{H}$ generally is not integrable.

An affine connection $\mathscr{H}$ is called *symmetric* if its connection symbols $\Gamma_j{}^i{}_k$ are symmetric with respect to their subscripts j and k. From (5.51), we see that this condition does not depend on the choice of the co-ordinate system, so that it defines an intrinsic property of the connection. Clearly, the local connections $\mathscr{H}_a$ are all symmetric, since the connection symbols of $\mathscr{H}_a$ vanish identically relative to the co-ordinate chart $(\mathscr{D}_a, \boldsymbol{\kappa}_a)$. Then from (5.53) the global connection $\mathscr{H}$ is also symmetric. Thus the general existence theorem, in fact, asserts that *the tangent bundle of any manifold can be equipped with a symmetric affine connection.* However, in later applications, we shall use affine connections that obey some additional restrictions. Since in some cases those additional restrictions cannot be satisfied by symmetric connections, we do not confine our treatment here to symmetric connections only.

If $\mathscr{H}$ is an arbitrary connection, then from (5.51) the quantities $C_{j\ k}^{\ i}$ defined by

$$C_{j\ k}^{\ i} \equiv \Gamma_{j\ k}^{\ i} - \Gamma_{k\ j}^{\ i} \tag{5.56}$$

satisfy the transformation rule of a tensor field. Thus we can define a tensor field $\mathbf{C}$, called the *torsion tensor* of $\mathscr{H}$, by requiring the $C_{j\ k}^{\ i}$ to be its components. A symmetric connection is called *torsion-free.*

Since the generators (5.44) of an affine connection have the same form as those of a Euclidean connection, the formula for the covariant derivative of a vector field relative to an affine connection is again given by (5.25). The only difference here is that the connection symbols of an affine connection need not vanish in any co-ordinate system, so that the simple formula (3.86) for the covariant derivative in terms of a Cartesian co-ordinate system has no counterpart in general. Indeed, if the manifold is not a Euclidean space, no Cartesian co-ordinate system can be defined for it. The parallel transports relative to an affine connection are linear isomorphisms, so that they give rise to parallel transports of the tensor spaces. The general formula for the covariant derivative of a tensor field relative to an affine connection is again given by (3.85).

Before closing this section, we mention here that the concept of a *metric* can be defined on a manifold in general. We pointed out in § 2 that an inner product on a vector space corresponds to a positive-definite and symmetric covariant tensor of order 2. On a manifold $\mathscr{M}$, we define a *Riemannian metric* g as a differentiable field of such covariant tensors on the tangent spaces. Using the technique of a partition of unity as before, we can prove that *upon any manifold, a Riemannian metric may be defined.* A manifold $\mathscr{M}$ equipped with a given metric g is called a *Riemannian manifold.*

A Riemannian manifold has a unique symmetric affine connection, called the *Riemannian connection*, relative to which the covariant derivative of the Riemannian metric vanishes identically. Indeed, the connection symbols of the Riemannian connection are given by the formula (3.88) with g_{ij} and g^{ij} now denoting the components of the tensor fields $\boldsymbol{g}$ and $\boldsymbol{g}^{-1}$. The Riemannian connection need not be integrable.

A Riemannian manifold $\mathscr{M}$ is called *locally Euclidean* if every point

$x \in \mathscr{M}$ can be covered by a co-ordinate chart relative to which the components of the Riemannian metric are constant fields. Such a co-ordinate chart is then called a *local Cartesian chart*, and it the components of the metric are equal to δ_{ij}, the chart is called a *local rectangular Cartesian chart*. Evidently, a Riemannian manifold is locally Euclidean if and only if its Riemannian connection is integrable. Further, the local horizontal sections are those having constant components with respect to any local Cartesian chart.

A *(globally) Euclidean manifold* is a locally Euclidean manifold that is diffeomorphic to a Euclidean space. Such a manifold is also called a *Euclidean point space*. Clearly, a Euclidean manifold can be equipped with a global Cartesian chart, and any such chart corresponds to a diffeomorphism of the manifold with a Cartesian space. As usual, we call the point having co-ordinates $(0, \ldots, 0)$ the *center* or the *origin*. The set of all global horizontal sections over a Euclidean manifold $\mathscr{M}$ forms an inner-product space $\mathscr{V}$, called the *translation space* of $\mathscr{M}$. The restriction map

$$|_x : \mathscr{V} \to \mathscr{M}_x \tag{5.57}$$

at any $x \in \mathscr{M}$ now defines the *Euclidean parallelism* on $\mathscr{M}$.

If we choose some point $x_0 \in \mathscr{M}$ as a reference origin, then we can define an isomorphism

$$\iota_{x_0} : \mathscr{M} \to \mathscr{V} \tag{5.58}$$

by choosing a Cartesian co-ordinate system (x^i) centered at x_0 and by making the assignment

$$\iota_{x_0}(x) = x^i \partial x_i \tag{5.59}$$

where the x^i are the co-ordinates of x. Clearly, this condition does not depend on the choice of the Cartesian co-ordinate system. We call the operation ι_{x_0} the *translation relative to* x_0 and denote it by

$$\iota_{x_0}(x) \equiv x - x_0 . \tag{5.60}$$

Then the subtraction operation $-$ is a mapping

$$- : \mathscr{M} \times \mathscr{M} \to \mathscr{V} \tag{5.61}$$

satisfying the following three basic conditions:

$$x - x = \mathbf{0} \quad \forall x \in \mathscr{M}, \tag{5.62}$$

$$(x-y)+(y-z) = x-z \quad \forall x, y, z \in \mathscr{M}, \tag{5.63}$$

and for every $\boldsymbol{x} \in \mathscr{M}$ and every $\boldsymbol{v} \in \mathscr{V}$ there exists a unique $\boldsymbol{y} \in \mathscr{M}$ such that

$$\boldsymbol{y} - \boldsymbol{x} = \boldsymbol{v}. \tag{5.64}$$

We denote the point $\boldsymbol{y}$ satisfying the last condition by

$$\boldsymbol{y} = \boldsymbol{x} + \boldsymbol{v} = \boldsymbol{v} = \boldsymbol{v} + \boldsymbol{x}. \tag{5.65}$$

Finally, we remark that an orientable Riemannian manifold can be equipped with a pair of canonical volume tensors $\pm\boldsymbol{V}$ by

$$\boldsymbol{V} \equiv \pm\sqrt{G}\,\partial x^1 \wedge \ldots \wedge \partial x^n, \tag{5.66}$$

where G denotes here the determinant of the matrix $[g_{ij}]$,

$$G \equiv \det\,[g_{ij}]. \tag{5.67}$$

We can remove the $\pm$ signs in (5.66) if an equivalence class of *positive* co-ordinate systems is assigned. In this case, we call $+\sqrt{G}\,\partial x^1 \wedge \ldots \wedge \partial x^n$ the (*positive*) *volume tensor* if (x^i) is a positive system. Then we can integrate a smooth function with respect to this volume tensor, as explained in the preceding section. A Euclidean manifold is orientable, of course.

6. Lie Groups

A Lie group is a manifold equipped with the structure of an abstract group whose group operation and inverse operation are differentiable with respect to the manifold structure. As usual, we write the group operation as a multiplication. Then for any point $\boldsymbol{x}$ belonging to a Lie group $\mathfrak{G}$, the *left multiplication* of $\boldsymbol{x}$ on $\mathfrak{G}$ is a diffeomorphism

$$\boldsymbol{L}_{\boldsymbol{x}} : \mathfrak{G} \to \mathfrak{G} \tag{6.1}$$

defined by

$$\boldsymbol{L}_{\boldsymbol{x}}(\boldsymbol{y}) = \boldsymbol{x}\boldsymbol{y} \quad \forall \boldsymbol{y} \in \mathfrak{G}. \tag{6.2}$$

Similarly, the *right multiplication* of $\boldsymbol{x}$ on $\mathfrak{G}$ is a diffeomorphism

$$\boldsymbol{R}_{\boldsymbol{x}} : \mathfrak{G} \to \mathfrak{G} \tag{6.3}$$

defined by

$$\boldsymbol{R}_{\boldsymbol{x}}(\boldsymbol{y}) = \boldsymbol{y}\boldsymbol{x} \quad \forall \boldsymbol{y} \in \mathfrak{G}. \tag{6.4}$$

The following rules are obvious:

$$L_e = R_e = \mathrm{id}_{\mathfrak{G}}, \tag{6.5}$$

$$L_{xy} = L_x L_y,\ R_{xy} = R_y R_x, \tag{6.6}$$

$$(L_x)^{-1} = L_{x^{-1}},\ (R_x)^{-1} = R_{x^{-1}}, \tag{6.7}$$

and

$$L_x R_y = R_y L_x, \tag{6.8}$$

for all $x, y \in \mathfrak{G}$. Here e denotes the identity element in $\mathfrak{G}$, and x^{-1} denotes the inverse element of x. Since the group operation need not be commutative, the operations L_x and R_x are not the same in general.

We denote the operation of taking the inverse of any element of $\mathfrak{G}$ by

$$J : \mathfrak{G} \to \mathfrak{G}. \tag{6.9}$$

That is,

$$J(x) = x^{-1} \quad \forall x \in \mathfrak{G}. \tag{6.10}$$

Then the left and the right multiplications are related by

$$L_x = J \circ R_{x^{-1}} \circ J, \tag{6.11}$$

and

$$R_x = J \circ L_{x^{-1}} \circ J. \tag{6.12}$$

Also, for any two elements $x, y \in \mathfrak{G}$, there exist unique $a, b \in \mathfrak{G}$ such that

$$y = L_a(x) = R_b(x). \tag{6.13}$$

Indeed, we have

$$a = yx^{-1},\ b = x^{-1}y. \tag{6.14}$$

Now suppose that v is a tangent vector at x. Then the induced linear maps L_{a*x} and R_{b*x} carry v into the vectors $L_{a*x}(v)$ and $R_{b*x}(v)$ at y, called respectively the *left* and the *right translations* of v. We call a vector field u *left-invariant* (respectively *right-invariant*) if its value $u(y)$ at any point y is the left translation (respectively the right translation) of its value $u(x)$ at any other point x, for all x and y. That is, a vector field u on $\mathfrak{G}$ is left invariant if and only if

$$L_{a*}(u) = u \tag{6.15}$$

for all $\boldsymbol{a} \in \mathfrak{G}$. Similarly, a vector field $\boldsymbol{v}$ is right-invariant if and only if

$$\boldsymbol{R}_{a*}(\boldsymbol{v}) = \boldsymbol{v} \tag{6.16}$$

for all $\boldsymbol{a} \in \mathfrak{G}$.

From the conditions (6.15) and (6.16) we see that every left-invariant field $\boldsymbol{u}$ has the representation

$$\boldsymbol{u}(\boldsymbol{x}) = \boldsymbol{L}_{x*e}(\boldsymbol{u}(\boldsymbol{e})) \quad \forall \boldsymbol{x} \in \mathfrak{G}, \tag{6.17}$$

and every right-invariant field $\boldsymbol{v}$ has the representation

$$\boldsymbol{v}(\boldsymbol{x}) = \boldsymbol{R}_{x*e}(\boldsymbol{v}(\boldsymbol{e})) \quad \forall \boldsymbol{x} \in \mathfrak{G}. \tag{6.18}$$

Further, from (6.11) or (6.12), $\boldsymbol{u}$ satisfies the condition (6.15) if and only if $\boldsymbol{J}_*(\boldsymbol{u})$ satisfies the condition (6.16). Thus the left-invariant fields and the right-invariant fields on $\mathfrak{G}$ are in one-to-one correspondence under the mapping $\boldsymbol{J}_*$.

We denote the collection of all left-invariant fields on $\mathfrak{G}$ by $\mathfrak{g}$. From (6.17), we see that $\mathfrak{g}$ forms a vector space, and the restriction map

$$|_e : \mathfrak{g} \to \mathfrak{G}_e \tag{6.19}$$

is a linear isomorphism, called the *standard representation* of $\mathfrak{g}$. In view of this representation, we have, of course,

$$\dim \mathfrak{g} = \dim \mathfrak{G}_e = \dim \mathfrak{G} = n. \tag{6.20}$$

From (6.17), we see that every left-invariant field is differentiable. In particular, we can determine a local integral curve $\boldsymbol{\lambda}_u(t)$ satisfying the initial condition

$$\boldsymbol{\lambda}_u(0) = \boldsymbol{e} \tag{6.21}$$

for any $\boldsymbol{u} \in \mathfrak{g}$. Now let $\boldsymbol{x}$ be an arbitrary point in $\mathfrak{G}$. Then in view of the fact that $\boldsymbol{u}$ is left-invariant, the left multiplication $\boldsymbol{L}_x$ carries the curve $\boldsymbol{\lambda}_u(t)$ into another integral curve

$$\boldsymbol{\lambda}_u(\boldsymbol{x}, t) \equiv \boldsymbol{x}\boldsymbol{\lambda}_u(t), \tag{6.22}$$

which satisfies the initial condition

$$\boldsymbol{\lambda}_u(\boldsymbol{x}, 0) = \boldsymbol{x}. \tag{6.23}$$

Consequently, every integral curve of $\boldsymbol{u}$ can be extended indefinitely for all $t \in \mathscr{R}$, and $\boldsymbol{\lambda}_u(t)$ becomes a 1-parameter group in $\mathfrak{G}$. That is, from

(6.22), we have

$$\boldsymbol{\lambda}_{\boldsymbol{u}}(t+s) = \boldsymbol{\lambda}_{\boldsymbol{u}}(\boldsymbol{\lambda}_{\boldsymbol{u}}(t), s) = \boldsymbol{\lambda}_{\boldsymbol{u}}(t)\boldsymbol{\lambda}_{\boldsymbol{u}}(s) \tag{6.24}$$

for all $t, s \in \mathscr{R}$.

The condition (6.22) means that the transformation group of $\mathfrak{G}$ generated by the vector field $\boldsymbol{u}$ is equal to the 1-parameter group $\boldsymbol{\lambda}_{\boldsymbol{u}}(t)$ acting on $\mathfrak{G}$ by right multiplication. Hence from (4.21), the Lie derivative of any vector field $\boldsymbol{z}$ on $\mathfrak{G}$ with respect to $\boldsymbol{u}$ can be expressed by

$$\mathscr{L}_{\boldsymbol{u}}\boldsymbol{z} = \lim_{t\to 0} \frac{1}{t}(\boldsymbol{R}_{\boldsymbol{\lambda}_{\boldsymbol{u}}(t)*}(\boldsymbol{z})-\boldsymbol{z}). \tag{6.25}$$

Now since the left multiplication commutes with the right multiplication (*cf.* (6.8)), the expression (6.25) implies that the Lie derivative of a left-invariant field with respect to another left-invariant field is again a left-invariant field. Consequently $\mathfrak{g}$ is a Lie algebra with respect to the bracket product. In this sense we call $\mathfrak{g}$ *the* (*left*) *Lie algebra* of $\mathfrak{G}$. Likewise, the collection $\bar{\mathfrak{g}}$ of all right-invariant fields is called *the* (*right*) *Lie algebra* of $\mathfrak{G}$.

Since $\boldsymbol{\lambda}_{\boldsymbol{u}}(t)$ forms a 1-parameter group for each $\boldsymbol{u} \in \mathfrak{g}$, its inverse is given by

$$\boldsymbol{J}(\boldsymbol{\lambda}_{\boldsymbol{u}}(t)) = \boldsymbol{\lambda}_{\boldsymbol{u}}(-t) = \boldsymbol{\lambda}_{-\boldsymbol{u}}(t). \tag{6.26}$$

Hence $\boldsymbol{J}_{*e}$ is the negation operator on $\mathfrak{G}_{\boldsymbol{e}}$:

$$\boldsymbol{J}_{*e}(\boldsymbol{w}) = -\boldsymbol{w} \quad \forall \boldsymbol{w} \in \mathfrak{G}_{\boldsymbol{e}}. \tag{6.27}$$

We put

$$\bar{\boldsymbol{u}} = \boldsymbol{J}_*(\boldsymbol{u}). \tag{6.28}$$

Then $\bar{\boldsymbol{u}} \in \bar{\mathfrak{g}}$ and

$$\bar{\boldsymbol{u}}(\boldsymbol{e}) = -\boldsymbol{u}(\boldsymbol{e}). \tag{6.29}$$

From (6.26), we have also

$$\boldsymbol{\lambda}_{\bar{\boldsymbol{u}}}(t) = \boldsymbol{J}(\boldsymbol{\lambda}_{\boldsymbol{u}}(t)) = \boldsymbol{\lambda}_{\boldsymbol{u}}(-t) = \boldsymbol{\lambda}_{-\boldsymbol{u}}(t). \tag{6.30}$$

As before, the transformation group of $\mathfrak{G}$ generated by $\bar{\boldsymbol{u}}$ is equal to the 1-parameter group $\boldsymbol{\lambda}_{\bar{\boldsymbol{u}}}(t)$ acting on $\mathfrak{G}$ by left multiplication, so that we can express the Lie derivative of any vector field $\boldsymbol{z}$ with respect to $\bar{\boldsymbol{u}}$ by

$$\mathscr{L}_{\bar{\boldsymbol{u}}}\boldsymbol{z} = \lim_{t\to 0} \frac{1}{t}(\boldsymbol{L}_{\boldsymbol{\lambda}_{\bar{\boldsymbol{u}}}(t)*}(\boldsymbol{z})-\boldsymbol{z}). \tag{6.31}$$

In particular, if z is equal to $\bar{v}$ for some $v \in \mathfrak{g}$, then

$$\begin{aligned}\mathscr{L}_{\bar{u}}\bar{v} &= \lim_{t\to 0} \frac{1}{t}(L_{\lambda_{\bar{u}}(t)*}(\bar{v})-\bar{v}) \\ &= J_*\left(\lim_{t\to 0}\frac{1}{t}(R_{\lambda_u(t)*}(v)-v)\right) = J_*(\mathscr{L}_u v). \qquad (6.32)\end{aligned}$$

Thus J_* is an isomorphism between the Lie algebras $\mathfrak{g}$ and $\bar{\mathfrak{g}}$. That is,

$$\overline{[u, v]} = [\bar{u}, \bar{v}]. \qquad (6.33)$$

for all $u, v \in \mathfrak{g}$.

From (6.29) and (6.30) we see that the 1-parameter group with a given initial tangent vector at e is uniquely determined by that tangent vector. Let w be a tangent vector at e. Then we denote the point corresponding to the parameter $t = 1$ on the 1-parameter group with initial tangent w by exp (w). In terms of this exponential map,

$$\exp : \mathfrak{G}_e \to \mathfrak{G}, \qquad (6.34)$$

the 1-parameter group determined by w can be expressed simply as $\exp(tw)$, $t \in \mathscr{R}$. Thus

$$\lambda_u(t) = \exp(tu(e)), \qquad (6.35)$$

and

$$\lambda_{\bar{u}}(t) = \exp(t\bar{u}(e)), \qquad (6.36)$$

for all $u \in \mathfrak{g}$ and $\bar{u} \in \bar{\mathfrak{g}}$.

From (6.35) or (6.36) we see that the exponential map is differentiable and that its induced map at the origin is the identity map, *viz*,

$$\exp_{*0}(w) = w \quad \forall w \in \mathfrak{G}_e. \qquad (6.37)$$

By the implicit function theorem, this condition implies that exp is a local diffeomorphism near the origin. In particular, if $\{e_i\}$ is a basis of $\mathfrak{G}_e$, then exp carries the Cartesian co-ordinate system induced by $\{e_i\}$ into a (local) co-ordinate system (x^i) near e by the condition

$$x = \exp(x^i e_i) \qquad (6.38)$$

for sufficiently small x^i. We call such a co-ordinate system (x^i) a *first canonical system* near e. Relative to (x^i), the 1-parameter group $\exp(tw)$ with initial tangent w has the co-ordinates

$$x^i = tw^i, \; i = 1, \ldots, n, \qquad (6.39)$$

where w^i is the ith component of $\boldsymbol{w}$ relative to $\boldsymbol{e}_i$, *viz*,

$$\boldsymbol{w} = w^i \boldsymbol{e}_i. \tag{6.40}$$

From (6.37), the natural basis of (x^i) at $\boldsymbol{e}$ coincides with $\boldsymbol{e}_i$,

$$\partial_{\boldsymbol{e}} x_i = \boldsymbol{e}_i, \quad i = 1, \ldots, n, \tag{6.41}$$

so that w^i is also the ith component of $\boldsymbol{w}$ in (x^i).

The Lie algebra $\mathfrak{g}$ gives rise to a parallelism on the Lie group in a natural way. We simply choose the sections corresponding to the vector fields in $\mathfrak{g}$ as the horizontal sections. The connection obtained in this way is called the *Cartan connection* of $\mathfrak{G}$. Clearly, the Cartan connection is an affine connection and is integrable.

Let D be the covariant derivative relative to the Cartan connection, and let $\{\boldsymbol{e}_i\}$ be a basis of $\mathfrak{g}$. Then by assumption

$$D\boldsymbol{e}_i = \boldsymbol{0}, \quad i = 1, \ldots, n. \tag{6.42}$$

From (5.25), this condition has the component form

$$\frac{\partial e_i^j}{\partial x^k} + \Gamma_{l\,k}^{\,j} e_i^l = 0 \tag{6.43}$$

relative to any co-ordinate system (x^i). Here the $\Gamma_{j\,k}^{\,i}$ are the connection symbols of the Cartan connection relative to (x^i). Since $\boldsymbol{e}_i$ is a basis, the component matrix $[e_i^{\,j}]$ is non-singular. Thus we can solve (6.43) for the connection symbols, namely

$$\Gamma_{j\,k}^{\,i} = -e^{-1}{}_{j}^{m} \frac{\partial e_m^i}{\partial x^k}. \tag{6.44}$$

This formula shows that the Cartan connection generally is not torsion-free, and the components of its torsion tensor are given by

$$\mathsf{C}_{j\,k}^{\,i} = e^{-1}{}_{k}^{m} \frac{\partial e_m^i}{\partial x^j} - e^{-1}{}_{j}^{m} \frac{\partial e_m^i}{\partial x^k}. \tag{6.45}$$

Since the torsion is a tensor field of type (1, 2), its contraction with any two vector fields $\boldsymbol{u}$ and $\boldsymbol{v}$ is a vector field $\mathsf{C}(\boldsymbol{u}, \boldsymbol{v})$ having components $\mathsf{C}_{j\,k}^{\,i} u^j v^k$. In particular, we have

$$(\mathsf{C}(e_i, e_j))^k = e_i^l \frac{\partial e_j^k}{\partial x^l} - e_j^l \frac{\partial e_i^k}{\partial x^l} = [e_i, e_j]^k. \tag{6.46}$$

Thus C corresponds to a tensor over $\mathfrak{g}$ with

$$\mathsf{C} : \mathfrak{g} \times \mathfrak{g} \to \mathfrak{g} \tag{6.47}$$

given by

$$\mathsf{C}(\boldsymbol{u}, \boldsymbol{v}) = [\boldsymbol{u}, \boldsymbol{v}] \tag{6.48}$$

for all $\boldsymbol{u}, \boldsymbol{v} \in \mathfrak{g}$.

So far, we have considered the vector fields that are invariant under the operation of left translations or the operation of right translations. Since the translations are linear isomorphisms, they give rise to induced isomorphisms of the tensor spaces also. Naturally, a tensor field is called *left-invariant* (respectively *right-invariant*) if it is invariant under the induced isomorphisms of the left translations (respectively right translations). Equivalently, a tensor field is left-invariant if and only if its covariant derivative with respect to the Cartan connection vanishes identically. For example, the torsion tensor of the Cartan connection is a left-invariant field. As before, every left-invariant field Ψ is differentiable and has the representation

$$\Psi(\boldsymbol{x}) = \otimes \boldsymbol{L}_{\boldsymbol{x}*\boldsymbol{e}}(\Psi(\boldsymbol{e})) \quad \forall \boldsymbol{x} \in \mathfrak{G}, \tag{6.49}$$

where $\otimes \boldsymbol{L}_{\boldsymbol{x}*}$ denotes the induced isomorphism of $\boldsymbol{L}_{\boldsymbol{x}*}$.

Let $\mathfrak{g}^*$ denote the collection of all left-invariant 1-forms on $\mathfrak{G}$. Then the contraction $\langle \boldsymbol{u}, \boldsymbol{w} \rangle$ is a constant function for any $\boldsymbol{u} \in \mathfrak{g}$ and $\boldsymbol{w} \in \mathfrak{g}^*$. Hence $\mathfrak{g}^*$ corresponds to the usual dual space of $\mathfrak{g}$. From (4.35), the exterior derivative of any left-invariant form is again left-invariant. Further, from (4.43), we have the *Maurer-Cartan relation*:

$$\partial \boldsymbol{w}(\boldsymbol{u}, \boldsymbol{v}) = -\tfrac{1}{2}\langle [\boldsymbol{u}, \boldsymbol{v}], \boldsymbol{w} \rangle \tag{6.50}$$

for any $\boldsymbol{u}, \boldsymbol{v} \in \mathfrak{g}$ and $\boldsymbol{w} \in \mathfrak{g}^*$.

We can extend the concept of a differential form to the concept of a *differential form with value in a Lie algebra* in an obvious way. Let $\mathscr{M}$ be a manifold. Then an r-form on $\mathscr{M}$ with value in $\mathfrak{g}$ is a differentiable field with value in $\mathfrak{g} \otimes \hat{\mathscr{M}}_{\boldsymbol{x}}^{r}$ at each $\boldsymbol{x} \in \mathscr{M}$. Let $\mathfrak{F}$ be such an r-form. Then it can be expressed in component form relative to a basis $\{\boldsymbol{e}_i\}$ of $\mathfrak{g}$ by

$$\mathfrak{F} = \boldsymbol{e}_i \otimes \mathfrak{F}^i, \tag{6.51}$$

where the components $\mathfrak{F}^i$ are ordinary r-forms on $\mathscr{M}$. We define the exterior derivative of $\mathfrak{F}$ by

$$\partial \mathfrak{F} \equiv \boldsymbol{e}_i \otimes \partial \mathfrak{F}^i. \tag{6.52}$$

Then $\partial ♂$ is an $(r+1)$-form on $\mathscr{M}$ with value in $\mathfrak{g}$, and that value is independent of the choice of $\{e_i\}$.

We define the exterior-bracket product of two forms with value in $\mathfrak{g}$ in a similar way. Let $♆$ be an s-form on $\mathscr{M}$ with value in $\mathfrak{g}$, say in component form

$$♆ = e_i \otimes ♆^i. \tag{6.53}$$

Then we define the exterior-bracket product $[♂ \wedge ♆]$ of $♂$ with $♆$ by

$$[♂ \wedge ♆] \equiv [e_i, e_j] \otimes (♂^i \wedge ♆^j). \tag{6.54}$$

Clearly, $[♂ \wedge ♆]$ is an $(r+s)$-form on $\mathscr{M}$ with value in $\mathfrak{g}$ and is independent of the choice of the basis e_i.

PROPOSITION. *The differential forms having values in a Lie algebra obey the following generalized Jacobi identities*:

$$[♂ \wedge ♆] = (-1)^{rs+1}[♆ \wedge ♂], \tag{6.55}$$

and

$$(-1)^{rt}[♂\wedge[♆\wedge☿]]+(-1)^{sr}[♆\wedge[☿\wedge♂]]+(-1)^{ts}[☿\wedge[♂\wedge♆]] = 0, \tag{6.56}$$

where $♂$ is any r-form, $♆$ is any s-form, and $☿$ is any t-form with value in $\mathfrak{g}$.

An important example of a differential form with value in a Lie algebra is the *canonical form* $\boldsymbol{\theta}$ of a Lie group $\mathfrak{G}$. We define $\boldsymbol{\theta}$ as a 1-form on $\mathfrak{G}$ with value in the Lie algebra $\mathfrak{g}$ of $\mathfrak{G}$ by the condition

$$\langle \boldsymbol{v}(\boldsymbol{x}), \boldsymbol{\theta}(\boldsymbol{x})\rangle = \boldsymbol{v} \quad \forall \boldsymbol{v} \in \mathfrak{g},\ \forall \boldsymbol{x} \in \mathfrak{G}. \tag{6.57}$$

In component form, $\boldsymbol{\theta}$ can be expressed by

$$\boldsymbol{\theta} = \boldsymbol{e}_i \otimes \boldsymbol{e}^i, \tag{6.58}$$

where $\{\boldsymbol{e}_i\}$ and $\{\boldsymbol{e}^i\}$ are the dual bases of $\mathfrak{g}$ and $\mathfrak{g}^*$. In terms of this canonical form $\boldsymbol{\theta}$, the Maurer-Cartan relation (6.50) takes the elegant form

$$\partial\boldsymbol{\theta} = -\tfrac{1}{2}[\boldsymbol{\theta}\wedge\boldsymbol{\theta}]. \tag{6.59}$$

Every Lie group is orientable, since from (6.49) it can be equipped with a left-invariant volume tensor that is non-zero everywhere. Let $\boldsymbol{V}$ be such a left-invariant volume tensor. Then the integral relative to $\boldsymbol{V}$ is also left-invariant in the sense that

$$\int f\mathbf{V} = \int \boldsymbol{L}_{\boldsymbol{a}}(f)\mathbf{V} \quad \forall \boldsymbol{a} \in \mathfrak{G}. \tag{6.60}$$

Here $\boldsymbol{L}_{\boldsymbol{a}}(f)$ is a function given by

$$\boldsymbol{L}_{\boldsymbol{a}}(f)|_{\boldsymbol{x}} = f(\boldsymbol{ax}) \quad \forall \boldsymbol{x} \in \mathfrak{G}. \tag{6.61}$$

A subgroup that is also a submanifold is called a *Lie subgroup*. Likewise, a subspace that is closed with respect to the bracket product of a Lie algebra is called a *Lie subalgebra*. From the Frobenius theorem, we have the following important relation between a Lie subgroup of $\mathfrak{G}$ and a Lie subalgebra of $\mathfrak{g}$:

PROPOSITION. *Any distribution on $\mathfrak{G}$ determined by a Lie subalgebra of $\mathfrak{g}$ is integrable; further, its maximal connected integral manifold passing through the identity element $\boldsymbol{e}$ is a connected Lie subgroup of $\mathfrak{G}$. Conversely, the Lie algebra of any connected Lie subgroup of $\mathfrak{G}$ can be extended into a Lie subalgebra of $\mathfrak{g}$; further, the Lie subgroup is the maximal connected integral manifold of that Lie subalgebra.*

From this proposition, we see that a one-to-one correspondence exists between the connected Lie subgroups of $\mathfrak{G}$ and the Lie subalgebras of $\mathfrak{g}$. Two Lie subgroups $\mathfrak{H}$ and $\bar{\mathfrak{H}}$ of $\mathfrak{G}$ are called *conjugate* if there exists an element $\boldsymbol{x} \in \mathfrak{G}$ such that

$$\bar{\mathfrak{H}} = \boldsymbol{x}\mathfrak{H}\boldsymbol{x}^{-1}. \tag{6.62}$$

An equivalence class of conjugate Lie subgroups is called a *type* of Lie subgroups. Since the conjugation operation

$$\boldsymbol{C}_{\boldsymbol{x}} : \mathfrak{G} \to \mathfrak{G} \tag{6.63}$$

defined by

$$\boldsymbol{C}_{\boldsymbol{x}} = \boldsymbol{L}_{\boldsymbol{x}} \circ \boldsymbol{R}_{\boldsymbol{x}^{-1}} = \boldsymbol{R}_{\boldsymbol{x}^{-1}} \circ \boldsymbol{L}_{\boldsymbol{x}} \tag{6.64}$$

is a diffeomorphism of $\mathfrak{G}$, $\mathfrak{H}$ and $\bar{\mathfrak{H}}$ are conjugate only if they are diffeomorphic. In particular, the Lie algebras $\mathfrak{h}$ and $\bar{\mathfrak{h}}$ of $\mathfrak{H}$ and $\bar{\mathfrak{H}}$ must have the same dimension.

The induced linear map $\boldsymbol{C}_{\boldsymbol{x}*}$ of $\boldsymbol{C}_{\boldsymbol{x}}$ is an automorphism of $\mathfrak{g}$. Indeed, for any $\boldsymbol{v} \in \mathfrak{g}$, we have

$$\begin{aligned} \boldsymbol{L}_{\boldsymbol{a}*}(\boldsymbol{C}_{\boldsymbol{x}*}(\boldsymbol{v})) &= \boldsymbol{L}_{\boldsymbol{a}*} \circ \boldsymbol{R}_{\boldsymbol{x}^{-1}*} \circ \boldsymbol{L}_{\boldsymbol{x}*}(\boldsymbol{v}), \\ &= \boldsymbol{L}_{\boldsymbol{a}*} \circ \boldsymbol{R}_{\boldsymbol{x}^{-1}*}(\boldsymbol{v}) = \boldsymbol{R}_{\boldsymbol{x}^{-1}*} \circ \boldsymbol{L}_{\boldsymbol{a}*}(\boldsymbol{v}), \\ &= \boldsymbol{R}_{\boldsymbol{x}^{-1}*}(\boldsymbol{v}) = \boldsymbol{R}_{\boldsymbol{x}^{-1}*} \circ \boldsymbol{L}_{\boldsymbol{x}*}(\boldsymbol{v}) = \boldsymbol{C}_{\boldsymbol{x}*}(\boldsymbol{v}), \end{aligned} \tag{6.65}$$

so that $\mathfrak{g}$ is closed under $\boldsymbol{C}_{x*}$. Further, from (6.25) for any $\boldsymbol{u}, \boldsymbol{v} \in \mathfrak{g}$ we have

$$[\boldsymbol{C}_{x*}(\boldsymbol{u}), \boldsymbol{C}_{x*}(\boldsymbol{v})] = [\boldsymbol{R}_{x^{-1}*}(\boldsymbol{u}), \boldsymbol{R}_{x^{-1}*}(\boldsymbol{v})], \\ = \boldsymbol{R}_{x^{-1}*}[\boldsymbol{u}, \boldsymbol{v}] = \boldsymbol{C}_{x*}[\boldsymbol{u}, \boldsymbol{v}], \tag{6.66}$$

so that $\boldsymbol{C}_{x*}$ is an isomorphism of Lie algebra. We denote the restriction of $\boldsymbol{C}_{x*}$ to $\mathfrak{g}$ by

$$\text{ad}\,(x) : \mathfrak{g} \to \mathfrak{g}, \tag{6.67}$$

called the *adjoint representation of* $\boldsymbol{x}$. The operator ad then maps $\mathfrak{G}$ into the general linear group $\mathfrak{GL}(\mathfrak{g})$ of $\mathfrak{g}$, *viz*,

$$\text{ad} : \mathfrak{G} \to : \mathfrak{GL}(\mathfrak{g}), \tag{6.68}$$

called the *adjoint representation* of $\mathfrak{G}$.

The general linear group of a vector space is an important Lie group for later applications, so that we consider its structure in more detail here. Since a linear transformation $\boldsymbol{L}$ of a vector space $\mathscr{V}$ is non-singular if and only if det $\boldsymbol{L}$ is non-zero, the general linear group $\mathfrak{GL}(\mathscr{V})$ of $\mathscr{V}$ is an open submanifold of the Euclidean space $\mathscr{L}(\mathscr{V};\mathscr{V})$. In particular,

$$\dim \mathfrak{GL}(\mathscr{V}) = n^2, \tag{6.69}$$

where n is the dimension of $\mathscr{V}$. Relative to the Euclidean parallelism on $\mathscr{L}(\mathscr{V};\mathscr{V})$ we can identify the tangent space $\mathfrak{GL}(\mathscr{V})_1$ of $\mathfrak{GL}(\mathscr{V})$ at the identity transformation $\mathbf{1}$ with the vector space $\mathscr{L}(\mathscr{V};\mathscr{V})$. Then relative to the standard representation of the Lie algebra $\mathfrak{gl}(\mathscr{V})$, we have

$$\mathfrak{gl}(\mathscr{V}) \cong \mathscr{L}(\mathscr{V};\mathscr{V}). \tag{6.70}$$

Using this representation of $\mathfrak{gl}(\mathscr{V})$, we can express the exponential map by the power series

$$\exp \boldsymbol{A} = \mathbf{1}+\boldsymbol{A}+\frac{1}{2!}\boldsymbol{A}^2+\ldots+\frac{1}{n!}\boldsymbol{A}^n+\ldots, \tag{6.71}$$

which converges absolutely for any $\boldsymbol{A} \in \mathfrak{gl}(\mathscr{V})$. Also, we can express the bracket product on $\mathfrak{gl}(\mathscr{V})$ by

$$[\boldsymbol{A}, \boldsymbol{B}] = \boldsymbol{AB}-\boldsymbol{BA}, \tag{6.72}$$

where the product on the right-hand side is the product of linear transformations in the usual sense. The series representation (6.71) implies the following two important formulae:

$$\exp(\boldsymbol{L}\boldsymbol{A}\boldsymbol{L}^{-1}) = \boldsymbol{L}(\exp \boldsymbol{A})\boldsymbol{L}^{-1}, \tag{6.73}$$

and

$$\det(\exp \boldsymbol{A}) = \exp(\operatorname{tr} \boldsymbol{A}), \tag{6.74}$$

for any $\boldsymbol{A} \in \mathfrak{gl}(\mathscr{V})$ and any $\boldsymbol{L} \in \mathfrak{GL}(\mathscr{V})$. Thus $\boldsymbol{A}, \bar{\boldsymbol{A}} \in \mathfrak{gl}(\mathscr{V})$ are conjugate in the sense of Lie algebra, and $\boldsymbol{L}, \bar{\boldsymbol{L}} \in \mathfrak{GL}(\mathscr{V})$ are conjugate in the sense of Lie groups if and only if they are conjugate in the sense of linear transformations in $\mathscr{L}(\mathscr{V}; \mathscr{V})$. Since the operation of conjugation corresponds to the transformation rule of the components of a linear transformation under a change of basis, $\boldsymbol{A}$ and $\bar{\boldsymbol{A}}$ are conjugate (respectively, $\boldsymbol{L}$ and $\bar{\boldsymbol{L}}$ are conjugate) if and only if the components of $\boldsymbol{A}$ and $\bar{\boldsymbol{A}}$ (respectively, $\boldsymbol{L}$ and $\bar{\boldsymbol{L}}$) are identical relative to some bases $\{\boldsymbol{e}_i\}$ and $\{\bar{\boldsymbol{e}}_i\}$. Thus within the context of the general linear group, a type of Lie subalgebras of $\mathfrak{gl}(\mathscr{V})$, or a type of Lie subgroups of $\mathfrak{GL}(\mathscr{V})$ corresponds respectively to a particular Lie subalgebra of $\mathfrak{gl}(n)$ or a particular Lie subgroup of $\mathfrak{GL}(n)$ associated with the Cartesian n-space $\mathscr{R}^n$.

From (6.74), we see that the Lie subalgebra $\mathfrak{sl}(\mathscr{V})$ of the Lie subgroup $\mathfrak{SL}(\mathscr{V})$ is characterized by

$$\boldsymbol{A} \in \mathfrak{sl}(\mathscr{V}) \Leftrightarrow \operatorname{tr}(\boldsymbol{A}) = 0. \tag{6.75}$$

The condition on the right-hand side clearly characterizes a Lie subalgebra, since from (6.72) we have

$$\operatorname{tr}([\boldsymbol{A}, \boldsymbol{B}]) = \operatorname{tr}(\boldsymbol{A}\boldsymbol{B}) - \operatorname{tr}(\boldsymbol{B}\boldsymbol{A}) = 0 \tag{6.76}$$

for all $\boldsymbol{A}, \boldsymbol{B} \in \mathfrak{gl}(\mathscr{V})$.

Now we introduce an inner product on $\mathscr{V}$. Then $\mathscr{L}(\mathscr{V}; \mathscr{V})$ is equipped with an operation of transposition

$$\boldsymbol{T} : \mathscr{L}(\mathscr{V}; \mathscr{V}) \to \mathscr{L}(\mathscr{V}; \mathscr{V}). \tag{6.77}$$

From (6.71), we can verify easily that

$$\exp(\boldsymbol{A}^T) = (\exp(\boldsymbol{A}))^T \tag{6.78}$$

for all $\boldsymbol{A} \in \mathfrak{gl}(\mathscr{V})$. Since we have also

$$\exp(-\boldsymbol{A}) = (\exp(\boldsymbol{A}))^{-1}, \tag{6.79}$$

we see that the Lie algebra $\mathfrak{o}(\mathscr{V})$ of the orthogonal group $\mathfrak{O}(\mathscr{V})$ is characterized by the condition

$$A \in \mathfrak{o}(\mathscr{V}) \Leftrightarrow A^T = -A. \tag{6.80}$$

The right-hand side again characterizes a Lie subalgebra, since from (6.72) we have

$$[A, B]^T = B^T A^T - A^T B^T = -(AB - BA) = -[A, B] \tag{6.81}$$

for all $A, B \in \mathfrak{o}(\mathscr{V})$.

Before closing this section, we mention two more useful results here: First, a theorem of CARTAN states that every closed subgroup of a Lie group is a Lie subgroup, and second, a theorem of LIE gives the complete classification of the types of Lie subalgebras of $\mathfrak{gl}(\mathscr{V})$ for a three-dimensional space $\mathscr{V}$. In the context of elasticity, as we shall see later, CARTAN's theorem implies that the symmetry group of the response function of a material is a Lie subgroup. Since the symmetry group is required to be a subgroup of the unimodular group, we need LIE's classification for the Lie subalgebras of $\mathfrak{sl}(\mathscr{V})$ only. Further, as explained before, the classification can be expressed in terms of the component matrices relative to a basis of $\mathscr{V}$.

Classification of the types of connected Lie subgroups of $\mathfrak{Sl}(3)$ and their corresponding Lie subalgebras

Reference number	Lie subgroup det $[L^i_j] = 1$	Lie subalgebra tr $[A^i_j] = 0$	dim.
1	$\begin{bmatrix} e^{\alpha a} & 0 & 0 \\ b & e^{\beta a} & 0 \\ c & d & e^{\gamma a} \end{bmatrix}$	$\begin{bmatrix} \alpha a & 0 & 0 \\ b & \beta a & 0 \\ c & d & \gamma a \end{bmatrix}$	5, 4, 3
2	$\begin{bmatrix} e^{\alpha a} & 0 & 0 \\ b & e^{\beta a} & 0 \\ c & 0 & e^{\gamma a} \end{bmatrix}$	$\begin{bmatrix} \alpha a & 0 & 0 \\ b & \beta a & 0 \\ c & 0 & \gamma a \end{bmatrix}$	4, 3, 2
3	$\begin{bmatrix} e^{\alpha a} & b & c \\ 0 & e^{\beta a} & 0 \\ 0 & 0 & e^{\gamma a} \end{bmatrix}$	$\begin{bmatrix} \alpha a & b & c \\ 0 & \beta a & 0 \\ 0 & 0 & \gamma a \end{bmatrix}$	4, 3, 2
4	$\begin{bmatrix} e^{\alpha a} & 0 & 0 \\ b & e^{\beta a} & 0 \\ 0 & 0 & e^{\gamma a} \end{bmatrix}$	$\begin{bmatrix} \alpha a & 0 & 0 \\ b & \beta a & 0 \\ 0 & 0 & \gamma a \end{bmatrix}$	3, 2, 1

5	$\begin{bmatrix} e^{\alpha a} & 0 & 0 \\ 0 & e^{\beta a} & 0 \\ 0 & 0 & e^{\gamma a} \end{bmatrix}$	$\begin{bmatrix} \alpha a & 0 & 0 \\ 0 & \beta a & 0 \\ 0 & 0 & \gamma a \end{bmatrix}$	2, 1, 0
6	$\begin{bmatrix} e^{\alpha a}\cos a & e^{\alpha a}\sin a & 0 \\ -e^{\alpha a}\sin a & e^{\alpha a}\cos a & 0 \\ b & c & e^{\beta a} \end{bmatrix}$	$\begin{bmatrix} \alpha a & a & 0 \\ -a & \alpha a & 0 \\ b & c & \beta a \end{bmatrix}$	4, 3
7	$\begin{bmatrix} e^{\alpha a}\cos a & e^{\alpha a}\sin a & b \\ -e^{\alpha a}\sin a & e^{\alpha a}\cos a & c \\ 0 & 0 & e^{\beta a} \end{bmatrix}$	$\begin{bmatrix} \alpha a & a & b \\ -a & \alpha a & c \\ 0 & 0 & \beta a \end{bmatrix}$	4, 3
8	$\begin{bmatrix} e^{\alpha a}\cos a & e^{\alpha a}\sin a & 0 \\ -e^{\alpha a}\sin a & e^{\alpha a}\cos a & 0 \\ 0 & 0 & e^{\beta a} \end{bmatrix}$	$\begin{bmatrix} \alpha a & a & 0 \\ -a & \alpha a & 0 \\ 0 & 0 & \beta a \end{bmatrix}$	2, 1
9	$\begin{bmatrix} a & b & c \\ d & e & f \\ g & h & i \end{bmatrix}$	$\begin{bmatrix} a & b & c \\ d & e & f \\ g & h & i \end{bmatrix}$	8
10	$\begin{bmatrix} a & b & c \\ d & e & f \\ 0 & 0 & g \end{bmatrix}$	$\begin{bmatrix} a & b & c \\ d & e & f \\ 0 & 0 & g \end{bmatrix}$	6
11	$\begin{bmatrix} a & b & 0 \\ c & d & 0 \\ e & f & g \end{bmatrix}$	$\begin{bmatrix} a & d & 0 \\ b & e & 0 \\ c & f & g \end{bmatrix}$	6
12	$\begin{bmatrix} a & b & c \\ d & e & f \\ 0 & 0 & 1 \end{bmatrix}$	$\begin{bmatrix} a & b & c \\ d & e & f \\ 0 & 0 & 0 \end{bmatrix}$	5
13	$\begin{bmatrix} a & d & 0 \\ b & e & 0 \\ c & f & 1 \end{bmatrix}$	$\begin{bmatrix} a & b & 0 \\ c & d & 0 \\ e & f & 0 \end{bmatrix}$	5
14	$\begin{bmatrix} a & b & 0 \\ c & d & 0 \\ 0 & 0 & e \end{bmatrix}$	$\begin{bmatrix} a & b & 0 \\ c & d & 0 \\ 0 & 0 & e \end{bmatrix}$	4
15	$\begin{bmatrix} a & b & 0 \\ c & d & 0 \\ 0 & 0 & 1 \end{bmatrix}$	$\begin{bmatrix} a & b & 0 \\ c & d & 0 \\ 0 & 0 & 0 \end{bmatrix}$	3

16	The proper orthogonal group $\mathfrak{SO}(3)$	$\begin{bmatrix} 0 & -a & -b \\ a & 0 & -c \\ b & c & 0 \end{bmatrix}$	3
17	$\begin{bmatrix} e^a & 0 & 0 \\ b & e^{-2a} & c \\ ae^a & 0 & e^a \end{bmatrix}$	$\begin{bmatrix} a & 0 & 0 \\ b & -2a & c \\ a & 0 & a \end{bmatrix}$	3
18	$\begin{bmatrix} e^a & b & 0 \\ 0 & e^{-2a} & 0 \\ ae^a & c & e^a \end{bmatrix}$	$\begin{bmatrix} a & b & 0 \\ 0 & -2a & 0 \\ a & c & a \end{bmatrix}$	3
19	$\exp\begin{bmatrix} 0 & a & b \\ b & c & 0 \\ a & 0 & -c \end{bmatrix}$	$\begin{bmatrix} 0 & a & b \\ b & c & 0 \\ a & 0 & -c \end{bmatrix}$	3
20	$\begin{bmatrix} 1 & 0 & a \\ ae^b & e^b & c \\ 0 & 0 & e^{-b} \end{bmatrix}$	$\begin{bmatrix} 0 & 0 & a \\ a & b & c \\ 0 & 0 & -b \end{bmatrix}$	3
21	$\begin{bmatrix} e^a & 0 & 0 \\ b & e^{-2a} & 0 \\ ae^a & 0 & e^a \end{bmatrix}$	$\begin{bmatrix} a & 0 & 0 \\ b & -2a & 0 \\ a & 0 & a \end{bmatrix}$	2
22	$\begin{bmatrix} e^a & 0 & 0 \\ 0 & e^{-2a} & 0 \\ ae^a & b & e^a \end{bmatrix}$	$\begin{bmatrix} a & 0 & 0 \\ 0 & -2a & 0 \\ a & b & a \end{bmatrix}$	2
23	$\begin{bmatrix} 1 & 0 & a \\ ae^b & e^b & \frac{1}{2}a^2e^b \\ 0 & 0 & e^{-b} \end{bmatrix}$	$\begin{bmatrix} 0 & 0 & a \\ a & b & 0 \\ 0 & 0 & -b \end{bmatrix}$	2
24	$\begin{bmatrix} 1 & a & 0 \\ 0 & 1 & 0 \\ a & b & 1 \end{bmatrix}$	$\begin{bmatrix} 0 & a & 0 \\ 0 & 0 & 0 \\ a & b & 0 \end{bmatrix}$	2
25	$\begin{bmatrix} e^a & 0 & 0 \\ 0 & e^{-2a} & 0 \\ ae^a & 0 & e^a \end{bmatrix}$	$\begin{bmatrix} a & 0 & 0 \\ 0 & -2a & 0 \\ a & 0 & a \end{bmatrix}$	1
26	$\begin{bmatrix} 1 & a & 0 \\ 0 & 1 & 0 \\ a & \frac{1}{2}a^2 & 1 \end{bmatrix}$	$\begin{bmatrix} 0 & a & 0 \\ 0 & 0 & 0 \\ a & 0 & 0 \end{bmatrix}$	1

In the above table $a, b, c, \ldots$ are arbitrary real numbers and α, β, γ are arbitrary real parameters which can take on either fixed values or arbitrary values, subject only to the restriction that det $[L^i_j] = 1$ and tr $[A^i_j] = 0$. For example, choosing $\alpha = \beta = \gamma = 0$ in family No. 1 yields a type of Lie subgroups and Lie subalgebras with components

$$\begin{bmatrix} 1 & 0 & 0 \\ b & 1 & 0 \\ c & d & 1 \end{bmatrix}, \quad \begin{bmatrix} 0 & 0 & 0 \\ b & 0 & 0 \\ c & d & 0 \end{bmatrix},$$

where b, c, d are arbitrary. Likewise, choosing $\alpha = 1, \beta = 2$, and $\gamma = -3$ yields the components

$$\begin{bmatrix} e^a & 0 & 0 \\ b & e^{2a} & 0 \\ c & d & e^{-3a} \end{bmatrix}, \quad \begin{bmatrix} a & 0 & 0 \\ b & 2a & 0 \\ c & d & -3a \end{bmatrix},$$

where a, b, c, d are arbitrary. Finally, if we let α, β, and γ be arbitrary to within the condition $\alpha+\beta+\gamma = 0$, we obtain the components

$$\begin{bmatrix} e^a & 0 & 0 \\ c & e^b & 0 \\ d & f & e^{-(a+b)} \end{bmatrix}, \quad \begin{bmatrix} a & 0 & 0 \\ b & c & 0 \\ d & f & -(a+c) \end{bmatrix},$$

where a, b, c, d, and f are arbitrary.

Choosing the parameters α, β, γ in this manner, we can determine ∞^1-types in each of the families No. 1–No. 8. The remaining families in the table contain no parameters, so that each of them is a single type.

CHAPTER II

Continuum Mechanics

Just as Chapter I has given in outline the background in "pure" mathematics upon which this book on elasticity will occasionally draw, Chapter II sketches the corresponding background in continuum mechanics. The scope of this chapter is necessarily broader than what is actually needed for elasticity. Its purpose is not only to set elasticity at its proper place in continuum mechanics as a whole, but also to make the book self-contained. Thus he whose interest lies in elasticity alone may pass over this chapter entirely, using it only for reference[1] as need may arise in his reading of the sequel.

1. Kinematics

In any theory of physics an *event world*, the seat of all such phenomena as may occur, is assumed at the outset, either implicitly or expressly. In classical mechanics the event world is taken to be the *Newtonian space-time*[2] $\mathscr{E}$, a 4-dimensional differentiable manifold formed by the disjoint union

$$\mathscr{E} = \bigcup_{s \in \mathscr{T}} \mathscr{E}_s, \tag{1.1}$$

[1] The material in §§ 1-6, along with many related topics, is treated in much greater detail, though from a point of yieuw now out of date, in Chapters B-F of Truesdell & Toupin 's article "The Classical Field Theories" in Flügge 's *Handbuch der Physik* Volume III(1, Berlin etc., Springer, 1960 (hereinafter cited as CFT.
The material in § 7 is developed in much greater detail, though again in a way which is no longer entirely modern, in Chapter C of NFTM.

[2] For a formal axiomatic construction of "neo-classical" space-time, see W. Noll, "The foundations of mechanics" (Lectures at Bressanone, 1965), pp. 159-200 of *Non-linear Continuum Theories*, ed. C. Truesdell & G. Grioli, Roma, Ed. Cremonese, 1966; corrected reprint, Carnegie Institute of Technology Mathematics Department Report No. 66-2. However, here we do not follow Noll's development or use his terminology. *Cf.* also the forthcoming book by C. Truesdell, *Introduction à la Mécanique Rationnelle des Milieux Continus*, Paris, Masson.

where $\mathscr{T}$ is a 1-dimensional oriented Euclidean manifold called the collection of *instants* s, and where $\mathscr{E}_s$ is a 3-dimensional oriented Euclidean manifold called the *instantaneous space* at s, the points $\boldsymbol{x}_s$ of which are called *places*. A point $(\boldsymbol{x}_s, s)$ of $\mathscr{E}$ is called an *event*. Since a given event is a point in one and only one instantaneous space $\mathscr{E}_s$, it is said to *occur* at the instant s and the place $\boldsymbol{x}_s$. Two events $(\boldsymbol{x}_s, s)$ and $(\bar{\boldsymbol{x}}_s, s)$ occurring at the same instant s are called *simultaneous*. Thus there is a time interval between any two events, but there is a distance only between simultaneous events.

At different instants s and $\bar{s}$, the instantaneous spaces $\mathscr{E}_s$ and $\mathscr{E}_{\bar{s}}$ are disjoint submanifolds of $\mathscr{E}$ which are not related by any particular canonical isomorphism. Thus there is no way to identify the places $\boldsymbol{x}_s$ and $\boldsymbol{x}_{\bar{s}}$ if $s \neq \bar{s}$, and $\mathscr{E}$ is not itself a product manifold. However it can be set into correspondence with the product manifold $\mathscr{R}^4 = \mathscr{R}^3 \times \mathscr{R}$ through a *frame of reference* on $\mathscr{E}$, defined as follows:

DEFINITION. A *frame of reference* φ on $\mathscr{E}$ is a Cartesian co-ordinate system $x^4(\cdot)$ on $\mathscr{T}$ and a Cartesian co-ordinate system $(x^i(\cdot, s))$, $i = 1, 2, 3$ on $\mathscr{E}_s$, for each $s \in \mathscr{T}$, such that the quadruple $(x^\alpha, \alpha = 1, 2, 3, 4)$ is a global co-ordinate system on $\mathscr{E}$.

Thus the time interval between the events whose co-ordinates are x^α and $\bar{x}^\alpha$ in φ is $|\bar{x}^4 - x^4|$; if the events are simultaneous, the distance between them is $(\sum_{i=1}^{3}(\bar{x}^i - x^i)^2)^{\frac{1}{2}}$.

The frame φ makes $\mathscr{E}$ a copy of $\mathscr{R}^4$:

$$\varphi : \mathscr{E} \to \mathscr{R}^4. \tag{1.2}$$

Of course, there are infinitely many frames of reference. In this section we shall lay down and develop and apply definitions in terms of some one frame, deferring until the next section comparison of effects of using different frames.

When we wish to enter into details about a frame of reference, we write (x^α) for φ, and we call $(x^i(\cdot, s))$ the *space co-ordinate system* φ_s on $\mathscr{E}_s$ and $x^4(\cdot)$ the *time co-ordinate system* on $\mathscr{T}$, relative to φ. Then the instant s_0 such that $x^4(s_0) = 0$ is called the *reference instant*, the origin $\boldsymbol{O}(s)$ of φ_s in $\mathscr{E}_s$ is called the *reference origin*, and the orthonormal basis $\{\boldsymbol{e}_i(s)\}$ of φ_s in the *instantaneous translation space* $\mathscr{V}_s$ of $\mathscr{E}_s$ is called the *reference basis* at s. It should be noted that, by the convention for oriented manifolds, the reference basis $\{\boldsymbol{e}_i(s)\}$ is positively oriented in $\mathscr{E}_s$, and the basis vector ∂x_4 is a positive vector for $\mathscr{T}$. In the physical interpretation, it is understood that a positive basis in $\mathscr{E}_s$ is *right-handed* in the conventional sense, and a positive vector for $\mathscr{T}$ points from the *past* into the *future*. Thus a frame of reference φ corresponds to an *observer*, who is equipped with a standard clock to measure the time and a standard yardstick to measure the spatial co-ordinates relative to an instantaneous spatial frame, at every instant s. While it is common in physics to speak of "the frame of reference of an observer", we shall use "frame of reference" and "observer" as synonymous, to denote in ordinary language the mathematical object φ.

Now let $\boldsymbol{\lambda} : (-\infty, \infty) \to \mathscr{E}$ be a smooth curve in $\mathscr{E}$, and let its representation relative to φ be $x^\alpha = \lambda^\alpha(t)$. Then $\boldsymbol{\lambda}$ is called a *world line* if $\lambda^4(t) = t+c$, where c is a constant. (In the next section, we shall see that in fact this condition does not depend on the choice of φ.) A world line is said to be *natural* with respect to the frame φ if $c = 0$. In most contexts for which world lines are useful, there is no loss in generality[1] if we restrict attention to natural world lines. Such a world line $\boldsymbol{\lambda}$ is characterized by the set of functions $\{\lambda^i(t) = \lambda^i(x^4)\}$ giving its *spatial co-ordinates* λ^i as functions of the *time* $t = x^4$. As usual, we can express the tangent vector of $\boldsymbol{\lambda}$ in component form by

$$\partial\boldsymbol{\lambda} = \dot{\lambda}^\alpha \partial x_\alpha = \dot{\lambda}^i \boldsymbol{e}_i + \boldsymbol{e}_4, \tag{1.3}$$

where $\boldsymbol{e}_4$ denotes the unit vector ∂x_4. We call the spatial projection of $\partial\boldsymbol{\lambda}$, namely,

$$\boldsymbol{v} = \dot{\lambda}^i \boldsymbol{e}_i, \tag{1.4}$$

the *velocity* of $\boldsymbol{\lambda}$ relative to the observer φ.

Relative to the assigned φ, a world line whose velocity vanishes for all time is said to be *at rest*; such a world line can be characterized by its spatial co-ordinates, which are constant: $\{x^i = \lambda^i(t) = \text{const.}\}$. Writing $\mathscr{E}_\varphi$ for the collection of all world lines at rest with respect to φ, we see that $\mathscr{E}_\varphi$ is a copy of any $\mathscr{E}_s$ and hence forms a 3-dimensional oriented Euclidean manifold in a natural way. We denote the canonical isomor-

---

[1] Generally the time is used as a variable with respect to which derivatives are calculated. The value of c is then immaterial. To make this point precise, we could eliminate translations along the time axis by defining a world line as the class of all curves in $\mathscr{E}$ which have the same parametric functions $\lambda^i(t)$ and are such that $\lambda^4(t) = t+c$ for some c. Such a complication seems unnecessary, however.

phism of $\mathscr{E}_\varphi$ with $\mathscr{E}_s$ by

$$\boldsymbol{i}_s : \mathscr{E}_\varphi \to \mathscr{E}_s, \tag{1.5}$$

and likewise for the translation spaces,

$$\boldsymbol{I}_s : \mathscr{V}_\varphi \to \mathscr{V}_s, \tag{1.6}$$

for each $s \in \mathscr{T}$. Then $\boldsymbol{I}_s$ is the induced linear map of $\boldsymbol{i}_s$. We call $\mathscr{E}_\varphi$ and $\mathscr{V}_\varphi$ the *absolute point space* and the *absolute translation space* of φ, while $\boldsymbol{i}_s$ and $\boldsymbol{I}_s$ are the *observations* or *observation maps* of φ on $\mathscr{E}_s$.

The observations specify an *absolute reference origin* $\boldsymbol{O}$ in $\mathscr{E}_\varphi$ and an *absolute spatial frame* $\{\boldsymbol{e}_i\}$ in $\mathscr{V}_\varphi$:

$$\boldsymbol{O} = i_s^{-1}(\boldsymbol{O}(s)),\ \boldsymbol{e}_i = \boldsymbol{I}_s^{-1}(\boldsymbol{e}_i(s)),\ i = 1, 2, 3, \tag{1.7}$$

independent of s. The co-ordinate system (x^i) determined by $\boldsymbol{O}$ and $\{\boldsymbol{e}_i\}$ on $\mathscr{E}_\varphi$ is called the *absolute space co-ordinate system.*

Using the observation maps, we can assign a smooth curve in $\mathscr{E}_\varphi$ to any world line in $\mathscr{E}$, and an absolute spatial vector or tensor over $\mathscr{V}_\varphi$ to any instantaneous spatial vector or tensor over some $\mathscr{V}_s$, and *vice versa*. In particular, the velocity of a world line in $\mathscr{E}$ is observed by φ as the tangent vector of the corresponding curve in $\mathscr{E}_\varphi$, which also may be called a world line without danger of confusion. In this sense, we define the *acceleration* of a world line $\boldsymbol{\lambda}$ relative to φ as the vector

$$\boldsymbol{a} \equiv \dot{\boldsymbol{v}} = \ddot{\lambda}^i \boldsymbol{e}_i. \tag{1.8}$$

Then $\boldsymbol{a}(t)$ can be regarded either as a spatial vector in $\mathscr{V}_s$ or as a spatial vector in $\mathscr{V}_\varphi$.

As we have said, frames of reference and observers are not intrinsic to $\mathscr{E}$. Consequently, the quantities determined by an observer φ, such as the spatial co-ordinates, the velocity, the acceleration, *etc.*, likewise are not intrinsic to a world line. In fact, if we assign two distinct frames of reference, say φ and $\bar{\varphi}$, then a given world line has two distinct sets of spatial co-ordinates, one for each frame. Hence two different observers φ and $\bar{\varphi}$ observe in general, different velocities and accelerations of the same world line. In the next section we shall consider the effects of a change of frame.

Having introduced the general mathematical structure of Newtonian space-time, we now consider *kinematics*, which is the mathematical theory of change with respect to time in an assigned frame of reference, and of the relations between such changes in different frames. In this book we are interested only in mathematical models for *deformable bodies*. Such a model is an oriented 3-dimensional differentiable manifold $\mathscr{B}$ which can be covered by a global co-ordinate chart. $\mathscr{B}$ is called the *body manifold* or simply *body*, and its elements are called *body-points*[1] and are denoted by light-face majuscules $X, Y, \ldots$ Any open submanifold of $\mathscr{B}$ is also a body; such a body is often called a *subbody* of $\mathscr{B}$. A *motion* of $\mathscr{B}$ is a mapping of its points into world lines. The distance between two body-points of $\mathscr{B}$ at the instant s is the distance between the points of intersection of their world lines with $\mathscr{E}_s$. Since $\mathscr{B}$ is deformable, this distance generally depends upon s for any pair of points in $\mathscr{B}$.

In continuum mechanics we assume that the mappings

$$\zeta_s : \mathscr{B} \to \mathscr{E}_s, \; s \in \mathscr{T}, \tag{1.9}$$

determined by the motion be orientation-preserving diffeomorphisms. The body point X is said to *occupy* the place $\zeta_s(X)$ in $\mathscr{E}_s$ at the instant s. By (1.9), there exists a diffeomorphism

$$\zeta_{s,\bar{s}} \equiv \zeta_{\bar{s}} \circ \zeta_s^{-1} : \zeta_s(\mathscr{B}) \to \zeta_{\bar{s}}(\mathscr{B}) \tag{1.10}$$

for any two instants s and $\bar{s}$. We call the mapping ζ_s (or, sometimes, its image $\zeta_s(\mathscr{B})$), the *abstract configuration* of $\mathscr{B}$ at s, and the mapping $\zeta_{s,\bar{s}}$ the *abstract deformation* of $\mathscr{B}$ from the configuration at s to the configuration at $\bar{s}$.

Now let φ be a given frame of reference. Then the motion of $\mathscr{B}$ is observed by φ as a 1-parameter family of diffeomorphisms

$$\chi_t : \mathscr{B} \to \mathscr{E}_\varphi, \; t \in \mathscr{R}. \tag{1.11}$$

Naturally, we call χ_t the *absolute configuration* of $\mathscr{B}$ at the time t, as observed by φ. The *absolute place* $\mathbf{x}$ occupied by a body-point $X \in \mathscr{B}$ at the time t can now be characterized by the spatial co-ordinates $x^i(t)$ of $\chi_t(X)$ in $\mathscr{E}_\varphi$.

Often it is convenient to introduce a *reference configuration* $\boldsymbol{\kappa}$ of $\mathscr{B}$ relative to φ. A reference configuration of $\mathscr{B}$ (not to be confused with a frame of reference on $\mathscr{E}$) is simply a particular absolute configuration $\boldsymbol{\kappa}$ of $\mathscr{B}$:

$$\boldsymbol{\kappa} : \mathscr{B} \to \mathscr{E}_\varphi. \tag{1.12}$$

[1] We use the name "body-point" for what in classical continuum mechanics is called a "particle".

In applications, $\boldsymbol{\kappa}$ is often chosen as χ_t for some particular t, say $t = 0$, so that $\boldsymbol{\kappa}$ is the absolute configuration of $\mathscr{B}$ at the initial time, but this special choice is not necessary. The spatial co-ordinates of a typical body-point $X \in \mathscr{B}$ in the reference configuration $\boldsymbol{\kappa}$ are denoted by $(X^A, A = 1, 2, 3)$ and are called the *referential co-ordinates*[1] of the body-point X in the reference configuration $\boldsymbol{\kappa}$. We shall denote $\boldsymbol{\kappa}(X)$, $\boldsymbol{\kappa}(Y)$, *etc.* by boldface majuscules $\mathbf{X}$, $\mathbf{Y}$, *etc.* Properly, so as to remind the reader that $\boldsymbol{\kappa}$ has been selected, we ought to write $\mathbf{X}_{\boldsymbol{\kappa}}$, *etc.*, but we do not do so.

Using a reference configuration $\boldsymbol{\kappa}$, we can then represent any absolute configuration χ_t of $\mathscr{B}$ by the mapping

$$\boldsymbol{\kappa}_t : \boldsymbol{\kappa}(\mathscr{B}) \to \chi_t(\mathscr{B}), \tag{1.13}$$

where, of course,

$$\boldsymbol{\kappa}_t = \chi_t \circ \boldsymbol{\kappa}^{-1}. \tag{1.14}$$

This mapping is called the *deformation* of $\mathscr{B}$ from $\boldsymbol{\kappa}$ to χ_t. In co-ordinate form, it is given by the equations

$$x^i = \kappa^i(\mathbf{X}, t) = \kappa^i(X^A, t), \quad i = 1, 2, 3, \tag{1.15}$$

and the functions κ^i are called the *deformation functions.*

Since a motion assigns a world line to a body-point, we can call the velocity and the acceleration of that world line the velocity and acceleration of the body-point relative to $\boldsymbol{\varphi}$. By considering all the body-points of $\mathscr{B}$, we thus obtain *vector fields* $\boldsymbol{v}$ and $\boldsymbol{a}$ defined on the configurations of $\mathscr{B}$. In component form relative to the absolute spatial frame,

$$v^i = \frac{\partial \kappa^i}{\partial t}, \tag{1.16}$$

and

$$a^i = \frac{\partial^2 \kappa^i}{\partial t^2}, \tag{1.17}$$

the domain of κ^i being the set of all referential co-ordinates X^A and times t, as above. Since the relation (1.15) is invertible for each fixed t, we can

---

[1] In the literature of hydrodynamics, the coordinates (x^i) and (X^A) are often called "Eulerian" and "Lagrangean", respectively. *Material co-ordinates* of X are its local co-ordinates with respect to some co-ordinate system on $\mathscr{B}$. Referential co-ordinates are possible material co-ordinates, and a global material co-ordinate system is referential. In this book we do not need material co-ordinates.

also use the spatial co-ordinates x^i and the time t as the independent variables in terms of which to describe the fields[1] $\boldsymbol{v}$ and $\boldsymbol{a}$. In this sense the acceleration field $\boldsymbol{a}$ is given by

$$a^i = \frac{\partial v^i}{\partial t} + \frac{\partial v^i}{\partial x^j} v^j \equiv \frac{dv^i}{dt}, \tag{1.18}$$

or, in vector notation,

$$\boldsymbol{a} = \frac{\partial \boldsymbol{v}}{\partial t} + (\operatorname{grad} \boldsymbol{v})\, \boldsymbol{v} \equiv \frac{d\boldsymbol{v}}{dt}, \tag{1.19}$$

where grad $\boldsymbol{v}$ is a tensor field (over $\mathscr{V}_\varphi$ or $\mathscr{V}_s$) called the *spatial velocity gradient* relative to $\boldsymbol{\varphi}$.

For a deformable body, we define three kinds of *deformation gradients*. First, the *abstract deformation gradient* of a configuration ζ_s of $\mathscr{B}$ is a two-point tensor field $\nabla\zeta_s$ on $\mathscr{B}$ whose value at any point $X \in \mathscr{B}$ corresponds to the induced linear maps

$$\zeta_{s*X} : \mathscr{B}_X \to \mathscr{V}_s, \tag{1.20}$$

and

$$\zeta^*_{sX} : \mathscr{V}_s \to \mathscr{B}^*_X. \tag{1.21}$$

Here $\mathscr{B}_X$ is the tangent space of $\mathscr{B}$ at X, and we have invoked the standard isomorphism between $\mathscr{V}_s$ and $\mathscr{V}^*_s$ via the inner product. Second, the

---

[1] In mathematical physics it often becomes cumbrous to denote a function and its value by different letters in all cases. Here the v^i are the spatial components of the field $\boldsymbol{v}$, regarded as functions of either (X^1, X^2, X^3, t) or (x^1, x^2, x^3, t) through the transformation (1.15) for appropriate domains in $\mathscr{E}_\varphi$ or $\mathscr{E}_s$.

In this book we shall follow a notation for differential invariants common in continuum mechanics. Let f be the common scalar value of the functions $\hat{f}(\boldsymbol{x})$ and $\check{f}(\boldsymbol{X})$ at time t:

$$f = \hat{f}(\boldsymbol{x}) = \check{f}(\boldsymbol{X}),$$

where $\boldsymbol{x}$ is the spatial position of the body-point X whose position in $\boldsymbol{\kappa}$ is $\boldsymbol{X}$. Then

$$\operatorname{grad} f \equiv \nabla\hat{f}, \quad \operatorname{Grad} f \equiv \nabla\check{f},$$

where ∇ is the gradient operator. Similarly, if $\boldsymbol{f}$ is a vector,

$$\operatorname{div} \boldsymbol{f} = \operatorname{tr}(\nabla\hat{\boldsymbol{f}}), \quad \operatorname{Div} \boldsymbol{f} = \operatorname{tr}(\nabla\check{\boldsymbol{f}}).$$

Of course, if functions and their values were always denoted by different letters, these notations would be superfluous.

absolute deformation gradient of a configuration χ_t of $\mathscr{B}$ relative to φ is a two-point tensor field $\nabla\chi_t$ on $\mathscr{B}$ whose value at X corresponds to the induced linear maps

$$\chi_{t*X} : \mathscr{B}_X \to \mathscr{V}_\varphi, \tag{1.22}$$

and

$$\chi_{tX}^{*} : \mathscr{V}_\varphi \to \mathscr{B}_X^{*}. \tag{1.23}$$

Finally, the gradient of the deformation κ_t of $\mathscr{B}$ from a reference configuration κ to χ_t is a tensor field on $\kappa(\mathscr{B})$ whose value at X corresponds to the induced linear maps

$$\boldsymbol{F}_\kappa \equiv \kappa_{t*\mathbf{X}} : \mathscr{V}_\varphi \to \mathscr{V}_\varphi, \tag{1.24}$$

and

$$\boldsymbol{F}_\kappa^T \equiv \kappa_{t\mathbf{X}}^{*} : \mathscr{V}_\varphi \to \mathscr{V}_\varphi. \tag{1.25}$$

The linear map $\boldsymbol{F}_\kappa$, often called the *gradient*, is the commonest in continuum mechanics. Here κ is the reference configuration in terms of which $\boldsymbol{F}_\kappa$ is defined, but when a single κ is selected and held fixed, $\boldsymbol{F}_\kappa$ is denoted simply by $\boldsymbol{F}$.

Using the chain rule, we can write down many obvious relations among these three kinds of deformation gradients; *e.g.* from (1.14)

$$\chi_{t*X} = \kappa_{t*\mathbf{X}} \circ \kappa_{*X}, \tag{1.26}$$

etc. Also, if μ is the mapping of κ onto another reference configuration $\bar{\kappa}$, that is,

$$\mu : \kappa(\mathscr{B}) \to \bar{\kappa}(\mathscr{B}), \tag{1.27}$$

then

$$\kappa_t = \bar{\kappa}_t \circ \mu, \tag{1.28}$$

and hence

$$\boldsymbol{F}_\kappa = \boldsymbol{F}_{\bar{\kappa}}\boldsymbol{P}, \tag{1.29}$$

where $\boldsymbol{P}$ is the gradient of μ, namely

$$\boldsymbol{P} = \mu_* = \bar{\kappa}_* \circ \kappa_*^{-1}. \tag{1.30}$$

The importance of the deformation gradients, of course, lies in their providing *first-order approximations* to the configurations and the deformations at a fixed time. For instance, if the deformation functions for κ_t are given by (1.15), then the matrix of components of the deformation gradient $\boldsymbol{F}$ in the absolute spatial frame is

$$[F^i_A] = \left[\frac{\partial \kappa^i}{\partial X^A}\right]. \tag{1.31}$$

Thus at a given body-point X_0 and a given time t_0, the values of the F^i_A at $\boldsymbol{\kappa}(X_0)$ and t_0 determine the functions $\kappa^i(X, t_0)$ to within an error that is of the first order in $(X^A - X^A_0)$ near (X_0, t_0) relative to the value $\kappa^i(X_0, t_0)$.

Since the deformation is orientation-preserving, the determinant of $[F^i_A]$ is necessarily positive. Thus we can determine the unique *polar decomposition* for $\boldsymbol{F}$, say

$$\boldsymbol{F} = \boldsymbol{R}\boldsymbol{U} = \boldsymbol{V}\boldsymbol{R}, \tag{1.32}$$

where $\boldsymbol{R}$ is a rotation and where $\boldsymbol{U}$ and $\boldsymbol{V}$ are positive-definite symmetric second-order tensors over $\mathscr{V}_\varphi$. According to this theorem (*cf.* § I.2), the first-order deformation at the body point X_0 and the time t_0 may be regarded as produced by pure stretches along a certain orthogonal triad, followed by a rotation, or as produced by that same rotation first, followed by the same stretches effected along the axes into which the original ones have been rotated. If

$$\boldsymbol{C} \equiv \boldsymbol{F}^T\boldsymbol{F} = \boldsymbol{U}^2 \tag{1.33}$$

and

$$\boldsymbol{B} \equiv \boldsymbol{F}\boldsymbol{F}^T = \boldsymbol{V}^2, \tag{1.34}$$

then clearly $\boldsymbol{C}$ determines $\boldsymbol{U}$ uniquely, and likewise $\boldsymbol{B}$ determines $\boldsymbol{V}$. We call $\boldsymbol{R}$ the *rotation tensor* of $\boldsymbol{F}$; $\boldsymbol{U}$ and $\boldsymbol{V}$, the *right* and *left stretch tensors* of $\boldsymbol{F}$; and $\boldsymbol{C}$ and $\boldsymbol{B}$, the *right* and *left Cauchy-Green tensors* of $\boldsymbol{F}$. If the deformations whose gradients are $\boldsymbol{F}$ and $\bar{\boldsymbol{F}}$ have the same right or left stretch tensors, the one deformation follows from the other by a suitable rotation. Specifically, if $\bar{\boldsymbol{U}} = \boldsymbol{U}$, then $\bar{\boldsymbol{F}} = \bar{\boldsymbol{R}}\boldsymbol{R}^T\boldsymbol{F}$, while if $\bar{\boldsymbol{V}} = \boldsymbol{V}$, then $\bar{\boldsymbol{F}} = \boldsymbol{F}\boldsymbol{R}^T\bar{\boldsymbol{R}}$. Loosely, the two deformations are said to "differ" by a rotation. Likewise, if $\boldsymbol{C} = \bar{\boldsymbol{C}}$, then $\boldsymbol{U} = \bar{\boldsymbol{U}}$, and again the deformations from which these tensors are calculated differ by a rotation.

More than this, since $\boldsymbol{F}$ is the gradient of a deformation, it is not difficult to prove that if $\boldsymbol{U}$ = const., then $\boldsymbol{R}$ = const., so that if $\boldsymbol{U} = \mathbf{1}$, the deformation is a rigid-body displacement. Consequently, if $\boldsymbol{U} = \bar{\boldsymbol{U}}$, the deformations corresponding to these Cauchy-Green tensors differ by a rigid displacement of the whole body. The same holds if $\boldsymbol{V} = \bar{\boldsymbol{V}}$.

Application of the chain rule yields the trivial but important formula

$$\dot{\boldsymbol{F}} = (\text{grad}\ \boldsymbol{v})\boldsymbol{F}, \tag{1.35}$$

which relates $\dot{\boldsymbol{F}}$ to the gradient of the velocity field, relative to the frame of reference φ.

Now we define the *relative deformation* $\chi_{t,u}$ as the deformation from χ_t to χ_u:

$$\chi_{t,u} \equiv \chi_u \circ \chi_t^{-1}. \tag{1.36}$$

The gradient $\boldsymbol{F}_{t,u}$ of $\chi_{t,u}$ is called the *relative deformation gradient* from time t to time u. Of course, we can determine $\chi_{t,u}$ and $\boldsymbol{F}_{t,u}$ by the identities

$$\chi_{t,u} = \boldsymbol{\kappa}_u \circ \boldsymbol{\kappa}_t^{-1} \tag{1.37}$$

and

$$\boldsymbol{F}_{t,u} = \boldsymbol{F}_{\boldsymbol{\kappa}}(u)\boldsymbol{F}_{\boldsymbol{\kappa}}(t)^{-1}, \tag{1.38}$$

where $\boldsymbol{\kappa}_t$ is defined by (1.14) in terms of some particular reference configuration $\boldsymbol{\kappa}$, with respect to which $\boldsymbol{F}_{\boldsymbol{\kappa}}(\cdot)$ also is calculated, but it should be noted that $\chi_{t,u}$ and $\boldsymbol{F}_{t,u}$, by their very definitions, are independent of the choice of that reference configuration.

Again, we have the unique polar decompositions of $\boldsymbol{F}_{t,u}$,

$$\boldsymbol{F}_{t,u} = \boldsymbol{R}_{t,u}\boldsymbol{U}_{t,u} = \boldsymbol{V}_{t,u}\boldsymbol{R}_{t,u}. \tag{1.39}$$

In particular, when $u = t$, we have

$$\boldsymbol{F}_{t,t} = \boldsymbol{R}_{t,t} = \boldsymbol{U}_{t,t} = \boldsymbol{V}_{t,t} = \boldsymbol{1}. \tag{1.40}$$

Here $\boldsymbol{1}$ denotes the identity map (considered as a second-order tensor as usual) of $\mathscr{V}_{\varphi}$. Of course, if we consider all the points of $\mathscr{B}$, then $\boldsymbol{F}_{t,u}$, for fixed times t and u, is a tensor field on $\chi_t(\mathscr{B})$, say in terms of the absolute spatial co-ordinates (x^i)

$$\boldsymbol{F}_{t,u} = \boldsymbol{F}_{t,u}(x^1, x^2, x^3). \tag{1.41}$$

Holding x^i and t fixed and differentiating this tensor field with respect to u, we then obtain from (1.39)

$$\dot{\boldsymbol{F}}_{t,u} \equiv \frac{d}{du}\boldsymbol{F}_{t,u} = \dot{\boldsymbol{R}}_{t,u}\boldsymbol{U}_{t,u} + \boldsymbol{R}_{t,u}\dot{\boldsymbol{U}}_{t,u}. \tag{1.42}$$

Notice that the operator d/du here has the same meaning as that given earlier in (1.18), *i.e.*, $d/du\, \boldsymbol{F}_{t,u}(x^1, x^2, x^3)$ denotes the time derivative of $\boldsymbol{F}_{t,u}(x^1, x^2, x^3)$ along the world line of the particular body point $X \in \mathscr{B}$ whose spatial co-ordinates at time t are (x^1, x^2, x^3). This operator, denoted also by a dot, which occurs frequently in kinematics, is often called the *material derivative*. In this same notation, (1.35) takes the form

$$\dot{\boldsymbol{F}}_{t,u} = (\operatorname{grad} \boldsymbol{v}(u))\boldsymbol{F}_{t,u}, \tag{1.43}$$

where $\boldsymbol{v}(u)$ is the spatial velocity field at time u.

Setting $u = t$ in (1.42) and using (1.40), we get

$$\dot{\boldsymbol{F}}_{t,t} \equiv \left.\left(\frac{d}{du}\boldsymbol{F}_{t,u}\right)\right|_{u=t} = \boldsymbol{W}(t)+\boldsymbol{D}(t), \tag{1.44}$$

where

$$\boldsymbol{W}(t) \equiv \dot{\boldsymbol{R}}_{t,t}, \quad \boldsymbol{D}(t) = \dot{\boldsymbol{U}}_{t,t}. \tag{1.45}$$

As these formulae suggest, $\boldsymbol{W}(t)$ is called the *spin*, and $\boldsymbol{D}(t)$ is called the *stretching* of the motion at the time t. Like the relative deformation gradient $\boldsymbol{F}_{t,u}$, the spin and the stretching are tensor fields on $\chi_t(\mathscr{B})$. The reader will show easily that $\boldsymbol{W}(t)$ is skew, while $\boldsymbol{D}(t)$ is symmetric. Also, by setting $u = t$ in (1.43) and then applying (1.40) and (1.44), we see that

$$\operatorname{grad} \boldsymbol{v}(t) = \boldsymbol{W}(t)+\boldsymbol{D}(t). \tag{1.46}$$

These results express the ***Cauchy-Stokes decomposition theorem***:

$$\begin{aligned} \boldsymbol{W} &= \tfrac{1}{2}\operatorname{grad} \boldsymbol{v}-\tfrac{1}{2}(\operatorname{grad} \boldsymbol{v})^T, \\ \boldsymbol{D} &= \tfrac{1}{2}\operatorname{grad} \boldsymbol{v}+\tfrac{1}{2}(\operatorname{grad} \boldsymbol{v})^T. \end{aligned} \tag{1.47}$$

In particular, the motion has zero spin if and only if the curl of the velocity field vanishes. Such a motion is said to be *irrotational*. Since $\boldsymbol{D}$ is a symmetric tensor, an irrotational motion may be regarded as a pure stretching along a suitable orthogonal triad. Likewise, the motion has zero stretching if and only if $\boldsymbol{D}(t) = \boldsymbol{0}$. It is easy to show that such a motion, which is called *rigid* at the time t, may be represented as the result of superimposing an instantaneous rotation upon a uniform velocity field. Thus, in general, from (1.45) and (1.46) we see that the velocity field near the place whose spatial co-ordinates are x_0^i may be regarded, with an error which is of the first order in $(x^i - x_0^i)$ at the given time t_0, as the result of superimposing a uniform translation, an instantaneous rotation, and a pure stretching along a suitable orthogonal triad.

The Cauchy-Stokes additive decomposition of the velocity gradient is the counterpart in terms of infinitesimal transformations of Cauchy's multiplicative decompositions (1.32) of the deformation gradient.

In analogy to (1.33), we define the *relative right Cauchy-Green tensor* $\boldsymbol{C}_{t,u}$ by

$$\boldsymbol{C}_{t,u} \equiv \boldsymbol{F}_{t,u}^{T}\boldsymbol{F}_{t,u}. \tag{1.48}$$

Then we can compute the successive material derivatives of $\boldsymbol{C}_{t,u}$ at $u = t$:

$$\boldsymbol{A}_n(t) \equiv \left(\frac{d^n}{du^n}\boldsymbol{C}_{t,u}\right)\Bigg|_{u=t}, \quad n = 0, 1, 2, 3, \ldots \tag{1.49}$$

The tensor field $\boldsymbol{A}_n(t)$ on $\chi_t(\mathscr{B})$ is known as the *Rivlin-Ericksen tensor* of *order n*. Of course,

$$\boldsymbol{A}_0(t) = \mathbf{1}. \tag{1.50}$$

From (1.48), (1.49), and (1.47),

$$\boldsymbol{A}_1(t) = \dot{\boldsymbol{F}}_{t,t} + \dot{\boldsymbol{F}}_{t,t}^{T} = 2\boldsymbol{D}(t). \tag{1.51}$$

In general,

$$\boldsymbol{A}_n(t) = \overset{(n)}{\boldsymbol{F}}_{t,t} + \binom{n}{1}\overset{(n-1)}{\boldsymbol{F}}_{t,t}\dot{\boldsymbol{F}}_{t,t}^{T} + \ldots + \binom{n}{i}\overset{(n-i)}{\boldsymbol{F}}_{t,t}\overset{(i)}{\boldsymbol{F}}{}_{t,t}^{T} + \ldots + \overset{(n)}{\boldsymbol{F}}{}_{t,t}^{T}, \tag{1.52}$$

where $\binom{n}{i}$ depotes the binomial coefficient.

2. Frame-Indifference

In the preceding section we have defined the concept of a frame of reference or observer φ on $\mathscr{E}$. Relative to a given frame of reference φ, we have introduced various kinematical quantities for a deformable body $\mathscr{B}$ in motion, *e.g.*, the velocity, the acceleration, the deformation gradients, *etc.* These quantities may be called *observable kinematical quantities* in the motion of $\mathscr{B}$, since the motion determines them uniquely relative to the given observer. More generally, let Φ denote the collection of all frames of reference on $\mathscr{E}$. (We shall characterize Φ in detail shortly.) Then we call an *observable quantity* a function f defined on Φ, say

$$f : \Phi \to \mathscr{S}, \tag{2.1}$$

where $\mathscr{S}$ is the set of values of $\boldsymbol{f}$. In the cases of commonest interest, $\mathscr{S}$ is a set of scalars, vectors, tensors, or functions of some kind. Naturally, for each $\varphi \in \Phi$ we call the value $\boldsymbol{f}(\varphi)$ *the value of* $\boldsymbol{f}$ *as observed by* φ.

For a given body $\mathscr{B}$, we write $\mathfrak{M}$ for the collection of all motions of $\mathscr{B}$. Then we define an *observable kinematical quantity* in the motion of $\mathscr{B}$ in general to be a function $\boldsymbol{g}$ on $\Phi \times \mathfrak{M}$, say

$$\boldsymbol{g} : \Phi \times \mathfrak{M} \to \mathscr{S}. \tag{2.2}$$

Of course, for each $\varphi \in \Phi$ and $\zeta \in \mathfrak{M}$ the value $\boldsymbol{g}(\varphi, \zeta)$ is called the *value of* $\boldsymbol{g}$ *as observed by* φ *in the motion* ζ. Since ζ is given by a set of abstract configurations (1.9) of $\mathscr{B}$, namely

$$\zeta = \{\zeta_s, s \in \mathscr{T}, \zeta_s : \mathscr{B} \to \mathscr{E}_s\}, \tag{2.3}$$

an observable kinematical quantity $\boldsymbol{g}$ in general can be regarded as a functional

$$\boldsymbol{g}(\varphi, \zeta) = \boldsymbol{G}(\zeta_s, \varphi). \tag{2.4}$$

In this sense, $\boldsymbol{g}$ is a special kind of observable quantity whose value observed by each φ is a functional $\boldsymbol{G}(\cdot, \varphi)$ defined on the set of motions $\mathfrak{M}$ for $\mathscr{B}$.

Now suppose that $\boldsymbol{f}$ be a given observable quantity, and let Ψ be a subset of Φ. Then $\boldsymbol{f}$ is said to be *frame-indifferent with respect to* Ψ if the value of $\boldsymbol{f}$ is invariant on Ψ. That is to say, for arbitrary members $\varphi, \bar{\varphi} \in \Psi$

$$\boldsymbol{f}(\varphi) = \boldsymbol{f}(\bar{\varphi}). \tag{2.5}$$

In particular, if $\boldsymbol{f}$ is frame-indifferent with respect to the whole set of frames Φ, then it is said simply to be *frame-indifferent*. On the other hand, if $\boldsymbol{f}$ is not frame-indifferent with respect to Φ, then it is said to be *not frame-indifferent*.

The mathematical meaning of frame-indifference is clear. In the physical interpretation, frame-indifference means *invariance under a change of observer*. Such invariance is an important concept in mechanics, since all mathematical models of material response are assumed to enjoy it. For example, in the crudest theory of elasticity, that for a simple spring, it is only natural to require that the elastic constant be a frame-indifferent property of the spring, independent of the observer. In this section we consider in general the condition of frame-indifference for various kinds of observable quantitities. In § 4 and § 7 we shall apply the condition to the theories of deformable bodies.

So as to be in a position to state conditions of frame-indifference explicitly, we must first show how to specify all frames of reference on $\mathscr{E}$ in terms of any given one. Let φ and $\bar{\varphi}$ be two frames of reference with reference origins $\boldsymbol{O}(s)$, $\bar{\boldsymbol{O}}(s) \in \mathscr{E}_s$, reference instants s_0, $\bar{s}_0 \in \mathscr{T}$, and reference bases $\{\boldsymbol{e}_i(s)\}$, $\{\bar{\boldsymbol{e}}_i(s)\} \in \mathscr{V}_s$. Then we can characterize $\bar{\boldsymbol{O}}(s)$ by its spatial co-ordinates $y^i(s)$ relative to φ, and likewise $\{\bar{\boldsymbol{e}}_i(s)\}$ by its components $\{Q_i^j(s)\}$ relative to φ:

$$\bar{\boldsymbol{O}}(s) - \boldsymbol{O}(s) = y^i(s)\boldsymbol{e}_i(s), \tag{2.6}$$

and

$$\bar{\boldsymbol{e}}_i(s) = Q_i^j(s)\boldsymbol{e}_j(s). \tag{2.7}$$

Here $[Q_i^j(s)]$ must be a proper-orthogonal matrix, since both $\{\boldsymbol{e}_i(s)\}$ and $\{\bar{\boldsymbol{e}}_i(s)\}$ are positive orthonormal bases for $\mathscr{V}_s$. Also we can characterize $\bar{s}_0$ by its absolute time t_0 relative to φ:

$$x^4(\bar{s}_0) = t_0, \tag{2.8}$$

or equivalently

$$\bar{s}_0 - s_0 = t_0\,\boldsymbol{e}_4. \tag{2.9}$$

Of course, the basis vectors $\boldsymbol{e}_4$ and $\bar{\boldsymbol{e}}_4$ for $x^4(\cdot)$ and $\bar{x}^4(\cdot)$ are identical, since they both are positive unit vectors for the 1-dimensional oriented Euclidean manifold $\mathscr{T}$. As before, the values of the functions $x^4(\cdot)$ and $\bar{x}^4(\cdot)$, which are the absolute times with respect to φ and $\bar{\varphi}$, respectively, will be denoted by t and $\bar{t}$.

Now let $\boldsymbol{e}$ be a typical point in $\mathscr{E}$, say $\boldsymbol{e} \in \mathscr{E}_s$. Then the co-ordinates $\varphi(\boldsymbol{e}) = (x^1, x^2, x^3, t)$ and $\bar{\varphi}(\boldsymbol{e}) = (\bar{x}^1, \bar{x}^2, \bar{x}^3, \bar{t})$ relative to φ and $\bar{\varphi}$ can be determined in the standard way:

$$\boldsymbol{e} - \boldsymbol{O}(s) = x^i\boldsymbol{e}_i(s), \quad \boldsymbol{e} - \bar{\boldsymbol{O}}(s) = \bar{x}^i\bar{\boldsymbol{e}}_i(s), \tag{2.10}$$

and

$$s - s_0 = t\boldsymbol{e}_4, \; s - \bar{s}_0 = \bar{t}\boldsymbol{e}_4. \tag{2.11}$$

By applying the basic relations (2.6), (2.7), and (2.9) to (2.10) and (2.11) we get the following co-ordinate transformation:

$$x^i = y^i(s) + Q_j^i(s)\bar{x}^j, \quad i = 1, 2, 3, \tag{2.12}$$

and

$$t = t_0 + \bar{t}. \tag{2.13}$$

Here, of course, we can express s as a function of the absolute time $\bar{t}$ relative to $\bar{\varphi}$ by inverting the equation $\bar{t} = \bar{x}^4(s)$, namely, $s = \bar{s}(\bar{t})$. Thus (2.12) and (2.13) give (x^1, x^2, x^3, t) as functions of $(\bar{x}^1, \bar{x}^2, \bar{x}^3, \bar{t})$. Alternatively, we may use the absolute time t relative to φ. We shall not encumber the notation by introducing new symbols for the functions of t or $\bar{t}$ so defined: $Q^i_j(s) = Q^i_j(\bar{s}(\bar{t})) = Q^i_j(s(t))$. When there is no danger of confusion, we shall write $Q^i_j(t)$ and $Q^i_j(\bar{t})$ for these functions. The notation $\dot{Q}^i_j$ will stand for the derivative of the first of these functions with respect to t or of the second with respect to $\bar{t}$, these two derivatives being equal as a consequence of (2.13). The notations $y^i(t)$, $\dot{y}^i$, $\ddot{y}^i$, and $\ddot{Q}^i_j$ are similarly defined.

Since both φ and $\bar{\varphi}$ are co-ordinate systems on $\mathscr{E}$, the co-ordinate transformation (2.12), (2.13) must be smooth. Hence y^i and Q^j_k are necessarily smooth functions on $\mathscr{T}$.

Now conversely, suppose there be given a triple $(y^i(\cdot), Q^j_k(\cdot), t_0)$ such that $y^i(\cdot)$ and $Q^j_k(\cdot)$ are smooth functions on $\mathscr{T}$, with $[Q^j_k(s)]$ being proper-orthogonal for each s, and let φ be a given frame of reference as before. Then there exists a unique frame of reference $\bar{\varphi}$ determined by the basic relations (2.6), (2.7), and (2.9). Moreover, (x^α) and $(\bar{x}^\alpha)$ are related by the co-ordinate transformation (2.12) and (2.13). Consequently, once a particular frame φ is given, all other frames are determined from it by the assignment of triples $(y^i(\cdot), Q^j_k(\cdot), t_0)$ which have the properties just stated. Thus we have exhibited the set of all frames of reference Φ in terms of any one of its elements φ. In this sense, (2.12) and (2.13) are the *general equations of a change of frame*. Commonly they are described as showing that any change of frame may be resolved into a time-dependent translation, a time-dependent rotation, and a shift of the origin of time, in succession.

Now let us consider an observable quantity $\boldsymbol{f}$ whose value $\boldsymbol{f}(\varphi)$ observed by each $\varphi \in \Phi$ is an event in $\mathscr{E}$, so that $\boldsymbol{f}(\varphi)$ is specified by the co-ordinates (x^α) relative to φ:

$$(x^1, x^2, x^3, x^4) = \varphi(\boldsymbol{f}(\varphi)). \tag{2.14}$$

Suppose that this observable quantity $\boldsymbol{f}$ be frame-indifferent. Then there exists an event $\boldsymbol{e} \in \mathscr{E}$ such that

$$\boldsymbol{e} = \boldsymbol{f}(\varphi) \quad \forall \varphi \in \Phi. \tag{2.15}$$

Thus if we write $(\bar{x}^\alpha)$ for the co-ordinates of $\boldsymbol{f}(\bar{\varphi})$ relative to $\bar{\varphi}$ as in (2.14), then (x^α) and $(\bar{x}^\alpha)$ must be related by the transformation law

(2.12) and (2.13). In this sense, we have shown that (2.12) and (2.13) express the *condition of frame-indifference* for an event-valued observable quantity.

Suppose now that φ and $\overline{\varphi}$ be two frames of reference related by (2.12) and (2.13), and let $\boldsymbol{\lambda} : (-\infty, \infty) \to \mathscr{E}$ be a curve in $\mathscr{E}$ whose co-ordinate forms in φ and $\overline{\varphi}$ are, respectively, $(\lambda^\alpha(t))$ and $(\overline{\lambda}^\alpha(t))$. Then from (2.13) the condition

$$\lambda^4(t) = t + c_\varphi \tag{2.16}$$

holds if and only if

$$\overline{\lambda}^4(t) = t + c_{\overline{\varphi}}, \tag{2.17}$$

where c_φ and $c_{\overline{\varphi}}$ are constants. Consequently, the *condition for a curve to be a world line*, given in the previous section as a definition, is independent of the choice of frame. Either of the constants c_φ, $c_{\overline{\varphi}}$ determines the other in terms of the constant t_0 in the change of frame (2.13):

$$c_{\overline{\varphi}} = c_\varphi - t_0 . \tag{2.18}$$

Thus we see that the observable quantity $\{c_\varphi, \varphi \in \Phi\}$ for a given world line is not frame-indifferent. In particular, if $c_\varphi = 0$, generally $c_{\overline{\varphi}} \neq 0$, so that if $\boldsymbol{\lambda}$ is natural with respect to the frame φ, it is not natural with respect to $\overline{\varphi}$ unless $t_0 = 0$ in the transformation (2.13) from φ to $\overline{\varphi}$.

Next we consider an observable quantity $\boldsymbol{f}$ whose value for each frame $\varphi \in \Phi$ is a world line: $\boldsymbol{f}(\varphi) = \boldsymbol{\lambda}_\varphi$. The co-ordinate representation of $\boldsymbol{\lambda}_\varphi$ in φ is

$$\varphi(\boldsymbol{\lambda}_\varphi(t)) = (\lambda^1(t), \lambda^2(t), \lambda^3(t), \lambda^4(t)), \ t \in \mathscr{R}, \tag{2.19}$$

where $\lambda^4(t)$ is given by (2.16). Now suppose that this observable quantity be frame-indifferent. Then from (2.5) $\boldsymbol{\lambda}_\varphi = \boldsymbol{\lambda}_\varphi$ for all $\varphi, \overline{\varphi} \in \Phi$. Such is the case, of course, if and only if $\boldsymbol{\lambda}_\varphi(t) = \boldsymbol{\lambda}_{\overline{\varphi}}(t)$ $\forall t$. Thus for each fixed t, $\{\boldsymbol{\lambda}_\varphi(t), \varphi \in \Phi\}$ is a frame-indifferent event-valued observable quantity. Consequently, the condition of frame-indifference for a world line can be read off from the previous condition (2.12) and (2.13), namely,

$$\lambda^i(t) = y^i(\overline{\lambda}^4(t)) + Q^i_j(\overline{\lambda}^4(t))\overline{\lambda}^j(t), \quad i = 1, 2, 3 \tag{2.20}$$

and

$$\lambda^4(t) = t_0 + \overline{\lambda}^4(t). \tag{2.21}$$

Here the notation is the same as before: $\overline{\varphi}$ is an arbitrary frame, and

$$\overline{\varphi}(\boldsymbol{\lambda}_{\overline{\varphi}}(t)) = (\overline{\lambda}^1(t), \overline{\lambda}^2(t), \overline{\lambda}^3(t), \overline{\lambda}^4(t)). \tag{2.22}$$

The triple $(y^i(\cdot),\ Q^j_k(\cdot),\ t_0)$ characterizes the change of frame from φ to $\bar{\varphi}$.

Next, let us consider an observable quantity f whose value for each observer φ is a spatial vector $f(\varphi) \in \mathscr{V}_{s_0}$ for some fixed instant s_0, so that $f(\varphi)$ can be specified by its components relative to the spatial frame of φ:

$$f(\varphi) = f^i e_i(s_0). \tag{2.23}$$

The condition of frame-indifference (2.5) takes the following form in terms of the components of f:

$$f^i e_i(s_0) = \bar{f}^i \bar{e}_i(s_0). \tag{2.24}$$

From the co-ordinate relation (2.7) for the change of frame, this condition is equivalent to the usual vector transformation rule:

$$f^i = Q^i_j(s_0)\bar{f}^j, \tag{2.25}$$

which therefore expresses a *necessary and sufficient condition that the vector f be frame-indifferent.*

By exactly the same argument, we can express the condition of frame-indifference for a tensor-valued observable quantity f by the tensor transformation rule. For example, consider the case when $f(\varphi)$ is a second-order tensor with co-ordinate representation

$$f(\varphi) = f^i_j e_i(s_0) \otimes e^j(s_0). \tag{2.26}$$

Then the condition of frame-indifference (2.5) leads to the following *necessary and sufficient transformation rule for a frame-indifferent second-order tensor*:

$$f^i_j = Q^i_k(s_0)\bar{f}^k_l Q^{T\,l}_j(s_0), \tag{2.27}$$

where

$$f(\bar{\varphi}) = \bar{f}^k_l e_k(s_0) \otimes e^l(s_0). \tag{2.28}$$

and where $Q^{T\,l}_j$ stands for the set of functions such that $[Q^{T\,l}{}_j(s_0)]$ is the transpose of the matrix $[Q^l_j(s_0)]$.

We remark that the terms *vector-valued* or *tensor-valued* observable quantities here refer to functions on Φ whose values are vectors or tensors over some instantaneous space $\mathscr{V}_s$, *not* over the absolute spaces of the various observers. Suppose that f be an observable quantity whose value $f(\varphi)$ for each fixed $\varphi \in \Phi$ is an absolute vector or an absolute tensor over the absolute space $\mathscr{V}_\varphi$. Then f cannot possibly be frame-indifferent, since for two distinct frames φ and $\bar{\varphi}$, the absolute spaces $\mathscr{V}_\varphi$ and $\mathscr{V}_{\bar{\varphi}}$ are themselves distinct.

We now show that the velocity and the acceleration of a fixed world line $\boldsymbol{\lambda} : (-\infty, \infty) \to \mathscr{E}$ are *not frame-indifferent.* By definition, the component forms of the velocity and the acceleration relative to φ and $\bar{\varphi}$ are

$$\boldsymbol{v} = \dot{\lambda}^i \boldsymbol{e}_i, \ \bar{\boldsymbol{v}} = \dot{\bar{\lambda}}^i \bar{\boldsymbol{e}}_i, \tag{2.29}$$

and

$$\boldsymbol{a} = \ddot{\lambda}^i \boldsymbol{e}_i, \ \bar{\boldsymbol{a}} = \ddot{\bar{\lambda}}^i \bar{\boldsymbol{e}}_i, \tag{2.30}$$

the bases $\{\boldsymbol{e}_i\}$ and $\{\bar{\boldsymbol{e}}_i\}$ being related by (2.7). Since $\boldsymbol{\lambda}$ is fixed, independent of frame to begin with, the components (λ^i) and $(\bar{\lambda}^i)$ must satisfy the condition (2.20) for all $t \in (-\infty, \infty)$. Differentiating that condition twice, we obtain

$$\dot{\lambda}^j = \dot{y}^j + Q^j_i \dot{\bar{\lambda}}^i + \dot{Q}^j_i \bar{\lambda}^i, \tag{2.31}$$

and

$$\ddot{\lambda}^j = \ddot{y}^j + Q^j_i \ddot{\bar{\lambda}}^i + 2\dot{Q}^j_i \dot{\bar{\lambda}}^i + \ddot{Q}^j_i \bar{\lambda}^i. \tag{2.32}$$

These equations show that $\boldsymbol{v}$ and $\boldsymbol{a}$ do not satisfy the general condition (2.25) for frame-indifference of a vector-valued observable quantity. Thus the velocity and the acceleration are not frame-indifferent.

The transformation laws (2.31) and (2.32) suggest the following

DEFINITION. Two frames φ and $\bar{\varphi}$ are said to be related by a *Galilean transformation* if $\dot{Q}^i_j = \ddot{y}^k = 0 \ \forall i, j, k$ and $\forall s$. Similarly, φ and $\bar{\varphi}$ are said to be related by a *rigid transformation* if $\dot{Q}^i_j = \dot{y}^k = 0 \ \forall i, j, k, s$.

Clearly, a rigid transformation is a Galilean transformation; moreover, transformations of both these kinds are equivalence relations on Φ. Their equivalence classes are called, respectively, *Galilean classes* and *rigid classes.* Then the formulae (2.31) and (2.32) imply the following

PROPOSITION. *The velocity of a world line is frame-indifferent with respect to any rigid class; the acceleration is frame-indifferent with respect to any Galilean class.*

The converse of this proposition holds in the sense that a rigid class and a Galilean class are maximal subsets in Φ with respect to which the velocity and the acceleration, respectively, of an arbitrary world line are frame-indifferent. From this proposition, we see that the absolute space[1] (and hence the absolute translation space) is frame-indifferent with respect to any rigid class. Consequently the observation maps (1.5) and (1.6) are also frame-indifferent with respect to such a class. Let us write Φ_0 for a typical rigid class in Φ. Then the (frame-indifferent) observation maps of its members may be called the *observation maps of* Φ_0. In this sense, Φ_0 gives rise to a representation[2] of $\mathscr{E}$ by the product manifold $\mathscr{E}_{\Phi_0} \times \mathscr{T}$, $(\mathscr{E}_{\Phi_0} \equiv \mathscr{E}_\varphi \;\; \forall \varphi \in \Phi_0)$,

$$\Phi_0 : \mathscr{E} \to \mathscr{E}_{\Phi_0} \times \mathscr{T}. \tag{2.33}$$

As we shall see later, in *dynamics*, which is a mathematical theory for the determination of motion, an important axiom asserts that there is a distinguished Galilean class in Φ, called the *inertial class*, relative to which the relations between forces and motions take their simplest forms.

Next we show that the Rivlin-Ericksen tensors (1.49), considered as tensors over the instantaneous spaces, are frame-indifferent. Since the Rivlin-Ericksen tensors are observable kinematical quantities, we must show that their values are independent of the choice of $\varphi \in \Phi$ at each instant $s_0 \in \mathscr{T}$, in each motion ζ of $\mathscr{B}$. Hence we choose an arbitrary frame of reference φ, and let $\chi_t : \mathscr{B} \to \mathscr{E}_\varphi$, $t \in (-\infty, \infty)$, characterize the motion of $\mathscr{B}$ as observed by φ. Then the relative deformation $\chi_{t,u}$ is given by the deformation functions[3] $(\chi_{t,u})^i(x^1, x^2, x^3)$, and the relative deformation gradient $\boldsymbol{F}_{t,u}$ has the component representation

$$[(F_{t,u})^i_j] = \left[\frac{\partial(\chi_{t,u})^i}{\partial x^j}\right]. \tag{2.34}$$

Hence by the definition (1.48) the relative Cauchy-Green tensor $\boldsymbol{C}_{t,u}$ is

---

[1] *Note.* The absolute space $\mathscr{E}_\varphi$ can be regarded as a function of φ, so frame-indifference here has its usual meaning: $\mathscr{E}_\varphi = \mathscr{E}_{\overline{\varphi}}$ if φ and $\overline{\varphi}$ belong to the same rigid class in $\boldsymbol{\Phi}$.

[2] NOLL, in the lectures cited in the previous section, calls such a representation "a frame of reference" on $\mathscr{E}$, which is, of course, an equivalence class of frames of reference in our sense. Thus the formulation developed here is essentially equivalent to NOLL's, though his is co-ordinate-free.

[3] So as to avoid confusion, we separate by parentheses component indices from such other subscripts or superscripts as a character may bear. Thus $(\chi_{t,u})^i$ is the *i*-component of $\chi_{t,u}$.

given by

$$[(C_{t,u})_{kl}] = [(F_{t,u})^i_j]^T[(F_{t,u})^i_j] = \left[\delta_{ij}\frac{\partial(\chi_{t,u})^i}{\partial x^k}\frac{\partial(\chi_{t,u})^j}{\partial x^l}\right]. \tag{2.35}$$

Let the component form of the tensor $\boldsymbol{A}_n(t)$ over $\mathscr{V}_{s(t)}$ be

$$\boldsymbol{A}_n(t) = (A_n(t))_{kl}\,\boldsymbol{e}^k(s(t)) \otimes \boldsymbol{e}^l(s(t)). \tag{2.36}$$

Then the components $A_n(t)_{kl}$ are given by

$$A_n(t)_{kl} = \left[\frac{d^n}{du^n}\left(\delta_{ij}\frac{\partial(\chi_{t,u})^i}{\partial x^k}\frac{\partial(\chi_{t,u})^j}{\partial x^l}\right)\right]_{u=t}. \tag{2.37}$$

The meaning of this equation is similar to that of (1.49).

Now if we choose another frame of reference $\bar{\boldsymbol{\varphi}}$ which is related to $\boldsymbol{\varphi}$ by (2.6), (2.7), and (2.9), then the new relative deformation $\bar{\boldsymbol{\chi}}_{\bar{t},\bar{u}}$ as observed by $\bar{\boldsymbol{\varphi}}$ is characterized by a new set of deformation functions, $(\bar{\chi}_{\bar{t},\bar{u}})^i(\bar{x}^1, \bar{x}^2, \bar{x}^3)$. We choose the times $\bar{t}$ and $\bar{u}$ relative to $\bar{\boldsymbol{\varphi}}$ in such a way that

$$\bar{s}(\bar{t}) = s(t) = s_0,\ \bar{s}(\bar{u}) = s(u) = s, \tag{2.38}$$

so that t, u and $\bar{t}$, $\bar{u}$ are the times of the same instants s_0, s relative to $\boldsymbol{\varphi}$ and $\bar{\boldsymbol{\varphi}}$. Then the transformation law (2.12) implies that

$$(\chi_{t,u})^i = y^i(s) + Q^i_j(s)(\bar{\chi}_{\bar{t},\bar{u}})^j, \tag{2.39}$$

where (x^i) and $(\bar{x}^i)$ are the arguments of $(\chi_{t,u})^i$ and $(\bar{\chi}_{\bar{t},\bar{u}})^i$, respectively, and where

$$x^i = y^i(s_0) + Q^i_j(s_0)\bar{x}^j. \tag{2.40}$$

Differentiating (2.39) with respect to x^k and using the chain rule based on (2.40), we then obtain

$$\frac{\partial(\bar{\chi}_{\bar{t},\bar{u}})^j}{\partial \bar{x}^m} = Q^{Tj}_i(s_0)\frac{\partial(\chi_{t,u})^i}{\partial x^k}Q^k_m(s_0). \tag{2.41}$$

If we now write down the component form of the Rivlin-Ericksen tensor $\boldsymbol{A}_n(t)$ over $\mathscr{V}_{\bar{s}(\bar{t})}$ as observed by $\bar{\boldsymbol{\varphi}}$:

$$\bar{\boldsymbol{A}}_n(\bar{t}) = \bar{A}_n(\bar{t})_{kl}\,\bar{\boldsymbol{e}}^k(\bar{s}(\bar{t})) \otimes \bar{\boldsymbol{e}}^l(\bar{s}(\bar{t})), \tag{2.42}$$

then as in (2.37)

$$\bar{A}_n(\bar{t})_{kl} = \left[\frac{d^n}{d\bar{u}^n}\left(\delta_{ij}\frac{\partial(\bar{\chi}_{\bar{t},\bar{u}})^i}{\partial\bar{x}^k}\frac{\partial(\bar{\chi}_{\bar{t},\bar{u}})^j}{\partial\bar{x}^l}\right)\right]_{\bar{u}=\bar{t}}. \tag{2.43}$$

But, substituting (2.41) into (2.43), we obtain

$$\begin{aligned}\bar{A}_n(\bar{t})_{pq} &= Q_p^k(s_0)\left[\frac{d^n}{du^n}\left(\delta_{ij}\frac{\partial(\chi_{t,u})^i}{\partial x^k}\frac{\partial(\chi_{t,u})^j}{\partial x^l}\right)\right]_{u=t} Q_q^l(s_0),\\ &= Q_p^k(s_0)A_n(t)_{kl}Q_q^l(s_0).\end{aligned} \tag{2.44}$$

Consequently $\bar{A}_n(\bar{t}) = A_n(t)$ as tensors over $\mathscr{V}_{s(t)} = \mathscr{V}_{\bar{s}(\bar{t})} = \mathscr{V}_{s_0}$, and the proof is complete.

We now treat frame-indifference for a general observable kinematical quantity. Consider the quantity $\boldsymbol{g}$ in (2.2) to be represented by the functional equation (2.4). Suppose that we specify the motion ζ given by (2.3) in terms of the spatial co-ordinates relative to φ, say

$$\varphi(\zeta) = \{\varphi(\zeta_s),\ s \in \mathscr{T},\ \varphi(\zeta_s) : \mathscr{B} \to \mathscr{R}^3\} \tag{2.45}$$

where $\varphi(\zeta_s)$ is characterized by the scalar fields χ^i:

$$\varphi(\zeta_s) = (\chi^1(\cdot, s), \chi^2(\cdot, s), \chi^3(\cdot, s)) \tag{2.46}$$

defined[1] on $\mathscr{B}$. Then we can represent $\boldsymbol{g}$ also by the equation

$$\boldsymbol{g}(\varphi, \zeta) = \underset{\substack{s\in\mathscr{T}\\ X\in\mathscr{B}}}{\boldsymbol{G}}(\chi^1(X, s), \chi^2(X, s), \chi^3(X, s), \varphi). \tag{2.47}$$

For the moment, let us consider the special case when $\boldsymbol{g}$ is vector-valued, say $\boldsymbol{g}(\varphi, \zeta) \in \mathscr{V}_{s_0}$ $\forall \varphi \in \Phi$, $\zeta \in \mathfrak{M}$, for some fixed instant s_0. Then we can reduce the representation further by using the component form of $\boldsymbol{G}$ relative to the instantaneous spatial frame $\{\boldsymbol{e}_i(s_0)\}$ of φ at s_0:

$$\boldsymbol{g}(\varphi, \zeta) = \underset{\substack{s\in\mathscr{T}\\ X\in\mathscr{B}}}{G^i}(\chi^1(X, s), \chi^2(X, s), \chi^3(X, s), \varphi)\boldsymbol{e}_i(s_0), \tag{2.48}$$

where the functionals G^i are scalar-valued. More generally, if $\boldsymbol{g}$ is tensor-valued, then a similar representation is possible; *e.g.*, if $\boldsymbol{g}(\varphi, \zeta)$ is a second-

---

[1] The value of the function $\chi^i(X, s)$ is the ith spatial co-ordinate of $\boldsymbol{\chi}_t(X)$ in $\mathscr{E}_\varphi$ at the instant $s(t)$:

$$x^i(t) = \chi^i(X, s(t)).$$

order tensor over $\mathscr{V}_{s_0}$ $\forall \boldsymbol{\varphi}, \zeta$, then

$$\boldsymbol{g}(\boldsymbol{\varphi}, \zeta) = \underset{\substack{s \in \mathscr{T} \\ X \in \mathscr{B}}}{\boldsymbol{G}^{ij}}(\chi^1(X, s), \chi^2(X, s), \chi^3(X, s), \boldsymbol{\varphi}) \boldsymbol{e}_i(s_0) \otimes \boldsymbol{e}_j(s_0). \tag{2.49}$$

We now remark that all the observable kinematical quantities we have defined explicitly so far are such that when cast into the standard representation in the component form as given by (2.48) or (2.49), the functionals whose values are their components, if considered as observable quantities themselves, are always frame-indifferent. In other words, the functional $G^i(\cdot, \boldsymbol{\varphi})$ in (2.48) and the functional $G^{ij}(\cdot, \boldsymbol{\varphi})$ in (2.49) do *not* depend on $\boldsymbol{\varphi}$. This basic fact is easy to understand; it merely reflects the condition that the *procedure* for specifying an observable kinematical quantity with respect to a frame is independent of that frame. For example, to specify the velocity of a world line with respect to a frame $\boldsymbol{\varphi}$, we first determine the spatial co-ordinates of that world line in $\boldsymbol{\varphi}$. Then we differentiate the functions giving its spatial co-ordinates so as to get the components of the velocity relative to the spatial frame of $\boldsymbol{\varphi}$, as shown in $(2.29)_1$. Of course, we follow exactly the same procedure in any other frame, say $\overline{\boldsymbol{\varphi}}$, as shown in $(2.29)_2$.

That the procedure for specifying an observable kinematical quantity with respect to a frame is independent of that frame, should not be confused with the condition that the kinematical quantity thus specified be itself frame-indifferent. Using the example just considered, we see that the spatial co-ordinates of the world line relative to distinct frames indeed are distinct sets of functions, and hence so are their derivatives. Thus the velocity, which is specified by the procedure, need not be frame-indifferent; indeed, we have shown above that it is not. However, the fact that the components of the velocity are the derivatives of the spatial co-ordinates, holds for all frames. Consequently, the functionals expressing the components of the velocity in terms of the spatial co-ordinates should not depend on the frame. It is not surprising that the procedure for specifying a kinematical quantity relative to a frame should always reduce to a representation in component form such as (2.48) and (2.49), since the natural way for an observer to describe a motion is through its spatial co-ordinates, as in (2.46); likewise, the natural way to specify a tensor for each motion relative to an observer is to use the component form relative to the instantaneous spatial frame belonging to that observer. Thus the representation in component form as given by (2.48) or (2.49) is typical, and the component functionals should not depend on the frame.

Now suppose that an observable kinematical quantity $\boldsymbol{g}$ satisfy this invariance condition in its representation by components. For example, let $\boldsymbol{g}$ be vector-valued and hence given by

$$\boldsymbol{g}(\varphi, \zeta) = \underset{\substack{s\in\mathcal{T} \\ X\in\mathcal{B}}}{G^i}(\chi^1(X, s), \chi^2(X, s), \chi^3(X, s))\boldsymbol{e}_i(s_0). \tag{2.50}$$

Then what condition on G^i is necessary and sufficient that the quantity $\boldsymbol{g}$ itself be frame-indifferent?

To answer this question is easy. Since the functionals G^i are frame-indifferent, for any other frame of reference $\overline{\varphi}$ the value of $\boldsymbol{g}(\overline{\varphi}, \zeta)$ is given by

$$\boldsymbol{g}(\overline{\varphi}, \zeta) = \underset{\substack{s\in\mathcal{T} \\ X\in\mathcal{B}}}{G^i}(\bar{\chi}^1(X, s), \bar{\chi}^2(X, s), \bar{\chi}^3(X, s))\bar{\boldsymbol{e}}_i(s_0). \tag{2.51}$$

From (2.25), $\boldsymbol{g}$ is frame-indifferent if and only if

$$\begin{aligned}\underset{\substack{s\in\mathcal{T} \\ X\in\mathcal{B}}}{G^k}&(\chi^1(X, s), \chi^2(X, s), \chi^3(X, s)) \\ &= \underset{\substack{s\in\mathcal{T} \\ X\in\mathcal{B}}}{G^l}(\bar{\chi}^1(X, s), \bar{\chi}^2(X, s), \bar{\chi}^3(X, s))Q^k_l(s_0),\end{aligned} \tag{2.52}$$

where $Q^k_l(s_0)$ is given by (2.7) and where χ^i and $\bar{\chi}^i$, being the spatial co-ordinates of the same motion ζ relative to the two frames φ and $\overline{\varphi}$, must be related by the transformation law (2.12) for each $X \in \mathcal{B}$ and $s \in \mathcal{T}$:

$$\chi^i(X, s) = y^i(s)+Q^i_j(s)\bar{\chi}^j(X, s), \quad i = 1, 2, 3. \tag{2.53}$$

Let us consider the transformation (2.53) from another point of view. We choose now to regard the fields $\bar{\chi}^i(X, s)$, $X \in \mathcal{B}$, $s \in \mathcal{T}$, as the spatial co-ordinates of *another motion* $\bar{\zeta}$ relative to the *same* frame of reference φ. In the motion ζ the body point $X \in \mathcal{B}$ occupies the place

$$\zeta_s(X) = \boldsymbol{o}(s)+\chi^i(X, s)\boldsymbol{e}_i(s) \tag{2.54}$$

in $\mathcal{E}_s$ at the instant s, while in the motion $\bar{\zeta}$, X occupies the place

$$\bar{\zeta}_s(X) = \boldsymbol{o}(s)+\bar{\chi}^i(X, s)\boldsymbol{e}_i(s). \tag{2.55}$$

Thus (2.53) expresses the abstract relation

$$\zeta_s(X) = \boldsymbol{y}(s)+\boldsymbol{Q}(s)(\bar{\zeta}_s(X)-\boldsymbol{o}(s))+\boldsymbol{o}(s), \tag{2.56}$$

where $\boldsymbol{y}(s)$ is a vector in $\mathscr{V}_s$ given by

$$\boldsymbol{y}(s) \equiv y^i(s)\boldsymbol{e}_i(s). \tag{2.57}$$

Since $\boldsymbol{Q}(s)$, as defined by (2.56), is proper-orthogonal (over $\mathscr{V}_s$), the relation (2.56) says that for each fixed s the place $\zeta_s(X)$ can be obtained from $\bar{\zeta}_s(X)$ by a rotation $\boldsymbol{Q}(s)$ centered at $\boldsymbol{o}(s)$, and then followed by a translation $\boldsymbol{y}(s)$. Since $\boldsymbol{y}(s)$ and $\boldsymbol{Q}(s)$ are independent of the body point X, the motions ζ and $\bar{\zeta}$ differ by a *rigid motion* in the common sense.

Returning to our study of frame-indifference, we recall that (2.52) is a definite mathematical restriction on the functionals G^i, so that we need not retain the physical interpretation used to motivate it. If, as we may, we choose to regard ζ and $\bar{\zeta}$ as different motions relative to the same frame φ, then for all $s \in \mathscr{T}$

$$\varphi(\bar{\zeta}_s) = (\bar{\chi}^1(\cdot, s), \bar{\chi}^2(\cdot, s), \bar{\chi}^3(\cdot, s)), \tag{2.58}$$

and the restriction (2.52) means that the values $\boldsymbol{g}(\varphi, \zeta)$ and $\boldsymbol{g}(\varphi, \bar{\zeta})$ must be related by the condition

$$\boldsymbol{g}(\varphi, \zeta) = \boldsymbol{Q}(s_0)\boldsymbol{g}(\varphi, \bar{\zeta}), \tag{2.59}$$

where, for any $s \in \mathscr{T}$, $\boldsymbol{Q}(s)$ denotes the tensor

$$\boldsymbol{Q}(s) \equiv Q_l^k(s)\boldsymbol{e}^l(s) \otimes \boldsymbol{e}_k(s), \tag{2.60}$$

the Q_l^k being the functions appearing in (2.53).

We summarize the result in the following

PROPOSITION. *A vector-valued kinematical quantity $\boldsymbol{g}$ given by the representation* (2.50) *is frame-indifferent if and only if the functionals G^i satisfy the identity* (2.52); *or, equivalently, relative to any one frame of reference φ, the functional $\boldsymbol{g}(\varphi, \cdot)$ obeys the identity* (2.59) *for each pair of motions ζ and $\bar{\zeta}$ that differ by a rigid motion.*

Similar results hold for other kinds of observable kinematical quantities.

The condition of frame-indifference (2.59) obtained here is not the same as the previous general condition (2.52). Applying (2.25) to the observable quantity $\boldsymbol{g}(\varphi, \zeta)$, we get the usual condition:

$$\boldsymbol{g}(\varphi, \cdot) = \boldsymbol{g}(\bar{\varphi}, \cdot) \quad \forall \varphi, \bar{\varphi} \in \Phi, \tag{2.61}$$

which is merely a special case of (2.5). The condition (2.59), however, is of entirely different nature. Here φ is fixed, but the two motions ζ and $\bar{\zeta}$ differ by a rigid motion. Thus (2.5), or equivalently (2.61), is an *invariance law* for $\boldsymbol{g}$ with respect to a *change of frame* in a given motion, while (2.59) is a *transformation law* for $\boldsymbol{g}$ with respect to a *change of motion* in a given frame. These two laws are equivalent if and only if the component functionals for $\boldsymbol{g}$ as shown in the general representation (2.48) are frame-indifferent, since in this case, and only in this case, both laws are characterized by the condition (2.52).

NOLL[1], calling two motions that differ by a rigid motion "equivalent motions", adopted the condition (2.59) as the *definition* of frame-indifference. The transformation (2.59) then characterizes the relation between the values of $\boldsymbol{g}$ in equivalent motions relative to the fixed frame φ. While the definition of equivalent motions is phrased in terms of a given observer φ, the relation itself does not depend upon φ, since if two motions differ by a rigid motion in one frame of reference, they do so in all. Moreover, the translation $\boldsymbol{y}(s)$ and the rotation $\boldsymbol{Q}(s)$ relating ξ_s and $\bar{\xi}_s$ in (2.56) are independent of φ. Thus the condition that ξ and $\bar{\xi}$ be equivalent motions is frame-indifferent.

It seems more natural to define frame-indifference in terms of a condition relative to a change of frame rather than a condition relative to a change of motion; moreover, the meaning of the word "indifference" should be brought out by an invariance law rather than by a transformation law. For this reason, we do not follow NOLL's definition here, although his condition of frame-indifference is mathematically equivalent to ours, as shown by the proposition above.

The condition of frame-indifference as given by (2.59) is valid for vector-valued kinematical quantities only. A parallel analysis, of course, can be carried out for a tensor-valued quantity. When $\boldsymbol{G}(\varphi, \zeta)$ is a second-order tensor, say the one given by (2.49) with component functionals independent of the frame, the resulting necessary and sufficient condition for frame-indifference is

$$\boldsymbol{G}(\varphi, \zeta) = \boldsymbol{Q}(s_0)\boldsymbol{G}(\varphi, \bar{\zeta})\boldsymbol{Q}^T(s_0). \tag{2.62}$$

The standard isomorphism $\mathscr{V}_s \to \mathscr{V}_s^*$ via the inner product is understood here.

So as to illustrate these ideas, we now use the condition (2.62) to show that the Rivlin-Ericksen tensors $\boldsymbol{A}_n(s_0)$ are frame-indifferent at each fixed $s_0 \in \mathscr{T}$. Relative to the fixed frame of reference φ, let us choose a fixed reference configuration $\boldsymbol{\kappa}$ of $\mathscr{B}$. As before, we denote the absolute configurations of ζ_s and $\bar{\zeta}_s$ by χ_t and $\bar{\chi}_t$, t being the absolute time of the

---

[1] P. 173 of the work cited on p. 81.

instant s. Now since ζ differs from $\bar{\zeta}$ by a rigid motion as shown in (2.56), we see that

$$\boldsymbol{F}(t) = \boldsymbol{Q}(t)\bar{\boldsymbol{F}}(t), \tag{2.63}$$

where $\boldsymbol{Q}(t)$ is defined by (2.56) but is now considered as a tensor over $\mathscr{V}_\varphi$, *i.e.*,

$$\boldsymbol{Q}(t) = Q_l^k(s(t))\boldsymbol{e}_k \otimes \boldsymbol{e}^l. \tag{2.64}$$

and where $\boldsymbol{F}(t)$ and $\bar{\boldsymbol{F}}(t)$ are the deformation gradients of $\boldsymbol{\chi}_t$ and $\bar{\boldsymbol{\chi}}_t$ relative to $\boldsymbol{\kappa}$ as usual. (Of course, at each fixed time t, $\boldsymbol{F}(t)$ and $\bar{\boldsymbol{F}}(t)$ are *tensor fields*, but $\boldsymbol{Q}(t)$ is a *constant tensor*.)

Now from the identity (1.38) we have

$$\boldsymbol{F}_{t_0,t} = \boldsymbol{F}(t)\boldsymbol{F}(t_0)^{-1}, \quad \bar{\boldsymbol{F}}_{t_0,t} = \bar{\boldsymbol{F}}(t)\bar{\boldsymbol{F}}(t_0)^{-1}, \tag{2.65}$$

where t_0 denotes the absolute time of the fixed instant s_0. From (2.63), we then get

$$\boldsymbol{F}_{t_0,t} = \boldsymbol{Q}(t)\bar{\boldsymbol{F}}_{t_0,t}\boldsymbol{Q}^T(t_0). \tag{2.66}$$

Consequently by (1.48)

$$\boldsymbol{C}_{t_0,t} = \boldsymbol{F}_{t_0,t}^T\boldsymbol{F}_{t_0,t} = \boldsymbol{Q}(t_0)\bar{\boldsymbol{C}}_{t_0,t}\boldsymbol{Q}^T(t_0), \tag{2.67}$$

and hence by (1.49)

$$\boldsymbol{A}_n(t_0) = \boldsymbol{Q}(t_0)\bar{\boldsymbol{A}}_n(t_0)\boldsymbol{Q}^T(t_0), \tag{2.68}$$

or equivalently,

$$\boldsymbol{A}_n(s_0) = \boldsymbol{Q}(s_0)\bar{\boldsymbol{A}}_n(s_0)\boldsymbol{Q}^T(s_0) \tag{2.69}$$

when the various tensors are regarded as spatial tensors over $\mathscr{V}_{s_0}$. Thus $\boldsymbol{A}_n$ satisfies the condition (2.62) for frame-indifference. (The tensor denoted by $\bar{\boldsymbol{A}}_n$ in (2.68) is the Rivlin-Ericksen tensor of the motion $\bar{\zeta}$ as observed by $\boldsymbol{\varphi}$, while in (2.42) $\bar{\boldsymbol{A}}_n$ denotes the Rivlin-Ericksen tensor of the motion ζ as observed by $\bar{\boldsymbol{\varphi}}$.)

3. Axioms of Dynamics and Thermodynamics in Inertial Frames

In the theory of kinematics we have defined Newtonian space-time, frames of reference, and body manifolds, the concepts in terms of which we may describe the motions of deformable bodies. In mechanics, of course, we are more interested in the determination of the motions. A mathematical theory for this purpose is called *dynamics*.

In this book we follow one of the classical approaches to dynamics, that developed by NEWTON, EULER, and many others, according to which motions are produced by the action of *forces* upon *masses*. Both force and mass, like body and motion, are mathematical concepts, the properties of which are to be set forth through axiomatic structures. Only very recently have axioms for systems of forces been proposed, by NOLL[1]. Since NOLL's axioms cover situations far more general than those needed in this book, we shall not present them, resting content instead with an informal treatment along traditional lines. For mass, however, the theory of measure, presumed known to the reader of this book, provides a fully mathematical structure.

In order to apply measure theory, we assume that every body $\mathscr{B}$ is equipped with a positive finite measure M, called the *mass distribution.* The measurable sets for M are the *Borel sets* of $\mathscr{B}$ as defined in measure theory. If $\mathscr{U}$ is a Borel set, then the value $M(\mathscr{U})$ is called the *mass of* $\mathscr{U}$. Since M is assigned directly to the body, of course, it is frame-indifferent and independent of the motion of $\mathscr{B}$.

In continuum mechanics we require also that M be absolutely continuous with respect to the volume measure induced by any volume tensor field on $\mathscr{B}$. Thus in any configuration, say $\zeta_s : \mathscr{B} \to \mathscr{E}_s$, there exists an integrable field $\rho(\cdot, s)$ on $\zeta_s(\mathscr{B})$ such that

$$M(\mathscr{U}) = \int_{\zeta_s(\mathscr{U})} \rho(\cdot, s) dV \tag{3.1}$$

for all Borel sets $\mathscr{U}$ in $\mathscr{B}$, where dV denotes the canonical volume element for the Euclidean manifold $\mathscr{E}_s$. We call the function ρ the *mass density* in the configuration ζ_s, and we commonly assume that it is positive and smooth. Like the existence of ρ, the smoothness of ρ is independent of the choice of the configuration. If it exists, the value of ρ at the configuration $\zeta_s(X)$ of the body point X is called the *mass density* of X in ζ_s, or, more simply, just the *density*.

Now let φ be a given frame of reference on $\mathscr{E}$. Then the abstract configuration ζ_s corresponds to an absolute configuration $\boldsymbol{\chi}_t : \mathscr{B} \to \mathscr{E}_\varphi$, where t is the absolute time of s. Since $\boldsymbol{\chi}_t$ is given by

$$\boldsymbol{\chi}_t = \boldsymbol{i}_s^{-1} \circ \zeta_s, \tag{3.2}$$

---

[1] P. 186 of the lectures cited above, p. 81. *Cf.* also the introductory book by TRUESDELL, cited above, p. 81.

where $\boldsymbol{i}_s$ is the observation map of φ, defined by (1.5), the condition (3.1) implies that there exists an integrable function $\rho_\varphi(\cdot, t)$ on $\chi_t(\mathscr{B})$ such that

$$M(\mathscr{U}) = \int_{\chi_t(\mathscr{U})} \rho_\varphi(\cdot, t) dV_\varphi . \tag{3.3}$$

Here dV_φ denotes the canonical volume element in $\mathscr{E}_\varphi$. Now since $\boldsymbol{i}_s$ is an isomorphism of the Euclidean manifolds $\mathscr{E}_\varphi$ and $\mathscr{E}_s$, it is measure-preserving with respect to dV_φ and dV. Hence, trivially, ρ_φ and ρ satisfy the relation

$$\rho(\zeta_s(X), s) = \rho_\varphi(\chi_t(X), t) \quad \forall X \in \mathscr{B}. \tag{3.4}$$

For this reason, we shall suppress the subscript φ in the notations ρ_φ and dV_φ.

In continuum mechanics we commonly need to integrate with respect to mass only over open subsets, called *subbodies* of the given body $\mathscr{B}$. Since, by definition, the mass of a body is invariant with respect to t,

$$\frac{d}{dt} \int_{\chi_t(\mathscr{U})} \rho \, dV = 0 \tag{3.5}$$

for any subbody $\mathscr{U}$ of $\mathscr{B}$. Therefore, if Ψ is a function which is smooth with respect to t along the world lines of the material points of $\mathscr{B}$,

$$\frac{d}{dt} \int_{\chi_t(\mathscr{U})} \rho \Psi \, dV = \int_{\chi_t(\mathscr{U})} \rho \dot{\Psi} \, dV, \tag{3.6}$$

no matter what be the frame φ. This trivial but important formula will be used freely henceforth without further remark[1]. As mentioned in Chapter 1, symbols of the planets, Ψ being one, are used to denote vector or tensor fields of unspecified order.

For a motion $\chi_t : \mathscr{B} \to \mathscr{E}_\varphi$, $t \in \mathscr{R}$, we define the *linear momentum*[2]

---

[1] In theories taking account of chemical reactions specific rules allowing for changes of mass are laid down. This book does not enter into such complications. However, we remark that the conditions for frame-indifference given in the next section must be replaced by more elaborate ones if variable mass is allowed to the body under consideration.

[2] More general concepts of linear momentum, rotational momentum, and kinetic energy, while now common in some parts of continuum mechanics, are not needed in this book.

of $\mathscr{B}$ relative to $\boldsymbol{\varphi}$ as the vector $\boldsymbol{L}(t) \in \mathscr{V}_{\varphi}$ given by

$$\boldsymbol{L}(t) \equiv \int_{\chi_t(\mathscr{B})} \rho(\cdot, t)\boldsymbol{v}(\cdot, t)\,dV, \tag{3.7}$$

where $\boldsymbol{v}$ is the velocity field. The linear momentum of an arbitrary subbody $\mathscr{U} \subset \mathscr{B}$ is defined in a similar way. As usual, we can regard $\boldsymbol{L}(t)$ as a vector in $\mathscr{V}_s$ also, via the observation map $\boldsymbol{I}_s$. Further, since $\boldsymbol{v}$ is not frame-indifferent, neither is $\boldsymbol{L}$.

We define the *rotational momentum* of $\mathscr{B}$ relative to $\boldsymbol{\varphi}$ as the skew tensor $\boldsymbol{H}(t)$ given by

$$\boldsymbol{H}(t) \equiv \int_{\chi_t(\mathscr{B})} \rho(\cdot, t)\boldsymbol{r}(\cdot) \wedge \boldsymbol{v}(\cdot, t)\,dV, \tag{3.8}$$

where $\boldsymbol{r}(\cdot)$ denotes the position vector of the field position $\boldsymbol{x}$ of integration in $\mathscr{E}_{\varphi}$, and where "$\wedge$" denotes the exterior product (*cf.* I.4.37)). We define the rotational momentum of a subbody of $\mathscr{B}$ in exactly the same way. Like $\boldsymbol{L}$, $\boldsymbol{H}$ is not frame-indifferent.

Finally, we define the *kinetic energy* of $\mathscr{B}$ relative to $\boldsymbol{\varphi}$ as the scalar $E(t)$ given by

$$E(t) \equiv \int_{\chi_t(\mathscr{B})} \tfrac{1}{2}\rho(\cdot, t)\boldsymbol{v}(\cdot, t) \cdot \boldsymbol{v}(\cdot, t)\,dV. \tag{3.9}$$

From its definition, E is non-negative. Like $\boldsymbol{L}$ and $\boldsymbol{H}$, it is not frame-indifferent.

As we have said, the mathematical model for the "cause" of motion, or, properly, for the departure of motion from some preferred motion, is provided by the action of a *force*. Specifically, let $\zeta_s : \mathscr{B} \to \mathscr{E}_s$, $s \in \mathscr{T}$, be a motion of $\mathscr{B}$. Then, by assumption, a *system of forces* acting on the configuration $\zeta_s(\mathscr{B})$ consists of two parts: a *body force* $\boldsymbol{B}$, which is a $\mathscr{V}_s$-valued finite measure defined on each configuration $\zeta_s(\mathscr{B})$, and a *contact force* $\boldsymbol{C}_{\mathscr{S}}$, which is a $\mathscr{V}_s$-valued finite measure defined on each piecewise smooth oriented surface $\mathscr{S}$ in $\zeta_s(\mathscr{B})$. In applications, we assume that the boundary of $\zeta_s(\mathscr{B})$ is a piecewise smooth surface in $\mathscr{E}_s$, oriented in the usual way, so that its positive normal points into the exterior of $\zeta_s(\mathscr{B})$. Then we require that there be a contact force defined on the boundary $\partial\zeta_s(\mathscr{B})$ also. The contact force $\boldsymbol{C}_{\mathscr{S}}$ is assumed to be a local *set function* in the following sense: Suppose that $\mathscr{S}_0$ is a measurable set belonging to two surfaces $\mathscr{S}$ and $\bar{\mathscr{S}}$. Then

$$\boldsymbol{C}_{\mathscr{S}}(\mathscr{S}_0) = \pm \boldsymbol{C}_{\bar{\mathscr{S}}}(\mathscr{S}_0), \tag{3.10}$$

where the sign is + if the orientation of $\bar{\mathscr{S}}$ agrees with that of $\mathscr{S}$ on $\mathscr{S}_0$, while otherwise the sign is −. Under this assumption, we can regard $\boldsymbol{C}$ as a set function defined for oriented piecewise smooth surfaces by the condition

$$\boldsymbol{C}(\mathscr{S}) \equiv \boldsymbol{C}_{\mathscr{S}}(\mathscr{S}). \tag{3.11}$$

Now suppose that $\mathscr{D}$ is an open set in $\zeta_s(\mathscr{B})$ with piecewise smooth boundary. Then the values $\boldsymbol{B}(\mathscr{D})$ and $\boldsymbol{C}(\partial\mathscr{D})$ are called the *body force* and the *contact force* acting on $\mathscr{D}$. Here $\partial\mathscr{D}$ is oriented in the usual way, as just specified. The sum of the body force and the contact force is called the *total force* acting on $\mathscr{D}$, which we denote by $\boldsymbol{F}(\mathscr{D})$:

$$\boldsymbol{F}(\mathscr{D}) = \boldsymbol{B}(\mathscr{D}) + \boldsymbol{C}(\partial\mathscr{D}). \tag{3.12}$$

In applications, we usually assume that $\boldsymbol{B}$ and $\boldsymbol{C}$ are absolutely continuous with respect to Euclidean volume in $\zeta_s(\mathscr{B})$ and to the surface measure on smooth surfaces in $\zeta_s(\mathscr{B})$, respectively. Then we can represent the body force and the contact force by the integrals

$$\boldsymbol{B}(\mathscr{D}) = \int_{\mathscr{D}} \rho \boldsymbol{b}\, dV, \tag{3.13}$$

and

$$\boldsymbol{C}(\partial\mathscr{D}) = \int_{\partial\mathscr{D}} \boldsymbol{t}\, dA, \tag{3.14}$$

where dA denotes the element of surface area on $\partial\mathscr{D}$. The fields $\boldsymbol{b}$ and $\boldsymbol{t}$ in the integrands are called the *body force density* and the *contact force density*. With the convention of sign indicated in (3.14), $\boldsymbol{t}$ is said to represent the "tension"; with the opposite convention, it would represent the "pressure" on $\partial\mathscr{D}$.

By their very definitions, the body force and the contact force, and hence also their densities, are frame-indifferent. However, if φ is a frame of reference, then via the observation map we can regard the body force and the contact force also as vectors in $\mathscr{V}_\varphi$, the translation space of $\mathscr{E}_\varphi$.

We define the *total moment* of the forces acting on $\mathscr{D}$ relative to $\boldsymbol{O}_\varphi$ as the skew tensor

$$\int_{\mathscr{D}} \rho \boldsymbol{r} \wedge \boldsymbol{b}\, dV + \int_{\partial\mathscr{D}} \boldsymbol{r} \wedge \boldsymbol{t}\, dA, \tag{3.15}$$

where $\boldsymbol{r}$ is the position vector of the field position of integration, as before. Of course, the moment can be regarded as a tensor either over $\mathscr{V}_{\varphi}$ or over $\mathscr{V}_s$. While it is not frame-indifferent, it depends upon φ only through the reference origin $\boldsymbol{O}(s)$.

Now we are ready to state the axioms of dynamics [1].

I. *There exists a frame of reference φ relative to which the linear momentum of any body $\mathscr{B}$ in a motion is independent of t provided that the total force acting on the configuration of $\mathscr{B}$ vanish for all t.*

A frame of reference φ specified by the above principle is called an *inertial frame*. The following proposition, proof of which is obvious in view of results in the previous section, characterizes the class of such frames.

PROPOSITION. *Let φ be an inertial frame. Then a frame $\overline{\varphi}$ is an inertial frame if and only if φ and $\overline{\varphi}$ are related by a Galilean transformation.*

A motion $\zeta_s : \mathscr{B} \to \mathscr{E}_s$, $s \in \mathscr{T}$, is said to be an *inertial motion* if the velocity of each body point of $\mathscr{B}$ is invariant in t with respect to any inertial frame. From the previous proposition, we see that this concept is well defined. Axiom I asserts that an inertial motion is a *preferred motion* in mechanics. Departure from a preferred motion is governed by the following axiom:

II. *Relative to an inertial frame φ, the rate of change of the linear momentum of a body $\mathscr{B}$ in a motion is equal to the total force acting on the configuration of $\mathscr{B}$, and the rate of change of the rotational momentum is equal to the total moment of the forces acting on the configuration of $\mathscr{B}$.*

Clearly Axiom II is consistent with Axiom I and is independent of the choice of the inertial frame. Since a subbody $\mathscr{U}$ of $\mathscr{B}$ is also a body, Axioms I and II apply to it also. Thus in terms of the integrals introduced by (3.7), (3.8), (3.13), (3.14), and (3.15), Axiom II is expressed as follows:

$$\frac{d}{dt}\int_{\chi_t(\mathscr{U})} \rho \boldsymbol{v}\, dV = \int_{\chi_t(\mathscr{U})} \rho \boldsymbol{a}\, dV = \int_{\chi_t(\mathscr{U})} \rho \boldsymbol{b}\, dV + \int_{\partial\chi_t(\mathscr{U})} \boldsymbol{t}\, dA, \tag{3.16}$$

and

$$\frac{d}{dt}\int_{\chi_t(\mathscr{U})} \rho \boldsymbol{r}\wedge\boldsymbol{v}\, dV = \int_{\chi_t(\mathscr{U})} \rho \boldsymbol{r}\wedge\boldsymbol{a}\, dV = \int_{\chi_t(\mathscr{U})} \rho \boldsymbol{r}\wedge\boldsymbol{b}\, dV + \int_{\partial\chi_t(\mathscr{U})} \boldsymbol{r}\wedge\boldsymbol{t}\, dA, \tag{3.17}$$

[1] Axiom I is intended to replace the rather vague First Law of NEWTON. Axiom II expresses EULER's First and Second Laws.

for all subbodies $\mathscr{U}$ of $\mathscr{B}$. The acceleration $\boldsymbol{a}$, like the velocity $\boldsymbol{v}$, is that of the motion of $\mathscr{B}$ with respect to an inertial frame.

The conditions (3.5), (3.16), and (3.17) are known as the *principle of balance of mass, the principle of balance of linear momentum, and the principle of balance of rotational momentum*, respectively. These principles are the ***axioms of dynamics***[1].

While in this book we are mainly interested in dynamics, for some problems we shall draw also upon certain principles of *thermodynamics*. Thermodynamics is an extension of dynamics in that it considers not only the parameters associated with motion but also those associated with heat and temperature. We shall need two such: *entropy* and *internal energy*. By definition, these thermodynamic parameters are finite (signed) measures defined on each configuration $\zeta_s(\mathscr{B})$ of $\mathscr{B}$ in a motion. If $\mathscr{D}$ is an open set in $\zeta_s(\mathscr{B})$, then the values of these measures are called the *entropy* and the *internal energy* in $\mathscr{D}$. We call the sum of the internal energy and the kinetic energy relative to a frame of reference $\boldsymbol{\varphi}$ the *total energy* relative to that frame. The internal energy and the entropy are frame-indifferent, but the total energy is not.

As before, we assume that the entropy and the internal energy are absolutely continuous with respect to the Euclidean volume dV, so that they can be represented by the integrals

$$\int_{\mathscr{D}} \rho\eta \, dV, \tag{3.18}$$

and

$$\int_{\mathscr{D}} \rho\varepsilon \, dV, \tag{3.19}$$

respectively, where the fields η and ε in the integrands are called the *entropy density* and the *internal energy density*.

Now the thermodynamic parameters, like the motion, also have certain preferred trends, which will be made precise in later paragraphs. The *causes* of the departure from these trends are then introduced into the mathematical structure. While forces, which are the causes of change of linear momentum and rotational momentum, contribute their share also

---

[1] If the action of impulses is to be taken into account, these principles need to be replaced by more general ones in which the linear momentum and rotational momentum are not always differentiable with respect to *t*. *Cf.* CFT, §§ 198 and 206.

to the change of total energy, there are other possible causes for that change, and these are represented mathematically by the concept of *energy supply*. Likewise, the concept of *entropy supply*, which may or may not be closely related to the energy supply, represents the causes of change of entropy. In thermodynamics these supplies take their places beside force (which is the supply of momentum).

Like force, which is the supply of linear momentum, by definition the energy supply is made up of two parts: a *body energy supply* R, which is a finite, signed measure on each configuration $\zeta_s(\mathscr{B})$, and a *contact energy supply* $Q_{\mathscr{S}}$, which is a finite, signed measure defined on each piecewise smooth oriented surface $\mathscr{S}$ in $\zeta_s(\mathscr{B})$. Again, we assume that $\partial\zeta_s(\mathscr{B})$ is a piecewise smooth surface which is oriented in the standard way and that there is a contact energy supply defined on $\partial\zeta_s(\mathscr{B})$. Also we assume that $Q_{\mathscr{S}}$ is a local set function as explained before, so that we can write the contact energy supply on $\mathscr{S}$ as $Q(\mathscr{S})$:

$$Q(\mathscr{S}) \equiv Q_{\mathscr{S}}(\mathscr{S}). \tag{3.20}$$

Naturally, for any open region $\mathscr{D}$ with piecewise smooth boundary we call the values $R(\mathscr{D})$ and $Q(\partial\mathscr{D})$ the *body energy supply* and the *contact energy supply* in $\mathscr{D}$. In physical interpretation, $R(\mathscr{D})$ is often associated with radiation, while $Q(\partial\mathscr{D})$ is associated with the conduction of heat. The body force $\boldsymbol{B}$ and the contact force $\boldsymbol{C}$ also give rise to energy supplies, these being, by definition, simply their powers, Thus the *energy supply through body force* relative to a frame of reference $\boldsymbol{\varphi}$ is

$$\int_{\mathscr{D}} \rho \boldsymbol{b} \cdot \boldsymbol{v}\, dV, \tag{3.21}$$

and the *energy supply through contact force* relative to $\boldsymbol{\varphi}$ is

$$\int_{\partial\mathscr{D}} \boldsymbol{t} \cdot \boldsymbol{v}\, dA, \tag{3.22}$$

where $\boldsymbol{v}$ is the velocity field of the motion relative to $\boldsymbol{\varphi}$. We assume further that R and Q are absolutely continuous, so that $R(\mathscr{D})$ and $Q(\mathscr{D})$ are given by the integrals

$$R(\mathscr{D}) = \int_{\mathscr{D}} \rho r\, dV, \tag{3.23}$$

and

$$Q(\mathscr{D}) = \int_{\partial\mathscr{D}} q\, dA, \tag{3.24}$$

where the fields r and q in the integrands are called the *body energy supply density* and the *contact energy supply density*, respectively. With the convention of sign indicated by (3.24), q is said to represent the "influx" of energy; with the opposite sign, it would represent the "efflux". The sum of the energy supplies of all kinds is called the *total energy supply* in $\mathscr{D}$.

Now we are ready to state the first thermodynamic axiom:

III. *Relative to an inertial frame* $\boldsymbol{\varphi}$, *the rate of change of the total energy of a body* $\mathscr{B}$ *in a motion is equal to the total energy supply in the configuration of* $\mathscr{B}$.

Since the motion is required to satisfy the axioms of dynamics, the reader will show easily that Axiom III holds in all inertial frames if and only if it holds in one. Under the assumptions (3.20), (3.21), (3.22), (3.23), and (3.24), Axiom III takes the form

$$\begin{aligned}\frac{d}{dt}\int_{\chi_t(\mathscr{U})} \rho(\varepsilon+\tfrac{1}{2}\boldsymbol{v}\cdot\boldsymbol{v})dV &= \int_{\chi_t(\mathscr{U})} \rho(\dot{\varepsilon}+\boldsymbol{a}\cdot\boldsymbol{v})dV \\ &= \int_{\chi_t(\mathscr{U})} \rho(r+\boldsymbol{b}\cdot\boldsymbol{v})dV + \int_{\partial\chi_t(\mathscr{U})} (q+\boldsymbol{t}\cdot\boldsymbol{v})dA,\end{aligned} \tag{3.25}$$

for all subbodies $\mathscr{U}$ of $\mathscr{B}$, provided the frame of reference be an inertial one. This condition is known as the *principle of balance of energy*.

Following a similar approach, we introduce the concept of *entropy supply*. For any region $\mathscr{D}$ in $\zeta_s(\mathscr{B})$, as before, we postulate existence of a *body entropy supply* and a *contact entropy supply* which can be represented by the integrals

$$\int_{\mathscr{D}} \rho h\, dV, \tag{3.26}$$

and

$$\int_{\partial\mathscr{D}} k\, dA, \tag{3.27}$$

respectively, where the fields h and k in the integrands are called the *body entropy supply density* and the *contact entropy supply density*, respectively. By their very definitions, these supplies and their densities are frame-indifferent.

The second thermodynamic axiom is

IV. *The rate of change of the total entropy of a body $\mathscr{B}$ in a motion cannot fall below the total entropy supply in the configuration of $\mathscr{B}$.*

In terms of the integral representations, this axiom takes the following form:

$$\frac{d}{dt}\int_{\chi_t(\mathscr{U})} \rho\eta\, dV = \int_{\chi_t(\mathscr{U})} \rho\dot{\eta}\, dV \geqq \int_{\chi_t(\mathscr{U})} \rho h\, dV + \int_{\partial\chi_t(\mathscr{U})} k\, dA \tag{3.28}$$

for all subbodies $\mathscr{U}$ of $\mathscr{B}$. The surplus of the entropy rate over the total entropy supply is called the *entropy production*, which is assumed to be absolutely continuous with respect to the volume measure also. Since (3.28) assserts an inequality, this axiom is known as the *principle of irreversibility*. Another common name for it is the CLAUSIUS-DUHEM *inequality*[1]. Each term in it is frame-indifferent, so that, unlike Axioms II and III, it is here stated in a form not restricted to inertial frames.

The energy supplies r and q are commonly assumed to stand in direct proportion to the entropy supplies h and k, the common ratio r/h and q/k being by further assumption a positive scalar field θ defined on the configuration:

$$\theta = \frac{r}{h} = \frac{q}{k}. \tag{3.29}$$

The field θ, necessarily frame-indifferent, is called the *absolute temperature* in the configuration. While more general relations between energy and

[1] While the "second law of thermodynamics" is universally acclaimed by physicists, there is disagreement among them as to what that law is, and the statements in physics books are not sufficiently concrete to enter a mathematical theory. The justice of the CLAUSIUS-DUHEM inequality is still a matter of dispute. The now standard motivation for it, due essentially to C. ECKART, is presented and explained in Lecture II of TRUESDELL's *Rational Thermodynamics*, New York *etc.*, McGraw Hill, 1969. In the form (3.28), with the specializations (3.29), it appeared first in CFT. In an important paper, W. A. DAY shows how to derive (3.28) for certain materials from a system of axioms including an assumption regarding the boundedness of useful work: "Thermodynamics based on a work axiom", *Arch. Rational Mech. Anal.* **31**, 1–34 (1968). See also DAY's Springer Tract in Natural Philosophy, *The Thermodynamics of Simple Materials with Fading Memory*, Berlin *etc.*, Springer, 1972.

entropy are sometimes considered in continuum mechanics, only the classical special case (3.29) will be used in this book[1].

The set of axioms obtained by adjoining the two axioms of thermodynamics to the three axioms of dynamics is called the ***axioms of thermomechanics***.

4. Frame-Indifferent Dynamical Principles

While each integral appearing in the principle of balance of mass (3.5) and the principle of irreversibility (3.28) is frame-indifferent, we have required that the frame of reference φ used in stating the principles of balance of linear momentum, balance of rotational momentum, and balance of energy, namely (3.16), (3.17), and (3.25), be an inertial one. When the motion is referred to an arbitrary frame of reference, these latter principles no longer retain the same simple, special forms. In this section we shall derive their general forms, relative to an arbitrary frame of reference, always assuming the principle of balance of mass as before.

First we consider change of frame relative to an inertial frame. Let φ be an inertial frame, and let $\bar{\varphi}$ be an arbitrary frame. Then as before we can characterize the relation between φ and $\bar{\varphi}$ by the co-ordinate transformations (2.12) and (2.13). Now suppose that $\zeta_s : \mathscr{B} \to \mathscr{E}_s$, $s \in \mathscr{T}$, is a given motion of $\mathscr{B}$. Then ζ_s can be represented in terms either of the co-ordinates (x^i) for φ or the co-ordinates $(\bar{x}^i)$ for $\bar{\varphi}$, the two absolute motions being, say,

$$\chi_t : \mathscr{B} \to \mathscr{E}_\varphi, \ t \in \mathscr{R}, \tag{4.1}$$

and

$$\bar{\chi}_{\bar{t}} : \mathscr{B} \to \mathscr{E}_{\bar{\varphi}}, \ \bar{t} \in \mathscr{R}. \tag{4.2}$$

By (2.30) and (2.32), the acceleration fields $\boldsymbol{a}$ and $\bar{\boldsymbol{a}}$, obtained from these two absolute motions at the absolute times t and $\bar{t}$ corresponding to the

[1] In their paper "An axiomatic foundation for continuum thermodynamics", *Arch. Rational Mech. Anal.* **26**, 83–117 (1967), M. E. GURTIN & W. O. WILLIAMS have set up an axiomatic structure for a theory of heat and temperature in which $r/h = \varphi > 0$, $q/h = \theta > 0$, but the conductive temperature θ and the radiative temperature φ need not be equal. A still more general possibility is allowed by I. MÜLLER, "On the entropy inequality", *ibid.* 118–141, who considers the effect of leaving q and h unrelated except by implication through a class of constitutive equations.

same instant s, differ by a vector field $\boldsymbol{i}$:

$$\boldsymbol{i} = \bar{\boldsymbol{a}} - \boldsymbol{a}, \tag{4.3}$$

where $\boldsymbol{i}$ has the following component form relative to φ:

$$\boldsymbol{i} = -(\ddot{y}^j + 2\dot{Q}^j_i \bar{v}^i + \ddot{Q}^j_i \bar{x}^i)\boldsymbol{e}_j. \tag{4.4}$$

Here $\bar{x}^i$ are the co-ordinates, and $\bar{v}^i$ are the components of the velocity $\bar{\boldsymbol{v}}$ of a typical body-point $X \in \mathscr{B}$ relative to the frame $\bar{\varphi}$.

PROPOSITION. *In a given motion of $\mathscr{B}$, the vector field* **i** *defined by* (4.4) *is independent of the choice of the inertial frame φ.*

To prove this proposition, we need only remark that since the acceleration $\boldsymbol{a}$ is invariant under change of inertial frame, so also is $\boldsymbol{i}$. (In fact, it is easy to show that $\boldsymbol{i}$ is a function only of the motion ζ and of the Galilean class to which $\bar{\varphi}$ belongs.)

Therefore, $\boldsymbol{i}$ is an observable kinematical quantity as defined in § 2, *viz*,

$$\boldsymbol{i} = \boldsymbol{i}(\bar{\varphi}, \zeta); \tag{4.5}$$

$\boldsymbol{i}$ is called the *inertial body force* in ζ relative to $\bar{\varphi}$. While by its very definition (4.4), $\boldsymbol{i}$ is not frame-indifferent – indeed,

$$\boldsymbol{i}(\varphi, \zeta) = \boldsymbol{0} \tag{4.6}$$

if the frame φ is inertial – *the field $\bar{\boldsymbol{a}} - \boldsymbol{i}$ is a frame-indifferent field whose value is the acceleration field in an inertial frame.* Here $\bar{\boldsymbol{a}}$ is the acceleration of the motion ζ with respect to the general frame $\bar{\varphi}$.

Returning to the principle of linear momentum in the form $(3.16)_2$, we see that each integral on the right-hand side is a frame-indifferent quantity. To express the principle itself in a general frame, we need only replace the left-hand side by that frame-indifferent quantity which reduces in an inertial frame to the rate of change of linear momentum. By (4.3), then, we must replace $\boldsymbol{a}$ by the frame-indifferent field $\bar{\boldsymbol{a}} - \boldsymbol{i}$. Dropping the bars and using φ to denote a general frame, we can express this result as follows:

PROPOSITION (*Clairaut's Principle*). *The principle of linear momentum* (3.16) *holds in a motion ζ of $\mathscr{B}$ if and only if relative to the arbitrary frame* *of reference* φ

$$\frac{d}{dt}\int_{\chi_t(\mathscr{U})} \rho \boldsymbol{v}\, dV = \int_{\chi_t(\mathscr{U})} \rho \boldsymbol{a}\, dV,$$
$$= \int_{\chi_t(\mathscr{U})} \rho \boldsymbol{b}^*\, dV + \int_{\partial\chi_t(\mathscr{U})} \boldsymbol{t}\, dA \tag{4.7}$$

for each subbody $\mathscr{U}$ *of* $\mathscr{B}$, *where the apparent body force density* $\boldsymbol{b}^*$ *is given by*

$$\boldsymbol{b}^* = \boldsymbol{b} + \boldsymbol{i}, \tag{4.8}$$

$\boldsymbol{i}$ *being the inertial body force in the motion* ζ *relative to* φ.

Evidently we can generalize the principle of balance of rotational momentum in a similar way.

PROPOSITION. *The principle of rotational momentum* (3.17) *holds in a motion* ζ *of* $\mathscr{B}$ *if and only if relative to the arbitrary frame of reference* φ

$$\frac{d}{dt}\int_{\chi_t(\mathscr{U})} \rho \boldsymbol{r} \wedge \boldsymbol{v}\, dV = \int_{\chi_t(\mathscr{U})} \rho \boldsymbol{r} \wedge \boldsymbol{a}\, dV,$$
$$= \int_{\chi_t(\mathscr{U})} \rho \boldsymbol{r} \wedge \boldsymbol{b}^*\, dV + \int_{\partial\chi_t(\mathscr{U})} \boldsymbol{r} \wedge \boldsymbol{t}\, dA. \tag{4.9}$$

From now on the two general expressions (4.7) and (4.9), the forms of which are independent of the frame of reference φ, will be called the *principles of balance of linear momentum and rotational momentum*, respectively.

Before taking up the frame-indifference of the principle of balance of energy, we establish the following.

LEMMA[1]. *Let* $\boldsymbol{f}$ *and* $\boldsymbol{t}$ *be frame-indifferent vector fields defined and integrable in a region* $\mathscr{D}$ *and on its boundary* $\partial\mathscr{D}$, *respectively. Then their total power in a motion with velocity field* $\mathbf{v}$, *namely,*

$$\int_{\mathscr{D}} \boldsymbol{f} \cdot \boldsymbol{v}\, dV + \int_{\partial\mathscr{D}} \boldsymbol{t} \cdot \boldsymbol{v}\, dA, \tag{4.10}$$

[1] While in this book only a three-dimensional Euclidean manifold is considered, the proof applies equally in n dimensions. It would suffice also to restrict attention to infinitesimal rotations.

is a frame-indifferent scalar if and only if

$$\int_{\mathscr{D}} \boldsymbol{f}\, dV + \int_{\partial\mathscr{D}} \boldsymbol{t}\, dA = \boldsymbol{0} \tag{4.11}$$

and

$$\int_{\mathscr{D}} \boldsymbol{r} \wedge \boldsymbol{f}\, dV + \int_{\partial\mathscr{D}} \boldsymbol{r} \wedge \boldsymbol{t}\, dA = \boldsymbol{0}. \tag{4.12}$$

To see this, we first note that the formula for transforming the velocity under change of frame, namely, (2.31), may be abbreviated as

$$\bar{\boldsymbol{v}} - \boldsymbol{v} = \boldsymbol{c} + \boldsymbol{\Omega}\boldsymbol{r}, \tag{4.13}$$

where $\boldsymbol{r}$ is the position vector of $\bar{\boldsymbol{x}}$, the position of X in the barred frame, and where

$$\boldsymbol{c} \equiv -\dot{y}^j \boldsymbol{e}_j,$$
$$\boldsymbol{\Omega} = -\boldsymbol{\Omega}^T \equiv -\dot{Q}^k_{\ m} Q^{Tm}_{\ \ \ l} \boldsymbol{e}_k \otimes \boldsymbol{e}^l. \tag{4.14}$$

The equality considered by the lemma is thus

$$\begin{aligned}\int_{\mathscr{D}} \boldsymbol{f}\cdot\boldsymbol{v}\, dV + \int_{\partial\mathscr{D}} \boldsymbol{t}\cdot\boldsymbol{v}\, dA &= \int_{\mathscr{D}} \boldsymbol{f}\cdot\bar{\boldsymbol{v}}\, dV + \int_{\partial\mathscr{D}} \boldsymbol{t}\cdot\bar{\boldsymbol{v}}\, dA = \int_{\mathscr{D}} \boldsymbol{f}\cdot\boldsymbol{v}\, dV + \\ &+ \int_{\partial\mathscr{D}} \boldsymbol{t}\cdot\boldsymbol{v}\, dA + \boldsymbol{c}\cdot\left[\int_{\mathscr{D}} \boldsymbol{f}\, dV + \int_{\partial\mathscr{D}} \boldsymbol{t}\, dA\right] \\ &+ \operatorname{tr}\left(\boldsymbol{\Omega}\left[\int_{\mathscr{D}} \boldsymbol{r}\wedge\boldsymbol{f}\, dV + \int_{\partial\mathscr{D}} \boldsymbol{r}\wedge\boldsymbol{t}\, dA\right]\right),\end{aligned} \tag{4.15}$$

where the second form follows because $\boldsymbol{c}$ and $\boldsymbol{\Omega}$ are, at each time, constant fields, independent of the position of integration. Thus the equality reduces to

$$\boldsymbol{c}\cdot\left[\int_{\mathscr{D}} \boldsymbol{f}\, dV + \int_{\partial\mathscr{D}} \boldsymbol{t}\, dA\right] + \operatorname{tr}\left(\boldsymbol{\Omega}\left[\int_{\mathscr{D}} \boldsymbol{r}\wedge\boldsymbol{f}\, dV + \int_{\partial\mathscr{D}} \boldsymbol{r}\wedge\boldsymbol{t}\, dA\right]\right) = 0. \tag{4.16}$$

Since the values of $\boldsymbol{c}$ and $\boldsymbol{\Omega}$ are arbitrary, (4.16) is equivalent to (4.11) and (4.12).

Let us now take $\mathscr{D}$ as being the set $\zeta_s(\mathscr{U})$; by (3.1) and (3.3), integration over this set is equivalent to integration over $\chi_t(\mathscr{U})$ and $\bar{\chi}_{\bar{t}}(\mathscr{U})$, where t and $\bar{t}$ are the absolute times of the same instant s relative to φ and $\bar{\varphi}$, respectively. If we take for $\boldsymbol{f}$ the frame-indifferent field $\rho(\boldsymbol{b}^* - \boldsymbol{a})$, where $\boldsymbol{b}^*$ is the apparent body force density (4.8), and interpret $\boldsymbol{t}$ as the contact force density acting on $\partial\zeta_s(\mathscr{U})$, the expression (4.10) becomes the *total power P* of all forces acting upon the subbody $\mathscr{U}$ in the configuration $\zeta_s(\mathscr{U})$:

$$P \equiv \int_{\zeta_s(\mathscr{U})} \rho(\boldsymbol{b}^* - \boldsymbol{a}) \cdot \boldsymbol{v}\, dV + \int_{\partial\zeta_s(\mathscr{U})} \boldsymbol{t} \cdot \boldsymbol{v}\, dA. \tag{4.17}$$

Moreover, (4.11) and (4.12) become the general, frame-indifferent principles of linear momentum and rotational momentum, (4.7) and (4.9), respectively. Interpretation of the lemma in this context gives an elegant result due to NOLL[1]:

PROPOSITION. *The total power P is frame-indifferent if and only if the principles of linear momentum and rotational momentum hold in all frames.*

As an immediate corollary of NOLL's theorem we obtain a statement of the principle of balance of energy with respect to an arbitrary frame.

PROPOSITION. *Granted the principles of balance of linear momentum and rotational momentum, the principle of balance of energy* (3.25) *holds in a motion ζ of $\mathscr{B}$ if and only if relative to an arbitrary frame of reference $\boldsymbol{\varphi}$*

$$\begin{aligned} \frac{d}{dt}\int_{\chi_t(\mathscr{U})} \rho(\varepsilon + \tfrac{1}{2}v^2)dV &= \int_{\chi_t(\mathscr{U})} \rho(\dot{\varepsilon} + \boldsymbol{a} \cdot \boldsymbol{v})dV, \\ &= \int_{\chi_t(\mathscr{U})} \rho(\boldsymbol{b}^* \cdot \boldsymbol{v} + r)dV + \int_{\partial\chi_t(\mathscr{U})} (\boldsymbol{t} \cdot \boldsymbol{v} + q)dA. \end{aligned} \tag{4.18}$$

To be sure of clarity, we remind the reader that the fields $\boldsymbol{b}$, $\boldsymbol{t}$, r, q and ε are frame-indifferent in virtue of their definitions in the preceding section, and that throughout this section the principle of balance of mass has been assumed.

5. Cauchy's Postulate and Fundamental Theorem

In the two preceding sections we have seen that the principles of balance (3.5), (4.7), (4.9), and (4.18) and the principle of irreversibility (3.28) all

---

[1] W. NOLL, "La mécanique classique, basée sur un axiome d'objectivité", *La Méthode Axiomatique . . . Colloque International* (1959), Paris, Gauthier-Villars, 1963.

have the form of an ***integral balance inequality***:

$$\frac{d}{dt}\int_{\chi_t(\mathscr{U})} \rho\Psi\, dV = \int_{\chi_t(\mathscr{U})} \rho\dot{\Psi}\, dV \geqq \int_{\chi_t(\mathscr{U})} \rho ♃\, dV + \int_{\partial\chi_t(\mathscr{U})} ♀\, dA, \tag{5.1}$$

or, equivalently,

$$\int_{\mathscr{D}} ♂\, dV \geqq \int_{\partial\mathscr{D}} ♀\, dA, \tag{5.2}$$

where ♂ denotes the field $\rho(\dot{\Psi} - ♃)$, and where $\mathscr{D}$ is an arbitrary bounded open domain in $\chi_t(\mathscr{B})$ with piecewise smooth boundary, since the subbody $\mathscr{U}$ in (5.1) and (5.2) is arbitrary. (Again we use signs of the planets to stand for general tensor fields of unspecified order.) Naturally, ♂ and ♀ are integrable fields over their domains. The field Ψ may be called an *intensity*; it and its time derivative are assumed to be integrable. The field ♀, defined on $\partial\mathscr{D}$, is the *contact supply density* of Ψ, while ♂ is the rate of growth of Ψ less the *body supply density* ♃, per unit mass. In what follows we shall indicate the domain $\mathscr{S}$ of ♀ by a subscript, writing $♀_{\mathscr{S}}$.

We note the following

LEMMA. *Let* $-\mathscr{S}$ *denote the surface having the same underlying set as* $\mathscr{S}$ *but with the opposite orientation. Then*

$$♀_{-\mathscr{S}}(\boldsymbol{x}) = -♀_{\mathscr{S}}(\boldsymbol{x}). \tag{5.3}$$

Proof. From measure theory, the surface integral in (5.2) is differentiable almost everywhere with respect to surface measure, so that

$$\lim_{\Delta\mathscr{A}\to\{\boldsymbol{x}\}} \frac{\int_{\Delta\mathscr{A}} ♀_{\mathscr{S}}\, dA}{\int_{\Delta\mathscr{A}} dA} = ♀_{\mathscr{S}}(\boldsymbol{x}) \tag{5.4}$$

for almost every $\boldsymbol{x}$ on $\mathscr{S}$, where $\Delta\mathscr{A}$ is a net of surface elements on $\mathscr{S}$ converging to the point $\boldsymbol{x}$; for example, $\Delta\mathscr{A}$ may be the intersection of $\mathscr{S}$ with a circular cylinder of radius Δr and with the normal to $\mathscr{S}$ at $\boldsymbol{x}$ as its axis. From the assumption that the surface integral in (5.2) is a local set-function in the sense exemplified by (3.10), we know that

$$\int_{\Delta\mathscr{A}} ♀_{\mathscr{S}}\, dA = -\int_{-\Delta\mathscr{A}} ♀_{-\mathscr{S}}\, dA. \tag{5.5}$$

Substitution into (5.4) yields (5.3).

The traditional treatment, deriving from CAUCHY himself, is different from ours here. Cauchy's postulate, which we are about to state as (5.6), is *assumed*, and then (5.3) is *proved* by appeal to the integral balance inequality. Thus the appropriate special case of (3.10) follows. Here we prove, rather than assume, Cauchy's postulate, but we do assume, rather than prove, the existence of a measure satisfying (3.10).

It turns out that if we make in regard to $♀_{\mathscr{S}}$ one fundamental assumption, known as *Cauchy's postulate*, then a major simplification of (5.2) follows. In this section we treat Cauchy's postulate and its several consequences in general. The specific cases that correspond to the specific relations of balance, which give rise to the *field equations* and the *jump conditions*, will be considered in the next section.

CAUCHY'S POSTULATE. *Suppose that two surfaces $\mathscr{S}$ and $\bar{\mathscr{S}}$ have a common normal direction at the place* $\mathbf{x}$. *Then*

$$♀_{\mathscr{S}}(\boldsymbol{x}) = ♀_{\bar{\mathscr{S}}}(\boldsymbol{x}). \tag{5.6}$$

In (3.10) we have already encountered a special case of this assumption.

Cauchy's postulate asserts that the supply $♀_{\mathscr{S}}(\boldsymbol{x})$ on any surface $\mathscr{S}$ depends upon that surface only through its positive unit normal $\boldsymbol{n}_{\mathscr{S}}(\boldsymbol{x})$:

$$♀_{\mathscr{S}}(\boldsymbol{x}) = ☿(\boldsymbol{x}, \boldsymbol{n}_{\mathscr{S}}(\boldsymbol{x})), \quad \boldsymbol{x} \in \mathscr{S}. \tag{5.7}$$

From (5.3) and (5.7), then,

$$☿(\cdot, -\boldsymbol{n}) = -☿(\cdot, \boldsymbol{n}). \tag{5.8}$$

We now show that Cauchy's postulate is a consequence of some general conditions on the fields ♂ and ♀.

PROPOSITION[1]. *Let $\mathscr{D}_0$ be a fixed compact region in $\mathscr{E}_\varphi$ such that the balance law* (5.2) *holds for every regular open domain $\mathscr{D}$ in the interior of $\mathscr{D}_0$. Suppose that the field ♂ is essentially bounded on $\mathscr{D}_0$, and suppose that ♀ is essentially and uniformly bounded on all piecewise smooth oriented surfaces $\mathscr{S}$ in the interior of $\mathscr{D}_0$. Then there exists a field $☿(\cdot, \boldsymbol{n})$ on the interior of $\mathscr{D}_0$ for each unit vector $\boldsymbol{n} \in \mathscr{V}_\varphi$, such that the representation formula* (5.7) *holds almost everywhere* (*with respect to the surface measure*) *on each $\mathscr{S}$.*

---

[1] Theorem IV of W. NOLL, "The foundations of classical mechanics in the light of recent advances in continuum mechanics", *The Axiomatic Method, with special reference to geometry and physics* (1957), Amsterdam, North-Holland Co., 1959, pp. 266–281.

Proof. Within $\mathscr{D}$, consider the region $\Delta\mathscr{D}$ bounded by the surfaces $\mathscr{S}$, $\bar{\mathscr{S}}$, and a circular cylinder of radius Δr and axis along the common normal to $\mathscr{S}$ and $\bar{\mathscr{S}}$ at $\boldsymbol{x}$, as shown shaded in the figure[1].

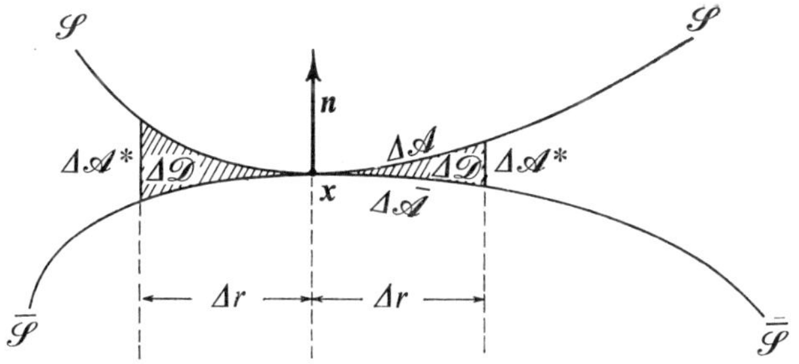

Let $\Delta\mathscr{A}$ and $\Delta\bar{\mathscr{A}}$ be the parts of $\mathscr{S}$ and $\bar{\mathscr{S}}$, respectively, that bound this region, and let $\Delta\mathscr{A}^*$ be the cylindrical part of the boundary. Then as $\Delta r^2 \to 0$,

$$
\begin{aligned}
\int_{\Delta\mathscr{A}} dA &= \pi\Delta r^2 + o(\Delta r^2), \\
\int_{\Delta\bar{\mathscr{A}}} dA &= \pi\Delta r^2 + o(\Delta r^2), \\
\int_{\Delta\mathscr{A}^*} dA &= 2\pi\Delta r O(\Delta r^2) = o(\Delta r^2), \\
\int_{\Delta\mathscr{D}} dV &= O(\Delta r^4) = o(\Delta r^3).
\end{aligned}
\tag{5.9}
$$

[1] In view of (5.3), we may modify the argument so as to cover the case when $\mathscr{S}$ and $\bar{\mathscr{S}}$ cross each other as shown below:

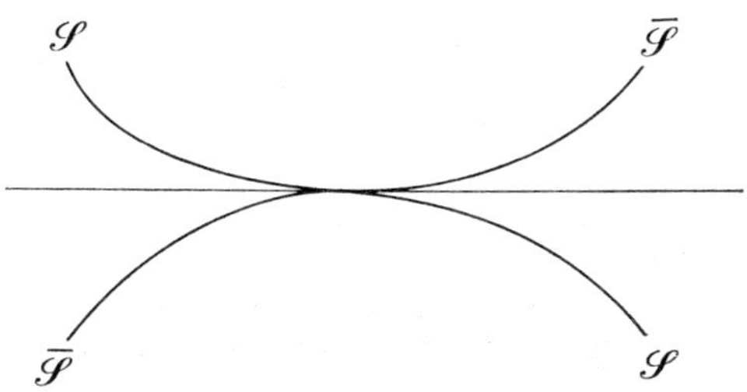

To establish these estimates, it suffices to consider the case when $\bar{\mathscr{S}}$ is the tangent plane to $\mathscr{S}$ at the place in question, since two applications of the results for this case yield the more general ones desired. Let (x, y) be Cartesian co-ordinates on $\bar{\mathscr{S}}$ with origin at the place called $\boldsymbol{x}$ up to now, and let z be the Cartesian co-ordinate normal to $\bar{\mathscr{S}}$. Then near the origin $\mathscr{S}$, since it is smooth, has the representation $z = f(x, y)$, and both $\partial f/\partial x$ and $\partial f/\partial y$ vanish at the origin. For a fixed radius Δr, it is possible to construct a paraboloid of revolution $\mathscr{S}^*$ that is tangent to the plane $\bar{\mathscr{S}}$ at the origin and lies entirely above $\mathscr{S}$, as shown in the figure. Formally, if we set

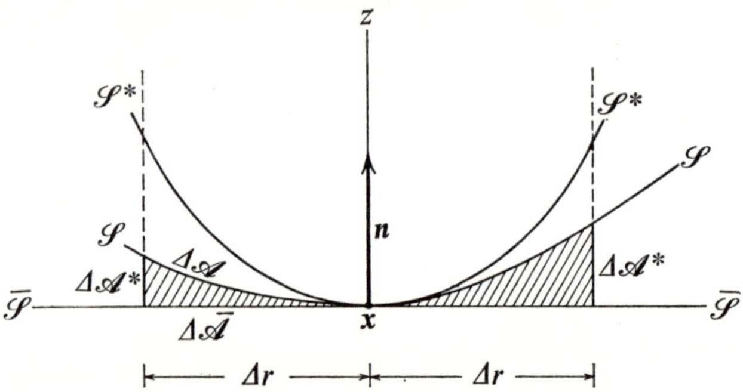

$$K \equiv \max\left(\frac{\partial^2 f}{\partial x^2}, \frac{\partial^2 f}{\partial x\,\partial y}, \frac{\partial^2 f}{\partial y^2}\right) \text{ when } x^2+y^2 \leqq \Delta r^2, \tag{5.10}$$

then such a paraboloid is given by the equation

$$z = \tfrac{1}{2}K(x^2+y^2). \tag{5.11}$$

The estimates (5.9) can now be read off from (5.11).

To complete the proof of the lemma, we now apply the balance relation (5.2) to the shaded region $\Delta\mathscr{D}$ in the previous figure, so obtaining

$$\int_{\Delta\mathscr{A}} \text{♀}_{\mathscr{S}}\,dA - \int_{\Delta\bar{\mathscr{A}}} \text{♀}_{\bar{\mathscr{S}}}\,dA + \int_{\Delta\mathscr{A}^*} \text{♀}_{\Delta\mathscr{A}^*}\,dA \leqq \int_{\Delta\mathscr{D}} \text{♂}\,dV, \tag{5.12}$$

where, of course, we have used (5.3). Since ♀ and ♂ are essentially bounded, (5.12) and the estimates $(5.9)_{3,4}$ yield[1]

$$\int_{\Delta\mathscr{A}} \text{♀}_{\mathscr{S}}\,dA - \int_{\Delta\bar{\mathscr{A}}} \text{♀}_{\bar{\mathscr{S}}}\,dA = o(\Delta r^2). \tag{5.13}$$

[1] When (5.2) and (5.12) are inequalities, the production is assumed to be absolutely continuous with an essentially bounded density relative to the volume measure.

If we now divide by $\pi \Delta r^2$, as $\Delta r \to 0$ we obtain

$$\lim_{\Delta r \to 0} \frac{\int_{\Delta \mathscr{A}} ♀_{\mathscr{S}} \, dA}{\pi \Delta r^2} = \lim_{\Delta r \to 0} \frac{\int_{\Delta \mathscr{A}} ♀_{\bar{\mathscr{S}}} \, dA}{\pi \Delta r^2}. \tag{5.14}$$

From $(5.9)_{1,2}$, we see that (5.14) implies (5.6) if (5.4) is valid at $\boldsymbol{x}$ for both $\mathscr{S}$ and $\bar{\mathscr{S}}$. Hence at any point $\boldsymbol{x} \in \mathscr{D}_0$ and for any unit vector $\boldsymbol{n}$, we define $☿(\boldsymbol{x}, \boldsymbol{n})$ by (5.7) if there exists a smooth surface $\mathscr{S}$ passing through $\boldsymbol{x}$, such that $\boldsymbol{n}_{\mathscr{S}}(\boldsymbol{x})$ is equal to $\boldsymbol{n}$ and (5.4) is valid for $\mathscr{S}$ at $\boldsymbol{x}$; otherwise, we define $☿(\boldsymbol{x}, \boldsymbol{n})$ arbitrarily. Then (5.7) is valid almost everywhere on each $\mathscr{S}$. Q.E.D.

If we assume the supply $♀_{\mathscr{S}}$ to be continuous, then the condition (5.4) holds at every $\boldsymbol{x}$; moreover, in this case the field $☿(\cdot, \boldsymbol{n})$ in (5.7) is continuous for each fixed unit vector $\boldsymbol{n}$. In general, if $\boldsymbol{w}$ is an arbitrary vector in $\mathscr{V}_\varphi$, then we define the field $☿(\cdot, \boldsymbol{w})$ by the conditions

$$☿(\cdot, \boldsymbol{w}) \equiv \|\boldsymbol{w}\| ☿\left(\cdot, \frac{\boldsymbol{w}}{\|\boldsymbol{w}\|}\right), \quad \boldsymbol{w} \neq \boldsymbol{0}, \tag{5.15}$$

and

$$☿(\cdot, \boldsymbol{0}) = 0. \tag{5.16}$$

Thus by definition $☿(\boldsymbol{x}, \boldsymbol{w})$ is linear in $\boldsymbol{w}$ on each 1-dimensional subspace of $\mathscr{V}_\varphi$, *i.e.*,

$$☿(\boldsymbol{x}, \alpha \boldsymbol{w}) = \alpha ☿(\boldsymbol{x}, \boldsymbol{w}) \tag{5.17}$$

for all $\alpha \in \mathscr{R}$. Here we have used also the condition (5.8), which is of course a reflection of (5.3).

It turns out that under very general conditions on the field $☿$, at any given place $\boldsymbol{x}$ the function $☿(\boldsymbol{x}, \cdot)$ is always linear on $\mathscr{V}_\varphi$. The classical version of this result is the celebrated ***Cauchy theorem***, proof of which, reproducing an argument created by CAUCHY himself, can be found in almost every textbook on continuum mechanics [1].

CAUCHY'S FUNDAMENTAL THEOREM. *Suppose that Cauchy's postulate* (5.6) *holds, and suppose that the field* $☿(\cdot, \boldsymbol{n})$ *is continuous for each* $\boldsymbol{n}$. *Then at each place* $\boldsymbol{x}$, $☿(\boldsymbol{x}, \cdot)$ *is a linear function on* $\mathscr{V}_\varphi$.

---

[1] A rigorous and elegant version of CAUCHY's proof, along with a different proof due to NOLL, is given by GURTIN in § 15 of "The Linear Theory of Elasticity", *Handbuch der Physik* VIa/2 (ed. C. TRUESDELL), 1–295, Berlin *etc.*, Springer, 1972.

The condition of continuity in the hypothesis of Cauchy's theorem can be replaced by the much weaker one of integrability, as follows.

PROPOSITION[1]. *Let* $\mathscr{D}_0$ *be the compact region in the previous proposition, and suppose that* $☿(\cdot, \boldsymbol{n})$ *be an integrable field on* $\mathscr{D}_0$ *for each* $\boldsymbol{n}$ *such that the representation formula* (5.7) *holds. Then there exists a field of linear functions* $☿(\cdot)$ *such that for each* $\boldsymbol{n}$, $☿(\boldsymbol{x}, \boldsymbol{n}) = \langle ☿(\boldsymbol{x}), \boldsymbol{n}\rangle$ *for almost every* $\boldsymbol{x} \in \mathscr{D}_0$ *(with respect to the volume measure on* $\mathscr{E}_{\boldsymbol{\varphi}}$*).*

We sketch the proof as follows: First we choose a sequence of C^∞ test functions ψ_ε, $\varepsilon > 0$, defined on the Euclidean manifold $\mathscr{E}_{\boldsymbol{\varphi}}$ in such a way that for each fixed ε the support of ψ_ε is contained in the ball $\mathscr{B}(\varepsilon, \boldsymbol{O})$ with radius ε centered at the origin $\boldsymbol{O}$ and such that

$$\int_{\mathscr{B}(\varepsilon, \boldsymbol{O})} \psi_\varepsilon dV = 1. \tag{5.18}$$

Next we define the convolutions $☿_\varepsilon(\cdot, \boldsymbol{n})$ and $♂_\varepsilon(\cdot)$ of $☿(\cdot, \boldsymbol{n})$ and $♂(\cdot)$ with respect to ψ_ε by

$$☿_\varepsilon(\boldsymbol{x}, \boldsymbol{n}) \equiv \int_{\mathscr{E}_{\boldsymbol{\varphi}}} \psi_\varepsilon(\boldsymbol{x}-\boldsymbol{y})☿(\boldsymbol{y}, \boldsymbol{n})dV_{\boldsymbol{y}}, \tag{5.19}$$

and

$$♂_\varepsilon(\boldsymbol{x}) = \int_{\mathscr{E}_{\boldsymbol{\varphi}}} \psi_\varepsilon(\boldsymbol{x}-\boldsymbol{y})♂(\boldsymbol{y})dV_{\boldsymbol{y}}, \tag{5.20}$$

where $\boldsymbol{x}$ is an arbitrary place in the interior of $\mathscr{D}_0$ which is at a distance greater than ε from the boundary of $\mathscr{D}_0$. We denote the set of all such points $\boldsymbol{x}$ as $\mathscr{D}_\varepsilon$. Since $☿(\cdot, \boldsymbol{n})$ is integrable and ψ_ε is C^∞, we can differentiate (5.19) and (5.20) with respect $\boldsymbol{x}$ under the integral sign. Thus the convolutions $☿_\varepsilon(\cdot, \boldsymbol{n})$ and $♂_\varepsilon(\cdot)$ are C^∞ fields on $\mathscr{D}_\varepsilon$; moreover, for any fixed ε_0

$$☿_\varepsilon(\cdot, \boldsymbol{n}) \to ☿(\cdot, \boldsymbol{n}) \tag{5.21}$$

and

$$♂_\varepsilon(\cdot) \to ♂(\cdot) \tag{5.22}$$

as $\varepsilon \to 0$ in the topology of the space of integrable functions $\mathfrak{L}(\mathscr{D}_{\varepsilon_0})$ over $\mathscr{D}_{\varepsilon_0}$.

---

[1] M. E. GURTIN, V. J. MIZEL, & W. O. WILLIAMS, "A note on Cauchy's stress theorem", *J. Math. Anal. Applic.* **22**, 398–401 (1968).

Now let $\mathscr{D}$ be a regular open domain in $\mathscr{D}_{\varepsilon 0}$. Then from (5.19), (5.20), (5.7), the balance law (5.2), and Fubini's theorem, we can assert that

$$\int_{\mathscr{D}} \mars_0 \, dV \geqq \int_{\partial \mathscr{D}} \mercury_\varepsilon \, dA, \tag{5.23}$$

where $\mercury_\varepsilon$ is given by

$$\mercury_\varepsilon(\cdot) \equiv \mercury_\varepsilon(\cdot, \boldsymbol{n}_{\partial \mathscr{D}}(\cdot)). \tag{5.24}$$

But since the fields $\mars_\varepsilon$ and $\venus_\varepsilon$ are continuous, by the classical version of Cauchy's theorem $\mercury_\varepsilon(\boldsymbol{x}, \cdot)$ is necessarily a linear function on $\mathscr{V}_\varphi$, *viz*

$$\mercury_\varepsilon(\boldsymbol{x}, \boldsymbol{n}) = \langle \mercury_\varepsilon(\boldsymbol{x}), \boldsymbol{n} \rangle, \tag{5.25}$$

where $\mercury_\varepsilon(\cdot)$ is a covector field on $\mathscr{D}_{\varepsilon 0}$. Now let $\mercury(\cdot)$ be the limit of $\mercury_\varepsilon(\cdot)$ as $\varepsilon \to 0$ in the topology of the space of integrable covector fields over $\mathscr{D}_{\varepsilon 0}$. Then for each $\boldsymbol{n}$, using the previous condition (5.21), in the limit as $\varepsilon \to 0$ in (5.25) we obtain

$$\mercury(\boldsymbol{x}, \boldsymbol{n}) = \langle \mercury(\boldsymbol{x}), \boldsymbol{n} \rangle \tag{5.26}$$

for almost every $\boldsymbol{x} \in \mathscr{D}_{\varepsilon 0}$. But since ε_0 is arbitrary, and since the volume of $\mathscr{D}_{\varepsilon 0}$ approaches that of $\mathscr{D}_0$ as $\varepsilon_0 \to 0$, we finally conclude that for each $\boldsymbol{n}$ (5.26) holds for almost every $\boldsymbol{x}$ in $\mathscr{D}_0$. Q.E.D.

6. Field Equations and Jump Conditions

CAUCHY's Fundamental Theorem is the cornerstone of continuum physics. It enables us to replace the integral relations of balance by partial differential equations, called *field equations*, at points interior to regions where the fields themselves are sufficiently smooth, and by *jump conditions* at points on many kinds of surfaces of discontinuity.

When $\mars$ in (5.2) is a tensor field of a certain order m, so also is $\venus$. The linear function $\mercury(\cdot, \boldsymbol{n})$ delivered by Cauchy's theorem is therefore of the form

$$\mercury(\cdot, \boldsymbol{n}) = \saturn(\cdot)\boldsymbol{n}, \tag{6.1}$$

where $\saturn$, the *flux density* corresponding to $\mercury$, is a tensor field of order $m+1$, defined throughout the interior of $\mathscr{D}_0$ and of course independent of the surface $\mathscr{S}$, and where the operation indicated is a linear one. Thus the relation of balance (5.2) becomes

$$\int_{\mathscr{D}} \mathfrak{s}\, dV \geqq \int_{\partial\mathscr{D}} \mathfrak{h}\boldsymbol{n}\, dA. \tag{6.2}$$

When $\partial\mathscr{D}$ and $\mathfrak{h}$ are smooth enough that the divergence theorem holds for the integral on the right-hand side, (6.2) becomes

$$\int_{\mathscr{D}} (\mathfrak{s} - \operatorname{div} \mathfrak{h})\, dV \geqq 0, \tag{6.3}$$

where "div" denotes the spatial divergence operator. In terms of components relative to the Cartesian co-ordinate system on $\mathscr{E}_{\varphi}$,

$$(\mathfrak{h}\boldsymbol{n})^{km\ldots pq} = \mathfrak{h}^{km\ldots pqr} n_r,$$
$$(\operatorname{div} \mathfrak{h})^{km\ldots pq} = \frac{\partial \mathfrak{h}^{km\ldots pqr}}{\partial x^r}. \tag{6.4}$$

Since (6.3) must hold for all regions $\mathscr{D}$ interior to $\mathscr{D}_0$, when $\mathfrak{s}$ and div $\mathfrak{h}$ are continuous (6.3) is equivalent to the *field condition*

$$\mathfrak{s} \geqq \operatorname{div} \mathfrak{h} \tag{6.5}$$

at each interior point of $\mathscr{D}_0$. In terms of the original variables occurring in (5.1), the condition (6.5) is

$$\rho\dot{\Psi} \geqq \operatorname{div} \mathfrak{h} + \rho\mathfrak{A}. \tag{6.6}$$

This is the ***general field condition*** of continuum physics.

In continuum physics there are also cases in which at certain points $\dot{\Psi}$ and div$\mathfrak{h}$ fail to exist, and hence the field condition (6.6) cannot hold. Typical problems of this kind occur in the theory of *wave propagation* (see Chapter VI). In that theory we consider an oriented smooth surface, say $\mathscr{S}_t$, and we assume that the integrands in (6.2) are smooth fields on both sides of $\mathscr{S}_t$ but may suffer jump discontinuities across it. Such a surface is called a *singular surface* in a configuration $\chi_t(\mathscr{B})$ of the body $\mathscr{B}$. That is to say, for any field Ψ under consideration, the limits $\Psi^+(\boldsymbol{x})$ and $\Psi^-(\boldsymbol{x})$ of Ψ taken from the positive side and the negative side of $\mathscr{S}_t$ exist for every point $\boldsymbol{x} \in \mathscr{S}_t$, but in general $\Psi^+(\boldsymbol{x})$ need not equal $\Psi^-(\boldsymbol{x})$. We define the *jump* $[\Psi]$ of Ψ across $\mathscr{S}_t$ as the difference of these limits:

$$[\Psi] \equiv \Psi^+ - \Psi^-. \tag{6.7}$$

We assume that $[\Psi]$ is a smooth field on $\mathscr{S}_t$.

In general, the singular surface $\mathscr{S}_t$, for all times $t \in \mathscr{R}$, need not be the locus of a single surface in the body $\mathscr{B}$. We say that the singular surface *propagates smoothly* in the body $\mathscr{B}$ if there exists a co-ordinate system (q^1, q^2) on $\mathscr{S}_t$ for each time t, such that a place $\boldsymbol{x} = (x^i) \in \mathscr{S}_t$ if and only if

$$x^i = f^i(q^1, q^2, t), \quad i = 1, 2, 3, \tag{6.8}$$

where f^i are smooth functions. Then we define the *normal speed* $u(\boldsymbol{x})$ at any point $\boldsymbol{x} \in \mathscr{S}_t$ as

$$u(\boldsymbol{x}) = \boldsymbol{c}(\boldsymbol{x}) \cdot \boldsymbol{n}_{\mathscr{S}_t}(\boldsymbol{x}), \tag{6.9}$$

where $\boldsymbol{c}(\boldsymbol{x})$ is the velocity of the fictitious "surface point" with fixed surface co-ordinates, *i.e.*,

$$\boldsymbol{c} = \frac{\partial f^i}{\partial t} \boldsymbol{e}_i . \tag{6.10}$$

This surface velocity, of course, is not intrinsic to the singular surface $\mathscr{S}_t$, since it depends on the choice of the surface co-ordinates (q^1, q^2) on $\mathscr{S}_t$ at different times t. However, we may prove[1] that the normal speed u is intrinsic to $\mathscr{S}_t$. This fact is more or less obvious, since a (time-dependent) change of surface co-ordinates affects only the tangential component of the surface velocity $\boldsymbol{c}$ in (6.10). The *intrinsic speeds* $U^\pm$ of $\mathscr{S}_t$ relative to the motion of $\mathscr{B}$ are defined as the differences

$$U^\pm = u - \boldsymbol{v}^\pm \cdot \boldsymbol{n}_{\mathscr{S}_t}, \tag{6.11}$$

where $\boldsymbol{v}$ is the velocity field of $\mathscr{B}$.

KOTCHINE'S THEOREM. *If $\partial(\rho\Psi)/\partial t$ and $\rho\text{♃}$ are essentially bounded in a sufficiently small neighborhood of the point $\boldsymbol{x}$ on $\mathscr{S}_t$, and if the jumps $[\rho\Psi]$, $[\boldsymbol{v}\cdot\boldsymbol{n}]$, and $[\text{♄}\boldsymbol{n}]$ are continuous functions of position on $\mathscr{S}_t$, then (5.1) implies at $\boldsymbol{x}$*

$$-[\rho\Psi U] \geqq [\text{☿}] = [\text{♄}]\boldsymbol{n}, \tag{6.12}$$

where $\boldsymbol{n}$ is the positive unit normal to $\mathscr{S}_t$ at $\boldsymbol{x}$.

The proof of the inequality $(6.12)_1$, which does not require use of the Cauchy theorem, is left to the reader[2]. The equality $(6.12)_2$, which merely repeats the result of the Cauchy theorem, converts the jump condition

---

[1] *Cf.* § 177 of CFT.

[2] *Cf.* § 193 of CFT.

into a boundary condition on the field ♄ on surfaces where ♄ fails to be differentiable and hence the field equation or inequality (6.6) is meaningless. The equation or inequality (6.12) is the ***general jump condition*** of continuum physics.

We now apply these results to the principles of balance of mass, linear momentum, rotational momentum, and energy and to the principle of irreversibility. Thus we first compare (5.1) with (3.5), (4.7), (4.9), (4.18), and (3.28), and then apply (6.1), giving to the field ♄ an appropriate notation and name in each case.

	♆	♃	♀	♄
Mass	1	0	0	0
Linear Momentum	$\boldsymbol{v}$	$\boldsymbol{b}^*$	$\boldsymbol{t}$	$\boldsymbol{T}$
Rotational Momentum	$\boldsymbol{r}\wedge\boldsymbol{v}$	$\boldsymbol{r}\wedge\boldsymbol{b}^*$	$\boldsymbol{r}\wedge\boldsymbol{t}$	$\boldsymbol{r}\wedge\boldsymbol{T}$
Energy	$\varepsilon+\frac{1}{2}v^2$	$r+\boldsymbol{b}^*\cdot\boldsymbol{v}$	$\boldsymbol{t}\cdot\boldsymbol{v}+q$	$\boldsymbol{T}^T\boldsymbol{v}+\boldsymbol{h}$
Entropy	η	r/θ	q/θ	$\boldsymbol{h}/\theta$

The value of the field $\boldsymbol{T}$ at $\boldsymbol{x}$ is called the *stress tensor* there. In view of Cauchy's theorem (6.1), it delivers the contact force density $\boldsymbol{t}$ upon the surface whose normal is $\boldsymbol{n}$ at $\boldsymbol{x}$:

$$\boldsymbol{t} = \boldsymbol{T}\boldsymbol{n}. \tag{6.13}$$

The value of the vector field $\boldsymbol{h}$ at $\boldsymbol{x}$ is the *heating flux* there. It delivers the contact supply of energy q in a similar way:

$$q = \boldsymbol{h}\cdot\boldsymbol{n}. \tag{6.14}$$

The integral principles assume the following forms in terms of $\boldsymbol{T}$ and $\boldsymbol{h}$:

Balance of mass (3.5):

$$\frac{d}{dt}\int_{\chi_t(\mathscr{U})} \rho\, dV = 0 \tag{6.15}$$

Balance of linear momentum (4.7):

$$\frac{d}{dt}\int_{\chi_t(\mathscr{U})} \rho\boldsymbol{v}\, dV = \int_{\chi_t(\mathscr{U})} \rho\boldsymbol{a}\, dV = \int_{\chi_t(\mathscr{U})} \rho\boldsymbol{b}^* dV + \int_{\partial\chi_t(\mathscr{U})} \boldsymbol{T}\boldsymbol{n}\, dA. \tag{6.16}$$

Balance of rotational momentum (4.9):

$$\frac{d}{dt}\int_{\chi_t(\mathscr{U})} \rho \boldsymbol{r} \wedge \boldsymbol{v}\, dV = \int_{\chi_t(\mathscr{U})} \rho \boldsymbol{r} \wedge \boldsymbol{a}\, dV$$

$$= \int_{\chi_t(\mathscr{U})} \rho \boldsymbol{r} \wedge \boldsymbol{b}^* dV + \int_{\partial\chi_t(\mathscr{U})} \boldsymbol{r} \wedge \boldsymbol{T}\boldsymbol{n}\, dA. \quad (6.17)$$

Balance of energy (4.18):

$$\frac{d}{dt}\int_{\chi_t(\mathscr{U})} \rho(\varepsilon + \tfrac{1}{2}v^2) dV = \int_{\chi_t(\mathscr{U})} \rho(\dot{\varepsilon} + \boldsymbol{a} \cdot \boldsymbol{v}) dV,$$

$$= \int_{\chi_t(\mathscr{U})} \rho(r + \boldsymbol{b}^* \cdot \boldsymbol{v}) dV + \int_{\partial\chi_t(\mathscr{U})} (\boldsymbol{h} \cdot \boldsymbol{n} + \boldsymbol{v} \cdot \boldsymbol{T}\boldsymbol{n}) dA. \quad (6.18)$$

Principle of irreversibility (3.28):

$$\frac{d}{dt}\int_{\chi_t(\mathscr{U})} \rho\eta\, dV = \int_{\chi_t(\mathscr{U})} \rho\dot{\eta}\, dV \geqq \int_{\chi_t(\mathscr{U})} \rho\frac{r}{\theta}\, dV + \int_{\partial\chi_t(\mathscr{U})} \frac{\boldsymbol{h} \cdot \boldsymbol{n}}{\theta}\, dA. \quad (6.19)$$

The corresponding field equations are as follows:

Balance of mass (D'ALEMBERT, EULER):

$$\dot{\rho} + \rho \operatorname{div} \boldsymbol{v} = 0. \quad (6.20)$$

Balance of linear momentum (EULER, CAUCHY):

$$\rho\boldsymbol{a} = \operatorname{div} \boldsymbol{T} + \rho\boldsymbol{b}^*. \quad (6.21)$$

Balance of rotational momentum (CAUCHY):

$$\boldsymbol{T} = \boldsymbol{T}^T. \quad (6.22)$$

Balance of energy (FOURIER, KIRCHHOFF, C. NEUMANN):

$$\rho\dot{\varepsilon} = \operatorname{tr}(\boldsymbol{T} \operatorname{grad} \boldsymbol{v}) + \operatorname{div} \boldsymbol{h} + \rho r. \quad (6.23)$$

Principle of irreversibility (CLAUSIUS, DUHEM, TRUESDELL & TOUPIN):

$$\rho\dot{\eta} \geqq \operatorname{div}\frac{\boldsymbol{h}}{\theta} + \rho\frac{r}{\theta}. \quad (6.24)$$

Of these, only (6.21) and (6.24) follow from the general principle (6.6) by trivial substitution. In fact, (6.20) does not follow from it at all, since, as we have remarked, balance of mass was assumed at various points in

§§ 3 and 4, and in particular at the first step. To obtain the classical equation (6.20), however, we need only differentiate (6.15) (*cf.* § 156 of CFT). The symmetry of the stress tensor, expressed by (6.22), follows by subtracting the moment of (6.21) from the field equation corresponding to (6.17). Likewise, (6.23) follows by subtracting from the field equation corresponding to (6.18) the inner product of (6.21) by $\boldsymbol{v}$. Thus (6.20) – (6.24) express in local form and collectively but not individually the principles of balance of mass, linear momentum, rotational momentum, and energy, and the principle of irreversibility. In particular, we notice that the body force density has been eliminated from all these equations except (6.21), where it contributes to $\boldsymbol{b}^*$. The field equations (6.21) and (6.22) are called ***Cauchy's first and second laws of motion.*** In the energy equation (6.23), each term is individually frame-indifferent. The only one for which this fact is not obvious is tr $(\boldsymbol{T} \operatorname{grad} \boldsymbol{v})$, which is called the *stress power*, since it is that part of the power of the stress which contributes to the internal energy. That it is frame-indifferent, follows because the condition expressed by equation (6.23) is frame-indifferent, as we have shown, in effect, in § 4, and because each other term is plainly frame-indifferent, but of course the reader may construct also a direct proof.

Now let $\boldsymbol{\kappa}$ be a fixed reference configuration of $\mathscr{B}$. Then we can express the various fields in (6.15)–(6.19) in terms of the referential co-ordinates (X^A) and the time t also. In this sense we have the following field equations and field inequality:

Balance of mass:

$$\rho = \frac{\rho_{\boldsymbol{\kappa}}}{\det \boldsymbol{F}}. \tag{6.25}$$

Balance of linear momentum:

$$\operatorname{Div} \boldsymbol{T}_{\boldsymbol{\kappa}} + \rho_{\boldsymbol{\kappa}} \boldsymbol{b}^* = \rho_{\boldsymbol{\kappa}} \boldsymbol{a}. \tag{6.26}$$

Balance of rotational momentum:

$$\boldsymbol{T}_{\boldsymbol{\kappa}} \boldsymbol{F}^T = \boldsymbol{F} \boldsymbol{T}_{\boldsymbol{\kappa}}^T \tag{6.27}$$

Balance of energy:

$$\rho_{\boldsymbol{\kappa}} \dot{\varepsilon} = \operatorname{tr} (\boldsymbol{T}_{\boldsymbol{\kappa}}^T \dot{\boldsymbol{F}}) + \operatorname{Div} \boldsymbol{h}_{\boldsymbol{\kappa}} + \rho_{\boldsymbol{\kappa}} r. \tag{6.28}$$

Principle of irreversibility

$$\rho_{\boldsymbol{\kappa}} \dot{\eta} \geqq \operatorname{Div} \frac{\boldsymbol{h}_{\boldsymbol{\kappa}}}{\theta} + \rho_{\boldsymbol{\kappa}} \frac{r}{\theta}. \tag{6.29}$$

In the above expressions, ρ_κ denotes the density field in the reference configuration κ, *i.e.*,

$$M(\mathscr{U}) = \int_{\kappa(\mathscr{U})} \rho_\kappa \, dV \tag{6.30}$$

for any subbody $\mathscr{U}$ of $\mathscr{B}$; $\boldsymbol{T}_\kappa$ denotes the *Piola-Kirchhoff stress tensor*, which is defined by the condition

$$\int_{\partial\kappa(\mathscr{U})} \boldsymbol{T}_\kappa \boldsymbol{N} \, dA_\kappa = \int_{\partial\chi_t(\mathscr{U})} \boldsymbol{T}\boldsymbol{n} \, dA, \tag{6.31}$$

where $\boldsymbol{N}$ is the positive unit normal to $\partial\kappa(\mathscr{U})$ at $\boldsymbol{X}$, so that

$$\boldsymbol{T}_\kappa = (\det \boldsymbol{F}) \, \boldsymbol{T}(\boldsymbol{F}^T)^{-1}; \tag{6.32}$$

and $\boldsymbol{h}_\kappa$ is the heat flux vector relative to κ, defined by the condition

$$\int_{\partial\kappa(\mathscr{U})} \boldsymbol{h}_\kappa \cdot \boldsymbol{N} \, dA_\kappa = \int_{\partial\chi_t(\mathscr{U})} \boldsymbol{h} \cdot \boldsymbol{n} \, dA, \tag{6.33}$$

so that

$$\boldsymbol{h}_\kappa = (\det \boldsymbol{F})\boldsymbol{F}^{-1}\boldsymbol{h}. \tag{6.34}$$

Here dA_κ denotes the element of area in κ. The other notations in (6.25)–(6.29) are the same as before, except that Div denotes the divergence operator with respect to position in κ, *viz*

$$\text{Div } \boldsymbol{h}_\kappa = \frac{\partial h_\kappa^A}{\partial X^A}, \tag{6.35}$$

and

$$(\text{Div } T_\kappa)^i = \frac{\partial T_\kappa^{iA}}{\partial X^A}. \tag{6.36}$$

Notice that the term $\dot{\boldsymbol{F}}$ in (6.28) can be written as Grad $\boldsymbol{v}$ also, since

$$\dot{F}^i_A = \frac{\partial}{\partial t}\left(\frac{\partial x^i}{\partial X^A}\right) = \frac{\partial}{\partial X^A}\left(\frac{\partial x^i}{\partial t}\right) = \frac{\partial v^i}{\partial X^A} = (\text{Grad } v)^i_A. \tag{6.37}$$

In just the same way, we may apply Kotchine's theorem (6.12) to each of the basic principles (6.15)–(6.19) so as to obtain the following *jump conditions* at a singular surface $\mathscr{S}_t$:

Balance of mass:

$$[\rho U] = 0. \tag{6.38}$$

Balance of linear momentum:

$$[\rho U \boldsymbol{v}] + [\boldsymbol{T}]\boldsymbol{n} = \boldsymbol{0}. \tag{6.39}$$

Balance of rotational momentum:

The jump condition is simply the moment of (6.39) and hence is satisfied if (6.39) is.

Balance of energy

$$[\rho U(\varepsilon + \tfrac{1}{2}v^2)] + [\boldsymbol{T}\boldsymbol{v} + \boldsymbol{h}] \cdot \boldsymbol{n} = 0. \tag{6.40}$$

Principle of irreversibility:

$$-[\rho U \eta] \geqq \left[\frac{\boldsymbol{h}}{\theta}\right] \cdot \boldsymbol{n}. \tag{6.41}$$

In each case, for the quantity in question we have assumed that the body supply is absolutely continuous[1], this assumption being implicit in Kotchine's theorem.

7. Constitutive Relations

In § 3 we have introduced the concepts of forces and energy supplies as mathematical models for the physical *causes* of motions of a deformable body, and we have set up in terms of them general axioms of dynamics and thermodynamics. These portray in the theory the physical properties *common* to most bodies in nature. In the physical world, however, we find that bodies are made up of various *materials*. Two bodies the same in size and shape but of two different materials certainly need not have the same motion even if subjected to exactly the same forces and energy supplies. For example, two springs of identical form and dimensions, the one of steel and the other of copper, suffer different elongations when stretched by equal forces. In the mathematical interpretation, this example shows that the dynamical principles alone should not suffice to determine the motion of a deformable body even when the forces and the energy supplies are known. In order that the theory furnish models for

---

[1] In particular, then, there are no singular sources on any surface or line in a region containing $\mathscr{S}_t$.

the *diversity* of materials, it must include a mathematical concept of material response. This concept is embodied in *constitutive relations.*

Constitutive relations of a degenerate kind occur even in elementary mechanics, in which each mass-point is assigned a mass, and each rigid body is assigned a mass density, from which, in view of its shape, its tensor of inertia with respect to any body-point may be determined. Although constitutive relations of this same kind hold in continuum theories, they are not usually mentioned or even named, since they are not typical. The constitutive relations of continuum thermodynamics relate the stress tensor, the flux of energy, the internal energy density, and the entropy density to the experience the body has undergone.

While NOLL[1] has sketched a formal system of constitutive axioms for purely mechanical theories of continuous media in his neo-classical spacetime, thermodynamic theories are still being treated from several different, only partially meshed and formalized standpoints. In this book we shall not need to call on anything like the generality of the modern thermodynamics of materials, yet it would be wrong to pass clean over this new domain of mathematical science, since it has furnished new foundations for the classical theory of elasticity. Therefore we shall merely describe and illustrate the modern approach to constitutive relations.

The basic principles are these:

I. Determinism
II. Local Action
III. Equipresence
IV. Universal Dissipation
V. Material Frame-Indifference
VI. Material Symmetry

In their present forms, at least, the first three are not mathematical but rather physical or metaphysical. In short, they are rules to guide us when we come to set up constitutive relations in the first place. They give us directions about the kind of assumptions to make. The fourth and fifth are specific and hence mathematical, capable of final and explicit statement as soon as a particular class of constitutive relations has been laid down for study. The last is physical, showing us one way to specialize constitutive relations to particular cases.

[1] On p. 196 of his C.I.M.E. Bressanone lectures, cited above on p. 81.

We shall now discuss these six principles in turn, illustrating each by applications to elasticity or to a somewhat more embracing theory.

Since constitutive relations are mathematical statements representing observed material response, we assume now and henceforth that a particular frame of reference φ is assigned, although, as we shall see later, the constitutive relations themselves must be frame-indifferent.

I. *Determinism.* Natural science assumes that natural phenomena can be not only *described* but also – to some extent – *predicted.* That is, past and present experience *determines* in some sense[1] the present response. The mathematical statements of this determination are called *constitutive relations.*

As we have seen in the previous section, the general laws common to all bodies in continuum physics have the form of a *principle of balance* relating an *intensity* Ψ_Γ, a *body supply density* $\mathfrak{Y}_\Gamma$, and a *flux* $\mathfrak{h}_\Gamma$; such a principle follows by substituting (6.1) into (5.1):

$$\int_{\chi_t(\mathscr{U})} \rho \dot{\Psi}_\Gamma \, dV \geqq \int_{\chi_t(\mathscr{U})} \rho \mathfrak{Y}_\Gamma \, dV + \int_{\partial\chi_t(\mathscr{U})} \mathfrak{h}_\Gamma \boldsymbol{n} \, dA, \tag{7.1}$$

where $\Gamma = 1, 2, \ldots, k$. For mechanics and thermodynamics, $k = 5$, and the specific names and notations for the five Ψ_Γ, $\mathfrak{Y}_\Gamma$, and $\mathfrak{h}_\Gamma$ are listed in the table on p. 130. "Experience" of a body $\mathscr{B}$ up to the time t is represented by the *motion* $\chi_u(Y)$ and the *temperature field* $\theta(\chi_u(Y))$, where $Y \in \mathscr{B}$ and $-\infty < u \leqq t$. Let us denote these two quantities by $\chi^t_{\mathscr{B}}$ and $\theta^t_{\mathscr{B}}$, respectively. Then we may express as follows the

Principle of Determinism (Noll)[2].

---

[1] The principle of determinism in analytical dynamics, with its notorious property of reversibility, is a degenerate and untypical special case. Also the classical ideas about well set boundary-value or initial-value problems are far too special to cover modern concepts of determinism. The objective of mathematical physics is to correlate phenomena. While, of course, nineteenth-century programs do succeed in correlating certain phenomena, they illustrate only the simplest kinds of correlation and do not afford typical examples.

[2] Noll's statement of 1957 concerns pure mechanics and refers to the stress alone. It has, of course, long antecedents in special cases in classical physics, notably in Boltzmann's theory of infinitesimal visco-elasticity, and a major special case for non-linear materials had been stated the year before by Green & Rivlin.

The form of the principle given here is appropriate for materials not subject to internal constraints. For a corresponding principle in constrained materials, see § 30 of NFTM, and for the case of an incompressible elastic material, see § III.1 below.

$$
\begin{aligned}
\neptune_\Gamma(\chi_t(X)) &= \hat{\neptune}_\Gamma(\chi^t_{\mathscr{B}}, \theta^t_{\mathscr{B}}; X),\\
\jupiter_\Gamma(\chi_t(X)) &= \hat{\jupiter}_\Gamma(\chi^t_{\mathscr{B}}, \theta^t_{\mathscr{B}}; X),\\
\saturn_\Gamma(\chi_t(X)) &= \hat{\saturn}_\Gamma(\chi^t_{\mathscr{B}}, \theta^t_{\mathscr{B}}; X),
\end{aligned}
\tag{7.2}
$$

where $\hat{\neptune}_\Gamma$, $\hat{\jupiter}_\Gamma$, $\hat{\saturn}_\Gamma$ *are functionals of their first two arguments and functions of their last one.* According to this principle, the present and past values of the motion of $\mathscr{B}$ and the temperature field over $\mathscr{B}$ determine the present values of the intensity $\neptune_\Gamma$, the body supply density $\jupiter_\Gamma$, and the flux $\saturn_\Gamma$, at each body point $X \in \mathscr{B}$.

We note that the specializing relations implied by the table on p. 130, for example $\neptune_1 = 1$, $\jupiter_1 = 0$, $\saturn_1 = 0$, $\neptune_2 = \boldsymbol{v}$, all are consistent with the principle of determinism as stated here.

The constitutive relations (7.2) must be consistent with the five principles of balance themselves. The principle of irreversibility requires a special analysis, which we shall postpone till later in this section, but we shall now dispose of the other four principles. Substituting (7.2) into the four principles of balance, we see that the functional $\hat{\jupiter}_\Gamma$ is determined by the functionals $\hat{\neptune}_\Gamma$ and $\hat{\saturn}_\Gamma$; further, that the value of the functional $\hat{\saturn}_2$, which gives the stress $\boldsymbol{T}$, must be a symmetric tensor field. Conversely, if these restrictions are satisfied, then (7.2) is consistent with the principles of balance. Henceforth we assume, then, that the functional $\hat{\jupiter}_\Gamma$ is determined and that the functional $\hat{\saturn}_2$ is restricted in the way we have just explained.

In the interpretation, indeed, the body supply density $\jupiter_\Gamma$ and the contact supply density $\saturn\boldsymbol{n}$ represent causes of two different kinds. The body supply density serves as a model for causes from without, like gravity and radiation. The contact supply density $\saturn\boldsymbol{n}$ expresses the reaction of contiguous bodies. In experimental physics, a body is given, and the experimenter changes the surroundings so as to perform tests upon it. While only a few particular body supply densities correspond to actual tests known today, more may be discovered tomorrow, and there is no ground for excluding any. This fact suggests the broad view that in the theory we should be able to adjust the functional $\hat{\jupiter}_\Gamma$ so as to yield any desired value $\jupiter_\Gamma$ at any particular place and time. That is, the functionals $\hat{\neptune}_\Gamma$ and $\hat{\saturn}_\Gamma$ are not themselves restricted so as to yield any particular class of functionals $\hat{\jupiter}_\Gamma$. On this understanding, the four principles of balance do not restrict the functionals $\hat{\neptune}_\Gamma$ and $\hat{\saturn}_\Gamma$ in any way besides requiring that the values of $\hat{\mathfrak{h}}_2$ be symmetric. Thus the principle of determinism itself is general enough to cover most cases studied in the thermodynamics of continua so far, but before illustrating it we shall state the next general principle, which severely limits the determinism allowed in the theory.

From the principles of balance and determinism, the constitutive relations for the fluxes $\mathfrak{h}_\Gamma$ and the intensities Ψ_Γ characterize the response of the body $\mathscr{B}$ completely; they are, then, said to define a *material body* $\mathscr{B}$, the body-points of which, now equipped with constitutive relations, are called *material points.*

II. *Local Action.* For a given material point X, the flux $\mathfrak{h}_\Gamma$ and the intensity Ψ_Γ at $\chi_t(X)$ are models for the interaction of material on material at X and the accumulated result of all past supplies at X (to within an additive constant, of course); both concepts are *local* in nature. Hence we assume that the functionals $\hat{\mathfrak{h}}_\Gamma$ and $\hat{\Psi}_\Gamma$ may depend upon the configurations $\chi_u(\mathscr{B})$ at all past times u, but *only within a neighborhood of* X. Consequently, action at a distance is excluded from the causes that can be represented by the contact supplies.

We say that $\chi^t_{\mathscr{B}}$ and $\theta^t_{\mathscr{B}}$ are *locally equivalent* to $\bar{\chi}^t_{\mathscr{B}}$ and $\bar{\theta}^t_{\mathscr{B}}$ at X if there exists a neighborhood of x over which $\chi^t_{\mathscr{B}}$ coincides with $\bar{\chi}^t_{\mathscr{B}}$ and $\theta^t_{\mathscr{B}}$ coincides with $\bar{\theta}^t_{\mathscr{B}}$. Then a mathematically explicit statement of the local restriction on $\hat{\mathfrak{h}}_\Gamma$ and $\hat{\Psi}_\Gamma$ is the

Principle of Local Action (Noll)[1]. *The intensity* Ψ_Γ *and the flux* $\mathfrak{h}_\Gamma$ *at* $\chi_t(X)$ *depend upon* $\chi^t_{\mathscr{B}}$ *and* $\theta^t_{\mathscr{B}}$ *only through their local equivalence class at* X.

The specializing relations implied by the table on p. 130 all are consistent with the principle of local action as just stated.

An example, the simple body. The principles of determinism and local action are obviously satisfied by the special case when the functionals $\hat{\Psi}_\Gamma$ and $\hat{\mathfrak{h}}_\Gamma$ depend upon $\chi^t_{\mathscr{B}}$ and $\theta^t_{\mathscr{B}}$ only through the histories of the motion and temperature fields and their gradients at X itself. That is, the functionals $\hat{\Psi}_\Gamma$ and $\hat{\mathfrak{h}}_\Gamma$ in (7.2) may be replaced[2] by functionals of

$$\nabla\chi_u(X),\ \theta(\chi_u(X)),\ \text{grad}\ \theta(\chi_u(X)),\ \text{where}\ -\infty < u \leqq t. \tag{7.3}$$

[1] All the classical theories of continua obey this principle. Noll seems to have been the first to impose it as a general restriction on constitutive relations.

[2] We leave it to the reader to show that the principle of material frame-indifference, stated below, makes it impossible for the positions $\chi_u(X)$ to appear in the list of independent variables.

A material point whose constitutive relations are of this form is a *simple point* in the sense of NOLL[1]. A body made up of simple points is a *simple body*.

In the practice of continuum mechanics constitutive relations are commonly stated in terms of a reference configuration $\boldsymbol{\kappa}$. The constitutive functionals themselves then depend upon $\boldsymbol{\kappa}$. While such dependence might seem to put an unnecessary complication into the mathematics, it reflects the fact that in nature we encounter bodies only in some configuration or other, and that when we try to describe the response of a real body, we always do so in terms of the configuration in which we find it or conjecture it once to have been. At any particular material point X, the response of the material will depend upon $\boldsymbol{\kappa}$ only through the induced local reference $\boldsymbol{\kappa}_{*X}$, but to simplify the writing we shall here omit the subscript "$*X$". To express the consitutive relations of a simple point in terms of $\boldsymbol{\kappa}$, we replace the independent variable $\nabla\boldsymbol{\chi}_u(X)$ by the deformation gradient $\boldsymbol{F}_{\boldsymbol{\kappa}}$, defined by (1.24). Likewise, we regard the temperature θ as a function of $\boldsymbol{X}$ and t,

$$\theta = \theta_{\boldsymbol{\kappa}}(\boldsymbol{X}, t) = \theta(\boldsymbol{\chi}_t(\boldsymbol{\kappa}^{-1}(\boldsymbol{X})), t), \tag{7.4}$$

and we set

$$\boldsymbol{g}_{\boldsymbol{\kappa}}(\boldsymbol{X}, t) \equiv \operatorname{Grad} \theta_{\boldsymbol{\kappa}} = \boldsymbol{F}_{\boldsymbol{\kappa}}^T \operatorname{grad} \theta. \tag{7.5}$$

As before, we write f^t for the restriction of a function $f(s)$ on $(-\infty, +\infty)$ to the interval $(-\infty, t]$. Then for a simple point, those constitutive relations that are not eliminable by means of the table on p. 130 are, explicitly,

$$\begin{aligned} \boldsymbol{T}_{\boldsymbol{\kappa}}(t) &= \tilde{\boldsymbol{T}}_{\boldsymbol{\kappa}}(\boldsymbol{F}^t_{\boldsymbol{\kappa}}, \theta^t_{\boldsymbol{\kappa}}, \boldsymbol{g}^t_{\boldsymbol{\kappa}}), \\ \boldsymbol{h}_{\boldsymbol{\kappa}}(t) &= \tilde{\boldsymbol{h}}_{\boldsymbol{\kappa}}(\boldsymbol{F}^t_{\boldsymbol{\kappa}}, \theta^t_{\boldsymbol{\kappa}}, \boldsymbol{g}^t_{\boldsymbol{\kappa}}), \\ \psi_{\boldsymbol{\kappa}}(t) &= \tilde{\psi}_{\boldsymbol{\kappa}}(\boldsymbol{F}^t_{\boldsymbol{\kappa}}, \theta^t_{\boldsymbol{\kappa}}, \boldsymbol{g}^t_{\boldsymbol{\kappa}}), \\ \eta_{\boldsymbol{\kappa}}(t) &= \tilde{\eta}_{\boldsymbol{\kappa}}(\boldsymbol{F}^t_{\boldsymbol{\kappa}}, \theta^t_{\boldsymbol{\kappa}}, \boldsymbol{g}^t_{\boldsymbol{\kappa}}), \end{aligned} \tag{7.6}$$

---

[1] See NOLL's fundamental memoir on continuum mechanics, "A mathematical theory of the mechanical behavior of continuous media", *Arch. Rational Mech. Anal.* **2**, 197–226 (1958), reprinted in *Continuum Mechanics II., The Rational Mechanics of Materials*, Internat. Sci. Rev. Ser. Vol. 8, ed. C. TRUESDELL, N.Y., Gordon & Breach, 1965, and in *Continuum Theory of Inhomogeneities in Simple Bodies*, Berlin *etc.*, Springer-Verlag, 1968. In this paper NOLL considers purely mechanical effects only. Dependence upon the temperature field and the present temperature gradient was introduced by COLEMAN, "Thermodynamics of materials with memory", *Arch. Rational Mech. Anal.* **17**, 1–46 (1964), and the history of the temperature gradient was considered soon afterward by others.

where $\boldsymbol{T}_\kappa$ and $\boldsymbol{h}_\kappa$ are defined in terms of $\boldsymbol{T}$, $\boldsymbol{h}$, and $\boldsymbol{F}_\kappa$ by (6.32) and (6.34); where $\tilde{\boldsymbol{T}}_\kappa$, $\tilde{\boldsymbol{h}}_\kappa$, $\tilde{\psi}_\kappa$, and $\tilde{\eta}_\kappa$ are functionals; where for brevity we have omitted the argument $\boldsymbol{X}$ for all functions and functionals appearing in (7.6); and where ψ, the *free-energy density*, is given by the definition

$$\psi \equiv \varepsilon - \eta\theta \tag{7.7}$$

and is introduced for formal convenience, so as to make the application of the principle of universal dissipation easier. In part IV of this section we shall illustrate that application in a major if untypical simple special case, that of the thermo-elastic material, which we now define.

A special case, the thermo-elastic point. The simple point may show effects of *memory* in that past values of the deformation and temperature fields may influence its present fluxes and intensities. In the special case when no memory is allowed, the point is called *thermo-elastic.* The constitutive functionals in (7.6) then reduce to functions of the present values of $\boldsymbol{F}_\kappa$, θ_κ, and $\boldsymbol{g}_\kappa$:

$$\begin{aligned} \boldsymbol{T}_\kappa &= \tilde{\boldsymbol{T}}_\kappa(\boldsymbol{F}_\kappa, \theta_\kappa, \boldsymbol{g}_\kappa), \\ \boldsymbol{h}_\kappa &= \tilde{\boldsymbol{h}}_\kappa(\boldsymbol{F}_\kappa, \theta_\kappa, \boldsymbol{g}_\kappa), \\ \psi_\kappa &= \tilde{\psi}_\kappa(\boldsymbol{F}_\kappa, \theta_\kappa, \boldsymbol{g}_\kappa), \\ \eta_\kappa &= \tilde{\eta}_\kappa(\boldsymbol{F}_\kappa, \theta_\kappa, \boldsymbol{g}_\kappa), \end{aligned} \tag{7.8}$$

where the arguments $\boldsymbol{X}$ and t are left unwritten throughout.

When the fields $\boldsymbol{F}_\kappa(\boldsymbol{X}, t)$, $\theta_\kappa(\boldsymbol{X}, t)$, and $\boldsymbol{g}_\kappa(\boldsymbol{X}, t)$ are independent of t, they are said to be *static* (with respect, of course, to the particular frame of reference φ). For static fields, obviously, the constitutive relations of any simple point reduce to those of a certain thermo-elastic point. Thus *in statics, a general simple point cannot be distinguished from a thermo-elastic point*, so that static thermo-elasticity exhausts the *statics of simple materials.* Since much of the development of thermo-elasticity and its special cases is restricted to statics, this obvious remark helps to explain the broad usefulness of what is only, it might seem, a very special material model. In time-dependent conditions, however, the effects of memory, which elasticity leaves altogether out of account, may be various and decisive.

III. *Equipresence.* Experiment is rarely if ever sufficient to determine a constitutive relation. In practice, a constitutive class is set down by hypothesis, and then experiment or preconception or both are called upon to find within that class some special case simple enough to manage in mathematical problems yet sufficient to cover the phenomena. While nowadays the constitutive classes commonly and successfully studied are far broader than what seemed ultimately general only two decades ago, specialization of one kind or another is always imposed by the theorist. The rule of equipresence restricts the arbitrariness of this procedure. It prohibits the theorist from prejudicing a special theory by guessing a classification of phenomena, setting them apart in categories of his own imagining. If some interaction of effects may conceivably occur, he shall not take upon himself to assume it will not. Of course, he may later *prove* it to be incompatible with some other general principle, such as the principle of dissipation, or with some other natural requirement, such as a particular material symmetry.

RULE OF EQUIPRESENCE (TRUESDELL). *A quantity present as an independent variable in one constitutive relation must be assumed so present in all.*

Clearly this rule is satisfied by the general constitutive relations (7.2) by the special constitutive relations (7.6) defining a simple point, and by the still more special constitutive relations (7.8) defining a thermoelastic point. These do not illustrate how the principle is used, since in fact they were suggested originally by application of it. We pause here to remark on a simple case, one that did arise in the development of thermoelasticity a few years ago.

The classical theory of *elasticity* rests on a purely mechanical constitutive relation, namely,

$$\boldsymbol{T}_\kappa = \tilde{\boldsymbol{T}}_\kappa(\boldsymbol{F}_\kappa), \tag{7.9}$$

in which the effects of temperature changes and gradients presently experienced by the material point X are left altogether out of account. The classical theory of *heat conduction* in rigid bodies rests on purely thermal constitutive relations, namely,

$$\begin{aligned} \boldsymbol{h} &= \tilde{\boldsymbol{h}}(\theta, \operatorname{grad} \theta), \\ \varepsilon &= \tilde{\varepsilon}(\theta) \end{aligned} \tag{7.10}$$

(which in fact might seem to violate the rule of equipresence). A mathematician wishing to construct a theory of the interaction of the effects deformation and temperature may well ask whether deformation affects $\boldsymbol{h}$ and ε, and whether temperature and its gradient affect $\boldsymbol{T}$. The rule of equipresence tells him to presume that the answer in both cases is yes unless he can prove otherwise.[1] In this way he is forced to set up (7.8) as his constitutive relations for the combined theory. In fact, it was in this way that (7.8) first was suggested. We shall see later that the principle of universal dissipation forces $\tilde{\boldsymbol{T}}_\kappa$, $\tilde{\psi}_\kappa$, and $\tilde{\eta}_\kappa$ to be independent of $\boldsymbol{g}_\kappa$ after all. This independence is thus made the content of a mathematical theorem, not just a guess.

Like many other developments in modern mechanics, the rule of equipresence eliminates a certain amount of unnecessary "physics" and gives a bigger role to mathematics.

IV. *Universal Dissipation.* In presenting the rule of determinism we have remarked that constitutive relations must be consistent with the general principles of balance of mass, linear momentum, rotational momentum, and energy and with the principle of irreversibility. We have taken the standpoint that constitutive relations for the body supply densities are assignable independently of the others and in a class broad enough to allow arbitrary values at any one place and time. In this way, the principles of balance impose no restriction at all upon the functionals $\hat{\mathfrak{P}}_\Gamma$ and $\hat{\mathfrak{h}}_\Gamma$, except that the value of $\hat{\mathfrak{h}}_2$ be a symmetric tensor. On the assumption that the four principles of balance are satisfied we may use (6.23) (to derive which, we recall, the three previously stated principles (6.20), (6.21), and (6.22) have been used) so as to eliminate r from (6.24). Indeed, identically,

$$\theta\left[\rho\dot{\eta}-\operatorname{div}\frac{\boldsymbol{h}}{\theta}-\rho\frac{r}{\theta}\right]=\rho\dot{\psi}-\rho\eta\dot{\theta}+\operatorname{tr}(\boldsymbol{T}\operatorname{grad}\boldsymbol{v})-\frac{\boldsymbol{h}}{\theta}\cdot\operatorname{grad}\theta, \quad (7.11)$$

where ψ is the free-energy density (7.7). Thus (6.24) becomes the *reduced dissipation inequality*:

$$\rho\dot{\psi}+\rho\eta\dot{\theta}-\operatorname{tr}(\boldsymbol{T}\operatorname{grad}\boldsymbol{v})+\theta^{-1}\boldsymbol{h}\cdot\operatorname{grad}\theta\leqq 0. \quad (7.12)$$

---

[1] Some recent statements of equipresence add at the end the redundant phrase, "unless its presence contradicts some other physical principle," which only blunts equipresence and obscures its nature. It is *a rule for setting up hypotheses*, not a divination of the outcome of a mathematical theory. In fact the rule of equipresence is merely one case of a general philosophical precept in mathematics: Do not *assume* something if you can *prove* it. It reflects also the suspicion, which in mathematical physics often has turned out to be just, that what cannot be proved from natural assumptions may not always be true.

It expresses the principle of irreversibility, on the assumption that the four principles of balance are satisfied. In terms of the fields (6.32), (6.34), (7.4), and (7.5), defined in terms of a reference configuration $\boldsymbol{\kappa}$, (7.12) assumes the form

$$\rho_\kappa \dot{\psi}_\kappa + \rho_\kappa \eta_\kappa \dot{\theta}_\kappa - \operatorname{tr}(\boldsymbol{T}_\kappa \dot{\boldsymbol{F}}_\kappa^T) + \theta_\kappa^{-1} \boldsymbol{h}_\kappa \cdot \boldsymbol{g}_\kappa \leqq 0. \tag{7.13}$$

In a fundamental memoir on thermo-elasticity COLEMAN & NOLL[1] proposed to interpret the principle of irreversibility as an *identical requirement* on constitutive equations. Thus the functionals $\hat{\mathfrak{P}}_\Gamma$ ane $\hat{\mathfrak{h}}_\Gamma$ occurring in the constitutive relations (7.2) must be of such a kind that (7.13) is satisfied *for all* histories $\chi_{\mathscr{B}}^t$ and $\theta_{\mathscr{B}}^t$ of the motion and temperature of $\mathscr{B}$. In this way (7.13) becomes a condition on the constitutive functionals, so that a definite mathematical problem results, *viz*, to find within a given class the most general $\hat{\mathfrak{P}}_\Gamma$ and $\hat{\mathfrak{h}}_\Gamma$ that satisfy (7.13) for every motion and temperature field. Formally, we impose the principle of

UNIVERSAL DISSIPATION (COLEMAN & NOLL). *All constitutive relations are such as to satisfy the reduced dissipation inequality for all arguments in their domains.*

We illustrate this principle in a major if untypically simple special case, that of the thermo-elastic material. Substitution of (7.8) into (7.13) yields

$$\rho_\kappa \frac{\partial \tilde{\psi}_\kappa}{\partial \boldsymbol{g}_\kappa} \cdot \dot{\boldsymbol{g}}_\kappa + \rho_\kappa \left(\tilde{\eta}_\kappa + \frac{\partial \tilde{\psi}_\kappa}{\partial \theta_\kappa} \right) \dot{\theta}_\kappa -$$

$$- \operatorname{tr}\left[\left(\tilde{\boldsymbol{T}}_\kappa - \rho_\kappa \frac{\partial \tilde{\psi}_\kappa}{\partial \boldsymbol{F}_\kappa} \right) \dot{\boldsymbol{F}}_\kappa^T \right] + \frac{\tilde{\boldsymbol{h}}_\kappa}{\theta_\kappa} \cdot \boldsymbol{g}_\kappa \leqq 0. \tag{7.14}$$

At each place and time this inequality is to hold for every history of deformation and temperature. Since we may construct such histories having any desired values and any desired derivatives at any given place and time, (7.14) becomes an identical inequality, to be satisfied for arbitrary $\dot{\boldsymbol{g}}_\kappa$, $\dot{\theta}_\kappa$, and $\dot{\boldsymbol{F}}_\kappa$. Hence it is equivalent to the following four relations:

(i)
$$\frac{\partial \tilde{\psi}_\kappa}{\partial \boldsymbol{g}_\kappa} = \boldsymbol{0}, \tag{7.15}$$

[1] B. D. COLEMAN & W. NOLL, "The thermodynamics of elastic materials with heat conduction and viscosity", *Arch. Rational Mech. Anal.* **13**, 167–178 (1963).

that is,

$$\psi = \psi_\kappa(X, t) = \tilde{\psi}_\kappa(F_\kappa, \theta_\kappa). \tag{7.16}$$

(ii)
$$\eta = \tilde{\eta}_\kappa(F_\kappa, \theta_\kappa) = -\frac{\partial\tilde{\psi}_\kappa(F_\kappa, \theta_\kappa)}{\partial\theta_\kappa}. \tag{7.17}$$

(iii)
$$T_\kappa = \tilde{T}_\kappa(F_\kappa, \theta_\kappa) = \rho_\kappa \frac{\partial\tilde{\psi}_\kappa(F_\kappa, \theta_\kappa)}{\partial F_\kappa}. \tag{7.18}$$

Thus the function $\tilde{\psi}_\kappa$ is a *thermodynamic potential* for the specific entropy η_κ and the Piola-Kirchhoff stress T_κ, and in particular these quantities are independent of g_κ. We have proved, then, that within the constitutive class defined by (7.8) the temperature gradient cannot affect the free energy, the stress, or the entropy that correspond to a given deformation gradient and temperature. This reduction illustrates what we said in regard to the rule of equipresence, namely, that certain interactions which that rule tells us to allow in laying down a constitutive class may subsequently be *proved* impossible as a consequence of other principles. Finally,

(iv)
$$h_\kappa \cdot g_\kappa \leqq 0, \tag{7.19}$$

or, equivalently,

$$h \cdot \operatorname{grad} \theta \leqq 0. \tag{7.20}$$

This last is *Fourier's inequality*, stating that h cannot subtend an acute angle from grad θ, or, in physical terms, the flux of energy never points toward a hot spot from a cold one.

The reduced dissipation inequality has been investigated and solved for far more general constitutive relations by COLEMAN. The celebrated *Coleman theorem* has two parts: The first, when specialized to a thermoelastic material, is the theorem of COLEMAN & NOLL we have just derived, while the second establishes the consistency of thermodynamics with GIBBS' thermostatics and obtains a theorem of minimum free energy. In this book we shall not take up thermostatics at all, so we shall not need the full strength of COLEMAN's theorem.[1]

---

[1] An abstract axiomatic structure sufficient to obtain COLEMAN's theorem under very general conditions has been set up by M. E. GURTIN, "On the thermodynamics of materials with memory", *Arch. Rational Mech. Anal.* **28**, 40–50 (1968). A treatment under slightly weaker hypotheses may be found in Chapter 3 of C. TRUESDELL's *Rational Thermodynamics*, New York *etc.*, McGraw-Hill, 1969.

The potential relations (7.17) and (7.18) are often found in another form. Let us suppose that (7.17) be invertible for θ_{κ}:

$$\theta = \hat{\theta}_{\kappa}(F_{\kappa}, \eta_{\kappa}). \tag{7.21}$$

Then from (7.7) we have

$$\varepsilon = \hat{\varepsilon}_{\kappa}(F_{\kappa}, \eta_{\kappa}) \equiv \tilde{\psi}_{\kappa}(F_{\kappa}, \hat{\theta}_{\kappa}(F_{\kappa}, \eta_{\kappa})) + \eta_{\kappa}\hat{\theta}_{\kappa}(F_{\kappa}, \eta_{\kappa}). \tag{7.22}$$

The reader will easily verify the potential relations

$$\theta = \hat{\theta}_{\kappa}(F_{\kappa}, \eta_{\kappa}) = \frac{\partial \hat{\varepsilon}_{\kappa}(F_{\kappa}, \eta_{\kappa})}{\partial \eta_{\kappa}}, \tag{7.23}$$

$$T_{\kappa} = \hat{T}_{\kappa}(F_{\kappa}, \eta_{\kappa}) = \rho_{\kappa} \frac{\partial \hat{\varepsilon}_{\kappa}(F_{\kappa}, \eta_{\kappa})}{\partial F_{\kappa}}. \tag{7.24}$$

Thus also the function $\hat{\varepsilon}_{\kappa}(F_{\kappa}, \eta_{\kappa})$ is a thermodynamic potential.

In terms of the Cauchy stress T, by (6.32) we may write the potential relations (7.18) and (7.24) in the forms

$$T = \rho \frac{\partial \tilde{\psi}(F, \theta)}{\partial F} F^T = \rho \frac{\partial \hat{\varepsilon}(F, \eta)}{\partial F} F^T, \tag{7.25}$$

where we have dropped the subscript κ throughout the two right-hand members. In order that T be a symmetric tensor, as it must be, the functions $\tilde{\psi}$ and $\hat{\varepsilon}$ must be such that

$$\begin{aligned} F\left(\frac{\partial \tilde{\psi}}{\partial F}\right)^T &= \frac{\partial \tilde{\psi}}{\partial F} F^T, \\ F\left(\frac{\partial \hat{\varepsilon}}{\partial F}\right)^T &= \frac{\partial \hat{\varepsilon}}{\partial F} F^T. \end{aligned} \tag{7.26}$$

We shall see below that these identities hold as a consequence of the principle of material frame-indifference. Thus (7.25) assumes the alternative form

$$T = \rho F \left(\frac{\partial \tilde{\psi}(F, \theta)}{\partial F}\right)^T = \rho F \left(\frac{\partial \hat{\varepsilon}(F, \eta)}{\partial F}\right)^T. \tag{7.27}$$

V. *Material frame-indifference.* Since the constitutive relations serve as a mathematical model for a material, they must not depend on the observer, or frame of reference. This idea is clear enough in its physical interpretation. For example, consider the constitutive relation for a simple elastic spring:

$$f = k\delta, \tag{7.28}$$

where f denotes the tensile force and δ denotes the elongation. Everybody takes it for granted that the spring constant k expresses a property of the spring, the same for all observers. For more complicated constitutive relations, such as (7.2), (7.6), or (7.8), which depend explicitly on the frame of reference $\boldsymbol{\varphi}$ in their very forms, frame-indifference is no longer so obvious. The basic idea of frame-indifference is that if a quantity is frame-indifferent with respect to a certain class of frames, any constitutive relation delivering that quantity should enjoy the same invariance. In continuum mechanics interest concentrates upon the contact supplies $\mathfrak{h}_\Gamma, \Gamma = 1, 2, \dots 5$, all of which are frame-indifferent by definition. Moreover, the intensities Ψ_Γ not already specified by the table on p. 130 also are frame-indifferent. Thus we lay down the following

Principle of Material Frame-Indifference[1]. *Constitutive relations for the contact supplies $\mathfrak{h}_\Gamma$ and for those intensities Ψ_Γ that are frame-indifferent must be invariant under changes of frame.*

In reference to the constitutive relations (7.2), this principle applies non-trivially to the functionals $\hat{\Psi}_4 - \frac{1}{2}v^2$ and $\hat{\Psi}_5$, whose values are the scalars ε and η; to the functional $\hat{\mathfrak{h}}_2$, whose value is the tensor $\boldsymbol{T}$; and to the functional $\hat{\mathfrak{h}}_4 - \boldsymbol{T}^T\boldsymbol{v}$ whose value is the vector $\boldsymbol{h}$. We may summarize as follows the four constitutive relations affected:

$$\mathfrak{S}(\chi_t(X)) = \hat{\mathfrak{S}}(\chi^t_{\mathscr{B}}, \theta^t_{\mathscr{B}}; X). \tag{7.29}$$

According to the principle of material frame-indifference, then, the functional $\mathfrak{S}(\cdot, \cdot; X)$ is a frame-indifferent observable kinematical quantity[2] as specified in § 2.

We now determine a necessary and sufficient condition for the functional $\hat{\mathfrak{S}}$ to satisfy this requirement. For definiteness, we carry through the analysis for the case when $\hat{\mathfrak{S}}$ is vector-valued, so that the component form of (7.29) relative to $\boldsymbol{\varphi}$ is

$$\mathfrak{S}(\chi_t(X)) = \hat{\mathfrak{S}}^k(\chi^t_{\mathscr{B}}, \theta^t_{\mathscr{B}}; X)\boldsymbol{e}_k. \tag{7.30}$$

---

[1] The principle of frame-indifference, in various forms and applications, is due to Poisson, Cauchy, Zaremba, Oldroyd, Noll, and others. *Cf.* §§ 18–19A of NFTM.

[2] We extend the concepts of § 2 in an obvious way so as to include dependence on $\theta^t_{\mathscr{B}}$ as well as $\chi^t_{\mathscr{B}}$.

As we explained more generally in § 2, we assume that the component functionals $\hat{\mathfrak{K}}^k$ are independent of the choice of frame. That is, all observers agree to follow the same *procedure* for finding $\hat{\mathfrak{K}}^k$. Then, with the same component functionals $\hat{\mathfrak{K}}^k$, in $\bar{\varphi}$ we have

$$\mathfrak{K}(\bar{\chi}_{\bar{t}}(X)) = \hat{\mathfrak{K}}^k(\bar{\chi}^{\bar{t}}_{\mathscr{B}}, \bar{\theta}^{\bar{t}}_{\mathscr{B}}; X)\bar{e}_k, \tag{7.31}$$

where the components of χ_t and $\bar{\chi}_{\bar{t}}$ are related by (2.53) (*cf.* (2.12)); where t and $\bar{t}$ are related by (2.13) and where

$$\bar{\theta}(\bar{\chi}_{\bar{t}}(X)) = \theta(\chi_t(X)). \tag{7.32}$$

The requirement of frame-indifference, of course, is

$$\mathfrak{K}(\chi_t(X)) = \mathfrak{K}(\bar{\chi}_{\bar{t}}(X)). \tag{7.33}$$

By (7.30), (7.31), and (2.7) we thus obtain the following necessary and sufficient condition:

$$\hat{\mathfrak{K}}^k(\chi^t_{\mathscr{B}}, \theta^t_{\mathscr{B}}; X) = Q^k_j(t)\hat{\mathfrak{K}}^j(\bar{\chi}^{\bar{t}}_{\mathscr{B}}, \theta^{\bar{t}}_{\mathscr{B}}; X), \tag{7.34}$$

to be satisfied identically in the motion χ_t, the proper-orthogonal matrix $[Q^j_i(t)]$, the three functions $y^i(t)$, and the constant t_0, these being the quantities that specify the transformations (2.13) and (2.53).

We recapitulate the assumptions made:

1. There exists an observable kinematical quantity $\hat{\mathfrak{K}}$ which specifies the value of the field $\mathfrak{K}$ at the place occupied in the frame φ by the body point X in any motion of $\mathscr{B}$.

2. The procedure for determining the components $\hat{\mathfrak{K}}^k$ of $\hat{\mathfrak{K}}$ in φ from the motion $\chi^t_{\mathscr{B}}$ and the temperature field $\theta^t_{\mathscr{B}}$ in $\mathscr{B}$ with respect to φ up to the time t is independent of φ.

Under these assumptions, (7.34) is a *necessary and sufficient condition* on the component functionals $\hat{\mathfrak{K}}^k$ in order that the value $\mathfrak{K}$ specified by $\hat{\mathfrak{K}}$ be a *frame-indifferent vector field.*

As we explained in general terms at the end of § 2, we can convert (7.34) into a condition on the functional $\hat{\mathfrak{K}}$ relative to a single frame of reference. We include the cases when $\mathfrak{K}$ is scalar-valued or tensor-valued, since these follow by parallel analyses. Moreover, we recall the concept of motions equivalent in the sense that they differ by a rigid-body motion, as explained in § 2, and we amplify it by the corresponding requirement (7.32) for the associated temperature fields. Thus we obtain the following

PROPOSITION.[1] *The constitutive relation* (7.29) *obeys the principle of material frame-indifference if and only if for scalar-valued* $\mathfrak{H}$,

$$\hat{\mathfrak{H}}(\chi^t_{\mathscr{B}}, \theta^t_{\mathscr{B}}; X) = \hat{\mathfrak{H}}(\bar{\chi}^{\bar{t}}_{\mathscr{B}}, \bar{\theta}^{\bar{t}}_{\mathscr{B}}; X); \tag{7.35}$$

for vector-valued $\mathfrak{H}$,

$$\hat{\mathfrak{H}}(\chi^t_{\mathscr{B}}, \theta^t_{\mathscr{B}}, X) = \boldsymbol{Q}(t)\hat{\mathfrak{H}}(\bar{\chi}^{\bar{t}}_{\mathscr{B}}, \bar{\theta}^{\bar{t}}_{\mathscr{B}}; X); \tag{7.36}$$

for tensor-valued $\mathfrak{H}$,

$$\hat{\mathfrak{H}}(\chi^t_{\mathscr{B}}, \theta^t_{\mathscr{B}}, X) = \boldsymbol{Q}(t)\hat{\mathfrak{H}}(\bar{\chi}^{\bar{t}}_{\mathscr{B}}, \bar{\theta}^{\bar{t}}_{\mathscr{B}}; X)\boldsymbol{Q}^T(t), \tag{7.37}$$

where χ, θ *and* $\bar{\chi}$, $\bar{\theta}$ *are any two members of the same equivalence class related by* (2.53) *and* (7.32).

For the simple body, whose points obey the constitutive relations (7.6), the requirement of frame-indifference may be given a more explicit form. If χ_t and $\bar{\chi}_{\bar{t}}$ belong to the same equivalence class, by (2.53) we have for the corresponding deformation gradients with respect to a particular reference configuration $\boldsymbol{\kappa}$

$$\boldsymbol{F}_{\boldsymbol{\kappa}}(X, s) = \boldsymbol{Q}(s)\bar{\boldsymbol{F}}_{\boldsymbol{\kappa}}(X, s), \quad s \in \mathscr{T}, \tag{7.38}$$

or, if we drop the body-point X from the notation but indicate the absolute times by t and $\bar{t}$, as usual,

$$\boldsymbol{F}_{\boldsymbol{\kappa}}(t) = \boldsymbol{Q}(t)\bar{\boldsymbol{F}}_{\boldsymbol{\kappa}}(\bar{t}) = \bar{\boldsymbol{Q}}(\bar{t})\bar{\boldsymbol{F}}_{\boldsymbol{\kappa}}(\bar{t}). \tag{7.39}$$

Likewise, the temperature gradients (7.5) satisfy

$$\boldsymbol{g}_{\boldsymbol{\kappa}}(t) = \bar{\boldsymbol{g}}_{\boldsymbol{\kappa}}(\bar{t}), \tag{7.40}$$

in the same abbreviated notation.[2]

Since the Cauchy stress $\boldsymbol{T}$ is a frame-indifferent tensor, by (6.32) we have the following transformation law for the Piola-Kirchhoff stress $\boldsymbol{T}_{\boldsymbol{\kappa}}$ under change of frame:

---

[1] While this proposition has antecedents going back a century, in the above general form it is due to NOLL.

[2] The spatial temperature gradient grad θ, of course, satisfies instead the transformation law

$$\operatorname{grad} \theta(t) = \bar{\boldsymbol{Q}}(\bar{t}) \operatorname{grad} \bar{\theta}(\bar{t}).$$

$$\begin{aligned} \bar{T}_\kappa &= (\det \bar{F}_\kappa)\bar{T}(\bar{F}_\kappa^T)^{-1}, \\ &= (\det F_\kappa)Q^T T Q((Q^T F_\kappa)^T)^{-1}, \\ &= Q^T T_\kappa . \end{aligned} \tag{7.41}$$

Thus when we specialize (7.37) to the case of a simple material, we obtain

$$\tilde{T}_\kappa(F_\kappa^t, \theta_\kappa^t, g_\kappa^t) = Q(t)\tilde{T}_\kappa((\bar{Q}^{\bar{t}})^T F_\kappa^{\bar{t}}, \theta_\kappa^{\bar{t}}, g_\kappa^{\bar{t}}), \tag{7.42}$$

where $\bar{Q}^{\bar{t}}$ is the restriction of $\bar{Q}(\bar{u})$ to the interval $(-\infty, \bar{t}]$. The times t and $\bar{t}$ differ by an arbitrary constant t_0 according to (2.13), and $Q(t)$ is an arbitrary function of t only whose values are proper-orthogonal tensors. The identity (7.42) must be satisfied by the functional $\tilde{T}_\kappa$ for all such t_0 and $Q(t)$ and for all non-singular tensor functions $F_\kappa(t)$, all positive scalar functions $\theta_\kappa(t)$, and all vector functions $g_\kappa(t)$. Similar identities must be satisfied by the functionals $\tilde{h}_k$, $\tilde{\psi}_\kappa$, and $\tilde{\eta}_\kappa$ occurring in (7.6).

Functional equations of this kind have been solved by NOLL[1], COLEMAN[2], and others, but we shall not need results of such generality.

A special case, the thermo-elastic body. A thermo-elastic point is defined by the special constitutive relations (7.8). For this case, (7.41) reduces to

$$\tilde{T}_\kappa(F_\kappa, \theta_\kappa, g_\kappa) = Q\tilde{T}_\kappa(Q^T F_\kappa, \theta_\kappa, g_\kappa), \tag{7.43}$$

where $\tilde{T}_\kappa$ is the function occurring in $(7.8)_1$. By the polar decomposition $(1.32)_1$,

$$\tilde{T}_\kappa(F_\kappa, \theta_\kappa, g_\kappa) = Q\tilde{T}_\kappa(Q^T R_\kappa U_\kappa, \theta_\kappa, g_\kappa). \tag{7.44}$$

Since this identity must hold for all rotations Q, for given F_κ, θ_κ, and g_κ, we may choose $Q = R_\kappa$ and obtain

$$\tilde{T}_\kappa(F_\kappa, \theta_\kappa, g_\kappa) = R_\kappa \tilde{T}_\kappa(U_\kappa, \theta_\kappa, g_\kappa) \tag{7.45}$$

as a necessary condition for frame-indifference. Conversely, this condition is evidently sufficient. Thus we obtain the following *reduced form* for the constitutive relation giving the stress at a thermo-elastic point:

$$T_\kappa = R_\kappa \tilde{T}_\kappa(U_\kappa, \theta_\kappa, g_\kappa). \tag{7.46}$$

The term "reduced" refers to the fact that the function $\tilde{T}_\kappa$, with its first argument restricted to positive-definite symmetric tensors, automati-

---

[1] W. NOLL, p. 216 of the work cited on p. 139.

[2] B. D. COLEMAN, p. 40 of the work cited on p. 139.

cally delivers a frame-indifferent tensor $\boldsymbol{T}_\kappa$. We may view (7.46) as being a frame-indifferent extension of $\tilde{\boldsymbol{T}}_\kappa(\cdot, \theta_\kappa, \boldsymbol{g}_\kappa)$ from the domain of positive-definite symmetric tensors to the domain of proper non-singular tensors. If we prefer to use the Cauchy stress $\boldsymbol{T}$, we may rewrite (7.46) in the form

$$\boldsymbol{T} = \boldsymbol{R}\tilde{\boldsymbol{T}}(\boldsymbol{U}, \theta, \boldsymbol{R}^T \text{grad}\, \theta)\boldsymbol{R}^T, \tag{7.47}$$

where $\tilde{\boldsymbol{T}}$ is a function dependent on κ (though it is not the function denoted before as $\tilde{\boldsymbol{T}}_\kappa$), and where we have dropped the subscript κ from $\boldsymbol{R}$ and $\boldsymbol{U}$. Similarly, a reduced form for the constitutive relation delivering the heating flux $\boldsymbol{h}$ is

$$\boldsymbol{h} = \boldsymbol{R}\tilde{\boldsymbol{h}}(\boldsymbol{U}, \theta, \boldsymbol{R}^T \text{grad}\, \theta), \tag{7.48}$$

as the reader will easily verify, while the free-energy density ψ and the entropy density η obey the reduced constitutive relations

$$\begin{aligned} \psi &= \tilde{\psi}(\boldsymbol{U}, \theta, \boldsymbol{R}^T \text{grad}\, \theta), \\ \eta &= \tilde{\eta}(\boldsymbol{U}, \theta, \boldsymbol{R}^T \text{grad}\, \theta). \end{aligned} \tag{7.49}$$

According to the principle of universal dissipation, as we have seen, the argument $\boldsymbol{R}^T$ grad θ in fact must drop out of (7.47) and (7.49).

There are many other reduced forms for the constitutive equations of a thermo-elastic point. They may be derived by transforming the ones we have given above or by other direct arguments reflecting the same principle of invariance.

In view of $(7.49)_1$, it is easy to see that $(7.26)_1$ is satisfied. Conversely, if $(7.26)_1$ is assumed, (7.49) without the argument $\boldsymbol{R}^T$grad θ follows. Similarly, $(7.49)_1$ and (7.47) are mathematically equivalent in virtue of $(7.25)_1$ or $(7.27)_1$. Thus we have a celebrated theorem due in principle to NOLL[1]:

PROPOSITION. *In a thermo-elastic body, on the assumption that the principles of linear momentum and universal dissipation are satisfied, any of these three statements implies the other two*:

1. *The principle of rotational momentum holds.*
2. *The stress tensor* $\boldsymbol{T}$ *is frame-indifferent.*
3. *The free-energy function* ψ *is frame-indifferent.*

---

[1] W. NOLL, "On the continuity of the solid and fluid states", *J. Rational Mech. Anal.* 4, 3–81 (1955), reprinted in *Continuum Mechanics II, Rational Mechanics of Materials*, vol. 8 of Internat. Sci. Rev. Ser., ed. C. TRUESDELL, N.Y., Gordon & Breach, 1965. See § 16. This work, of course, was too early to take account directly of the principle of universal dissipation.

VI. *Material symmetry.* The constitutive relations (7.6) for a simple point are expressed in terms of a particular reference configuration $\boldsymbol{\kappa}$. They may be expressed equally well in terms of any other reference configuration $\bar{\boldsymbol{\kappa}}$. Let us write a typical member of (7.6) in the form

$$\Psi_{\boldsymbol{\kappa}}(t) = \tilde{\Psi}_{\boldsymbol{\kappa}}(\boldsymbol{F}^t_{\boldsymbol{\kappa}}, \theta^t_{\boldsymbol{\kappa}}, \boldsymbol{g}^t_{\boldsymbol{\kappa}}). \tag{7.50}$$

Then, equally,

$$\Psi_{\bar{\boldsymbol{\kappa}}}(t) = \tilde{\Psi}_{\bar{\boldsymbol{\kappa}}}(\boldsymbol{F}^t_{\bar{\boldsymbol{\kappa}}}, \theta^t_{\bar{\boldsymbol{\kappa}}}, \boldsymbol{g}^t_{\bar{\boldsymbol{\kappa}}}). \tag{7.51}$$

By the chain rule (1.29) and its counterparts for $\theta_{\boldsymbol{\kappa}}$ and $\boldsymbol{g}_{\boldsymbol{\kappa}}$, (7.50) may be written also in terms of $\boldsymbol{F}_{\bar{\boldsymbol{\kappa}}}$, $\theta_{\bar{\boldsymbol{\kappa}}}$, and $\boldsymbol{g}_{\bar{\boldsymbol{\kappa}}}$:

$$\Psi_{\bar{\boldsymbol{\kappa}}}(t) = \tilde{\Psi}_{\boldsymbol{\kappa}}(\boldsymbol{F}^t_{\bar{\boldsymbol{\kappa}}}\boldsymbol{P}, \theta^t_{\bar{\boldsymbol{\kappa}}}, \boldsymbol{P}^T\boldsymbol{g}^t_{\bar{\boldsymbol{\kappa}}}), \tag{7.52}$$

where $\boldsymbol{P}$ is the gradient of the deformation from the configuration $\boldsymbol{\kappa}$ to the configuration $\bar{\boldsymbol{\kappa}}$, namely, $\boldsymbol{P} = \bar{\boldsymbol{\kappa}}_* \circ \boldsymbol{\kappa}_*^{-1}$. Comparison of (7.52) and (7.51) shows that the functional $\tilde{\Psi}_{\bar{\boldsymbol{\kappa}}}$ may be expressed in terms of the functional $\tilde{\Psi}_{\boldsymbol{\kappa}}$ as follows:

$$\tilde{\Psi}_{\bar{\boldsymbol{\kappa}}}(\boldsymbol{F}^t_{\bar{\boldsymbol{\kappa}}}, \theta^t_{\bar{\boldsymbol{\kappa}}}, \boldsymbol{g}^t_{\bar{\boldsymbol{\kappa}}}) = \tilde{\Psi}_{\boldsymbol{\kappa}}(\boldsymbol{F}^t_{\bar{\boldsymbol{\kappa}}}\boldsymbol{P}, \theta^t_{\bar{\boldsymbol{\kappa}}}, \boldsymbol{P}^T\boldsymbol{g}^t_{\bar{\boldsymbol{\kappa}}}). \tag{7.53}$$

Thus if the response of a simple point with respect to one reference configuration is set down, it can be determined easily with respect to any other reference configuration.

Suppose now $\boldsymbol{P}$ be any proper non-singular tensor such that

$$\tilde{\Psi}_{\boldsymbol{\kappa}}(\boldsymbol{F}^t_{\boldsymbol{\kappa}}, \theta^t_{\boldsymbol{\kappa}}, \boldsymbol{g}^t_{\boldsymbol{\kappa}}) = \tilde{\Psi}_{\boldsymbol{\kappa}}(\boldsymbol{F}^t_{\boldsymbol{\kappa}}\boldsymbol{P}, \theta^t_{\boldsymbol{\kappa}}, \boldsymbol{P}^T\boldsymbol{g}^t_{\boldsymbol{\kappa}}) \tag{7.54}$$

for all histories $\boldsymbol{F}^t_{\boldsymbol{\kappa}}$, $\theta^t_{\boldsymbol{\kappa}}$, and $\boldsymbol{g}^t_{\boldsymbol{\kappa}}$. Then by (7.53), $\boldsymbol{P}$ is the gradient of the deformation from $\boldsymbol{\kappa}$ to another reference configuration $\bar{\boldsymbol{\kappa}}$ that is indistinguishable from it by any property of the functional $\tilde{\Psi}_{\boldsymbol{\kappa}}$. If the functional $\tilde{\Psi}_{\boldsymbol{\kappa}}$ is regarded as expressing the outcome of all possible experiments described by the theory, we may say that $\boldsymbol{\kappa}$ and $\bar{\boldsymbol{\kappa}}$ are *indistinguishable by any experiment.*

The reader will easily supply a simple argument to show that the set of all such $\boldsymbol{P}$ forms a group, the *symmetry group*[1] of the functional $\tilde{\Psi}_{\boldsymbol{\kappa}}$ at X. Since $\tilde{\Psi}_{\boldsymbol{\kappa}}$ stands here for any one of the functionals $\tilde{\boldsymbol{T}}_{\boldsymbol{\kappa}}$, $\tilde{\boldsymbol{h}}_{\boldsymbol{\kappa}}$, $\tilde{\psi}_{\boldsymbol{\kappa}}$, and $\tilde{\eta}_{\boldsymbol{\kappa}}$, there are four such groups. The largest group contained in all four is the *symmetry group* $\mathfrak{G}_{\boldsymbol{\kappa}}$ for $\boldsymbol{\kappa}$ of the material point X whose place in $\boldsymbol{\kappa}$ is

---

[1] In this precise form, the concept of symmetry group is due to NOLL, who called it the *isotropy group. Cf.* p. 213 of the work cited on p. 139.

$\boldsymbol{X}$. The symmetry group $\mathfrak{G}_{\kappa}$ is the group of all static deformation gradients at X relative to $\boldsymbol{\kappa}$ such that no subsequent experiment on the body in that neighborhood can determine whether or not a deformation having that gradient has occurred.

We can also express the symmetry of a simple point X independently of any reference configuration. Writing a typical constitutive relation for X now in the form

$$\Psi(t) = \tilde{\Psi}(\boldsymbol{\chi}_*^t, \theta^t, \boldsymbol{g}^t), \tag{7.55}$$

where the arguments of $\tilde{\Psi}$ are those given in (7.3), we say that a linear automorphism

$$\boldsymbol{A} : \mathscr{B}_X \to \mathscr{B}_X \tag{7.56}$$

is an *abstract material automorphism* of X for the response functional $\tilde{\Psi}$ if

$$\tilde{\Psi}(\boldsymbol{\chi}_*^t \circ \boldsymbol{A},\ \theta^t,\ \boldsymbol{g}^t) = \tilde{\Psi}(\boldsymbol{\chi}_*^t, \theta^t,\ \boldsymbol{g}^t) \tag{7.57}$$

for all $\boldsymbol{\chi}_*^t$, θ^t, and $\boldsymbol{g}^t$. The reader will verify easily that this condition is independent of the frame of reference $\boldsymbol{\varphi}$. The collection of all such automorphisms forms a subgroup of the general linear group $\mathfrak{GL}(\mathscr{B}_X)$, called the *abstract symmetry group of the functional* $\tilde{\Psi}$ at X. Again there is one such group for each of the functionals $\tilde{\boldsymbol{T}}_{\kappa}$, $\tilde{\boldsymbol{h}}_{\kappa}$, $\tilde{\psi}_{\kappa}$, and $\tilde{\eta}_{\kappa}$; the intersection $\mathfrak{G}$ of these four groups may be called the *abstract symmetry group of the simple point X*.

Now suppose that $\boldsymbol{\kappa}$ is a reference configuration as before. Then the functionals $\tilde{\Psi}$ and $\tilde{\Psi}_{\kappa}$ are related by

$$\tilde{\Psi}_{\kappa}(\boldsymbol{F}_{\kappa}^t, \theta_{\kappa}^t, \boldsymbol{g}_{\kappa}^t) = \tilde{\Psi}(\boldsymbol{\chi}_*^t, \theta^t, \boldsymbol{g}^t), \tag{7.58}$$

where the arguments of $\tilde{\Psi}$ and $\tilde{\Psi}_{\kappa}$ are related by (7.4), (7.5), and

$$\boldsymbol{F}_{\kappa}^t = \boldsymbol{\chi}_*^t \circ \boldsymbol{\kappa}_*^{-1}. \tag{7.59}$$

Using these relations, we see that a tensor $\boldsymbol{P}$ satisfies the identity (7.54) for $\tilde{\Psi}_{\kappa}$ if and only if the tensor $\boldsymbol{A}$ given by

$$\boldsymbol{A} = \boldsymbol{\kappa}_*^{-1} \circ \boldsymbol{P} \circ \boldsymbol{\kappa}_* \tag{7.60}$$

satisfies (7.57) for $\tilde{\Psi}$. In particular, the symmetry group $\mathfrak{G}_{\kappa}$ for the functional $\tilde{\Psi}_{\kappa}$ relative to any reference configuration $\boldsymbol{\kappa}$ can be determined from the abstract symmetry group $\mathfrak{G}$ by

$$\mathfrak{G}_{\kappa} = \boldsymbol{\kappa}_* \circ \mathfrak{G} \circ \boldsymbol{\kappa}_*^{-1}. \tag{7.61}$$

From this relation we see that if $\bar{\kappa}$ is a reference configuration obtained from κ by a deformation whose gradient is $\boldsymbol{P}$, *i.e.*, $\boldsymbol{P} = \bar{\kappa}_* \circ \kappa_*^{-1}$, then the symmetry group $\mathfrak{G}_{\bar{\kappa}}$ is related to the symmetry group $\mathfrak{G}_\kappa$ by NOLL's rule:

$$\mathfrak{G}_{\bar{\kappa}} = \boldsymbol{P}\mathfrak{G}_\kappa \boldsymbol{P}^{-1}. \tag{7.62}$$

Particular kinds of material symmetry correspond to particular concepts of physical behavior. Thus the theorist himself, in designing a mathematical model for a particular physical material, will wish to consider only those constitutive functionals that have a particular symmetry group. Formally, we lay down the

RULE OF MATERIAL SYMMETRY[1]. *Material symmetry of a body is specified by the symmetry groups of its material points.*

If $\mathfrak{G}$ contains an element that is not unimodular, the stress can remain unchanged when the body is compressed or expanded in a certain way, and such behavior corresponds to no known physical object. Thus it is reasonable to consider in the theory only those materials whose symmetry groups are subgroups of the proper unimodular group:

$$\mathfrak{G}_\kappa \subset \mathfrak{SL}(3). \tag{7.63}$$

By (7.62), (7.63) implies that $\mathfrak{G}_{\bar{\kappa}} \subset \mathfrak{SL}(3)$ for all $\bar{\kappa}$. There are also mathematical reasons for laying down (7.63) as an assumption[2].

A *solid* material point is one such that for some reference configuration, no non-orthogonal deformation is indiscernible. That is, there exists a κ such that

$$\mathfrak{G}_\kappa \subset \mathfrak{O}, \tag{7.64}$$

where $\mathfrak{O}$ is the orthogonal group. A reference configuration such that (7.64) holds is said to be *undistorted* for the solid in question. For a given solid, only certain particular reference configurations have this property. In terms of the abstract symmetry group $\mathfrak{G}$, the condition (7.64) means that on the tangent space $\mathscr{B}_X$ there exists an inner product *invariant* with

---

[1] The rule of material symmetry derives from work of CAUCHY and the physicists of the last century who studied crystalline optics. Like the rule of equipresence, it is a rule of guidance in setting up theories, not a mathematical axiom.

[2] M. E. GURTIN & W. O. WILLIAMS, "On the inclusion of the complete symmetry group in the unimodular group", *Arch. Rational Mech. Anal.* **23**, 163–172 (1966).

respect to the material automorphisms in $\mathfrak{G}$. Such an inner product is called an *intrinsic inner product* at X. An *isotropic* material particle is one such that for some reference configuration $\boldsymbol{\kappa}$, all rotations are indiscernible:

$$\mathfrak{G}_{\kappa} \supset \mathfrak{S}\mathfrak{O}(3). \tag{7.65}$$

A reference configuration such that (7.65) holds is called *undistorted* for the isotropic material in question. In general, only certain particular reference configurations are undistorted for a given isotropic material, but it is not difficult to prove that for an isotropic solid any reference configuration undistorted as just defined is also undistorted in the earlier sense introduced for solids, whether isotropic or not. Since the rotation group is maximal in the special linear group[1], (7.65) implies that

$$\mathfrak{G}_{\kappa} = \mathfrak{S}\mathfrak{O}(3) \text{ or } \mathfrak{G}_{\kappa} = \mathfrak{S}\mathfrak{L}(3). \tag{7.66}$$

A *fluid crystal* may be defined as a non-solid, and a *fluid* as a fluid crystal that has the same isotropy group with respect to all configurations:

$$\mathfrak{G}_{\kappa} = \mathfrak{G}_{\bar{\kappa}} \tag{7.67}$$

for all $\boldsymbol{\kappa}$ and all $\bar{\boldsymbol{\kappa}}$. Since the special linear group is simple[2], (7.67) is satisfied if and only if $\mathfrak{G}_{\kappa}$ = the trivial group or $\mathfrak{G}_{\kappa}$ = the special linear group. Thus we have the following easy theorems[3]:

1. *The symmetry group of a fluid is the special linear group, for all reference configurations.*

2. *All isotropic materials are fluids or solids; all fluids are isotropic; an isotropic material is a fluid if and only if all configurations are undistorted.*

---

[1] A short elementary proof is given by W. Noll, "Proof of the maximality of the orthogonal group in the unimodular group", *Arch. Rational Mech. Anal.* **18**, 100–102 (1965).

[2] *Cf.* § II.2 of the book by J. Dieudonné, *La geométrie des groupes classiques*, Berlin-Göttingen-Heidelberg, Springer, 1955.

[3] The definition of "isotropic" is due in principle to Cauchy; the definition of "solid" is Noll's, and the definition of "fluid" is Truesdell's, derived from but modifying that of Noll. The existence of "fluid crystals" in the mathematical sense specified here seems first to have been seen by Noll. For a more extensive treatment of material symmetry, see §§ 32–33 bis of NFTM and Lecture 6 of Truesdell's *Elements of Continuum Mechanics*, Springer, 1965, as well as §§ IV. 14–17 of Truesdell's *Introduction à la Mécanique Rationnelle des Milieux Continus*, Paris, Masson, in press.

3. *A fluid crystal is isotropic if and only if it is a fluid.*

4. *The only materials that have the same symmetry groups with respect to all reference configurations are fluids and triclinic solids, the last being the materials that have the trivial symmetry group.*

8. Representations for Constitutive Relations

In the previous section we required a constitutive relation to satisfy the principle of material frame-indifference and the rule of material symmetry. While the condition based upon the principle of material frame-indifference was solved for some major special cases, that based upon the rule of material symmetry was only stated, not solved for any case. In fact, we did not even prove that these two kinds of conditions are consistent. In this section we give an explicit general representation[1] for all constitutive relations subjected to conditions of both kinds. From this general representation we can deduce representation formulae for the constitutive relations defining special classes of materials, such as thermoelastic materials having a given symmetry.

Our analysis begins at the concept of a transformation group. Let $\mathscr{D}$ be a set, and let $\mathfrak{G}$ be a group. Then $\mathfrak{G}$ is called a *transformation group* of $\mathscr{D}$ by *right multiplication* if an operation

$$\boldsymbol{R} : \mathscr{D} \times \mathfrak{G} \to \mathscr{D} \tag{8.1}$$

is assigned, such that the following two conditions are satisfied:

$$\boldsymbol{R}(\boldsymbol{x}, \mathbf{1}_{\mathfrak{G}}) = \boldsymbol{x}, \tag{8.2}$$

and

$$\boldsymbol{R}(\boldsymbol{R}(\boldsymbol{x}, \boldsymbol{G}_1), \boldsymbol{G}_2) = \boldsymbol{R}(\boldsymbol{x}, \boldsymbol{G}_1 \boldsymbol{G}_2) \tag{8.3}$$

for all $\boldsymbol{x} \in \mathscr{D}$ and all $\boldsymbol{G}_1, \boldsymbol{G}_2 \in \mathfrak{G}$. Here $\mathbf{1}_{\mathfrak{G}}$ denotes the identity element of $\mathfrak{G}$, and $\boldsymbol{G}_1 \boldsymbol{G}_2$ denotes the group product of $\boldsymbol{G}_2$ by $\boldsymbol{G}_1$. For brevity, we write

$$\boldsymbol{x}\boldsymbol{G} \equiv \boldsymbol{R}(\boldsymbol{x}, \boldsymbol{G}). \tag{8.4}$$

Now let $\mathscr{D}$ and $\mathscr{W}$ be two given sets, and let $\mathfrak{G}$ be a given group, which

---

[1] C.-C. Wang, "A general representation theorem for constitutive relations", *Arch. Rational Mech. Anal.*, **32**, 1–25 (1969).

is a transformation group on $\mathscr{D}$ and $\mathscr{W}$ by right multiplication. Then we say that a mapping

$$\Psi : \mathscr{D} \to \mathscr{W} \tag{8.5}$$

obeys the *generalized symmetry condition* relative to $\mathfrak{G}$ if it satisfies the identity

$$\Psi(\boldsymbol{x}\boldsymbol{G}) = \Psi(\boldsymbol{x})\boldsymbol{G} \quad \forall \boldsymbol{x} \in \mathscr{D}, \tag{8.6}$$

for all $\boldsymbol{G} \in \mathfrak{G}$.

From known examples, we see that the symmetry condition on most deterministic constitutive relations can be regarded as a special case of the generalized symmetry condition. Specifically, for the constitutive relation (7.50), $\mathscr{D}$ is the function space consisting of all triples $(\boldsymbol{F}^t_{\boldsymbol{\kappa}}, \theta^t_{\boldsymbol{\kappa}}, \boldsymbol{g}^t_{\boldsymbol{\kappa}})$, where $\boldsymbol{F}^t_{\boldsymbol{\kappa}}$, $\theta^t_{\boldsymbol{\kappa}}$, and $\boldsymbol{g}^t_{\boldsymbol{\kappa}}$ have their usual meanings as explained before, $\mathscr{W}$ is a space of scalars, vectors, or tensors, and $\mathfrak{G}$ is the relative symmetry group $\mathfrak{G}_{\boldsymbol{\kappa}}$. The operations of an element $\boldsymbol{P} \in \mathfrak{G}_{\boldsymbol{\kappa}}$ on $\mathscr{D}$ and $\mathscr{W}$, of course, are defined by

$$\boldsymbol{R}((\boldsymbol{F}^t_{\boldsymbol{\kappa}}, \theta^t_{\boldsymbol{\kappa}}, \boldsymbol{g}^t_{\boldsymbol{\kappa}}), \boldsymbol{P}) \equiv (\boldsymbol{F}^t_{\boldsymbol{\kappa}}\boldsymbol{P}, \theta^t_{\boldsymbol{\kappa}}, \boldsymbol{P}^T\boldsymbol{g}^t_{\boldsymbol{\kappa}}), \tag{8.7}$$

and

$$\boldsymbol{R}(\Psi_{\boldsymbol{\kappa}}, \boldsymbol{P}) \equiv \Psi_{\boldsymbol{\kappa}} \tag{8.8}$$

for all $(\boldsymbol{F}^t_{\boldsymbol{\kappa}}, \theta^t_{\boldsymbol{\kappa}}, \boldsymbol{g}^t_{\boldsymbol{\kappa}}) \in \mathscr{D}$ and all $\Psi_{\boldsymbol{\kappa}} \in \mathscr{W}$. Clearly, these operations satisfy the conditions (8.2) and (8.3). Further, relative to these operations, the symmetry condition (7.54) becomes a special case of (8.6). In a similar way, we can visualize the abstract symmetry condition (7.57) as a special case of (8.6).

Now we proceed to solve the condition (8.6) in general. To this end, we introduce the following equivalence relation on $\mathscr{D}$:

$$\boldsymbol{x} \sim \boldsymbol{y} \Leftrightarrow \boldsymbol{y} = \boldsymbol{x}\boldsymbol{G} \text{ for some } \boldsymbol{G} \in \mathfrak{G}. \tag{8.9}$$

An equivalence class $\mathscr{X}$ with respect to this equivalence relation is called a *left coset* of $\mathfrak{G}$ in $\mathscr{D}$. Clearly, if $\boldsymbol{x} \in \mathscr{X}$, then in the set-theoretical sense

$$\mathscr{X} = \boldsymbol{x}\mathfrak{G}. \tag{8.10}$$

Let $\boldsymbol{x}$ be a fixed point in $\mathscr{D}$. Then we define a subgroup $\mathfrak{G}_{\boldsymbol{x}}$ in $\mathfrak{G}$ for $\boldsymbol{x}$ by the condition

$$\boldsymbol{G} \in \mathfrak{G}_{\boldsymbol{x}} \Leftrightarrow \boldsymbol{G}\boldsymbol{x} = \boldsymbol{x}. \tag{8.11}$$

We call $\mathfrak{G}_{\boldsymbol{x}}$ the *invariant group* of $\boldsymbol{x}$ in $\mathfrak{G}$. For the special case given by (8.7), the invariant group is a trivial group, consisting in the identity element alone, for all points in the domain $\mathscr{D}$. In general, however, the invariant group need not be a trivial group.

PROPOSITION. *The invariant group satisfies the following transformation law over each left coset,*

$$\mathfrak{G}_{xG} = G^{-1}\mathfrak{G}_x G \tag{8.12}$$

for all $G \in \mathfrak{G}$. *In particular, if* $\mathfrak{G}_x$ *is the trivial group, then* $\mathfrak{G}_y$ *is also the trivial group for all* $y \sim x$.

The proof is obvious.

Now let G be an arbitrary element of the invariant group $\mathfrak{G}_x$ for x. Then the condition (8.6) implies that the value $\Psi(x)$ must be invariant also under G. Thus we introduce a subset $\mathscr{W}_x$ in $\mathscr{W}$ for each fixed point $x \in \mathscr{D}$ by the condition

$$w \in \mathscr{W}_x \Leftrightarrow wG = w \quad \forall G \in \mathfrak{G}_x. \tag{8.13}$$

We call $\mathscr{W}_x$ the *admissible range* of x in $\mathscr{W}$. If the invariant group $\mathfrak{G}_x$ for x happens to be the trivial group, then from (8.2) the admissible range $\mathscr{W}_x$ coincides with the whole set $\mathscr{W}$. In general, however, the admissible range $\mathscr{W}_x$ need not even be a non-empty set. Of course, if $\mathscr{W}_x = \emptyset$, then the condition (8.6) has no solution.

PROPOSITION. *The admissible range satisfies the following transformation law over each left coset,*

$$\mathscr{W}_{xG} = \mathscr{W}_x G \tag{8.14}$$

for all $G \in \mathfrak{G}$. *In particular,* $\mathscr{W}_x$ *is empty or non-empty if and only if* $\mathscr{W}_y$ *is empty or non-empty, respectively, for all* $y \sim x$. *Thus the condition of having an empty or non-empty admissible range is a property of a left coset.*

The proof is again obvious.

Now we are ready to give a general representation for the condition (8.6).

PROPOSITION. *A function* Ψ *in* (8.5) *satisfies the generalized symmetry condition* (8.6) *relative to a given symmetry group* $\mathfrak{G}$ *if and only if it has the form*

$$\Psi(x) = \Psi(x_{\mathscr{X}})G, \tag{8.15}$$

where $\mathbf{x}_{\mathscr{X}}$ *is a fixed representative element for the left coset* $\mathscr{X}$ *containing* $\mathbf{x}$*, so that*

$$\mathbf{x} = \mathbf{x}_{\mathscr{X}} \mathbf{G}, \tag{8.16}$$

and where the value $\Psi(\mathbf{x}_{\mathscr{X}})$ *satisfies the condition*

$$\Psi(\mathbf{x}_{\mathscr{X}}) \in \mathscr{W}\mathbf{x}_{\mathscr{X}} \tag{8.17}$$

for all $\mathscr{X}$. *In particular, if* $\mathscr{W}\mathbf{x}_{\mathscr{X}} = \emptyset$, *then* (8.6) *has no solution on* $\mathscr{X}$.

Proof. Necessity is obvious, since both (8.15) and (8.17) are direct consequences of (8.6).

To prove sufficiency, notice first that the condition (8.17) ensures that Ψ be well defined by (8.15). Further, for any fixed element $\hat{\mathbf{G}} \in \mathfrak{G}$, we have

$$\mathbf{x}\hat{\mathbf{G}} = \mathbf{x}_{\mathscr{X}} \mathbf{G}\hat{\mathbf{G}}, \tag{8.18}$$

so that from (8.15) the value $\Psi(\mathbf{x}\hat{\mathbf{G}})$ is given by

$$\Psi(\mathbf{x}\hat{\mathbf{G}}) = \Psi(\mathbf{x}_{\mathscr{X}})\mathbf{G}\hat{\mathbf{G}}. \tag{8.19}$$

Comparing this value with the value $\Psi(\mathbf{x})$ given by (8.15), we see that (8.6) is satisfied. Q.E.D.

Following exactly the same idea, we next generalize the condition based on the principle of material frame-indifference. For this purpose, we introduce the concept of a transformation group by left multiplication. Let $\mathscr{D}$ be an a set, and let $\mathfrak{Q}$ be a group, as before. Then $\mathfrak{Q}$ is called a *transformation group* of $\mathscr{D}$ by *left multiplication* if an operation

$$\mathbf{L} : \mathfrak{Q} \times \mathscr{D} \to \mathscr{D} \tag{8.20}$$

is assigned, such that the following two conditions are satisfied:

$$\mathbf{L}(\mathbf{1}_{\mathfrak{Q}}, \mathbf{x}) = \mathbf{x}, \tag{8.21}$$

and

$$\mathbf{L}(\mathbf{Q}_1, \mathbf{L}(\mathbf{Q}_2, \mathbf{x})) = \mathbf{L}(\mathbf{Q}_1 \mathbf{Q}_2, \mathbf{x}), \tag{8.22}$$

for all $\mathbf{x} \in \mathscr{D}$ and all $\mathbf{Q}_1, \mathbf{Q}_2 \in \mathfrak{Q}$. As usual, $\mathbf{1}_{\mathfrak{Q}}$ denotes the identity element of $\mathfrak{Q}$, and $\mathbf{Q}_1 \mathbf{Q}_2$ denotes the group product of $\mathbf{Q}_2$ by $\mathbf{Q}_1$ in $\mathfrak{Q}$. For brevity we write

$$\mathbf{Q}\mathbf{x} = \mathbf{L}(\mathbf{Q}, \mathbf{x}). \tag{8.23}$$

Now let Ψ be a mapping given by (8.5), and let $\mathfrak{Q}$ be a transformation group of $\mathscr{D}$ and $\mathscr{W}$ by left multiplication. Then we say that Ψ obeys the *generalized condition of frame-indifference* relative to $\mathfrak{Q}$ if it satisfies the identity

$$\Psi(\mathbf{Q}\mathbf{x}) = \mathbf{Q}\Psi(\mathbf{x}) \quad \forall \mathbf{x} \in \mathscr{D}, \tag{8.24}$$

for all $\mathbf{Q} \in \mathfrak{Q}$.

The conditions (8.6) and (8.24), of course, are equivalent; either can be regarded as a special case of the other. To make (8.24) a special case of (8.6), we introduce the following concept: We define the *adjoint group* $\mathfrak{Q}^*$ of the group $\mathfrak{Q}$ as the group made up of the same elements as $\mathfrak{Q}$ but the group product $*$ for $\mathfrak{Q}^*$ is that for $\mathfrak{Q}$ in reversed order, *i.e.*,

$$\mathbf{Q}_1 * \mathbf{Q}_2 \equiv \mathbf{Q}_2 \mathbf{Q}_1 \tag{8.25}$$

for all $\mathbf{Q}_1$, $\mathbf{Q}_2$ in $\mathfrak{Q}$. Using this concept, we obtain immediately the following

PROPOSITION. *Let $\mathfrak{Q}$ be a transformation group of $\mathscr{D}$ by left multiplication. Then $\mathfrak{Q}^*$ is a transformation group of $\mathscr{D}$ by right multiplication relative to the operation*

$$\mathbf{R} : \mathscr{D} \times \mathfrak{Q}^* \to \mathscr{D} \tag{8.26}$$

given by

$$\mathbf{R}(\mathbf{x}, \mathbf{Q}) \equiv \mathbf{L}(\mathbf{Q}, \mathbf{x}). \tag{8.27}$$

From this proposition we see that (8.24) can be regarded as a special case of (8.6) with $\mathfrak{G} = \mathfrak{Q}^*$. Hence we can read off a representation for (8.24) from (8.15). Specifically, we define a *right coset* $\mathscr{Y}$ of $\mathfrak{Q}$ in $\mathscr{D}$ as an equivalence class with respect to the equivalence relation

$$\mathbf{x} \approx \mathbf{y} \Leftrightarrow \mathbf{y} = \mathbf{Q}\mathbf{x} \text{ for some } \mathbf{Q} \in \mathfrak{Q}. \tag{8.28}$$

For each fixed element $\mathbf{x} \in \mathscr{D}$ we define the *invariant group* $\mathfrak{Q}_{\mathbf{x}}$ for $\mathbf{x}$ as the subgroup of $\mathfrak{Q}$ given by the condition

$$\mathbf{Q} \in \mathfrak{Q}_{\mathbf{x}} \Leftrightarrow \mathbf{Q}\mathbf{x} = \mathbf{x}. \tag{8.29}$$

Then $\mathfrak{Q}_{\mathbf{x}}$ satisfies the transformation law

$$\mathfrak{Q}_{\mathbf{Q}\mathbf{x}} = \mathbf{Q}\mathfrak{Q}_{\mathbf{x}}\mathbf{Q}^{-1} \tag{8.30}$$

for all $\mathbf{Q} \in \mathfrak{Q}$. Further, we define the *admissible range* $\mathscr{W}_{\mathbf{x}}^*$ for $\mathbf{x}$ as the subset of $\mathscr{W}$ satisfying the condition

$$\mathbf{w} \in \mathscr{W}_{\mathbf{x}}^* \Leftrightarrow \mathbf{Q}\mathbf{w} = \mathbf{w} \quad \forall \mathbf{Q} \in \mathfrak{Q}_{\mathbf{x}}. \tag{8.31}$$

Then $\mathscr{W}^*_x$ obeys the transformation law

$$\mathscr{W}^*_{Qx} = Q\mathscr{W}^*_x. \tag{8.32}$$

Having defined $\mathscr{Y}$, $\mathfrak{Q}_x$, and $\mathscr{W}^*_x$, we can give immediately the following representation theorem for the condition (8.24):

PROPOSITION. *A function* Ψ *in* (8.5) *satisfies the generalized condition of frame-indifference relative to a given transformation group* $\mathfrak{Q}$ *of* $\mathscr{D}$ *and* $\mathscr{W}$ *if and only if it has the form*

$$\Psi(\boldsymbol{x}) = Q\Psi(\boldsymbol{x}_{\mathscr{Y}}), \tag{8.33}$$

where $\boldsymbol{x}_{\mathscr{Y}}$ *is a fixed representative element for the right coset* $\mathscr{Y}$ *containing* $\boldsymbol{x}$, *so that*

$$\boldsymbol{x} = Q\boldsymbol{x}_{\mathscr{Y}}, \tag{8.34}$$

and where the value $\Psi(\boldsymbol{x}_{\mathscr{Y}})$ *satisfies the condition*

$$\Psi(\boldsymbol{x}_{\mathscr{Y}}) \in \mathscr{W}^*_{\boldsymbol{x}_{\mathscr{Y}}} \tag{8.35}$$

for all $\mathscr{Y}$. *In particular, if* $\mathscr{W}^*_{\boldsymbol{x}_{\mathscr{Y}}} = \emptyset$, *then* (8.24) *has no solution on* $\mathscr{Y}$.

The reader will verify easily that for the special case (7.43) the transformation group $\mathfrak{Q}$ is the proper orthogonal group such that

$$L(Q, (\boldsymbol{F}_\kappa, \theta_\kappa, \boldsymbol{g}_\kappa)) \equiv (Q\boldsymbol{F}_\kappa, \theta_\kappa, \boldsymbol{g}_\kappa) \tag{8.36}$$

for all $(\boldsymbol{F}_\kappa, \theta_\kappa, \boldsymbol{g}_\kappa)$ in the domain $\mathscr{D}$ of $\tilde{\boldsymbol{T}}_\kappa$, and

$$L(Q, \boldsymbol{T}_\kappa) \equiv Q\boldsymbol{T}_\kappa \tag{8.37}$$

for all $\boldsymbol{T}_\kappa$ in the range of $\tilde{\boldsymbol{T}}_\kappa$. Further, for this special case the invariant group is the trivial group for all points in $\mathscr{D}$, so that the restriction (8.35) becomes redundant. From the polar decomposition theorem we can choose the triple $(\boldsymbol{U}_\kappa, \theta_\kappa, \boldsymbol{g}_\kappa)$ as a representative element for the right coset $\mathscr{Y}$ containing $(\boldsymbol{F}_\kappa, \theta_\kappa, \boldsymbol{g}_\kappa)$. Then (8.33) implies that (7.45) is a representation for (7.43).

A representation for a constitutive relation, of course, must be a general solution both for the condition of material symmetry and for the condition of material frame-indifference. Following the generalizations of the two kinds of condition, we now consider a function Ψ in (8.5) subjected to both the conditions (8.6) and (8.24) relative to the two given transformation groups $\mathfrak{G}$ and $\mathfrak{Q}$ as before. Observing the applications of those two conditions to specific constitutive relations, we require also that the actions of $\mathfrak{G}$ on $\mathscr{D}$ and $\mathscr{W}$ commute with the actions of $\mathfrak{Q}$ on $\mathscr{D}$ and $\mathscr{W}$. Hence we write

$$\boldsymbol{QxG} \equiv (\boldsymbol{Qx})\boldsymbol{G} = \boldsymbol{Q}(\boldsymbol{xG}), \tag{8.38}$$

and

$$\boldsymbol{QwG} \equiv (\boldsymbol{Qw})\boldsymbol{G} = \boldsymbol{Q}(\boldsymbol{wG}), \tag{8.39}$$

for all $\boldsymbol{x} \in \mathscr{D}$, $\boldsymbol{w} \in \mathscr{W}$, $\boldsymbol{Q} \in \mathfrak{Q}$, and $\boldsymbol{G} \in \mathfrak{G}$. Further, we say that a function Ψ in (8.5) obeys a *combined condition* of material symmetry and material frame-indifference if it satisfies the identity

$$\Psi(\boldsymbol{QxG}) \equiv \boldsymbol{Q}\Psi(\boldsymbol{x})\boldsymbol{G} \quad \forall \boldsymbol{x} \in \mathscr{D}, \tag{8.40}$$

for all $\boldsymbol{Q} \in \mathfrak{Q}$ and all $\boldsymbol{G} \in \mathfrak{G}$.

PROPOSITION. *A function* Ψ *in* (8.5) *obeys* (8.40) *if and only if it obeys both* (8.6) *and* (8.24).

Now we seek a general representation for Ψ satisfying the condition (8.40). For this purpose we introduce the following concept: Let $\mathfrak{G}_1$ and $\mathfrak{G}_2$ be any two groups. Then the *product group* $\mathfrak{G}_1 \times \mathfrak{G}_2$ is the group consisting in all ordered pairs $(\boldsymbol{G}_1, \boldsymbol{G}_2)$, where $\boldsymbol{G}_1 \in \mathfrak{G}_1$, and $\boldsymbol{G}_2 \in \mathfrak{G}_2$, such that the group product is given by

$$(\boldsymbol{G}_1, \boldsymbol{G}_2)(\bar{\boldsymbol{G}}_1, \bar{\boldsymbol{G}}_2) = (\boldsymbol{G}_1\bar{\boldsymbol{G}}_1, \boldsymbol{G}_2\bar{\boldsymbol{G}}_2). \tag{8.41}$$

PROPOSITION. *Let* $\mathfrak{G}_1$ *and* $\mathfrak{G}_2$ *be two transformation groups of* $\mathscr{D}$ *by right multiplication such that the action of* $\mathfrak{G}_1$ *commutes with the action of* $\mathfrak{G}_2$. *Then* $\mathfrak{G}_1 \times \mathfrak{G}_2$ *is a transformation group of* $\mathscr{D}$ *by right multiplication relative to the operation*

$$\boldsymbol{R}(\boldsymbol{x}, (\boldsymbol{G}_1, \boldsymbol{G}_2)) \equiv (\boldsymbol{xG}_1)\boldsymbol{G}_2 = (\boldsymbol{xG}_2)\boldsymbol{G}_1. \tag{8.42}$$

The proof is obvious.

For the combined condition (8.40), the sets $\mathscr{D}$ and $\mathscr{W}$ are equipped with two transformation groups $\mathfrak{Q}^*$ and $\mathfrak{G}$ by right multiplication. Because by hypothesis, the actions of $\mathfrak{Q}^*$ commute with the actions of $\mathfrak{G}$, we may apply the preceding proposition and therefore regard the product group $\mathfrak{Q}^* \times \mathfrak{G}$ as a transformation group of $\mathscr{D}$ and $\mathscr{W}$ by right multiplication. Then the combined condition (8.40) becomes a special case of the condition (8.6) relative to $\mathfrak{Q}^* \times \mathfrak{G}$, so that we can read off a representation for Ψ satisfying (8.40) from the representation (8.15).

To make this idea precise, we introduce as before an equivalence relation $\stackrel{\approx}{\sim}$ on $\mathscr{D}$:

$$\boldsymbol{x} \stackrel{\approx}{\sim} \boldsymbol{y} \Leftrightarrow \boldsymbol{y} = \boldsymbol{QxG} \text{ for some } \boldsymbol{Q} \in \mathfrak{Q} \text{ and } \boldsymbol{G} \in \mathfrak{G}. \tag{8.43}$$

An equivalence class $\mathscr{L}$ with respect to this equivalence relation is called a *combined coset* of $\mathfrak{Q}$ and $\mathfrak{G}$ in $\mathscr{D}$. Since in using the general representation (8.15) it is important to know precisely the character of each combined coset $\mathscr{L}$, we give the following

PROPOSITION[1]. *A combined coset $\mathscr{L}$ of $\mathfrak{Q}$ and $\mathfrak{G}$ in $\mathscr{D}$ is an equivalence class of left cosets $\mathscr{X}$ of $\mathfrak{G}$ in $\mathscr{D}$ relative to the equivalence relation*

$$\mathscr{X} \approx \overline{\mathscr{X}} \Leftrightarrow \overline{\mathscr{X}} = \boldsymbol{Q}\mathscr{X} \text{ for some } \boldsymbol{Q} \in \mathfrak{Q}, \tag{8.44}$$

or, equivalently, a combined coset is an equivalence class of right cosets $\mathscr{Y}$ of $\mathfrak{Q}$ in $\mathscr{D}$ relative to the equivalence relation

$$\mathscr{Y} \sim \overline{\mathscr{Y}} \Leftrightarrow \overline{\mathscr{Y}} = \mathscr{Y}\boldsymbol{G} \text{ for some } \boldsymbol{G} \in \mathfrak{G}. \tag{8.45}$$

Thus we can represent each $\mathscr{L}$ schematically by the following sketch:

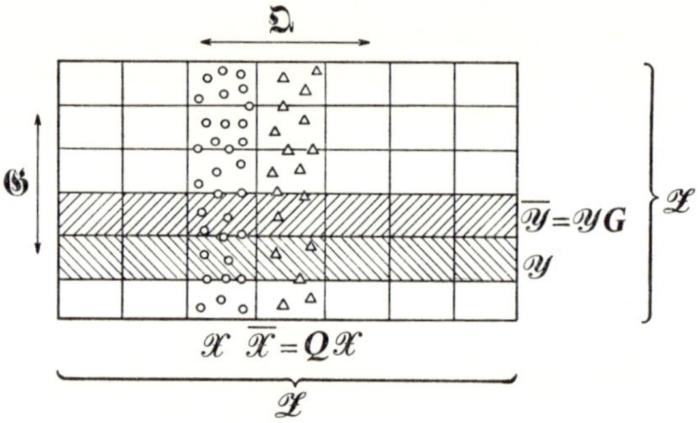

(8.46)

From the preceding proposition we see that each combined coset $\mathscr{L}$ is made up of generic units $\mathscr{K}$ of the form

$$\mathscr{K} = \mathscr{X} \cap \mathscr{Y}, \tag{8.47}$$

where $\mathscr{X}$ and $\mathscr{Y}$ are contained in $\mathscr{L}$. This condition means that $\boldsymbol{x}$ and $\boldsymbol{y}$ belong to the same generic unit $\mathscr{K}$ if and only if they belong to the

---

[1] C.–C. WANG, p. 11 of the work cited on p. 155.

same left coset $\mathscr{X}$ and the same right coset $\mathscr{Y}$, *i.e.*,

$$\mathbf{x}, \mathbf{y} \in \mathscr{K} \Leftrightarrow \mathbf{x} \sim \mathbf{y} \text{ and } \mathbf{x} \approx \mathbf{y}, \tag{8.48}$$

where the equivalence relations on the right-hand side are as defined by (8.9) and (8.28).

Next we define as before the notion of invariant group $\mathfrak{D}_{\mathbf{x}}$ for any fixed point $\mathbf{x}$

$$(\mathbf{Q}, \mathbf{G}) \in \mathfrak{D}_{\mathbf{x}} \Leftrightarrow \mathbf{x} = \mathbf{Q}\mathbf{x}\mathbf{G}. \tag{8.49}$$

Then $\mathfrak{D}_{\mathbf{x}}$ is a subgroup of the product group $\mathfrak{Q}^* \times \mathfrak{G}$; however, in general $\mathfrak{D}_{\mathbf{x}}$ is not a product group itself. On the product group $\mathfrak{Q}^* \times \mathfrak{G}$ we can define the *projections* of any $(\mathbf{Q}, \mathbf{G})$ in $\mathfrak{Q}^*$ and $\mathfrak{G}$ to be the elements $\mathbf{Q}$ and $\mathbf{G}$, respectively. We denote the $\mathfrak{Q}^*$-projection of $\mathfrak{D}_{\mathbf{x}}$ by $\hat{\mathfrak{Q}}_{\mathbf{x}}^*$ and the $\mathfrak{G}$-projection of $\mathfrak{D}_{\mathbf{x}}$ by $\hat{\mathfrak{G}}_{\mathbf{x}}$. Then $\hat{\mathfrak{G}}_{\mathbf{x}}^*$ and $\hat{\mathfrak{Q}}_{\mathbf{x}}^*$ are subgroups of $\mathfrak{G}$ and $\mathfrak{Q}$, and $\mathfrak{D}_{\mathbf{x}}$ is a subgroup of $\hat{\mathfrak{Q}}_{\mathbf{x}}^* \times \hat{\mathfrak{G}}_{\mathbf{x}}$. $\mathfrak{D}_{\mathbf{x}}$ is a *diagonal subgroup* of $\hat{\mathfrak{Q}}_{\mathbf{x}}^* \times \hat{\mathfrak{G}}_{\mathbf{x}}$, since the projections of $\mathfrak{D}_{\mathbf{x}}$ are onto $\hat{\mathfrak{Q}}_{\mathbf{x}}^*$ and $\hat{\mathfrak{G}}_{\mathbf{x}}$.

PROPOSITION. *The projections $\hat{\mathfrak{Q}}_{\mathbf{x}}$ and $\hat{\mathfrak{G}}_{\mathbf{x}}$ satisfy the following transformation laws on $\mathscr{L}$:*

$$\hat{\mathfrak{Q}}_{\mathbf{Q}\mathbf{x}\mathbf{G}} = \mathbf{Q}\hat{\mathfrak{Q}}_{\mathbf{x}}\mathbf{Q}^{-1}, \quad \hat{\mathfrak{G}}_{\mathbf{Q}\mathbf{x}\mathbf{G}} = \mathbf{G}^{-1}\hat{\mathfrak{G}}_{\mathbf{x}}\mathbf{G}. \tag{8.50}$$

In particular, $\hat{\mathfrak{Q}}_{\mathbf{x}}$ depends only on the left coset $\mathscr{X} \ni \mathbf{x}$, and $\hat{\mathfrak{G}}_{\mathbf{x}}$ depends only on the right coset $\mathscr{Y} \ni \mathbf{x}$. Consequently $\hat{\mathfrak{Q}}_{\mathbf{x}}$ and $\hat{\mathfrak{G}}_{\mathbf{x}}$ are invariant on each generic unit $\mathscr{K}$ in $\mathscr{L}$ (cf. (8.46)). *The diagonal subgroup $\mathfrak{D}_{\mathbf{x}}$, however, is not fixed on $\mathscr{K}$; rather, it satisfies the transformation law*

$$(\bar{\mathbf{Q}}, \bar{\mathbf{G}}) \in \mathfrak{D}_{\mathbf{Q}\mathbf{x}\mathbf{G}} \Leftrightarrow (\mathbf{Q}^{-1}\bar{\mathbf{Q}}\mathbf{Q},\ \mathbf{G}\bar{\mathbf{G}}\mathbf{G}^{-1}) \in \mathfrak{D}_{\mathbf{x}}. \tag{8.51}$$

PROPOSITION. *The invariant groups $\mathfrak{G}_{\mathbf{x}}$ and $\mathfrak{Q}_{\mathbf{x}}$ given by* (8.11) *and* (8.29) *are the maximal "vertical" and "horizontal" subgroups of the diagonal group $\mathfrak{D}_{\mathbf{x}}$ in the product group $\hat{\mathfrak{Q}}_{\mathbf{x}}^* \times \hat{\mathfrak{G}}_{\mathbf{x}}$.*

Next we introduce the admissible range $\mathscr{W}_{\mathbf{x}}$ for each fixed $\mathbf{x} \in \mathscr{D}$ as before by the condition

$$\mathbf{w} \in \hat{\mathscr{W}}_{\mathbf{x}} \Leftrightarrow \mathbf{Q}\mathbf{w}\mathbf{G} = \mathbf{w} \quad \forall (\mathbf{Q}, \mathbf{G}) \in \mathfrak{D}_{\mathbf{x}}. \tag{8.52}$$

PROPOSITION. *$\hat{\mathscr{W}}_{\mathbf{x}}$ is contained in the intersection $\mathscr{W}_{\mathbf{x}} \cap \mathscr{W}_{\mathbf{x}}^*$, where $\mathscr{W}_{\mathbf{x}}$ and $\mathscr{W}_{\mathbf{x}}^*$ are given by* (8.13) *and* (8.31). *Further, $\hat{\mathscr{W}}_{\mathbf{x}}$ satisfies the transformation law*

$$\hat{\mathscr{W}}_{\mathbf{Q}\mathbf{x}\mathbf{G}} = \mathbf{Q}\hat{\mathscr{W}}_{\mathbf{x}}\mathbf{G} \tag{8.53}$$

for all $\boldsymbol{Q} \in \mathfrak{Q}$ *and* $\mathbf{G} \in \mathfrak{G}$. *In particular, if* $\hat{\mathscr{W}}_{\mathbf{x}} = \emptyset$ *for some* $\mathbf{x} \in \mathscr{Z}$, *then* $\hat{\mathscr{W}}_{\bar{\mathbf{x}}} = \emptyset$ *for all* $\bar{\mathbf{x}} \in \mathscr{Z}$.

Now we are ready to state the general representation theorem, the proof of which is a consequence of (8.15) and (8.16).

Representation Theorem.[1] *A function* Ψ *of the form* (8.5) *satisfies the conditions* (8.6) *and* (8.24) *relative to the given transformation groups* $\mathfrak{G}$ *and* $\mathfrak{Q}$ *if and only if it is of the form*

$$\Psi(\boldsymbol{x}) = \boldsymbol{Q}\Psi(\boldsymbol{x}_{\mathscr{Z}})\boldsymbol{G}, \tag{8.54}$$

where $\mathbf{x}_{\mathscr{Z}}$ *is a representative element for the combined coset* $\mathscr{Z}$ *containing* $\boldsymbol{x}$, *so that*

$$\boldsymbol{x} = \boldsymbol{Q}\boldsymbol{x}_{\mathscr{Z}}\boldsymbol{G}, \tag{8.55}$$

and where $\Psi(\boldsymbol{x}_{\mathscr{Z}})$ *is a function of the representative element only such that*

$$\Psi(\boldsymbol{x}_{\mathscr{Z}}) \in \hat{\mathscr{W}}_{\boldsymbol{x}_{\mathscr{Z}}} \tag{8.56}$$

for all $\mathscr{Z}$.

In the next chapter we shall make use of this general theorem to obtain representations for the response functions of elastic materials and the stored-energy functions of hyperelastic materials.

It should be noted that the representation (8.54) has a very simple meaning: $\Psi(\boldsymbol{x}_{\mathscr{Z}})$ is an arbitrary $\hat{\mathscr{W}}_{\boldsymbol{x}_{\mathscr{Z}}}$-valued function defined for the representative elements of the various combined cosets of $\mathfrak{Q}^* \times \mathfrak{G}$ in $\mathscr{D}$, and then that function is extended to the whole domain $\mathscr{D}$ by the combined condition (8.40).

---

[1] C.–C. Wang, p. 13 of the work cited on p. 155.

CHAPTER III

Elastic and Hyperelastic Points

1. The Constitutive Equation of an Elastic Point

We select once and for all a frame of reference φ on the event world $\mathscr{E}$, so that $\mathscr{E}$ is represented in a definite way by the product manifold $\mathscr{R}^4 = \mathscr{R}^3 \times \mathscr{R}$ (*cf.* § II.1). In the preceding chapter we denoted the absolute space and the absolute translation space of φ by $\mathscr{E}_\varphi$ and $\mathscr{V}_\varphi$, respectively. Since φ is now fixed, we may suppress the subscript φ in the notation and write those spaces simply as $\mathscr{E}$ and $\mathscr{V}$, which we shall refer to sometimes as the *physical space* and the *physical translation space*, respectively. The letters $\mathfrak{L}$, $\mathfrak{U}$, and $\mathfrak{O}$ will denote groups over $\mathscr{V}$: the proper general linear group and its special linear and rotational subgroups, respectively. Also, whenever we use a co-ordinate system, it will be the absolute space co-ordinate system (x^i) and the time t associated with φ; moreover, we shall suppress the term "absolute", used before in connection with the fixed frame φ.

Now let $\mathscr{B}$ be a body consisting in body-points $X, Y, \ldots$. Then an (absolute) configuration of $\mathscr{B}$ is an orientation-preserving diffeomorphism of $\mathscr{B}$ into $\mathscr{E}$, say

$$\chi : \mathscr{B} \to \mathscr{E}. \tag{1.1}$$

Similarly, an (absolute) local configuration of a body-point X is an orientation-preserving isomorphism of the tangent space $\mathscr{B}_X$ and $\mathscr{V}$, say

$$\mu_X : \mathscr{B}_X \to \mathscr{V}. \tag{1.2}$$

Of course, the induced linear map χ_{*X} is a local configuration of X. Indeed, for a given X, any local configuration μ_X can be realized as an induced local configuration of some configuration χ, *viz*

$$\mu_X = \chi_{*X}. \tag{1.3}$$

A given field of local configurations on $\mathscr{B}$, however, is not an induced field unless it satisfies certain compatibility conditions.

As explained in § II.7, the response of $\mathscr{B}$ is characterized by a set of constitutive relations. If for every configuration χ of $\mathscr{B}$ the stress tensor $\boldsymbol{T}$ at $\chi(X)$ is uniquely determined by the induced local configuration χ_{*X} at X:

$$\boldsymbol{T} = \boldsymbol{G}(\chi_{*X}, X), \tag{1.4}$$

then X is called an *elastic point* of $\mathscr{B}$. The function $\boldsymbol{G}$ is called the *response function*, and the relation (1.4) itself is called the *constitutive equation* of X. Naturally, if all body-points of $\mathscr{B}$ are elastic, then $\mathscr{B}$ is called an *elastic body*. In this chapter we consider the theory of elastic points; the rest of the book concerns the theory of elastic bodies.

In order that the balance of linear momentum imply the balance of rotational momentum, it is necessary and sufficient that the stress tensor $\boldsymbol{T}$ be symmetric, as is asserted by (II.6.22). That such shall be the case in elasticity, it is necessary and sufficient that *the value of* **G**, *at every argument, be a symmetric tensor.* We assume henceforth that $\mathfrak{G}$ is of this kind.

The constitutive equation (1.4) must satisfy the principle of material frame-indifference (§ II.7). Since (1.4) is a special case of the constitutive relations (II. 7.8)$_1$ for a thermo-elastic body-point, we can read off the restriction on (1.4) from (II. 7.43). We leave it to the reader to verify that that relation when applied to the function $\boldsymbol{G}$ is equivalent to the identity

$$\boldsymbol{G}(\boldsymbol{Q}\boldsymbol{\mu}_X, X) = \boldsymbol{Q}\boldsymbol{G}(\boldsymbol{\mu}_X, X)\boldsymbol{Q}^T \tag{1.5}$$

for all $\boldsymbol{\mu}_X$ and all rotations $\boldsymbol{Q}$ over $\mathscr{V}$. Alternatively, he may obtain (1.5) by a simple direct argument employing the concept of frame-indifference.

Now let $\boldsymbol{v}_X$ be a fixed local configuration of X. Then we can factor any local configuration $\boldsymbol{\mu}_X$ through $\boldsymbol{v}_X$ as follows:

$$\mathscr{B}_X \xrightarrow{\boldsymbol{v}_X} \mathscr{V} \xrightarrow{\boldsymbol{F}} \mathscr{V}, \qquad \boldsymbol{\mu}_X : \mathscr{B}_X \to \mathscr{V} \tag{1.6}$$

so that

$$\boldsymbol{\mu}_X = \boldsymbol{F} \circ \boldsymbol{v}_X, \tag{1.7}$$

where $\boldsymbol{F}$ is an isomorphism of $\mathscr{V}$ given by

$$\boldsymbol{F} = \boldsymbol{\mu}_X \circ \boldsymbol{\nu}_X^{-1}. \tag{1.8}$$

We call $\boldsymbol{\nu}_X$ a *local reference configuration* of X, and $\boldsymbol{F}$ the *deformation gradient* from $\boldsymbol{\nu}_X$ to $\boldsymbol{\mu}_X$ (*cf.* (II.1.24)). Since $\boldsymbol{F}$ is uniquely determined by $\boldsymbol{\mu}_X$ through (1.8), and conversely $\boldsymbol{\mu}_X$ is uniquely determined by $\boldsymbol{F}$ through (1.7), the constitutive equation (1.4) can be written also in the form

$$\boldsymbol{T} = \boldsymbol{H}(\boldsymbol{F}), \tag{1.9}$$

where the dependence of $\boldsymbol{H}$ on X and on the local reference configuration $\boldsymbol{\nu}_X$ has been suppressed in the notation, and where $\boldsymbol{H}(\boldsymbol{F})^T = \boldsymbol{H}(\boldsymbol{F})$. We call $\boldsymbol{H}$ the *response function of X relative to* $\boldsymbol{\nu}_X$.

The constitutive equation (1.9) is a special case of $(\text{II}.7.8)_1$ and hence also of $(\text{II}.7.6)_1$. In this sense the elastic point is a special case of the thermo-elastic point, which in turn is a special simple point. In another sense, however, the elastic material does not fall within the theories of Chapter II, since no counterparts of the further constitutive relations, namely, $(\text{II}.7.8)_{2,3,4}$ and $(\text{II}.7.6)_{2,3,4}$, are assumed. The theory of elasticity, with some exceptions which we shall remark later, is a purely mechanical one. Such restrictions on the function $\tilde{\boldsymbol{T}}_{\boldsymbol{\kappa}}$ in $(\text{II}.7.8)_1$ as follow from the use of principles concerning the *other* constitutive relations, namely, $(\text{II}.7.8)_{2,3,4}$, do not necessarily hold in elasticity. In particular, the theorem of the thermodynamic potential, expressed by (II.7.18), need not apply to elasticity, since in elasticity no free-energy function $\psi_{\boldsymbol{\kappa}}$ need be introduced, and the principle of universal dissipation need not be involved.

The physical meaning of the deformation gradient $\boldsymbol{F}$ can be visualized in the following way. As suggested by the general remark (1.3), suppose that the local reference configuration $\boldsymbol{\nu}_X$ is the induced local configuration of a reference configuration $\boldsymbol{\kappa}$, *viz*,

$$\boldsymbol{\kappa} : \mathscr{B} \to \mathscr{E}, \tag{1.10}$$

and

$$\boldsymbol{\kappa}_{*X} = \boldsymbol{\nu}_X \tag{1.11}$$

(*cf.* (II.1.12)). Then the composite map $\boldsymbol{\chi} \circ \boldsymbol{\kappa}^{-1}$ is the deformation of $\mathscr{B}$ from $\boldsymbol{\kappa}$ to $\boldsymbol{\chi}$:

$$\boldsymbol{\chi} \circ \boldsymbol{\kappa}^{-1} : \boldsymbol{\kappa}(\mathscr{B}) \to \boldsymbol{\chi}(\mathscr{B}). \tag{1.12}$$

By the chain rule, we have

$$(\boldsymbol{\chi} \circ \boldsymbol{\kappa}^{-1})_* = \boldsymbol{\chi}_* \circ \boldsymbol{\kappa}_*^{-1}. \tag{1.13}$$

In particular, at the place $\boldsymbol{X} = \boldsymbol{\kappa}(X)$

$$(\boldsymbol{\chi} \circ \boldsymbol{\kappa}^{-1})_{*\boldsymbol{X}} = \boldsymbol{\chi}_{*X} \circ \boldsymbol{\kappa}_{*X}^{-1} = \boldsymbol{\mu}_X \circ \boldsymbol{\nu}_X^{-1} = \boldsymbol{F}. \tag{1.14}$$

Thus $\boldsymbol{F}$ is the gradient of the deformation from $\boldsymbol{\kappa}$ to $\boldsymbol{\chi}$ at $\boldsymbol{\kappa}(X)$.

From (1.8), the response functions $\boldsymbol{H}$ and $\boldsymbol{G}$ are related by

$$\boldsymbol{H}(\boldsymbol{\mu}_X \circ \boldsymbol{\nu}_X^{-1}) = \boldsymbol{G}(\boldsymbol{\mu}_X, X). \tag{1.15}$$

By (1.5), then, $\boldsymbol{H}$ must satisfy the equation

$$\boldsymbol{H}(\boldsymbol{QF}) = \boldsymbol{QH}(\boldsymbol{F})\boldsymbol{Q}^T \quad \forall \boldsymbol{Q} \in \mathfrak{O}, \ \forall \boldsymbol{F} \in \mathfrak{L}. \tag{1.16}$$

To obtain a general solution of this equation, we use the method already presented more generally in Part V of § II.7 and § II.8. Let

$$\boldsymbol{F} = \boldsymbol{RU} \tag{1.17}$$

be the polar decomposition of $\boldsymbol{F}$, where $\boldsymbol{R}$ is a rotation and where $\boldsymbol{U}$ is positive-definite and symmetric (*cf.* (I.2.41) or (II.1.32)). Then the condition (1.16) implies that

$$\boldsymbol{H}(\boldsymbol{F}) = \boldsymbol{H}(\boldsymbol{RU}) = \boldsymbol{RH}(\boldsymbol{U})\boldsymbol{R}^T. \tag{1.18}$$

Thus $\boldsymbol{H}(\boldsymbol{F})$ is determined for all arguments as soon as its restriction to positive-definite and symmetric arguments is specified. Conversely, let $\boldsymbol{H}(\boldsymbol{U})$ be a function defined for positive-definite and symmetric tensors only. Then the extension $\boldsymbol{H}(\boldsymbol{F})$ defined by the equation (1.18) for arbitrary non-singular tensors $\boldsymbol{F} = \boldsymbol{RU}$ always satisfies the restriction (1.16). To see this result, consider the polar decomposition of any tensor $\boldsymbol{QF}$:

$$\boldsymbol{QF} = \boldsymbol{Q}(\boldsymbol{RU}) = (\boldsymbol{QR})\boldsymbol{U}. \tag{1.19}$$

From (1.18), we get

$$\boldsymbol{H}(\boldsymbol{QF}) = (\boldsymbol{QR})\boldsymbol{H}(\boldsymbol{U})(\boldsymbol{QR})^T = \boldsymbol{Q}(\boldsymbol{RH}(\boldsymbol{U})\boldsymbol{R}^T)\boldsymbol{Q}^T = \boldsymbol{QH}(\boldsymbol{F})\boldsymbol{Q}^T. \tag{1.20}$$

Thus we have proved the following

PROPOSTION (RICHTER). *In order that the constitutive relation of an elastic point be frame-indifferent, it is necessary and sufficient that the response function* $\boldsymbol{H}(\boldsymbol{F})$ *be of the form* (1.18), *where* $\boldsymbol{H}(\boldsymbol{U})$ *is a function of positive-definite and symmetric tensors, and where the polar decomposition of* $\boldsymbol{F}$ *is* $\boldsymbol{RU}$.

A deformation in which $\boldsymbol{R} = \boldsymbol{1}$ is said to be a *pure strain*. RICHTER's theorem asserts that the restriction of the response function to pure strains determines it uniquely for all deformations. This fact may be presented in more detail as follows. Let $\boldsymbol{F}$ and $\bar{\boldsymbol{F}}$ be two non-singular tensors belonging to the same right coset of the rotation group $\mathfrak{O}$ in the proper general linear group $\mathfrak{L}$ over $\mathscr{V}$:

$$\bar{\boldsymbol{F}}\boldsymbol{F}^{-1} \in \mathfrak{O}, \tag{1.21}$$

say

$$\bar{\boldsymbol{F}} = \boldsymbol{Q}\boldsymbol{F}. \tag{1.22}$$

Then the identity (1.16) requires that

$$\boldsymbol{H}(\bar{\boldsymbol{F}}) = \boldsymbol{Q}\boldsymbol{H}(\boldsymbol{F})\boldsymbol{Q}^T. \tag{1.23}$$

Thus the restriction of $\boldsymbol{H}$ on any right coset of $\mathfrak{O}$ in $\mathfrak{L}$ is completely specified by the value of $\boldsymbol{H}$ at any one tensor belonging to that right coset. From the polar decomposition (1.17), the tensors $\boldsymbol{F}$ and $\boldsymbol{U}$ obviously belong to the same right coset. Further, from the polar decomposition (1.19) two tensors $\boldsymbol{F}$ and $\bar{\boldsymbol{F}}$ belong to the same right coset if and only if their polar decompositions share the same positive-definite and symmetric tensor $\boldsymbol{U}$. Consequently each right coset of $\mathfrak{O}$ in $\mathfrak{L}$ contains one and only one representative positive-definite and symmetric tensor $\boldsymbol{U}$. If the value $\boldsymbol{H}(\boldsymbol{U})$ of $\boldsymbol{H}$ at that representative tensor $\boldsymbol{U}$ is assigned, then the function $\boldsymbol{H}$ is completely specified on the whole right coset through the relation (1.23), or, explicitly, through the relation (1.18). For different right cosets, of course, we may assign the values of $\boldsymbol{H}$ at the representative tensors $\boldsymbol{U}$ independently. Thus the identity (1.16) requires that $\boldsymbol{H}$ be given for all positive-definite and symmetric tensors $\boldsymbol{U}$, which represent the right cosets of $\mathfrak{O}$ in $\mathfrak{L}$.

We define the *right Cauchy-Green tensor* $\boldsymbol{C}$ as follows (*cf.* (II. 1.33)):

$$\boldsymbol{C} \equiv \boldsymbol{U}^2. \tag{1.24}$$

Then $\boldsymbol{C}$ is positive-definite and symmetric; $\boldsymbol{C}$ and $\boldsymbol{U}$ determine each other uniquely; and also

$$\boldsymbol{C} = \boldsymbol{F}^T\boldsymbol{F}. \tag{1.25}$$

Consequently, we can write the constitutive equation (1.9) also in various other forms:

$$\boldsymbol{T} = \boldsymbol{R}\boldsymbol{K}(\boldsymbol{C})\boldsymbol{R}^T, \tag{1.26}$$

or

$$\boldsymbol{T} = \boldsymbol{F}\boldsymbol{J}(\boldsymbol{C})\boldsymbol{F}^T, \tag{1.27}$$

where

$$K(C) = H(C^{\frac{1}{2}}), \tag{1.28}$$

and

$$J(C) = C^{-\frac{1}{2}}H(C^{\frac{1}{2}})C^{-\frac{1}{2}}. \tag{1.29}$$

The values $K(C)$ and $J(C)$ are symmetric tensors for every C.

In general, the constitutive equation for an elastic point need not permit the stress ever to vanish. That is, the equation $H(F) = 0$ need have no solution F at X. If such an F exists, the local configuration ν_0 to which it corresponds is called a *natural state* for the elastic point X. If X has a natural state, we may select it as the reference configuration, so that

$$H(1) = 0. \tag{1.30}$$

In some important special cases no natural state exists, while in others there are such states. If (1.30) is satisfied, it follows from (1.18) that $H(R) = 0$ for all rotations R. That is, *the local configuration obtained by any rotation of a natural state is also a natural state.*

The material points described by the classical linear theory of elasticity are not special cases of elastic points, since the constitutive relations in linear elasticity do not obey the principle of material frame-indifference (§ II.7). Hence, strictly speaking, the linearly elastic point is not a model at all for a material point. Since, however, the linear formula for the stress can be obtained as a mathematical approximation to infinitely many possible true constitutive relations, linear elasticity has a clear and definite if degenerate status in the framework of the exact theory.

To identify the classical linear equation as the linearization of the exact constitutive equation of a certain elastic point, we first assume that point has a natural state ν_0; then use the constitutive equation (1.27); and finally assume that the function J is differentiable near $C = 1$. By the Taylor series expansion at $C = 1$,

$$T = 0 + \mathsf{L}(E) + o(E), \tag{1.31}$$

where

$$E = \tfrac{1}{2}(F + F^T) - 1 \tag{1.32}$$

and where $\mathsf{L}(E)$ is linear in E. In component form,

$$[\mathsf{L}(E)]_{ij} = \mathsf{C}_{ijkl}E^{kl}, \tag{1.33}$$

and

$$\mathsf{C}_{ijkl} = 2\left.\frac{\partial J_{ij}}{\partial C^{kl}}\right|_{C=1}. \tag{1.34}$$

The constants C_{ijkl} are the classical *elasticities*; by (1.34), they satisfy the symmetry conditions

$$\mathsf{C}_{ijkl} = \mathsf{C}_{jikl} = \mathsf{C}_{ijlk}. \tag{1.35}$$

The linear formula

$$T = \mathsf{L}(E) \tag{1.36}$$

without the error term $\boldsymbol{o}(\boldsymbol{E})$ defines the *classical linear theory*. Since this equation is not a special case of (1.27), there is no special case of the exact theory in which the error term $\boldsymbol{o}(\boldsymbol{E})$ vanishes identically. In fact, the linearized constitutive equation (1.36) obeys the principle of material frame-indifference only to within infinitesimal rotations. We cannot, then, expect the classical linear theory to yield any result close to that of the exact theory if the deformation gradient from the natural state involves either a large rotation or a large pure stretch. Whether or not the classical linear theory does indeed yield accurate results for small strain is still a partly open question. What is presently known about the status of solutions of the linearized differential equations of equilibrium with respect to the solutions of the exact equations in cases when the strains and rotations do indeed come out to be small according to these latter, is presented below in Chapter VII.

Unlike the classical linearly elastic material, the *classical elastic fluid* (also known as the *compressible perfect fluid* or *piezotropic fluid*) is based traditionally on the following two constitutive assumptions:

1. The fluid cannot support any shear stress. Thus the stress tensor in any configuration is always a hydrostatic pressure,

$$\boldsymbol{T} = -p\mathbf{1}. \tag{1.37}$$

2. The pressure in any configuration depends only on the density in that configuration

$$p = p(\rho). \tag{1.38}$$

(Almost invariably the function p is assumed to be positive-valued on $(0, \infty)$.)

From these two assumptions we can write the constitutive equation for the classical elastic fluid as

$$\boldsymbol{T} = h(\det \boldsymbol{F})\mathbf{1}, \tag{1.39}$$

or equivalently

$$\boldsymbol{T} = f(\det \boldsymbol{C})\mathbf{1}. \tag{1.40}$$

Since this equation is a special case of (1.26), we see that the classical elastic fluid satisfies the principle of material frame-indifference, as indeed is evident directly.

Not only are the classical constitutive assumptions 1 and 2 formal and unmotivated except as a convenient summary of various mathematical statements about fluids which go back some centuries, but also they do not really cover even all the old concepts of fluidity. For instance, in a shearing motion the classical Newtonian viscous fluid can support shear stress, and hence it does not satisfy Assumption 1. In fact, the elastic fluid is, as its name suggests, a fluid of a special kind. In § II.7, Part VI, we have given the general definition of fluid current in continuum mechanics today, and in the next section we shall prove that the classical elastic fluid results, as, again, its name suggests, when we demand that a fluid be an elastic material in the sense of this section.

Our definition of an elastic point does not apply to body points which are subject to *internal constraints*, such as incompressibility or inextensibility. In general, if the local configuration of a body point X in any motion is required to be a member of a certain proper subset $\mathscr{C}$ of the set of all orientation-preserving isomorphisms between $\mathscr{B}_X$ and $\mathscr{V}$, then X is called a *body-point with internal constraint*, and $\mathscr{C}$ is called the set of *admissible local configurations* of X. For example, if $\mathscr{C}$ is the set of all local configurations $\boldsymbol{\mu}_X$ such that the mass density in $\boldsymbol{\mu}_X$ is a given constant, then X is an *incompressible body-point*. It is natural to think of a constraint as being a constitutive relation and hence to require that it satisfy the principle of material frame-indifference. Thus the set $\mathscr{C}$ must satisfy the following restriction:

$$\boldsymbol{\mu}_X \in \mathscr{C} \Rightarrow \boldsymbol{Q} \circ \boldsymbol{\mu}_X \in \mathscr{C} \quad \forall \boldsymbol{Q} \in \mathfrak{O}. \tag{1.41}$$

Now let $\boldsymbol{\nu}_X$ be a fixed local reference configuration (which may or may not belong to $\mathscr{C}$). Then we can express $\boldsymbol{\mu}_X$ in terms of the deformation gradient $\boldsymbol{F}$, as shown in (1.6). In particular, there exists a subset $\mathscr{C}_{\nu} \subset \mathfrak{L}$ such that

$$\boldsymbol{F} \in \mathscr{C}_{\nu} \Leftrightarrow \boldsymbol{F} \circ \boldsymbol{\nu}_X \in \mathscr{C}. \tag{1.42}$$

Naturally, we call $\mathscr{C}_{\nu}$ the set of all *admissible deformation gradients of X relative to* $\boldsymbol{\nu}_X$. From (1.41), we see that $\mathscr{C}_{\nu}$ must satisfy the following restriction:

$$\boldsymbol{F} \in \mathscr{C}_{\nu} \Leftrightarrow \boldsymbol{QF} \in \mathscr{C}_{\nu} \quad \forall \boldsymbol{Q} \in \mathfrak{O}. \tag{1.43}$$

Thus $\mathscr{C}_{\nu}$ can be regarded as the disjoint union of a set of right cosets of $\mathfrak{O}$ in $\mathfrak{L}$. For the constraint of incompressibility,

$$\mathscr{C}_{\nu} = \{\boldsymbol{F} \mid \det \boldsymbol{F} = k\}, \tag{1.44}$$

where k is a given positive constant, which is equal to 1 if $\boldsymbol{\nu}_X$ is itself an admissible local configuration in $\mathscr{C}$. Clearly the set $\mathscr{C}_{\nu}$ given by (1.44) satisfies the condition (1.43).

To include body-points with internal constraints, we replace our constitutive assumption (1.4) by a more general one due to ERICKSEN & RIVLIN. If the stress tensor in any admissible local configuration $\boldsymbol{\mu}_X$ of X is determined by $\boldsymbol{\mu}_X$ to within a stress $\boldsymbol{N}$ that does no work in any local motion of X compatible with the constraint:

$$\boldsymbol{T} = \boldsymbol{N} + \boldsymbol{G}(\boldsymbol{\mu}_X, X), \tag{1.45}$$

then X is called an *elastic point*, $\boldsymbol{G}$ is called the *response function*, and the relation (1.45) itself is called the *constitutive equation* of X. The function

G, of course, is now defined on the set $\mathscr{C}$ only, and the value $G(\mu_X, X)$ is called the *determinate stress* in μ_X. The principle of material frame-indifference requires that G satisfy (1.5) on the set $\mathscr{C}$.

As before, we can write the constitutive equation relative to a local reference configuration ν_X also:

$$T = N + H(F), \tag{1.46}$$

where H is defined on the set $\mathscr{C}_\nu$ and is called the *response function of X relative to* ν_X. In terms of the deformation gradient F relative to ν_X, the indeterminate stress N must satisfy the condition

$$\operatorname{tr}(N\dot{F}F^{-1}) = 0 \tag{1.47}$$

for any smooth curve $F = F(t)$ in $\mathscr{C}_\nu$. The left-hand side of this equation, of course, is the stress power of N in the local motion characterized by $F = F(t)$. *Cf.* § II.6.

For an incompressible elastic point, the tangent vector $\dot{F}F^{-1}$ of any smooth curve $F = F(t)$ in the set (1.44) is a tensor with trace zero (*cf.* (I.6.75)). From (1.47), if we decompose N into the sum

$$N = \tfrac{1}{3}(\operatorname{tr} N)\mathbf{1} + N_D, \tag{1.48}$$

where $\operatorname{tr} N_D = 0$, then

$$\operatorname{tr}(N_D \dot{F}F^{-1}) = 0. \tag{1.49}$$

But since the trace of N_D is zero, the condition (1.49) is satisfied for all smooth curves $F = F(t)$ in $\mathscr{C}_\nu$ if and only if N_D vanishes. Thus N must be a hydrostatic stress:

$$N = -p\mathbf{1}, \tag{1.50}$$

and the constitutive equation (1.46) becomes

$$T = -p\mathbf{1} + H(F), \tag{1.51}$$

where p is not determined by the deformation. For simplicity, we may choose the local reference configuration ν_X to be an admissible one, and then H is defined on the special linear group $\mathfrak{U}$ over $\mathscr{V}$.

In this book we do not consider any internal constraint other than incompressibility[1].

---

[1] The solution of (1.47) for other constraints is given in § 30 of NFTM.

2. The Symmetry of an Elastic Point

As has been remarked in § II.7, the symmetry of a material point is specified by the symmetry group of the constitutive relation defining that point. To define the symmetry group of the constitutive relation (1.4) for an elastic point X, let an orientation-preserving linear automorphism

$$\boldsymbol{A} : \mathscr{B}_X \to \mathscr{B}_X \tag{2.1}$$

be called an *abstract material automorphism* of X if

$$\boldsymbol{G}(\boldsymbol{\mu}_X \circ \boldsymbol{A}, X) = \boldsymbol{G}(\boldsymbol{\mu}_X, X) \tag{2.2}$$

for all local configurations $\boldsymbol{\mu}_X$. Then the collection of all abstract material automorphisms forms the *abstract symmetry group* $\mathfrak{G}$, which is a subgroup of $\mathfrak{GL}(\mathscr{B}_X)$.

In the physical interpretation, a local configuration $\boldsymbol{\mu}_X$ represents a local experiment at X; the outcome of the experiment is the value $\boldsymbol{G}(\boldsymbol{\mu}_X, X)$. An automorphism $\boldsymbol{A}$ of $\mathscr{B}_X$ transforms any local configuration $\boldsymbol{\mu}_X$ into the local configuration $\boldsymbol{\mu}_X \circ \boldsymbol{A}$. The condition (2.2) requires that an abstract material isomorphism of X be an automorphism $\boldsymbol{A}$ such that the outcome in any local experiment is invariant under the transformation induced by $\boldsymbol{A}$. In other words, that transformation leaves the response of X invariant in all local experiments. The reader will verify easily that $\mathfrak{G}$ is frame-indifferent.

Now let $\boldsymbol{v}$ be a local reference configuration of X, and let (1.9) be the constitutive equation relative to $\boldsymbol{v}$. Then an automorphism $\boldsymbol{P}$ of $\mathscr{V}$ is called a *material automorphism relative to* $\boldsymbol{v}$ if

$$\boldsymbol{H}(\boldsymbol{F}\boldsymbol{P}) = \boldsymbol{H}(\boldsymbol{F}) \tag{2.3}$$

for all $\boldsymbol{F}$ in $\mathfrak{L}$. The collection of all such relative material automorphisms forms the *relative symmetry group* $\mathfrak{G}_{\boldsymbol{v}}$, which is a subgroup of $\mathfrak{L}$. Since $\boldsymbol{H}$ and $\boldsymbol{G}$ are related by (1.15), the relative symmetry group $\mathfrak{G}_{\boldsymbol{v}}$ can be obtained from the abstract symmetry group $\mathfrak{G}$ by (*cf.* (II. 7.61))

$$\mathfrak{G}_{\boldsymbol{v}} = \boldsymbol{v} \circ \mathfrak{G} \circ \boldsymbol{v}^{-1}. \tag{2.4}$$

In particular, if $\bar{\boldsymbol{v}}$ is another local reference configuration, then

$$\mathfrak{G}_{\bar{\boldsymbol{v}}} = \bar{\boldsymbol{v}} \circ \mathfrak{G} \circ \bar{\boldsymbol{v}}^{-1} = \boldsymbol{P}\mathfrak{G}_{\boldsymbol{v}}\boldsymbol{P}^{-1}, \tag{2.5}$$

where $\boldsymbol{P}$ is the deformation gradient from $\boldsymbol{v}$ to $\bar{\boldsymbol{v}}$, *viz*,

$$\boldsymbol{P} = \bar{\boldsymbol{v}} \circ \boldsymbol{v}^{-1}. \tag{2.6}$$

The transformation law $(2.5)_2$ means that the relative symmetry groups are inner-isomorphic within $\mathfrak{L}$.

In the physical interpretation the meaning of the condition (2.3) is similar to that of (2.2). Here $\boldsymbol{F}$ represents a local experiment relative to the reference configuration $\boldsymbol{v}$. Thus the transformation $\boldsymbol{P}$ is a relative material automorphism if it leaves the relative response function $\boldsymbol{H}$ invariant. Another way to visualize the meaning of a relative material automorphism is as follows: Two local reference configurations $\boldsymbol{v}$ and $\bar{\boldsymbol{v}}$ of X are called *isomorphic* if the response functions relative to them are identical, *viz*,

$$\boldsymbol{H}_{\boldsymbol{v}}(\boldsymbol{F}) = \boldsymbol{H}_{\bar{\boldsymbol{v}}}(\boldsymbol{F}) \quad \forall \boldsymbol{F} \in \mathfrak{L}. \tag{2.7}$$

Then the reader will verify easily that $\boldsymbol{P}$ is a material automorphism relative to $\boldsymbol{v}$ if and only if it is the deformation gradient from $\boldsymbol{v}$ to a local reference configuration $\bar{\boldsymbol{v}}$ which is isomorphic to $\boldsymbol{v}$. Thus $\boldsymbol{P}$ is given by (2.6), where $\boldsymbol{v}$ and $\bar{\boldsymbol{v}}$ satisfy (2.7). Clearly if $\boldsymbol{v}$ and $\bar{\boldsymbol{v}}$ are isomorphic, then $\mathfrak{G}_{\boldsymbol{v}}$ and $\mathfrak{G}_{\bar{\boldsymbol{v}}}$ are identical. The converse is false: If $\mathfrak{G}_{\boldsymbol{v}} = \mathfrak{G}_{\bar{\boldsymbol{v}}}$, then $\boldsymbol{v}$ and $\bar{\boldsymbol{v}}$ need not be isomorphic. Indeed, $\mathfrak{G}_{\boldsymbol{v}} = \mathfrak{G}_{\bar{\boldsymbol{v}}}$ whenever $\boldsymbol{v}$ and $\bar{\boldsymbol{v}}$ differ by a uniform dilatation, that is, when $\boldsymbol{P} = c\mathbf{1}$.

A relative response function $\boldsymbol{H}$ having a given symmetry group $\mathfrak{G}_{\boldsymbol{v}}$ can be visualized in the following way (*cf.* § II. 8): Suppose that $\boldsymbol{F}$ and $\bar{\boldsymbol{F}}$ are two tensors belonging to the same left coset of $\mathfrak{G}_{\boldsymbol{v}}$ in $\mathfrak{L}$, *i.e.*,

$$\boldsymbol{F}^{-1}\bar{\boldsymbol{F}} \in \mathfrak{G}_{\boldsymbol{v}}, \tag{2.8}$$

say

$$\bar{\boldsymbol{F}} = \boldsymbol{F}\boldsymbol{P}, \tag{2.9}$$

where $\boldsymbol{P} \in \mathfrak{G}_{\boldsymbol{v}}$. Then the condition (2.3) requires that the values of $\boldsymbol{H}$ at $\boldsymbol{F}$ and at $\bar{\boldsymbol{F}}$ be the same. In other words, $\boldsymbol{H}$ is invariant on each left coset of $\mathfrak{G}$ in $\mathfrak{L}$. Conversely, if $\boldsymbol{H}$ is invariant on each left coset of a subgroup $\mathfrak{H}$ in $\mathfrak{L}$, then the condition (2.3) holds for every member $\boldsymbol{P} \in \mathfrak{H}$, and hence $\mathfrak{H}$ is contained in $\mathfrak{G}_{\boldsymbol{v}}$. Therefore $\mathfrak{G}_{\boldsymbol{v}}$ is the maximal subgroup of $\mathfrak{L}$ such that $\boldsymbol{H}$ is invariant on each left coset of $\mathfrak{G}_{\boldsymbol{v}}$ in $\mathfrak{L}$. From this analysis we see a simple way to describe a general solution of the condition (2.3).

Namely, let us denote a typical left coset of $\mathfrak{G}_{\boldsymbol{v}}$ in $\mathfrak{L}$ by $\mathscr{X}$. Then $\mathfrak{L}$ can be regarded as the disjoint union of the left cosets

$$\mathfrak{L} = \bigcup_{\mathscr{X} \in \mathfrak{L}/\mathfrak{G}_{\mathbf{v}}} \mathscr{X}. \tag{2.10}$$

If $\boldsymbol{F}_{\mathscr{X}} \in \mathscr{X}$, by (2.3) we see that

$$\boldsymbol{H}(\boldsymbol{F}) = \boldsymbol{H}(\boldsymbol{F}_{\mathscr{X}}) \quad \forall \boldsymbol{F} \in \mathscr{X}, \tag{2.11}$$

where $\boldsymbol{F}$ has the decomposition

$$\boldsymbol{F} = \boldsymbol{F}_{\mathscr{X}} \boldsymbol{P} \tag{2.12}$$

for some $\boldsymbol{P} \in \mathfrak{G}_{\mathbf{v}}$. The value $\boldsymbol{H}(\boldsymbol{F}_{\mathscr{X}})$ is arbitrary.

To illustrate the meaning of the solution (2.11), we consider the special case when $\mathfrak{G}_{\mathbf{v}} = \mathfrak{O}$. By the polar decomposition $(\text{II}.1.32)_2$, two tensors $\boldsymbol{F}$ and $\bar{\boldsymbol{F}}$ belong to the same left coset $\mathscr{X}$ of $\mathfrak{O}$ in $\mathfrak{L}$ if and only if they have the same left stretch tensor, say

$$\boldsymbol{F} = \boldsymbol{V}\boldsymbol{R}, \quad \bar{\boldsymbol{F}} = \boldsymbol{V}\bar{\boldsymbol{R}}, \tag{2.13}$$

where $\boldsymbol{R}, \bar{\boldsymbol{R}} \in \mathfrak{O}$. We can choose, then, the tensor $\boldsymbol{V}$ as the representative element for $\mathscr{X}$. Relative to this choice of representative element, the general solution (2.11) reduces to

$$\boldsymbol{H}(\boldsymbol{F}) = \boldsymbol{H}(\boldsymbol{V}), \tag{2.14}$$

or equivalently,

$$\boldsymbol{H}(\boldsymbol{F}) = \bar{\boldsymbol{H}}(\boldsymbol{B}), \tag{2.15}$$

where $\boldsymbol{B}$ is the left Cauchy-Green tensor, and where

$$\bar{\boldsymbol{H}}(\boldsymbol{B}) \equiv \boldsymbol{H}(\boldsymbol{B}^{\frac{1}{2}}). \tag{2.16}$$

$\boldsymbol{H}(\boldsymbol{V})$ and $\bar{\boldsymbol{H}}(\boldsymbol{B})$ are functions of positive-definite and symmetric tensors only.

Since (2.11) delivers no more than a general solution of the condition (2.3) for material symmetry, a response function $\boldsymbol{H}$ determined by it need not satisfy the condition (1.16) for material frame-indifference. A representation formula for the response functions must be a general solution of both (2.3) and (1.16); we shall derive such a representation formula in general in the next section. Here we consider some special cases first:

Case 1. Elastic Fluids

By definition (*cf.* § II.7) the relative symmetry group of a fluid material point is the special linear group $\mathfrak{U}$. Hence the response function $\boldsymbol{H}$ satisfies the conditions

$$\boldsymbol{H}(\boldsymbol{F}) = \boldsymbol{H}(\boldsymbol{FP}) \quad \forall \boldsymbol{P} \in \mathfrak{U}, \tag{2.17}$$

and

$$\boldsymbol{H}(\boldsymbol{QF}) = \boldsymbol{QH}(\boldsymbol{F})\boldsymbol{Q}^T \quad \forall \boldsymbol{Q} \in \mathfrak{O}. \tag{1.16}$$

We wish to determine an explicit general solution for such $\boldsymbol{H}$.

Let $\boldsymbol{F}$ and $\bar{\boldsymbol{F}}$ be any two elements of $\mathfrak{L}$ such that

$$\det \boldsymbol{F} = \det \bar{\boldsymbol{F}}. \tag{2.18}$$

Then we can write

$$\bar{\boldsymbol{F}} = \boldsymbol{FP}, \tag{2.19}$$

where

$$\boldsymbol{P} = \boldsymbol{F}^{-1}\bar{\boldsymbol{F}}. \tag{2.20}$$

From (2.18), we see that $\boldsymbol{P} \in \mathfrak{U}$, so that $\boldsymbol{F}$ and $\bar{\boldsymbol{F}}$ belong to the same left coset $\mathscr{X}$ of $\mathfrak{U}$ in $\mathfrak{L}$. Conversely, if $\boldsymbol{F}$ and $\bar{\boldsymbol{F}}$ are related by (2.19) with $\boldsymbol{P} \in \mathfrak{U}$, then the condition (2.18) must hold. Consequently the value of $\det \boldsymbol{F}$ characterizes the left coset $\mathscr{X} \ni \boldsymbol{F}$ completely. From (2.11), we can express $\boldsymbol{H}$ in the form

$$\boldsymbol{H}(\boldsymbol{F}) = \hat{\boldsymbol{H}}(\det \boldsymbol{F}) = \tilde{\boldsymbol{H}}(\rho), \tag{2.21}$$

where ρ is the mass density in the configuration characterized by $\boldsymbol{F}$, and where $\hat{\boldsymbol{H}}$ and $\tilde{\boldsymbol{H}}$ are functions defined on $(0, \infty)$.

We now show that the function $\tilde{\boldsymbol{H}}$ is necessarily of the form

$$\tilde{\boldsymbol{H}}(\rho) = -p(\rho)\mathbf{1}, \tag{2.22}$$

i.e., for an elastic fluid point the stress in any configuration is a hydrostatic pressure (which may be negative), and this pressure is determined by the density alone. To prove this fact, we notice that

$$\tilde{\boldsymbol{H}}(\rho) = \boldsymbol{H}((\det \boldsymbol{F})^{\frac{1}{3}}\mathbf{1}). \tag{2.23}$$

From (2.17) we have

$$\boldsymbol{H}((\det \boldsymbol{F})^{\frac{1}{3}}\,\mathbf{1}\boldsymbol{Q}) = \boldsymbol{H}((\det \boldsymbol{F})^{\frac{1}{3}}\,\boldsymbol{Q}) = \tilde{\boldsymbol{H}}(\rho) \quad \forall \boldsymbol{Q} \in \mathfrak{O}, \tag{2.24}$$

since $\mathfrak{O}$ is a subgroup of $\mathfrak{U}$. On the other hand, from (1.16) we have

$$\boldsymbol{H}((\det \boldsymbol{F})^{\frac{1}{3}}\,\boldsymbol{Q}\mathbf{1}) = \boldsymbol{H}((\det \boldsymbol{F})^{\frac{1}{3}}\,\boldsymbol{Q}) = \boldsymbol{Q}\tilde{\boldsymbol{H}}(\rho)\boldsymbol{Q}^T. \tag{2.25}$$

Combining the last two equations shows that

$$\tilde{\boldsymbol{H}}(\rho)\boldsymbol{Q} = \boldsymbol{Q}\tilde{\boldsymbol{H}}(\rho) \tag{2.26}$$

for all $\boldsymbol{Q} \in \mathfrak{O}$. By a proposition in § I.2, this condition requires that $\tilde{\boldsymbol{H}}(\rho)$ be of the form (2.22).

Now we claim that

$$\boldsymbol{H}(\boldsymbol{F}) = -p(\rho)\mathbf{1} \tag{2.27}$$

is a representation formula for the response function of an elastic fluid. The results (2.21) and (2.22), already proved, show that (2.27) is a necessary condition. Conversely, (2.27) clearly satisfies the condition (2.17), since (2.21) is a general solution of that condition. Also, (2.27) satisfies the condition (1.16), since for any $\boldsymbol{Q} \in \mathfrak{O}$

$$p(\rho)\mathbf{1} = \boldsymbol{Q}[p(\rho)\mathbf{1}]\boldsymbol{Q}^T. \tag{2.28}$$

Thus the proof is complete.

From the representation formula (2.27) we see that an elastic fluid point as treated here is the same as the classical (piezotropic) compressible perfect fluid point, defined in the previous section. While from EULER's day until recently the constitutive equation (2.27) was taken as the *definition* of a fluid, its proof as a theorem within the general classification of § II.7 is due to NOLL (1958).

From (2.27) and (1.51), the constitutive equation for an incompressible elastic fluid point is

$$\boldsymbol{T} = -p\mathbf{1}, \tag{2.29}$$

where p is an undetermined hydrostatic pressure.

This constitutive equation, of course, is the same as that commonly taken as the definition of the classical incompressible perfect fluid, and similar remarks may be made regarding its proof here as a theorem.

Case 2. *Isotropic Elastic Solids.*

By definition (*cf.* § II. 7) for an isotropic elastic point there exists an undistorted reference configuration such that the relative symmetry group is the rotational group $\mathfrak{O}$. Hence the response function $\boldsymbol{H}$ satisfies the following two conditions:

$$\boldsymbol{H}(\boldsymbol{F}) = \boldsymbol{H}(\boldsymbol{F}\boldsymbol{P}) \quad \forall \boldsymbol{P} \in \mathfrak{O}, \tag{2.30}$$

and

$$\boldsymbol{H}(\boldsymbol{Q}\boldsymbol{F}) = \boldsymbol{Q}\boldsymbol{H}(\boldsymbol{F})\boldsymbol{Q}^T \quad \forall \boldsymbol{Q} \in \mathfrak{O}, \tag{1.16}$$

the reference configuration being undistorted. We proceed to find a general solution for such $\boldsymbol{H}$.

Let $\boldsymbol{F}$ and $\bar{\boldsymbol{F}}$ be two tensors in $\mathfrak{L}$ such that

$$\bar{\boldsymbol{F}} = \boldsymbol{F}\boldsymbol{Q} \tag{2.31}$$

for some rotation $\boldsymbol{Q} \in \mathfrak{O}$. Then the condition (2.30) requires that the value of $\boldsymbol{H}$ at $\boldsymbol{F}$ be the same as the value of $\boldsymbol{H}$ at $\bar{\boldsymbol{F}}$. On the other hand, if $\boldsymbol{F}$ and $\bar{\boldsymbol{F}}$ are not related by (2.31) for any $\boldsymbol{Q}$, then the condition (2.30) does not restrict the values $\boldsymbol{H}(\boldsymbol{F})$ and $\boldsymbol{H}(\bar{\boldsymbol{F}})$ in any way. Thus to solve the condition (2.30), we must characterize equivalence classes of tensors satisfying the condition (2.31). We claim that $\boldsymbol{F}$ and $\bar{\boldsymbol{F}}$ satisfy (2.31) if and only if

$$\boldsymbol{B} = \bar{\boldsymbol{B}}, \tag{2.32}$$

where, for any tensor $\boldsymbol{F}$, $\boldsymbol{B}$ denotes the left Cauchy-Green tensor as before, *viz*,

$$\boldsymbol{B} \equiv \boldsymbol{F}\boldsymbol{F}^T. \tag{2.33}$$

First, suppose that (2.31) holds. Then from (2.33) we have

$$\bar{\boldsymbol{B}} = \boldsymbol{F}\boldsymbol{Q}(\boldsymbol{F}\boldsymbol{Q})^T = \boldsymbol{F}\boldsymbol{Q}\boldsymbol{Q}^T\boldsymbol{F}^T = \boldsymbol{F}\boldsymbol{F}^T = \boldsymbol{B}, \tag{2.34}$$

which is (2.32). Conversely, suppose that (2.32) holds. Then from (2.33) again we have

$$(\boldsymbol{F}^{-1}\bar{\boldsymbol{F}})(\boldsymbol{F}^{-1}\bar{\boldsymbol{F}})^T = \boldsymbol{1}. \tag{2.35}$$

Consequently, since both $\boldsymbol{F}$ and $\bar{\boldsymbol{F}}$ are proper transformations, $\boldsymbol{F}^{-1}\bar{\boldsymbol{F}}$ is a rotation, and (2.31) holds. From (2.32), a general solution for (2.30) is

$$\boldsymbol{H}(\boldsymbol{F}) = \bar{\boldsymbol{H}}(\boldsymbol{B}). \tag{2.36}$$

Now we determine the condition on the function $\bar{\boldsymbol{H}}$ in order that the condition (1.16) be satisfied. From (2.33), the left Cauchy-Green tensor for the tensor $\boldsymbol{Q}\boldsymbol{F}$ is $\boldsymbol{Q}\boldsymbol{B}\boldsymbol{Q}^T$, where $\boldsymbol{B}$ is the left Cauchy-Green tensor for $\boldsymbol{F}$. Hence from (1.16) and (2.36), the condition on the function $\bar{\boldsymbol{H}}$ is

$$\bar{\boldsymbol{H}}(\boldsymbol{Q}\boldsymbol{B}\boldsymbol{Q}^T) = \boldsymbol{Q}\bar{\boldsymbol{H}}(\boldsymbol{B})\boldsymbol{Q}^T. \tag{2.37}$$

In the theory of invariants a function satisfying a condition such as (2.37) is called an *isotropic function*. The condition (1.16), then, requires that $\bar{\boldsymbol{H}}$ be an isotropic function.

Before solving (2.37) we remark first that that condition requires every proper vector of $\boldsymbol{B}$ to be a proper vector of $\overline{\boldsymbol{H}}(\boldsymbol{B})$. This fact is obvious, since we can choose the tensor $\boldsymbol{Q}$ in (2.37) as a rotation of 180° in the direction of any proper vector of $\boldsymbol{B}$. Then from the commutation theorem for symmetric tensors and orthogonal tensors (*cf.* § I. 2), $\boldsymbol{B}$ is invariant under such a tensor $\boldsymbol{Q}$, *viz*,

$$\boldsymbol{Q}\boldsymbol{B}\boldsymbol{Q}^T = \boldsymbol{B}. \tag{2.38}$$

Hence from (2.37) $\overline{\boldsymbol{H}}(\boldsymbol{B})$ must also be invariant under $\boldsymbol{Q}$. By the converse of the commutation theorem, the axis of $\boldsymbol{Q}$ is a proper vector of $\overline{\boldsymbol{H}}(\boldsymbol{B})$. From this remark, we see that there are three possibilities:

1. $\boldsymbol{B}$ has only one distinct proper number, say

$$\boldsymbol{B} = b\mathbf{1}. \tag{2.39}$$

Then the remark implies that $\overline{\boldsymbol{H}}(\boldsymbol{B})$ has only one distinct proper number also. Hence

$$\overline{\boldsymbol{H}}(\boldsymbol{B}) = \varphi_0\mathbf{1}, \tag{2.40}$$

where φ_0 is a scalar-valued function of $\boldsymbol{B}$.

2. $\boldsymbol{B}$ has only two distinct proper numbers, say

$$\boldsymbol{B} = a\mathbf{1}+b\boldsymbol{e}\otimes\boldsymbol{e}, \tag{2.41}$$

where $\boldsymbol{e}$ is a unit vector, and where $b \neq 0$. Then the remark implies that $\overline{\boldsymbol{H}}(\boldsymbol{B})$ has at most two distinct proper numbers; moreover, $\overline{\boldsymbol{H}}(\boldsymbol{B})$ has the form

$$\overline{\boldsymbol{H}}(\boldsymbol{B}) = \varphi_0\mathbf{1}+\varphi_1\boldsymbol{B}, \tag{2.42}$$

where φ_0 and φ_1 are scalar-valued functions of $\boldsymbol{B}$. Here we have used the fact that $\{\mathbf{1}, \boldsymbol{B}\}$ and $\{\mathbf{1}, \boldsymbol{e}\otimes\boldsymbol{e}\}$, where $\boldsymbol{B}$ is related to $\boldsymbol{e}\otimes\boldsymbol{e}$ by (2.41), generate the same subspace in the tensor space.

3. $\boldsymbol{B}$ has three distinct proper numbers, say

$$\boldsymbol{B} = a\boldsymbol{e}\otimes\boldsymbol{e}+b\boldsymbol{f}\otimes\boldsymbol{f}+c\boldsymbol{g}\otimes\boldsymbol{g} \tag{2.43}$$

where $\{\boldsymbol{e}, \boldsymbol{f}, \boldsymbol{g}\}$ is an orthonormal basis, and where a, b, c are unequal. Then $\overline{\boldsymbol{H}}(\boldsymbol{B})$ must be of the form

$$\overline{\boldsymbol{H}}(\boldsymbol{B}) = \varphi_0\mathbf{1}+\varphi_1\boldsymbol{B}+\varphi_2\boldsymbol{B}^2, \tag{2.44}$$

since the sets $\{\mathbf{1}, \boldsymbol{B}, \boldsymbol{B}^2\}$ and $\{\boldsymbol{e} \otimes \boldsymbol{e}, \boldsymbol{f} \otimes \boldsymbol{f}, \boldsymbol{g} \otimes \boldsymbol{g}\}$, where $\boldsymbol{B}$ satisfies (2.43), generate the same subspace in the tensor space. The coefficients φ_0, φ_1, and φ_2 are again scalar-valued functions of $\boldsymbol{B}$.

Now since

$$\boldsymbol{Q}\mathbf{1}\boldsymbol{Q}^T = \mathbf{1}, \quad \boldsymbol{Q}\boldsymbol{B}^2\boldsymbol{Q}^T = (\boldsymbol{Q}\boldsymbol{B}\boldsymbol{Q}^T)^2, \tag{2.45}$$

the condition (2.37) implies further that the coefficients φ_0, φ_1, and φ_2 in (2.40), (2.42), and (2.44) are isotropic functions of $\boldsymbol{B}$, *viz*,

$$\varphi_\Gamma(\boldsymbol{Q}\boldsymbol{B}\boldsymbol{Q}^T) = \varphi_\Gamma(\boldsymbol{B}), \ \Gamma = 0, 1, 2, \ \forall \boldsymbol{Q} \in \mathfrak{O}. \tag{2.46}$$

From the commutation theorem for (2.38) we know that a general solution of (2.46) is

$$\varphi_\Gamma = \varphi_\Gamma(I_{\boldsymbol{B}}, II_{\boldsymbol{B}}, III_{\boldsymbol{B}}), \quad \Gamma = 0, 1, 2, \tag{2.47}$$

where $I_{\boldsymbol{B}}$, $II_{\boldsymbol{B}}$, and $III_{\boldsymbol{B}}$ denote the three principal invariants of $\boldsymbol{B}$. Combining the results (2.40), (2.42), (2.44), and (2.47), we see that $\bar{\boldsymbol{H}}(\boldsymbol{B})$ can be written in the form (2.44) with φ_0, φ_1, and φ_2 given by (2.47). From (2.45), it is clear that (2.44) and (2.47) also suffice for the condition (2.37). Thus we have proved the following

PROPOSITION (RIVLIN & ERICKSEN).[1] *A function $\bar{\boldsymbol{H}}(\boldsymbol{B})$ from symmetric tensors into symmetric tensors is isotropic if and only if it has a representation of the form* (2.44), *where the coefficients* φ_0, φ_1, and φ_2 *are of the form* (2.47).

Interpretation of this algebraic theorem in terms of the response function $\bar{\boldsymbol{H}}$ of an elastic point satisfying the conditions (2.30) and (1.16) leads to the following

PROPOSITION[2]. *The constitutive equation of an elastic point has the form* (2.44) *with coefficients given by* (2.47) *if and only if that point is isotropic and the local reference configuration is undistorted.*

---

[1] Under the restriction to polynomial functions, this proposition is a special case of a theorem in classical invariant theory. The rediscovery of this special case in the context of continuum mechanics by REINER and PRAGER in 1945 was important in the early days of the renascence of rational mechanics.

[2] Major special cases of this theorem may be found, under more or less superfluous additional conditions, in the literature of elasticity theory from the work of FINGER (1894) onward, leading to the first fairly general treatments by REINER and RICHTER in 1948. As stated here, the theorem is due to RIVLIN & ERICKSEN (1955) but is expressed in the terms introduced later by NOLL.

There are several alternative forms of representations for the response function $\boldsymbol{H}$. By the Hamilton-Cayley theorem (*cf.* § I.1), $\boldsymbol{B}$ satisfies the equation

$$\boldsymbol{B}^3 - I_{\boldsymbol{B}}\boldsymbol{B}^2 + II_{\boldsymbol{B}}\boldsymbol{B} - III_{\boldsymbol{B}}\mathbf{1} = \mathbf{0}. \tag{2.48}$$

Hence

$$\boldsymbol{B}^2 = I_{\boldsymbol{B}}\boldsymbol{B} - II_{\boldsymbol{B}}\mathbf{1} + III_{\boldsymbol{B}}\boldsymbol{B}^{-1}. \tag{2.49}$$

Substituting this result into (2.44), we obtain

$$\boldsymbol{H}(\boldsymbol{F}) = \psi_0\mathbf{1} + \psi_1\boldsymbol{B} + \psi_{-1}\boldsymbol{B}^{-1}, \tag{2.50}$$

where ψ_0, ψ_1, and ψ_{-1} are functions of $I_{\boldsymbol{B}}$, $II_{\boldsymbol{B}}$, and $III_{\boldsymbol{B}}$ given in terms of φ_0, φ_1, and φ_2 by the relations

$$\psi_0 = \varphi_0 - II_{\boldsymbol{B}}\varphi_2, \tag{2.51}$$

$$\psi_1 = \varphi_1 + I_{\boldsymbol{B}}\varphi_2, \tag{2.52}$$

and

$$\psi_{-1} = III_{\boldsymbol{B}}\varphi_2. \tag{2.53}$$

Other alternative forms include

$$\boldsymbol{H}(\boldsymbol{F}) = \eta_0\mathbf{1} + \eta_1\boldsymbol{V} + \eta_2\boldsymbol{V}^2, \tag{2.54}$$

and

$$\boldsymbol{H}(\boldsymbol{F}) = \zeta_0\mathbf{1} + \zeta_1\boldsymbol{V} + \zeta_{-1}\boldsymbol{V}^{-1}, \tag{2.55}$$

where $\boldsymbol{V}$ is the left stretch tensor of $\boldsymbol{F}$, namely, the unique positive-definite and symmetric square root of $\boldsymbol{B}$, and where η_Γ and ζ_Γ are functions of the three principal invariants of $\boldsymbol{V}$. These two last alternative forms are rarely useful, however, since $\boldsymbol{V}$ is not a rational function of $\boldsymbol{F}$.

Now suppose that the material is incompressible. Then the same analysis yields a representation for the response function:

$$\boldsymbol{T} = -p\mathbf{1} + \varphi_1\boldsymbol{B} + \varphi_2\boldsymbol{B}^2 \tag{2.56}$$

with alternatives analogous to (2.50), (2.54), and (2.55). Here $\boldsymbol{B}$ is required to be unimodular, and φ_1 and φ_2 are functions of the first two invariants of $\boldsymbol{B}$. Of course, the third invariant of $\boldsymbol{B}$ now has the value 1.

It should be noted that the representations (2.44), (2.50), (2.54), and (2.55) all are special cases of the general solution (1.26) for the condition of material frame-indifference. From their definitions $\boldsymbol{B}$ and $\boldsymbol{C}$ are related by

$$\boldsymbol{B} = \boldsymbol{R}\boldsymbol{C}\boldsymbol{R}^{\mathrm{T}}, \tag{2.57}$$

so that

$$B^2 = RC^2R^T; \tag{2.58}$$

moreover, corresponding invariants of B and C are equal. Consequently for an isotropic material the function K in (1.26) is given by

$$K(C) = \varphi_0 \mathbf{1} + \varphi_1 C + \varphi_2 C^2, \tag{2.59}$$

among other possibilities.

We can linearize the representation (2.44) about a natural state, taking that state as the undistorted reference configuration for the representation. The result is

$$T = \lambda(\operatorname{tr} E)\mathbf{1} + 2\mu E + o(E), \tag{2.60}$$

where E is the classical strain tensor given by (1.32). The Lamé constants λ and μ are given by

$$\lambda = 2\left(\frac{\partial}{\partial I} + 2\frac{\partial}{\partial II} + \frac{\partial}{\partial III}\right)(\varphi_0 + \varphi_1 + \varphi_2)\bigg|_{(3,3,1)} \tag{2.61}$$

and

$$\mu = \varphi_1(3, 3, 1) + 2\varphi_2(3, 3, 1), \tag{2.62}$$

where we have suppressed the subscript B in the symbols for the invariants. In deducing the formulae (2.61) and (2.62), we have used the identities

$$\frac{\partial I}{\partial B} = \mathbf{1}, \tag{2.63}$$

$$\frac{\partial II}{\partial B} = I\mathbf{1} - B, \tag{2.64}$$

and

$$\frac{\partial III}{\partial B} = B^2 - IB + II\mathbf{1}, \tag{2.65}$$

which are easy to establish.

Case 3. A Type of Elastic Fluid Crystals[1].

We consider now a type of anisotropic fluid crystals which possess an intrinsic orientation characterized by a fixed line in the tangent space of

---

[1] C.-C. Wang, "A general theory of subfluids", *Arch. Rational Mech. Anal.* **20**, 1–40 (1965). *Cf.* also B. D. Coleman, "Simple liquid crystals", *Arch. Rational Mech. Anal.* **20**, 41–58 (1965). Also other types of elastic fluid crystals are introduced and studied in these papers.

the body-points. Relative to an orthonormal basis such that the third basis vector is located on that intrinsic line, the relative symmetry group is made up of tensors $\boldsymbol{P}$ whose components are given by

$$[P_{ij}] = \begin{bmatrix} a & b & 0 \\ c & d & 0 \\ e & f & g \end{bmatrix}, \tag{2.66}$$

where $a, \ldots, g$ are arbitrary real numbers subjected to the condition that

$$\det [P_{ij}] = 1. \tag{2.67}$$

Denoting the relative symmetry group as always by $\mathfrak{G}_{\boldsymbol{v}}$, where $\boldsymbol{v}$ is the reference configuration, we require the relative response function to satisfy the conditions

$$\boldsymbol{H}(\boldsymbol{FP}) = \boldsymbol{H}(\boldsymbol{F}) \quad \forall \boldsymbol{P} \in \mathfrak{G}_{\boldsymbol{v}}, \tag{2.3}$$

and

$$\boldsymbol{H}(\boldsymbol{QF}) = \boldsymbol{QH}(\boldsymbol{F})\boldsymbol{Q}^T \quad \forall \boldsymbol{Q} \in \mathfrak{O}. \tag{1.16}$$

We proceed to derive a representation for these two conditions for the given $\mathfrak{G}_{\boldsymbol{v}}$.

We consider first the value $\boldsymbol{H}(\boldsymbol{F})$ for

$$\boldsymbol{F} = c\mathbf{1}, \tag{2.68}$$

where c is a positive constant. From (2.66), (2.3), and (1.16), $\boldsymbol{H}(c\mathbf{1})$ must commute with all rotations $\hat{\boldsymbol{Q}}$ of the form

$$[\hat{Q}_{ij}] = \begin{bmatrix} \cos\theta, & -\sin\theta & 0 \\ \sin\theta, & \cos\theta & 0 \\ 0 & 0 & 1 \end{bmatrix}, \tag{2.69}$$

since $\hat{\boldsymbol{Q}} \in \mathfrak{G}_{\boldsymbol{v}} \cap \mathfrak{O}$. Using the commutation theorem cited several times in this section, we see that $\boldsymbol{H}(c\mathbf{1})$ is necessarily of the form

$$\boldsymbol{H}(c\mathbf{1}) = \varphi_0 \mathbf{1} + \varphi_1 \boldsymbol{e}_3 \otimes \boldsymbol{e}_3, \tag{2.70}$$

where $\boldsymbol{e}_3$ is the third basis vector, which is in the preferred direction of the material point.

Now let $\boldsymbol{F}$ be an arbitrary deformation gradient. Then the reader will verify easily that $\boldsymbol{F}$ can be factorized in the following way:

$$\boldsymbol{F} = (\det \boldsymbol{F})^{\frac{1}{3}}\boldsymbol{RP}, \tag{2.71}$$

where $\boldsymbol{P} \in \mathfrak{G}_{\mathbf{v}}$, and where the rotation $\boldsymbol{R}$ maps $\boldsymbol{e}_3$ onto the unit vector $\boldsymbol{i}_3$ given by

$$\boldsymbol{i}_3 = \frac{\boldsymbol{F}\boldsymbol{e}_3}{\|\boldsymbol{F}\boldsymbol{e}_3\|}. \tag{2.72}$$

In the physical interpretation, $\boldsymbol{i}_3$ is a unit vector in the preferred direction of the body-point in the present configuration. Substituting (2.71) into (2.3) and then using (1.16), we obtain

$$\boldsymbol{H}(\boldsymbol{F}) = \boldsymbol{R}\boldsymbol{H}((\det \boldsymbol{F})^{\frac{1}{3}}\mathbf{1})\boldsymbol{R}^T. \tag{2.73}$$

Then from (2.70) $\boldsymbol{H}(\boldsymbol{F})$ is necessarily of the form

$$\boldsymbol{H}(\boldsymbol{F}) = \varphi_0 \mathbf{1} + \varphi_1 \boldsymbol{i}_3 \otimes \boldsymbol{i}_3, \tag{2.74}$$

where φ_0 and φ_1 depend upon $\det \boldsymbol{F}$, or, equivalently, upon the density ρ.

The form (2.74), which we have shown to be a consequence of the conditions (2.3) and (1.16), is also sufficient that those conditions hold. This fact is more or less obvious. First, the preferred direction in the local configuration $\boldsymbol{F}\mathbf{v}$ is the same as that in the local configuration $\boldsymbol{F}\boldsymbol{P}\mathbf{v}$ for all $\boldsymbol{P}$. We can verify this directly from the formula (2.72). Thus (2.74) satisfies the condition (2.3). Second, the preferred directions in the local configurations $\boldsymbol{F}\mathbf{v}$ and $\boldsymbol{Q}\boldsymbol{F}\mathbf{v}$ are related by

$$\frac{\boldsymbol{Q}\boldsymbol{F}\boldsymbol{e}_3}{\|\boldsymbol{Q}\boldsymbol{F}\boldsymbol{e}_3\|} = \boldsymbol{Q}\frac{\boldsymbol{F}\boldsymbol{e}_3}{\|\boldsymbol{F}\boldsymbol{e}_3\|}, \tag{2.75}$$

since $\boldsymbol{Q} \in \mathfrak{O}$ and hence preserves the norm $\| \ \|$ on $\mathscr{V}$. Substituting (2.75) into (2.74), we see that that form satisfies also the condition (1.16). Thus we have proved that (2.74) is a representation for the relative response function $\boldsymbol{H}$ of this type of fluid crystals.

The three cases above furnish examples of explicit representations for the response functions of three kinds of elastic materials. We do not attempt to work out any further examples here, especially since it is not presently possible to exhibit explicit representations for all types which are of interest, but in the next section we shall prove a general though not explicit representation theorem for the response function of an elastic point having any assigned symmetry group.

It should be noted that the classification of the connected Lie subgroups of the special linear group is known (*cf.* § I.6). If we require that the response function of an elastic point be continuous, then its symmetry group is necessarily a closed Lie subgroup. Even in this case, however, that group need not be connected.

3. Representation for the Response Function of an Elastic Point

We have seen representations for the response functions of three specific types of elastic points in the previous section. In this section we construct a representation for the response function in general. Our treatment[1], while applying the technique developed in § II.8, is self-contained. We proceed to find a general solution for the following two conditions:

$$\boldsymbol{H}(\boldsymbol{F}\boldsymbol{P}) = \boldsymbol{H}(\boldsymbol{F}) \quad \forall \boldsymbol{P} \in \mathfrak{G}_{\nu} \tag{2.3}$$

and

$$\boldsymbol{H}(\boldsymbol{Q}\boldsymbol{F}) = \boldsymbol{Q}\boldsymbol{H}(\boldsymbol{F})\boldsymbol{Q}^{T} \quad \forall \boldsymbol{Q} \in \mathfrak{O}. \tag{1.16}$$

Since the reference configuration ν is held fixed throughout this section, we suppress the subscript and denote the relative symmetry group by $\mathfrak{G}$.

As in the more general theory developed in § II.8, we introduce an equivalence relation $\sim$ on the domain $\mathfrak{L}$ of the response function $\boldsymbol{H}$,

$$\boldsymbol{F} \sim \bar{\boldsymbol{F}} \Leftrightarrow \bar{\boldsymbol{F}} = \boldsymbol{F}\boldsymbol{P} \quad \text{for some } \boldsymbol{P} \in \mathfrak{G}. \tag{3.1}$$

An equivalence class $\mathscr{X}$ with respect to this equivalence relation is a left coset of $\mathfrak{G}$ in $\mathfrak{L}$ in the sense of group theory. Let $\boldsymbol{F}_{\mathscr{X}}$ be a representative element for $\mathscr{X}$, so that for each $\boldsymbol{F} \in \mathscr{X}$ there exists a $\boldsymbol{P} \in \mathfrak{G}$ such that

$$\boldsymbol{F} = \boldsymbol{F}_{\mathscr{X}}\boldsymbol{P}. \tag{2.12}$$

Then a general solution for (2.3) is

$$\boldsymbol{H}(\boldsymbol{F}) = \boldsymbol{H}(\boldsymbol{F}_{\mathscr{X}}), \tag{2.11}$$

as we have seen in the preceding section.

In order that the condition (1.16) be satisfied also, the function $\boldsymbol{H}(\boldsymbol{F}_{\mathscr{X}})$ cannot be arbitrary. So as to determine the restriction on $\boldsymbol{H}(\boldsymbol{F}_{\mathscr{X}})$

---

[1] C.-C. WANG, "On the response functions of elastic materials", *Arch. Rational Mech. Anal.* **33**, 331–342 (1969).

from (1.16), we introduce the group $\mathfrak{D}_{\mathscr{X}}$ for each left coset $\mathscr{X}$:

$$\mathfrak{D}_{\mathscr{X}} \equiv (\boldsymbol{F}_{\mathscr{X}} \mathfrak{G} \boldsymbol{F}_{\mathscr{X}}^{-1}) \cap \mathfrak{D}. \tag{3.2}$$

The reader will verify easily that this group depends only on the left coset $\mathscr{X}$ and is independent of the choice of the representative element $\boldsymbol{F}_{\mathscr{X}} \in \mathscr{X}$.

In the physical interpretations the group

$$\mathfrak{G}_{\mathscr{X}} \equiv \boldsymbol{F}_{\mathscr{X}} \mathfrak{G} \boldsymbol{F}_{\mathscr{X}}^{-1} \tag{3.3}$$

is the symmetry group relative to the configuration $\boldsymbol{F}_{\mathscr{X}} \boldsymbol{v}$ (*cf.* Noll's rule (II. 7.62)). Clearly, $\mathfrak{G}_{\mathscr{X}}$ depends only on the left coset $\mathscr{X}$, since the local configurations $\boldsymbol{F}$ belonging to a fixed left coset $\mathscr{X}$ are pairwise materially isomorphic. From (3.2), $\mathfrak{D}_{\mathscr{X}}$ is simply that part of $\mathfrak{G}_{\mathscr{X}}$ contained in the rotational group $\mathfrak{D}$.

The group $\mathfrak{D}_{\mathscr{X}}$ is important, since from (1.16) the value $\boldsymbol{H}(\boldsymbol{F}_{\mathscr{X}})$ must commute with all elements of $\mathfrak{D}_{\mathscr{X}}$,

$$\boldsymbol{Q}\boldsymbol{H}(\boldsymbol{F}_{\mathscr{X}}) = \boldsymbol{H}(\boldsymbol{F}_{\mathscr{X}})\boldsymbol{Q} \quad \forall \boldsymbol{Q} \in \mathfrak{D}_{\mathscr{X}}. \tag{3.4}$$

For this reason, we introduce the subspace $\mathscr{S}_{\mathscr{X}}$ in the space of symmetric tensors by the condition

$$\boldsymbol{S} \in \mathscr{S}_{\mathscr{X}} \Leftrightarrow \boldsymbol{S}\boldsymbol{Q} = \boldsymbol{Q}\boldsymbol{S} \quad \forall \boldsymbol{Q} \in \mathfrak{D}_{\mathscr{X}}. \tag{3.5}$$

Clearly, $\mathscr{S}_{\mathscr{X}}$ is a non-trivial subspace, since it always contains the tensor $\mathbf{1}$.

Next we introduce an equivalence relation $\approx$ among the left cosets of $\mathfrak{G}$ in $\mathfrak{L}$ (*cf.* (II. 8.44)),

$$\mathscr{X} \approx \overline{\mathscr{X}} \Leftrightarrow \overline{\mathscr{X}} = \boldsymbol{Q}\mathscr{X} \text{ for some } \boldsymbol{Q} \in \mathfrak{D}, \tag{3.6}$$

where the equation on the right-hand side holds in the sense of set theory. An equivalence class $\mathscr{Z}$ with respect to $\approx$ is called a *combined coset* of $\mathfrak{G}$ and $\mathfrak{D}$ in $\mathfrak{L}$. We may regard $\mathscr{Z}$ as a subset of $\mathfrak{L}$.

PROPOSITION. *The groups $\mathfrak{G}_{\mathscr{X}}$ and $\mathfrak{D}_{\mathscr{X}}$ and the subspace $\mathscr{S}_{\mathscr{X}}$ obey the transformation rules*

$$\mathfrak{G}_{\boldsymbol{Q}\mathscr{X}} = \boldsymbol{Q}\mathfrak{G}_{\mathscr{X}}\boldsymbol{Q}^T, \; \mathfrak{D}_{\boldsymbol{Q}\mathscr{X}} = \boldsymbol{Q}\mathfrak{D}_{\mathscr{X}}\boldsymbol{Q}^T, \; \mathscr{S}_{\boldsymbol{Q}\mathscr{X}} = \boldsymbol{Q}\mathscr{S}_{\mathscr{X}}\boldsymbol{Q}^T \tag{3.7}$$

on each combined coset $\mathscr{Z}$.

The proof is left to the reader.

From these transformation rules, we see that on each combined coset $\mathscr{Z}$ the groups $\mathfrak{D}_{\mathscr{X}}$, $\mathscr{X} \in \mathscr{Z}$, are inner-isomorphic in $\mathfrak{D}$ and the subspaces $\mathscr{S}_{\mathscr{X}}$, $\mathscr{X} \in \mathscr{Z}$, have the same dimension. In view of (3.4), we set

$$n_{\mathscr{Z}} \equiv \dim \mathscr{S}_{\mathscr{X}}, \mathscr{X} \in \mathscr{Z}, \tag{3.8}$$

and we call $n_{\mathscr{Z}}$ the *degree of freedom* of the response function on the combined coset $\mathscr{Z}$.

Now we adopt the following convention with regard to the choice of the representative element $\boldsymbol{F}_{\mathscr{X}}$ for $\mathscr{X}$ belonging to a fixed combined coset $\mathscr{Z}$: We require that

$$\boldsymbol{F}_{\overline{\mathscr{X}}} = \boldsymbol{Q}_{\mathscr{X}\overline{\mathscr{X}}} \boldsymbol{F}_{\mathscr{X}} \tag{3.9}$$

for any $\mathscr{X}$, $\overline{\mathscr{X}}$ in $\mathscr{Z}$. Here $\boldsymbol{Q}_{\mathscr{X}\overline{\mathscr{X}}}$ is any particular rotation such that

$$\overline{\mathscr{X}} = \boldsymbol{Q}_{\mathscr{X}\overline{\mathscr{X}}} \mathscr{X}. \tag{3.10}$$

Under this convention, a general solution for the conditions (2.3) and (1.16) is given by the following

REPRESENTATION THEOREM (WANG). *A response function* $\boldsymbol{H}$ *satisfies* (2.3) *and* (1.16) *if and only if it has the form*

$$\boldsymbol{H}(\boldsymbol{F}) = \boldsymbol{H}(\boldsymbol{F}_{\mathscr{X}}), \tag{2.11}$$

where the representative elements $\boldsymbol{F}_{\mathscr{X}}$ *satisfy the convention* (3.9) *and where the function* $\boldsymbol{H}(\boldsymbol{F}_{\mathscr{X}})$ *satisfies the conditions*

$$\boldsymbol{H}(\boldsymbol{F}_{\mathscr{X}}) \in \mathscr{S}_{\mathscr{X}} \tag{3.11}$$

and

$$\boldsymbol{H}(\boldsymbol{F}_{\overline{\mathscr{X}}}) = \boldsymbol{Q}_{\mathscr{X}\overline{\mathscr{X}}} \boldsymbol{H}(\boldsymbol{F}_{\mathscr{X}}) \boldsymbol{Q}^T_{\mathscr{X}\overline{\mathscr{X}}}. \tag{3.12}$$

Proof. In the preceding section we have shown that (2.11) is a general solution of (2.3). From (3.4) and (3.9), we see that (3.11) and (3.12) are necessary for (1.16). Conversely, if $\boldsymbol{Q}$ is any rotation, then the left coset $\mathscr{X}$ of $\boldsymbol{F}$ and the left coset $\overline{\mathscr{X}}$ of $\boldsymbol{QF}$ are equivalent in the sense of (3.6). Comparing the right-hand side of (3.6) with (3.10) and using the condition (2.8) for a left coset, we see that $\boldsymbol{Q}^T_{\mathscr{X}\overline{\mathscr{X}}}\boldsymbol{Q}$ belongs to the group $\mathfrak{D}_{\mathscr{X}}$ defined by (3.2). Hence from (3.11) and (3.5) the tensor $\boldsymbol{Q}_{\mathscr{X}\overline{\mathscr{X}}}$ in (3.12) can be replaced by the tensor $\boldsymbol{Q}$. Now substituting (2.11) into (3.12), we obtain (1.16). Q.E.D.

The reader will easily show that this theorem follows also from the general representation theorem in § II.8 as the special case appropriate to elasticity.

It should be noted that a general solution for the conditions (3.11) and (3.12) is given by

$$H(F_{\mathscr{X}}) = Q_{\mathscr{X}_{\mathfrak{X}}\mathscr{X}} H(F_{\mathscr{X}_{\mathfrak{X}}}) Q^T_{\mathscr{X}_{\mathfrak{X}}\mathscr{X}} \tag{3.13}$$

and

$$H(F_{\mathscr{X}_{\mathfrak{X}}}) \in \mathscr{S}_{\mathscr{X}_{\mathfrak{X}}}, \tag{3.14}$$

where $\mathscr{X}_{\mathfrak{X}}$ is a representative left coset in $\mathfrak{X}$. Combining (3.13) and (2.11), we can write the representation also in the form

$$H(F) = QH(F_{\mathscr{X}_{\mathfrak{X}}})Q^T, \tag{3.15}$$

where $F \in \mathfrak{X}$, so that there exist tensors $Q \in \mathfrak{O}$ and $P \in \mathfrak{G}$ such that

$$F = QF_{\mathscr{X}_{\mathfrak{X}}}P. \tag{3.16}$$

Then the relation between the special case (3.15) and the general formula (II. 8.54) is apparent.

The meaning of the representation (3.15) is clear. First, the function $H(F_{\mathscr{X}_{\mathfrak{X}}})$, having the degree of freedom $n_{\mathfrak{X}}$ at each $F_{\mathscr{X}_{\mathfrak{X}}}$, is arbitrary. Then the formula (3.15) extends the function $H(F_{\mathscr{X}_{\mathfrak{X}}})$ to the entire domain $\mathfrak{L}$ via the equivalence relation (3.16), which characterizes the combined cosets $\mathfrak{X}$.

We leave it to the reader to fix the positions of the specific representations (2.27), (2.44), and (2.74) within the context of the general one just derived.

4. The Constitutive Equation of a Hyperelastic Point

As we explained in § II.7, a constitutive equation is a mathematical model for the mechanical behavior of a material point. While such mathematical models ought to be general enough to allow for the great variety of materials occurring in nature, they all must be consistent with the basic axioms of mechanics. The constitutive equation for an elastic point, treated in the preceding three sections, is indeed consistent with those axioms, which include the principles of balance of mass, balance of linear momentum, and balance of rotational momentum.

The axioms of mechanics can be regarded as a subset of the axioms of thermodynamics[1], which extend and supplement those of mechanics by adjoining the principle of balance of energy and the principle of irreversibility. Likewise, we may seek to regard an elastic point in mechanics as some special case of a thermo-elastic point as defined in § II.7, but it is not instantly clear whether or not the constitutive equation of an elastic point is consistent with the two further axioms added to the mechanical ones in order to define the doctrine of thermodynamics. This problem can be made precise and solved in the following way[2].

First, we must specify the position of the constitutive equation (1.9) with respect to the constitutive equations (II. 7.8). There are four obvious ways to reduce a thermo-elastic point to an elastic one of a special kind:

Case 1 (Isothermal process). Suppose that the temperature at the body-point be held fixed throughout the motion of that point. From (II. 7.18), the stress tensor is a function of the deformation gradient only. Thus in an isothermal process a thermo-elastic point reduces to an elastic one.

Case 2 (Isentropic process). Suppose that the entropy at the body-point be fixed. From (II. 7.24), the stress tensor is again a function of the deformation gradient only. Thus also in an isentropic process a thermo-elastic point reduces to an elastic one.

Case 3. Suppose that in the constitutive equation $\boldsymbol{T}_\kappa = \tilde{\boldsymbol{T}}_\kappa(\boldsymbol{F}_\kappa, \theta_\kappa)$ of a thermo-elastic point the function $\tilde{\boldsymbol{T}}_\kappa$ is independent of its second argument. Such a thermo-elastic point is *ipso facto* an elastic point.

Case 4. Suppose that in the constitutive equation $\boldsymbol{T}_\kappa = \hat{\boldsymbol{T}}_\kappa(\boldsymbol{F}_\kappa, \eta_\kappa)$ of a thermo-elastic point the function $\hat{\boldsymbol{T}}_\kappa$ is independent of its second argument. Such a thermo-elastic point, likewise, is *ipso facto* an elastic point.

From the standpoint of § II.7, Cases 1 and 2 are merely formally parallel, not equivalent, since a thermo-elastic point subjected to an iso-

---

[1] Since for the most part the analysis in the present chapter is confined within pure mechanics and is self-contained, we remark to the reader that the above paragraphs are an exception. Without loss of continuity, he may skip over the remarks on thermodynamics and pass at once to the definition of "hyperelastic point" and "stored-energy function" on p. 192.

[2] In principle, the argument here goes back to KIRCHHOFF, PLANCK, VOIGT, and others in the last century. For the form given here *cf.* "the fully recoverable case" in CFT, § 256A, and "perfect materials" in NFTM, §§ 80 and 82.

thermal process need not have a constant entropy, nor need a thermo-elastic point in an isentropic process have a constant temperature. Moreover, the relative response function $\boldsymbol{H}$ in the elastic constitutive equation (1.9) that results for an isothermal process for a given thermo-elastic point is not generally the same as the one that results when the same thermo-elastic point is subjected to an isentropic process. Cases 3 and 4, however, are equivalent, since the condition for either one implies the other, as we shall see. While physical relevance for Cases 3 and 4 may be limited, all four cases are consistent with the mathematical theory.

Now we determine the restriction the axioms of thermodynamics impose upon the response functions of such thermo-elastic points as reduce to elastic ones in any of the four cases just described. We have shown in § II.7 that the principle of irreversibility requires the response function of a thermo-elastic particle to be given by (II. 7.27), whatever be the process. By specializing the first and second forms, respectively, in (II. 7.25) we see that in an isothermal process or in an isentropic process

$$\boldsymbol{T} = \rho \boldsymbol{F} \left(\frac{\partial \hat{\sigma}}{\partial \boldsymbol{F}} \right)^T, \tag{4.1}$$

where $\hat{\sigma}$ is a function of $\boldsymbol{F}$:

$$\sigma = \hat{\sigma}(\boldsymbol{F}), \tag{4.2}$$

but the function $\hat{\sigma}$ is not generally the same in the two cases. Specifically, in an isothermal process

$$\hat{\sigma}(\boldsymbol{F}) = \tilde{\psi}(\boldsymbol{F}, \theta), \tag{4.3}$$

while in an isentropic process

$$\hat{\sigma}(\boldsymbol{F}) = \hat{\varepsilon}(\boldsymbol{F}, \eta). \tag{4.4}$$

A fixed local reference configuration $\boldsymbol{v}$ relative to which the deformation gradient $\boldsymbol{F}$ is computed is understood here. This simple analysis shows that in Cases 1 and 2 the axioms of thermodynamics limit the response function to the form given by (4.1).

Next we consider Case 3. By hypothesis the stress tensor is independent of the temperature. Then from (II. 7.18) the free-energy function $\tilde{\psi}$ satisfies the condition

$$\frac{\partial^2 \tilde{\psi}}{\partial \boldsymbol{F}\, \partial \theta} = \boldsymbol{0}. \tag{4.5}$$

Integrating this equation, we obtain

$$\tilde{\psi}(\boldsymbol{F}, \theta) = \hat{\sigma}(\boldsymbol{F}) + \bar{\psi}(\theta). \tag{4.6}$$

Substituting this result into (II. 7.27) shows that (4.1) holds also in this case. Further, from (II. 7.17), the entropy is given by

$$\eta = -\frac{d\bar{\psi}}{d\theta}. \tag{4.7}$$

As before, we assume that the entropy relation is invertible with respect to the temperature, so that $\theta = \bar{\theta}(\eta)$, say. By (4.7) and (II. 7.7), the internal energy is given by

$$\varepsilon = \hat{\sigma}(\boldsymbol{F}) + \bar{\varepsilon}(\eta), \tag{4.8}$$

where $\bar{\varepsilon}(\eta) = \eta\bar{\theta}(\eta) + \bar{\psi}(\bar{\theta}(\eta))$. Substituting this result into (II. 7.24), we see that the stress is independent of the entropy as well as of the temperature.

In Case 4 the same argument may be applied in reverse order, starting from the hypothesis that the stress is independent of the entropy. Thus Cases 3 and 4 are equivalent.

We have shown, then, that in all four of the cases considered the constitutive equation of elasticity must reduce to the special form (4.1) in order to fall within the framework of thermodynamics.

In the theory of elasticity itself, the special case (4.1) defines a *hyperelastic point*. The function $\hat{\sigma}$ is called the *stored-energy function*, and its value σ is called the *stored-energy density*.

So as to find the physical meaning of the stored-energy function, we may differentiate (4.2) with respect to t and by use of the chain rule show that

$$\rho\dot{\hat{\sigma}} = \operatorname{tr}(\boldsymbol{T}\dot{\boldsymbol{F}}\boldsymbol{F}^{-1}) = \operatorname{tr}(\boldsymbol{T} \operatorname{grad} \boldsymbol{v}) \tag{4.9}$$

in any motion. Comparing this equation with the energy equation (II.6.23), we see that σ is that part of ε accumulated by the effects of mechanical work alone. More specifically, (II. 6.23) and (4.9) imply that

$$\rho\frac{d}{dt}(\varepsilon - \hat{\sigma}) = \operatorname{div} \boldsymbol{h} + \rho r. \tag{4.10}$$

In particular, if the thermal energy supplies div $\boldsymbol{h}$ and r are zero, then $\hat{\sigma}$ differs from ε only by an additive constant, for each body-point.

Mathematically, however, the thermal energy supplies, like the thermodynamical variables, never enter into the theory of elasticity, which is a purely mechanical theory stated within the framework of the axioms of dynamics alone. We may choose to regard (4.9) as asserting the balance of purely mechanical energy, but that equation is really an identity whose solution[1] is the constitutive equation (4.1). The true equation of balance of total energy is obtained by inserting in (4.10) the solution $\sigma = \hat{\sigma}(\boldsymbol{F})$ of (4.1).

The constitutive equation (4.1) must satisfy the principle of material frame-indifference. From (1.16) and (4.1) we have

$$\boldsymbol{Q}^T \left.\frac{\partial \hat{\sigma}}{\partial F}\right|_{\boldsymbol{QF}} = \left.\frac{\partial \hat{\sigma}}{\partial \boldsymbol{F}}\right|_{\boldsymbol{F}} \quad \forall \boldsymbol{Q} \in \mathfrak{O}. \tag{4.11}$$

Since (4.2) can be regarded as a constitutive equations also, we assume that it too must obey the principle of material frame-indifference. Thus

$$\hat{\sigma}(\boldsymbol{QF}) = \hat{\sigma}(\boldsymbol{F}) \quad \forall \boldsymbol{Q} \in \mathfrak{O}. \tag{4.12}$$

It turns out that the conditions (4.11) and (4.12) are equivalent, as was first proved by NOLL.

Indeed, by differentiating (4.12) with respect to $\boldsymbol{F}$ we obtain (4.11). Conversely, integrating (4.11), we obtain

$$\hat{\sigma}(\boldsymbol{QF}) = \hat{\sigma}(\boldsymbol{F}) + \hat{\sigma}(\boldsymbol{Q}) - \hat{\sigma}(\boldsymbol{1}). \tag{4.13}$$

Thus it remains to show that

$$\hat{\sigma}(\boldsymbol{Q}) = \hat{\sigma}(\boldsymbol{1}) \quad \forall \boldsymbol{Q} \in \mathfrak{O}. \tag{4.14}$$

To see this, we set $\boldsymbol{F} = \overline{\boldsymbol{Q}}$ in (4.13) and integrate the resulting equation

---

[1] The argument here goes back, in principle though not in correct detail, to G. GREEN (1839) and was first given precisely by KIRCHHOFF and KELVIN. For this reason hyperelastic materials are sometimes called "Green-elastic", while elastic materials are called "Cauchy-elastic".

In the looser literature one sometimes reads that "thermodynamics proves that all elastic materials are hyperelastic". This false claim, to the extent it has any concrete foundation at all, rests on one of two misinterpretations of the results given in our text above: first, that (4.9) is the equation of balance of energy, or, second, that the four special cases analysed exhaust all of elasticity or the part there of that is compatible with thermo-elasticity. The reader may verify, for example, that in some but not all iso-energetic deformations a thermo-elastic point is hyperelastic.

over the rotational group with respect to a left-invariant volume element $d\mu_{\bar{Q}}$ (*cf.* § I.6):

$$\int_{\mathfrak{D}} \hat{\sigma}(Q\bar{Q})d\mu_{\bar{Q}} = \int_{\mathfrak{D}} \hat{\sigma}(\bar{Q})d\mu_{\bar{Q}} + (\hat{\sigma}(Q) - \hat{\sigma}(\mathbf{1}))\int_{\mathfrak{D}} d\mu_{\bar{Q}}. \tag{4.15}$$

From the invariance of $d\mu_{\bar{Q}}$, the left-hand side is equal to the first term on the right-hand side in (4.15). Since $\mathfrak{D}$ is compact, the integral over $\mathfrak{D}$ is finite and non-zero, so

$$(\hat{\sigma}(Q) - \hat{\sigma}(\mathbf{1}))\int_{\mathfrak{D}} d\mu_{\bar{Q}} = 0, \tag{4.16}$$

which implies (4.14).

We shall give a different proof of (4.14) in a later section.

By an argument of just the same kind as that applied to the response function to derive (1.18), we get a general solution[1] of (4.12):

$$\hat{\sigma}(F) = \hat{\sigma}(U), \tag{4.17}$$

or, alternatively,

$$\hat{\sigma}(F) = \bar{\sigma}(C), \tag{4.18}$$

where $\bar{\sigma}$ is a function of positive-definite and symmetric tensors.

Of course the constitutive equation (4.1) must satisfy (II. 6.22), expressing the local balance of rotational momentum: The value of the response function must be a symmetric tensor. It turns out that this condition is also equivalent to the condition (4.12). Indeed, differentiating the general solution (4.18) with respect to F, we see that

$$\rho F\left(\frac{\partial\hat{\sigma}}{\partial F}\right)^T = \rho F\left(\frac{\partial\bar{\sigma}}{\partial C} + \left(\frac{\partial\bar{\sigma}}{\partial C}\right)^T\right)F^T = 2\rho F\frac{\partial\bar{\sigma}}{\partial C}F^T. \tag{4.19}$$

Thus (4.12) implies that the values of response function are symmetric tensors. Conversely, by the energy relation (4.9), for any local motion of the form

$$F(t) = Q(t)F_0, \quad Q(t) \in \mathfrak{D}, \tag{4.20}$$

we have

$$\rho\dot{\sigma} = \operatorname{tr}(T\dot{Q}Q^{-1}) = 0, \tag{4.21}$$

---

[1] The solution itself is due to G. GREEN; several derivations of it, varying in strictness and in unnecessary additional conditions, were given in the last century and this; the simple proof we expect our readers to see at a glance here is due to NOLL.

since $\boldsymbol{T}$ is symmetric and $\dot{\boldsymbol{Q}}\boldsymbol{Q}^{-1}$ is skew. Consequently $\hat{\sigma}$ is invariant on each smooth curve of the form (4.20) in $\mathfrak{L}$. This condition implies (4.12) immediately, since $\mathfrak{D}$ is connected. We summarize the foregoing results in a

PROPOSITION (NOLL). *The following three conditions are equivalent:*

1. *The value* $\boldsymbol{T}$ *of the response function in* (4.1) *is always a symmetric tensor.*

2. *The response function obeys the principle of material frame-indifference.*

3. *The stored-energy function obeys the principle of material frame-indifference.*

Hyperelastic materials, which are defined by constitutive equations of the form (4.1), form only a proper subset of elastic materials. We can characterize this subset by the usual condition of integrability. From (4.1), $\boldsymbol{H}(\boldsymbol{F})$ is the relative response function of a hyperelastic point if and only if the function $\rho^{-1}\boldsymbol{H}(\boldsymbol{F})(\boldsymbol{F}^{-1})^T$, where ρ as a function of $\boldsymbol{F}$ is given by (II. 6.25), is the gradient of a scalar function $\hat{\sigma}(\boldsymbol{F})$. A necessary and sufficient condition for this, of course, is that the gradient of $\rho^{-1}\boldsymbol{H}(\boldsymbol{F})$ $(\boldsymbol{F}^{-1})^T$ with respect to $\boldsymbol{F}$ be a symmetric tensor with respect to the variable $\boldsymbol{F}$. From (4.9), an equivalent integral condition is

$$\oint \frac{1}{\rho} \operatorname{tr}(\boldsymbol{H}(\boldsymbol{F})\dot{\boldsymbol{F}}\boldsymbol{F}^{-1})dt = 0 \tag{4.22}$$

for all closed circuits

$$\boldsymbol{F} = \boldsymbol{F}(t) \tag{4.23}$$

in $\mathfrak{L}$. If the integrability condition (4.22) holds, then the stored-energy function $\hat{\sigma}$ is given by

$$\hat{\sigma}(\boldsymbol{F}) = \int_{\mathbf{1}}^{\boldsymbol{F}} \frac{1}{\rho} \operatorname{tr}(\boldsymbol{H}(\boldsymbol{F})\dot{\boldsymbol{F}}\boldsymbol{F}^{-1})dt + \hat{\sigma}(\mathbf{1}), \tag{4.24}$$

where the integral is taken along any smooth path joining $\mathbf{1}$ to $\boldsymbol{F}$, and where $\hat{\sigma}(\mathbf{1})$ is an arbitrary constant.

We now consider the linearization of the constitutive equation (4.1) about a natural state $\boldsymbol{\nu}_0$, which we choose to be the reference configuration. First, comparing the representation (4.19) with the previous result (1.7), we see that for a hyperelastic point the function $\boldsymbol{J}(\boldsymbol{C})$ is given by

$$\boldsymbol{J}(\boldsymbol{C}) = 2\rho \frac{\partial \bar{\sigma}}{\partial \boldsymbol{C}}, \tag{4.25}$$

where ρ can be expressed as a function of C in view of (II.6.25):

$$\rho = \frac{\rho_0}{(\det \boldsymbol{C})^{\frac{1}{2}}}, \tag{4.26}$$

ρ_0 being the density in the natural state $\boldsymbol{v}_0$. Now from (1.34), the classical elasticities C_{ijkl} are given by

$$C_{ijkl} = 4\rho_0 \left.\frac{\partial^2 \bar{\sigma}}{\partial C^{ij} \partial C^{kl}}\right|_{\boldsymbol{C}=\boldsymbol{1}}. \tag{4.27}$$

Consequently, in addition to the previously derived symmetry conditions (1.35), we have the following further one:

$$C_{ijkl} = C_{klij}. \tag{4.28}$$

Conversely, this last is also the condition of integrability for (4.25) within the framework of the linearized theory. Thus (4.28) is a criterion for a linear hyperelastic point. This criterion is called *Green's condition.*

We define an *incompressible hyperelastic point* by the constitutive equation

$$\boldsymbol{T} = -p\boldsymbol{1} + \rho \boldsymbol{F} \left(\frac{\partial \hat{\sigma}}{\partial \boldsymbol{F}}\right)^T, \tag{4.29}$$

where p is an undetermined hydrostatic pressure as explained in § 1. If we choose for the reference configuration $\boldsymbol{v}$ an admissible configuration, then the domain of the response function is the special linear group $\mathfrak{U}$. Of course, we do not expect the response function to depend upon the behavior of the stored-energy function $\hat{\sigma}$ outside the domain $\mathfrak{U}$. This condition is reflected mathematically by the fact that if $\hat{\sigma}_1$ and $\hat{\sigma}_2$ are two stored-energy functions which agree on the special linear group $\mathfrak{U}$, then the response functions based on them can differ from each other by at most a hydrostatic pressure.

5. The Symmetry of a Hyperelastic Point

Since a hyperelastic point is a special kind of elastic point, its symmetry can be characterized by the symmetry group of the response function as defined in § 2. Specifically, relative to a fixed reference configuration $\boldsymbol{v}$ a tensor $\boldsymbol{P}$ is a material automorphism if the response function $\boldsymbol{H}$ obeys the transformation rule (2.3) with respect to $\boldsymbol{P}$. For a hyperelastic point, the response function is given by (4.1). Then in terms of the stored-energy function $\hat{\sigma}$ the transformation (2.3) takes the form

$$\left.\frac{\partial\hat{\sigma}}{\partial \boldsymbol{F}}\right|_{\boldsymbol{F}} = \left.\frac{\partial\hat{\sigma}}{\partial \boldsymbol{F}}\right|_{\boldsymbol{FP}} \boldsymbol{P}^T, \tag{5.1}$$

where we have used the condition that $\boldsymbol{P}$ is unimodular, so that the density in the configuration $\boldsymbol{Fv}$ is equal to that in $\boldsymbol{FPv}$.

The conditions (5.1) and (4.11) are formally identical. Just as (4.13) is the integral of (4.11), the integral of (5.1) is TRUESDELL's relation

$$\hat{\sigma}(\boldsymbol{FP}) = \hat{\sigma}(\boldsymbol{F})+\hat{\sigma}(\boldsymbol{P})-\hat{\sigma}(\boldsymbol{1}). \tag{5.2}$$

However, the argument leading to (4.12) does not apply here so as to prove that (5.2) is equivalent to

$$\hat{\sigma}(\boldsymbol{FP}) = \hat{\sigma}(\boldsymbol{F}), \tag{5.3}$$

since, unlike the rotational group $\mathfrak{O}$, the symmetry group is not compact in general. In fact, (5.3) is not always a consequence of (5.2). Thus we are led to the following

DEFINITION. A tensor $\boldsymbol{P}$ satisfying the identity (5.3) is called a *material automorphism for* $\hat{\sigma}$ relative to $\boldsymbol{v}$. The group $\hat{\mathfrak{G}}_{\boldsymbol{v}}$ of all such material automorphisms is called the *energy symmetry group of the hyperelastic point*.

The reader will easily define an *abstract material automorphism for* $\hat{\sigma}$ and prove the following transformation rule:

$$\hat{\mathfrak{G}}_{\boldsymbol{Kv}} = \boldsymbol{K}\hat{\mathfrak{G}}_{\boldsymbol{v}}\boldsymbol{K}^{-1} \tag{5.4}$$

for all $\boldsymbol{K} \in \mathfrak{L}$. From now on we employ some one reference configuration $\boldsymbol{v}$, and we suppress the subscript $\boldsymbol{v}$ in the notations $\hat{\mathfrak{G}}_{\boldsymbol{v}}$ and $\mathfrak{G}_{\boldsymbol{v}}$.

Before going into the details of the solutions for the conditions (5.2) and (5.3), we prove a few preliminary results.

A subgroup $\hat{\mathfrak{G}}$ of a subgroup $\mathfrak{G}$ of the linear group $\mathfrak{L}$ is called *frame-indifferent* if for all $\boldsymbol{K} \in \mathfrak{L}$

$$\boldsymbol{K}\hat{\mathfrak{G}}\boldsymbol{K}^{-1} \cap \mathfrak{O} = \boldsymbol{K}\mathfrak{G}\boldsymbol{K}^{-1} \cap \mathfrak{O} \tag{5.5}$$

in the sense of set theory.

PROPOSITION (TRUESDELL). *The energy symmetry group $\hat{\mathfrak{G}}$ of a hyperelastic point is a frame-indifferent subgroup of the symmetry group $\mathfrak{G}$ of that point.*

Indeed, the fact that $\hat{\mathfrak{G}}$ is a subgroup of $\mathfrak{G}$ is obvious, since (5.3) clearly implies (5.2). Moreover, if $\boldsymbol{P}$ is a rotation, (4.14) shows that the conditions (5.2) and (5.3) become the same. Thus

$$\hat{\mathfrak{G}} \cap \mathfrak{D} = \mathfrak{G} \cap \mathfrak{D}. \tag{5.6}$$

The final result (5.5) then follows from the transformation rules (5.4) and (2.5).

COROLLARY (TRUESDELL). *For a hyperelastic solid,* $\hat{\mathfrak{G}} = \mathfrak{G}$.

As we shall see, $\hat{\mathfrak{G}}$ and $\mathfrak{G}$ coincide also for a hyperelastic fluid, but for a hyperelastic fluid crystal $\hat{\mathfrak{G}}$ and $\mathfrak{G}$ may differ.

Now we derive representations for the stored-energy functions of some special types of hyperelastic points. A general representation for the stored-energy function is given in the next section.

Case 1. Hyperelastic Fluids.

By definition $\mathfrak{G}$ is equal to the special linear group $\mathfrak{U}$, and we have already established the representation (2.22). In order for the fluid to be hyperelastic, by (4.1) it is thus necessary and sufficient that there exist a function $\hat{\sigma}(\boldsymbol{F})$ such that

$$-p(\rho)\mathbf{1} = \rho \boldsymbol{F} \left(\frac{\partial \hat{\sigma}}{\partial \boldsymbol{F}}\right)^T. \tag{5.7}$$

For given $p(\rho)$, this equation can always be solved, and its general solution is

$$\hat{\sigma}(\boldsymbol{F}) = f(\rho), \tag{5.8}$$

where

$$f(\rho) = \int \frac{p(\rho)}{\rho^2}\, d\rho. \tag{5.9}$$

This solution shows not only that $\hat{\mathfrak{G}} = \mathfrak{U}$ but also that *every elastic fluid is hyperelastic*. While the argument presumes the fluid compressible, the two results obtained by it are valid trivially for incompressible fluids also.

Case 2. Hyperelastic Isotropic Solids.

For a hyperelastic solid, we have proved above that $\mathfrak{G} = \hat{\mathfrak{G}}$; if the solid is also isotropic, the conditions that $\hat{\sigma}$ must satisfy in order to be a stored-energy function are

$$\hat{\sigma}(\boldsymbol{FQ}) = \hat{\sigma}(\boldsymbol{F}) \quad \forall \boldsymbol{Q} \in \mathfrak{O} \tag{5.10}$$

and

$$\hat{\sigma}(\boldsymbol{QF}) = \hat{\sigma}(\boldsymbol{F}) \quad \forall \boldsymbol{Q} \in \mathfrak{O}, \tag{4.12}$$

the former being the specialization of (5.3) to the present case, and the reference configuration being undistorted. Clearly, a general solution of (5.10) is

$$\hat{\sigma}(\boldsymbol{F}) = f(\boldsymbol{B}). \tag{5.11}$$

Then from (4.12) f is an isotropic function of $\boldsymbol{B}$, *viz*,

$$f(\boldsymbol{QBQ}^{\mathsf{T}}) = f(\boldsymbol{B}). \tag{5.12}$$

In § 2 we have shown that a general solution of (5.12) is given by

$$\hat{\sigma}(\boldsymbol{F}) = h(I_{\boldsymbol{B}}, II_{\boldsymbol{B}}, III_{\boldsymbol{B}}). \tag{5.13}$$

Substituting this representation formula for the stored-energy function into the constitutive relation (4.1), we obtain the following

PROPOSITION (FINGER). *The response function of a hyperelastic isotropic point has the form*

$$\boldsymbol{T} = \bar{\boldsymbol{H}}(\boldsymbol{B}) = 2\rho\left(III_{\boldsymbol{B}}\frac{\partial h}{\partial III_{\boldsymbol{B}}}\boldsymbol{1} + \left[\frac{\partial h}{\partial I_{\boldsymbol{B}}} + I_{\boldsymbol{B}}\frac{\partial h}{\partial II_{\boldsymbol{B}}}\right]\boldsymbol{B} - \frac{\partial h}{\partial II_{\boldsymbol{B}}}\boldsymbol{B}^2\right) \tag{5.14}$$

if and only if the reference configuration is undistorted.

This result shows that for a hyperelastic isotropic solid the coefficients φ_0, φ_1, and φ_2 in the representation (2.44) are determined from h in the following specific way:

$$\varphi_0 = 2\rho\, III_{\boldsymbol{B}}\frac{\partial h}{\partial III_{\boldsymbol{B}}}, \tag{5.15}$$

$$\varphi_1 = 2\rho\left(\frac{\partial h}{\partial I_{\boldsymbol{B}}} + I_{\boldsymbol{B}}\frac{\partial h}{\partial II_{\boldsymbol{B}}}\right), \tag{5.16}$$

and

$$\varphi_2 = -2\rho\frac{\partial h}{\partial II_{\boldsymbol{B}}}. \tag{5.17}$$

Here we have required that the coefficients φ_0, φ_1, and φ_2 in (2.44) be smooth and hence unique.

In general, the simultaneous equations (5.15)–(5.17) cannot be integrated to obtain the function h unless the functions φ_0, φ_1, and φ_2 satisfy certain conditions of integrability. Thus, in contrast to the case of an elastic fluid, not every isotropic elastic solid is hyperelastic. Conditions of integrability on φ_0, φ_1, and φ_2 are easily obtained from (5.15)–(5.17):

$$\frac{\partial}{\partial II_{\boldsymbol{B}}}\left(\frac{\varphi_0}{\rho}\right) + III_{\boldsymbol{B}}\frac{\partial}{\partial III_{\boldsymbol{B}}}\left(\frac{\varphi_2}{\rho}\right) = 0, \tag{5.18}$$

$$III_{\boldsymbol{B}}\frac{\partial}{\partial III_{\boldsymbol{B}}}\left(\frac{\varphi_1}{\rho}\right) - I_{\boldsymbol{B}}\frac{\partial}{\partial II_{\boldsymbol{B}}}\left(\frac{\varphi_0}{\rho}\right) - \frac{\partial}{\partial I_{\boldsymbol{B}}}\left(\frac{\varphi_0}{\rho}\right) = 0, \tag{5.19}$$

and

$$\frac{\partial}{\partial II_{\boldsymbol{B}}}\left(\frac{\varphi_1}{\rho}\right) + \frac{\partial}{\partial I_{\boldsymbol{B}}}\left(\frac{\varphi_2}{\rho}\right) + I_{\boldsymbol{B}}\frac{\partial}{\partial II_{\boldsymbol{B}}}\left(\frac{\varphi_2}{\rho}\right) = 0. \tag{5.20}$$

Here

$$\rho = \frac{\rho_{\boldsymbol{v}}}{\sqrt{III_{\boldsymbol{B}}}}, \tag{5.21}$$

where $\rho_{\boldsymbol{v}}$ is the density in the reference configuration $\boldsymbol{v}$.

For an incompressible point, the general solution of (5.12) is

$$\hat{\sigma}(\boldsymbol{F}) = h(I_{\boldsymbol{B}}, II_{\boldsymbol{B}}), \tag{5.22}$$

since $III_{\boldsymbol{B}} = 1$. Thus we obtain the following

PROPOSITION (RIVLIN). *The constitutive equation of an incompressible isotropic hyperelastic point has the form*

$$\boldsymbol{T} = -p\mathbf{1} + 2\rho\left(\frac{\partial h}{\partial I_{\boldsymbol{B}}} + I_{\boldsymbol{B}}\frac{\partial h}{\partial II_{\boldsymbol{B}}}\right)\boldsymbol{B} - 2\rho\frac{\partial h}{\partial II_{\boldsymbol{B}}}\boldsymbol{B}^2 \tag{5.23}$$

if and only if the reference configuration is undistorted.

Hence the coefficients φ_1 and φ_2 in (2.56) are related to the stored-energy function h by

$$\varphi_1 = 2\rho\left(\frac{\partial h}{\partial I_{\boldsymbol{B}}} + I_{\boldsymbol{B}}\frac{\partial h}{\partial II_{\boldsymbol{B}}}\right), \tag{5.24}$$

and

$$\varphi_2 = -2\rho \frac{\partial h}{\partial II_B}. \tag{5.25}$$

An alternative form of (5.23) is

$$\boldsymbol{T} = -p\mathbf{1} + 2\rho \frac{\partial h}{\partial I_B} \boldsymbol{B} - 2\rho \frac{\partial h}{\partial II_B} \boldsymbol{B}^{-1}. \tag{5.26}$$

In general the simultaneous equations (5.24) and (5.25) are not integrable. The condition of integrability for them is

$$\frac{\partial}{\partial II_B}(\varphi_1 + I_B \varphi_2) + \frac{\partial}{\partial I_B} \varphi_2 = 0. \tag{5.27}$$

In the linearized theory every isotropic elastic material is hyperelastic, since the elastic constants in the constitutive relation (2.60) satisfy trivially Green's condition (4.28).

Case 3. A Type of Hyperelastic Fluid Crystals.

We consider hyperelastic fluid crystals[1] belonging to Case 3 of § 2. The relative symmetry group $\mathfrak{G}$ of the response function is given by (2.66). We claim that the energy symmetry group $\hat{\mathfrak{G}}$ for this type of fluid crystals must be different from $\mathfrak{G}$.

To see this, suppose that the two symmetry groups $\hat{\mathfrak{G}}$ and $\mathfrak{G}$ were the same. Then the stored-energy function $\hat{\sigma}$ would satisfy the following two conditions

$$\hat{\sigma}(\boldsymbol{F}\boldsymbol{P}) = \hat{\sigma}(\boldsymbol{F}) \quad \forall \boldsymbol{P} \in \mathfrak{G} \tag{5.28}$$

and

$$\hat{\sigma}(\boldsymbol{Q}\boldsymbol{F}) = \hat{\sigma}(\boldsymbol{F}) \quad \forall \boldsymbol{Q} \in \mathfrak{O}. \tag{4.12}$$

But from the factorization (2.71) these two conditions would imply

$$\hat{\sigma}(\boldsymbol{F}) = h(\rho). \tag{5.29}$$

Therefore $\hat{\mathfrak{G}} = \mathfrak{U}$, and hence $\mathfrak{G} = \mathfrak{U}$, contradicting the hypothesis that $\mathfrak{G}$ is given by (2.66) and is different from $\mathfrak{U}$.

The correct conditions on $\hat{\sigma}$, of course, are

$$\hat{\sigma}(\boldsymbol{F}\boldsymbol{P}) = \hat{\sigma}(\boldsymbol{F}) + \hat{\sigma}(\boldsymbol{P}) - \hat{\sigma}(\mathbf{1}) \quad \forall \boldsymbol{P} \in \mathfrak{G} \tag{5.30}$$

---

[1] C.-C. WANG, p. 14 of the work cited on p. 183.

and

$$\hat{\sigma}(\boldsymbol{QF}) = \hat{\sigma}(\boldsymbol{F}) \quad \forall \boldsymbol{Q} \in \mathfrak{D}. \tag{4.12}$$

We proceed to derive a general solution of these two conditions. First we prove that $\hat{\mathfrak{G}}$ consists of tensors $\hat{\boldsymbol{P}}$ having the component form

$$[\hat{P}_{ij}] = \begin{bmatrix} a & b & 0 \\ c & d & 0 \\ e & f & \pm 1 \end{bmatrix}, \tag{5.31}$$

where the basis is that for the component form (2.66), and where $a, \ldots, f$ are real numbers satisfying the condition

$$\det [\hat{P}_{ij}] = 1. \tag{5.32}$$

The conditions (5.31) and (5.32) mean that $\hat{\boldsymbol{P}}$ is a unimodular tensor such that $\hat{\boldsymbol{P}}\boldsymbol{e}_3 = +\boldsymbol{e}_3$ or $\hat{\boldsymbol{P}}\boldsymbol{e}_3 = -\boldsymbol{e}_3$. By direct multiplication of component matrices it can be shown that the group $\hat{\mathfrak{G}}$ consisting of tensors $\hat{\boldsymbol{P}}$ is generated by the collection of tensors whose component matrices are

$$[(\hat{P}_0)_{ij}] = \begin{bmatrix} \pm 1 & 0 & 0 \\ 0 & \pm 1 & 0 \\ 0 & 0 & \pm 1 \end{bmatrix}, \quad \det [\hat{\boldsymbol{P}}_0] = 1, \tag{5.33}$$

$$[(\hat{P}_1)_{ij}] = \begin{bmatrix} 1 & 0 & 0 \\ a & 1 & 0 \\ 0 & 0 & 1 \end{bmatrix}, \quad a \in \mathscr{R}, \tag{5.34}$$

$$[(\hat{P}_2)_{ij}] = \begin{bmatrix} 1 & a & 0 \\ 0 & 1 & 0 \\ 0 & 0 & 1 \end{bmatrix}, \quad a \in \mathscr{R}, \tag{5.35}$$

$$[(\hat{P}_3)_{ij}] = \begin{bmatrix} 1 & 0 & 0 \\ 0 & 1 & 0 \\ a & 0 & 1 \end{bmatrix}, \quad a \in \mathscr{R}, \tag{5.36}$$

and

$$[(\hat{P}_4)_{ij}] = \begin{bmatrix} 1 & 0 & 0 \\ 0 & 1 & 0 \\ 0 & a & 1 \end{bmatrix}, \quad a \in \mathscr{R}. \tag{5.37}$$

Since the tensors $\hat{\boldsymbol{P}}_0$ are orthogonal and belong to $\hat{\mathfrak{G}}$, from (5.5) we see that $\hat{\boldsymbol{P}}_0 \in \hat{\mathfrak{G}}$. Next we prove that $\hat{\boldsymbol{P}}_1 \in \hat{\mathfrak{G}}$. From (5.34), we have

$$[(\hat{P}_1^{-1})_{ij}] = \begin{bmatrix} 1 & 0 & 0 \\ -a & 1 & 0 \\ 0 & 0 & 1 \end{bmatrix}, \tag{5.38}$$

so that

$$\hat{\boldsymbol{P}}_1^{-1} = \hat{\boldsymbol{Q}}_0 \hat{\boldsymbol{P}}_1 \hat{\boldsymbol{Q}}_0^T, \tag{5.39}$$

where

$$[(\hat{Q}_0)_{ij}] = \begin{bmatrix} 1 & 0 & 0 \\ 0 & -1 & 0 \\ 0 & 0 & -1 \end{bmatrix}. \tag{5.40}$$

Comparing (5.40) and (5.33), we see that $\hat{\boldsymbol{Q}}_0 \in \hat{\mathfrak{G}}$. Now from the condition (5.30) and (4.12), we have

$$\hat{\sigma}(\hat{\boldsymbol{P}}_1^{-1}) = \hat{\sigma}(\hat{\boldsymbol{Q}}_0 \hat{\boldsymbol{P}}_1 \hat{\boldsymbol{Q}}_0^T) = \hat{\sigma}(\hat{\boldsymbol{Q}}_0 \hat{\boldsymbol{P}}_1) = \hat{\sigma}(\hat{\boldsymbol{P}}_1). \tag{5.41}$$

On the other hand, from (5.30)

$$\hat{\sigma}(\mathbf{1}) = \hat{\sigma}(\hat{\boldsymbol{P}}_1^{-1}\hat{\boldsymbol{P}}_1) = \hat{\sigma}(\hat{\boldsymbol{P}}_1^{-1}) + \hat{\sigma}(\hat{\boldsymbol{P}}_1) - \hat{\sigma}(\mathbf{1}). \tag{5.42}$$

Combining these two equations, we obtain

$$\hat{\sigma}(\hat{\boldsymbol{P}}_1) = \hat{\sigma}(\mathbf{1}). \tag{5.43}$$

Hence from (5.3) and (5.30) $\hat{\boldsymbol{P}}_1 \in \hat{\mathfrak{G}}$. The same argument shows that $\hat{\boldsymbol{P}}_2$, $\hat{\boldsymbol{P}}_3$, and $\hat{\boldsymbol{P}}_4$ also belong to $\hat{\mathfrak{G}}$. Thus all tensors $\hat{\boldsymbol{P}}$ of the form (5.31) are elements of $\hat{\mathfrak{G}}$.

This result implies that the value $\hat{\sigma}(\boldsymbol{P})$ for $\boldsymbol{P}$ depends only on the absolute value of the $\langle 33 \rangle$ component of $\boldsymbol{P}$, because any two tensors $\boldsymbol{P}_1$ and $\boldsymbol{P}_2$ which are of the form (2.66) and whose $\langle 33 \rangle$ components have the same absolute value differ only by a factor $\hat{\boldsymbol{P}}$ of the form (5.31), say $\boldsymbol{P}_2 = \boldsymbol{P}_1 \hat{\boldsymbol{P}}$. Indeed, $\hat{\boldsymbol{P}} = \boldsymbol{P}_1^{-1}\boldsymbol{P}_2$, since the $\langle 33 \rangle$ component of any tensor $\boldsymbol{P}$ of the form (2.66) is simply the reciprocal of the $\langle 33 \rangle$ component of $\boldsymbol{P}^{-1}$. By (2.66) the square of P_{33} is equal to the $\langle 33 \rangle$ component of the Cauchy-Green tensor $\boldsymbol{C}$ of $\boldsymbol{P}$, so

$$\hat{\sigma}(\boldsymbol{P}) = h(C_{33}). \tag{5.44}$$

We proceed to determine the nature of the function h.

Substituting (5.44) into (5.30), we see that the function h satisfies the condition

$$h(xy) = h(x) + h(y) - h(1) \tag{5.45}$$

for all x, $y \in (0, \infty)$, which is the domain of h. To solve this equation we introduce the transformation

$$\xi = \log x, \tag{5.46}$$

and we put

$$\bar{h}(\xi) = h(x) - h(1). \tag{5.47}$$

Then the condition (5.45) becomes

$$\bar{h}(\xi + \eta) = \bar{h}(\xi) + \bar{h}(\eta) \tag{5.48}$$

for all $\xi, \eta \in \mathscr{R}$. Since $\bar{h}$ is continuous, a general solution of (5.48) is

$$\bar{h}(\xi) = k\xi \tag{5.49}$$

for some constant k.

The constant k cannot be zero, since otherwise from (5.47) we should infer that $\hat{\mathfrak{G}} = \mathfrak{G}$, which we have shown already to be untrue. Inverting the transformation (5.46), we obtain

$$h(x) = h(1) + k \log x. \tag{5.50}$$

Since this equation implies that $\boldsymbol{P} \in \mathfrak{G}$ is in $\hat{\mathfrak{G}}$ only if $|P_{33}| = 1$, it confirms our claim that $\hat{\mathfrak{G}}$ consists of proper unimodular tensors $\boldsymbol{P}$ given by (5.31).

Substituting (5.50) into (5.44) and using the factorization (2.71), we finally obtain a representation for the stored-energy function,

$$\hat{\sigma}(\boldsymbol{F}) = f(\rho) + k \log C_{33}. \tag{5.51}$$

We can compute the response function immediately from this representation by the constitutive eqation (4.1). The result is

$$\boldsymbol{H}(\boldsymbol{F}) = \varphi_0 \mathbf{1} + 2k\rho \boldsymbol{i}_3 \otimes \boldsymbol{i}_3, \tag{5.52}$$

where $\boldsymbol{i}_3$ is given by (2.72). Comparing this equation with (2.74), we see that for this type of fluid crystal the condition for hyperelasticity is that the response coefficient φ_1 in (2.74) be a homogeneous function of ρ of degree 1.

Thus far we have supposed that the fluid crystal is not subject to any internal constraint. For an incompressible hyperelastic point having the same type of crystalline symmetry, the constitutive equation reduces to

$$\boldsymbol{T} = -p\mathbf{1} + \tau \boldsymbol{i}_3 \otimes \boldsymbol{i}_3, \tag{5.53}$$

where p is an undetermined hydrostatic pressure, and where τ is a non-zero material constant which gives the amount of the uniaxial tension in the preferred direction $\boldsymbol{i}_3$. Further, every incompressible elastic point of this type is hyperelastic.

Before closing this section let us make some comments about the procedure leading toward the general solution (5.51). If we rewrite (5.30) in the form

$$\hat{\sigma}(\boldsymbol{F}\boldsymbol{P})-\hat{\sigma}(\boldsymbol{F}) = \hat{\sigma}(\boldsymbol{P})-\hat{\sigma}(\boldsymbol{1}), \tag{5.54}$$

we see that the variation of $\hat{\sigma}$ on any left coset $\mathscr{X}$ of $\mathfrak{G}$ is the same as that on the trivial coset $\mathfrak{G}$ itself. In other words the restriction of $\hat{\sigma}$ on any $\mathscr{X}$ differs from that on $\mathfrak{G}$ by an additive constant only. Hence to solve (5.30) in general, it suffices to solve it only on $\mathfrak{G}$. Now from (5.3) we know that $\hat{\sigma}$ is constant on each left coset of $\hat{\mathfrak{G}}$ in $\mathfrak{G}$. This condition is expressed by (5.44) for the special type of materials considered here. The constant values of $\hat{\sigma}$ on different cosets of $\hat{\mathfrak{G}}$ in $\mathfrak{G}$ are not arbitrary, however. Indeed, (5.30) implies that the function h, which characterizes the restriction of $\hat{\sigma}$ on $\mathfrak{G}$, must satisfy the condition (5.45), which can be solved and the solution is (5.50). The general solution of (5.30) is then found by substituting (5.50) into (5.54). In the next section we shall apply the same procedure to solve the condition (5.30) in general for arbitrary $\mathfrak{G}$.

6. Representation for the Stored-Energy Function of a Hyperelastic Point

In the preceding section we have found representations for the stored-energy functions of three specific types of hyperelastic materials. In this section we derive a representation[1] for stored-energy functions in general. Thus we seek a general solution for the simultaneous identities

$$\hat{\sigma}(\boldsymbol{F}\boldsymbol{P}) = \hat{\sigma}(\boldsymbol{F})+\hat{\sigma}(\boldsymbol{P})-\hat{\sigma}(\boldsymbol{1}) \tag{5.2}$$

and

$$\hat{\sigma}(\boldsymbol{Q}\boldsymbol{F}) = \hat{\sigma}(\boldsymbol{F}) \tag{4.12}$$

---

[1] C.–C. Wang, "On the stored-energy functions of hyperelastic materials", *Arch. Rational Mech. Anal.* **23**, 1–14 (1966). The problem had been set by preliminary work of Wang, "On the symmetries of certain hyperelastic simple liquid crystals", *Proc. Nat. Acad. Sci. U.S.A.* **55**, 468–471 (1966).

for all $Q \in \mathfrak{O}$ and all $P \in \mathfrak{G}$, where $\mathfrak{G}$ is some assigned group. As before, we denote by $\hat{\mathfrak{G}}$ the energy symmetry group. We first prove some preliminary results.

PROPOSITION. *$\hat{\mathfrak{G}}$ is a normal subgroup of $\mathfrak{G}$.*

Proof. If $P \in \mathfrak{G}$ and $\hat{P} \in \hat{\mathfrak{G}}$, we must show that $P\hat{P}P^{-1} \in \hat{\mathfrak{G}}$. Since $P^{-1} \in \mathfrak{G}$, from (5.2) we have

$$\hat{\sigma}(PP^{-1}) = \hat{\sigma}(P)+\hat{\sigma}(P^{-1})-\hat{\sigma}(1). \tag{6.1}$$

Thus

$$\hat{\sigma}(P)+\hat{\sigma}(P^{-1}) = 2\hat{\sigma}(1). \tag{6.2}$$

Similarly,

$$\hat{\sigma}(FP\hat{P}P^{-1}) = \hat{\sigma}(FP\hat{P})+\hat{\sigma}(P^{-1})-\sigma(1). \tag{6.3}$$

But $\hat{P} \in \hat{\mathfrak{G}}$, so that

$$\hat{\sigma}(FP\hat{P}) = \hat{\sigma}(FP). \tag{6.4}$$

Applying (5.2) again, we get

$$\hat{\sigma}(FP) = \hat{\sigma}(F)+\hat{\sigma}(P)-\hat{\sigma}(1). \tag{6.5}$$

Now combining (6.2)–(6.5), we obtain

$$\hat{\sigma}(FP\hat{P}P^{-1}) = \hat{\sigma}(F). \tag{6.6}$$

Thus $P\hat{P}P^{-1} \in \hat{\mathfrak{G}}$. Q.E.D.

PROPOSITION. *The quotient group $\mathfrak{G}/\hat{\mathfrak{G}}$ is an Abelian group.*

Proof. For an arbitrary element $P \in \mathfrak{G}$ let $\{P\}$ denote the $\hat{\mathfrak{G}}$-coset containing P. Then we must show that

$$\{P_1\}\{P_2\} = \{P_2\}\{P_1\} \tag{6.7}$$

for all $P_1, P_2 \in \mathfrak{G}$. This condition amounts to

$$P_1 P_2 P_1^{-1} P_2^{-1} \in \hat{\mathfrak{G}}, \tag{6.8}$$

which does hold in virtue of four successive applications of (5.2).

COROLLARY. *If $\mathfrak{G}$ is a non-Abelian simple group, then $\hat{\mathfrak{G}} = \mathfrak{G}$. In particular, $\hat{\mathfrak{G}} = \mathfrak{G}$ if $\mathfrak{G} = \mathfrak{O}$ or if $\mathfrak{G} = \mathfrak{U}$.*

This corollary reaffirms results obtained in § 5 by more special arguments, namely, that $\hat{\mathfrak{G}} = \mathfrak{G}$ for any isotropic point referred to an undistorted configuration. In the parallel analysis for the condition (4.13) the counterpart of this corollary implies the condition (4.12), which we proved before by integrating the condition (4.13) over the rotational group $\mathfrak{O}$.

So far we have discovered three important properties of the symmetry group $\hat{\mathfrak{G}}$ within the symmetry group $\mathfrak{G}$:

1. $\hat{\mathfrak{G}}$ is a frame-indifferent subgroup of $\mathfrak{G}$.
2. $\hat{\mathfrak{G}}$ is a normal subgroup of $\mathfrak{G}$.
3. $\mathfrak{G}/\hat{\mathfrak{G}}$ is an Abelian group.

We claim that the collection of all subgroups $\tilde{\mathfrak{G}}$ satisfying these three conditions has an infimum in $\mathfrak{G}$ in the sense of set theory. This fact is a consequence of the following

PROPOSITION. *Suppose that two subgroups $\tilde{\mathfrak{G}}_1$ and $\tilde{\mathfrak{G}}_2$ satisfy the three conditions just stated. Then so does the subgroup $\tilde{\mathfrak{G}}_1 \cap \tilde{\mathfrak{G}}_2$.*

The proof is left to the reader.

We denote the intersection of all subgroups[1] $\tilde{\mathfrak{G}}$ satisfying the three conditions as $\tilde{\mathfrak{G}}_0$, which is, of course, the infimum of $\tilde{\mathfrak{G}}$. This infimum $\tilde{\mathfrak{G}}_0$ may or may not be realized as the symmetry group of some stored-energy function for the type of hyperelastic materials under consideration. Since $\mathfrak{G}$ and $\hat{\mathfrak{G}}$ are both closed[2] Lie subgroups of $\mathfrak{U}$, a necessary condition on $\hat{\mathfrak{G}}$ is

$$\dim \mathfrak{G} \geqq \dim \hat{\mathfrak{G}} \geqq \dim \mathfrak{G} - 1. \tag{6.9}$$

Indeed, from (5.2) $\boldsymbol{P} \in \mathfrak{G}$ is in $\hat{\mathfrak{G}}$ if and only if $\hat{\sigma}(\boldsymbol{P}) = \hat{\sigma}(\boldsymbol{1})$. Thus $\hat{\mathfrak{G}}$ is the locus of a scalar equation $\hat{\sigma}(\boldsymbol{P}) - \hat{\sigma}(\boldsymbol{1}) = 0$ on $\mathfrak{G}$. The infimum $\tilde{\mathfrak{G}}_0$, however, may have dimension lower than dim $\mathfrak{G} - 1$.

Next we introduce a function σ_0 defined on the group $\mathfrak{G}$ by

$$\sigma_0(\boldsymbol{P}) = \hat{\sigma}(\boldsymbol{P}) - \hat{\sigma}(\boldsymbol{1}). \tag{6.10}$$

Clearly this function takes a constant value on each coset of $\hat{\mathfrak{G}}$ in $\mathfrak{G}$, so that it can be regarded as a function on the quotient group $\mathfrak{G}/\hat{\mathfrak{G}}$, *viz*,

$$\sigma_0(\{\boldsymbol{P}\}) = \sigma_0(\boldsymbol{P}). \tag{6.11}$$

Then we have the following

---

[1] The set of subgroups $\tilde{\mathfrak{G}}$ is non-empty, since it always contains the group $\mathfrak{G}$ itself.

[2] This follows because the stored-energy function and the response function are assumed to be continuous.

PROPOSITION. σ_0 *is an additive function on the Abelian group* $\mathfrak{G}/\hat{\mathfrak{G}}$.

Proof. We must show that

$$\sigma_0(\{P_1 P_2\}) = \sigma_0(\{P_1\}) + \sigma_0(\{P_2\}) \tag{6.12}$$

for all $P_1, P_2 \in \mathfrak{G}$. From (6.10) this equation amounts to

$$\hat{\sigma}(P_1 P_2) = \hat{\sigma}(P_1) + \hat{\sigma}(P_2) - \hat{\sigma}(1), \tag{6.13}$$

but this relation is a direct consequence of the condition (5.2). Q.E.D.

Since $\hat{\mathfrak{G}}$ satisfies the three conditions given on the preceding page, and since $\tilde{\mathfrak{G}}_0$ is the intersection of all subgroups satisfying these conditions, certainly $\tilde{\mathfrak{G}}_0$ is contained in $\hat{\mathfrak{G}}$. Consequently we can regard σ_0 as an additive function on $\mathfrak{G}/\tilde{\mathfrak{G}}_0$ also. Conversely, if σ_0 is an additive function on $\mathfrak{G}/\tilde{\mathfrak{G}}_0$, then it is *ipso facto* an additive function on $\mathfrak{G}/\hat{\mathfrak{G}}$, where $\hat{\mathfrak{G}}$ is the locus of $\sigma_0(P) = 0$ in $\mathfrak{G}$. However, unlike $\tilde{\mathfrak{G}}_0$, $\hat{\mathfrak{G}}$ is not unique.

As in § 3, let $\mathscr{X}$ denote a left coset of $\mathfrak{G}$ in $\mathfrak{L}$, and let $\mathscr{Z}$ denote a combined coset of $\mathfrak{G}$ and $\mathfrak{O}$. Also, let $F_{\mathscr{X}}$ be a fixed representative element in $\mathscr{X}$ such that the condition (3.9) is satisfied, and let $\mathscr{X}_{\mathscr{Z}}$ be a representative left coset in $\mathscr{Z}$. Then a general solution for the simultaneous identities (5.2) and (4.12) is given by the following

REPRESENTATION THEOREM (WANG). *A stored-energy function* $\hat{\sigma}$ *satisfies the conditions* (5.2) *and* (4.12) *if and only if it has the form*

$$\hat{\sigma}(F) = \sigma_0(P) + \sigma_1(F), \tag{6.14}$$

where P *is the right* $\mathfrak{G}$-*factor of* F *with respect to* $F_{\mathscr{X}}$ *in the decomposition* (2.12), σ_0 *is a function on* $\mathfrak{G}$ *which can be regarded as an additive function on* $\mathfrak{G}/\tilde{\mathfrak{G}}_0$, *and* σ_1 *is a frame-indifferent function whose symmetry group is* $\mathfrak{G}$, i.e.,

$$\sigma_1(F) = \sigma_1(F_{\mathscr{X}_{\mathscr{Z}}}). \tag{6.15}$$

Proof. Necessity is obvious. From the decomposition (3.16) and the identities (5.2) and (4.12), we have

$$\hat{\sigma}(F) = \hat{\sigma}(QF_{\mathscr{X}_{\mathscr{Z}}}P) = \hat{\sigma}(F_{\mathscr{X}_{\mathscr{Z}}}) + \hat{\sigma}(P) - \hat{\sigma}(1). \tag{6.16}$$

Thus σ_0 is given by (6.10) and σ_1 is given by

$$\sigma_1(F) = \hat{\sigma}(F_{\mathscr{X}_{\mathscr{Z}}}). \tag{6.17}$$

Clearly, σ_0 and σ_1 have the properties stated in the theorem.

Conversely, suppose that $\hat{\sigma}$ is of the form (6.14). Then we must show that the conditions (5.2) and (4.12) are satisfied. We prove (5.2) first.

Let $\boldsymbol{P}^*$ be an arbitrary element in $\mathfrak{G}$. Then for any tensor $\boldsymbol{F} \in \mathfrak{L}$, the tensors $\boldsymbol{F}$ and $\boldsymbol{F}\boldsymbol{P}^*$ belong to the same left coset of $\mathfrak{G}$ in $\mathfrak{L}$, say $\mathscr{X}$. Then from (2.12)

$$\boldsymbol{F} = \boldsymbol{F}_{\mathscr{X}}\boldsymbol{P}, \;\; \boldsymbol{F}\boldsymbol{P}^* = \boldsymbol{F}_{\mathscr{X}}(\boldsymbol{P}\boldsymbol{P}^*). \tag{6.18}$$

Hence from (6.15) and (6.14)

$$\hat{\sigma}(\boldsymbol{F}\boldsymbol{P}^*) - \hat{\sigma}(\boldsymbol{F}) = \sigma_0(\boldsymbol{P}\boldsymbol{P}^*) - \sigma_0(\boldsymbol{P}). \tag{6.19}$$

Now since σ_0 is an additive function on $\mathfrak{G}/\tilde{\mathfrak{G}}_0$,

$$\sigma_0(\boldsymbol{P}\boldsymbol{P}^*) - \sigma_0(\boldsymbol{P}) = \sigma_0(\boldsymbol{P}^*). \tag{6.20}$$

Thus it remains to prove that

$$\sigma_0(\boldsymbol{P}^*) = \hat{\sigma}(\boldsymbol{P}^*) - \sigma(\mathbf{1}). \tag{6.21}$$

Clearly $\mathfrak{G}$ forms a left coset by itself. Hence from (6.14)

$$\hat{\sigma}(\boldsymbol{P}^*) = \sigma_0(\bar{\boldsymbol{P}}) + \sigma_1(\boldsymbol{P}^*), \tag{6.22}$$

and

$$\hat{\sigma}(\mathbf{1}) = \sigma_0(\boldsymbol{F}_{\mathfrak{G}}^{-1}) + \sigma_1(\mathbf{1}). \tag{6.23}$$

where $\boldsymbol{F}_{\mathfrak{G}}$ is the representative element of $\mathfrak{G}$, and where

$$\boldsymbol{P}^* = \boldsymbol{F}_{\mathfrak{G}}\bar{\boldsymbol{P}}. \tag{6.24}$$

Subtracting (6.23) from (6.22) and using (6.15), we obtain

$$\hat{\sigma}(\boldsymbol{P}^*) - \hat{\sigma}(\mathbf{1}) = \sigma_0(\bar{\boldsymbol{P}}) - \sigma_0(\boldsymbol{F}_{\mathfrak{G}}^{-1}). \tag{6.25}$$

From the additivity of σ_0 and the factorization (6.24), we see that the right-hand side of (6.25) is equal to $\sigma_0(\boldsymbol{P}^*)$. Thus (6.21) holds, and the proof of (5.2) is complete.

Next we prove (4.12). Let $\boldsymbol{Q}^*$ be an arbitrary element of $\mathfrak{O}$. Then the left coset $\mathscr{X}$ containing $\boldsymbol{F}$ and the left coset $\mathscr{X}^*$ containing $\boldsymbol{Q}^*\boldsymbol{F}$ belong to the same combined coset $\mathscr{Z}$. Hence

$$\boldsymbol{F} = \boldsymbol{Q}_{\mathscr{X}_{\mathscr{Z}}\mathscr{X}}\boldsymbol{F}_{\mathscr{X}_{\mathscr{Z}}}\boldsymbol{P}, \;\; \boldsymbol{Q}^*\boldsymbol{F} = \boldsymbol{Q}_{\mathscr{X}_{\mathscr{Z}}\mathscr{X}^*}\boldsymbol{F}_{\mathscr{X}_{\mathscr{Z}}}\boldsymbol{P}^*, \tag{6.26}$$

where $\boldsymbol{P}, \boldsymbol{P}^* \in \mathfrak{G}$. From (6.14) and (6.15),

$$\hat{\sigma}(\boldsymbol{Q}^*\boldsymbol{F}) - \hat{\sigma}(\boldsymbol{F}) = \sigma_0(\boldsymbol{P}^*) - \sigma_0(\boldsymbol{P}). \tag{6.27}$$

Thus it remains to prove that

$$\sigma_0(\boldsymbol{P}^*)-\sigma_0(\boldsymbol{P}) = 0. \tag{6.28}$$

From (6.26), we have

$$\boldsymbol{F}_{\mathcal{X}\mathfrak{X}}\boldsymbol{P}^*\boldsymbol{P}^{-1}\boldsymbol{F}_{\mathcal{X}\mathfrak{X}}^{-1} = \boldsymbol{Q}_{\mathcal{X}\mathfrak{X}\mathcal{X}^*}^T\boldsymbol{Q}^*\boldsymbol{Q}_{\mathcal{X}\mathfrak{X}\mathcal{X}}. \tag{6.29}$$

Since $\boldsymbol{P}^*\boldsymbol{P}^{-1} \in \mathfrak{G}$, the condition (5.5) implies that the tensor $\boldsymbol{P}^*\boldsymbol{P}^{-1}$ belongs to all frame-indifferent subgroups of $\mathfrak{G}$. Hence

$$\boldsymbol{P}^*\boldsymbol{P}^{-1} \in \tilde{\mathfrak{G}}_0. \tag{6.30}$$

From the additivity of σ_0 on $\mathfrak{G}/\tilde{\mathfrak{G}}_0$, this condition implies that

$$0 = \sigma_0(\boldsymbol{P}^*\boldsymbol{P}^{-1}) = \sigma_0(\boldsymbol{P}^*)-\sigma_0(\boldsymbol{P}). \tag{6.31}$$

Thus (6.28) holds, and the proof is complete.

As in § 3, we leave the positions of the specific representations (5.8), (5.13), and (5.51) within the framework of the general representation to the reader.

7. Inequalities in Elasticity, I. General Background

In the preceding sections we have considered the restrictions imposed upon the response functions of elastic points or hyperelastic points by the general theory of constitutive relations, mainly through the principle of material frame-indifference and the rule of material symmetry. Response functions delimited only by these principles, however, are still too general in that nothing regarding the direction of response is said. This excessive generality is recognized and excised in the classical linearized theory of isotropic elastic materials, where the two elasticities, λ and μ in (2.60), are not left entirely arbitrary constants, as they would be if no further limitation were imposed. In order that (2.60) describe a response that is physically reasonable, the inequalities

$$\mu > 0, \quad 3\lambda+2\mu > 0 \tag{7.1}$$

are usually laid down as further axioms. These inequalities may be motivated by several different kinds of argument. For example, $(7.1)_1$ is necessary and sufficient that in some one simple shear (*cf.* § IV.3, in the next chapter), and hence in all simple shears, the shear stress be directed in the same sense as the shear effected, and $(7.1)_2$ is necessary and sufficient that tension be needed to produce some one uniform dilatation, and hence any uniform dilatation, while pressure be needed to produce any uniform condensation, both of which requirements seem to be confirmed by experience with bodies in nature. In this section we first summarize other implications of the classical inequalities (7.1). Then we shall consider the more general problem of finding restrictions upon the response functions within the framework of the non-linear theory of elasticity. This problem was examined in detail by TRUESDELL & TOUPIN[1]; see also NFTM §§ 51–53.

Using "C" as a mnemonic for "classical", we shall call (7.1) the *C-inequalities*. Within the context of the classical linearized theory, they arise naturally in connection with the following lines of thought.

A. Statics. Besides the above-mentioned implications in special strains, in any strain whatever $(7.1)_1$ is necessary and sufficient that the shearing stress on any plane shall have the same direction as the shear of that plane, while $(7.1)_2$ is necessary and sufficient that the mean tension have the same sign as the increment in volume.

B. Work theorem. The C-inequalities are necessary and sufficient that the work done in any non-rigid local deformation be positive.

C. Stability. The C-inequalities are sufficient but not necessary in order that for every homogeneous body the deformation that satisfies the conditions of equilibrium shall correspond to a lesser total stored energy than does any other deformation having the same boundary values. Necessary and sufficient conditions are given by the weaker inequalities.

$$\mu > 0, \quad \lambda + 2\mu > 0. \tag{7.2}$$

D. Uniqueness theorems. The C-inequalities are sufficient in order that the mixed boundary-value problem of equilibrium, for smooth enough boundaries and boundary data, have a solution that is unique, at least to within a rigid displacement. Moreover, the C-inequalities or their opposites are also necessary in order that, for a general region, *both* the place

[1] C. TRUESDELL & R. A. TOUPIN, "Static grounds for inequalities in finite strain of elastic materials", *Arch. Rational Mech. Anal.* **12**, 1–33 (1963).

This problem was proposed in a lecture by TRUESDELL, "Das ungelöste Hauptproblem der endlichen Elastizitätstheorie", *Z. angew. Math. Mech.* **36**, 97–103 (1956). English translation in *Continuum Mechanics III, Foundations of Elasticity Theory*, Int. Sci. Rev. Ser. Vol. VIII, ed. C. TRUESDELL, New York, Gordon & Breach, 1965.

boundary-value problem and the traction boundary-value problem have unique solutions, although they are not necessary for general uniqueness of solution of either of these problems by itself. They are sufficient but not necessary for the uniqueness of solution of the equations of motion corresponding to assigned initial displacements and velocities upon the boundary surface.

E. Wave speeds.[1] The C-inequalities are sufficient but not necessary in order that the squared speeds of waves[2] corresponding to all amplitudes satisfying the propagation condition be positive. Necessary and sufficient conditions, again, are the less restrictive inequalities (7.2).

In the general theory, whether a full set of restrictions such as to ensure physically reasonable behavior in any possible deformation may be established once and for all, is a matter still under discussion. Here we shall follow TRUESDELL & TOUPIN[3], who have explored implications of most of the proposed restrictions – in particular, the inequality of COLEMAN & NOLL and the condition of strong ellipticity. As we shall see in the next section, the former requires that the response function be monotone under a certain restriction, while the latter is necessary and sufficient that the acoustic tensor[4] for acceleration waves or waves of higher order be positive-definite, for every direction of propagation. For hyperelastic materials, the two conditions require that the stored-energy function be restrictedly convex, and that the squared speeds corresponding to all amplitudes satisfying the propagation condition for acceleration waves or waves of higher order be positive. Finally, for the limiting case of isotropic linear elasticity, the two conditions reduce to (7.1) and (7.2), respectively. Thus in this limiting case the Coleman-Noll inequality is stronger than the condition of strong ellipticity. In finite strain, however, neither implies the other in general.

A one-parameter family of inequalities generalizing the Coleman-Noll inequality has been formulated and studied recently by HILL, and a par-

---

[1] The theory of acceleration waves propagating in elastic bodies is presented in Chapter VI.

[2] *I.e.* propagating singular surfaces of first or higher order. Waves of first order are commonly called *shock waves*; those of second or higher order, *acceleration waves* or *higher-order waves*.

[3] *Cf.* TRUESDELL & TOUPIN, *op. cit. ante*.

[4] *Cf.* (VI. 3.34).

ticular one of them has been recommended by him as a general constitutive inequality for elastic materials.

In the next section we shall state the Coleman-Noll inequality, the condition of strong ellipticity, and the Hill inequalities, and then in § 9 we shall deduce and analyse their various consequences, especially for isotropic elastic or hyperelastic materials.

8. Inequalities in Elasticity, II. Statement of Major Inequalities

A. The Generalized Coleman-Noll Inequality

A transformation f of some inner-product space into itself is called *monotone* (*increasing*) if

$$(f(x)-f(y))\cdot(x-y)>0 \tag{8A.1}$$

for every pair of distinct points x, y. When the dimension of the space is 1, this concept of monotonicity clearly reduces to the usual increasing monotonicity for real functions of one real variable. TRUESDELL & TOUPIN[1], generalizing a proposal of COLEMAN & NOLL, set up for study as a possible *a priori* inequality in elasticity the requirement that the transformation from deformation gradient to Piola-Kirchhoff stress tensor be monotone with respect to pairs of deformations differing from one another by a pure stretch. We shall call such a condition of restricted monotonicity the *generalized Coleman-Noll condition*, or briefly the *GCN condition*.

Let the constitutive relation for the Piola-Kirchhoff stress of an elastic point relative to a reference configuration $\boldsymbol{\kappa}$ be

$$\boldsymbol{T}_{\boldsymbol{\kappa}}=\boldsymbol{A}(\boldsymbol{F}). \tag{8A.2}$$

Then from (8A.1), the GCN condition is the inequality[2]

$$\operatorname{tr}\{[\boldsymbol{A}(\boldsymbol{F}^*)-\boldsymbol{A}(\boldsymbol{F})][\boldsymbol{F}^{*T}-\boldsymbol{F}^T]\}>0 \tag{8A.3}$$

---

[1] *Cf.* NFTM, § 52. COLEMAN & NOLL themselves, also motivated by their own earlier work on thermostatics, proposed the same inequality somewhat later and called it the "weakened thermostatic inequality".

[2] For the space of second-order tensors, the inner product of $\boldsymbol{A}$ and $\boldsymbol{B}$ is given by $\operatorname{tr}(\boldsymbol{A}\boldsymbol{B}^T)$. *Cf.* (I.2.31).

provided that

$$F^* = SF, \tag{8A.4}$$

where S is positive-definite, symmetric, and not equal to $\mathbf{1}$.

Clearly, some restriction such as (8A.4) is needed here, since (8A.3) cannot hold for arbitrary F and F^*. To see this, we need only choose $F = \mathbf{1}$ and $F^* = Q$, where Q is a rotation. From (II.6.32), $A(F)$ is related to $H(F)$ by

$$A(F) = (\det F)H(F)(F^T)^{-1}. \tag{8A.5}$$

Hence if the reference configuration is a natural state, $A(\mathbf{1}) = \mathbf{0}$, and from (1.16) and (8A.5), also $A(Q) = \mathbf{0}$. Thus for this choice of F and F^*, (8A.3) cannot hold. The condition (8A.4) serves to forbid the particular choice of F and F^* we have just considered. Of course, it does not follow that this particular supplementary condition is the only one under which it would make sense to lay down the basic inequality (8A.3) as an axiom. The specific restriction (8A.4) can be justified only indirectly by verifying that it leads to acceptable consequences.

The GCN condition may be laid down as an *a priori* requirement on all F in the domain of A, or as a characteristic distinguishing some such F from others, say only for those related by positive-definite, symmetric S satisfying $\|S-\mathbf{1}\| < \varepsilon$ for some given ε. For simplicity, we shall not pursue the refinement due to this weaker requirement. In the one special case we now consider, consequences of the general requirement or the weaker requirement are obviously the same.

To apply the GCN condition to the classical linearized theory, we choose $F = \mathbf{1}$, $F^* = \mathbf{1}+E$, where E represents an infinitesimal strain. Then (8A.3) reduces to the requirement that

$$\operatorname{tr}(T^*E) = \operatorname{tr}\{\mathsf{L}(E)E\} > 0 \tag{8A.6}$$

provided that $E \neq \mathbf{0}$. As is usual in the classical theory, the reference configuration is here assumed to be a natural state. The condition (8A.6) means that the fourth-order tensor of elasticity L (*cf.* (1.36)) is positive-definite with respect to the space of symmetric second-order tensors. This is the usual *a priori* inequality of the linearized theory. Thus the GCN condition includes and generalizes all the requirements customarily imposed in the classical infinitesimal theory.[1] In particular, for an isotropic material GCN $\Rightarrow$ C.

In finite strain the GCN condition implies the condition of uniqueness

$$A(\boldsymbol{SF}) \neq A(\boldsymbol{F}) \tag{8A.7}$$

for any $\boldsymbol{F}$ and any positive-definite, symmetric $\boldsymbol{S}$ other than $\mathbf{1}$. This means that if a pure stretch be superimposed upon an arbitrary state of strain, the Piola-Kirchhoff stress tensor cannot remain unchanged thereby. It does not follow from the GCN condition, however, that the Cauchy stress tensor is necessarily changed by a pure stretch.

For an elastic body[2], the GCN condition implies the following

UNIQUENESS THEOREM (TRUESDELL & TOUPIN). *If all points of an elastic body obey the GCN condition, then a given mixed boundary-value problem of place and traction cannot have two solutions that differ from each other by a convex potential deformation.*

As usual, we say that two configurations χ and $\bar{\chi}$ of a body *differ by a potential deformation* if the deformation gradient from χ to $\bar{\chi}$ is a symmetric tensor at every point. That is, locally, there exists a *displacement potential* $\varphi = \varphi(x^i)$ such that the deformation functions

$$\bar{x}^i = \bar{x}^i(x^j) \tag{8A.8}$$

from $\chi(\mathscr{B})$ to $\bar{\chi}(\mathscr{B})$ are given by

$$\bar{x}^i = \frac{\partial \varphi}{\partial x^i}. \tag{8A.9}$$

A potential deformation is *convex* if its deformation gradient is positive-definite and symmetric at every point. This means that the Hessian form of the displacement potential is positive-definite.

---

[1] Here by the "classical infinitesimal theory" we mean the theory of CAUCHY, in which there need be no stored-energy function. The reduction to the theory of GREEN, which follows by assuming also the existence of a stored-energy function, neither affects the *a priori* inequality nor is affected by it.

[2] The theory of the elastic body is presented in Chapters IV and V. Here we assume the usual condition of smoothness sufficient for the validity of the divergence theorem.

To prove the uniqueness theorem, let χ and χ^* be two equilibrium configurations of the elastic body $\mathscr{B}$ subject to the same body force $\boldsymbol{b}$ and the same surface traction $\boldsymbol{t}$ or displacement $\boldsymbol{u}$ on the boundary of $\mathscr{B}$ in a reference configuration $\boldsymbol{\kappa}$. Then from (II. 6.26), the Piola-Kirchhoff stress tensors $\boldsymbol{T}_\kappa$ and $\boldsymbol{T}_\kappa^*$ for χ and χ^*, respectively, must satisfy the equation of equilibrium:

$$\text{Div}\, \boldsymbol{T}_\kappa + \rho_\kappa \boldsymbol{b} = \boldsymbol{0} \tag{8A.10}$$

and

$$\text{Div}\, \boldsymbol{T}_\kappa^* + \rho_\kappa \boldsymbol{b} = \boldsymbol{0} \tag{8A.11}$$

in $\boldsymbol{\kappa}(\mathscr{B})$ and the boundary conditions

$$\boldsymbol{u} = \boldsymbol{u}^* \quad \text{or} \quad \boldsymbol{t}_\kappa = \boldsymbol{t}_\kappa^* \tag{8A.12}$$

on $\partial\boldsymbol{\kappa}(\mathscr{B})$. Now taking the inner product of Div $(\boldsymbol{T}_\kappa - \boldsymbol{T}_\kappa^*)$ with $\boldsymbol{u} - \boldsymbol{u}^*$, integrating the result by parts, and then using the divergence theorem together with the boundary conditions (8A.12), we obtain

$$\int_{\kappa(\mathscr{B})} \text{tr}\, \{(\boldsymbol{T}_\kappa^* - \boldsymbol{T}_\kappa)(\boldsymbol{F}^* - \boldsymbol{F})^T\} dV_\kappa = 0. \tag{8A.13}$$

Hence if χ and χ^* differ by a convex potential deformation, then $\boldsymbol{F}$ and $\boldsymbol{F}^*$ are related by (8A.4) with $\boldsymbol{S}$ being the positive-definite tensor field grad(grad φ). Hence from (8A.3) the integrand on the left-hand side of (8A.13) is positive, so that that equation cannot hold. Thus the theorem is proved.

This argument is essentially a generalization to finite elasticity of the one invented by KIRCHHOFF to prove his uniqueness theorem in the classical infinitesimal theory. We leave it to the reader to see that when specialized to the infinitesimal theory this same proof does indeed yield KIRCHHOFF's theorem, in the statement of which "convex potential deformation" is replaced by "non-rigid deformation".

Next, we relate the GCN condition to a differential inequality involving the gradient of the response function $\boldsymbol{A}$. We set

$$\mathsf{A}_k{}^A{}_m{}^B = \mathsf{A}_k{}^A{}_m{}^B(\boldsymbol{F}) \equiv \frac{\partial A_k{}^A}{\partial F^m{}_B}. \tag{8A.14}$$

The fourth-order tensor $\boldsymbol{\mathsf{A}}$ with components $\mathsf{A}_k{}^A{}_m{}^B$ is the gradient of $\boldsymbol{A}$ with respect to $\boldsymbol{F}$. We claim that the GCN condition implies the differential inequality[1]

$$\mathsf{A}_k{}^A{}_m{}^B F^p{}_A F^q{}_B D^k_p D^m_q \geqq 0 \tag{8A.15}$$

[1] *Cf.* NFTM, § 52.

for all symmetric tensors $\boldsymbol{D}$. To see this we consider the straight line segment

$$\boldsymbol{F}(\tau) = (\mathbf{1}+\tau\boldsymbol{D})\boldsymbol{F}, \quad \boldsymbol{D} \neq \mathbf{0}. \tag{8A.16}$$

Since $\mathbf{1}+\tau\boldsymbol{D}$ is positive-definite and symmetric for sufficiently small τ, we may substitute $\boldsymbol{F}(\tau)$ for $\boldsymbol{F}^*$ in (8A.3), obtaining

$$\tau[f(\tau)-f(0)] > 0, \tag{8A.17}$$

where $f(\tau)$ is the stress power along the straight line segment, *viz*,

$$f(\tau) = \operatorname{tr}\{(\boldsymbol{DF})^T\boldsymbol{A}((\mathbf{1}+\tau\boldsymbol{D})\boldsymbol{F})\}. \tag{8A.18}$$

Now from (8A.17), we have $\dot{f}(0) \geqq 0$. But from (8A.18), $\dot{f}(0)$ is equal to the right-hand side of (8A.15). Thus (8A.15) is proved.

We define components of the fourth-order tensor B by

$$\mathsf{B}_k{}^p{}_m{}^q \equiv \frac{1}{\det \boldsymbol{F}} \mathsf{A}_k{}^A{}_m{}^B F^p{}_A F^q{}_B. \tag{8A.19}$$

In terms of B the condition (8A.15) becomes

$$\mathsf{B}^{kmpq} D_{km} D_{pq} \geqq 0 \tag{8A.20}$$

for all symmetric tensors $\boldsymbol{D}$. We shall call the strict inequality corresponding to (8A.20), *viz*,

$$\mathsf{B}^{kmpq} D_{km} D_{pq} > 0 \tag{8A.21}$$

for all non-zero symmetric $\boldsymbol{D}$, the *GCN$^+$ condition.*

The distinction between the GCN condition and the GCN$^+$ condition can be seen most easily from the restrictions they impose upon the stress power f, defined by (8A.18). Since any $\boldsymbol{F}$ and $\boldsymbol{F}^*$ related by (8A.4) can be joined in the domain of $\boldsymbol{A}$ by a straight line segment of the form (8A.16) with $\boldsymbol{D} = \boldsymbol{S}-\mathbf{1}$ and $\tau \in [0, 1]$, *the GCN condition is necessary and sufficient that the stress power $f(\tau)$ be a monotone increasing function of τ on all straight line segments of the form* (8A.16). On the other hand, *the GCN$^+$ condition is necessary and sufficient that the initial slope $\dot{f}(0)$ of the stress power $f(\tau)$ be strictly positive on all straight line segments of the form* (8A.16)[1]. In fact, we can even assert the following more symmetrical condition[1]: *The GCN$^+$ condition is necessary and sufficient*

[1] *Cf.* NFTM, § 52, and M. Hayes, "Static implications of the strong-ellipticity condition", *Arch. Rational Mech. Anal.* **33**, 181–191 (1969).

that the slope $\dot{f}(\tau)$ of the stress power $f(\tau)$ be strictly positive on all straight line segments of the form (8A.16). The sufficiency is obvious, since $\dot{f}(\tau) > 0$ for any τ clearly implies $\dot{f}(0) > 0$ for $\tau = 0$. Conversely, for any τ_0 such that $\mathbf{1}+\tau_0\boldsymbol{D}$ is positive-definite, we can rewrite the straight line segment (8A.16) as

$$\bar{\boldsymbol{F}}(\bar{\tau}) = (\mathbf{1}+\bar{\tau}\bar{\boldsymbol{D}})\boldsymbol{F}(\tau_0), \tag{8A.22}$$

where

$$\bar{\tau} = \tau-\tau_0, \; \bar{\boldsymbol{D}} = \boldsymbol{D}(\mathbf{1}+\tau_0\boldsymbol{D})^{-1} = (\mathbf{1}+\tau_0\boldsymbol{D})^{-1}\boldsymbol{D}. \tag{8A.23}$$

For definiteness, we denote the stress power along the straight line segment (8A.22) by $\bar{f}(\bar{\tau})$, *i.e.*,

$$\bar{f}(\bar{\tau}) = \operatorname{tr}\{(\bar{\boldsymbol{D}}\boldsymbol{F}(\tau_0))^T\boldsymbol{A}((\mathbf{1}+\bar{\tau}\bar{\boldsymbol{D}})\boldsymbol{F}(\tau_0))\}. \tag{8A.24}$$

Since (8A.22) is merely (8A.16) after a change of parameter $(8\text{A}.23)_1$, $\bar{f}(\bar{\tau})$ and $f(\tau)$ are related by

$$\bar{f}(\bar{\tau}) = f(\bar{\tau}+\tau_0). \tag{8A.25}$$

In particular, we have

$$\dot{\bar{f}}(0) = \dot{f}(\tau_0). \tag{8A.26}$$

Now applying the first necessary and sufficient condition for GCN$^+$ to the straight line (8A.22), we see that $\dot{\bar{f}}(0) > 0$. From (8A.26), this means $\dot{f}(\tau_0) > 0$. But since τ_0 is arbitrary, the second necessary and sufficient condition for GCN$^+$ is proved.

Having proved the second, more symmetrical necessary and sufficient condition for GCN$^+$, we can assert immediately that

$$\text{GCN}^+ \Rightarrow \text{GCN}, \tag{8A.27}$$

since $\dot{f} > 0$ clearly implies that f is a monotone increasing function. The converse of (8A.27) is not true in general, of course.

So far, we have considered the GCN condition for elastic materials in general. For hyperelastic materials, we now show that the GCN condition reduces to the requirement that the stored-energy function $\hat{\sigma}$ be a *restricted convex function*. Specifically, the condition on $\hat{\sigma}$ is

$$\hat{\sigma}(\boldsymbol{F}^*)-\hat{\sigma}(\boldsymbol{F})-\operatorname{tr}\{(\boldsymbol{F}^*-\boldsymbol{F})\hat{\sigma}_{\boldsymbol{F}}(\boldsymbol{F})^T\} > 0 \tag{8A.28}$$

for all $\boldsymbol{F}$ and $\boldsymbol{F}^*$ satisfying (8A.4). The sufficiency is obvious, since from

(8A.5) and (4.1) we can rewrite (8A.28) as

$$\hat{\sigma}(\boldsymbol{F}^*)-\hat{\sigma}(\boldsymbol{F})-\frac{1}{\rho_{\kappa}}\operatorname{tr}\{(\boldsymbol{F}^*-\boldsymbol{F})\boldsymbol{A}(\boldsymbol{F})^T\}>0. \tag{8A.29}$$

Now interchanging $\boldsymbol{F}$ and $\boldsymbol{F}^*$ and adding the result to (8A.29), we obtain (8A.3). Conversely, for any $\boldsymbol{F}$ and $\boldsymbol{F}^*$ related by (8A.4) integration along the straight line (8A.16) with $\boldsymbol{D}=\boldsymbol{S}-\mathbf{1}$ and $\tau\in[0,1]$, followed by application of (8A.3), shows that

$$\begin{aligned}&\hat{\sigma}(\boldsymbol{F}^*)-\hat{\sigma}(\boldsymbol{F})-\operatorname{tr}\{(\boldsymbol{F}^*-\boldsymbol{F})\hat{\sigma}_{\boldsymbol{F}}(\boldsymbol{F})^T\}\\&\qquad=\frac{1}{\rho_{\kappa}}\int_0^1\operatorname{tr}\{(\boldsymbol{F}(\tau)-\boldsymbol{F})^T(\boldsymbol{A}(\boldsymbol{F}(\tau))-\boldsymbol{A}(\boldsymbol{F}))\}\frac{d\tau}{\tau}>0.\end{aligned} \tag{8A.30}$$

Thus for a hyperelastic material the conditions (8A.28) and (8A.3) are equivalent[1].

TRUESDELL & TOUPIN called the condition (8A.28) the *Coleman-Noll condition* or, briefly, the *C-N condition.* For a hyperelastic material having a natural state, which we shall choose to be the reference configuration as usual, the C-N condition implies

$$\hat{\sigma}(\boldsymbol{S})>\hat{\sigma}(\mathbf{1}) \tag{8A.31}$$

for all positive-definite, symmetric tensors $\boldsymbol{S}$ other than $\mathbf{1}$. Hence in terms of the reduced stored-energy function $\bar{\sigma}$ defined by (4.18) we have

$$\bar{\sigma}(\boldsymbol{C})>\bar{\sigma}(\mathbf{1}) \tag{8A.32}$$

for all positive-definite, symmetric $\boldsymbol{C}$ other than $\mathbf{1}$. Thus $\bar{\sigma}$ has a strict minimum at a natural state. Further, any two natural states can differ at most by a rotation. In view of (8A.32), we see that the C–N condition reduces in the infinitesimal theory the statement that the stored-energy function is a positive-definite quadratic form in the classical strain tensor $\boldsymbol{E}$. In particular, we see again that in isotropic linear elasticity C–N $\Rightarrow$ C.

---

[1] In our text we regard the GCN inequality as being imposed on the response function $\boldsymbol{A}$ at all of its arguments. TRUESDELL & TOUPIN, on the contrary, considered the inequality as a local one, imposed only in some part of the domain. To justify their proof by integration along straight lines, they included the assumption that the domain of $\boldsymbol{A}$ was convex. While sufficient for the proof, this assumption is not only too strong but also inappropriate, since the usual domain of $\boldsymbol{A}$ is not convex. The appropriate assumption for TRUESDELL & TOUPIN's form of the theorem would be that the domain of $\boldsymbol{A}$ contain all straight lines joining any two of its points that differ by a pure stretch.

B. The Strong Ellipticity Condition and Infinitesimal Stability.

For an elastic body, from (8A.2) and (II. 6.26) the requirement that the differential equations of equilibrium be strongly elliptic is[1]

$$\mathsf{A}_k{}^A{}_m{}^B u^k u^m v_A v_B > 0 \tag{8B.1}$$

for all non-vanishing vectors $\boldsymbol{u}$ and $\boldsymbol{v}$. From (8A.19), this condition is equivalent to

$$\mathsf{B}^{kmpq} u_k v_m u_p v_q > 0 \tag{8B.2}$$

for arbitrary non-vanishing $\boldsymbol{u}$ and $\boldsymbol{v}$. TRUESDELL & TOUPIN called (8B.1) or (8B.2) the *strong ellipticity condition* or briefly the *S-E condition.* As mentioned in the preceding section, this S–E condition is necessary and sufficient that the acoustic tensor for acceleration waves or higher-order waves be positive-definite, for every direction of propagation. Since in the infinitesimal theory the S–E condition reduces to the condition (7.2), which is too weak to assure physically reasonable response in all circumstances, we do not lay it down as a postulate, but we propose it for study, and we shall now determine its various consequences.

First, comparing (8B.2) with (8A.21), we see that the S–E condition can be expressed in the form (8A.17) also, provided that the tensors $\boldsymbol{D}$ there be replaced by tensors of rank 1. Since GCN$^+$ and S–E are stated in terms of different sets of tensors $\boldsymbol{D}$, the requirements for GCN$^+$ and S–E are not the same.

Second, replacing "$>$" by "$\geqq$" in (8B.1) or (8B.2), we obtain the *Hadamard condition* or briefly the *H condition*:

$$\mathsf{A}_k{}^A{}_m{}^B u^k u^m v_A v_B \geqq 0 \tag{8B.3}$$

for all vectors $\boldsymbol{u}$ and $\boldsymbol{v}$. Unlike the GCN condition, which was put forward as possibly sufficient to ensure physically reasonable response of a material in all circumstances, the H condition turns out to be a necessary condition for stability in any equilibrium configuration of an elastic body subject to given body force and mixed boundary condition.

TRUESDELL & NOLL called an equilibrium configuration $\boldsymbol{\kappa}$ subject to given conditions *infinitesimally stable* if the work done in every infinitesimal deformation compatible with the boundary conditions is not less than that required to effect the same infinitesimal deformation subject to

---

[1] *Cf.* NFTM, §§ 44, 52, 68, 71.

"dead loads", *i.e.*, at the same state of stress as in the given configuration $\boldsymbol{\kappa}$, which in stability theory is often called "the ground state". Since for an infinitesimal deformation

$$F = \mathbf{1}+H \tag{8B.4}$$

and

$$T_{\boldsymbol{\kappa}} = T_0+\mathsf{A}_0(H), \tag{8B.5}$$

where $T_0 = A(\mathbf{1})$ and $\mathsf{A}_0 = \mathsf{A}(\mathbf{1})$, we see that the local work done less the work done by the dead loads is

$$\operatorname{tr}\{(T_{\boldsymbol{\kappa}}-T_0)(F-\mathbf{1})^T\} = \operatorname{tr}\{\mathsf{A}_0(H)H^T\}. \tag{8B.6}$$

Hence the ground state $\boldsymbol{\kappa}$ is infinitesimally stable if

$$\int_{\boldsymbol{\kappa}(\mathscr{B})} \operatorname{tr}\{\mathsf{A}_0(H)H^T\}dV_{\boldsymbol{\kappa}} \geqq 0 \tag{8B.7}$$

for gradients H of all displacement fields compatible with the given boundary conditions. Our remark about (8B.3) is then contained in the following theorem, due essentially to HADAMARD[1]: *In order that a configuration* $\boldsymbol{\kappa}$ *of an elastic body* $\mathscr{B}$ *be infinitesimally stable for boundary conditions of place or traction or both mixed, it is necessary that the Hadamard condition* (8B.3) *hold at each point of* $\boldsymbol{\kappa}(\mathscr{B})$.

We shall give a proof of this theorem in § VII.8, in the course of our discussion of stability in general.

Like the GCN$^+$ condition, the S–E condition can be expressed as a restriction upon the stress power along some straight lines in the domain of A, as we shall now see. For the GCN$^+$ condition the straight lines used

[1] The formulation is that of TRUESDELL & NOLL, HADAMARD's having been expressed differently and limited to the boundary-value problem of place for hyperelastic materials. In HADAMARD's proof there is a gap, as was remarked by DUHEM. The first correct proof of HADAMARD's assertion was published by CATTANEO in 1946, although, as ERICKSEN has pointed out to us and as has been thereupon explained in detail by FICHERA, a theorem of GRAVES entails it as a corollary. *Cf.* L. M. GRAVES, "The Weierstrass condition for multiple integral variation problems", *Duke Math. J.* **5**, 656–660 (1939), and G. FICHERA, "Sulla propagazione delle onde in un mezzo elastico", *Papers in Honor of the* 80th *Birthday of N.I. Muskhelisvili*, Tbilissi, in press, where FICHERA gives also his own proof of HADAMARD's assertion. The theorem as stated in our text above is due to NOLL, and his proof may be found in § 68bis of NFTM. Our proof of it in § VII.8 is a slight improvement of his.

were of the form (8A.16). For the S–E condition, we need straight lines of the form

$$\boldsymbol{F}(\tau) = (\mathbf{1}+\tau \boldsymbol{u} \otimes \boldsymbol{v})\boldsymbol{F}, \tag{8B.8}$$

where $\boldsymbol{u}$ and $\boldsymbol{v}$ are arbitrary unit vectors, and where τ belongs to some interval $[0, \tau^*]$ such that

$$1+\tau^* \boldsymbol{u} \cdot \boldsymbol{v} > 0. \tag{8B.9}$$

This condition ensures that

$$\det(\mathbf{1}+\tau \boldsymbol{u} \otimes \boldsymbol{v}) = 1+\tau \boldsymbol{u} \cdot \boldsymbol{v} > 0 \tag{8B.10}$$

for all $\tau \in [0, \tau^*]$. We shall explain later the physical meaning of a deformation whose gradient has the form $\mathbf{1}+\tau \boldsymbol{u} \otimes \boldsymbol{v}$. Here we consider first the relation between the S–E condition and the stress power $f(\tau)$ on the straight line (8B.8). It should be noted that (8B.8) can be expressed in the form (8A.16) provided that $\boldsymbol{D}$ be a tensor of rank 1. Hence the argument here is formally the same as the previous argument in connection with the GCN^+ condition. Specifically, we can assert that *the S–E condition is necessary and sufficient that the initial slope $\dot{f}(0)$ of the stress power $f(\tau)$ be strictly positive on all straight lines of the form* (8B.8)[1].

To prove this, notice first that the stress power $f(\tau)$ along the straight line (8B.8) is given by

$$f(\tau) = \operatorname{tr}\{[(\boldsymbol{u} \otimes \boldsymbol{v})\boldsymbol{F}]^T \boldsymbol{A}((\mathbf{1}+\tau \boldsymbol{u} \otimes \boldsymbol{v})\boldsymbol{F})\}. \tag{8B.11}$$

This equation is the same as (8A.18) provided that $\boldsymbol{D}$ be replaced by $\boldsymbol{u} \otimes \boldsymbol{v}$. As before, we differentiate (8B.11) with respect to τ and then evaluate the result at $\tau = 0$, obtaining

$$\dot{f}(0) = \mathsf{A}_k{}^A{}_m{}^B F^p{}_A F^q{}_B u^k v_p u^m v_q . \tag{8B.12}$$

Since $\boldsymbol{F}$ is non-singular, comparing (8B.12) with (8B.1), we have proved our assertion regarding the S–E condition.

As before, we can state also a second, more symmetrical condition: *The S–E condition is necessary and sufficient that the slope $\dot{f}(\tau)$ of the stress power $f(\tau)$ be strictly positive on all straight lines of the form* (8B.8). Sufficiency is again obvious. To prove necessity, choosing any fixed $\tau_0 \in [0, \tau^*]$ as before, we can rewrite the straight line (8B.8) as

$$\bar{\boldsymbol{F}}(\bar{\tau}) = (\mathbf{1}+\bar{\tau} \boldsymbol{u} \otimes \boldsymbol{v})\boldsymbol{F}(\tau_0) \tag{8B.13}$$

---

[1] *Cf.* HAYES, *op. cit.* p. 217.

where

$$\bar{\tau} = \frac{\tau - \tau_0}{1 + \tau_0 \boldsymbol{u} \cdot \boldsymbol{v}}. \tag{8B.14}$$

As before, we denote the stress power along the straight line (8B.13) by $\bar{f}(\bar{\tau})$. From (8B.14), we have

$$\frac{d}{d\bar{\tau}} \bar{\boldsymbol{F}}(\bar{\tau}) = (1 + \tau_0 \boldsymbol{u} \cdot \boldsymbol{v}) \frac{d}{d\tau} \boldsymbol{F}(\tau). \tag{8B.15}$$

Hence the stress powers $\bar{f}(\bar{\tau})$ and $f(\tau)$ are related by

$$\bar{f}(\bar{\tau}) = (1 + \tau_0 \boldsymbol{u} \cdot \boldsymbol{v}) f(\bar{\tau}(1 + \tau_0 \boldsymbol{u} \cdot \boldsymbol{v}) + \tau_0). \tag{8B.16}$$

Differentiating (8B.16) with respect to $\bar{\tau}$ and evaluating the derivatives at $\bar{\tau} = 0$ yields

$$\bar{f}'(0) = (1 + \tau_0 \boldsymbol{u} \cdot \boldsymbol{v})^2 \dot{f}(\tau_0), \tag{8B.17}$$

where $\bar{f}'(\bar{\tau})$ denotes the derivative of $\bar{f}(\bar{\tau})$ with respect to $\bar{\tau}$. Now applying the first necessary and sufficient condition for S–E to the straight line (8B.13), we see that $\bar{f}'(0) > 0$. From (8B.17) this means $\dot{f}(\tau_0) > 0$. But since τ_0 is arbitrary, the second necessary and sufficient condition for S–E is proved.

In view of the similarity between the S–E condition and the GCN$^+$ condition, we may regard the S–E condition as the strict differential inequality associated with a certain *integrated strong ellipticity condition*, just as the GCN$^+$ condition is the strict differential inequality associated with the GCN condition. Obviously, this integrated strong ellipticity condition has the form (8A.3) also, except that now, in contrast with (8A.4), $\boldsymbol{F}$ and $\boldsymbol{F}^*$ must be related by

$$\boldsymbol{F}^* = \boldsymbol{P}\boldsymbol{F}, \tag{8B.18}$$

where $\boldsymbol{P}$ has the form

$$\boldsymbol{P} = \mathbf{1} + \tau \boldsymbol{u} \otimes \boldsymbol{v} \tag{8B.19}$$

with τ satisfying the condition $(8B.10)_2$. For definiteness, let us denote this integrated strong ellipticity condition as the *ISE condition.* Then we can assert that *the ISE condition is necessary and sufficient that the stress power $f(\tau)$ be a monotone increasing function of τ on all straight lines of the form* (8B.8), and also

$$\text{ISE} \Rightarrow \text{H}, \tag{8B.20}$$

and

$$\text{S–E} \Rightarrow \text{ISE}, \tag{8B.21}$$

where the converses of both implications fail in general. The proofs of these statements are parallel to those given in connection with the C–N and GCN conditions in subsection 8A. For the implication (8B.20), we need only to remind the reader that the H condition is the counterpart of the condition (8A.15).

Having discussed the formal similarity between S–E and GCN$^+$ and between ISE and GCN, we now turn to the physical meaning of the deformation gradient $\boldsymbol{P}$ given by (8B.19). HAYES[1] defines a *pure extension* as a homogeneous deformation of the form

$$x^i = X^i + \tau u^i(v_j X^j), \tag{8B.22}$$

where u^i, v^j are components of unit vectors $\boldsymbol{u}$, $\boldsymbol{v}$, and where τ is a non-negative number satisfying the condition $(8\text{B}.10)_2$. Differentiating (8B.22) with respect to X^A, we see that the deformation gradient of a pure extension has the form (8B.19). Thus we may say that while the GCN condition requires the Piola-Kirchhoff stress to be monotone with respect to deformations differing by a pure stretch, the S–E condition requires the same with respect to deformations differing by a pure extension.

To interpret pure extensions, without loss of generality we may choose $\boldsymbol{v}$ to be the basis vector $\boldsymbol{e}_1$ and $\boldsymbol{u}$ to be a unit vector on the (1, 2)-plane of a Cartesian co-ordinate system induced by the orthonormal basis $\{\boldsymbol{e}_i\}$. Then we have

$$\boldsymbol{u} = \sin\theta\,\boldsymbol{e}_1 + \cos\theta\,\boldsymbol{e}_2 \tag{8B.23}$$

and (8B.22) becomes

$$x^1 = X^1 + \tau X^1 \sin\theta,\quad x^2 = X^2 + \tau X^1 \cos\theta,\quad x^3 = X^3. \tag{8B.24}$$

Such a deformation can be illustrated by the adjoining sketch:

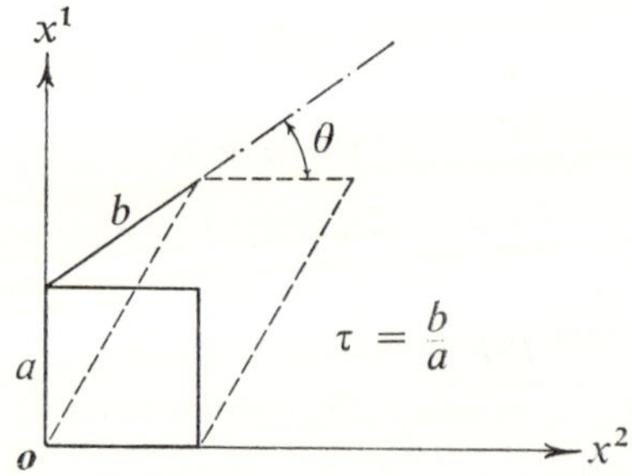

[1] *Cf.* HAYES *op. cit.* p. 217.

where the solid line denotes the reference configuration, while the dotted line denotes the actual configuration. The parameter τ, as given by the ratio of b to a, may be called the *amount of extension*. From this sketch we see that there are three simple special cases[1]:

$$\left.\begin{array}{l}\text{Simple shear}\\ \text{Simple tension}\\ \text{Simple compression}\end{array}\right\} \Leftrightarrow \left\{\begin{array}{l}\theta = 0\\ \theta = \pi/2\\ \theta = -\pi/2\end{array}\right. . \tag{8B.25}$$

Further, in each case τ reduces to the amount of shear, extension, or compression, respectively, as usually defined.

The supplementary condition (8B.18), like the condition (8A.4), also rules out the undesirable case in which $\boldsymbol{F}^* = \boldsymbol{QF}$ for a rotation $\boldsymbol{Q}$. However, since a pure extension in general is not a symmetric tensor, the two kinds of supplementary conditions, (8A.4) or (8B.18), are distinct. Likewise, straight lines of the forms (8A.16) or (8B.8) are different. For a straight line of the form (8A.16), each $\boldsymbol{S}(\tau)$ is a positive-definite, symmetric tensor; further, all $\boldsymbol{S}(\tau)$ share a principal basis, and their proper numbers $s_1(\tau)$, $s_2(\tau)$, $s_3(\tau)$ satisfy the condition that the ratios of the form

$$\frac{s_a(\tau)-1}{s_b(\tau)-1} \tag{8B.26}$$

are independent of τ for all $a, b = 1, 2, 3$, provided that the denominator do not vanish. On the other hand, for a straight line of the form (8B.8), each $\boldsymbol{P}(\tau)$ is a pure extension; further, all $\boldsymbol{P}(\tau)$ share the same unit vectors $\boldsymbol{u}$, $\boldsymbol{v}$, and their determinant $d(\tau)$ satisfies the condition

$$\frac{d(\tau)-1}{d(\bar{\tau})-1} = \frac{\tau}{\bar{\tau}} \tag{8B.27}$$

for all τ and $\bar{\tau}$, provided that $\bar{\tau} \neq 0$.

C. *Hill's Inequalities*

We have seen in the preceding two parts of this section that the GCN and the S-E inequalities are restrictions on the stress power on some particular *families of straight lines* in the domain of $\boldsymbol{A}$. HILL[2] has recently proposed

---

[1] *Cf.* HAYES, *op. cit.* p. 217.

[2] R. HILL, "On constitutive inequalities for simple materials – I", *J. Mech. Phys. Solids* **16**, 229–242 (1968).

some inequalities which, as we shall see now, are restrictions on the stress power on some particular *families of curves* in the domain of $\boldsymbol{A}$. It turns out that the family of straight lines of the form (8A.16) is included in HILL's families of curves; further, for that particular family HILL's inequality reduces to the GCN$^+$ inequality. In this sense, HILL's inequalities are generalizations of the GCN$^+$ condition. The family of straight lines of the form (8B.8), however, is not included in HILL's families, so that HILL's inequalities are not generalizations of the S–E condition. We shall explain the mathematical statements of HILL's inequalities first. Their motivations we shall give later.

To define HILL's inequalities, we start from the concept of *generalized strains* and their conjugate *generalized stresses.* In linear algebra, a real-valued function $g(x)$ of a real variable x gives rise to an isotropic, symmetric tensor-valued function $\boldsymbol{g}(\boldsymbol{A})$ of a symmetric tensor variable $\boldsymbol{A}$ by the spectral form

$$\boldsymbol{g}(\boldsymbol{A}) = g(a_i)\boldsymbol{e}_i \otimes \boldsymbol{e}_i, \tag{8C.1}$$

where

$$\boldsymbol{A} = a_i \boldsymbol{e}_i \otimes \boldsymbol{e}_i \tag{8C.2}$$

is a spectral form of $\boldsymbol{A}$. Here i is summed[1] from 1 to the dimension of the underlying inner-product space. Since the function g is arbitrary, the transformation $\boldsymbol{g}$ on the space of symmetric tensors, generally, is non-linear[2]. In particular, a straight line of the form

$$\boldsymbol{A}(\tau) = \boldsymbol{A}_0 + \tau\dot{\boldsymbol{A}} \tag{8C.3}$$

in the domain of $\boldsymbol{g}$ is transformed by $\boldsymbol{g}$ generally into a curve:

$$\boldsymbol{g}(\boldsymbol{A}(\tau)) = \boldsymbol{g}(\boldsymbol{A}_0 + \tau\dot{\boldsymbol{A}}) \tag{8C.4}$$

in the range of $\boldsymbol{g}$. Conversely, if $\boldsymbol{g}$ is invertible, a straight line of the form

$$\boldsymbol{B}(\tau) = \boldsymbol{B}_0 + \tau\dot{\boldsymbol{B}} \tag{8C.5}$$

in the range of $\boldsymbol{g}$ is transformed by $\boldsymbol{g}^{-1}$ into a curve:

$$\boldsymbol{g}^{-1}(\boldsymbol{B}(\tau)) = \boldsymbol{g}^{-1}(\boldsymbol{B}_0 + \tau\dot{\boldsymbol{B}}) \tag{8C.6}$$

in the domain of $\boldsymbol{g}$. As we shall see, HILL's families of curves are precisely those that can be expressed in the form (8C.6) relative to some particular

[1] The summation convection becomes awkward for spectral forms. This exception will effect eqs. (8C.1), (8C.2), (8C.11), (8C.12), (8C.13), (8C.15) *etc.*

[2] For this reason, we do not denote $\boldsymbol{g}$ by a symbol for a fourth-order tensor.

g – that is, families of pre-images of straight lines, each with respect to a particular g. Of course, this condition is meaningful only if we have reduced the function $\boldsymbol{A}$ to a symmetric tensor-valued function, since the transformation $\boldsymbol{g}$ is defined for symmetric tensors only. Such a reduction of $\boldsymbol{A}$ will be carried out as usual with the help of the principle of material frame-indifference.

More specifically, for any deformation gradient $\boldsymbol{F}$, HILL[1] in effect defines the *generalized strain tensors* $\boldsymbol{E}_g$ and $\bar{\boldsymbol{E}}_g$ relative to a function g by

$$\boldsymbol{E}_g \equiv \boldsymbol{g}(\boldsymbol{U}),\ \bar{\boldsymbol{E}}_g \equiv \boldsymbol{g}(\boldsymbol{V}), \tag{8C.7}$$

provided that g be a monotone increasing smooth function such that

$$g(1) = 0, \quad g'(1) = 1. \tag{8C.8}$$

Here $\boldsymbol{U}$ and $\boldsymbol{V}$ denote, as usual, the right stretch tensor and the left stretch tensor of $\boldsymbol{F}$. The condition $(8C.8)_1$ means simply that $\boldsymbol{E}_g$ and $\bar{\boldsymbol{E}}_g$ vanish at the reference configuration, as is only natural for a strain measure. Moreover, a positive proper number of $\boldsymbol{E}_g$ or $\bar{\boldsymbol{E}}_g$ corresponds to an extension, while a negative one corresponds to a compression. As we shall see, the condition $(8C.8)_2$ means that the generalized strain rates $\dot{\boldsymbol{E}}_g(0)$ and $\dot{\bar{\boldsymbol{E}}}_g(0)$ are equal to the stretching tensor $\boldsymbol{D}(0)$ at the reference configuration along any smooth curve $\boldsymbol{F} = \boldsymbol{F}(\tau)$ passing through $\mathbf{1}$ at $\tau = 0$. Since we already know $\boldsymbol{D}(0) = \dot{\boldsymbol{U}}(0) = \dot{\boldsymbol{V}}(0) = \dot{\boldsymbol{E}}(0)$ at the reference configuration, the preceding condition implies that both $\boldsymbol{E}_g$ and $\bar{\boldsymbol{E}}_g$ are close to the classical strain tensor $\boldsymbol{E}$ when $\boldsymbol{F}$ is sufficiently close to $\mathbf{1}$. This remark justifies the use of the name "generalized strain tensors" for $\boldsymbol{E}_g$ and $\bar{\boldsymbol{E}}_g$.

Since the function g is assumed to be monotone, the transformation $\boldsymbol{g}$ is invertible; further, since $\boldsymbol{U}$ and $\boldsymbol{V}$ are positive-definite, in order to define $\boldsymbol{E}_g$ and $\bar{\boldsymbol{E}}_g$ the function g need be specified on the positive real axis $(0, \infty)$ only. Since $\boldsymbol{g}$ is isotropic, (II.1.32) gives

$$\boldsymbol{E}_g = \boldsymbol{R}^T \bar{\boldsymbol{E}}_g \boldsymbol{R},\ \bar{\boldsymbol{E}}_g = \boldsymbol{R}\boldsymbol{E}_g \boldsymbol{R}^T, \tag{8C.9}$$

where $\boldsymbol{R}$ is the rotation of $\boldsymbol{F}$ as usual. Hence $\boldsymbol{E}_g$ is related to $\bar{\boldsymbol{E}}_g$ as $\boldsymbol{U}$ is related to $\boldsymbol{V}$.

We proceed now to prove that

$$\dot{\boldsymbol{E}}_g(0) = \dot{\bar{\boldsymbol{E}}}_g(0) = \dot{\boldsymbol{U}}(0) = \dot{\boldsymbol{V}}(0) = \dot{\boldsymbol{E}}(0) = \boldsymbol{D}(0) \tag{8C.10}$$

[1] HILL did not give any name to $\boldsymbol{E}_g$ or to $\bar{\boldsymbol{E}}_g$.

along all curves $\boldsymbol{F} = \boldsymbol{F}(\tau)$ with $\boldsymbol{F}(0) = \mathbf{1}$. Let

$$\boldsymbol{U}(\tau) = u_i(\tau)\boldsymbol{e}_i(\tau) \otimes \boldsymbol{e}_i(\tau) \tag{8C.11}$$

be a spectral form of $\boldsymbol{U}(\tau)$. Then from (8C.7) a spectral form of $\boldsymbol{E}_g(\tau)$ is

$$\boldsymbol{E}_g(\tau) = g(u_i(\tau))\boldsymbol{e}_i(\tau) \otimes \boldsymbol{e}_i(\tau). \tag{8C.12}$$

Differentiating (8C.11) and (8C.12) with respect to τ yields

$$\dot{\boldsymbol{U}} = \{\dot{u}_i\delta_{ij} + (u_i - u_j)w_{ij}\}\boldsymbol{e}_i \otimes \boldsymbol{e}_j, \tag*{(8C.13)$_1$}$$

and

$$\dot{\boldsymbol{E}}_g = \{g'(u_i)\dot{u}_i\delta_{ij} + (g(u_i) - g(u_j))w_{ij}\}\boldsymbol{e}_i \otimes \boldsymbol{e}_j, \tag*{(8C.13)$_2$}$$

where

$$w_{ij} = \dot{\boldsymbol{e}}_i \cdot \boldsymbol{e}_j, \tag{8C.14}$$

so that

$$\dot{\boldsymbol{e}}_i = w_{ij}\boldsymbol{e}_j, \quad w_{ij} = -w_{ji}. \tag{8C.15}$$

Since $g'(1) = 1$ and $\boldsymbol{U}(0) = \mathbf{1}$, from (8C.13) we obtain

$$\dot{\boldsymbol{E}}_g(0) = \dot{\boldsymbol{U}}(0). \tag{8C.16}$$

Exactly the same argument shows also that

$$\dot{\hat{\boldsymbol{E}}}_g(0) = \dot{\boldsymbol{V}}(0). \tag{8C.17}$$

Since in § II.1 we have proved (8C.10)$_{3,4,5}$, from (8C.16) and (8C.17) the proof of (8C.10) is complete.

Next, HILL[1] defines the *generalized stress* $\boldsymbol{T}_g$ relative to g as a symmetric tensor such that on any smooth curve $\boldsymbol{F} = \boldsymbol{F}(\tau)$ the stress power $f(\tau)$ satisfies the condition

$$f(\tau) = \operatorname{tr}(\boldsymbol{T}_g(\tau)\dot{\boldsymbol{E}}_g(\tau)). \tag{8C.18}$$

We shall show that for any elastic material $\boldsymbol{T}_g$ exists uniquely and is a function of $\boldsymbol{E}_g$.

To see that f is a linear form in $\dot{\boldsymbol{E}}_g$ at each fixed $\boldsymbol{E}_g$, we notice first that the transformation $\boldsymbol{g}$ from $\boldsymbol{U}$ to $\boldsymbol{E}_g$ is a diffeomorphism. Hence as usual we have

$$\dot{\boldsymbol{E}}_g = [\nabla \boldsymbol{g}](\dot{\boldsymbol{U}}) \tag{8C.19}$$

---

[1] *Cf.* HILL, *op. cit.* p. 225. HILL did not give any name to $\boldsymbol{T}_g$.

on any smooth curve $\boldsymbol{U}(\tau)$ passing through $\boldsymbol{U}$. Here $\nabla \boldsymbol{g}$ is a fourth-order tensor and is an automorphism on the space of symmetric tensors. Since the stress power f is given by[1]

$$f = \operatorname{tr}(\dot{\boldsymbol{F}}\boldsymbol{A}(\boldsymbol{F})^T) = \frac{\rho_\kappa}{\rho}\operatorname{tr}(\boldsymbol{T}\boldsymbol{D}), \tag{8C.20}$$

by use of the polar decomposition (II.1.32) of $\boldsymbol{F}$ we can rewrite (8C.20) in the form

$$f = \frac{\rho_\kappa}{\rho}\operatorname{tr}(\boldsymbol{U}^{-1}\boldsymbol{R}^T\boldsymbol{T}\boldsymbol{R}\dot{\boldsymbol{U}}), \tag{8C.21}$$

which shows that f is a linear form[2] in $\dot{\boldsymbol{U}}$; further, from the reduced form (1.18) the combination $\boldsymbol{R}^T\boldsymbol{T}\boldsymbol{R}$ is a function of $\boldsymbol{U}$ only, so that linear form in $\dot{\boldsymbol{U}}$ is a function of $\boldsymbol{U}$. Now since $\boldsymbol{U}$ is uniquely determined by $\boldsymbol{E}_g$, and since by the inverse of (8C.19) $\dot{\boldsymbol{U}}$ is linear in $\dot{\boldsymbol{E}}_g$, from (8C.21) we see that f is a linear form in $\dot{\boldsymbol{E}}_g$ also; further, that linear form is a function of $\boldsymbol{E}_g$. Thus $\boldsymbol{T}_g$ exists, and we have

$$\boldsymbol{T}_g = \hat{\boldsymbol{T}}_g(\boldsymbol{E}_g). \tag{8C.22}$$

In view of the basic condition (8C.18), we say that $(\boldsymbol{T}_g, \boldsymbol{E}_g)$ forms a pair of *conjugate variables*. Sometimes we also say that $\boldsymbol{T}_g$ is the *generalized stress tensor conjugate* to $\boldsymbol{E}_g$, and that $\boldsymbol{E}_g$ is the *generalized strain tensor conjugate* to $\boldsymbol{T}_g$. We shall also refer to (8C.22) as the *generalized constitutive relation* and $\hat{\boldsymbol{T}}_g(\boldsymbol{E}_g)$ the *generalized response function.* It should be noted that the principle of material frame-indifference is satisfied by (8C.22), since that equation is derived from the reduced form (1.18).

Since g is arbitrary, the interpretation of $\boldsymbol{T}_g$ in terms of force and area is not simple in general. In the reference configuration, however, $\boldsymbol{T}_g$ reduces to the Cauchy stress $\boldsymbol{T}$ in that configuration for all g. Indeed, from (8C.10) the conditions (8C.18) and (8C.21) become the same at $\tau = 0$ with $\boldsymbol{F}(0) = \boldsymbol{1}$. This remark serves to justify somewhat more the use of the name "generalized stress tensor" for $\boldsymbol{T}_g$.

---

[1] Here as in the preceding subsections f is the stress power per unit volume of the reference configuration κ. Since $\operatorname{tr}(\boldsymbol{T}\boldsymbol{D})$ gives only the stress power per unit volume of the actual configuration, we need the factor ρ_κ/ρ on the right-hand side of (8C.20).

[2] For an anisotropic material, the stress power is not a linear form in $\dot{\boldsymbol{V}}$, so we cannot define a generalized stress conjugate to $\bar{\boldsymbol{E}}_g$ by a condition similar to (8C.18) with $\dot{\boldsymbol{E}}_g(\tau)$ replaced by $\dot{\bar{\boldsymbol{E}}}_g(\tau)$.

Assuming $\hat{T}_g(E_g)$ to be a smooth function as usual, we have

$$\dot{T}_g = [\nabla \hat{T}_g](\dot{E}_g). \tag{8C.23}$$

For brevity we denote the value of $\nabla \hat{T}_g$ at $E_g = \mathbf{0}$ by E_g. This fourth-order tensor is the *generalized elasticity tensor* relative to g at the reference configuration. In his paper HILL proposes to consider and interpret the assertion that E_g be positive-definite over the space of symmetric tensors at every reference configuration, *i.e.*,

$$\operatorname{tr}\{[\mathsf{E}_g](D)D\} > 0 \tag{8C.24}$$

for all symmetric tensors D other than $\mathbf{1}$. This assertion we shall call HILL's *inequality relative to g*, or more briefly *the H_g inequality*. As we shall see, the GCN^+ conditon is just the H_{g_0} inequality, where g_0 is the linear function

$$g_0(x) = x-1. \tag{8C.25}$$

Clearly we can interpret the H_g inequality in terms of the initial slope $\dot{f}(0)$ on any straight line of the form

$$E_g = \tau D, \tag{8C.26}$$

where D is an arbitrary symmetric tensor. In fact, by exactly the same argument as before, we can assert that *the H_g inequality is necessary and sufficient that the initial slope $\dot{f}(0)$ of the stress power $f(\tau)$ be strictly positive on all straight lines of the form* (8C.26). It is understood that the reference configuration, used implicitly in this condition, is arbitrary.

Before going any further, we remark first that a straight line (8C.26) in the E_g-space, generally, does not correspond to a straight line in the U-space, the only exception being the case when g is equal to the linear function g_0 defined by (8C.25), since in that case we have

$$E_{g_0} = U-\mathbf{1}, \tag{8C.27}$$

so that (8C.26) corresponds to the straight line

$$U(\tau) = \mathbf{1}+\tau D, \tag{8C.28}$$

which is the same as (8A.16) with $F = \mathbf{1}$. The requirement that $F = \mathbf{1}$, of course, can be removed by a change of reference configuration, since the reference configuration for any H_g inequality is arbitrary. Thus we have shown that H_{g_0} and GCN^+, since both are equivalent to the requirement that $\dot{f}(0) > 0$ in the same family of motions $F = F(\tau)$, are the same condition.

The function $g_0(x)$ defined by (8C.25) is the only linear function satisfying the conditions (8C.8). Consequently g is not linear if $g \neq g_0$, and hence straight lines of the form (8C.26) in the $\boldsymbol{E}_g$-space correspond to curves of the form

$$\boldsymbol{U}(\tau) = \boldsymbol{g}^{-1}(\tau\boldsymbol{D}) \tag{8C.29}$$

in the $\boldsymbol{U}$-space. *The* H_g *inequality requires that* $\dot{f}(0) > 0$ *on all curves of the form* (8C.29). However, the H_g inequality is not necessarily different from the GCN$^+$ condition even if $g \neq g_0$. In fact, we shall prove that H_g and $\mathrm{H}_{\tilde{g}}$ are the same if

$$g''(1) = \tilde{g}''(1). \tag{8C.30}$$

Hence even if the function g is non-linear, H_g may still be the same as the GCN$^+$ condition; for example, g may be the function $(x-1)^3+x-1$.

In order to determine the dependence of H_g on g, we express $g(x)$ in Taylor series about $x = 1$

$$g(x) = (x-1)+\alpha(x-1)^2+\beta(x-1)^3+\ldots, \tag{8C.31}$$

where conditions (8C.8) have been used, and where $2\alpha = g''(1)$, $6\beta = g'''(1)\ldots$. From $(8C.1)_1$ and (8C.31), $\boldsymbol{E}_g$ is given by

$$\boldsymbol{E}_g = (\boldsymbol{U}-\boldsymbol{1})+\alpha(\boldsymbol{U}-\boldsymbol{1})^2+\beta(\boldsymbol{U}-\boldsymbol{1})^3+\ldots. \tag{8C.32}$$

Hence on any smooth curve $\boldsymbol{U}(\tau)$, $\dot{\boldsymbol{E}}_g$ and $\dot{\boldsymbol{U}}$ are related by

$$\dot{\boldsymbol{E}}_g = \dot{\boldsymbol{U}}+\alpha(\dot{\boldsymbol{U}}(\boldsymbol{U}-\boldsymbol{1})+(\boldsymbol{U}-\boldsymbol{1})\dot{\boldsymbol{U}})+\beta(\dot{\boldsymbol{U}}(\boldsymbol{U}-\boldsymbol{1})^2+(\boldsymbol{U}-\boldsymbol{1})\dot{\boldsymbol{U}}(\boldsymbol{U}-\boldsymbol{1}) \\ +(\boldsymbol{U}-\boldsymbol{1})^2\dot{\boldsymbol{U}})+\ldots. \tag{8C.33}$$

Now from the basic condition (8C.18) we have

$$\operatorname{tr}(\boldsymbol{T}_g\dot{\boldsymbol{E}}_g) = \operatorname{tr}(\boldsymbol{T}_{g_0}\dot{\boldsymbol{U}}), \tag{8C.34}$$

where g_0 is the linear function defined by (8C.25). Substituting (8C.33) into (8C.34), we get

$$\boldsymbol{T}_{g_0} = \boldsymbol{T}_g+\alpha(\boldsymbol{T}_g(\boldsymbol{U}-\boldsymbol{1})+(\boldsymbol{U}-\boldsymbol{1})\boldsymbol{T}_g)+\beta(\boldsymbol{T}_g(\boldsymbol{U}-\boldsymbol{1})^2+(\boldsymbol{U}-\boldsymbol{1})\boldsymbol{T}_g(\boldsymbol{U}-\boldsymbol{1}) \\ +(\boldsymbol{U}-\boldsymbol{1})^2\boldsymbol{T}_g)+\ldots, \tag{8C.35}$$

since $\dot{\boldsymbol{U}}$ is arbitrary. This condition implies that $\boldsymbol{T}_{g_0} = \boldsymbol{T}_g = \boldsymbol{T}$ at $\boldsymbol{U} = \boldsymbol{1}$, a result we already remarked. Differentiating (8C.35) with respect to

U at $U = \mathbf{1}$ and using (8C.10), we obtain

$$\mathsf{E}_{g_0}(D) = \mathsf{E}_g(D) + \alpha(TD + DT). \tag{8C.36}$$

From (8C.36), for any two functions g and $\tilde{g}$ we have

$$(\mathsf{E}_g - \mathsf{E}_{\tilde{g}})(D) = -\frac{m}{2}(TD + DT), \tag{8C.37}$$

where

$$m = 2(\alpha - \tilde{\alpha}) = g''(1) - \tilde{g}''(1). \tag{8C.38}$$

In component form (8C.37) is[1]

$$\mathsf{E}_{ijkl} - \tilde{\mathsf{E}}_{ijkl} = -\frac{m}{4}(T_{ik}\delta_{jl} + T_{il}\delta_{jk} + T_{jk}\delta_{il} + T_{jl}\delta_{ik}). \tag{8C.39}$$

This condition shows that if $m = 0$, or if $T = \mathbf{0}$, then E_g and $\mathsf{E}_{\tilde{g}}$ are the same tensor, and hence H_g and $\mathrm{H}_{\tilde{g}}$ are the same condition. From (8C.39) all E_g are the same at a natural state, and they reduce there to the classical elasticity tensor L. *Hence each H_g inequality includes and generalizes all the requirements customarily imposed in the classical linearized theory.*

Having proved that E_g, and hence H_g, depends only on the second derivative $g''(1)$, we see that HILL's inequalities can be exhausted by a one-parameter family of inequalities

$$\{\mathrm{H}_m \equiv \mathrm{H}_{g_m},\ m \in \mathscr{R}\}, \tag{8C.40}$$

where $\{g_m,\ m \in \mathscr{R}\}$ is any one-parameter family of monotone increasing smooth functions that satisfy (8C.8) and are such that

$$g_m''(1) = m \quad \forall m. \tag{8C.41}$$

HILL has noticed the following one-parameter family of this kind:

$$g_m(x) = \begin{cases} (x^{m+1} - 1)/(m+1) & \text{if } m \neq -1, \\ \log x & \text{if } m = -1, \end{cases} \tag{8C.42}$$

for $x \in (0, \infty)$. In particular, $g_0(x)$ is the linear function denoted similarly and defined by (8C.25).

[1] *Cf.* HILL, *op. cit.* p. 225.

The generalized strain tensors $\boldsymbol{E}_m \equiv \boldsymbol{E}_{g_m}$ and $\bar{\boldsymbol{E}}_m \equiv \bar{\boldsymbol{E}}_{g_m}$ have occasionally been used in other contexts[1]; for some particular values of m, they are well known. For example, when $m = 1$,

$$\boldsymbol{E}_1 = \tfrac{1}{2}(\boldsymbol{U}^2 - \mathbf{1}) = \tfrac{1}{2}(\boldsymbol{C} - \mathbf{1}) \tag{8C.43}$$

is the *Green-St. Venant strain tensor*[2]; when $m = -3$,

$$\bar{\boldsymbol{E}}_{-3} = \tfrac{1}{2}(\mathbf{1} - \boldsymbol{V}^{-2}) = \tfrac{1}{2}(\mathbf{1} - \boldsymbol{B}^{-1}) \tag{8C.44}$$

is the *Almansi-Hamel strain tensor*[2]; and when $m = -1$,

$$\boldsymbol{E}_{-1} = \log \boldsymbol{U} = \tfrac{1}{2} \log \boldsymbol{C} \tag{8C.45$_1$}$$

and

$$\bar{\boldsymbol{E}}_{-1} = \log \boldsymbol{V} = \tfrac{1}{2} \log \boldsymbol{B} \tag{8C.45$_2$}$$

are the *logarithmic strain tensors of Hencky*[2]. Of course, when $m = 0$, $\boldsymbol{E}_0$ is given by (8C.27), while $\bar{\boldsymbol{E}}_0$ is given by

$$\bar{\boldsymbol{E}}_0 = \boldsymbol{V} - \mathbf{1}. \tag{8C.46}$$

For some particular values of m, we can even derive explicit formulae for the corresponding generalized stress tensors. In doing so, we denote $\boldsymbol{T}_{g_m}$ more briefly by $\boldsymbol{T}_m$. By (8C.27), (8C.21) implies that

$$\boldsymbol{T}_0 = \frac{\rho_\kappa}{2\rho}(\boldsymbol{U}^{-1}\boldsymbol{R}^T\boldsymbol{T}\boldsymbol{R} + \boldsymbol{R}^T\boldsymbol{T}\boldsymbol{R}\boldsymbol{U}^{-1}), \tag{8C.47}$$

where of course we refer to the basic condition (8C.18), which defines the generalized stress. By (8C.42)

$$\boldsymbol{E}_{-2} = (\mathbf{1} - \boldsymbol{U}^{-1}), \tag{8C.48}$$

so that

$$\dot{\boldsymbol{E}}_{-2} = -(\dot{\boldsymbol{U}^{-1}}) = \boldsymbol{U}^{-1}\dot{\boldsymbol{U}}\boldsymbol{U}^{-1}. \tag{8C.49}$$

By (8C.49), (8C.21) gives

$$f = \frac{\rho_\kappa}{\rho} \operatorname{tr}(\boldsymbol{R}^T\boldsymbol{T}\boldsymbol{R}\boldsymbol{U}\dot{\boldsymbol{E}}_{-2}), \tag{8C.50}$$

---

[1] For example, by T. C. Doyle & J. L. Ericksen, "Nonlinear elasticity", *Adv. Appl. Mech.* **4**, 53–115 (1956).

[2] *Cf.* C. Truesdell, "The mechanical foundations of elasticity and fluid dynamics", *J. Rational Mech. Anal.* **1**, 125–300 (1952). Corrected reprint, *Continuum Mechanics* I, Intl. Sci. Rev. Ser. Vol. 8, New York, Gordon & Breach (1966). *Cf.* also CFT, §§ 33 and 33A.

which simply means that

$$T_{-2} = \frac{\rho_{\kappa}}{2\rho}(R^T TRU + UR^T TR). \tag{8C.51}$$

Now by (8C.43),

$$2\dot{E}_1 = (U^2)^{\cdot} = U\dot{U} + \dot{U}U, \tag{8C.52}$$

so that by (8C.21)

$$f = \frac{\rho_{\kappa}}{\rho}\operatorname{tr}(U^{-1}R^T TRU^{-1}\dot{E}_1), \tag{8C.53}$$

which means, as before, that

$$T_1 = \frac{\rho_{\kappa}}{\rho}U^{-1}R^T TRU^{-1} = \frac{\rho_{\kappa}}{\rho}F^{-1}T(F^{-1})^T. \tag{8C.54}$$

Finally, by (8C.42),

$$E_{-3} = \tfrac{1}{2}(1 - U^{-2}), \tag{8C.55}$$

so that

$$2\dot{E}_{-3} = -(\dot{U}^{-2}) = U^{-1}\dot{U}U^{-2} + U^{-2}\dot{U}U^{-1}. \tag{8C.56}$$

Then we can rewrite (8C.21) as

$$f = \frac{\rho_{\kappa}}{\rho}\operatorname{tr}(UR^T TRU\dot{E}_{-3}), \tag{8C.57}$$

which means also that

$$T_{-3} = \frac{\rho_{\kappa}}{\rho}UR^T TRU = \frac{\rho_{\kappa}}{\rho}F^T TF. \tag{8C.58}$$

Except in these few cases, the relation between T_m, $R^T TR$, and U is not simple.

So far, we have considered HILL's inequalities for an elastic material in general. Now if the material is hyperelastic, the stress power f in any motion must satisfy the energy equation:

$$\rho_{\kappa}\dot{\sigma} = f, \tag{8C.59}$$

where σ denotes the stored energy, which is given by a function of U or, equivalently, of E_g for any g. From (8C.59) and (8C.18) we see that the generalized constitutive relation for a hyperelastic material has the form

$$T_g = \rho_\kappa \frac{\partial \tilde{\sigma}}{\partial E_g}, \tag{8C.60}$$

where $\tilde{\sigma}$ denotes the stored energy taken as a function of E_g. Of course, (8C.22) reduces to this form locally if and only if $\nabla \hat{T}_g$ is symmetric over the space of symmetric second-order tensors. In particular, for a hyperelastic material, the generalized elasticity tensor E_g satisfies the symmetry condition

$$\mathsf{E}_{ijkl} = \mathsf{E}_{klij} \tag{8C.61}$$

in addition to its usual symmetry conditions

$$\mathsf{E}_{ijkl} = \mathsf{E}_{jikl} = \mathsf{E}_{ijlk}. \tag{8C.62}$$

From (8C.39) we see that[1] the condition (8C.61) holds relative to any one g if and only if it holds relative to all g.

Now from (8C.24) and (8C.60), we see that *for a hyperelastic material the H_g inequality is necessary and sufficient that the Hessian form of the stored-energy function relative to E_g be positive-definite at each reference configuration.*

In his paper HILL proposes to select, out of the infinite family $\{H_m, m \in \mathscr{R}\}$, the inequality H_{-1} as a possible *a priori* condition in elasticity. His argument is based upon a possible interpretation of the H_m inequality for incompressible elastic materials, as we shall now see. We recall that for an elastic material subject to internal constraints the stress tensor is determined by a constitutive relation only to within an arbitrary symmetric tensor that does no work in any motion compatible with the constraints. For an incompressible material, the undetermined part of the Cauchy stress is an arbitrary hydrostatic stress. Likewise, the undetermined part of the generalized stress T_m is an arbitrary symmetric tensor N_m such that

$$\operatorname{tr}(N_m \dot{E}_m) = 0 \tag{8C.63}$$

in all isochoric motions. For example, when $m = 0$, we obtain from (8C.47)

$$N_0 = -pU^{-1}, \tag{8C.64}$$

---

[1] *Cf.* HILL, *op. cit.* p. 225.

and when $m = 1$, we obtain from (8C.54)

$$N_1 = -pU^{-2}, \tag{8C.65}$$

etc., where p is arbitrary. The form of N_m for arbitrary m is not simple, however.

Now straight lines of the form (8C.26), even if tr$D$ is required to vanish, generally need not satisfy the condition of incompressibility. (For that matter, straight lines of the form (8A.16) need not be unimodular either.) Therefore, for an incompressible material the stress power is generally not defined on those straight lines. Hence the previous interpretation of HILL's inequalities in terms of the stress power cannot be applied directly to incompressible materials. However, as HILL remarks, in view of (8C.23) H_m requires also that

$$\operatorname{tr}(\dot{T}_m(0)D(0)) > 0 \tag{8C.66}$$

for all symmetric tensors $D(0)$. Here $\dot{T}_m(0)$ is the generalized stress rate on *any smooth curve* in the E_m-space with $E_m(0) = \mathbf{0}$ and $\dot{E}_m(0) = D(0)$. This interpretation frees H_m from any dependence on straight lines. Of course, it is this interpretation of H_m that HILL proposes to apply to *all* elastic materials – compressible or incompressible.

More specifically, HILL expresses the generalized constitutive relation for an incompressible elastic material in the form

$$T_m = N_m + \hat{T}_m(E_m), \tag{8C.67}$$

where $\hat{T}_m(E_m)$ is a generalized response function that gives the determinate generalized stress. Then he defines the H_m inequality as the requirement that

$$\operatorname{tr}(\dot{N}_m D) + \operatorname{tr}(\mathsf{E}_m(D)D) > 0 \tag{8C.68}$$

for all symmetric D such that $\operatorname{tr} D = 0$. Here $\dot{N}_m$ is the rate of N_m at $\tau = 0$ in any curve $E_m = E_m(\tau)$ compatible with the condition of incompressibility and satisfying the initial conditions $E_m(0) = \mathbf{0}$ and $\dot{E}_m(0) = D$ as usual. Now since $N_m(\tau)$ is still arbitrary to some extent, it is not clear whether or not the inequality (8C.68) can be satisfied by any generalized response function $\hat{T}_m(E_m)$. Indeed, as we shall see, HILL shows that (8C.68) *is possible only if* $m = -1$. This fact provides HILL's reason for rejecting all other H_m – including H_0, which is the GCN^+ condition.

To prove HILL's theorem, we differentiate (8C.63) with respect to τ and evaluate the result at $\tau = 0$, obtaining

$$\operatorname{tr}(\dot{\boldsymbol{N}}_m(0)\boldsymbol{D}(0)) - p(0)\operatorname{tr}(\ddot{\boldsymbol{E}}_m(0)) = 0. \tag{8C.69}$$

Here we have used the fact that

$$\boldsymbol{N}_m(0) = -p(0)\mathbf{1}, \tag{8C.70}$$

since the generalized stress reduces to the Cauchy stress at the reference configuration. Now differentiating the condition that the motion be isochoric:

$$\operatorname{tr}(\dot{\boldsymbol{U}}(\tau)\boldsymbol{U}(\tau)^{-1}) = 0, \tag{8C.71}$$

we obtain

$$\operatorname{tr}(\ddot{\boldsymbol{E}}_0(0)) \equiv \operatorname{tr}(\ddot{\boldsymbol{U}}(0)) = \operatorname{tr}(\boldsymbol{D}(0)^2). \tag{8C.72}$$

Hence from (8C.33) in the case in which $g = g_m$ we have

$$\operatorname{tr}(\ddot{\boldsymbol{E}}_m(0)) = (m+1)\operatorname{tr}(\boldsymbol{D}(0)^2). \tag{8C.73}$$

Substituting (8C.73) into (8C.69) yields

$$\operatorname{tr}(\dot{\boldsymbol{N}}_m(0)\boldsymbol{D}(0)) = p(0)(m+1)\operatorname{tr}(\boldsymbol{D}(0)^2). \tag{8C.74}$$

Now since $\operatorname{tr}(\boldsymbol{D}(0)^2)$ is positive-definite with respect to $\boldsymbol{D}(0)$, and since $p(0)$ is arbitrary, the formula (8C.74) shows clearly that the inequality (8C.68) is impossible for $\hat{\boldsymbol{T}}_m(\boldsymbol{E}_m)$ except when $m = -1$. This result supports HILL's choice of H_{-1} as a possible *a priori* condition in elasticity.

It seems questionable that an *a priori* condition for compressible elastic materials should be selected entirely by an argument based upon some possible interpretation of that condition for certain corresponding incompressible elastic materials[1]. If we choose to interpret the H_m inquality for incompressible materials as the requirement that

$$\operatorname{tr}\{(\dot{\boldsymbol{T}}_m - \dot{\boldsymbol{N}}_m)\boldsymbol{D}\} = \operatorname{tr}\{\mathsf{E}_m(\boldsymbol{D})\boldsymbol{D}\} > 0 \tag{8C.75}$$

---

[1] To the best of our knowledge, at present there is no theorem giving a precise status to the solutions according to the theory of incompressible bodies as limits of some appropriately selected family of solutions for certain corresponding compressible ones. Since the entire stress is determined by the deformation for any compressible material, yet in the limit of incompressibility, we must presume, an additional arbitrary hydrostatic stress suddenly appears, the point is by no means a trivial one. Also there is no unambiguous definition of the particular incompressible material to which a given compressible one should "reduce" in such a passage to the limit.

for all non-vanishing symmetric tensors $\boldsymbol{D}$ such that $\operatorname{tr} \boldsymbol{D} = 0$, so that the inequality becomes purely a condition of convexity on the determinate generalized response function, which is, after all, the function that characterizes the incompressible material, then HILL's argument in favor of the H_{-1} inequality no longer applies. The condition (8C.75) is not inconsistent with the condition (8C.24), since for a compressible material $\boldsymbol{N}_m \equiv \mathbf{0}$.

Well aware of the shoals to be avoided if any rational theory connecting compressible with incompressible materials is to be constructed, TRUESDELL & TOUPIN and COLEMAN & NOLL proposed the GCN and GCN$^+$ conditions for study in relation to unconstrained elastic materials only. Despite some mysterious claims that those inequalities "do not hold" for incompressible materials, we cannot even divine the interpretations of them for incompressible materials that would be necessary prerequisites for giving such claims meaning, let alone truth, and in NFTM TRUESDELL & NOLL expressly refused to lay down even for tentative study any candidate for the status of an *a priori* inequality for incompressible elastic materials. HILL's analysis shows that an interpretation of his own extension of the GCN$^+$ condition in the form (8C.68) with $m = 0$ is unacceptable for incompressible materials. A positive test of any *a priori* condition, of course, must be derived from consequences of that condition in all circumstances. As we shall see in the next section, the GCN condition does lead to an excessively strong restriction when applied to elastic fluids. On the other hand, we shall see also that the H_{-1} inequality cannot be satisfied by any elastic fluid whatever. Thus HILL's proposal is liable to objections at least as strong as those levelled by him and others at all inequalities proposed so far, and TRUESDELL's "main open problem in non-linear elasticity" of 1955 remains open today, despite the amount of valuable information which has been accumulated in researches devoted to it.

9. Inequalities in Elasticity, III. Specialization to Isotropic Materials

A. Specialization of the GCN Condition

We now consider isotropic materials. In § 2 we have shown that relative to an undistorted reference configuration $\boldsymbol{\kappa}$ of an isotropic elastic material point the Cauchy stress $\boldsymbol{T}$ is given by an isotropic function of the left stretch tensor $\boldsymbol{V}$. Hence $\boldsymbol{T}$ commutes with $\boldsymbol{V}$. Now with the help of the usual polar decomposition $\boldsymbol{F} = \boldsymbol{V}\boldsymbol{R}$, we can express the relation between $\boldsymbol{T}$ and $\boldsymbol{T}_\kappa$ in the form

$$\boldsymbol{T}_\kappa = (J\boldsymbol{T}\boldsymbol{V}^{-1})\boldsymbol{R} = \frac{\rho_\kappa}{\rho}\boldsymbol{T}\boldsymbol{V}^{-1}\boldsymbol{R}. \tag{9A.1}$$

$J\boldsymbol{T}\boldsymbol{V}^{-1}$ is a symmetric tensor which has a principal basis in common with $\boldsymbol{T}$ and $\boldsymbol{V}$. Since the proper numbers T_a of $J\boldsymbol{T}\boldsymbol{V}^{-1}$ are given by

$$T_a = v_1 v_2 v_3 \frac{t_a}{v_a}, \quad a = 1, 2, 3, \tag{9A.2}$$

which shows that they are equal to the normal tensions on the principal co-ordinate planes per unit area in the reference configuration, we call the T_a the *principal forces*. Like the principal stress t_a, the principal force T_a is the value of a function of the principal stretches (v_1, v_2, v_3). We shall see that the GCN condition implies that the transformation from principal stretches to principal forces be a monotone transformation. That is,

$$\sum_{a=1}^{3}(T_a^* - T_a)(v_a^* - v_a) > 0 \tag{9A.3}$$

if $v_a^* \neq v_a$ for some a. We call this inequality the *GCN_0 condition*.

To prove the GCN_0 condition, we choose $\boldsymbol{F}$ and $\boldsymbol{F}^*$ in such a way that

$$\boldsymbol{F} = \boldsymbol{V}\boldsymbol{R},\ \boldsymbol{F}^* = \boldsymbol{V}^*\boldsymbol{R},\ \boldsymbol{V}\boldsymbol{V}^* = \boldsymbol{V}^*\boldsymbol{V}, \tag{9A.4}$$

so that (8A.4) is satisfied with

$$\boldsymbol{S} = \boldsymbol{V}^*\boldsymbol{V}^{-1}. \tag{9A.5}$$

Then, using (9A.1), we find that

$$(\boldsymbol{T}_\kappa^* - \boldsymbol{T}_\kappa)(\boldsymbol{F}^{*T} - \boldsymbol{F}^T) = (J^*\boldsymbol{T}^*\boldsymbol{V}^{*-1} - J\boldsymbol{T}\boldsymbol{V}^{-1})(\boldsymbol{V}^* - \boldsymbol{V}). \tag{9A.6}$$

Now since the symmetric tensors $\boldsymbol{V}$, $\boldsymbol{V}^*$, $J\boldsymbol{T}\boldsymbol{V}^{-1}$, and $J^*\boldsymbol{T}^*\boldsymbol{V}^{*-1}$ have a principal basis in common, by taking the trace of (9A.6) and using (8A.3) we obtain (9A.3). Thus GCN_0 is a consequence of GCN for isotropic elastic materials.

We shall now deduce various consequences of the GCN_0 condition itself.

1. The tension-extension (T–E) inequalities

The condition that in order to lengthen a cube of isotropic material along one principal direction while its faces parallel to that direction are kept fixed, the tensile force must be increased, but to shorten it, the tensile force must be reduced, may be expressed in either of the following equivalent forms:

$$(T_a^* - T_a)(v_a^* - v_a) > 0, \quad (t_a^* - t_a)(v_a^* - v_a) > 0 \tag{9A.7}$$

provided $v_b^* = v_b$ if $b \neq a$ but $v_a^* \neq v_a$. We shall refer to (9A.7) as the *T–E inequalities*. Clearly $(9A.7)_1$ is a special case of (9A.3). Thus $GCN_0 \Rightarrow$ T–E.

2. The invertibility of force-stretch (IFS) condition.

The condition that if given pairs of opposing normal forces of magnitudes T_a are applied to the faces of a unit cube of isotropic elastic material, one and only one pure homogeneous deformation will result, requires that the equations

$$T_a = \hat{T}_a(v_1, v_2, v_3), \quad a = 1, 2, 3, \tag{9A.8}$$

be invertible. We shall refer to this requirement as the *IFS condition*. Clearly $GCN_0 \Rightarrow$ IFS, since if $v_a^* \neq v_a$ for any a but $T_a^* = T_a$ for all a, then (9A.3) cannot hold.

3. The extension-tension (E–T) inequalities

Granted that the IFS condition be satisfied, the condition that if one pair of opposing normal forces on a homogeneously strained block with faces perpendicular to the principal axes are increased in magnitude, while the remaining pairs are kept fixed, the block will lengthen in the corresponding direction, may be expressed in the form

$$(T_a^* - T_a)(v_a^* - v_a) > 0, \quad a = 1, 2, 3, \tag{9A.9}$$

if $T_b^* = T_b$ for $b \neq a$ but $T_a^* \neq T_a$. We shall refer to this condition as the *E–T inequalities*. Obviously, (9A.9) is a special case of (9A.3). Thus $GCN_0 \Rightarrow$ E–T.

4. The ordered-forces (O-F) inequalities

Consider again a block of isotropic elastic material supposed to be in equilibrium subject to pairs of equal and oppositely directed normal forces acting upon its faces. The condition that the greater stretch will occur in the direction of the greater force requires that

$$(T_a - T_b)(v_a - v_b) > 0, \quad a, b = 1, 2, 3, \tag{9A.10}$$

if $v_a \neq v_b$. We shall refer to (9A.10) as the *O–F inequalities.* To see that O–F is a consequence of GCN_0, we choose the v_a^* as a permutation of the v_a. Then (9A.3) implies

$$\sum_{a=1}^{3} (T_{p(a)} - T_a)(v_{p(a)} - v_a) > 0 \tag{9A.11}$$

provided that the permutation p not be the identity permutation and that $v_a \neq v_b$ if $a \neq b$. Permutation of any two of the v_a while leaving the other unchanged reduces (9A.11) to (9A.10).

5. The pressure-compression (P–C) inequality

Relative to a natural state of an isotropic elastic material, the condition that the volume should be decreased by pressure but increased by tension can be generalized by the inequality[1]

$$\sum_{a=1}^{3} T_a(v_a - 1) > 0 \tag{9A.12}$$

provided not all $v_a = 1$. We shall refer to this condition as the *P–C inequality*. Setting $v_a^* = 1$, we obtain $T_a^* = 0$, because the reference configuration is assumed to be a natural state. Then (9A.3) reduces to (9A.12). Thus we have shown that $GCN_0 \Rightarrow$ P–C.

Because the transformation (9A.8) is monotone, the Hessian matrix

$$[J_{ab}] \equiv \left[\frac{\partial \hat{T}_a}{\partial v_b}\right] \tag{9A.13}$$

is positive semi-definite; *i.e.*,

$$J_{ab}\, u^a u^b \geqq 0 \tag{9A.14}$$

for all vectors $\boldsymbol{u}$. We shall call the strict inequality corresponding to (9A.14), *viz*

$$J_{ab}\, u^a u^b > 0 \tag{9A.15}$$

[1] For some other possible generalizations, *cf.* NFTM, § 51.

for all non-vanishing $\boldsymbol{u}$, the *GCN_0^+ condition.* The usual argument based on integration along a straight line shows that the differential inequality (9A.15) implies the integral inequality (9A.3), *i.e.*, $\text{GCN}_0^+ \Rightarrow \text{GCN}_0$.

We now show that GCN_0^+ is a consequence of the general differential inequality GCN^+. As before, we consider the stress power f on any straight line of the form (8A.16). Here we choose the symmetric tensor $\boldsymbol{D}$ in such a way that

$$\boldsymbol{V}\boldsymbol{D} = \boldsymbol{D}\boldsymbol{V}, \tag{9A.16}$$

where $\boldsymbol{V}$ is the left stretch tensor of $\boldsymbol{F}$. By (9A.1) and (9A.16), (8A.18) implies that

$$f(\tau) = \text{tr}\,\{(\boldsymbol{D}\boldsymbol{V})^T(J(\tau)\boldsymbol{T}(\tau)\boldsymbol{V}(\tau)^{-1})\} = \sum_{a=1}^{3} d_a v_a T_a(\tau), \tag{9A.17}$$

where $T_a(\tau)$ is a principal force corresponding to $\boldsymbol{F}(\tau)$, *i.e.*,

$$T_a(\tau) = \hat{T}_a((1+\tau d_1)v_1,\ (1+\tau d_2)v_2,\ (1+\tau d_3)v_3), \tag{9A.18}$$

and where $J(\tau)$, $\boldsymbol{T}(\tau)$ and $\boldsymbol{V}(\tau)$ are defined similarly. From (9A.17) and (9A.18) it follows that

$$\dot{f}(0) = \sum_{a,b=1}^{3} (d_a v_a)(d_b v_b)\frac{\partial \hat{T}_a}{\partial v_b}. \tag{9A.19}$$

As remarked in the preceding section, the GCN^+ condition requires that $\dot{f}(0) > 0$. Since the proper numbers d_a of $\boldsymbol{D}$ are arbitrary, (9A.19) shows that the inequality $\dot{f}(0) > 0$ is equivalent to (9A.15). Thus we have shown that $\text{GCN}^+ \Rightarrow \text{GCN}_0^+$.

Note. Since the preceding proof makes use of straight lines of the form (8A.16) with $\boldsymbol{D}$ satisfying the condition (9A.16) only, it shows GCN_0^+ to be necessary but not sufficient for GCN^+ when the material is isotropic. In fact, we shall prove in Subsection 9C (*cf.* (9C.38) and (9C.42)) that for an isotropic elastic material GCN^+ is equivalent to GCN_0^+ and the conditions $A_\Gamma > 0$, $\Gamma = 1, 2, 3$, where

$$A_1 \equiv \frac{(t_2-t_3)(v_2^2+v_3^2)}{2(v_2^2-v_3^2)} - \tfrac{1}{4}(t_2+t_3), \tag{9A.20}$$

etc. Here $\boldsymbol{V}$ is assumed to be non-degenerate. In a case when $\boldsymbol{V}$ is degenerate, say with $v_2 = v_3 = v$, we may compute the ratio $(t_2-t_3)/(v_2-v_3)$

on the right-hand side of (9A.20) by the relation[1]

$$\lim_{v_2, v_3 \to v} \left(\frac{t_2 - t_3}{v_2 - v_3}\right) = \left(\frac{\partial \hat{t}_2}{\partial v_2} - \frac{\partial \hat{t}_3}{\partial v_2}\right)\Bigg|_{(v_1, v, v)}. \tag{9A.21}$$

Unfortunately, the full consequences of the inequalities $A_\Gamma > 0$ are not presently known, so we do not consider those inequalities in detail here.

Clearly, the GCN_0^+ condition has the following consequences.

1⁺. The T–E⁺ inequalities.

The strict differential inequalities corresponding to the T–E inequalities (9A.7) are

$$\frac{\partial \hat{T}_a}{\partial v_a} > 0, \quad \frac{\partial \hat{t}_a}{\partial v_a} > 0, \tag{9A.22}$$

which we shall call the *T–E⁺ inequalities*. Clearly, (9A.15) implies (9A.22), *i.e.*, $\mathrm{GCN}_0^+ \Rightarrow \mathrm{T-E}^+$.

2⁺. The IFS⁺ condition.

The strict differential condition corresponding to the IFS condition is

$$\det [J_{ab}] \neq 0, \tag{9A.23}$$

which we shall call the *IFS⁺ condition*. Since (9A.15) implies $\det [J_{ab}] > 0$, we see that $\mathrm{GCN}_0^+ \Rightarrow \mathrm{IFS}^+$.

3⁺. The E–T⁺ inequalities.

The strict differential inequalities corresponding to the E–T inequalities are

$$\frac{\partial \hat{v}_a}{\partial T_a} > 0, \quad a = 1, 2, 3, \tag{9A.24}$$

which we shall call the *E–T⁺ inequalities*. Since (9A.15) requires the inverse Hessian matrix

$$[J_{ab}]^{-1} = \left[\frac{\partial \hat{v}_a}{\partial T_b}\right] \tag{9A.25}$$

be positive-definite also, it clearly implies (9A.24). *I.e.*, $\mathrm{GCN}_0^+ \Rightarrow \mathrm{E-T}^+$.

---

[1] For more details about the degenerate cases, see § VI.4.

The preceding three consequences, while necessary, are generally not sufficient for the GCN_0^+ condition.

Next we consider specialization of the GCN condition to isotropic hyperelastic materials. We recall first that for a hyperelastic material the GCN condition is equivalent to the C–N condition. Also, in § 5 we proved that for an isotropic hyperelastic material the stored-energy function is a function of the principal stretches provided that the reference configuration be undistorted. We now show that the C–N condition requires the stored-energy function

$$\sigma = \hat{\sigma}(v_1, v_2, v_3) \tag{9A.26}$$

be a convex function, *i.e.*,

$$\hat{\sigma}(v_1^*, v_2^*, v_3^*) - \hat{\sigma}(v_1, v_2, v_3) - \sum_{a=1}^{3} (v_a^* - v_a) \frac{\partial \hat{\sigma}}{\partial v_a} > 0 \tag{9A.27}$$

if $v_a^* \neq v_a$ for some a. We shall refer to this inequality as the *C–N_0 condition.*

To prove that for an isotropic hyperelastic material C–N_0 is a consequence of C–N, we observe first that from (5.14) and (9A.2) the principal force T_a is given by

$$T_a = \rho_\kappa \frac{\partial \hat{\sigma}}{\partial v_a}. \tag{9A.28}$$

Therefore, integration along the straight line[1]

$$v_a(\tau) = v_a + \tau(v_a^* - v_a), \quad \tau \in [0, 1], \tag{9A.29}$$

followed by application of GCN_0 shows that

$$\hat{\sigma}(v_1^*, v_2^*, v_3^*) - \hat{\sigma}(v_1, v_2, v_3) - \sum_{a=1}^{3} (v_a^* - v_a) \frac{\partial \hat{\sigma}}{\partial v_a}$$

$$= \frac{1}{\rho_\kappa} \int_0^1 \sum_{a=1}^{3} \{(v_a(\tau) - v_a)[\hat{T}_a(v_1(\tau), v_2(\tau), v_3(\tau)) - \hat{T}_a(v_1, v_2, v_3)]\} \frac{d\tau}{\tau} > 0. \tag{9A.30}$$

[1] Since a symmetric tensor $V(\tau)$ is positive-definite if and only if its proper numbers $v_a(\tau)$ are positive, the straight line segment defined by (9A.29) is contained in the domain of $\hat{\sigma}$.

Now since for an isotropic hyperelastic material GCN_0 is a consequence of C–N, we have proved that C–N $\Rightarrow$ C–N_0. In fact, under the same assumption, C–N_0 is equivalent to GCN_0, since we can derive (9A.3) by adding (9A.27) to the same inequality with v_a^* and v_a interchanged.

Now returning to general hyperelastic materials, we can define the *C–N^+ condition* formally as the GCN^+ condition specialized to hyperelastic materials. Of course, we can express the C–N^+ condition also by a requirement that the Hessian form corresponding to the C–N condition (8A.28) be positive-definite. Then clearly C–N^+ $\Rightarrow$ C–N, since we have $GCN^+ \Rightarrow$ GCN in general. Likewise, we can define the *C–N_0^+ condition* formally as the GCN_0^+ condition specialized to hyperelastic materials. Then C–N_0^+ is also the strict differential inequality corresponding to C–N_0. Further, we have C–N^+ $\Rightarrow$ C–N_0^+, since $GCN^+ \Rightarrow GCN_0^+$ in general.

From (9A.28), the Hessian matrix $[J_{ab}]$ defined by (9A.13) is symmetric if and only if the material is hyperelastic. Here we have used the fact that the domain of $\hat{\sigma}$, namely, $(0, \infty)^3$, is simply connected. Now a symmetric Hessian matrix is positive-definite if and only if its principal minors are positive. This means that C–N_0^+ holds if and only if

$$\frac{\partial \hat{T}_a}{\partial v_a} > 0, \quad \frac{\partial \hat{v}_a}{\partial T_a} \det [J_{bc}] > 0, \; \det [J_{bc}] > 0, \quad a = 1, 2, 3. \tag{9A.31}$$

Hence C–N_0^+ is equivalent to T–E^+, E–T^+, IFS^+, and C when the reference configuration is an undistorted natural state. Here the C inequalities ensure that $\det [J_{bc}]$ is positive at the reference state; by continuity and IFS^+, $\det [J_{bc}]$ then remains positive for all states.

B. Specialization of the S–E Condition

As in Subsection 8B we consider the stress power $f(\tau)$ given by (8B.11) on straight lines of the form (8B.8), *i.e.*,

$$\boldsymbol{F}(\tau) = (\mathbf{1} + \tau \boldsymbol{u} \otimes \boldsymbol{w})\boldsymbol{F}, \tag{9B.1}$$

where $\boldsymbol{u}$ and $\boldsymbol{w}$ are unit vectors, and where τ belongs to some interval $[0, \tau^*]$ such that

$$1 + \tau^* \boldsymbol{u} \cdot \boldsymbol{w} > 0. \tag{9B.2}$$

We have replaced the notation $\boldsymbol{v}$ used before by the notation $\boldsymbol{w}$ so as to reserve the notation v_a for a principal stretch. No ambiguity will arise about the notation u_i, which means here a component of the unit vector

$\boldsymbol{u}$, since in this subsection we shall not use either the right stretch tensor $\boldsymbol{U}$ or its proper numbers.

For definiteness, let a spectral form of $\boldsymbol{V}$ be

$$\boldsymbol{V} = v_1 \boldsymbol{e}_1 \otimes \boldsymbol{e}_1 + v_2 \boldsymbol{e}_2 \otimes \boldsymbol{e}_2 + v_3 \boldsymbol{e}_3 \otimes \boldsymbol{e}_3 . \tag{9B.3}$$

Then we may choose $\boldsymbol{u} = \boldsymbol{w} = \boldsymbol{e}_1$. By $(8B.25)_2$ the tensor $(\mathbf{1} + \tau \boldsymbol{e}_1 \otimes \boldsymbol{e}_1)$ represents the deformation gradient of a simple extension of amount τ in the direction of $\boldsymbol{e}_1$. From (9A.1) and (9B.3) it follows that the stress power is given by

$$f(\tau) = \operatorname{tr} \{(\boldsymbol{e}_1 \otimes \boldsymbol{e}_1)\boldsymbol{V}(J(\tau)\boldsymbol{T}(\tau)\boldsymbol{V}(\tau)^{-1})\} = v_1 T_1(\tau), \tag{9B.4}$$

where $T_a(\tau)$ is a principal force corresponding to $\boldsymbol{F}(\tau)$, and $J(\tau)$, $\boldsymbol{V}(\tau)$, $\boldsymbol{T}(\tau)$ are defined similarly. Since a spectral form for $\boldsymbol{V}(\tau)$ is given by

$$\begin{aligned} \boldsymbol{V}(\tau) &= (\mathbf{1} + \tau \boldsymbol{e}_1 \otimes \boldsymbol{e}_1)(v_1 \boldsymbol{e}_1 \otimes \boldsymbol{e}_1 + v_2 \boldsymbol{e}_2 \otimes \boldsymbol{e}_2 + v_3 \boldsymbol{e}_3 \otimes \boldsymbol{e}_3) \\ &= (1+\tau)v_1 \boldsymbol{e}_1 \otimes \boldsymbol{e}_1 + v_2 \boldsymbol{e}_2 \otimes \boldsymbol{e}_2 + v_3 \boldsymbol{e}_3 \otimes \boldsymbol{e}_3 , \end{aligned} \tag{9B.5}$$

by substituting (9B.5) into (9A.8) we obtain

$$T_a(\tau) = \hat{T}_a((1+\tau)v_1 , v_2 , v_3). \tag{9B.6}$$

Hence it follows from (9B.4) that

$$\dot{f}(0) = v_1^2 \frac{\partial \hat{T}_1}{\partial v_1} . \tag{9B.7}$$

Now since the S–E condition requires $\dot{f}(0) > 0$, from (9B.7) we have proved that[1] S–E $\Rightarrow$ T–E$^+$.

Next, we may choose $\boldsymbol{u} = \boldsymbol{e}_2$ and $\boldsymbol{w} = \boldsymbol{e}_1$. Then by $(8B.25)_1$ the tensor $(\mathbf{1} + \tau \boldsymbol{e}_2 \otimes \boldsymbol{e}_1)$ is the deformation gradient of a simple shear of amount τ in the direction of $\boldsymbol{e}_2$ on the (1,2)-plane of the co-ordinate axes $\{\boldsymbol{e}_i\}$. For this particular choice of $\boldsymbol{u}$ and $\boldsymbol{w}$, the stress power is given by

$$f(\tau) = J \operatorname{tr} \{(\boldsymbol{e}_2 \otimes \boldsymbol{e}_1)(\mathbf{1} - \tau \boldsymbol{e}_2 \otimes \boldsymbol{e}_1)\boldsymbol{T}(\tau)\} = J T_{12}. \tag{9B.8}$$

Here J does not depend on τ, since simple shear is a unimodular deformation, and T_{12} is the shearing stress, *i.e.*, the shear stress in the direction of $\boldsymbol{e}_2$ on the co-ordinate plane perpendicular to $\boldsymbol{e}_1$. As we shall see in

---

[1] The proof here is essentially that of Hayes, *op. cit.* p. 217. Another proof can be read off from the formula (VI.5.16) for the speed of longitudinal principal waves in an isotropic elastic body.

§ IV.3, the Cauchy stress tensor $\boldsymbol{T}(\tau)$ corresponding to the deformation gradient $\boldsymbol{F}(\tau) = (\mathbf{1}+\tau \boldsymbol{e}_2 \otimes \boldsymbol{e}_1)\boldsymbol{V}\boldsymbol{R}$ has the component matrix

$$[T_{ij}] = \begin{bmatrix} T_{11} & T_{12} & 0 \\ T_{12} & T_{22} & 0 \\ 0 & 0 & T_{33} \end{bmatrix} \tag{9B.9}$$

relative to $\{\boldsymbol{e}_i\}$. Further, the normal stresses T_{ii} are values of certain even functions of τ, while the shear stress T_{12} is the value of an odd function of τ, and both functions depend on v_1, v_2, v_3 as parameters. In particular, T_{12} vanishes when $\tau = 0$.

Now since the S–E condition requires that $\dot{f}(\tau) > 0$, from (9B.8) it requires also that the *principal shear modulus*

$$\mu_{12} = \hat{\mu}_{12}(\tau; v_1, v_2, v_3) \equiv \frac{T_{12}}{\tau} \tag{9B.10}$$

be strictly positive for all v_1, v_2, v_3, and τ. In § IV.3 we shall consider in detail some properties of the special case when $v_1 = v_2 = v_3 = 1$. The corresponding function $\hat{\mu}_{12}$, which we may denote by

$$\hat{\mu}(\tau) \equiv \hat{\mu}_{12}(\tau; 1, 1, 1), \tag{9B.11}$$

is called the *generalized shear modulus*. Certain inequalities equivalent to the requirement that

$$\hat{\mu}(\tau) > 0 \tag{9B.12}$$

for all τ we shall call the *B–E inequalities*, as we now explain.

6. Baker-Ericksen (B–E) inequalities.

As we shall see in § IV.3, (9B.12) is equivalent to the condition that the greater principal stress shall occur always in the direction of the greater principal stretch:

$$(t_a - t_b)(v_a - v_b) > 0 \tag{9B.13}$$

provided $v_a \neq v_b$. We shall call this condition the *Baker-Ericksen inequalities*. The analysis leading to (9B.12) shows that[1] S–E ⇒ B–E.

---

[1] The proof here is essentially that of HAYES, *op. cit.* p. 217. Another proof can be read off from the formula (VI. 5.18) for the speeds of transverse principal waves in isotropic elastic body. TRUESDELL has attributed the implications S–E ⇒ T–E$^+$ and S–E ⇒ B–E in essence to unpublished work of ERICKSEN & TOUPIN in 1954–1955.

C. Specialization of Hill's Inequalities.

Relative to an undistorted reference configuration of an isotropic elastic material, the Cauchy stress $\boldsymbol{T}$ is given by an isotropic function of $\boldsymbol{V}$. Under the same assumption, the generalized stress $\boldsymbol{T}_g$ for any g is also given by an isotropic function of its conjugate generalized strain $\boldsymbol{E}_g$. To see this, we prove first that the stress power f can be regarded as a linear form in $\dot{\boldsymbol{V}}$. Indeed, since $\boldsymbol{T}$ commutes with $\boldsymbol{V}$, from (II.1.32) and (8C.21) we have

$$f = \frac{\rho_\kappa}{\rho} \operatorname{tr} (\boldsymbol{V}^{-1}\boldsymbol{T}\dot{\boldsymbol{V}}), \tag{9C.1}$$

where $(\rho_\kappa/\rho)\boldsymbol{V}^{-1}\boldsymbol{T}$ is the symmetric tensor whose proper numbers are the principal forces T_a. Since $(\rho_\kappa/\rho)\boldsymbol{V}^{-1}\boldsymbol{T}$ is an isotropic function of $\boldsymbol{V}$, (9C.1) shows not only that f is a linear form in $\dot{\boldsymbol{V}}$ but that that linear form is also an isotropic function of $\boldsymbol{V}$. Now from $(8\text{C}.7)_2$, we have

$$\dot{\bar{\boldsymbol{E}}}_g = [\nabla \boldsymbol{g}](\dot{\boldsymbol{V}}), \tag{9C.2}$$

where $\nabla \boldsymbol{g}$ is the same isomorphism as in (8C.19). Substituting the inverse of (9C.2) into (9C.1), we obtain

$$f = \operatorname{tr} (\bar{\boldsymbol{T}}_g \dot{\bar{\boldsymbol{E}}}_g), \tag{9C.3}$$

where $\bar{\boldsymbol{T}}_g$ is an isotropic function of $\boldsymbol{V}$, or, equivalently, of $\bar{\boldsymbol{E}}_g$, since $\bar{\boldsymbol{E}}_g$ is an isotropic function of $\boldsymbol{V}$, and conversely. Naturally we call $\bar{\boldsymbol{T}}_g$ *the generalized stress tensor conjugate* to $\bar{\boldsymbol{E}}_g$.

As remarked before, we can define $\bar{\boldsymbol{T}}_g$ relative to an undistorted reference configuration of an isotropic material only. If the material is not isotropic, or if the reference configuration is not undistorted, the stress power cannot be regarded as a linear form in $\dot{\boldsymbol{V}}$, and hence, likewise, in $\dot{\bar{\boldsymbol{E}}}_g$.

We now show that the functional relation

$$\bar{\boldsymbol{T}}_g = \hat{\bar{\boldsymbol{T}}}_g(\bar{\boldsymbol{E}}_g) \tag{9C.4}$$

is the same as the functional relation (8C.22) between $\boldsymbol{T}_g$ and $\boldsymbol{E}_g$, so that (9C.4) is also a generalized constitutive relation. To see that $\hat{\bar{\boldsymbol{T}}}_g$ and $\hat{\boldsymbol{T}}_g$ are the same function, we differentiate the basic relation (8C.9) between $\boldsymbol{E}_g$ and $\bar{\boldsymbol{E}}_g$ with respect to τ along any curve $\boldsymbol{F}(\tau)$ and then substitute the result into (9C.3), obtaining

$$f = \operatorname{tr} (\boldsymbol{R}\boldsymbol{T}_g \boldsymbol{R}^T \dot{\bar{\boldsymbol{E}}}_g). \tag{9C.5}$$

Comparing (9C.5) with (9C.3), we have

$$\hat{\hat{T}}_g(\bar{E}_g) = R\hat{T}_g(E_g)R^T. \tag{9C.6}$$

Now since $\hat{\hat{T}}_g(\bar{E}_g)$ is an isotropic function, (9C.6) and (8C.9) give

$$\hat{\hat{T}}_g(E_g) = \hat{T}_g(E_g) \tag{9C.7}$$

for all E_g. Thus $\hat{T}_g$ and $\hat{\hat{T}}_g$ are the same function. In particular, we have proved the opening remark that $\hat{T}_g(E_g)$ is an isotropic function, always assuming the reference configuration to be undistorted, of course.

Under the same assumption, we can even determine $\bar{T}_m$, and hence T_m, explicitly for all m. Indeed, let

$$V = v_i \bar{e}_i \otimes \bar{e}_i \tag{9C.8}$$

be a spectral form[1] of V. Then

$$T = t_i \bar{e}_i \otimes \bar{e}_i \tag{9C.9}$$

is a corresponding spectral form of T. Differentiating (9C.8) along any smooth curve $V = V(\tau)$, we obtain as before

$$\dot{V} = \{\dot{v}_i \delta_{ij} + (v_i - v_j)\bar{w}_{ij}\}\bar{e}_i \otimes \bar{e}_j. \tag{9C.10}$$

Now since

$$\bar{E}_m = g_m(V) = g_m(v_i)\bar{e}_i \otimes \bar{e}_i, \tag{9C.11}$$

we have also

$$\begin{aligned}\dot{\bar{E}}_m &= \{g_m'(v_i)\dot{v}_i \delta_{ij} + (g_m(v_i) - g_m(v_j))\bar{w}_{ij}\}\bar{e}_i \otimes \bar{e}_j, \\ &= \{v_i^m \dot{v}_i \delta_{ij} + (g_m(v_i) - g_m(v_j))\bar{w}_{ij}\}\bar{e}_i \otimes \bar{e}_j.\end{aligned} \tag{9C.12}$$

Here we have used the definition (8C.42) to compute the function g_m. Substituting (9C.8)–(9C.10) into (9C.1) yields

$$f = \frac{\rho_\kappa}{\rho} \frac{t_i \dot{v}_i}{v_i}. \tag{9C.13}$$

Now from (9C.12) the right-hand side of (9C.13) is equal to $(\rho_\kappa/\rho)\,\mathrm{tr}(TV^{-(m+1)}\dot{\bar{E}}_m)$, so we have also

$$f = \frac{\rho_\kappa}{\rho} \mathrm{tr}\,(TV^{-(m+1)}\dot{\bar{E}}_m). \tag{9C.14}$$

---

[1] *Cf.* Footnote 1, p. 226.

Comparing this equation with (9C.3), we conclude that

$$\bar{\boldsymbol{T}}_m = \frac{\rho_{\boldsymbol{\kappa}}}{\rho} \boldsymbol{T}\boldsymbol{V}^{-(m+1)}, \tag{9C.15}$$

or equivalently

$$\boldsymbol{T}_m = \frac{\rho_{\boldsymbol{\kappa}}}{\rho} \boldsymbol{R}^T\boldsymbol{T}\boldsymbol{R}\boldsymbol{U}^{-(m+1)} \tag{9C.16}$$

for all m. In particular, when $m = -1$ we have

$$\bar{\boldsymbol{T}}_{-1} = \frac{\rho_{\boldsymbol{\kappa}}}{\rho} \boldsymbol{T}, \quad \boldsymbol{T}_{-1} = \frac{\rho_{\boldsymbol{\kappa}}}{\rho} \boldsymbol{R}^T\boldsymbol{T}\boldsymbol{R}, \tag{9C.17}$$

and when $m = 0, 1, -2$, or -3, (9C.16) reduces to special cases of the previous formulae (8C.47), (8C.54), (8C.51), and (8C.58), respectively. From (9C.15), the *principal generalized stresses* t_{ma}, $a = 1, 2, 3$, are related as follows to the principal stresses t_a and the principal stretches v_a:

$$t_{ma} = v_1 v_2 v_3 \frac{t_a}{v_a^{m+1}}. \tag{9C.18}$$

Thus, in particular, t_{0a} is the principal force T_a.

Next, we consider the H_m inequality for isotropic elastic materials. Since the reference configuration must be undistorted in order that (9C.15) can be applied, a slight modification of the previous analysis of H_m – which refers to an arbitrary reference configuration – is needed here. We observe first that without loss of generality we can assume that the gradient of the deformation from some undistorted reference configuration to an arbitrary configuration be a pure stretch $\boldsymbol{V}$, since a rotation sends one undistorted reference configuration into another one. For definiteness, we denote the undistorted reference configuration by $\boldsymbol{\kappa}$ and the possibly distorted configuration $\boldsymbol{V} \circ \boldsymbol{\kappa}$ by $\boldsymbol{\kappa}_0$. We shall now consider the H_m inequality at the configuration $\boldsymbol{\kappa}_0$.

We recall that the H_m inequality holds at $\boldsymbol{\kappa}_0$ if and only if the initial slope $\dot{f}(0)$ of the stress power $f(\tau)$ is strictly positive on all straight lines of the form

$$\boldsymbol{E}_m(\tau) = \tau \boldsymbol{D} \tag{9C.19}$$

relative to $\boldsymbol{\kappa}_0$. Here the stress power f is measured per unit volume of $\boldsymbol{\kappa}_0$, of course. As remarked in Subsection 8C, a straight line of the form (9C.19) in the $\boldsymbol{E}_m$-space, generally, corresponds to a curve:

$$\boldsymbol{U}(\tau) = \boldsymbol{g}_m^{-1}(\tau\boldsymbol{D}) \tag{9C.20}$$

in the $\boldsymbol{U}$-space. Here $\boldsymbol{U}(\tau)$ is a pure stretch from $\boldsymbol{\kappa}_0$.

Now we can calculate $\dot{f}(0)$ in two different ways: First, from (8C.18) and (9C.19) we have

$$\dot{f}(0) = \operatorname{tr}(\dot{\boldsymbol{T}}_m(0)\boldsymbol{D}) = \operatorname{tr}(\mathsf{E}_m(\boldsymbol{D})\boldsymbol{D}), \tag{9C.21}$$

where $\boldsymbol{T}_m$ is the generalized stress tensor conjugate to $\boldsymbol{E}_m$ with respect to $\boldsymbol{\kappa}_0$. Second, since (9C.20) corresponds to a curve

$$\boldsymbol{F}(\tau) = \boldsymbol{U}(\tau)\boldsymbol{V} \tag{9C.22}$$

relative to $\boldsymbol{\kappa}$, we have

$$f = \frac{\rho_0}{\rho}\operatorname{tr}(\boldsymbol{T}\dot{\boldsymbol{F}}\boldsymbol{F}^{-1}) = \frac{\rho_0}{\rho_{\kappa}}\operatorname{tr}\left(\frac{\rho_{\kappa}}{\rho}\boldsymbol{T}\dot{\boldsymbol{U}}\boldsymbol{U}^{-1}\right), \tag{9C.23}$$

where ρ_0 denotes the density in $\boldsymbol{\kappa}_0$. Substituting $(9C.17)_1$ into (9C.23), we have also

$$f = \frac{\rho_0}{\rho_{\kappa}}\operatorname{tr}(\overline{\boldsymbol{T}}_{-1}\dot{\boldsymbol{U}}\boldsymbol{U}^{-1}), \tag{9C.24}$$

where $\overline{\boldsymbol{T}}_{-1}$ is referred to the undistorted reference configuration $\boldsymbol{\kappa}$. Differentiating (9C.24) with respect to τ, we have

$$\dot{f}(0) = \frac{\rho_0}{\rho_{\kappa}}\operatorname{tr}(\dot{\overline{\boldsymbol{T}}}_{-1}(0)\boldsymbol{D} + \overline{\boldsymbol{T}}_{-1}(0)\ddot{\boldsymbol{U}}(0) - \overline{\boldsymbol{T}}_{-1}(0)\boldsymbol{D}^2). \tag{9C.25}$$

Of course, (9C.25) is more convenient than (9C.21), since $\overline{\boldsymbol{T}}_{-1}$ is referred to $\boldsymbol{\kappa}$ while $\boldsymbol{T}_m$ is referred to $\boldsymbol{\kappa}_0$.

To determine the right-hand side of (9C.25), we write $\boldsymbol{D}$ and $\boldsymbol{U}(\tau)$ in spectral forms[1]:

$$\boldsymbol{D} = d_i \boldsymbol{f}_i \otimes \boldsymbol{f}_i, \quad \boldsymbol{U}(\tau) = u_i(\tau)\boldsymbol{f}_i \otimes \boldsymbol{f}_i. \tag{9C.26}$$

Then from (9C.19) and (8C.42) we have

$$\begin{aligned} &\frac{1}{m+1}(u_i(\tau)^{m+1} - 1) = \tau d_i \quad \text{if } m \neq -1, \\ &\log u_i(\tau) = \tau d_i \qquad\qquad\qquad \text{if } m = -1, \end{aligned} \tag{9C.27}$$

[1] *Cf.* Footnote 1, p. 226.

for all τ. Differentiating (9C.27) with respect to τ twice yields

$$\ddot{u}_i(0) = -md_i^2, \tag{9C.28}$$

which means

$$\ddot{U}(0) = -m\boldsymbol{D}^2. \tag{9C.29}$$

Next, from (9C.22) the left stretch tensor $\boldsymbol{V}(\tau)$ of $\boldsymbol{F}(\tau)$ relative to $\boldsymbol{\kappa}$ is given by

$$\boldsymbol{V}(\tau)^2 = \boldsymbol{U}(\tau)\boldsymbol{V}^2\boldsymbol{U}(\tau). \tag{9C.30}$$

Differentiating this equation with respect to τ once gives

$$\boldsymbol{V}\dot{\boldsymbol{V}}(0)+\dot{\boldsymbol{V}}(0)\boldsymbol{V} = \boldsymbol{D}\boldsymbol{V}^2+\boldsymbol{V}^2\boldsymbol{D}. \tag{9C.31}$$

Now we denote a spectral form of $\boldsymbol{V}(\tau)$ by

$$\boldsymbol{V}(\tau) = v_i(\tau)\bar{\boldsymbol{e}}_i(\tau) \otimes \bar{\boldsymbol{e}}_i(\tau), \tag{9C.32}$$

so that as before

$$\dot{\boldsymbol{V}}(0) = \{\dot{v}_i(0)\delta_{ij}+(v_i-v_j)\overline{w}_{ij}(0)\}\bar{\boldsymbol{e}}_i \otimes \bar{\boldsymbol{e}}_j. \tag{9C.33}$$

Then the component form of (9C.31) relative to $\{\bar{\boldsymbol{e}}_i\}$ implies that

$$\dot{v}_1(0) = v_1 d_{11}, \quad (v_1-v_2)\overline{w}_{12}(0) = \frac{v_1^2+v_2^2}{v_1+v_2} d_{12} \tag{9C.34}$$

and two similar pairs of formulae give the remaining two diagonal components and two off-diagonal components of $\dot{\boldsymbol{V}}(0)$. Here d_{ij} denotes a component of $\boldsymbol{D}$ relative to the principal basis $\{\bar{\boldsymbol{e}}_i\}$ of $\boldsymbol{V}$. Now since $\overline{\boldsymbol{T}}_{-1}(\tau)$ is the value of an isotropic function of $\boldsymbol{V}(\tau)$, a spectral form of $\overline{\boldsymbol{T}}_{-1}(\tau)$ is

$$\overline{\boldsymbol{T}}_{-1}(\tau) = t_{-1i}(\tau)\bar{\boldsymbol{e}}_i(\tau) \otimes \bar{\boldsymbol{e}}_i(\tau), \tag{9C.35}$$

where $t_{-1i}(\tau)$ is given by a function of $v_j(\tau)$, $j = 1, 2, 3$. Differentiating (9C.35) with respect to τ as before, we obtain

$$\dot{\overline{\boldsymbol{T}}}_{-1}(0) = \left\{\frac{\partial \hat{t}_{-1i}}{\partial v_k}\dot{v}_k(0)\delta_{ij}+(t_{-1i}-t_{-1j})\overline{w}_{1j}(0)\right\} \bar{\boldsymbol{e}}_i \otimes \bar{\boldsymbol{e}}_j, \tag{9C.36}$$

where the partial derivative $\partial\hat{t}_{-1i}/\partial v_k$ is evaluated at the argument (v_1, v_2, v_3) corresponding to $\boldsymbol{V}$ and where

$$t_{-1i} \equiv t_{-1i}(0) = \frac{\rho_{\boldsymbol{\kappa}}}{\rho_0} t_i, \tag{9C.37}$$

t_i being a principal stress, that is, a proper number of $\boldsymbol{T}$, the Cauchy stress in the configuration $\boldsymbol{\kappa}_0$.

Now substituting (9C.29), (9C.34), and (9C.36) into (9C.25), we finally obtain[1]

$$\begin{aligned}\frac{\rho_\kappa}{\rho_0}\,\mathring{f}(0) = &\left(v_1\frac{\partial t_{-11}}{\partial v_1}-(m+1)t_{-11}\right)d_{11}^2+\dots\\ &+\left(v_2\frac{\partial t_{-11}}{\partial v_2}+v_1\frac{\partial t_{-12}}{\partial v_1}\right)d_{11}\,d_{22}+\dots\\ &+\left[2(t_{-11}-t_{-12})\frac{v_1^2+v_2^2}{v_1^2-v_2^2}-(m+1)(t_{-11}+t_{-12})\right]d_{12}^2+\dots\end{aligned} \tag{9C.38}$$

where we have assumed that v_1, v_2, and v_3 are unequal. In a degenerate case, say when $v_1 = v_2 = v$, we can compute the ratio $(t_{-11}-t_{-12})/(v_1-v_2)$ by[2]

$$\lim_{v_1,v_2\to v}\left(\frac{t_{-11}-t_{-12}}{v_1-v_2}\right)=\left(\frac{\partial\hat{t}_{-11}}{\partial v_1}-\frac{\partial\hat{t}_{-12}}{\partial v_1}\right)\Bigg|_{(v,v,v_3)}. \tag{9C.39}$$

Since the H_m inequality requires $\mathring{f}(0) > 0$ for all symmetric tensors $\boldsymbol{D}$ other than $\boldsymbol{0}$, (9C.38) implies, first, that the coefficients of the shear components d_{12}^2, *etc.*, on its right-hand side be strictly positive, and second, that the quadratic form in the normal components d_{11}^2, *etc.*, be positive-definite. More specifically, we put

$$A_{m1} = \frac{(t_2-t_3)(v_2^2+v_3^2)}{2(v_2^2-v_3^2)}-\frac{m+1}{4}(t_2+t_3), \tag{9C.40}$$

etc. Then from (9C.37) the H_m inequality implies first

$$A_{m\Gamma} > 0,\quad \Gamma = 1, 2, 3. \tag{9C.41}$$

Of course, when $m = 0$, $A_{0\Gamma}$ is equal to A_Γ, defined by (9A.20). Taking $v_1 > v_2 > v_3$ in a non-degenerate case, we can rewrite (9C.41) as

$$\begin{aligned}&[(1-m)v_1^2+(3+m)v_2^2]t_{-11} > [(3+m)v_1^2+(1-m)v_2^2]t_{-12},\\ &[(1-m)v_2^2+(3+m)v_3^2]t_{-12} > [(3+m)v_2^2+(1-m)v_3^2]t_{-13},\\ &[(1-m)v_3^2+(3+m)v_1^2]t_{-13} > [(3+m)v_3^2+(1-m)v_1^2]t_{-11}.\end{aligned} \tag{9C.42}$$

[1] *Cf.* R. Hill, "Constitutive inequalities for isotropic elastic solids under finite strain", *Proc. Roy. Soc. Lond.* **A314**, 457–472 (1970).

[2] For more details about the degenerate cases, see § VI.4.

On the other hand, in a degenerate case, say when $v_1 = v_2 = v$, from (9C.39) we see that (9C.41) reduces to

$$v\left(\frac{\partial \hat{t}_{-11}}{\partial v_1} - \frac{\partial \hat{t}_{-12}}{\partial v_1}\right)\bigg|_{(v,\, v,\, v_3)} > (m+1)t_{-11}, \tag{9C.43}$$

etc.

Next, from (9C.15) and $(9C.17)_1$ the principal generalized stress t_{ma} of $\bar{\boldsymbol{T}}_m$ is related to t_{-1a} and v_a by

$$t_{ma} = \frac{t_{-1a}}{v_a^{m+1}}, \quad a = 1, 2, 3. \tag{9C.44}$$

Likewise, from (8C.42) the principal generalized stretch e_{ma} of $\bar{\boldsymbol{E}}_m$ is related to the principal stretch v_a by

$$e_{ma} = \frac{1}{m+1}\,(v_a^{m+1} - 1). \tag{9C.45}$$

Hence from $(9C.34)_1$ we have

$$\dot{e}_{ma} = v_a^m \dot{v}_a = v_a^{m+1} d_{aa}, \quad a = 1, 2, 3. \tag{9C.46}$$

Now using (9C.44) and (9C.46), we see that the quadratic form in the normal components of $\boldsymbol{D}$ on the right-hand side of (9C.38) can be written more compactly as

$$\dot{t}_{ma}\,\dot{e}_{ma} = \frac{\partial \hat{t}_{ma}}{\partial e_{mb}}\,\dot{e}_{ma}\,\dot{e}_{mb}\,. \tag{9C.47}$$

Hence the H_m inequality requires also that the Hessian matrix for the transformation from e_{ma} to t_{mb}:

$$\left[\frac{\partial \hat{t}_{ma}}{\partial e_{mb}}\right], \tag{9C.48}$$

be positive-definite at every point in the e_m-space.

Now by exactly the same argument as before, we see that the positive-definiteness of $[\partial \hat{t}_{ma}/\partial e_{mb}]$ is the strict differential inequality corresponding to the requirement that the transformation from e_{ma} to t_{mb} be monotone. That is, we can assert the inequality

$$\sum_{a=1}^{3} (t_{ma}^* - t_{ma})(e_{ma}^* - e_{ma}) > 0 \tag{9C.49}$$

if $e^*_{ma} \neq e_{ma}$ for some a. This inequality generalizes the GCN_0 condition, since when $m = 0$ it reduces to (9A.3). As before, we have the following five specific consequences of the positive-definiteness of $[\partial \hat{t}_{ma}/\partial e_{mb}]$:

1^+_m. *The generalized T–E⁺ inequalities.*

$$\frac{\partial \hat{t}_{ma}}{\partial e_{ma}} > 0, \quad a = 1, 2, 3. \tag{9C.50}$$

Clearly, these are the strict differential inequalities of the *generalized T–E inequalities*, which require that

$$(t^*_{ma} - t_{ma})(e^*_{ma} - e_{ma}) > 0, \quad a = 1, 2, 3, \tag{9C.51}$$

provided $e^*_{ma} \neq e_{ma}$ and $e^*_{mb} = e_{mb}$ if $b \neq a$. When $m = 0$, (9C.50) and (9C.51) reduce to $(9A.22)_1$ and $(9A.7)_1$, respectively.

2^+_m. *The generalized IFS⁺ condition.*

$$\det\left[\frac{\partial \hat{t}_{ma}}{\partial e_{mb}}\right] \neq 0. \tag{9C.52}$$

This is the strict differential condition of the *generalized IFS condition*, which requires the transformation from e_{ma} to t_{mb} be invertible. When $m = 0$, (9C.52) reduces to (9A.23).

3^+_m. *The generalized E–T⁺ inequalities.*

Under 2^+_m, we have

$$e_{ma} = \hat{e}_{ma}(t_{m1}, t_{m2}, t_{m3}). \tag{9C.53}$$

Then

$$\frac{\partial \hat{e}_{ma}}{\partial t_{ma}} > 0, \quad a = 1, 2, 3. \tag{9C.54}$$

Clearly, these are the strict differential inequalities corresponding to the *generalized E–T inequalities*, which require (9C.51) provided that $t^*_{ma} \neq t_{ma}$ and $t^*_{mb} = t_{mb}$ if $b \neq a$. When $m = 0$, (9C.54) reduces to (9A.24).

4_m. *The generalized ordered-stresses inequalities.*

$$(t_{ma} - t_{mb})(e_{ma} - e_{mb}) > 0 \text{ if } e_{ma} \neq e_{mb}. \tag{9C.55}$$

If we choose $e_{m1} > e_{m2} > e_{m3}$, or, equivalently $v_1 > v_2 > v_3$, then (9C.55) implies $t_{m1} > t_{m2} > t_{m3}$. From (9C.44) this means also that

$$\frac{t_{-11}}{v_1^{m+1}} > \frac{t_{-12}}{v_2^{m+1}} > \frac{t_{-13}}{v_3^{m+1}}. \tag{9C.56}$$

In particular, when $m = 0$, (9C.56) is equivalent to the ordered-forces inequalities:

$$t_{01} > t_{02} > t_{03}, \tag{9C.57}$$

which are the same as (9A.10), and when $m = -1$, (9C.56) reduces to the ordered-stresses inequalities

$$t_1 > t_2 > t_3, \tag{9C.58}$$

which are the same as the Baker-Ericksen inequalities (9B.13).

HILL[1] remarked that when $m = 1$, -1, or -3, the inequalities (9C.42) are equivalent to the inequalities (9C.56). Hence H_1, H_{-1}, *and* H_{-3} *are necessary and sufficient that the Hessian matrices* $[\partial \hat{t}_{ma}/\partial e_{mb}]$ *with the corresponding values of m be positive-definite.* For all other values of m, including 0, the inequalities (9C.56) and the positive-definiteness of $[\partial \hat{t}_{ma}/\partial e_{mb}]$ are distinct conditions.

5_m. *The generalized pressure-compression inequality.* When the reference configuration is a natural state,

$$\sum_{a=1}^{3} t_{ma}\, e_{ma} > 0. \tag{9C.59}$$

When $m = 0$, (9C.59) reduces to (9A.12).

Next, we consider consequences of the H_m inequality for an isotropic hyperelastic material. From (5.12), the stored-energy σ is now given by an isotropic function of V, or, equivalently, of $\bar{E}_m$. Then (8C.60) implies

$$t_{ma} = \rho_\kappa \frac{\partial \tilde{\sigma}}{\partial e_{ma}}, \tag{9C.60}$$

where $\tilde{\sigma}$ is the function of e_{m1}, e_{m2}, e_{m3} whose value is the stored energy. When $m = 0$, this equation reduces to (9A.28). From (9C.60) we see that $[\partial \hat{t}_{ma}/\partial e_{ma}]$ is positive-definite if and only if the Hessian form $[\partial^2 \tilde{\sigma}/\partial e_{ma} \partial e_{mb}]$ of $\tilde{\sigma}$ relative to e_{ma} is positive-definite at every point

[1] *Cf.* HILL, *op. cit.* p., 253.

in the e_m-space. Hence H_m implies that $\tilde{\sigma}$ is a strictly convex function of e_{m1}, e_{m2}, e_{m3}. In particular, we have the integrated inequality

$$\tilde{\sigma}(e_{m1}^*, e_{m2}^*, e_{m3}^*)-\tilde{\sigma}(e_{m1}, e_{m2}, e_{m3})-\sum_{a=1}^{3}(e_{ma}^*-e_{ma})\frac{\partial\tilde{\sigma}}{\partial e_{ma}}>0 \quad (9\mathrm{C}.61)$$

if $e_{ma}^* \neq e_{ma}$, or equivalently $v_a^* \neq v_a$, for some a. When $m = 0$, (9C.61) reduces to (9A.27). Now by exactly the same argument as before, we see that $[\partial^2\tilde{\sigma}/\partial e_{ma}\partial e_{mb}]$ is positive-definite if and only if for the same m the generalized T–E$^+$ inequalities, the generalized E–T$^+$ inequalities, the generalized IFS$^+$ condition, and the C inequalities hold, where for these last we assume that the reference configuration is an undistorted natural state. Of course, only when $m = 1, -1$, or -3 is the positive-definiteness of $[\partial^2\tilde{\sigma}/\partial e_{ma}\partial e_{mb}]$ equivalent to the H_m inequality. For other values of m, H_m implies also the inequalities (9C.42), which then restrict the function $\tilde{\sigma}$ in virtue of (9C.60), (9C.42) and (9C.44).

D. *Closing Remarks.*

We remark first that to the limited extent that experimental data are available, they seem to support the following *E inequalities.*

7. *Empirical (E) inequalities*

In terms of the representation formula (2.50), the E-inequalities are

$$\psi_0 \leqq 0, \ \psi_1 > 0, \ \psi_{-1} \leqq 0. \quad (9\mathrm{D}.1)$$

From (2.50), it can be shown easily that E ⇒ O–F & B–E. Beyond these implications, no theoretical motivation has been found for the E-inequalities.

In Subsections A through C we have presented some physically reasonable consequences of the major inequalities stated in § 8. There are, however, also some unreasonable consequences, especially for elastic fluids. First, since the S–E condition implies the B–E inequalities, it cannot possibly be satisfied by any elastic fluid, because the constitutive equation of such a fluid is

$$\boldsymbol{T} = -\bar{p}\left(\frac{1}{\rho_\kappa}\det \boldsymbol{F}\right)\mathbf{1} = -p(\rho)\mathbf{1} \quad (9\mathrm{D}.2)$$

according to which the shear stress vanishes for every $\boldsymbol{F}$. Second, for an elastic fluid we claim that GCN is equivalent to the following two conditions[1]:

i) $\bar{p}(\upsilon) > 0$ for all υ, or equivalently $p(\rho) > 0$ for all ρ, where $\upsilon \equiv \rho^{-1}$.
ii) $\upsilon^{\frac{2}{3}}\bar{p}(\upsilon)$ is a strictly decreasing function of υ, or equivalently $\rho^{-\frac{2}{3}}p(\rho)$ is a strictly increasing function of ρ.

While Condition i) seems to be in accord with experience, Condition ii) implies that

$$\upsilon \frac{d\bar{p}}{d\upsilon} \geqq -\tfrac{2}{3}\bar{p}, \tag{9D.3}$$

or equivalently

$$\rho \frac{dp}{d\rho} \geqq \tfrac{2}{3}p. \tag{9D.4}$$

It is known that such a condition on the compressibility cannot be satisfied near a critical point of a fluid capable of change of phase, *e.g.* a van der Waals fluid.

Before proving Conditions i) and ii), we recall first that every elastic fluid is hyperelastic; further, the stored-energy function can be represented by

$$\sigma = \bar{\sigma}\left(\frac{1}{\rho_\kappa} \det \boldsymbol{F}\right) = \tilde{\sigma}(\rho), \tag{9D.5}$$

where $\bar{\sigma}$ is given by

$$\bar{\sigma}(\upsilon) = -\int \bar{p}(\upsilon)d\upsilon, \quad \frac{d\bar{\sigma}}{d\upsilon} = -\bar{p}(\upsilon), \tag{9D.6}$$

and $\tilde{\sigma}$ is given by

$$\tilde{\sigma}(\rho) = \int \frac{p(\rho)}{\rho^2} d\rho, \quad \frac{d\tilde{\sigma}}{d\rho} = \frac{p(\rho)}{\rho^2}. \tag{9D.7}$$

As remarked before, when the material is hyperelastic, the GCN condition reduces to the C–N condition. From (9D.5) and (8A.28), the C–N condition for an elastic fluid has the form

$$\bar{\sigma}(\upsilon^*) - \bar{\sigma}(\upsilon) + \upsilon\bar{p}(\upsilon)\,\mathrm{tr}\,(\boldsymbol{S} - \boldsymbol{1}) > 0 \tag{9D.8}$$

---

[1] B. D. Coleman, "Mechanical and thermodynamical admissibility of stress-strain functions", *Arch. Rational Mech. Anal.* **9**, 172–186 (1962).

for all υ^* and υ satisfying the condition

$$\upsilon^* = \upsilon \det \boldsymbol{S}. \tag{9D.9}$$

Here, as before, $\boldsymbol{S}$ is an arbitrary positive-definite, symmetric tensor other than $\mathbf{1}$. We shall now prove that Conditions i) and ii) are necessary and sufficient for the inequality (9D.8).

To prove that Condition i) is necessary for (9D.8), we choose $\boldsymbol{S}$ to be an arbitrary unimodular, positive-definite, symmetric tensor $\boldsymbol{S}_0$ different from $\mathbf{1}$. Then (9D.9) implies that $\upsilon^* = \upsilon$, and hence (9D.8) reduces to

$$\upsilon \bar{p}(\upsilon) \operatorname{tr}(\boldsymbol{S}_0 - \mathbf{1}) > 0. \tag{9D.10}$$

Using the fact that the arithmetic mean is greater than the geometric mean, we have

$$\operatorname{tr}(\boldsymbol{S}_0 - \mathbf{1}) > 0. \tag{9D.11}$$

Hence (9D.10) implies $\bar{p}(\upsilon) > 0$.

To prove that condition ii) is necessary for (9D.8), we choose $\boldsymbol{S}$ to be an arbitrary multiple $\alpha^{\frac{1}{3}}\mathbf{1}$ of $\mathbf{1}$. Then (9D.9) implies $\upsilon^* = \alpha\upsilon$, and hence (9D.8) reduces to

$$\bar{\sigma}(\alpha\upsilon) - \bar{\sigma}(\upsilon) + 3\upsilon\bar{p}(\upsilon)(\alpha^{\frac{1}{3}} - 1) > 0. \tag{9D.12}$$

Since α and υ are arbitrary, we have also

$$\bar{\sigma}(\upsilon) - \bar{\sigma}(\alpha\upsilon) + 3\alpha\upsilon\bar{p}(\alpha\upsilon)(\alpha^{-\frac{1}{3}} - 1) > 0. \tag{9D.13}$$

Now adding (9D.12) and (9D.13), we obtain

$$3\upsilon^{\frac{1}{3}}(\alpha^{\frac{1}{3}} - 1)[\upsilon^{\frac{2}{3}}\bar{p}(\upsilon) - (\alpha\upsilon)^{\frac{2}{3}}\bar{p}(\alpha\upsilon)] > 0, \tag{9D.14}$$

which implies that $\upsilon^{\frac{2}{3}}\bar{p}(\upsilon)$ is a strictly decreasing function of υ.

Conversely, to prove that Conditions i) and ii) are sufficient for (9D.8), we observe that the left-hand side of (9D.12) is given by

$$\begin{aligned}
&\bar{\sigma}(\alpha\upsilon) - \bar{\sigma}(\upsilon) + 3\upsilon\bar{p}(\upsilon)(\alpha^{\frac{1}{3}} - 1) \\
&\quad = \int_1^{\alpha} \left[\frac{d\bar{\sigma}(\tau\upsilon)}{d\tau} + \upsilon\bar{p}(\upsilon)\tau^{-\frac{2}{3}} \right] d\tau \qquad (9D.15) \\
&\quad = \int_1^{\alpha} [\upsilon^{\frac{2}{3}}\bar{p}(\upsilon) - (\tau\upsilon)^{\frac{2}{3}}\bar{p}(\tau\upsilon)] \frac{\upsilon^{\frac{1}{3}}}{\tau^{\frac{2}{3}}}\, d\tau
\end{aligned}$$

when $\alpha \geqq 1$, and likewise by

$$\bar{\sigma}(\alpha v)-\bar{\sigma}(v)+3v\bar{p}(v)(\alpha^{\frac{1}{3}}-1)$$
$$=\int_{\alpha}^{1}[(\tau v)^{\frac{2}{3}}\bar{p}(\tau v)-v^{\frac{2}{3}}\bar{p}(v)]\frac{v^{\frac{1}{3}}}{\tau^{\frac{2}{3}}}\,d\tau \qquad (9D.16)$$

when $\alpha \leqq 1$. Hence we can recover the inequality (9D.12) from Condition ii). Now by using the fact that the arithmetic mean is greater than the geometric mean, we see that

$$\operatorname{tr}(\boldsymbol{S}-\mathbf{1}) > 3(\alpha^{\frac{1}{3}}-1), \qquad (9D.17)$$

where α is equal to the determinant of $\boldsymbol{S}$ as before. Then by Condition i) we have

$$v\bar{p}(v)\operatorname{tr}(\boldsymbol{S}-1) > 3v\bar{p}(v)(\alpha^{\frac{1}{3}}-1). \qquad (9D.18)$$

Adding (9D.12) and (9D.18), we recover (9D.8). Thus the proof is complete.

Now let us return to an isotropic elastic material in general. From $(9C.17)_1$, we see that the generalized stress $\bar{\boldsymbol{T}}_{-1}$ relative to an undistorted reference configuration is proportional to the Cauchy stress $\boldsymbol{T}$. Differentiating $(9C.17)_1$ with respect to τ along any smooth curve $\boldsymbol{F}=\boldsymbol{F}(\tau)$ with $\boldsymbol{F}(0)=\mathbf{1}$, we obtain

$$\dot{\boldsymbol{T}}_{-1}(0) = -p(\operatorname{tr}\boldsymbol{D})\mathbf{1}+\mathsf{L}(\boldsymbol{D}), \qquad (9D.19)$$

where p denotes the pressure in the reference configuration, $\boldsymbol{D}$ denotes the stretching tensor of $\boldsymbol{F}(\tau)$ at $\tau=0$, and L denotes the elasticity tensor at the reference configuration. From (2.60) we have

$$\mathsf{L}(\boldsymbol{D}) = \lambda(\operatorname{tr}\boldsymbol{D})\mathbf{1}+2\mu\boldsymbol{D}, \qquad (9D.20)$$

where λ and μ are the Lamé constants relative to the reference configuration. Since in general we have

$$\dot{\boldsymbol{T}}_{-1}(0) = \mathsf{E}_{-1}(\boldsymbol{D}), \qquad (9D.21)$$

from (9D.19), (9D.20), and (8C.38) we obtain

$$\dot{\boldsymbol{T}}_{m}(0) = \mathsf{E}_{m}(\boldsymbol{D}) = (\lambda-p)(\operatorname{tr}\boldsymbol{D})\mathbf{1}+(2\mu+(m+1)p)\boldsymbol{D}. \qquad (9D.22)$$

In particular, when $m=0$,

$$\dot{\boldsymbol{T}}_{0}(0) = \mathsf{E}_{0}(\boldsymbol{D}) = (\lambda-p)(\operatorname{tr}\boldsymbol{D})\mathbf{1}+(2\mu+p)\boldsymbol{D}. \qquad (9D.23)$$

Now from (9D.22) the H_m inequality holds at the reference configuration if and only if

$$0 < \operatorname{tr}(\mathsf{E}_m(\boldsymbol{D})\boldsymbol{D}) = (\lambda-p)(\operatorname{tr}\boldsymbol{D})^2+(2\mu+(m+1)p)\operatorname{tr}\boldsymbol{D}^2 \quad (9D.24)$$

for all symmetric tensors $\boldsymbol{D}$ other than $\mathbf{0}$. It is well known in the classical theory of linear elasticity that a necessary and sufficient condition for (9D.24) is

$$2\mu+(m+1)p > 0, \quad 3(\lambda-p)+(2\mu+(m+1)p) > 0. \quad (9D.25)$$

Thus the H_m inequality implies the following lower bounds for the shear modulus and the bulk modulus at an undistorted state[1]:

$$\mu > -\left(\frac{m+1}{2}\right)p, \; \kappa \equiv \lambda+\tfrac{2}{3}\mu > \left(\frac{2-m}{3}\right)p. \quad (9D.26)$$

In particular, when $m = 0$, the H_0 inequality, which is the GCN^+ condition, implies[2]

$$\mu > -\tfrac{1}{2}p, \quad \kappa > \tfrac{2}{3}p. \quad (9D.27)$$

Since the GCN^+ condition is the strict differential condition for the GCN condition, we must replace "$>$" by "$\geqq$" in (9D.24)–(9D.27) if the GCN condition is imposed, *i.e.*, the GCN condition implies the following weaker lower bounds:

$$\mu \geqq -\tfrac{1}{2}p, \quad \kappa \geqq \tfrac{2}{3}p, \quad (9D.28)$$

where equality holds only on a nowhere dense set of values of ρ. In particular, if the material is an elastic fluid, then $(9D.28)_2$ reduces to the previous Condition ii), but $(9D.28)_1$ is just a bit weaker than the previous Condition i).

Now we return to the general lower bounds (9D.26) induced by H_m. We notice that when $m = -1$, the H_{-1} inequality implies that

$$\mu > 0, \quad \kappa > p. \quad (9D.29)$$

That H_{-1} implies $\mu > 0$, we have already observed in Subsection 9C, where we showed that $H_{-1} \Rightarrow$ B–E. While HILL at first proposed the inequalities H_m, $m \in \mathscr{R}$, for elastic materials in general but later emphasized

---

[1] *Cf.* HILL, § 3 of *op. cit.* p. 225, § 7 of *op. cit.* p. 253.

[2] § 6 of B. D. COLEMAN & W. NOLL, "Material symmetry and thermostatic inequalities in finite elastic deformations", *Arch. Rational Mech. Anal.* **15**, 87–111 (1964).

them for isotropic elastic solids only, if we were to apply H_m to an elastic fluid we should find from $(9D.29)_1$ that H_{-1} was impossible. HILL, however, favors the H_{-1} inequality on the grounds of an argument which we have explained at the end of § 8. He notes[1] also that $(9D.26)_1$ could be violated for any $m < -1$ at sufficiently high pressures and for any $m > -1$ at sufficiently large tensions. This remark again supports his choice $m = -1$.

In summary, while the GCN condition or the more general H_m conditions serves to forbid certain physically unreasonable response, and while the S–E condition serves to ensure certain plausible consequences, *none of these can be general.* All of them are too restrictive to serve as *a priori* inequalities for all elastic materials. Some *general constitutive inequality* there must be, but the problem of finding it remains open.

---

[1] *Cf.* HILL, § 4 of *op. cit.* p. 225.

CHAPTER IV

Homogeneous Elastic Bodies

1. The Concept of an Elastic Body

In the preceding chapter we have derived and explained the constitutive equation for an elastic body-point. A body that consists of elastic points alone is called an *elastic body*.

The mathematical structure of an elastic body is determined by the constitutive equations of the material points that make it up. Clearly, that structure is much more complex than what the constitutive equation of a single point expresses, since there are infinitely many points in any body.

The simplest kind of elastic body is the homogeneous one, defined as follows: $\mathscr{B}$ is a *homogeneous*[1] elastic body if there exists a reference configuration $\boldsymbol{\kappa}$ for $\mathscr{B}$ such that the response function relative to the induced local reference configuration $\boldsymbol{\kappa}_*$ is independent of the body-point. A reference configuration $\boldsymbol{\kappa}$ satisfying this condition is called a *homogeneous* configuration of $\mathscr{B}$.

In the physical interpretation, a homogeneous configuration $\boldsymbol{\kappa}$ determines a co-ordinate system on $\mathscr{B}$ (by lifting back to it the reference co-ordinate system on the absolute space $\mathscr{E}$) in which the relative response function is independent of the co-ordinates. We regard $\boldsymbol{\kappa}$ as a configuration in which all body-points of $\mathscr{B}$ are in the same state as far as mechanical response is concerned. In this chapter we consider the theory of homogeneous elastic bodies, which until recently came near to exhausting not only the applications but also the non-trivial mathematics of the general theory of elasticity, once past the local formulation.

The scope of this book, however, is not limited to homogeneous bodies. In the next chapter we consider the theory of a more general class called

---

[1] Of course, the concept of homogeneity may be formulated in theories more general than elasticity, and the same goes for all the rest of the contents of this section.

smooth, materially uniform elastic bodies. While, as we shall see later, all homogeneous bodies are trivially smooth and materially uniform, a smooth materially uniform body need not be homogeneous. In order to fix the position of homogeneous bodies within the more general class, we give now precise definitions of material uniformity and smoothness. A reader whose interest lies entirely in the theory of homogeneous bodies may skip the rest of this section and go on now to the next.

Not requiring henceforth in this section that the elastic body $\mathscr{B}$ be homogeneous, we call two points X and Y of $\mathscr{B}$ *materially isomorphic*[1] if there exists an isomorphism

$$\boldsymbol{I}(X, Y) : \mathscr{B}_X \to \mathscr{B}_Y \tag{1.1}$$

such that

$$\boldsymbol{G}(\boldsymbol{\eta}_Y \circ \boldsymbol{I}(X, Y), X) = \boldsymbol{G}(\boldsymbol{\eta}_Y, Y) \tag{1.2}$$

for all local configurations $\boldsymbol{\eta}_Y$ of Y. Here $\boldsymbol{G}$ denotes the response function as defined in (III. 1.4). We call $\boldsymbol{I}(X, Y)$ a *material isomorphism* from X to Y. Of course, $\boldsymbol{I}(X, Y)^{-1}$ is then a material isomorphism from Y to X.

In the physical interpretation, the constitutive equation (III. 1.4) gives the outcome – the stress tensor – in a local experiment characterized by a local configuration of an elastic point. Then the condition (1.2) means that the outcome in any particular local experiment $\boldsymbol{\eta}_Y$ at Y is the same as that in the corresponding experiment $\boldsymbol{\eta}_Y \circ \boldsymbol{I}(X, Y)$ at X, and conversely. Thus the existence of the material isomorphism $\boldsymbol{I}(X, Y)$ means that the mechanical response of the body-point Y is indistinguishable from that of X by any experiment whatever. We encountered a major special case of a material isomorphism before, though we did not so name it. Indeed, if $Y = X$, the condition (1.2) for $\boldsymbol{I}(X, X)$ reduces to (III. 2.2), which defines a material automorphism of X. Thus a material automorphism of X is a material isomorphism of X with itself.

The following proposition gives the general relation between material isomorphisms and symmetry groups.

Proposition (Noll). *Let $\mathfrak{G}_X$ and $\mathfrak{G}_Y$ denote the symmetry groups of X and Y, and let $\boldsymbol{I}(X, Y)$ be a material isomorphism from X to Y. Then*

$$\mathfrak{G}_Y = \boldsymbol{I}(X, Y) \circ \mathfrak{G}_X \circ \boldsymbol{I}(X, Y)^{-1}. \tag{1.3}$$

[1] This concept was introduced by Noll. He made the condition (1.12) part of the definition. *Cf.* NFTM, § 27.

Conversely, let $\bar{\mathbf{I}}(X, Y)$ *be another material isomorphism from X to Y. Then*

$$\mathbf{I}(X, Y)^{-1} \circ \bar{\mathbf{I}}(X, Y) \in \mathfrak{G}_X, \tag{1.4}$$

and

$$\bar{\mathbf{I}}(X, Y) \circ \mathbf{I}(X, Y)^{-1} \in \mathfrak{G}_Y. \tag{1.5}$$

Consequently the collection

$$\mathbf{I}(X, Y) \circ \mathfrak{G}_X, \tag{1.6}$$

which, in view of the relation (1.3), *is the same as the collection*

$$\mathfrak{G}_Y \circ \mathbf{I}(X, Y), \tag{1.7}$$

gives all material isomorphisms from X to Y.

This proposition is a straightforward consequence of the definitions (1.2) and (III. 2.2).

Material isomorphism can be characterized also in terms of the relative response functions, as shown by the following

PROPOSITION (NOLL). *X and Y are materially isomorphic if and only if there exist local reference configurations* $\mathbf{v}_X$ *and* $\mathbf{v}_Y$ *such that the corresponding relative response functions of X and Y are the same*:

$$\mathbf{H}(\mathbf{F}, X) = \mathbf{H}(\mathbf{F}, Y) \quad \forall \mathbf{F} \in \mathfrak{L}. \tag{1.8}$$

Further, in this case the mapping $\mathbf{I}(X, Y)$ *given by*

$$\mathbf{I}(X, Y) \equiv \mathbf{v}_Y^{-1} \circ \mathbf{v}_X \tag{1.9}$$

is a material isomorphism from X to Y. Conversely, if $\mathbf{I}(X, Y)$ *is a material isomorphism from X to Y, and if* $\mathbf{v}_X$ *is any local reference configuration of X, then the local reference configuration* $\mathbf{v}_Y$ *given by*

$$\mathbf{v}_Y \equiv \mathbf{v}_X \circ \mathbf{I}(X, Y)^{-1} \tag{1.10}$$

fulfils the condition (1.8) *with respect to* $\mathbf{v}_X$.

This proposition is a direct consequence of the definitions (1.2) and (III. 1.15).

The condition (1.8) brings out clearly the physical meaning of material isomorphism: With respect to appropriate references, the mechanical response of X is the same as that of Y. Specifically, we may bring neighborhoods of X and Y into configurations such that the stress required to effect any given deformation of those neighborhoods is the same for each. We call the references $\mathbf{v}_X$ and $\mathbf{v}_Y$ used in the condition (1.8) *materially isomorphic reference configurations* for X and Y. The conditions (1.9) and (1.10) characterize such reference configurations. From (1.8), if $\mathbf{v}_X$ and $\mathbf{v}_Y$ are materially isomorphic, then their corresponding relative symmetry groups $\mathfrak{G}_{\mathbf{v}_X}$ and $\mathfrak{G}_{\mathbf{v}_Y}$ are the same:

$$\mathfrak{G}_{\mathbf{v}_X} = \mathfrak{G}_{\mathbf{v}_Y}. \tag{1.11}$$

Until recently the mass densities $\rho_{\mathbf{v}_X}$ and $\rho_{\mathbf{v}_Y}$ were required to be the same in any materially isomorphic reference configurations $\mathbf{v}_X$ and $\mathbf{v}_Y$:

$$\rho_{\mathbf{v}_X} = \rho_{\mathbf{v}_Y}. \tag{1.12}$$

Since we have assumed that the symmetry group is contained in the unimodular group, by (1.6) we see that the condition (1.12) is consistent with the conditions (1.9) or (1.10). However, in static problems, since the inertia vanishes, a condition such as (1.12) serves no purpose, so we shall impose it only in problems referring to motions.

A body $\mathscr{B}$ is *materially uniform*[1] if its points are pairwise materially isomorphic. In the physical interpretation a body is materially uniform if and only if it consists of one material alone, at least as far as concerns its mechanical response.

Now let $\mathscr{B}$ be a materially uniform elastic body. Then clearly we can generalize the concept of materially isomorphic reference configurations to more than two points in $\mathscr{B}$. Specifically, let $\mathscr{U}$ be a subbody of $\mathscr{B}$, and let $\mathbf{v}$ be a field of local reference configurations on $\mathscr{U}$. Then $\mathbf{v}$ is called a *material reference*[2] for $\mathscr{U}$ if the values of $\mathbf{v}$ are pairwise materially isomorphic. Thus an elastic body is materially uniform if and only if it can be equipped with a global material reference.

The material reference for $\mathscr{U}$ is not unique. From (1.9), a field of local reference configuration on $\mathscr{U}$ is a material reference if and only if $\mathbf{v}_Y^{-1} \circ \mathbf{v}_X$ is a material isomorphism from X to Y for all $X, Y \in \mathscr{U}$.

---

[1] This concept is due to NOLL, who was the first to see that a uniform body need not be homogeneous, and hence also the first to construct a precise and explicit mechanical theory of what are called in the physics of the solid state "dislocations". *Cf.* NFTM, §§ 27 and 34, and "Materially uniform simple bodies with inhomogeneities", *Arch. Rational Mech. Anal.* **27**, 1–32 (1967), reprinted in *Continuum Theory of Inhomogeneities in Simple Bodies*, New York, Springer, 1968.

[2] *Cf.* NFTM, § 34, and NOLL, p. 8 of the work cited in the preceding foonote.

Thus a representation for a material reference is given by

$$\boldsymbol{v}_Y = \boldsymbol{v}_X \circ \boldsymbol{I}(X, Y)^{-1}, \quad Y \in \mathscr{U}, \tag{1.13}$$

where $\boldsymbol{v}_X$ is an arbitrary local reference configuration of a fixed point $X \subset \mathscr{U}$ and where $\boldsymbol{I}(X, Y)$ is an arbitrary material isomorphism from X to an arbitrary point $Y \in \mathscr{U}$. An alternative representation is

$$\boldsymbol{v} = \boldsymbol{P}\boldsymbol{\mu}\boldsymbol{A}, \tag{1.14}$$

where $\boldsymbol{\mu}$ is a fixed material reference, $\boldsymbol{P}$ is an arbitrary tensor in $\mathfrak{L}$, and $\boldsymbol{A}$ is a tensor field with value in the symmetry group, *i.e.*, $\boldsymbol{A}(X) \in \mathfrak{G}_X$ $\forall X \in \mathscr{U}$.

From (1.8), the response function relative to a material reference $\boldsymbol{v}$ is independent of the body-point. We put

$$\boldsymbol{H}_{\boldsymbol{v}}(\boldsymbol{F}) \equiv \boldsymbol{H}(\boldsymbol{F}, \boldsymbol{X}) \quad \forall X \in \mathscr{U}, \tag{1.15}$$

and we call $\boldsymbol{H}_{\boldsymbol{v}}$ the *response function of* $\mathscr{U}$ *relative to* $\boldsymbol{v}$. Likewise, from (1.11), the symmetry group relative to $\boldsymbol{v}$ is independent of the body-point. We put

$$\mathfrak{G}_{\boldsymbol{v}} = \mathfrak{G}_{\boldsymbol{v}X} \quad \forall X \in \mathscr{U}, \tag{1.16}$$

and we call $\mathfrak{G}_{\boldsymbol{v}}$ *the symmetry group of* $\mathscr{U}$ *relative to* $\boldsymbol{v}$.

PROPOSITION (NOLL). *The relative response function and the relative symmetry group of* $\mathscr{U}$ *satisfy the following transformation rules*:

$$\boldsymbol{H}_{\bar{\boldsymbol{v}}}(\boldsymbol{F}) = \boldsymbol{H}_{\boldsymbol{v}}(\boldsymbol{F}\boldsymbol{P}), \tag{1.17}$$

and

$$\mathfrak{G}_{\bar{\boldsymbol{v}}} = \boldsymbol{P}\mathfrak{G}_{\boldsymbol{v}}\boldsymbol{P}^{-1}, \tag{1.12}$$

where $\boldsymbol{P}$ *can be any of the tensors given by*

$$\boldsymbol{P} = \bar{\boldsymbol{v}}_X \circ \boldsymbol{v}_X^{-1} \tag{1.19}$$

for any point $X \in \mathscr{U}$.

The proof of these transformation rules is the same as that given above for their counterparts for a single body-point.

A material reference $\boldsymbol{v}$ for $\mathscr{U}$ need not be the induced field of a reference configuration of $\mathscr{U}$. In fact, $\boldsymbol{v}$ need not even be a smooth field. This consideration leads us to a precise concept of smoothness in elasticity.

A material reference is called a *material chart* if it is a smooth field. Then $\mathscr{B}$ is called a *smooth*[1] materially uniform body if for every point X in $\mathscr{B}$ there exists a subbody $\mathscr{U} \ni X$ which can be equipped with a material chart. In the physical interpretation, this condition requires that the mechanical response of the body-points be distributed smoothly over the body.

While any material reference $\mathbf{v}$ for $\mathscr{U}$ can be extended in many ways to a global material reference for $\mathscr{B}$, a material chart need not have any smooth global extension. Thus the condition of smoothness is strictly local, as it should be. In general, a smooth materially uniform body need not possess any global material chart. Some examples illustrating this fact will be given in the next chapter. The lack of a global material chart for a smooth materially uniform body $\mathscr{B}$ in general leads us to the concept of a *material atlas*, which is similar to the co-ordinate atlas of a differentiable manifold, as we proceed to explain.

Let $\mathbf{v}$ be a material chart for $\mathscr{U}$, and let $\bar{\mathbf{v}}$ be a material chart for $\bar{\mathscr{U}}$. Then we say that $\mathbf{v}$ and $\bar{\mathbf{v}}$ are *compatible* if their corresponding relative response functions $\boldsymbol{H}_{\mathbf{v}}$ and $\boldsymbol{H}_{\bar{\mathbf{v}}}$ are the same. Suppose that the subbodies $\mathscr{U}$ and $\bar{\mathscr{U}}$ overlap. Then a necessary and sufficient condition that $\mathbf{v}$ and $\bar{\mathbf{v}}$ be compatible is

$$\mathbf{v}_X^{-1} \circ \bar{\mathbf{v}}_X \in \mathfrak{G}_X \tag{1.20}$$

for any one body-point X (and hence for all body-points) in $\mathscr{U} \cap \bar{\mathscr{U}}$. A maximal collection of mutually compatible material charts is called a *material atlas* for $\mathscr{B}$.

We denote a material atlas for $\mathscr{B}$ by

$$\mathfrak{A} = \{(\mathscr{U}_\alpha, \mathbf{v}_\alpha), \quad \alpha \in I\}, \tag{1.21}$$

where I denotes an index set and where $\mathbf{v}_\alpha$ denotes a material chart in $\mathfrak{A}$ for the subbody $\mathscr{U}_\alpha$. By the conditions of smoothness and maximality, the collection $\{\mathscr{U}_\alpha, \alpha \in I\}$ must form an open covering of $\mathscr{B}$:

$$\bigcup_{\alpha \in I} \mathscr{U}_\alpha = \mathscr{B}. \tag{1.22}$$

[1] From this point onward in this section, the development follows C.–C. WANG, "On the geometric structures of simple bodies, a mathematical foundation for the theory of continuous distributions of dislocations", *Arch. Rational Mech. Anal.* **27**, 33–94 (1967), reprinted in *Continuum Theory of Inhomogeneities in Simple Bodies*, New York, Springer, 1968.

We now prove the existence of a material atlas for $\mathscr{B}$. Let $\{\mathscr{U}_\beta, \beta \in J\}$ be an open covering of $\mathscr{B}$ such that each $\mathscr{U}_\beta$ can be equipped with a material chart $\boldsymbol{\mu}_\beta$. By the condition of smoothness on $\mathscr{B}$, such an open covering exists. We choose a particular point $X_\beta \in \mathscr{U}_\beta$ for each β and a particular point $X \in \mathscr{B}$. Now let $\boldsymbol{v}$ be a local reference configuration for X, and let $\boldsymbol{I}(X, X_\beta)$ be a material isomorphism from X to X_β. We set

$$\boldsymbol{P}_\beta \equiv \boldsymbol{\mu}_\beta(X_\beta) \circ \boldsymbol{I}(X, X_\beta) \circ \boldsymbol{v}^{-1}, \tag{1.23}$$

and

$$\boldsymbol{v}_\beta \equiv \boldsymbol{P}_\beta^{-1} \circ \boldsymbol{\mu}_\beta. \tag{1.24}$$

In view of the representation (1.14), we see that each $\boldsymbol{v}_\beta$ is a material chart for $\mathscr{U}_\beta$. All the $\boldsymbol{v}_\beta$ are mutually compatible, since from (1.23) the response function $\boldsymbol{H}_\beta$ relative to $\boldsymbol{v}_\beta$ is the same as the response function $\boldsymbol{H}_{\boldsymbol{v}}$ for the fixed point X relative to the fixed local reference configuration $\boldsymbol{v}$:

$$\begin{aligned}\boldsymbol{H}_\beta(\boldsymbol{F}) = \boldsymbol{G}(\boldsymbol{F} \circ \boldsymbol{v}_\beta(X_\beta), X_\beta) &= \boldsymbol{G}(\boldsymbol{F} \circ \boldsymbol{v} \circ \boldsymbol{I}(X, X_\beta)^{-1}, X_\beta) \\ &= \boldsymbol{G}(\boldsymbol{F} \circ \boldsymbol{v}, X) = \boldsymbol{H}_{\boldsymbol{v}}(\boldsymbol{F}, X) \quad \forall \beta \in J.\end{aligned} \tag{1.25}$$

Consequently the collection

$$\mathfrak{A}_0 = \{(\mathscr{U}_\beta, \boldsymbol{v}_\beta), \quad \beta \in J\} \tag{1.26}$$

forms a material pre-atlas for $\mathscr{B}$. We can maximize $\mathfrak{A}_0$ in the standard way, obtaining a unique material atlas $\mathfrak{A}$ containing $\mathfrak{A}_0$. Thus the existence of a material atlas for $\mathscr{B}$ is proved.

This existence theorem has a trivial converse: If $\mathscr{B}$ can be equipped with a material atlas $\mathfrak{A}$ satisfying the condition (1.22), then $\mathscr{B}$ is a smooth materially uniform body.

As before, the compatibility condition for the material charts in a material atlas implies that the relative response functions $\boldsymbol{H}_\alpha$ are the same for all $\alpha \in I$. We put

$$\boldsymbol{H}_{\mathfrak{A}}(\boldsymbol{F}) \equiv \boldsymbol{H}_\alpha(\boldsymbol{F}) \quad \forall \alpha \in I, \tag{1.27}$$

and we call $\boldsymbol{H}_{\mathfrak{A}}$ the *response function of $\mathscr{B}$ relative to the material atlas* $\mathfrak{A}$. Likewise, we put

$$\mathfrak{G}_{\mathfrak{A}} \equiv \mathfrak{G}_{\boldsymbol{v}_\alpha} \quad \forall \alpha \in I, \tag{1.28}$$

and we call $\mathfrak{G}_{\mathfrak{A}}$ the *symmetry group of $\mathscr{B}$ relative to* $\mathfrak{A}$. Of course, $\mathfrak{G}_{\mathfrak{A}}$

is the symmetry group of $\boldsymbol{H}_{\mathfrak{A}}$:

$$\boldsymbol{P} \in \mathfrak{G}_{\mathfrak{A}} \Leftrightarrow \boldsymbol{H}_{\mathfrak{A}}(\boldsymbol{FP}) = \boldsymbol{H}_{\mathfrak{A}}(\boldsymbol{F}) \quad \forall \boldsymbol{F} \in \mathfrak{L}. \tag{1.29}$$

The maximality condition on a material atlas now implies the following criterion: Two material atlases $\mathfrak{A}$ and $\mathfrak{B}$ for $\mathscr{B}$ are the same if and only if

$$\boldsymbol{H}_{\mathfrak{A}}(\boldsymbol{F}) = \boldsymbol{H}_{\mathfrak{B}}(\boldsymbol{F}) \quad \forall \boldsymbol{F} \in \mathfrak{L}. \tag{1.30}$$

Hence each material atlas $\mathfrak{A}$ for $\mathscr{B}$ is characterized by a distinct relative response function $\boldsymbol{H}_{\mathfrak{A}}$ of $\mathscr{B}$.

Since a relative response function of $\mathscr{B}$ is nothing but a copy of a relative response function of a particular body-point X in $\mathscr{B}$, the collection of all possible relative response functions for $\mathscr{B}$ is well defined. We can exhaust this collection by the relative response function $\boldsymbol{H}_{\mathfrak{A}}$ for all possible material atlases for $\mathscr{B}$. To see this fact, let $\boldsymbol{P}$ be an arbitrary tensor in $\mathfrak{L}$. Then for any fixed material atlas $\mathfrak{A}$ given by (1.21), the collection

$$\boldsymbol{P}\mathfrak{A} = \{(\mathscr{U}_\alpha, \boldsymbol{P} \circ \boldsymbol{v}_\alpha), \quad \alpha \in I\} \tag{1.31}$$

is also a material atlas for $\mathscr{B}$. We leave it to the reader to verify that $\boldsymbol{P}\mathfrak{A}$ is compatible and maximal. From the transformation rules (1.17) and (1.18), we see that

$$\boldsymbol{H}_{\boldsymbol{P}\mathfrak{A}}(\boldsymbol{F}) = \boldsymbol{H}_{\mathfrak{A}}(\boldsymbol{FP}) \quad \forall \boldsymbol{F} \in \mathfrak{L}, \tag{1.32}$$

and

$$\mathfrak{G}_{\boldsymbol{P}\mathfrak{A}} = \boldsymbol{P}\mathfrak{G}_{\mathfrak{A}}\boldsymbol{P}^{-1}. \tag{1.33}$$

Since $\boldsymbol{P}$ is arbitrary, the transformation rule (1.32) shows that the relative response function $\boldsymbol{H}_{\mathfrak{A}}$ for arbitrary $\mathfrak{A}$ is exhaustive.

Now since a material atlas is characterized by its corresponding relative response function for $\mathscr{B}$, the transformation rule (1.32) shows that the operation (1.31) sending one material atlas onto another is also exhaustive. Thus we have proved a representation theorem for material atlases:

PROPOSITION. *Let $\mathfrak{A}$ and $\mathfrak{B}$ be any two material atlases for $\mathscr{B}$. Then there exists a tensor $\boldsymbol{P} \in \mathfrak{L}$ such that*

$$\mathfrak{B} = \boldsymbol{P}\mathfrak{A}. \tag{1.34}$$

Of course, if one material atlas $\mathfrak{A}$ for $\mathscr{B}$ is given, then (1.34) characterizes all material atlases for $\mathscr{B}$. For two given material atlases $\mathfrak{A}$ and $\mathfrak{B}$, the tensor $\boldsymbol{P}$ that occurs in the relation (1.34) is not unique. A choice for $\boldsymbol{P}$ can be determined in the following way: Let X be any particular point in $\mathscr{B}$, and let $(\mathscr{V}, \boldsymbol{\tau})$ and $(\mathscr{U}, \boldsymbol{v})$ be any material charts in $\mathfrak{B}$ and $\mathfrak{A}$, respectively, such that

$$X \in \mathscr{U} \bigcap \mathscr{V}. \tag{1.35}$$

Then the tensor $\boldsymbol{P}$ given by

$$\boldsymbol{P} = \boldsymbol{\tau}_X \circ \boldsymbol{v}_X^{-1} \tag{1.36}$$

satisfies (1.34). From (1.32) and the criterion (1.30) we see that

$$A \in \mathfrak{G}_{\mathfrak{A}} \Leftrightarrow \mathfrak{A} = A\mathfrak{A}. \tag{1.37}$$

Consequently if $\boldsymbol{P}$ is any tensor satisfying (1.34), then all others are given by the collection $\boldsymbol{P}\mathfrak{G}_{\mathfrak{A}}$.

So far we have considered smooth materially uniform elastic bodies in general. Homogeneous bodies, which we defined and discussed at the beginning of this section, are special cases. We now develop the concept of homogeneity within the more general one of uniformity.

We say that a material chart $\boldsymbol{v}$ for $\mathscr{U}$ is a *homogeneous chart* if $\boldsymbol{v}$ is the induced field of a configuration $\boldsymbol{\kappa}$ of $\mathscr{U}$, *viz*

$$\boldsymbol{v}_X = \boldsymbol{\kappa}_{*X} \quad \forall X \in \mathscr{U}. \tag{1.38}$$

In this case $\boldsymbol{\kappa}$ is called a *homogeneous configuration* of $\mathscr{U}$. We call $\mathscr{B}$ a *locally homogeneous* body if for every body-point X in $\mathscr{B}$ there exists a (local) homogeneous material chart covering X. As usual, such material charts need not be defined globally on $\mathscr{B}$. Thus a locally homogeneous body need not be homogeneous in the sense defined at the beginning of this section. Of course, a homogeneous body is always locally homogeneous, since by definition a homogeneous body can be equipped with a global homogeneous material chart covering all its points. Thus we have the following two implications:

homogeneity $\Rightarrow$ local homogeneity $\Rightarrow$ smooth uniformity,

but neither can be reversed. Any body that is not homogeneous is called *inhomogeneous*, even if it happens to be smooth and locally homogeneous. In the next chapter we shall present and study examples of inhomogeneous but locally homogeneous bodies. Of course, there are inhomogeneous bodies that are neither smooth nor uniform, a trivial example being the union of two disjoint homogeneous bodies $\mathscr{B}$ and $\mathscr{C}$ such that no point of $\mathscr{B}$ is materially isomorphic to any point of $\mathscr{C}$.

The classification of material points into solids, fluids, and fluid crystals can be applied also to materially uniform bodies without any ambiguity. Also by mere extension of the definitions for elastic points we can define an *incompressible elastic body*, a *hyperelastic body*, an *isotropic elastic body, etc.* In the remaining sections of this chapter we consider the theory of homogeneous elastic bodies. The theory of smooth materially uniform bodies in general is reserved for the next chapter.

2. Static Universal Solutions

Throughout the rest of this chapter, $\mathscr{B}$ will denote a homogeneous elastic body, and $\boldsymbol{\kappa}$ will denote a homogeneous reference configuration of $\mathscr{B}$. Relative to $\boldsymbol{\kappa}$, the constitutive equation takes the form

$$\boldsymbol{T} = \boldsymbol{H}(\boldsymbol{F}), \tag{2.1}$$

where the response function $\boldsymbol{H}$ is independent of the body-point. The value of the response function gives the stress tensor in any configuration $\boldsymbol{\chi}$ whose deformation gradient relative to $\boldsymbol{\kappa}$ is $\boldsymbol{F}$, *viz*,

$$\boldsymbol{F} = \boldsymbol{\chi}_* \circ \boldsymbol{\kappa}_*^{-1}, \tag{2.2}$$

and

$$\boldsymbol{T}(X) = \boldsymbol{H}(\boldsymbol{\chi}_{*X} \circ \boldsymbol{\kappa}_{*X}^{-1}) \quad \forall \boldsymbol{x} \in \boldsymbol{\chi}(\mathscr{B}), \tag{2.3}$$

where $\boldsymbol{x}$ denotes the position of the body-point X in the configuration $\boldsymbol{\chi}$:

$$\boldsymbol{x} = \boldsymbol{\chi}(X). \tag{2.4}$$

In the reference configuration $\boldsymbol{\kappa}$ the body-point X occupies the position $\boldsymbol{X}$, where

$$\boldsymbol{X} = \boldsymbol{\kappa}(X). \tag{2.5}$$

Then the deformation from $\boldsymbol{\kappa}$ to $\boldsymbol{\chi}$ is a mapping which sends the point $\boldsymbol{X} \in \boldsymbol{\kappa}(\mathscr{B})$ into the point $\boldsymbol{x} \in \boldsymbol{\chi}(\mathscr{B})$.

The simplest kind of deformation is a *homogeneous* one, defined by the condition that the deformation gradient $\boldsymbol{F}$ be a constant tensor field:

$$\boldsymbol{F}(\boldsymbol{x}) = \boldsymbol{F} \quad \forall \boldsymbol{x} \in \boldsymbol{\chi}(\mathscr{B}). \tag{2.6}$$

In this case the deformation can be given explicitly by

$$\boldsymbol{x} = \boldsymbol{F}(\boldsymbol{X} - \boldsymbol{o}) + \boldsymbol{c}, \tag{2.7}$$

where $\boldsymbol{o}$ denotes the reference origin and $\boldsymbol{c}$ denotes an arbitrary fixed place. A homogeneous deformation may be regarded as a linear transformation $\boldsymbol{F}$ followed by a rigid translation. In a homogeneous deformation every pair of non-coincident straight lines in $\boldsymbol{\kappa}(\mathscr{B})$ is mapped onto a pair of non-coincident straight lines in $\boldsymbol{\chi}(\mathscr{B})$.

Since any homogeneous configuration may be selected as the reference configuration $\boldsymbol{\kappa}$, by setting $\boldsymbol{F} = \mathbf{1}$ in (2.1) we see that the stress $\boldsymbol{T}$ in any homogeneous configuration of $\mathscr{B}$ is a constant tensor. In particular, since a homogeneous deformation always maps one homogeneous configuration onto another[1], the stress $\boldsymbol{T}$ required to effect such a deformation is a constant tensor, and, still more particularly,

$$\operatorname{div} \boldsymbol{T} = \mathbf{0}. \tag{2.8}$$

This trivial analysis shows that every configuration $\boldsymbol{\chi}(\mathscr{B})$ of $\mathscr{B}$ which follows by a homogeneous deformation from $\boldsymbol{\kappa}$ is a configuration of static equilibrium provided that the boundary $\partial\boldsymbol{\chi}(\mathscr{B})$ be supported by the surface forces

$$\boldsymbol{t} = \boldsymbol{T}\boldsymbol{n}, \tag{2.9}$$

where $\boldsymbol{n}$ denotes the outward unit normal of $\partial\boldsymbol{\chi}(\mathscr{B})$.

Such equilibrium configurations are most important from the physical standpoint, since, first, it is relatively easy to apply forces of some desired kind to a boundary, and second, the field equation of equilibrium is satisfied, no matter what be the elastic body selected for the experiment. The shape of the body and the particular response function $\boldsymbol{H}$ enter the problem only through the boundary condition (2.9). Finally, and most important, if we know the boundary forces required for all homogeneous deformations[2], then we can determine the response function $\boldsymbol{H}$.

It is interesting and important to learn whether there be any other deformation that shares the first and second properties just stated for homogeneous deformations. More specifically, we say that a configuration $\boldsymbol{\chi}$ is *controllable* if $\boldsymbol{\chi}(\mathscr{B})$ can be maintained in equilibrium by surface forces

---

[1] The converse, in general, is false: Two homogeneous configurations of a homogeneous elastic body need not differ by a homogeneous deformation. For example, in the case of a homogeneous elastic fluid any isochoric deformation, homogeneous or not, maps a homogeneous configuration onto another one.

[2] In fact, the data for pure stretches suffice, since the response function must have the reduced form (III.1.18).

exerted on the boundary $\partial\chi(\mathscr{B})$ only. Let $\mathfrak{G}$ be a particular group. Then a deformation $\chi \circ \kappa^{-1}$ is called a *static universal solution with respect to* $\mathfrak{G}$ if $\chi(\mathscr{B})$ is controllable for all choices of response functions whose symmetry groups relative to κ contain $\mathfrak{G}$.

The group $\mathfrak{G}$ is an important factor in the condition for a universal solution. Indeed, the term *universal* refers to the demand that the equation of equilibrium be satisfied for *every* response function $\boldsymbol{H}$ whose symmetry group contains the given group $\mathfrak{G}$. In general, let $\mathfrak{S}_{\mathfrak{G}}$ denote the collection of all universal solutions with respect to $\mathfrak{G}$ for a given homogeneous elastic body $\mathscr{B}$ relative to a fixed homogeneous reference configuration κ. Then

$$\mathfrak{G}_1 \supset \mathfrak{G}_2 \Rightarrow \mathfrak{S}_{\mathfrak{G}_1} \supset \mathfrak{S}_{\mathfrak{G}_2}. \tag{2.10}$$

Thus $\mathscr{B}$ has the biggest class of universal solutions if it is made up of fluid points, and the smallest if it is made up of triclinic solid points, for which $\mathfrak{G} = \{\mathbf{1}\}$. That is to say

$$\mathfrak{S}_{\{\mathbf{1}\}} \subset \mathfrak{S}_{\mathfrak{G}} \subset \mathfrak{S}_{\mathfrak{U}} \quad \forall \mathfrak{G}. \tag{2.11}$$

Now let $\mathfrak{S}_0$ denote the collection of all homogeneous deformations relative to κ. Then what we proved before amounts to the relation

$$\mathfrak{S}_0 \subset \mathfrak{S}_{\mathfrak{G}}. \tag{2.12}$$

We proceed to determine the static universal solutions with respect to various groups $\mathfrak{G}$.

Case 1. Homogeneous Fluid Body.

For a homogeneous fluid body we claim that $\mathfrak{S}_{\mathfrak{U}}$ consists in all deformations such that $\det \boldsymbol{F}$ is uniform, *viz*,

$$\det \boldsymbol{F}(\boldsymbol{x}) = d > 0 \quad \forall \boldsymbol{x} \in \chi(\mathscr{B}), \tag{2.13}$$

so that

$$\mathfrak{S}_0 \subsetneqq \mathfrak{S}_{\mathfrak{U}}. \tag{2.14}$$

Indeed, from (III. 2.27) the consititutive equation of $\mathscr{B}$ relative to κ has the reduced form

$$\boldsymbol{T} = f(\det \boldsymbol{F})\mathbf{1}. \tag{2.15}$$

Substituting this representation into the field equation (2.8), we obtain

$$\frac{df}{d(\det \boldsymbol{F})} \operatorname{grad}(\det \boldsymbol{F}) = \mathbf{0}. \tag{2.16}$$

For a static universal solution in $\mathfrak{S}_{\mathfrak{U}}$, the field equation must be satisfied for all choices of response functions having the form (2.15). Thus f is an arbitrary function in (2.16). Consequently the governing equation for $\boldsymbol{F}$ is

$$\operatorname{grad}(\det \boldsymbol{F}) = \mathbf{0}. \tag{2.17}$$

The general solution of this equation is given by (2.13). Thus, as asserted, every static universal solution is a deformation such that $\det \boldsymbol{F}$ is uniform, and conversely.

Not every tensor field $\boldsymbol{F}$ satisfying the condition (2.13) can be integrated to yield a deformation relative to $\boldsymbol{\kappa}$. A consequence of the condition of integrability for (2.2) is that the curvature tensor $\mathbf{R}$ based on $\boldsymbol{B}^{-1}$ be equal to zero:

$$\mathbf{R}(\boldsymbol{x}) = \mathbf{0} \tag{2.18}$$

for all $\boldsymbol{x}$ in the domain of $\boldsymbol{F}$. As usual, $\boldsymbol{B}$ is the left Cauchy-Green tensor of $\boldsymbol{F}$:

$$\boldsymbol{B} = \boldsymbol{F}\boldsymbol{F}^T. \tag{2.19}$$

Hence a universal solution in $\mathfrak{S}_{\mathfrak{U}}$ corresponds to an integrable field $\boldsymbol{F}$ satisfying (2.13).

Case 2. Homogeneous Isotropic Solid Body.

We claim that $\mathfrak{S}_{\mathfrak{O}}$ is equal to $\mathfrak{S}_0$. That is,

$$\mathfrak{S}_0 = \mathfrak{S}_{\{\mathbf{1}\}} = \mathfrak{S}_{\mathfrak{G}} = \mathfrak{S}_{\mathfrak{O}} \quad \forall \mathfrak{G} \subset \mathfrak{O}. \tag{2.20}$$

Since the symmetry group is $\mathfrak{O}$, the reference configuration $\boldsymbol{\kappa}$ is undistorted. Hence the constitutive equation of $\mathscr{B}$ has the reduced form (*cf.* (III. 2.50))

$$\boldsymbol{T} = \psi_0 \mathbf{1} + \psi_1 \boldsymbol{B} + \psi_{-1} \boldsymbol{B}^{-1}, \tag{2.21}$$

where ψ_0, ψ_1, and ψ_{-1} are regarded as functions of the three principal invariants I_a of $\boldsymbol{B}^{-1}$. Substituting (2.21) into (2.8), we obtain

$$\frac{\partial \psi_0}{\partial I_a} \operatorname{grad} I_a + \frac{\partial \psi_1}{\partial I_a} \boldsymbol{B} \operatorname{grad} I_a + \frac{\partial \psi_{-1}}{\partial I_a} \boldsymbol{B}^{-1} \operatorname{grad} I_a + \psi_1 \operatorname{div} \boldsymbol{B} + \psi_{-1} \operatorname{div} \boldsymbol{B}^{-1} = \mathbf{0}, \tag{2.22}$$

where the index a is summed from 1 to 3. Since ψ_0, ψ_1, and ψ_{-1} are arbitrary functions, the governing equations for $\boldsymbol{B}$ are

$$\operatorname{grad} I_a = \mathbf{0}, \quad a = 1, 2, 3, \tag{2.23}$$

$$\operatorname{div} \boldsymbol{B} = \mathbf{0}, \tag{2.24}$$

and

$$\operatorname{div} \boldsymbol{B}^{-1} = \boldsymbol{0}. \tag{2.25}$$

Further, the condition (2.18) must also be satisfied. In component form that condition reads

$$\tfrac{1}{2}[(B^{-1})_{km,pq}+(B^{-1})_{qp,km}-(B^{-1})_{kp,mq}-(B^{-1})_{qm,kp}] \\ +B^{rs}(A_{qpr}A_{kms}-A_{qmr}A_{kps}) = 0, \tag{2.26}$$

where

$$2A_{kmp} \equiv (B^{-1})_{kp,m}+(B^{-1})_{mp,k}-(B^{-1})_{km,p}. \tag{2.27}$$

The comma denotes partial differentiation with respect to (x^1, x^2, x^3):

$$(B^{-1})_{km,p} \equiv \frac{\partial(B^{-1})_{km}}{\partial x^p}. \tag{2.28}$$

Summing (2.26) with respect to the pairs of indices (k, m) and (p, q) and then using the conditions (2.23)–(2.25), we obtain

$$B^{rs}A_{kpr}A_{kps} = 0, \tag{2.29}$$

which can be written as

$$(V^{rs}A_{kpr})^2 = 0, \tag{2.30}$$

where $\boldsymbol{V}$ is the left stretch tensor:

$$\boldsymbol{V} = \boldsymbol{B}^{\frac{1}{2}}. \tag{2.31}$$

Since $\boldsymbol{V}$ is positive-definite, (2.30) implies that

$$A_{kpr} = 0. \tag{2.32}$$

But from (2.27), we have

$$(B^{-1})_{ps,k} = A_{kps}+A_{skp}. \tag{2.33}$$

Hence $\boldsymbol{B}$ must be a constant tensor field.

The condition (2.32) implies that $\boldsymbol{F}$ is a constant tensor also, since from the transformation law for the Christoffel symbols A_{kpr}, (2.32) holds if and only if the co-ordinates (x^1, x^2, x^3) and (X^1, X^2, X^3) are related by an affine transformation (2.7). Thus (2.20) is proved.

The result (2.20) is a weaker version of the celebrated

ERICKSEN THEOREM: *All universal solutions for a hyperelastic homo-*

geneous isotropic body are homogeneous deformations from an undistorted configuration.

That is, the only universal solutions are those corresponding to the deformation of one homogeneous undistorted configuration into another. More specifically, let $\tilde{\tilde{\mathfrak{S}}}_{\mathfrak{G}}$ denote the collection of all deformations $\boldsymbol{\chi} \circ \boldsymbol{\kappa}^{-1}$ such that $\boldsymbol{\chi}(\mathscr{B})$ is controllable for all hyperelastic response functions possessing the symmetry group $\mathfrak{G}$. Then ERICKSEN's theorem asserts that

$$\mathfrak{S}_0 = \tilde{\tilde{\mathfrak{S}}}_{\{\mathbf{1}\}} = \tilde{\tilde{\mathfrak{S}}}_{\mathfrak{O}}. \tag{2.34}$$

Of course, in general

$$\mathfrak{S}_0 \subset \tilde{\tilde{\mathfrak{S}}}_{\mathfrak{G}} \subset \tilde{\tilde{\mathfrak{S}}}_{\mathfrak{U}}, \tag{2.35}$$

and

$$\mathfrak{S}_{\mathfrak{G}} \subset \tilde{\tilde{\mathfrak{S}}}_{\mathfrak{G}}, \tag{2.36}$$

but clearly

$$\mathfrak{S}_{\mathfrak{U}} = \tilde{\tilde{\mathfrak{S}}}_{\mathfrak{U}}, \tag{2.37}$$

since all elastic fluids are hyperelastic. ERICKSEN's theorem (2.34) implies then

$$\mathfrak{S}_{\mathfrak{G}} = \tilde{\tilde{\mathfrak{S}}}_{\mathfrak{G}} \quad \forall \mathfrak{G} \subset \mathfrak{O}. \tag{2.38}$$

To prove ERICKSEN's theorem (2.34), we write the constitutive equation of $\mathscr{B}$ relative to $\boldsymbol{\kappa}$ as

$$T_k^m = -2\rho_0 \sqrt{I_3}\, \frac{\partial \tilde{\sigma}}{\partial I_a}\, \frac{\partial I_a}{\partial (B^{-1})_{mp}}\, (B^{-1})_{kp}, \tag{2.39}$$

where ρ_0 is a constant and where $\tilde{\sigma}$ is independent of the body-point. We can choose ρ_0 as the density of any particular body-point X in the reference configuration $\boldsymbol{\kappa}$. Then $\tilde{\sigma}$ is the stored-energy function of X relative to $\boldsymbol{\kappa}$. Since $\mathscr{B}$ is homogeneous, (2.39) is then valid for all other body-points also. However, if the density of another body-point Y in $\boldsymbol{\kappa}$ is not equal to ρ_0, then $\tilde{\sigma}$ is not the stored-energy density of Y relative to $\boldsymbol{\kappa}$ in the usual sense[1]. We regard $\tilde{\sigma}$ as a function of I_1, I_2, I_3.

Substituting (2.39) into the field equation (2.8), we obtain

$$\frac{\partial \tilde{\sigma}}{\partial I_a}\left(\sqrt{I_3}\, \frac{\partial I_a}{\partial (B^{-1})_{mp}}\, (B^{-1})_{kp}\right)_{,m} + \sqrt{I_3}\, \frac{\partial^2 \tilde{\sigma}}{\partial I_a\, \partial I_b}\, I_{b,m}\, \frac{\partial I_a}{\partial (B^{-1})_{mp}}\, (B^{-1})_{kp} = 0. \tag{2.40}$$

---

[1] Since we assume that the body force vanishes, the inertia of $\mathscr{B}$ has no position in statics. This fact is plain from the field equation (2.8).

Since this condition is to hold for arbitrary $\tilde{\sigma}$, we obtain the governing equations

$$\left(\sqrt{I_3}\,\frac{\partial I_a}{\partial (B^{-1})_{mp}}\,B^{-1}_{kp}\right)_{,m}=0, \tag{2.41}$$

and

$$\left(I_{b,m}\frac{\partial I_a}{\partial (B^{-1})_{mp}}+I_{a,m}\frac{\partial I_b}{\partial (B^{-1})_{mp}}\right)(B^{-1})_{kp}=0,\quad a,b=1,2,3. \tag{2.42}$$

Of course, as before, (2.18) also must hold.

Setting $a=3$ in (2.41) and using the formula for the derivative of a determinant, we see that

$$0=\left(\sqrt{I_3}\,\frac{\partial I_3}{\partial (B^{-1})_{mp}}\,(B^{-1})_{kp}\right)_{,m}=(I_3^{\frac{3}{2}})_{,k}. \tag{2.43}$$

Thus I_3 must be a constant field as in (2.13). Next setting $a=1$ in (2.41) and using the formula for the derivative of the trace, we obtain

$$0=\sqrt{I_3}\left(\frac{\partial I_1}{\partial (B^{-1})_{mp}}\,(B^{-1})_{kp}\right)_{,m}=\sqrt{I_3}\,(B^{-1})^{m}_{k,m}. \tag{2.44}$$

Thus (2.25) must hold. Finally, setting $a=3$ and $b=1$ in (2.42) yields the condition

$$0=\left(I_{1,m}\frac{\partial I_3}{\partial (B^{-1})_{mp}}+I_{3,m}\frac{\partial I_1}{\partial (B^{-1})_{mp}}\right)(B^{-1})_{kp}=I_3 I_{1,k}. \tag{2.45}$$

which shows that I_1 is uniform.

Now we can proceed in exactly the same way as before to obtain (2.32). Thus ERICKSEN's theorem (2.34) is proved.

The cases of a fluid and an isotropic solid, which we have just treated, give the two extremes for the sets $\mathfrak{S}_{\mathfrak{G}}$. An intermediate set between the extremes, of course, must be some $\mathfrak{S}_{\mathfrak{G}}$ in which $\mathfrak{G}$ is a symmetry group for a type of fluid crystals. We consider next

Case 3. Homogeneous Fluid Crystal Body with $\mathfrak{G}$ Given by (III. 2.66).

For this type of fluid crystal, the constitutive equation has the reduced form

$$\boldsymbol{T}=\varphi_0(\det \boldsymbol{F})\mathbf{1}+\varphi_1(\det \boldsymbol{F})\boldsymbol{i}\otimes\boldsymbol{i}, \tag{2.46}$$

where $\boldsymbol{i}$ is given by (III. 2.72). Substituting (2.46) into (2.8), we obtain

$$\frac{d\varphi_0}{d(\det \boldsymbol{F})}\operatorname{grad}(\det \boldsymbol{F})+\frac{d\varphi_1}{d(\det \boldsymbol{F})}(\operatorname{grad}(\det \boldsymbol{F})\cdot \boldsymbol{i})\boldsymbol{i} \\ +\varphi_1(\operatorname{grad}\boldsymbol{i})\boldsymbol{i}+\varphi_1(\operatorname{div}\boldsymbol{i})\boldsymbol{i} = \mathbf{0}. \tag{2.47}$$

Since φ_0 and φ_1 are arbitrary, the governing equations are

$$\operatorname{grad}(\det \boldsymbol{F}) = \mathbf{0}, \tag{2.17}$$

and

$$(\operatorname{grad}\boldsymbol{i}+(\operatorname{div}\boldsymbol{i})\mathbf{1})\,\boldsymbol{i} = \mathbf{0}. \tag{2.48}$$

Of course, $\boldsymbol{F}$ is also subjected to the condition (2.18), and $\boldsymbol{i}$ is a unit vector:

$$\boldsymbol{i}\cdot\boldsymbol{i} = 1. \tag{2.49}$$

The general solution of (2.17) is given as before by (2.13). We proceed to analyse equations (2.48) and (2.49).

Taking the gradient of (2.49), we obtain

$$(\operatorname{grad}\boldsymbol{i})^T\boldsymbol{i} = \mathbf{0}. \tag{2.50}$$

Hence

$$\boldsymbol{i}\cdot(\operatorname{grad}\boldsymbol{i})\boldsymbol{i} = 0. \tag{2.51}$$

Now taking the inner product of (2.48) with $\boldsymbol{i}$ and then using (2.49) and (2.51), we show that $\boldsymbol{i}$ is a solenoidal field:

$$\operatorname{div}\boldsymbol{i} = 0. \tag{2.52}$$

Substitution of (2.52) into (2.48) yields

$$(\operatorname{grad}\boldsymbol{i})\boldsymbol{i} = \mathbf{0}, \tag{2.53}$$

which implies that $\boldsymbol{i}$ is invariant on each of its integral curves. Consequently the integral curves of $\boldsymbol{i}$ are all straight lines. Finally, from (2.50) and (2.53), we see that

$$[(\operatorname{grad}\boldsymbol{i})-(\operatorname{grad}\boldsymbol{i})^T]\boldsymbol{i} = \mathbf{0}. \tag{2.54}$$

Thus

$$(\operatorname{curl}\boldsymbol{i})\wedge\boldsymbol{i} = \mathbf{0}, \tag{2.55}$$

which means that $\boldsymbol{i}$ is either a screw field[1] or a lamellar field.

---

[1] *Cf.* CFT, App. § 34.

A necessary and sufficient condition[1] for a screw field $\boldsymbol{i}$ to be solenoidal is that the surfaces characterized by the condition

$$(\text{curl}\, \boldsymbol{i}) \cdot (\text{curl}\, \boldsymbol{i}) = \text{const.} \tag{2.56}$$

be vector sheets[1] of $\boldsymbol{i}$. Since the integral curves of $\boldsymbol{i}$ are all straight lines, the vector sheets given by (2.56) must be ruled surfaces.

The reader will verify easily that a solution $\boldsymbol{i}_0$ for the system is given by[2]

$$(i_0^1, i_0^2, i_0^3) = (0, \cos\theta(x^1), \sin\theta(x^1)), \tag{2.57}$$

where θ is an arbitrary angle depending on x^1 in a Cartesian co-ordinate system (x^1, x^2, x^3). For this solution, the vector sheets given by (2.56) are parallel planes perpendicular to the x^1-axis, and the integral curves of $\boldsymbol{i}_0$ are parallel lines on each of the vector planes. Of course, any vector field $\boldsymbol{i}$ obtained from a rigid deformation of $\boldsymbol{i}_0$ is also a solution.

3. A Central Example: Simple Shear of a Rectangular Block

A simple shear[3] is a homogeneous deformation characterized by the following components of deformation:

$$\begin{aligned} x^1 &= X^1 + KX^2, \\ x^2 &= X^2, \\ x^3 &= X^3, \end{aligned} \tag{3.1}$$

where the constant K, which is called the *amount of shear*, is equal to the tangent of the angle of shear θ in the (1, 2) co-ordinate plane, as shown in the adjoining sketch:

---

[1] CFT, App. § 34.

[2] When some regularity conditions at infinity are satisfied, the solution (2.57) is complete for the system (2.49), (2.52), and (2.55), as has been shown by A. W. Marris & C.-C. Wang "Solenoidal screw fields of constant magnitude", *Arch. Rational Mech. Anal.* **39**, 70–88 (1970). A complete geometric description of solutions of that system without any regularity conditions but with possible singularities such as sinks or sources has been given recently by W.-L. Yin, "Construction of solenoidal rectilinear vector fields of unit magnitude", *ibid.* **43**, 321–324 (1971).

[3] For a more detailed treatment see § 45 of CFT.

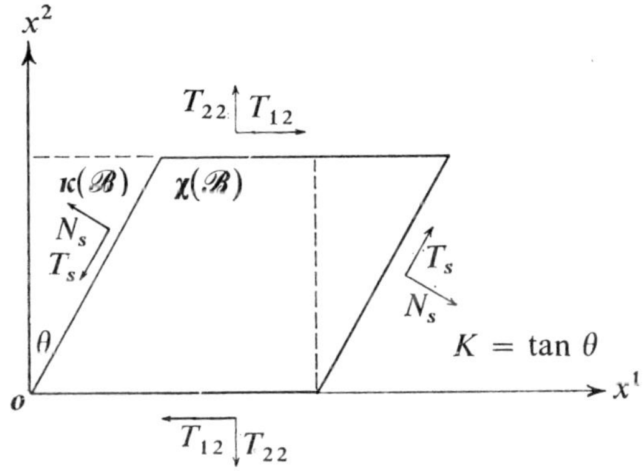

The components of the deformation gradient are easily computed:

$$[F^i_{\ A}] = \begin{bmatrix} 1 & K & 0 \\ 0 & 1 & 0 \\ 0 & 0 & 1 \end{bmatrix}. \tag{3.2}$$

Then the components of the left Cauchy-Green tensor $\boldsymbol{B}$ are given by

$$[B^{ij}] = \begin{bmatrix} 1+K^2 & K & 0 \\ K & 1 & 0 \\ 0 & 0 & 1 \end{bmatrix}, \tag{3.3}$$

so that

$$I_{\boldsymbol{B}} = II_{\boldsymbol{B}} = 3+K^2,\ III_{\boldsymbol{B}} = 1. \tag{3.4}$$

The inverse of $\boldsymbol{B}$ is given by

$$[(B^{-1})_{ij}] = \begin{bmatrix} 1 & -K & 0 \\ -K & 1+K^2 & 0 \\ 0 & 0 & 1 \end{bmatrix}, \tag{3.5}$$

and the invariants of $\boldsymbol{B}^{-1}$ are the same as the corresponding invariants of $\boldsymbol{B}$.

Since a simple shear is a homogeneous deformation, the stress tensor in $\chi(\mathscr{B})$ is a constant tensor. As before, we consider the following three cases:

Case 1. Fluid Body.

From the constitutive equation (2.15) for a fluid body, we see that the stress tensor in $\chi(\mathscr{B})$ is a hydrostatic stress:

$$\boldsymbol{T} = f(1)\mathbf{1}. \tag{3.6}$$

Further, that stress is the same as the stress in the reference configuration $\boldsymbol{\kappa}$, so that one and the same stress suffices to maintain any amount of shear. This simple result shows clearly that the principle of determinism for the stress tensor does not imply its converse, since for an elastic fluid body the deformation cannot be determined by the stress alone.

Case 2. Isotropic Solid Body[1]

The constitutive equation (2.21) for an isotropic body implies that the component matrix of the stress tensor is

$$[T_{ij}] = (\hat{\psi}_0+\hat{\psi}_1+\hat{\psi}_{-1})\begin{bmatrix}1&0&0\\0&1&0\\0&0&1\end{bmatrix}+(\hat{\psi}_1-\hat{\psi}_{-1})K\begin{bmatrix}0&1&0\\1&0&0\\0&0&0\end{bmatrix}$$

$$+\hat{\psi}_1 K^2\begin{bmatrix}1&0&0\\0&0&0\\0&0&0\end{bmatrix}+\hat{\psi}_{-1} K^2\begin{bmatrix}0&0&0\\0&1&0\\0&0&0\end{bmatrix}, \qquad (3.7)$$

where

$$\hat{\psi}_\Gamma = \hat{\psi}_\Gamma(K^2) = \psi_\Gamma(3+K^2, 3+K^2, 1), \quad \Gamma = 0, 1, -1. \qquad (3.8)$$

In particular, the shear stress T_{12} is related to the amount of shear by

$$T_{12} = (\hat{\psi}_1-\hat{\psi}_{-1})K. \qquad (3.9)$$

We define the *shear modulus* $\hat{\mu}$ as the function

$$\hat{\mu} = \hat{\mu}(K^2) = \hat{\psi}_1(K^2)-\hat{\psi}_{-1}(K^2). \qquad (3.10)$$

Then

$$T_{12} = \hat{\mu}K. \qquad (3.11)$$

As far as the mathematical theory is concerned, it is possible[2] for there

---

[1] The analysis of simple shear in a homogeneous isotropic elastic solid is due principally to RIVLIN (1948). For more detail, see § 54 of NFTM.

[2] In § II.7 we have remarked that every isotropic elastic material is a fluid or a solid, and that the constitutive equation (2.15) is necessary and sufficient for a fluid. In terms of the representation (2.21), then, a necessary and sufficient condition for an isotropic point to be a solid point is merely that ψ_1 and ψ_{-1} not vanish identically and ψ_0 not reduce to a function of I_3 alone. While $\hat{\mu}(K^2) = 0$ for a fluid, the converse is false, and it is certainly possible that $\psi_1 = \psi_{-1}$ in a solid. Indeed, the theory includes elastic solids that cannot support shear stress at all, namely, those whose constitutive equation is of the form $\boldsymbol{T} = \psi_0(I_1, I_2, I_3)\mathbf{1}$, where ψ_0 does not reduce to a function of I_3 alone.

to be an isotropic elastic solid in which $\hat{\mu}(K^2) = 0$ for all K, or that $\hat{\mu}(0) = 0$. In applications, however, these solids are of little interest, and indeed we expect that[1] $\hat{\mu}(K^2) > 0$, at least for small K, so that $\hat{\mu}(0) > 0$.

Since $\hat{\mu}$ is an even function of K while T_{12} is an odd function of K,

$$T_{12}|_{K=0} = 0, \quad \left.\frac{d\hat{\mu}}{dK}\right|_{K=0} = 0, \tag{3.12}$$

provided that $\hat{\mu}$ be differentiable at $K = 0$. If $\hat{\mu}$ is smooth near $K = 0$, then for sufficiently small K

$$\hat{\mu}(K^2) = \mu_0 + \mu_1 K^2 + o(K^2), \tag{3.13}$$

where

$$\mu_0 = \hat{\mu}(0), \tag{3.14}$$

and

$$\mu_1 = \left.\frac{d\hat{\mu}}{dK^2}\right|_{K=0}. \tag{3.15}$$

Since

$$\mu_0 = \lim_{K \to 0} \frac{T_{12}}{K}, \tag{3.16}$$

we see that in the classical linearized theory of isotropic solids, which is defined by (III. 2.60), the constant μ_0 is the shear modulus of the body at the reference configuration, assumed to be a natural state. We may call μ_0 the *shear constant*. By (3.11) and (3.13),

$$T_{12} = \mu_0 K + \mu_1 K^3 + o(K^3). \tag{3.17}$$

Thus the departure of the shear stress from the value $\mu_0 K$ given by the classical linearized theory is of third or higher order in K, always on the assumption that $\hat{\mu}$ be smooth near $K = 0$.

The normal stresses can be read off from (3.7) also:

$$\begin{aligned} T_{11} &= \hat{\psi}_0 + \hat{\psi}_1 + \hat{\psi}_{-1} + \hat{\psi}_1 K^2, \\ T_{22} &= \hat{\psi}_0 + \hat{\psi}_1 + \hat{\psi}_{-1} + \hat{\psi}_{-1} K^2, \\ T_{33} &= \hat{\psi}_0 + \hat{\psi}_1 + \hat{\psi}_{-1}. \end{aligned} \tag{3.18}$$

In general, these normal stresses do not vanish, and they are even func-

[1] Restrictions such as this so as to ensure physically reasonable response for elastic materials have been discussed in general in §§ IV. 7–9.

tions of K. Thus a simple shear cannot be effected by a shear stress alone. Further, if the reference configuration is a natural state, so that

$$(\hat{\psi}_0+\hat{\psi}_1+\hat{\psi}_{-1})|_{K=0} = 0, \tag{3.19}$$

and if the functions $\hat{\psi}_0$, $\hat{\psi}_1$, and $\hat{\psi}_{-1}$ are smooth near $K = 0$, then

$$\begin{aligned} T_{11} &= O(K^2) = o(K), \\ T_{22} &= O(K^2) = o(K), \\ T_{33} &= O(K^2) = o(K). \end{aligned} \tag{3.20}$$

That is, the normal stresses are all of second or higher order in K for sufficiently small K.

The result (3.20) shows that in the classical linearized theory, defined by (III.2.60), the normal stresses in simple shear are of the order of terms neglected. In that theory, consequently, a simple shear is maintained by the action of shear stress alone. As we have pointed out before, the classical theory is not a special case but rather an approximation, and so we should not be surprised when a result it predicts turns out to be strictly impossible. Such is the case here. In the general theory of isotropic elastic solids, shear stresses applied to boundaries *never* suffice to maintain a simple shear, no matter how small, since by (3.18) and (3.9) the normal stresses cannot all vanish unless the shear stress T_{12} vanishes also. Comparison of (3.20) with (3.17) shows that for small amounts of shear, the dominant effect of non-linearity is generally the presence of the normal stresses of order K^2, since the correction to the classical linear relation $T_{12} = \mu_0 K$ is of order K^3 or higher.

Not only are the normal stresses not equal to zero, but in general they cannot even equal one another. Indeed, by subtracting $(3.18)_2$ from $(3.18)_1$ and using (3.9), we obtain RIVLIN's universal relation:

$$T_{11}-T_{22} = KT_{12}. \tag{3.21}$$

Hence when $K \neq 0$, a necessary and sufficient condition that $T_{11} = T_{22}$ is $T_{12} = 0$. As we have seen, the condition $T_{12} = 0$ is satisfied in all elastic fluids for all shears, but in elastic solids it is unnatural at any amount of shear K, or at least for small K. The fact that in general $T_{11} \neq T_{22}$ is called the *Poynting effect*.

Of course, the right-hand side of (3.21) is of the order neglected in the classical linearized theory, so there we obtain $T_{11} = T_{22}$ as an approximation. Indeed, as shown already, $T_{11} = 0$, $T_{22} = 0$. The effects originally observed by POYNTING (1905) are more complicated than the simple fact we have chosen to name after him. Accordng to POYNTING, the faces of a block which is subject to the shear stress called for by the linearized theory will draw together or spread apart rather than effecting a simple shear, and torsion effects a change of length proportional to the square of the twist. The name is usually applied also to the effect we have derived, since simple shear fails to occur in the experiment because the unequal normal stresses called for by (3.21) are not supplied. *Cf.* NFTM, §§ 54, 56, 57, and 66.

The relation (3.21) does not depend on the response function $\boldsymbol{H}$. Formulae of this kind, which are called *universal relations*, are most important since they reflect a type of material symmetry directly in terms of stress and deformation. As we shall see later, the universal relation (3.21) follows from applying to the case of simple shear the fact that for an isotropic elastic point a principal basis of $\boldsymbol{B}$ is a principal basis also of $\boldsymbol{T}$.

If we assume that the reference configuration $\boldsymbol{\kappa}$ is a natural state, we may set

$$T_{33} = \hat{\tau}(K^2)K^2 \tag{3.22}$$

and require in view of (3.19) that $\hat{\tau}$ be bounded near $K = 0$. Then by (3.18) the mean stress in a simple shear is given by

$$\tfrac{1}{3}(T_{11}+T_{22}+T_{33}) = (\hat{\tau}+\tfrac{1}{3}(\hat{\psi}_1+\hat{\psi}_{-1}))K^2, \tag{3.23}$$

which is also of the second order in K at $K = 0$, and generally it does not vanish. This fact is called the *Wertheim-Kelvin effect*.

In the classical linearized theory, defined by (III.2.60), the mean stress vanishes since each normal stress is zero. The Wertheim-Kelvin effect is thus characteristic of finite deformation.

The effect found by WERTHEIM (1858) in his experiments on torsion and the related effect inferred in a semi-heuristic theory by KELVIN (1867) are more complicated than the simple fact we have chosen to name after them. WERTHEIM demonstrated that torsion of a hollow cylinder effects a change of volume proportional to the square of the twist. According to KELVIN, the volume of a block which is subject to the shear stress called for by the linearized theory ought to increase in proportion to the square of the amount of shear, so that a deformation more complicated than simple shear results. *Cf.* NFTM, §§ 54 and 66, for the work of KELVIN and later students if this problem; the experiments of WERTHEIM are described and analysed by J. F. BELL, "The experimental foundations of solid mechanics", FLÜGGE's *Handbuch der Physik*, Vol. VIa/1, ed. C. TRUESDELL, Berlin *etc.*, Springer-Verlag, to appear in 1973.

More generally, the term *Wertheim-Kelvin effect* is applied to any case in which the mean stress is zero in the classical linear theory but generally not zero in finite deformation, and a *Poynting effect* has a similar meaning with respect to inequality of individual stress components.

Next we determine the stress vector on the boundary of the rectangular block. As remarked above, the stress vector is given by (2.9). Referring again to the sketch of the deformation, we see that the components of the unit normals of the slanted planes are given by

$$
\begin{aligned}
n^1 &= \pm(1+K^2)^{-\frac{1}{2}}, \\
n^2 &= \mp K(1+K^2)^{-\frac{1}{2}}, \\
n^3 &= 0.
\end{aligned}
\tag{3.24}
$$

Substituting these components into (2.9), we get

$$T_s = \frac{T_{12}}{1+K^2}, \tag{3.25}$$

and

$$N_s = T_{22} - KT_s = T_{22} - \frac{KT_{12}}{1+K^2}, \tag{3.26}$$

where T_s and N_s are the tangential and normal stresses on the slanted faces. In particular, if $K \neq 0$,

$$|T_s| < |T_{12}|, \tag{3.27}$$

and

$$N_s < T_{22}, \tag{3.28}$$

where for the inequality (3.28) we have assumed that $\hat{\mu}(K^2) > 0$, which, as we have remarked before, while physically natural, is not a consequence of the mathematical theory. The inequalities (3.27) and (3.28) express other aspects of the Poynting and Wertheim-Kelvin effects.

Since a simple shear is a homogeneous deformation, it can be regarded as a pure rotation followed by a pure stretch. The reader will verify easily that the principal stretches v_i are given in terms of the proper numbers b_i of $\boldsymbol{B}$ and of the amount of shear K by the relations

$$
\begin{aligned}
v_1^2 &= b_1 = 1+\tfrac{1}{2}K^2 + K\sqrt{1+\tfrac{1}{4}K^2}, \\
v_2^2 &= b_2 = \frac{1}{b_1} = 1+\tfrac{1}{2}K^2 - K\sqrt{1+\tfrac{1}{4}K^2}, \\
v_3^2 &= b_3 = 1,
\end{aligned}
\tag{3.29}
$$

so that

$$v_1 - v_2 = K,$$

$$b_1 - b_2 = 2K\sqrt{1+\tfrac{1}{4}K^2}. \tag{3.30}$$

The principal basis $\{\boldsymbol{f}_1, \boldsymbol{f}_2, \boldsymbol{f}_3\}$ common to $\boldsymbol{V}$, $\boldsymbol{B}$, and $\boldsymbol{T}$ is shown in the adjoining sketch.

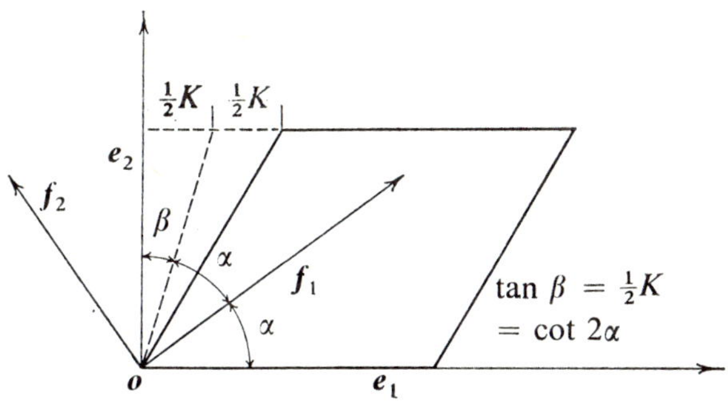

Here

$$\boldsymbol{f}_1 = (\cos\alpha)\boldsymbol{e}_1 + (\sin\alpha)\boldsymbol{e}_2,$$

$$\boldsymbol{f}_2 = -(\sin\alpha)\boldsymbol{e}_1 + (\cos\alpha)\boldsymbol{e}_2, \tag{3.31}$$

$$\boldsymbol{f}_3 = \boldsymbol{e}_3,$$

and $\alpha \in (0, \frac{1}{2}\pi)$ since otherwise $\boldsymbol{F}$ would be singular.

We now prove the interpretation for the universal relation (3.21) which we stated just before (3.22).

PROPOSITION. *Let the bases* $\{\boldsymbol{e}_i\}$ *and* $\{\boldsymbol{f}_i\}$ *be related by* (3.31). *Then a symmetric tensor* $\boldsymbol{A}$ *has the spectral form*

$$\boldsymbol{A} = a_1 \boldsymbol{f}_1 \otimes \boldsymbol{f}_1 + a_2 \boldsymbol{f}_2 \otimes \boldsymbol{f}_2 + a_3 \boldsymbol{f}_3 \otimes \boldsymbol{f}_3 \tag{3.32}$$

if and only if its components A_{ij} *relative to* $\{\boldsymbol{e}_k\}$ *satisfy the relations*

$$A_{13} = A_{23} = 0, \tag{3.33}$$

and

$$A_{11} - A_{22} = KA_{12}. \tag{3.34}$$

Proof. If (3.32) holds, from (3.31) we find at once that

$$\begin{aligned} A_{11} &= a_1 \cos^2 \alpha + a_2 \sin^2 \alpha, \\ A_{22} &= a_1 \sin^2 \alpha + a_2 \cos^2 \alpha, \\ A_{12} &= (a_1 - a_2) \sin \alpha \cos \alpha, \\ A_{33} &= a_3, \\ A_{13} &= A_{23} = 0. \end{aligned} \tag{3.35}$$

Thus

$$\begin{aligned} A_{11} - A_{22} &= (a_1 - a_2)(\cos^2 \alpha - \sin^2 \alpha), \\ &= (a_1 - a_2) \sin 2\alpha \cot 2\alpha, \\ &= A_{12} K, \end{aligned} \tag{3.36}$$

so that necessity is proved.

Conversely, suppose that (3.32) and (3.34) hold. Then we must show that f_1, f_2, and f_3 are proper vectors of A. By (3.33) and (3.31),

$$\begin{aligned} A_{1j} f_1^j &= A_{11} \cos \alpha + A_{12} \sin \alpha, \\ A_{2j} f_1^j &= A_{12} \cos \alpha + A_{22} \sin \alpha, \end{aligned} \tag{3.37}$$

and

$$A_{3j} f_1^j = 0.$$

But now the condition (3.34) implies that

$$\frac{A_{11} \cos \alpha + A_{12} \sin \alpha}{\cos \alpha} = \frac{A_{12} \cos \alpha + A_{22} \sin \alpha}{\sin \alpha}. \tag{3.38}$$

Hence f_1 is a proper vector of A. By the same argument we can prove that f_2 and f_3 are proper vectors of A. Therefore A has the spectral form (3.32), so that sufficiency is proved. Q.E.D.

Since the spectral form (3.32) gives the universal restriction on the stresses corresponding to a simple shear of an isotropic elastic body (in fact, for that matter, the spectral form (3.32) gives the universal restriction on the stress at an isotropic elastic point in any deformation provided that $\{f_i\}$ be a principal basis for the left Cauchy-Green tensor of that deformation), the preceding proposition implies that the three universal relations

$$T_{11} - T_{22} = K T_{12} \tag{3.21}$$

$$T_{13} = T_{23} = 0, \tag{3.39}$$

and

$$T_{11}, T_{22}, T_{33} \text{ are even functions of } K, \tag{3.40}$$

are *complete* in the sense that in a simple shear any universal relation must be a consequence of these three.

Referring again to the principal basis $\{f_i\}$, we see that the spectral form of the stress tensor is

$$\boldsymbol{T} = t_1 \boldsymbol{f}_1 \otimes \boldsymbol{f}_1 + t_2 \boldsymbol{f}_2 \otimes \boldsymbol{f}_2 + t_3 \boldsymbol{f}_3 \otimes \boldsymbol{f}_3, \tag{3.41}$$

where the principal stresses are given by

$$\begin{aligned} t_1 &= \hat{\psi}_0 + \hat{\psi}_1 b_1 + \hat{\psi}_{-1} b_1^{-1}, \\ t_2 &= \hat{\psi}_0 + \hat{\psi}_1 b_2 + \hat{\psi}_{-1} b_2^{-1}, \\ t_3 &= \hat{\psi}_0 + \hat{\psi}_1 + \hat{\psi}_{-1}. \end{aligned} \tag{3.42}$$

Hence, since $b_1 b_2 = 1$,

$$t_1 - t_2 = (\hat{\psi}_1 - \hat{\psi}_{-1})(b_1 - b_2). \tag{3.43}$$

Thus another representation for the shear modulus $\hat{\mu}$, defined by (3.10), is

$$\hat{\mu} = \frac{t_1 - t_2}{b_1 - b_2}. \tag{3.44}$$

Case 3. Fluid Crystal Body.

We consider simple shears for the fluid crystal body introduced in the Case 3 of § III.2. Since the body is anisotropic, the relative position of the preferred axis with respect to the plane of shear is important. If in the reference configuration $\boldsymbol{\kappa}$ the preferred axis is parallel to $\boldsymbol{e}_1$ or $\boldsymbol{e}_3$, then the actual configuration $\boldsymbol{\chi}$ is materially isomorphic to the reference configuration. The analysis of this case is trivial and is similar to that of Case 1. The situation is slightly more complex if the preferred axis is parallel to $\boldsymbol{e}_2$ in $\boldsymbol{\kappa}$. The stress tensor in the actual configuration is then given by

$$\boldsymbol{T} = -p\mathbf{1} + \tau \boldsymbol{i} \otimes \boldsymbol{i}, \tag{3.45}$$

where p and τ are two constants defined by the relations

$$p = -\varphi_0(1), \quad \tau = \varphi_1(1), \tag{3.46}$$

φ_0 and φ_1 being the functions in (2.46), and where $\boldsymbol{i}$ is a unit vector given by

$$\boldsymbol{i} = \sin\theta\, \boldsymbol{e}_1 + \cos\theta\, \boldsymbol{e}_2, \tag{3.47}$$

θ being the angle of shear as shown in the last sketch. Also $\tau \neq 0$, for otherwise the constitutive equation (3.45) would reduce to that of an elastic fluid. Thus in order to effect a simple shear of this kind of subfluid in this particular configuration, it is always necessary to apply a suitable non-zero shear stress.

Substituting (3.47) into (3.45), we find for the components of $\boldsymbol{T}$ the formulae

$$\begin{aligned} T_{11} &= -p+\tau\sin^2\theta, \\ T_{22} &= -p+\tau\cos^2\theta, \\ T_{33} &= -p, \\ T_{12} &= \tau\sin\theta\cos\theta, \\ T_{13} &= T_{23} = 0. \end{aligned} \tag{3.48}$$

Consequently the shear modulus is given by

$$\hat{\mu} = \hat{\mu}(K^2) = \frac{T_{12}}{K} = \frac{\tau}{1+K^2}, \tag{3.49}$$

which is an even function of K such that

$$\hat{\mu} > 0 \Leftrightarrow \tau > 0. \tag{3.50}$$

Further,

$$\tau = \hat{\mu}(0) = \mu_0 . \tag{3.51}$$

If $\tau > 0$, then μ_0 is the maximum value of the function $\hat{\mu}$, which is a monotonically decreasing function as $K^2 \to \infty$.

The boundary forces in $\boldsymbol{\chi}$ can be determined easily in this case. Referring to the previous sketch of the deformation, we have

$$N_s = -p, \tag{3.52}$$

and

$$T_s = 0. \tag{3.53}$$

The canonical form (3.45), of course, gives implicitly the universal restriction on the stress tensor in a simple shear. In terms of the referential components of $\boldsymbol{T}$, we have the following complete set of universal relations:

$$T_{13} = T_{23} = 0, \tag{3.54}$$

$$\frac{T_{11}-T_{33}}{T_{12}} = K, \tag{3.55}$$

$$\frac{T_{22}-T_{33}}{T_{12}} = \frac{1}{K}, \tag{3.56}$$

$$T_{11}+T_{22} = \text{constant independent of } K, \tag{3.57}$$

and

$$T_{33} = \text{constant independent of } K. \tag{3.58}$$

We leave the proof of the completeness of this set to the reader.

Finally, we remark that the results found in the analysis of simple shear cannot offer any condition for the existence of a stored-energy function, since the simple shears, regarded as a subset of the domain $\mathfrak{L}$ of the response function, cannot form any non-trivial loop.

4. Static Universal Solutions for Some Incompressible Bodies

In the last two sections we considered a body $\mathscr{B}$ not subject to any internal constraint, so that the constitutive equation of $\mathscr{B}$ relative to $\boldsymbol{\kappa}$ was (2.1). We turn now to incompressible elastic bodies, for which by (III. 1.51) the constitutive equation is

$$\boldsymbol{T} = -p\mathbf{1}+\boldsymbol{H}(\boldsymbol{F}), \tag{4.1}$$

where p is a hydrostatic stress which is not determined by $\boldsymbol{F}$. For simplicity we choose $\boldsymbol{\kappa}$ to be an admissible configuration of $\mathscr{B}$. Then the response function $\boldsymbol{H}$ is defined on the special linear group $\mathfrak{U}$.

As before, we can introduce the concept of a universal solution for incompressible bodies whose symmetry groups relative to $\boldsymbol{\kappa}$ contain a given group $\mathfrak{G}$. We denote the collection of all such universal solutions by $\hat{\mathfrak{S}}_{\mathfrak{G}}$. Clearly $\hat{\mathfrak{S}}_{\mathfrak{G}}$ contains all isochoric deformations in the previous collection $\mathfrak{S}_{\mathfrak{G}}$, since we can choose p to be zero. In particular, all homogeneous isochoric deformations are universal solutions for all homogeneous incompressible elastic bodies[1]. On the other hand, $\hat{\mathfrak{S}}_{\mathfrak{G}}$ is not a subset of $\mathfrak{S}_{\mathfrak{G}}$, as we shall now see in the following three specific cases.

Case 1. Incompressible Fluid Body.

An incompressible fluid body provides a trivial case, since the stress

[1] For development of the properties of these deformations, including the Poynting effect in isotropic bodies, see § 55 of NFTM.

tensor is entirely unrelated to the deformation:

$$\boldsymbol{T} = -p\mathbf{1}, \tag{4.2}$$

so that $\hat{\mathfrak{S}}_{\mathfrak{u}}$ consists of all isochoric deformations of $\boldsymbol{\kappa}$.

Case 2. Incompressible Isotropic Solid Body.

By (III. 2.56) the stress tensor for an incompressible isotropic body is given by

$$\boldsymbol{T} = -p\mathbf{1}+\varphi_1\boldsymbol{B}+\varphi_2\boldsymbol{B}^2, \tag{4.3}$$

or alternatively

$$\boldsymbol{T} = -p\mathbf{1}+\psi_1\boldsymbol{B}+\psi_{-1}\boldsymbol{B}^{-1}. \tag{4.4}$$

Since $III_{\boldsymbol{B}} = 1$, we have

$$I_{\boldsymbol{B}} = II_{\boldsymbol{B}^{-1}} \equiv I,\ II_{\boldsymbol{B}} = I_{\boldsymbol{B}^{-1}} \equiv II, \tag{4.5}$$

say, and the response coefficients φ_1, φ_2, ψ_1, and ψ_{-1} are functions of I and II. Substituting (4.4) into the field equation (2.8), we obtain

$$\operatorname{div}(\psi_1\boldsymbol{B}+\psi_{-1}\boldsymbol{B}^{-1}) = \operatorname{grad} p. \tag{4.6}$$

Now p is not related to the deformation otherwise than through this condition. Thus if $\boldsymbol{B}$ is given, we are free to adjust p at will. In order for a scalar field p satisfying (4.6) to exist, it is necessary that

$$\operatorname{curl}(\operatorname{div}(\psi_1\boldsymbol{B}+\psi_{-1}\boldsymbol{B}^{-1})) = \mathbf{0}. \tag{4.7}$$

For a simply connected body, this local condition is also sufficient for the existence of a satisfactory field p, but for a multiply connected body generally it is not.

If we set

$$\boldsymbol{A} = \operatorname{div}(\psi_1\boldsymbol{B}+\psi_{-1}\boldsymbol{B}^{-1}), \tag{4.8}$$

we can express the gradient of $\boldsymbol{A}$ in component form as

$$\begin{aligned} A_{k,p} = {}& a^1_{kp}\frac{\partial^2\psi_1}{\partial I^2}+a^2_{kp}\frac{\partial^2\psi_1}{\partial I\,\partial II}+a^3_{kp}\frac{\partial^2\psi_1}{\partial II^2}+a^4_{kp}\frac{\partial\psi_1}{\partial I}\\ &+a^5_{kp}\frac{\partial\psi_1}{\partial II}+a^6_{kp}\psi_1+a^7_{kp}\frac{\partial^2\psi_{-1}}{\partial I^2}+a^8_{kp}\frac{\partial^2\psi_{-1}}{\partial I\,\partial II}\\ &+a^9_{kp}\frac{\partial^2\psi_{-1}}{\partial II^2}+a^{10}_{kp}\frac{\partial\psi_{-1}}{\partial I}+a^{11}_{kp}\frac{\partial\psi_{-1}}{\partial II}+a^{12}_{kp}\psi_{-1}, \end{aligned} \tag{4.9}$$

where

$$\begin{aligned} a^1_{kp} &= I_{,m} I_{,p} B^m_k, \\ a^2_{kp} &= (I_{,m} II_{,p} + II_{,m} I_{,p}) B^m_k, \\ a^3_{kp} &= II_{,m} II_{,p} B^m_k, \\ a^4_{kp} &= I_{,p} B^m_{k,m} + (I_{,m} B^m_k)_{,p}, \\ a^5_{kp} &= II_{,p} B^m_{k,m} + (II_{,m} B^m_k)_{,p}, \\ a^6_{kp} &= B^m_{k,mp}, \end{aligned} \tag{4.10}$$

and where $a^7_{kp}, \ldots, a^{12}_{kp}$ are the same as $a^1_{kp}, \ldots, a^6_{kp}$, respectively, except that $\boldsymbol{B}$ is replaced by $\boldsymbol{B}^{-1}$. In order that $\boldsymbol{B}$ correspond to a universal solution, the condition (4.7) must hold for all choices of ψ_1 and ψ_{-1}. Hence from (4.7) the governing equations for the field $\boldsymbol{B}$ are

$$a^\Delta_{kp} = a^\Delta_{pk}, \quad \Delta = 1, \ldots, 12. \tag{4.11}$$

Further, $\boldsymbol{B}$ must satisfy the constraint

$$\det \boldsymbol{B} = 1 \tag{4.12}$$

and the compatibility condition (2.18).

The governing equations (4.11), (4.12), and (2.18) form a highly overdetermined system for the tensor field $\boldsymbol{B}$. So far, the general solution of this system has not been found. After several important infinite classes of particular solutions had been discovered by trial by RIVLIN (1949) and others, ERICKSEN[1] attacked the general problem. In a deep analysis he showed that four major families of inhomogeneous deformations, together with the isochoric homogeneous deformations, came near to exhausting all solutions, there being only two very restricted further possibilities left open. To find the solutions for these last and thus to determine, finally, a general solution of (4.7), remains a major unsolved problem of the theory of elasticity. For a decade it was guessed that the class of solutions in the two further possibilities was empty, and for one of

[1] J. L. ERICKSEN, "Deformations possible in every isotropic, incompressible, perfectly elastic body", *Z. angew. Math. Phys.* **5**, 466–489 (1954). Reprinted in *Continuum Mechanics IV, Problems of Non-linear Elasticity*, Intl. Sci. Rev. Ser. Vol. 8, ed. C. TRUESDELL, New York, Gordon & Breach 1965.

these classes such was proved to be the case very recently by MARRIS & SHIAU[1]. The other class left open by ERICKSEN was that in which the principal stretches are constant but the local rotation $\boldsymbol{R}$ does not reduce to $\mathbf{1}$. In 1965 the conjecture that this class, likewise, was empty was disproved by the discovery[2] that it includes a fifth family of universal deformations. For certain classes of deformations, however, the known families of solutions have been proved[3] to be complete.

Most of the remainder of this section will be used to set forth and verify the five presently known families of inhomogeneous universal solutions[4]. In the next section we shall study in detail some important examples.

We first present as lemmas some purely static results for certain simple systems of stresses. To this end it is convenient to use the *physical components* of a vector or a tensor with respect to an orthogonal curvilinear co-ordinate system (y^1, y^2, y^3) on the absolute space $\mathscr{E}$, these components being those with respect to an orthonormal basis in the co-ordinate directions. In a Cartesian co-ordinate system, these components coincide with the contravariant and covariant components, but in other co-ordinate systems, they do not. The indices denoting physical components are set in brackets $\langle\ \rangle$; for example, the physical components of the stress tensor $\boldsymbol{T}$ are $T_{\langle ij\rangle}$.

The equations of equilibrium in terms of the physical components relative to the commonest co-ordinate systems are listed on the following page.

---

[1] A. W. MARRIS & J. F. SHIAU, "Universal deformations in isotropic incompressible hyperelastic materials when the deformation tensor has equal proper values", *Arch. Rational Mech. Anal.* **36**, 135–160 (1970).

[2] M. SYNGH & A. C. PIPKIN, "Note on Ericksen's problem", *Z. angew. Math. Phys.* **16**, 706–709 (1965). W. W. KLINGBEIL & R. T. SHIELD, "On a class of solutions in plane finite elasticity", *Z. angew. Math. Phys.* **17**, 489–511 (1966) (see § 5).

[3] R. L. FOSDICK & K. W. SCHULER, "On Ericksen's problem for plane deformations with uniform transverse stretch", *Int. J. Engng. Sci.* **7**, 217–233 (1969). W. C. MÜLLER, "Some further results on the Ericksen-problem for deformations with constant strain invariants", *Z. angew. Math. Phys.* **21**, 633–636 (1970). C. B. KAFADAR, "On Ericksen's problem", *Arch. Rational Mech. Anal.* **47**, 15–27 (1972).

[4] We follow the method of NFTM, §§ 56–57, where more detail regarding the first four families may be found.

1). *Rectangular Cartesian System* (x, y, z).

$$\begin{aligned}
\frac{\partial T_{\langle xx\rangle}}{\partial x}+\frac{\partial T_{\langle xy\rangle}}{\partial y}+\frac{\partial T_{\langle xz\rangle}}{\partial z}&=0,\\
\frac{\partial T_{\langle yx\rangle}}{\partial x}+\frac{\partial T_{\langle yy\rangle}}{\partial y}+\frac{\partial T_{\langle yz\rangle}}{\partial z}&=0,\\
\frac{\partial T_{\langle zx\rangle}}{\partial x}+\frac{\partial T_{\langle zy\rangle}}{\partial y}+\frac{\partial T_{\langle zz\rangle}}{\partial z}&=0.
\end{aligned} \tag{4.13}$$

2). *Cylindrical System* (r, θ, z).

$$\begin{aligned}
\frac{\partial T_{\langle rr\rangle}}{\partial r}+\frac{1}{r}\frac{\partial T_{\langle r\theta\rangle}}{\partial \theta}+\frac{\partial T_{\langle rz\rangle}}{\partial z}+\frac{T_{\langle rr\rangle}-T_{\langle \theta\theta\rangle}}{r}&=0,\\
\frac{\partial T_{\langle r\theta\rangle}}{\partial r}+\frac{1}{r}\frac{\partial T_{\langle \theta\theta\rangle}}{\partial \theta}+\frac{\partial T_{\langle \theta z\rangle}}{\partial z}+\frac{2}{r}T_{\langle r\theta\rangle}&=0,\\
\frac{\partial T_{\langle rz\rangle}}{\partial r}+\frac{1}{r}\frac{\partial T_{\langle \theta z\rangle}}{\partial \theta}+\frac{\partial T_{\langle zz\rangle}}{\partial z}+\frac{1}{r}T_{\langle rz\rangle}&=0.
\end{aligned} \tag{4.14}$$

3). *Spherical System* (r, θ, φ).

$$\begin{aligned}
&\frac{\partial T_{\langle rr\rangle}}{\partial r}+\frac{1}{r}\frac{\partial T_{\langle r\theta\rangle}}{\partial \theta}+\frac{1}{r\sin\theta}\frac{\partial T_{\langle r\varphi\rangle}}{\partial \varphi}\\
&\qquad+\frac{1}{r}[2T_{\langle rr\rangle}-T_{\langle \theta\theta\rangle}-T_{\langle \varphi\varphi\rangle}+\cot\theta\, T_{\langle r\theta\rangle}]=0,\\
&\frac{\partial T_{\langle r\theta\rangle}}{\partial r}+\frac{1}{r}\frac{\partial T_{\langle \theta\theta\rangle}}{\partial \theta}+\frac{1}{r\sin\theta}\frac{\partial T_{\langle \theta\varphi\rangle}}{\partial \varphi}\\
&\qquad+\frac{1}{r}[3T_{\langle r\theta\rangle}+\cot\theta\,(T_{\langle \theta\theta\rangle}-T_{\langle \varphi\varphi\rangle})]=0,\\
&\frac{\partial T_{\langle r\varphi\rangle}}{\partial r}+\frac{1}{r}\frac{\partial T_{\langle \theta\varphi\rangle}}{\partial \theta}+\frac{1}{r\sin\theta}\frac{\partial T_{\langle \varphi\varphi\rangle}}{\partial \varphi}\\
&\qquad+\frac{1}{r}[3T_{\langle r\varphi\rangle}+2\cot\theta\, T_{\langle \theta\varphi\rangle}]=0.
\end{aligned} \tag{4.15}$$

Now we consider the following static problems.

1). *Planar Problems.* If the physical components of the determinate stress $\boldsymbol{T}+p\mathbf{1}$ with respect to a Cartesian system (x, y, z) depend on x alone, the equations of equilibrium (4.13) reduce to

$$\begin{aligned} \frac{\partial T_{\langle xx\rangle}}{\partial x} &= 0, \\ \frac{\partial T_{\langle xy\rangle}}{\partial x} - \frac{\partial p}{\partial y} &= 0, \\ \frac{\partial T_{\langle xz\rangle}}{\partial x} - \frac{\partial p}{\partial z} &= 0. \end{aligned} \tag{4.16}$$

If further $T_{\langle xy\rangle}$ and $T_{\langle xz\rangle}$ vanish, then all quantities $T_{\langle xx\rangle}$, $T_{\langle yy\rangle}$, $T_{\langle zz\rangle}$, $T_{\langle yz\rangle}$, and p are functions of x only. Moreover $T_{\langle xx\rangle}$ is constant. Thus a class of particular solutions of (4.13) is given by

$$\begin{aligned} T_{\langle xx\rangle} &= -p_0, \\ T_{\langle yy\rangle} &= f(x), \\ T_{\langle zz\rangle} &= g(x), \\ T_{\langle yz\rangle} &= h(x), \\ T_{\langle xy\rangle} &= T_{\langle xz\rangle} = 0, \end{aligned} \tag{4.17}$$

where f, g, and h are arbitrary.

2). *Cylindrical Problems.* If the physical components of the determinate stress with respect to a cylindrical co-ordinate system (r, θ, z) are functions of r alone, the equations of equilibrium (4.14) reduce to

$$\begin{aligned} \frac{\partial T_{\langle rr\rangle}}{\partial r} + \frac{1}{r}(T_{\langle rr\rangle} - T_{\langle\theta\theta\rangle}) &= 0, \\ \frac{\partial}{\partial r}(r^2 T_{\langle r\theta\rangle}) - r\frac{\partial p}{\partial \theta} &= 0, \\ \frac{\partial}{\partial r}(r T_{\langle rz\rangle}) - r\frac{\partial p}{\partial z} &= 0. \end{aligned} \tag{4.18}$$

If further $T_{\langle r\theta\rangle}$ and $T_{\langle rz\rangle}$ vanish, then all the quantities $T_{\langle rr\rangle}$, $T_{\langle\theta\theta\rangle}$, $T_{\langle zz\rangle}$, $T_{\langle\theta z\rangle}$, and p are functions of r only. Moreover, $T_{\langle rr\rangle}$ and $T_{\langle\theta\theta\rangle}$ are related by $(4.18)_1$. Thus a class of particular solutions of (4.14) is given by

$$\begin{aligned}
T_{\langle rr\rangle} &= f(r) = -\int \frac{T_{\langle rr\rangle}-T_{\langle\theta\theta\rangle}}{r}\,dr,\\
T_{\langle\theta\theta\rangle} &= \frac{d}{dr}(rf(r)),\\
T_{\langle zz\rangle} &= g(r),\\
T_{\langle\theta z\rangle} &= h(r),\\
T_{\langle r\theta\rangle} &= T_{\langle rz\rangle} = 0,
\end{aligned}\tag{4.19}$$

where f, g, and h are arbitrary. When we come to apply these purely static results to a particular incompressible body, such as an elastic one, we notice that the normal stress difference $T_{\langle rr\rangle}-T_{\langle\theta\theta\rangle}$ equals the difference of the corresponding components of the determinate stress and hence is delivered by the constitutive equation.

In another type of cylindrical problem we assume that the physical components of the determinate stress $\boldsymbol{T}+p\mathbf{1}$ are constants. Then the equations of equilibrium (4.14) reduce not only to (4.18) but to the still simpler forms

$$\begin{aligned}
\frac{\partial p}{\partial r} &= \frac{T_{\langle rr\rangle}-T_{\langle\theta\theta\rangle}}{r},\\
\frac{\partial p}{\partial\theta} &= 2T_{\langle r\theta\rangle},\\
\frac{\partial p}{\partial z} &= \frac{1}{r}\,T_{\langle rz\rangle}.
\end{aligned}\tag{4.20}$$

If, further, $T_{\langle rz\rangle}$ vanishes, then the system (4.20) is integrable, and its general solution is

$$p = (T_{\langle rr\rangle}-T_{\langle\theta\theta\rangle})\log r+2\theta T_{\langle r\theta\rangle}+p_0. \tag{4.21}$$

Thus a class of particular solutions of (4.14) is given by

$$\begin{aligned}
T_{\langle rr\rangle} &= -p+a,\\
T_{\langle\theta\theta\rangle} &= -p+b,\\
T_{\langle zz\rangle} &= -p+c,\\
T_{\langle r\theta\rangle} &= d\\
T_{\langle rz\rangle} &= T_{\langle\theta z\rangle} = 0,
\end{aligned}\tag{4.22}$$

where a, b, c, and d are arbitrary constants, and where p is given by (4.21). As before, the normal stress difference $T_{\langle rr \rangle} - T_{\langle \theta\theta \rangle}$ is delivered by the constitutive equation.

3). *Spherical Problems.* If the physical components of the determinate stress with respect to a spherical co-ordinate system (r, θ, φ) are functions of r only, and if also

$$T_{\langle r\theta \rangle} = T_{\langle r\varphi \rangle} = T_{\langle \theta\varphi \rangle} = 0 \tag{4.23}$$

and

$$T_{\langle \theta\theta \rangle} = T_{\langle \varphi\varphi \rangle}, \tag{4.24}$$

then the equations of equilibrium (4.15) reduce to

$$\begin{aligned} \frac{\partial T_{\langle rr \rangle}}{\partial r} &= -\frac{2}{r}(T_{\langle rr \rangle} - T_{\langle \theta\theta \rangle}), \\ \frac{\partial p}{\partial \theta} &= 0, \\ \frac{\partial p}{\partial \varphi} &= 0. \end{aligned} \tag{4.25}$$

Hence $T_{\langle rr \rangle}$, $T_{\langle \theta\theta \rangle}$, $T_{\langle \varphi\varphi \rangle}$, and p are functions of r only. For the particular solutions of (4.15) that correspond to these assumptions,

$$\begin{aligned} T_{\langle rr \rangle} &= f(r) = -2\int \frac{T_{\langle rr \rangle} - T_{\langle \theta\theta \rangle}}{r}\, dr, \\ T_{\langle \theta\theta \rangle} &= \frac{1}{2r}\frac{d}{dr}(r^2 f(r)), \\ T_{\langle \varphi\varphi \rangle} &= \frac{1}{2r}\frac{d}{dr}(r^2 f(r)), \\ T_{\langle r\theta \rangle} &= T_{\langle r\varphi \rangle} = T_{\langle \theta\varphi \rangle} = 0, \end{aligned} \tag{4.26}$$

where f is an arbitrary function. Again the normal-stress difference $T_{\langle rr \rangle} - T_{\langle \theta\theta \rangle}$ is delivered by the constitutive equation.

These purely static results enable us to find easily certain universal solutions. We first construct isochoric deformations of $\boldsymbol{\kappa}$ such that, in virtue of the constitutive equation, the stress system in the actual configuration is of the form (4.17), (4.19), (4.22), or (4.26), for all choices of the response coefficients ψ_1 and ψ_{-1}. Since corresponding pressure fields have been exhibited for these cases already, these deformations are static universal solutions for $\mathscr{B}$ relative to $\boldsymbol{\kappa}$. All five known families of inhomogeneous universal solutions fall under one or another of these cases, as we shall see now.

Family 1. This family consists in deformations such that

$$r = \sqrt{2AX}, \quad \theta = BY + CZ, \quad z = DY + EZ \tag{4.27}$$

relative to a Cartesian co-ordinate system (X, Y, Z) in the reference configuration $\boldsymbol{\kappa}$ and a cylindrical co-ordinate system (r, θ, z) in the actual configuration $\boldsymbol{\chi}$. In order that these deformations be isochoric, it is necessary and sufficient that the constants $A, \ldots, E$ satisfy

$$A(BE - CD) = 1. \tag{4.28}$$

We claim that the stress tensor in $\boldsymbol{\chi}(\mathscr{B})$ has the form (4.19) relative to (r, θ, z) for all choices of ψ_1 and ψ_{-1}.

Indeed, from (4.27) the matrix of components of the deformation gradient $\boldsymbol{F}$ relative to (X, Y, Z) and (r, θ, z) is

$$[F^k_A] = \begin{bmatrix} \dfrac{A}{r} & 0 & 0 \\ 0 & B & C \\ 0 & D & E \end{bmatrix}. \tag{4.29}$$

Then the contravariant components of $\boldsymbol{B}$ in the cylindrical co-ordinate system (r, θ, z) are given by

$$B^{km} = F^k_A F^m_B \delta^{AB}. \tag{4.30}$$

Thus

$$[B^{km}] = \begin{bmatrix} \dfrac{A^2}{r^2} & 0 & 0 \\ 0 & B^2 + C^2 & BD + CE \\ 0 & BD + CE & D^2 + E^2 \end{bmatrix}. \tag{4.31}$$

Since the covariant components of $\boldsymbol{B}^{-1}$ are related to the contravariant components of $\boldsymbol{B}$ by

$$(B^{-1})_{km} B^{ml} = \delta^l_k, \tag{4.32}$$

we see that

$$[(B^{-1})_{km}] = \begin{bmatrix} \frac{r^2}{A^2} & 0 & 0 \\ 0 & A^2(D^2+E^2) & -A^2(BD+CE) \\ 0 & -A^2(BD+CE) & A^2(B^2+C^2) \end{bmatrix}. \qquad (4.33)$$

From (4.31) and (4.33) the principal invariants of $\boldsymbol{B}$ are

$$\begin{aligned} I = I_{\boldsymbol{B}} &= B^{km}g_{km} = \frac{A^2}{r^2} + (B^2+C^2)r^2 + D^2 + E^2, \\ II = II_{\boldsymbol{B}} = I_{\boldsymbol{B}^{-1}} &= (B^{-1})_{km}g^{km} + \frac{r^2}{A^2} + \frac{A^2(D^2+E^2)}{r^2} + A^2(B^2+C^2), \end{aligned} \qquad (4.34)$$

where g^{km} and g_{km} are the components of the metric tensor in the cylindrical co-ordinate system (r, θ, z). The physical components of $\boldsymbol{B}$ and $\boldsymbol{B}^{-1}$ in (r, θ, z) are given by

$$B_{\langle km \rangle} = \sqrt{g_{kk}g_{mm}}\,B^{km} \quad \text{(unsummed)},$$

and

$$B^{-1}_{\langle km \rangle} = \sqrt{g^{kk}g^{mm}}(B^{-1})_{km} \quad \text{(unsummed)}. \qquad (4.35)$$

Thus

$$[B_{\langle km \rangle}] = \begin{bmatrix} \frac{A^2}{r^2} & 0 & 0 \\ 0 & r^2(B^2+C^2) & r(BD+CE) \\ 0 & r(BD+CE) & D^2+E^2 \end{bmatrix}, \qquad (4.36)$$

and

$$[B^{-1}_{\langle km \rangle}] = \begin{bmatrix} \frac{r^2}{A^2} & 0 & 0 \\ 0 & \frac{1}{r^2}A^2(D^2+E^2) & -\frac{1}{r}A^2(BD+CE) \\ 0 & -\frac{1}{r}A^2(BD+CE) & A^2(B^2+C^2) \end{bmatrix}. \qquad (4.37)$$

Now substituting (4.36), (4.37), and (4.34) into the constitutive equation (4.4), we see that the stress is of the form (4.19) provided that p be a function of r alone. Consequently the deformations in this family are universal solutions. To within an additive constant hydrostatic pressure, the physical components of the stress tensor relative to (r, θ, z) are given by

$$
\begin{aligned}
T_{\langle rr\rangle} &= -\int \frac{T_{\langle rr\rangle}-T_{\langle\theta\theta\rangle}}{r}\,dr,\\
&= -\int\left\{\left[\frac{A^2}{r^2}-r^2(B^2+C^2)\right]\bar{\psi}_1+\left[\frac{r^2}{A^2}-\frac{A^2}{r^2}(D^2+E^2)\right]\bar{\psi}_{-1}\right\}\frac{dr}{r},\\
&= \frac{1}{2}\int\left(\frac{dI}{dr}\bar{\psi}_1-\frac{dII}{dr}\bar{\psi}_{-1}\right)dr,\\
T_{\langle\theta\theta\rangle} &= T_{\langle rr\rangle}-\left[\frac{A^2}{r^2}-r^2(B^2+C^2)\right]\bar{\psi}_1-\left[\frac{r^2}{A^2}-\frac{A^2}{r^2}(D^2+E^2)\right]\bar{\psi}_{-1},\\
T_{\langle zz\rangle} &= T_{\langle rr\rangle}-\left[\frac{A^2}{r^2}-(D^2+E^2)\right]\bar{\psi}_1-\left[\frac{r^2}{A^2}-A^2(B^2+C^2)\right]\bar{\psi}_{-1},\\
T_{\langle\theta z\rangle} &= (BD+CE)\left[r\bar{\psi}_1-\frac{A^2}{r^2}\bar{\psi}_{-1}\right], \qquad (4.38)\\
T_{\langle\theta r\rangle} &= T_{\langle zr\rangle} = 0,
\end{aligned}
$$

where we have set $\bar{\psi}_\Gamma(r) \equiv \psi_\Gamma(I, II)$, $\Gamma = 1, -1$, with I and II given as functions of r by (4.34). Thus $\boldsymbol{T}$ is indeed of the form (4.19) for all choices of ψ_1 and ψ_{-1}.

Family 2. We employ a cylindrical co-ordinate system (R, Θ, Z) in the reference configuration $\boldsymbol{\kappa}$ and a rectangular Cartesian co-ordinate system (x, y, z) in the actual configuration $\boldsymbol{\chi}$. The components of deformation are

$$x = \tfrac{1}{2}AR^2, \quad y = B\Theta+CZ, \quad z = D\Theta+EZ, \qquad (4.39)$$

where $A, \ldots, E$ are constants satisfying the condition (4.28). Following the same procedure as for the previous family, we can prove that the stress is of the form (4.17) for all choice of ψ_1 and ψ_{-1}. Thus the deformations in this family are universal solutions.

Family 3. We use a cylindrical co-ordinate system in both the reference configuration and the actual one. The components of deformation are

$$r = \sqrt{AR^2+B}, \quad \theta = C\Theta+DZ, \quad z = E\Theta+FZ, \qquad (4.40)$$

where $A, \ldots, F$ are constants satisfying the condition

$$A(CF-DE) = 1. \tag{4.41}$$

It is easy to show that relative to (r, θ, z) the stress is of the form (4.19) for all choice of ψ_1 and ψ_{-1}. Thus these deformations form a family of universal solutions.

Family 4. We use a spherical co-ordinate system in both the reference configuration and the actual one. The components of deformation are

$$r = (\pm R^3 + A)^{\frac{1}{3}}, \quad \theta = \pm\Theta, \quad \varphi = \Phi, \tag{4.42}$$

where A is an arbitrary constant. It is easy to show that for all choices of ψ_1 and ψ_{-1} the stress is of the form (4.26) relative to (r, θ, φ). Thus these deformations form a family of universal solutions.

Family 5. We use a cylindrical co-ordinate system in both the reference configuration and the actual one. The components of deformation are

$$r = AR, \; \theta = B \log R + C\Theta, \; z = DZ, \tag{4.43}$$

where $A, \ldots, D$ are constants satisfying the conditions

$$A^2CD = 1. \tag{4.44}$$

It is easy to show that for all choices of ψ_1 and ψ_{-1} the stress is of the form (4.22) relative to (r, θ, z). Thus these deformations form a family of universal solutions.

Case 3. A Type of Incompressible Fluid Crystal Body[1].

We consider the type of fluid crystals treated in Case 3 of § 2, but we now require that the body be incompressible. The constitutive equation is

$$\boldsymbol{T} = -p\mathbf{1} + \tau \boldsymbol{i} \otimes \boldsymbol{i}, \tag{4.45}$$

where τ is a constant, and where $\boldsymbol{i}$ is the preferred axis in the actual configuration and is given by (III. 2.72). Substituting (4.45) into the field equation (2.8), we obtain

$$\tau[(\operatorname{grad} \boldsymbol{i})\boldsymbol{i} + \boldsymbol{i} \operatorname{div} \boldsymbol{i}] = \operatorname{grad} p. \tag{4.46}$$

[1] C.-C. Wang, "Universal solutions and universal relations for some types of elastic fluid crystals", *Q. Appl. Math.* **28**, 57–68 (1970).

As before, an integrability condition for the existence of p is

$$\tau \operatorname{curl} [(\operatorname{grad} \boldsymbol{i})\boldsymbol{i}+\boldsymbol{i} \operatorname{div} \boldsymbol{i}] = \mathbf{0}. \tag{4.47}$$

For a universal solution the constant τ is arbitrary. Thus the governing equation for the vector field $\boldsymbol{i}$ is

$$\operatorname{curl} [(\operatorname{grad} \boldsymbol{i})\boldsymbol{i}+\boldsymbol{i} \operatorname{div} \boldsymbol{i}] = \mathbf{0}. \tag{4.48}$$

Of course, $\boldsymbol{i}$ must also be a unit vector:

$$\boldsymbol{i} \cdot \boldsymbol{i} = 1. \tag{2.49}$$

Since τ is not equal to zero (for otherwise the fluid crystal would reduce to a fluid), the equations (4.47) and (4.48) are equivalent. Thus any static solution for a specific fluid crystal belonging to this type is necessarily a universal solution.

As remarked at the beginning of § 4, an isochoric universal solution for a body with no internal constraint is a universal solution for a corresponding incompressible body. Indeed, the vector field $\boldsymbol{i}_0$ given by (2.57) is a solution of (4.48). There are, however, solutions of (4.48) and (2.49) that do not satisfy the previous conditions (2.52) and (2.55). We give three families of such universal solutions here.

Family 1. The deformations are such that relative to a cylindrical coordinate system (r, θ, z) in the configuration χ the physical components of $\boldsymbol{i}$ are

$$i_{\langle r\rangle} = \cos\alpha, \quad i_{\langle\theta\rangle} = \sin\alpha, \quad i_{\langle z\rangle} = 0, \tag{4.49}$$

where α is an arbitrary angle. Clearly the stress tensor corresponding to the solution (4.49) has the form (4.22). Thus the body is in equilibrium in the configuration χ under the action of the pressure field (4.21). Two special cases of (4.49) are

I). *Plane Radial Distribution.*

Setting $\alpha = 0$ in (4.49), we get

$$i_{\langle r\rangle} = 1, \quad i_{\langle\theta\rangle} = i_{\langle z\rangle} = 0. \tag{4.50}$$

The curl of this vector field is zero, but the divergence is not.

II) *Circular Distribution.*

Setting $\alpha = \frac{1}{2}\pi$ in (4.49), we get

$$i_{\langle r\rangle} = i_{\langle z\rangle} = 0,\ i_{\langle\theta\rangle} = 1. \tag{4.51}$$

This vector field is solenoidal but is not a screw field.

Family 2. The deformations are such that relative to a cylindrical coordinate system (r, θ, z) in χ the physical components of $\boldsymbol{i}$ are

$$i_{\langle r\rangle} = 0,\ i_{\langle\theta\rangle} = \cos\alpha(r),\ i_{\langle z\rangle} = \sin\alpha(r), \tag{4.52}$$

where α is an arbitrary function of r. In χ the stress tensor has the form (4.19) relative to (r, θ, z). Thus the deformations form a family of universal solutions. The vector field $\boldsymbol{i}$ given by (4.52) is solenoidal but is not a screw field. The circular distribution II) in the previous family is also a special case of this family with $\alpha(r) = 0$.

Family 3. The deformations are such that relative to a spherical coordinate system (r, θ, φ) in χ the physical components of $\boldsymbol{i}$ are

$$i_{\langle\theta\rangle} = i_{\langle\varphi\rangle} = 0,\ i_{\langle r\rangle} = 1. \tag{4.53}$$

We call this field $\boldsymbol{i}$ a *central radial distribution*. In χ the stress has the physical components given by (4.26) relative to (r, θ, φ). Thus the deformations are universal solutions. Clearly in this case $\operatorname{curl}\boldsymbol{i} = \mathbf{0}$, but $\operatorname{div}\boldsymbol{i} \neq 0$.

5. Examples of Static Universal Solutions for Incompressible Isotropic Bodies[1]

Example 1. Pure Torsion of an Isotropic Solid Circular Cylinder.

The special case of Family 3 of Case 2 in the previous section when

$$A = 1,\ B = 0,\ C = 1,\ E = 0,\ F = 1 \tag{5.1}$$

in (4.40), so that the components of deformation are

$$r = R,\ \theta = \Theta + DZ,\ z = Z, \tag{5.2}$$

corresponds to pure torsion of a solid cylinder. The parameter D is the

---

[1] Most of the results given in this section are due to RIVLIN (1947–1949). For more details, further cases, and references, see § 57 of NFTM.

angle of twist per unit length in the z-direction. The matrix of components of the deformation gradient $\boldsymbol{F}$ is

$$[F^i_{\ A}] = \begin{bmatrix} 1 & 0 & 0 \\ 0 & 1 & D \\ 0 & 0 & 1 \end{bmatrix}, \tag{5.3}$$

which shows that the deformation is a local simple shear of amount Dr. The left Cauchy-Green tensor of the deformation has the component matrix

$$[B^{km}] = \begin{bmatrix} 1 & 0 & 0 \\ 0 & \dfrac{1}{r^2}+D^2 & D \\ 0 & D & 1 \end{bmatrix}, \tag{5.4}$$

which has the inverse matrix

$$[(B^{-1})_{km}] = \begin{bmatrix} 1 & 0 & 0 \\ 0 & r^2 & -Dr^2 \\ 0 & -Dr^2 & 1+D^2r^2 \end{bmatrix}. \tag{5.5}$$

Then the matrices of the physical components of $\boldsymbol{B}$ and $\boldsymbol{B}^{-1}$ are

$$[B_{\langle km\rangle}] = \begin{bmatrix} 1 & 0 & 0 \\ 0 & 1+D^2r^2 & Dr \\ 0 & Dr & 1 \end{bmatrix} \tag{5.6}$$

and

$$[B^{-1}_{\langle km\rangle}] = \begin{bmatrix} 1 & 0 & 0 \\ 0 & 1 & -Dr \\ 0 & -Dr & 1+D^2r^2 \end{bmatrix}. \tag{5.7}$$

Hence

$$I_{\boldsymbol{B}} = II_{\boldsymbol{B}} = 3+D^2r^2. \tag{5.8}$$

Putting (5.6)–(5.8) into (4.4) and (4.19) shows that the physical components of the stress tensor relative to (r, θ, z) are

$$\begin{aligned} T_{\langle rr\rangle} &= D^2\int r\hat{\psi}_1(D^2r^2)dr, \\ T_{\langle\theta\theta\rangle} &= T_{\langle rr\rangle}+D^2r^2\hat{\psi}_1(D^2r^2), \\ T_{\langle zz\rangle} &= T_{\langle rr\rangle}+D^2r^2\hat{\psi}_{-1}(D^2r^2), \\ T_{\langle\theta z\rangle} &= Dr\,\hat{\mu}(D^2r^2), \\ T_{\langle r\theta\rangle} &= T_{\langle rz\rangle} = 0, \end{aligned} \tag{5.9}$$

where $\hat{\psi}_1$, $\hat{\psi}_{-1}$, and $\hat{\mu}$ are given by (3.8) and (3.10). As in a simple shear, the normal stresses $T_{\langle zz\rangle}$ and $T_{\langle\theta\theta\rangle}$ do not vanish. This fact is the Poynting effect for this class of solutions.

Let a be the radius of the circular cylinder. Then the total twisting moment τ is found as follows from $(5.9)_4$:

$$\tau = 2\pi\int_0^a r^2 T_{\langle\theta z\rangle}\,dr = 2\pi D\int_0^a r^3\hat{\mu}(D^2r^2)dr. \tag{5.10}$$

Thus, as in the classical linear theory, the torsional stiffness of the cylinder is completely determined by the shear modulus $\hat{\mu}$. The resultant axial force, however, is not related in any simple way to any normal traction in simple shear. Under the natural boundary condition

$$T_{\langle rr\rangle}|_{r=a} = 0, \tag{5.11}$$

the radial stress is given by

$$T_{\langle rr\rangle} = -D^2\int_r^a r\hat{\psi}_1(D^2r^2)dr, \tag{5.12}$$

and the resultant axial force is

$$N = -\pi D^2\int_0^a r^3(\psi_1(D^2r^2)-2\hat{\psi}_{-1}(D^2r^2))dr. \tag{5.13}$$

Such experimental data as there are suggest that the functions ψ_1 and ψ_{-1} satisfy the conditions

$$\psi_1 > 0,\ \psi_{-1} \leqq 0, \tag{5.14}$$

which are called the *E-inequalities*[1]. If these conditions hold, then $N < 0$: The resultant axial force needed so as to effect a pure torsion is always a pressure, not a tension. Hence is suggested an explanation for the Poynting effect in its original form: An incompressible circular cylinder, altogether free on its outer surface and free of tension or pressure on its ends, elongates when it is twisted. An exact solution such as to model this effect can be found, of course, only if we consider torsional deformation together with an axial extension and a radial contraction. This kind of deformation furnishes our next example.

---

[1] *Cf.* § III. 9D.

Example 2. Torsion, Extension, and Inflation.

If in Family 3 of Case 2 in the previous section we choose

$$B = 0,\ C = 1,\ E = 0, \tag{5.15}$$

in (4.40), so that the components of deformation are

$$r = \sqrt{A}R, \quad \theta = \Theta + DZ, \quad z = FZ, \tag{5.16}$$

where

$$AF = 1, \tag{5.17}$$

the deformation represents an axial stretch F followed by a twist of amount D/F. If a is the radius of the undeformed cylinder, $a\sqrt{A}$ is that of the deformed one. In order that the deformation be isochoric, the axial stretch is accompanied by radial expansion in the linear ratio $\sqrt{A}$.

The matrix of components of the deformation gradient $\boldsymbol{F}$ is

$$[F^i_{\ A}] = \begin{bmatrix} \sqrt{A} & 0 & 0 \\ 0 & 1 & D \\ 0 & 0 & F \end{bmatrix}. \tag{5.18}$$

Then the matrix of components of the left Cauchy-Green tensor $\boldsymbol{B}$ is

$$[B^{km}] = \begin{bmatrix} A & 0 & 0 \\ 0 & \dfrac{A}{r^2} + D^2 & DF \\ 0 & DF & F^2 \end{bmatrix}, \tag{5.19}$$

and the matrix inverse to this one is given by

$$[(B^{-1})_{km}] = \begin{bmatrix} \dfrac{1}{A} & 0 & 0 \\ 0 & \dfrac{r^2}{A} & -Dr^2 \\ 0 & -Dr^2 & A^2 + AD^2r^2 \end{bmatrix}. \tag{5.20}$$

The matrices of the physical components of $\boldsymbol{B}$ and $\boldsymbol{B}^{-1}$ are

$$[B_{\langle km\rangle}] = \begin{bmatrix} A & 0 & 0 \\ 0 & A + D^2r^2 & DFr \\ 0 & DFr & F^2 \end{bmatrix} \tag{5.21}$$

and

$$[(B^{-1})_{\langle km\rangle}] = \begin{bmatrix} \dfrac{1}{A} & 0 & 0 \\ 0 & \dfrac{1}{A} & -Dr \\ 0 & -Dr & A^2+AD^2r^2 \end{bmatrix}. \tag{5.22}$$

Thus

$$I_{\boldsymbol{B}} = 2A+F^2+D^2r^2,$$

and (5.23)

$$II_{\boldsymbol{B}} = \frac{2}{A}+A^2+AD^2r^2.$$

Since $I_{\boldsymbol{B}} \neq II_{\boldsymbol{B}}$ unless $A = F = 1$, this deformation is not a local simple shear except when it reduces to pure torsion.

Substituting (5.21)–(5.23) into (4.4) and (4.19), we obtain the physical components of the stress tensor relative to (r, θ, z):

$$\begin{aligned} T_{\langle rr\rangle} &= D^2\int r\psi_1\,dr, \\ T_{\langle\theta\theta\rangle} &= T_{\langle rr\rangle}+D^2r^2\psi_1, \\ T_{\langle zz\rangle} &= T_{\langle rr\rangle}+(F^2-A)\psi_1+\left(A^2+AD^2r^2-\frac{1}{A}\right)\psi_{-1}, \\ T_{\langle\theta z\rangle} &= Dr(F\psi_1-\psi_{-1}), \\ T_{\langle r\theta\rangle} &= T_{\langle rz\rangle} = 0, \end{aligned} \tag{5.24}$$

where the arguments of ψ_1 and ψ_{-1} are $(I_{\boldsymbol{B}}, II_{\boldsymbol{B}})$, given by (5.23).

As before, we can compute the total twisting moment τ by the integral

$$\tau = 2\pi\int_0^{a\sqrt{A}} r^2T_{\langle\theta z\rangle}\,dr = \frac{2\pi D}{F}\int_0^a\left(\psi_1-\frac{1}{F}\psi_{-1}\right)R^3dR. \tag{5.25}$$

Now the torsional stiffness is no longer determined by the shear modulus $\hat{\mu}$. Again, we choose the boundary condition that the mantle of the cylinder is free of stress:

$$T_{\langle rr\rangle}|_{r=a\sqrt{A}} = 0. \tag{5.26}$$

Then the radial stress is given by

$$T_{\langle rr\rangle} = -D^2\int_r^{a\sqrt{A}} r\psi_1\,dr = -AD^2\int_R^a R\psi_1\,dR, \tag{5.27}$$

and the resultant axial force is

$$N = 2\pi\left(F-\frac{1}{F^2}\right)\int_0^a\left(\psi_1-\frac{1}{F}\psi_{-1}\right)R\,dR-\frac{\pi D^2}{F^2}\int_0^a\left(\psi_1-\frac{2}{F}\psi_{-1}\right)R^3dR. \tag{5.28}$$

Setting $N = 0$ gives an equation relating the twist D and the extension F in a torsional deformation free of resultant axial pressure. While the solution of this equation is not presently known, we can easily find an approximate one for small D. Writing (5.28) in the form

$$2\left(F-\frac{1}{F^2}\right)\int_0^a\left(\psi_1-\frac{1}{F}\psi_{-1}\right)R\,dR = \frac{D^2}{F^2}\int_0^a\left(\psi_1-\frac{2}{F}\psi_{-1}\right)R^3dR, \tag{5.29}$$

and assuming that ψ_1 and ψ_{-1} are smooth as usual, we see that (5.29) characterizes a smooth relation between D and F. Of course, that relation is satisfied in the reference configuration, in which $D = 0$ and $F = 1$. Thus we expect that $F-1$ will be small if D/F is small. In fact, after some calculations we can show that

$$\Pi \equiv \lim_{D\to 0}\frac{F-1}{(D/F)^2} = \tfrac{1}{3}\lim_{D\to 0}\frac{F^3-1}{D^2} = \tfrac{1}{12}a^2\,\frac{\bar{\psi}_1-2\bar{\psi}_{-1}}{\bar{\psi}_1-\bar{\psi}_{-1}}, \tag{5.30}$$

where $\bar{\psi}_1$ and $\bar{\psi}_{-1}$ are the constants

$$\bar{\psi}_1 \equiv \psi_1(3, 3), \quad \bar{\psi}_{-1} \equiv \psi_{-1}(3, 3). \tag{5.31}$$

The ratio Π, which gives the approximate solution

$$F^3 = 1+3\Pi D^2+\ldots, \tag{5.32}$$

is called the *Poynting modulus* of the rod. By use of (3.10) we find for Π the following expression in terms of the response coefficients:

$$\Pi = \tfrac{1}{12}a^2\left(1-\frac{\bar{\psi}_{-1}}{\mu_0}\right), \tag{5.33}$$

where μ_0 is the shear constant at the reference configuration. Since $\bar{\psi}_{-1}$ is not determined uniquely by the shear constant μ_0, the Poynting modulus cannot be determined by effects covered by the classical linear theory alone. Further, if ψ_1 and ψ_{-1} obey the E-inequalities (5.14), then Π is positive and is never less than the value $a^2/12$.

Returning now to the general formulae (5.25) and (5.28), we consider the response to small twist D when the stretch F is held fixed. The limiting value of the torsional stiffness for the stretched cylinder is

$$K = K(F) \equiv \lim_{D\to 0} \frac{\tau}{D/F} = \tfrac{1}{2}\pi a^4 \left(\tilde{\psi}_1 - \frac{1}{F}\tilde{\psi}_{-1}\right), \tag{5.34}$$

and the limiting value of the axial force is

$$\tilde{N} = \tilde{N}(F) \equiv \lim_{D\to 0} N = \pi a^2 \left(F - \frac{1}{F^2}\right)\left(\tilde{\psi}_1 - \frac{1}{F}\tilde{\psi}_{-1}\right), \tag{5.35}$$

where $\tilde{\psi}_1$ and $\tilde{\psi}_{-1}$ are the values of ψ_1 and ψ_{-1} at the argument $(2/F + F^2,\ 2F + 1/F^2)$. Taking the ratio of K and $\tilde{N}$ yields RIVLIN's universal relation:

$$\frac{\tilde{N}(F)}{K(F)} = \frac{2}{a^2}\left(F - \frac{1}{F^2}\right). \tag{5.36}$$

A complete set of universal relations for this example, however, is not known.

Example 3. Eversion of a Circular Isotropic Tube.

Again considering Family 3 of Case 2 in the previous section, in order to represent eversion of a circular tube we choose

$$C = 1,\ D = 0,\ E = 0 \tag{5.37}$$

in (4.40), so that the deformation becomes

$$r = \sqrt{AR^2 + B}, \quad \theta = \Theta, \quad z = FZ, \tag{5.38}$$

a purely radial extension. As before, the constants A and F are related by (5.17). We let the inner and the outer radii of the tube in the reference configuration be a_1 and a_2, and we consider only the case when the tube is turned inside out. From $(5.38)_1$ we see that conditions necessary and sufficient for this case are

$$0 < \sqrt{Aa_2^2 + B} < \sqrt{Aa_1^2 + B}, \tag{5.39}$$

which require that A be negative but Aa_2^2+B be positive:

$$A < 0 < Aa_2^2+B. \tag{5.40}$$

We can determine the stress tensor in the same way as before. The radial stress $T_{\langle rr\rangle}$ is

$$T_{\langle rr\rangle} = -\int \left(\frac{A^2R^2}{r^2} - \frac{r^2}{R^2}\right)\left(\psi_1 - \frac{1}{A^2}\psi_{-1}\right)\frac{dr}{r}, \tag{5.41}$$

where the arguments of ψ_1 and ψ_{-1} are

$$\begin{aligned} I_{\boldsymbol{B}} &= \frac{A^2R^2}{r^2} + \frac{r^2}{R^2} + F^2, \\ II_{\boldsymbol{B}} &= \frac{r^2}{A^2R^2} + \frac{R^2}{r^2} + A^2. \end{aligned} \tag{5.42}$$

As boundary condition we choose to require that the everted tube be free both inside and outside. Then (5.41) implies

$$\int_{a_1}^{a_2} \left[\frac{R^2}{(R^2+B/A)^2} - \frac{1}{R^2}\right]\left(\psi_1 - \frac{1}{A^2}\psi_{-1}\right) R\,dR = 0. \tag{5.43}$$

Similarly, if the everted tube is free of resultant axial force, then

$$\int_{a_1}^{a_2} \left[\left(\frac{A^2R^2}{AR^2+B} - \frac{2}{A^2} + \frac{AR^2+B}{R^2}\right)\psi_1 + \left(\frac{AR^2+B}{A^2B^2} - 2A^2 + \frac{R^2}{AR^2+B}\right)\psi_{-1}\right] R\,dR = 0. \tag{5.44}$$

The conditions (5.43) and (5.44) relate the parameters A and B in an eversion of an infinitely long tube free to extend. Whether or not these equations have any solution A, B satisfying the condition (5.40), is not presently known in general. However, for a *Mooney material*, which is defined as a material for which ψ_1 and ψ_{-1} are constants satisfying the E-inequalities (5.14), it can be shown that a unique solution for eversion exists. Further, for a *neo-Hookean material*, which is a Mooney material with $\psi_{-1} \equiv 0$, it can be shown that

$$-1 < A < 0. \tag{5.45}$$

Hence, from (5.17)

$$-\infty < F < -1, \tag{5.46}$$

which means that if we evert a long tube of neo-Hookean material, we always lengthen it.

Example 4. Eversion of an Isotropic Spherical Shell.

To represent eversion of a spherical shell, in Family 4 of Case 2 in the previous section we choose the negative sign in (4.42). Thus the components of deformation are

$$r = (A-R^3)^{\frac{1}{3}}, \quad \theta = -\Theta, \quad \varphi = \Phi. \tag{5.47}$$

If we let a_1 and a_2 be the inner and outer radii in the reference configuration, (5.47) represents an eversion if and only if

$$0 < A-a_2^3 < A-a_1^3, \tag{5.48}$$

which requires that A be greater than a_2^3.

From (5.47), the deformation gradient has the matrix of components

$$[F^i_A] = \begin{bmatrix} -\dfrac{R^2}{r^2} & 0 & 0 \\ 0 & -1 & 0 \\ 0 & 0 & 1 \end{bmatrix}. \tag{5.49}$$

Then

$$[B_{\langle km\rangle}] = \operatorname{diag}\left(\frac{R^4}{r^4}, \frac{r^2}{R^2}, \frac{r^2}{R^2}\right), \tag{5.50}$$

and

$$[B^{-1}_{\langle km\rangle}] = \operatorname{diag}\left(\frac{r^4}{R^4}, \frac{R^2}{r^2}, \frac{R^2}{r^2}\right), \tag{5.51}$$

so that

$$I_{\boldsymbol{B}} = \frac{R^4}{r^4} + \frac{2r^2}{R^2}, \quad II_{\boldsymbol{B}} = \frac{r^4}{R^4} + \frac{2R^2}{r^2}. \tag{5.52}$$

Substituting (5.50)–(5.52) into (4.4) and $(4.26)_1$ yields the radial stress:

$$T_{\langle rr\rangle} = 2\int \left[\left(\frac{r^2}{R^2} - \frac{R^4}{r^4}\right)\psi_1 + \left(\frac{R^2}{r^2} - \frac{r^4}{R^4}\right)\psi_{-1}\right]\frac{dr}{r}. \tag{5.53}$$

The boundary condition that the inner and the outer surfaces be free in the everted state is

$$\int_{a_1}^{a_2} \left[\left(\frac{r^2}{R^2} - \frac{R^4}{r^4}\right)\psi_1 + \left(\frac{R^2}{r^2} - \frac{r^4}{R^4}\right)\psi_{-1}\right]\frac{R^2 dR}{r^3} = 0, \tag{5.54}$$

where r is given by $(5.47)_1$. The parameter A is to be found as a solution of (5.54). While a general solution is not presently known, one can be proved to exist for a Mooney material.

Example 5. Pure Bending.

Four kinds of pure bending are included in the families of deformations of incompressible isotropic bodies which we derived in the previous section.

I. *Bending of a Rectangular Block into a Sector of a Hollow Circular Cylinder.*

The special case of (4.27) in which

$$C = 0,\ D = 0,\ E = 1 \tag{5.55}$$

corresponds to bending of a rectangular block into a sector of a hollow circular cylinder. The components of deformation are

$$r = \sqrt{2AX}, \quad \theta = BY, \quad z = Z, \tag{5.56}$$

where

$$AB = 1. \tag{5.57}$$

This deformation carries the co-ordinate planes of X, Y, and Z in a Cartesian co-ordinate system into the co-ordinate surfaces or r, θ, and z, respectively, in a cylindrical co-ordinate system.

II. *Bending of a Sector of a Hollow Circular Cylinder into a Rectangular Block.*

A deformation in which a sector of a hollow cylinder is bent into a rectangular block is the inverse of the previous kind and is a special case of (4.39) with the same restriction (5.55). Thus the components of deformation are

$$x = \tfrac{1}{2}AR^2,\ y = B\Theta,\ z = Z, \tag{3.58}$$

where (5.57) holds. Clearly, this deformation carries the co-ordinate surfaces of R, Θ, and Z into the co-ordinate planes of x, y, and z, respectively.

III. *Bending of a Sector of a Hollow Circular Cylinder into Another Sector with a Different Curvature.*

The deformation in which a sector of one hollow circular cylinder is bent into a sector of another may be regarded as a composition of Case II with Case I. It is given by the special case of (4.40) in which

$$D = 0, \; E = 0, \; F = 1. \tag{5.59}$$

The components of deformation are

$$r = \sqrt{AR^2 + B}, \; \theta = C\Theta, \; z = Z, \tag{5.60}$$

where

$$AC = 1. \tag{5.61}$$

The co-ordinate surfaces of R, Θ, and Z are carried into those of r, θ, and z, respectively. The curvature is reversed if the parameter A is negative, as has been explained in Example 3.

IV. *Bending of a Sector of a Hollow Sphere into Another Sector with a Different Curvature*[1].

The deformation of a sector of one hollow sphere into a sector of another is a special case of (4.43) with

$$B = 0, \; A = D. \tag{5.62}$$

In a cylindrical co-ordinate system the components of deformation are

$$r = AR, \quad \theta = C\Theta, \quad z = AZ, \tag{5.63}$$

where

$$A^3C = 1. \tag{5.64}$$

But in a spherical co-ordinate system the components of deformation become

$$r = AR, \quad \theta = \Theta, \quad \varphi = C\Phi. \tag{5.65}$$

Thus the deformation carries the co-ordinate surfaces of R, Θ, and Φ into those of r, θ, and φ, respectively. Since A must be positive, the curvature of the spherical sector cannot be reversed in the bending.

---

[1] R. R. Huilgol, "Dislocation of a spherical sector in finite elasticity", *Z. angew. Math. Mech.* **48**, 203 (1968).

6. Dynamic Universal Solutions for Some Incompressible Elastic Bodies

In this section we consider a class of exact solutions of the dynamical equations of incompressible elastic bodies. The solutions are special in that they correspond to *quasi-equilibrated motions*[1], defined by the condition that at each time t the instantaneous configuration be a possible configuration of equilibrium for the body, subject to surface traction alone. As before, we assume that the body is homogeneous and incompressible, and that its density is uniform in a homogeneous configuration. If for a given body we have any family of static solutions depending on a real parameter s throughout some interval of $\mathscr{R}$, by taking s as the time we obtain, trivially, a quasi-equilibrated motion of the same body, but in general it will not satisfy the dynamical equations unless particular body forces be supplied. We now find a necessary and sufficient condition that a quasi-equilibrated motion be controllable, that is, a dynamically possible motion for the given body, subject to surface forces alone.

Let χ_t, $t \in (-\infty, \infty)$, be a motion of $\mathscr{B}$ relative to an admissible homogeneous reference configuration $\boldsymbol{\kappa}$, in which we assume that the density field is constant. Then χ_t is dynamically possible for $\mathscr{B}$ subject to zero body force if the dynamical equation

$$\operatorname{div} \boldsymbol{T} = \rho \boldsymbol{a} \tag{6.1}$$

is satisfied at each time t. Here $\boldsymbol{a}$ denotes the acceleration of the motion. Since the body is incompressible, the motion must be isochoric. Further, the stress tensor $\boldsymbol{T}$ is given by

$$\boldsymbol{T} = -p\mathbf{1} + \boldsymbol{H}(\boldsymbol{F}), \tag{4.1}$$

where $\boldsymbol{H}(\boldsymbol{F})$ is the determinate stress. Substituting (4.1) into (6.1), we see that χ_t is dynamically possible if and only if there exists a single-valued scalar field p at each time t such that

$$\operatorname{grad} p = \operatorname{div} \boldsymbol{H}(\boldsymbol{F}) - \rho \boldsymbol{a}. \tag{6.2}$$

---

[1] The concept of quasi-equilibrated motion was introduced and first analysed by TRUESDELL. *Cf.* NFTM, § 61. The treatment in the text above incorporates some additions and corrections made by C.-C. WANG, "Universal solutions for incompressible laminated bodies", *Arch. Rational Mech. Anal.* **29**, 161–192 (1968), reprinted in *Continuum Theory of Inhomogeneities in Simple Bodies*, New York, Springer-Verlag, 1968.

Now suppose that the motion χ_t be quasi-equilibrated. Then we can determine a pressure field $\bar{p}$ such that the stress field $\bar{\boldsymbol{T}}$ given by

$$\bar{\boldsymbol{T}} = -\bar{p}\mathbf{1}+\boldsymbol{H}(\boldsymbol{F}) \tag{6.3}$$

satisfies the equation of equilibrium:

$$\text{grad}\,\bar{p} = \text{div}\,\boldsymbol{H}(\boldsymbol{F}). \tag{6.4}$$

Subtracting (6.2) from (6.4) yields a necessary and sufficient condition for a quasi-equilibrated motion to be controllable:

$$\boldsymbol{a} = \text{grad}\,\zeta, \tag{6.5}$$

where ζ is a single-valued acceleration-potential given by

$$\zeta = \frac{\bar{p}-p}{\rho}. \tag{6.6}$$

The importance of the class of quasi-equlibrated motions lies in the fact that every dynamic universal solution relative to some particular symmetry group always belongs to this class. We have the following[1]

Proposition (Wang). *An isochoric motion χ_t is a dynamic universal solution for some particular type of incompressible elastic body subject to zero body force if and only if the acceleration $\boldsymbol{a}$ is a lamellar field with a single-valued (possibly time-dependent) potential, and the determinate stress field is equilibrated at each time t by a suitable pressure field.*

The sufficiency has been proved by (6.5). To prove the necessity, we remark that the condition of material frame-indifference (III. 1.16) and the condition of material symmetry (III. 2.3) relative to a particular symmetry group $\mathfrak{G}$ are both closed with respect to the operation of scalar multiplication on the response function $\boldsymbol{H}$. Consequently, for a dynamic universal solution relative to $\mathfrak{G}$, since (6.2) holds, so must the relation

$$\text{grad}\,p^* = \text{div}\,\alpha\boldsymbol{H}(\boldsymbol{F})-\rho\boldsymbol{a} \tag{6.7}$$

for each constant α, where p^* is a single-valued pressure field. Subtracting (6.7) from (6.2) yields immediately the necessary conditions (6.4) and

---

[1] C.-C. Wang. "Proof that all dynamic universal solutions are quasi-equilibrated", *Z. angew. Math. Mech.* **50**, 311–314 (1970).

(6.5), with possible solutions for $\bar{p}$ and ζ given by

$$\bar{p} = \frac{p - p^*}{1-\alpha}, \; \zeta = \frac{\alpha p - p^*}{(1-\alpha)\rho}. \tag{6.7}$$

In view of this proposition, we see that the problem of finding dynamic universal solutions relative to any particular symmetry group is equivalent to the problem of finding dynamically possible quasi-equilibrated motions whose instantaneous configurations can be obtained by static universal solutions.

We proceed now to determine some families of quasi-equilibrated solutions for particular kinds of bodies. In the next section we shall study in detail some important examples.

Case 1. Incompressible Fluid Body.

As in statics, the case of an incompressible fluid body is a trivial one, since the determinate stress vanishes identically in all motions. Hence all isochoric motions are quasi-equilibrated, and those satisfying (6.5) are dynamically possible.

The condition (6.5) by itself, first obtained by D'ALEMBERT and EULER, has been studied extensively since the eighteenth century, and much of classical hydrodynamics is neither more nor less than the analysis of it in special cases. In particular, according to KELVIN's theorem, it is a necessary and sufficient condition that the circulation of velocity around every reducible circuit of body-points remain constant in time; it is equivalent to two of the three hydrodynamic theorems of HELMHOLTZ; and it implies several theorems of Bernoullian type. *Cf.* §§ 105–138 of CFT. In the cases that follow now, all these theorems hold, but only certain kinds of classical hydrodynamic motions satisfy the further conditions imposed by the nature of the material response.

Case 2. Incompressible Isotropic Body.

In the six families of static universal solutions for incompressible isotropic bodies discussed in §§ 4–5, we may replace the constants A, B, ... by functions of time $A(t)$, $B(t)$ ... and so obtain quasi-equilibrated motions. It then becomes a definite and easy problem to find those choices of the functions that satisfy (6.5).

A more general class of quasi-equilibrated solutions, differing from those we are about to set forth by an arbitrary rigid motion, can be considered also. To represent this general class, six more functions of t are needed. However, since analysis of the restrictions imposed by (6.5) upon these superimposed rigid motions is rather complex, we consider here only the simplest cases.

Family 0. Homogeneous Motions.

At each time t the instantaneous deformation is homogeneous. Hence the motion is given by (2.7) with $\boldsymbol{F}$ and $\boldsymbol{c}$ being functions of t. Since the motion must be isochoric, $\boldsymbol{F}$ has to be unimodular for all t. It is easy to see that (6.5) holds if and only if

$$\ddot{\boldsymbol{F}}^T = \boldsymbol{F}^T\ddot{\boldsymbol{F}}\boldsymbol{F}^{-1}. \tag{6.9}$$

When this condition is satisfied, the acceleration-potential ζ is given by

$$\zeta = \tfrac{1}{2}(\ddot{\boldsymbol{F}}\boldsymbol{F}^{-1}\boldsymbol{x})\cdot\boldsymbol{x}+(\ddot{\boldsymbol{c}}-\ddot{\boldsymbol{F}}\boldsymbol{F}^{-1}\boldsymbol{c})\cdot\boldsymbol{x}+\Phi(t) \tag{6.10}$$

where $\boldsymbol{x}$ is written for the position vector $\boldsymbol{x}-\boldsymbol{o}$, and where $\Phi(t)$ is an arbitrary function of t.

Family 1.

As suggested by (4.27), the components of deformation are

$$r^2 = 2AX+B, \quad \theta = CY+DZ, \quad z = EY+FZ, \tag{6.11}$$

where $A, \ldots, F$ are functions of t satisfying the condition

$$A(CF-DE) = 1 \tag{4.41}$$

for all t. We proceed to determine the restrictions on the functions $A, \ldots, F$ due to the condition (6.5).

First we compute the time derivatives of the components of deformation, obtaining

$$\begin{aligned}
\dot{r} &= \frac{\dot{A}}{2A}r+\left(\frac{\dot{B}}{2}-\frac{B\dot{A}}{2A}\right)\frac{1}{r},\\
\dot{\theta} &= \theta A(F\dot{C}-E\dot{D})+zA(C\dot{D}-D\dot{C}),\\
\dot{z} &= \theta A(F\dot{E}-E\dot{F})+zA(C\dot{F}-D\dot{E}).\\
\ddot{r} &= \left[\frac{\ddot{A}}{2A}-\left(\frac{\dot{A}}{2A}\right)^2\right]r+\left[\frac{\dot{A}}{A}\left(\frac{B\dot{A}}{A}-\dot{B}\right)-\left(\frac{B\ddot{A}}{A}-\ddot{B}\right)\right]\frac{1}{2r}\\
&\qquad -\left(\frac{B\dot{A}}{A}-\dot{B}\right)^2\frac{1}{4r^3},
\end{aligned} \tag{6.12}$$

$$\begin{aligned}
\ddot{\theta} &= \theta A(F\ddot{C}-E\ddot{D})+zA(C\ddot{D}-D\ddot{C}),\\
\ddot{z} &= \theta A(F\ddot{E}-E\ddot{F})+zA(C\ddot{F}-D\ddot{E}).
\end{aligned}$$

From these we can determine the covariant components of the acceleration field relative to (r, θ, z) by the formulae

$$\begin{aligned} a_r &= \ddot{r} - r\dot{\theta}^2, \\ a_\theta &= r^2\ddot{\theta} + 2r\dot{r}\dot{\theta}, \\ a_z &= \ddot{z}. \end{aligned} \tag{6.13}$$

The condition (6.5) requires that

$$\operatorname{curl} \boldsymbol{a} = \boldsymbol{0}. \tag{6.14}$$

Relative to (r, θ, z), this condition can be expressed as

$$\frac{\partial a_r}{\partial \theta} = \frac{\partial a_\theta}{\partial r}, \quad \frac{\partial a_r}{\partial z} = \frac{\partial a_z}{\partial r}, \quad \frac{\partial a_\theta}{\partial z} = \frac{\partial a_z}{\partial \theta}. \tag{6.15}$$

Substituting (6.12) and (6.13) into (6.15) yields the governing equations

$$C\dot{D} - D\dot{C} = 0, \tag{6.16}$$

$$F\ddot{E} - E\ddot{F} = 0, \tag{6.17}$$

and

$$F\ddot{C} - E\ddot{D} + (F\dot{C} - E\dot{D})\left(AF\dot{C} - AE\dot{D} + \frac{\dot{A}}{A}\right) = 0. \tag{6.18}$$

We wish to solve this system of differential equations together with the condition (4.41) for isochoric motion.

First, we notice that the function B does not appear in the governing equations, so that it is arbitrary. Second, the function C must vanish either always or never. Indeed, if $C \neq 0$, then the integral of (6.16) is

$$D = kC, \tag{6.19}$$

where k is a constant. Hence suppose that C, while not identically zero, vanishes for some t. Then the set

$$\mathscr{U} \equiv \{t : C(t) \neq 0\} \tag{6.20}$$

is open and satisfies the condition

$$\phi \subsetneqq \mathscr{U} \subsetneqq \mathscr{R}. \tag{6.21}$$

Thus there exists a point t_0 with

$$C(t_0) = 0, \tag{6.22}$$

while t_0 is also a limit point of $\mathscr{U}$, *i.e.*,

$$C(t) \neq 0 \tag{6.23}$$

on an open interval with t_0 as an end point. But since the integral (6.19) holds on that interval, continuity requires that

$$D(t_0) = 0, \tag{6.24}$$

which together with (6.20) clearly contradicts the condition (4.41). Hence the hypothesis (6.23) is false. Thus we have shown that there are only two possibilities:

I) $\mathscr{U} = \mathscr{R}$, or equivalently $C \neq 0$ for all t. In this case the integral (6.19) holds for all t. Then (4.41) reduces to

$$AC(F-kE) = 1, \tag{6.25}$$

and (6.18) reduces to

$$\ddot{C}+\dot{C}\left(\frac{\dot{C}}{C}+\frac{\dot{C}}{A}\right) = 0. \tag{6.26}$$

An integral of this differential equation is

$$\dot{C} = \frac{k_1}{AC}, \tag{6.27}$$

where k_1 is a constant. The differential equation (6.17) has an obvious integral also:

$$F\dot{E}-E\dot{F} = k_2, \tag{6.28}$$

where k_2 is a constant.

Now we consider the following three possibilities:

Ia). $F \equiv 0$ for all t. In this subcase $k_2 = 0$, and the equation (6.28) is satisfied for all E. Further, the condition (6.25) reduces to

$$-kACE = 1. \tag{6.29}$$

Thus the constant k must be non-zero. The complete solution for this subcase is

$$B,\ E = \text{arbitrary functions of } t, \text{ but } E \neq 0,$$

$$\begin{aligned} F &\equiv 0, \\ C &= k'\left(\int_0^t E\,dt + k''\right) \neq 0, \\ D &= -k'''\left(\int_0^t E\,dt + k''\right) \neq 0, \\ A &= \frac{1}{k'''E\left(\int_0^t E\,dt + k''\right)}, \end{aligned} \tag{6.30}$$

where k', k'', and k''' are arbitrary constants.

Ib). $F(t_0) \neq 0$ for some t_0. We consider the solution for t near t_0. Integrating (6.28) yields

$$E = F\left(k' + k_2 \int_0^t \frac{1}{F^2}\,dt\right). \tag{6.31}$$

We put

$$\xi = \frac{1}{AC} = F\left[1 - k\left(k' + k_2 \int_0^t \frac{1}{F^2}\,dt\right)\right]. \tag{6.32}$$

Then the complete solution for t near t_0 is

$$B,\ F = \text{arbitrary functions of } t, \text{ but } F \neq 0,$$

$$\begin{aligned} E &= F\left(k' + k_2 \int_{t_0}^t \frac{1}{F^2}\,dt\right) \neq 0, \\ C &= k''\left(\int_{t_0}^t \xi\,dt + k'''\right) \neq 0, \\ D &= kC = kk''\left(\int_{t_0}^t \xi\,dt + k'''\right), \\ A &= \frac{1}{C\xi}, \end{aligned} \tag{6.33}$$

where k, k', k'', k''', and k_2 are arbitrary constants.

Ic). $F(t_0) = 0$ for some t_0, but $F \not\equiv 0$. We claim that t_0 must be an isolated root of F. To see this, we put as before

$$\mathscr{U} \equiv \{t : F(t) \neq 0\}. \tag{6.34}$$

Then (6.21) holds. Thus we can choose t_0 in such a way that

$$F(t_0) = 0, \tag{6.35}$$

but

$$F(t) \neq 0 \tag{6.36}$$

on an interval with t_0 as an end point. Now suppose that

$$k_2 = 0. \tag{6.37}$$

Then we can integrate (6.28) as before and so show that

$$E = \bar{k}F \tag{6.38}$$

on the interval for which (6.36) holds. By continuity, we get

$$E(t_0) = 0, \tag{6.39}$$

contradicting the condition (6.25). Hence the assumption (6.37) is false. Having proved that $k_2 \neq 0$, we obtain from (6.28)

$$\dot{F}(t_0) \neq 0. \tag{6.30}$$

Thus t_0 is an isolated root of F.

Now we consider the solution for t near t_0. By (6.25), $E(t_0)$ and k do not vanish. Hence we can integrate (6.28) to get

$$F = -k_2 E \int_{t_0}^{t} \frac{dt}{E^2}. \tag{6.41}$$

As before, we put

$$\xi = \frac{1}{AC} = -E\left(k + k_2 \int_{t_0}^{t} \frac{dt}{E^2}\right). \tag{6.42}$$

Then the complete solution is

$$\begin{aligned} B, E &= \text{arbitrary functions of } t, \text{ but } E \neq 0, \\ F &= -k_2 E \int_{t_0}^{t} \frac{dt}{E^2} \neq 0, \\ C &= k_1 \left(\int_{t_0}^{t} \xi\, dt + k_0\right) \neq 0, \\ D &= kC = kk_1 \left(\int_{t_0}^{t} \xi\, dt + k_0\right) \neq 0, \\ A &= \frac{1}{C\xi}, \end{aligned} \tag{6.43}$$

where k, k_0, k_1, and k_2 are arbitrary constants.

II). $C \equiv 0$ for all t. In this case the governing equations are

$$ADE = -1, \tag{6.44}$$

$$F\dot{E} - E\dot{F} = k_2, \tag{6.45}$$

and

$$\ddot{D} + \dot{D}\left(\frac{\dot{A}}{A} + \frac{\dot{D}}{D}\right) = 0. \tag{6.46}$$

The last equation can be integrated at once, yielding

$$\dot{D} = \frac{k'''}{AD}. \tag{6.47}$$

Again there are three possibilities:

IIa). $F \equiv 0$ for all t. The complete solution for this subcase is exactly the same as that of Subcase Ia) except that now $C \equiv 0$.

IIb). $F(t_0) \neq 0$ for some t_0. The solutions for B, E, and F are the same as those of the Subcase Ib) except that $E \neq 0$, and the solutions for A and D are the same as those of Subcase Ia).

IIc). $F(t_0) = 0$ for some t_0, but $F \not\equiv 0$. Again t_0 must be an isolated root of F. The solutions for B, E, and F are the same as those of Subcase Ic), and the solutions for A and D are the same as those of Subcase Ia).

In general a solution may belong to different subcases in different intervals of time.

The acceleration-potential ζ for this family, in all cases, is given by

$$\zeta = \frac{1}{2}\left[\frac{\ddot{A}}{2A} - \left(\frac{\dot{A}}{2A}\right)^2\right] r^2 + \frac{1}{2}\left[\frac{\dot{A}}{A}\left(\frac{B\dot{A}}{A} - \dot{B}\right) - \left(\frac{B\ddot{A}}{A} - \ddot{B}\right)\right] \log r$$
$$+ \frac{1}{8}\left(\frac{B\dot{A}}{A} - \dot{B}\right)^2 \frac{1}{r^2} - \tfrac{1}{2}A^2(F\dot{C} - E\dot{D})^2 r^2\theta^2 + \tfrac{1}{2}(A\dot{B} - B\dot{A})(F\dot{C} - E\dot{D})\theta^2$$
$$+ \tfrac{1}{2}A(C\ddot{F} - D\ddot{E})z^2 + \Phi(t), \tag{6.48}$$

where $\Phi(t)$ is an arbitrary function of t.

Family 2.

The components of deformation are

$$x = \tfrac{1}{2}AR^2, \; y = B\Theta + CZ, \; z = D\Theta + EZ, \tag{4.39}$$

where $A, \ldots, E$ are functions of t satisfying the condition

$$A(BE-CD) = 1. \tag{4.28}$$

The components of the acceleration field are

$$\begin{aligned} a_x &= \ddot{x} = \frac{\ddot{A}}{A}x, \\ a_y &= \ddot{y} = yA(E\ddot{B}-D\ddot{C})+zA(B\ddot{C}-C\ddot{B}), \\ a_z &= \ddot{z} = yA(E\ddot{D}-D\ddot{E})+zA(B\ddot{E}-C\ddot{D}). \end{aligned} \tag{6.49}$$

Then the restriction due to the condition (6.5) is

$$B\ddot{C}-C\ddot{B} = E\ddot{D}-D\ddot{E}. \tag{6.50}$$

This differential equation can be integrated at once, yielding the integral

$$B\dot{C}-C\dot{B} = E\dot{D}-D\dot{E}+k, \tag{6.51}$$

where k is a constant. We consider the following two cases:

I). $B(t_0) \neq 0$ for some t_0. In this case the complete solution is

$$\begin{aligned} &B, D, E = \text{arbitrary functions of } t, \text{ but } B \neq 0 \text{ and} \\ &E, D \text{ do not vanish simultaneously}, \\ &C = B\left[k'+\int_{t_0}^{t}\frac{1}{B^2}(E\dot{D}-D\dot{E}+k)dt\right], \\ &\text{where } k' \text{ is a constant such that } E(t_0)-k'D(t_0) \neq 0, \\ &A = \frac{1}{BE-CD}. \end{aligned} \tag{6.52}$$

II). $B(t_0) = 0$ for some t_0. In this case

$$C(t_0)D(t_0) \neq 0. \tag{6.53}$$

The complete solution near t_0 is

$$\begin{aligned} &C, D, E = \text{arbitrary functions of } t, \text{ but } CD \neq 0, \\ &B = -C\int_{t_0}^{t}\frac{1}{C^2}(E\dot{D}-D\dot{E}+k)dt, \\ &A = \frac{1}{BE-CD}. \end{aligned} \tag{6.54}$$

The root t_0 need not be isolated.

Again a solution for this family may belong to different cases on different time intervals. The acceleration-potential is given by

$$\zeta = \frac{1}{2}\frac{\ddot{A}}{A}x^2 + \tfrac{1}{2}A(E\ddot{B} - D\ddot{C})y^2 + A(B\ddot{C} - C\ddot{B})yz$$
$$+ \tfrac{1}{2}A(B\ddot{E} - C\ddot{D})z^2 + \Phi(t). \qquad (6.55)$$

Family 3.

The components of deformation are

$$r^2 = AR^2 + B,\ \theta = C\Theta + DZ,\ z = E\Theta + FZ, \qquad (6.56)$$

where $A, \ldots, F$ are functions of t satisfying the condition (4.41). In terms of the functions $A, \ldots, F$, the velocity and the acceleration have the same components as those of Family 1. Thus the solution for the condition (6.5) is also the same as that of Family 1.

Family 4.

The components of deformation are

$$r^3 = \pm R^3 + A,\ \theta = \pm\Theta,\ \varphi = \Phi, \qquad (6.57)$$

where A is a function of t. The reader will verify easily that the components of the acceleration field are

$$a_r = \frac{\ddot{A}}{3r^2} - \frac{2\dot{A}^2}{9r^5},$$
$$a_\theta = a_\varphi = 0. \qquad (6.58)$$

Further, the condition (6.5) does not restrict the function A in any way, and the acceleration-potential ζ is given by

$$\zeta = -\frac{\ddot{A}}{3r} + \frac{\dot{A}^2}{18r^4} + \Phi(t). \qquad (6.59)$$

Family 5.

The components of deformation are

$$r = AR,\ \theta = B\log R + C\Theta,\ z = DZ, \qquad (6.60)$$

where $A, \ldots, D$ are functions of t such that

$$A^2CD = 1. \qquad (6.61)$$

The components of the acceleration field are

$$a_r = \frac{\ddot{A}}{A} r - \left[\dot{B} \log \frac{r}{A} + \frac{\dot{C}}{C}\left(\theta - B \log \frac{r}{A}\right)\right]^2 r,$$

$$a_\theta = \left[\ddot{B} \log \frac{r}{A} + \frac{\ddot{C}}{C}\left(\theta - B \log \frac{r}{A}\right)\right] r^2$$

$$+2 \frac{\dot{A}}{A}\left[\dot{B} \log \frac{r}{A} + \frac{\dot{C}}{C}\left(\theta - B \log \frac{r}{A}\right)\right] r^2, \tag{6.62}$$

$$a_z = \frac{\ddot{D}}{D} z.$$

Now the condition (6.5) implies that

$$\ddot{B} + \dot{B}\left(\frac{\dot{C}}{C} + 2\frac{\dot{A}}{A}\right) = 0, \tag{6.63}$$

$$\ddot{C} + \dot{C}\left(\frac{\dot{C}}{C} + 2\frac{\dot{A}}{A}\right) = 0, \tag{6.64}$$

$$\ddot{B} - B\frac{\ddot{C}}{C} + 2\frac{\dot{A}\dot{B}}{A} - 2\frac{\dot{A}\dot{C}B}{AC} = 0. \tag{6.65}$$

We wish to solve this system of differential equations together with the condition (6.61).

First, the equation (6.64) can be integrated at once, yielding

$$A^2 C\dot{C} = k_1. \tag{6.66}$$

Next subtracting (6.65) from (6.63), we get

$$\frac{\dot{B}\dot{C}}{C} + \frac{B\ddot{C}}{C} + 2\frac{\dot{A}\dot{C}B}{AC} = 0, \tag{6.67}$$

which can be integrated, yielding

$$A^2 B\dot{C} = k_2. \tag{6.68}$$

We consider the following two possibilities:

I). $k_1 = 0$. The complete solution for this case is

$$C = k \neq 0,$$

$$A = \text{an arbitrary non-zero function of } t,$$

$$B = k'\int_0^t \frac{dt}{A^2} + k'', \tag{6.69}$$

$$D - \frac{1}{kA^2},$$

where k, k', and k'' are arbitrary constants.

II). $k_1 \neq 0$. In this case (6.66) and (6.68) imply that B is proportional to C, and the compete solution is

$$C = \text{an arbitrary non-zero function of } t, \text{ but } \dot{C} \neq 0,$$

$$A = \left(\frac{k_1}{C\dot{C}}\right)^{\frac{1}{2}}, \tag{6.70}$$

$$B = k'C,$$

$$D = \frac{1}{A^2C},$$

where k_1 and k' are arbitrary constants.

The acceleration-potential for this family is

$$\zeta = \frac{r^2}{2}\left(\frac{\ddot{A}}{A} - \alpha^2 + \alpha\beta - \frac{\beta}{2}\right) + \frac{1}{2}\frac{\dot{D}}{D}z^2 + \Phi(t), \tag{6.71}$$

where

$$\alpha = \left(\dot{B} - \frac{B\dot{C}}{C}\right)\log\frac{r}{A} + \frac{\dot{C}}{C}\theta, \tag{6.72}$$

$$\beta = \dot{B} - \frac{B\dot{C}}{C}. \tag{6.73}$$

Case 3. Incompressible Fluid Crystal Body.

There are numerous families of quasi-equilibrated solutions for the type of incompressible fluid crystal body treated as Case 3 in § 4. We give only a few examples here.

Family 1. Flow Between Coaxial Circular Cylinders

For flow between coaxial circular cylinders, the components of deformation are

$$r = R, \quad \theta = \Theta + f(R, t), \quad z = Z + g(R, t), \tag{6.74}$$

where f and g are unknown functions that must be determined by the condition (6.5). We assume that the initial distribution of the preferred axis $\boldsymbol{i}$ is given by (4.52). From (6.74), the matrix of components of the deformation gradient $\boldsymbol{F}$ is

$$[F_{\langle iA\rangle}] = \begin{bmatrix} 1 & 0 & 0 \\ R\dfrac{\partial f}{\partial R} & 1 & 0 \\ \dfrac{\partial g}{\partial R} & 0 & 1 \end{bmatrix}. \tag{6.75}$$

Hence the motion is isochoric and preserves $\boldsymbol{i}$:

$$\det \boldsymbol{F} = 1, \qquad \boldsymbol{F}\boldsymbol{i} = \boldsymbol{i}. \tag{6.76}$$

In particular, the motion is quasi-equilibrated.

Next we consider the restrictions on f and g due to the condition (6.5). We first compute the components of the acceleration field:

$$\begin{aligned} a_r &= -r\dot{f}^2, \\ a_\theta &= r^2\ddot{f}, \\ a_z &= \ddot{g}. \end{aligned} \tag{6.77}$$

Hence the governing equations are

$$\frac{\partial(r^2\ddot{f})}{\partial r} = 0, \tag{6.78}$$

$$\frac{\partial \ddot{g}}{\partial r} = 0. \tag{6.79}$$

These differential equations can be integrated at once, yielding the following complete solution:

$$f(R, t) = \frac{A(t)}{R^2} + B(R)t + C(R), \tag{6.80}$$

and

$$g(R, t) = D(R)t + E(R) + F(t), \tag{6.81}$$

where $A, \ldots, F$ are arbitrary functions of the arguments indicated.

In the physical interpretation of this solution, the meaning of the functions $C(R)$ and $E(R)$ is clear: $C(R)$ gives an initial rigid rotation about the z-axis, and $E(R)$ gives an initial rigid translation in the direction of z. The meaning of the remaining four functions A, B, D, and F can be seen from the components of the velocity field:

$$v_{\langle r\rangle} = 0, \quad v_{\langle\theta\rangle} = \frac{\dot{A}(t)}{r} + rB(r), \quad v_{\langle z\rangle} = D(r) + \dot{F}(t). \tag{6.82}$$

Family 2. Twisting and Radial Expansion Flow.

The components of deformation are

$$r = f(R, t), \quad \theta = \Theta + g(R, t), \quad z = h(Z, t). \tag{6.83}$$

The matrix of components of the deformation gradient is

$$[F_{\langle iA\rangle}] = \begin{bmatrix} \dfrac{\partial f}{\partial R} & 0 & 0 \\ f\dfrac{\partial g}{\partial R} & \dfrac{f}{R} & 0 \\ 0 & 0 & \dfrac{\partial h}{\partial z} \end{bmatrix}. \tag{6.84}$$

Then the condition for isochoric motion is

$$\frac{f}{R}\frac{\partial f}{\partial R}\frac{\partial h}{\partial z} = 1. \tag{6.85}$$

Since the variables are separated in this equation, the solution is easily found to be

$$f(R, t) = \sqrt{A(t)R^2 + B(t)}, \tag{6.86}$$

and

$$h(Z, t) = \frac{1}{A(t)}Z + C(t), \tag{6.87}$$

where A, B, and C are arbitrary functions of t, with $A \neq 0$.

For this family of motions we assume that the initial distribution of the preferred axis $\boldsymbol{i}$ is again given by (4.52). However, unlike Family 1, this family does not preserve the field $\boldsymbol{i}$. But from (6.84), (6.86), and (6.87), the instantaneous distribution of the preferred axis always has the form (4.52), *i.e.*,

$$i_{\langle r\rangle} = 0, \quad i_{\langle\theta\rangle} = \cos\alpha(r, t), \quad i_{\langle z\rangle} = \sin\alpha(r, t). \tag{6.88}$$

Thus the motion is quasi-equilibrated.

Next we determine the restrictions on the functions g, A, B, and C due to the condition (6.5). From (6.13) and (6.83), a_r is a function of r and t only, while a_z is a function of z and t. Hence the restriction due to (6.5) is

$$\frac{\partial a_\theta}{\partial r} = 0. \tag{6.89}$$

In terms of g, A, and B this condition takes the form

$$(AR^2+B)\ddot{g}+(\dot{A}R^2+\dot{B})\dot{g} = \text{a function of } t. \tag{6.90}$$

This differential equation can be integrated at once, yielding the complete solution:

$$\begin{aligned} A,\ B,\ C &= \text{arbitrary functions of } t \text{ with } A \neq 0,\\ g(R, t) &= \int \frac{D(t)+E(R)}{A(t)R^2+B(t)}\,dt+F(R), \end{aligned} \tag{6.91}$$

where

$$\begin{aligned} D &= \text{an arbitrary function of } t,\\ E,\ F &= \text{arbitrary functions of } R. \end{aligned} \tag{6.92}$$

Family 3.

Suppose that the initial distribution of the preferred axis $\boldsymbol{i}$ is given by the static universal solution (4.49). Then the deformation (6.60) (Family 5 of Case 2) carries that distribution into another one of the same kind. Thus (6.60) gives also a family of quasi-equilibrated motions for this type of fluid crystals; moreover, the analysis given before for the restriction due to the dynamical equation remains unchanged in this case, since it concerns only the acceleration field of the motion.

Family 4.

Suppose that the initial distribution of the preferred axis $\boldsymbol{i}$ is given by the static universal solution (4.53). Then the deformation (6.57) (Family 4 of Case 2) preserves that distribution. Hence (6.57) gives also a family of dynamic universal solutions for this type of fluid crystals.

7. Examples of Dynamic Universal Solutions for Incompressible Isotropic Bodies

Example 1. Radial Oscillation of a Hollow Circular Cylinder.

To analyse radial oscillation of a cylindrical tube[1], in Family 3 of Case 2 of the previous section we consider the special case when

$$r^2 = R^2 - R_1^2 + r_1^2, \ \theta = \Theta, \ z = Z, \tag{7.1}$$

where R_1 and r_1 are the inner radii in the reference and the actual configurations, respectively. In a radial oscillation r_1 is a function of t. From (6.48) the acceleration potential for this motion is

$$\zeta = (r_1 \dot{r}_1)^{\cdot} \log r + \frac{r_1^2 \dot{r}_1^2}{2r^2}. \tag{7.2}$$

To within an additive constant, the pressure field can be determined by use of (6.6). Combining this result with the determinate stress for the actual configuration yields the complete stress system. In particular,

$$T_{\langle rr \rangle} = \rho \left[(r_1 \ddot{r}_1 + \dot{r}_1^2) \log r + \frac{r_1^2 \dot{r}_1^2}{2r^2} \right] + \int \frac{1}{r} \left(\frac{r^2}{R^2} - \frac{R^2}{r^2} \right) (\psi_1 - \psi_{-1}) dr, \tag{7.3}$$

where the arguments of ψ_1 and ψ_{-1} are

$$I = II = 1 + \frac{r^2}{R^2} + \frac{R^2}{r^2}. \tag{7.4}$$

We set

$$x = r_1(t)/R_1, \tag{7.5}$$

$$\xi = r^2/R^2, \tag{7.6}$$

$$\delta = \frac{R_2^2}{R_1^2} - 1 > 0, \tag{7.7}$$

and

$$f(x, \delta) = \frac{1}{\rho R_1^2} \int_{(\delta + x^2)/(\delta + 1)}^{x^2} (1 + \xi) \hat{\mu} \left(\xi + \frac{1}{\xi} - 2 \right) \frac{d\xi}{\xi^2}, \tag{7.8}$$

---

[1] The results are due to KNOWLES. We follow here the simpler and more general treatment given in NFTM, § 62.

where $\hat{\mu}$ is the generalized shear modulus, defined by (3.10), and where R_2 denotes the outer radius of the cylinder in the reference configuration. Substituting (7.5)–(7.8) into (7.3) and denoting the difference of the pressures on the inner and the outer surfaces by $P(t)$, we obtain the differential equation

$$x \log\left(1+\frac{\delta}{x^2}\right)\ddot{x}+\left[\log\left(1+\frac{\delta}{x^2}\right)-\frac{\delta}{\delta+x^2}\right]\dot{x}^2+f(x,\delta)=\frac{2P}{\rho R_1^2}. \tag{7.9}$$

Under a given loading $P(t)$ and an initial condition, (7.9) determines the motion. We consider the following two cases:

I. Free Oscillation.

To treat free oscillation, we set $P(t) \equiv 0$. Then from (7.9) we obtain at once the following "integral of energy":

$$\tfrac{1}{2}\dot{x}^2x^2 \log\left(1+\frac{\delta}{x^2}\right)+F(x,\delta)=C, \tag{7.10}$$

where C is a constant of integration, and where

$$F(x,\delta)=\int_1^x \lambda f(\lambda,\delta)d\lambda. \tag{7.11}$$

The "energy constant" C can be determined from the initial conditions, say

$$x(0)=x_0,\quad \dot{x}(0)=\dot{x}_0. \tag{7.12}$$

Then (7.10) can be written as

$$\tfrac{1}{2}\dot{x}^2x^2 \log\left(1+\frac{\delta}{x^2}\right)+F(x,\delta)=\tfrac{1}{2}\dot{x}_0^2x_0^2 \log\left(1+\frac{\delta}{x_0^2}\right)+F(x_0,\delta). \tag{7.13}$$

Since the "kinetic energy" term $\frac{1}{2}\dot{x}^2x^2 \log(1+\delta/x^2)$ in this equation is positive-definite in $\dot{x}$, the solution $x(t)$ must satisfy the condition

$$F(x(t),\delta) < C \tag{7.14}$$

for all t except at those values of t at which $\dot{x}$ vanishes. Thus, as in all oscillatory motions, the form of $x(t)$ depends critically on the "potential energy" $F(x, \delta)$.

We pause here to remark that in the derivation of the differential equation (7.9) and its integral (7.10) we did not require the material to be hyperelastic. The integral (7.10), however, seems like a statement of conservation of energy, with $F(x, \delta)$ playing the role of the "potential energy" or the "stored energy". The reason for this result is not hard to see. Like the simple shears treated in § 3, the deformations (7.1) depend upon only one parameter at each body-point; moreover, the one-parameter family of deformations cannot form a nontrivial loop in the domain of all admissible deformation gradients. Consequently the response of a general elastic material in the motion (7.1) cannot be distinguished from that of a general hyperelastic material.[1]

We now examine the general character of the function $F(x, \delta)$. From (7.8) and (7.11), if we assume that $\hat{\mu}$ is positive, then the integrand in (7.11) is positive when $x > 1$ and is negative when $0 < x < 1$. Thus for each fixed δ the function $F(x, \delta)$ is monotonically decreasing when $0 < x < 1$, zero when $x = 1$, and monotonically increasing when $x > 1$, as shown in the adjoining sketch:

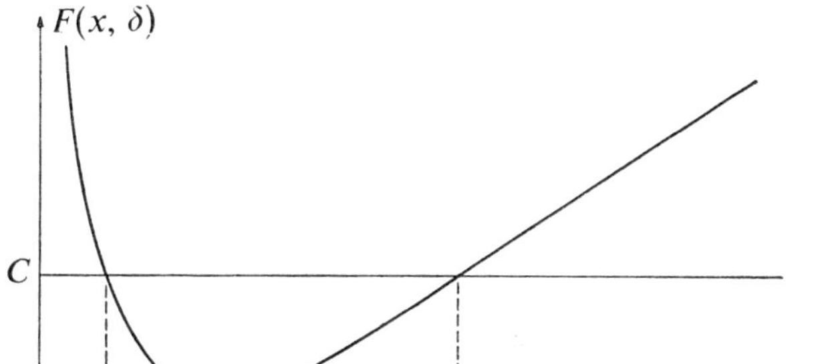

In order that the motion be an oscillation, the equation

$$F(x, \delta) = C \tag{7.14}$$

must have exactly two roots, say at $x = a$ and $x = b$, where $0 < a < 1 < b$. In this case the period τ is given by

$$\tau = 2\int_a^b \frac{dx}{\dot{x}} = 2\int_a^b \left[\frac{x^2 \log\left(1+\dfrac{\delta}{x}\right)}{2C-2F(x, \delta)}\right]^{\frac{1}{2}} dx, \tag{7.15}$$

[1] This fact seems to have been recognized first by TRUESDELL & NOLL, when they generalized the analysis of KNOWLES for this problem from hyperelastic to elastic materials.

and the two roots b and a are the maximum and the minimum values, respectively, of x in the motion.

In general we cannot assert that the equation (7.14) has two roots for each given value of C. Indeed, the condition $\hat{\mu} > 0$ suffices that there be two roots when C is sufficiently small. Further, if $\hat{\mu}$ is bounded below by a positive number, then $F(x, \delta) \to \infty$ when $x \to +0$ or when $x \to +\infty$, and then any initial value gives rise to an oscillatory motion.

By (7.7), the parameter δ characterizes the thickness of the wall relative to the radius of the cylinder, and the limiting case of a thin shell is obtained as $\delta \to 0$. In that limit the "energy equation" (7.13) reduces to

$$\dot{x}^2 + W(x^2) = \dot{x}_0^2 + W(x_0^2), \tag{7.16}$$

where

$$W(x^2) = \frac{1}{\rho R_1^2} \int_0^{x^2+1/x^2-2} \hat{\mu}(\lambda)d\lambda. \tag{7.17}$$

Since W satisfies the identity

$$W(x^2) = W\left(\frac{1}{x^2}\right), \tag{7.18}$$

the two roots a and b of the equation

$$W(x^2) = \dot{x}_0^2 + W(x_0^2) \tag{7.19}$$

are related by

$$ab = 1, \tag{7.20}$$

which means that the geometric mean of the actual maximum and minimum radii attained by the shell in any free oscillation is the radius of the shell in the reference configuration. Also the formula for the period τ is simplified:

$$\tau = 2\int_a^{1/a} [\dot{x}_0^2 + W(x_0^2) - W(x^2)]^{-\frac{1}{2}} dx, \tag{7.21}$$

where a is the smaller root of (7.19).

II. Oscillation Induced by an Impulse

To treat the motion resulting from a pressure impulse, we assume that the loading is

$$\frac{2P(t)}{\rho R_1^2} = \begin{cases} 0 & \text{when } t \leqq 0, \\ p & \text{when } t > 0, \end{cases} \tag{7.22}$$

where p is a constant, and that the initial condition is

$$x_0 = 1, \; \dot{x}_0 = 0. \tag{7.23}$$

In this case, integration of (7.9) yields

$$\tfrac{1}{2}\dot{x}^2 x^2 \log\left(1+\frac{\delta}{x^2}\right) + F(x, \delta) = \tfrac{1}{2}p(x^2-1), \; t > 0. \tag{7.24}$$

Then the motion is an oscillation if and only if the equation

$$F(x, \delta) = \tfrac{1}{2}p(x^2-1) \tag{7.25}$$

has a positive root a other than 1. Further, a must be greater or less than 1 according as p is positive or negative, as shown in the adjoining sketch:

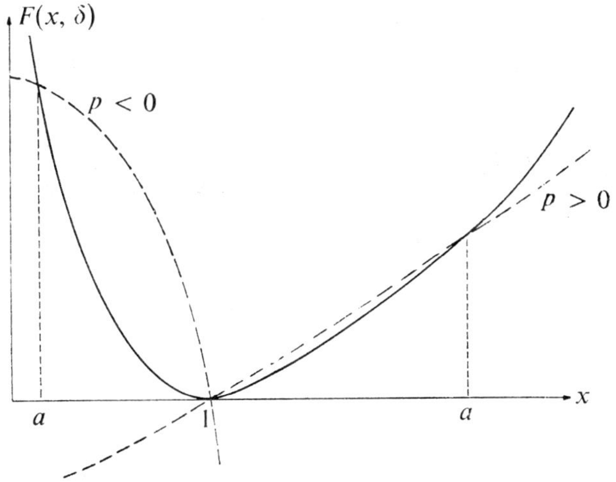

Now from (7.11), regarding δ as a parameter, we see that

$$F'(x, \delta) = xf(x, \delta), \tag{7.26}$$

so the equation

$$f(x, \delta) = p \tag{7.27}$$

must have a root b between $x = 1$ and $x = a$. Further, by (7.9) that root characterizes a static solution with respect to the static loading of the amount p. Thus we have shown that in an oscillatory motion induced by a pressure impulse, the equilibrium radius lies between the maximum and the minimum radii of the motion. The period of oscillation is given by

$$\tau = 2\left|\int_1^a x\left[\frac{\log\left(1+\dfrac{\delta}{x^2}\right)}{p(x^2-1)-2F(x,\delta)}\right]^{\frac{1}{2}} dx\right|. \tag{7.28}$$

Again we cannot assert that in general the equation (7.25) has a root a other than 1. However, if $\hat{\mu} > 0$ and if p is sufficiently small, then such a root does exist.

Example 2. Radial Oscillation of a Hollow Spherical Shell

To analyse the radial oscillation of a sphere[1], in Family 4 of Case 2 in the previous section we consider the special case when

$$r^3 = R^3 - R_1^3 + r_1^3,\ \theta = \Theta,\ \varphi = \Phi. \tag{7.29}$$

From (6.59), the acceleration-potential is

$$\zeta = -(r\ddot{r}_1 + \tfrac{3}{2}\dot{r}_1^2). \tag{7.30}$$

Then the stress system can be determined in the standard way as explained in the previous example. In particular

$$T_{\langle rr\rangle} = -\rho(r\ddot{r}_1 + \tfrac{3}{2}\dot{r}_1^2) + 2\int\left(\frac{r^2}{R^2} - \frac{R^4}{r^4}\right)\left(\psi_1 - \frac{r^2}{R^2}\psi_{-1}\right)\frac{dr}{r}, \tag{7.31}$$

where the arguments of ψ_1 and ψ_{-1} are

$$I = \frac{R^4}{r^4} + \frac{2r^2}{R^2},\ II = \frac{r^4}{R^4} + \frac{2R^2}{r^2}. \tag{7.32}$$

As before, we set

$$x = \frac{r_1(t)}{R_1}, \tag{7.33}$$

$$\delta = \frac{R_2^3}{R_1^3} - 1 > 0, \tag{7.34}$$

$$\xi = \frac{r^3}{R^3}, \tag{7.35}$$

---

[1] We follow the analysis of NFTM, § 62.

and

$$g(x, \delta) = \frac{4}{3\rho R_1^3} \int_{(\delta+x^3)/(\delta+1)}^{x^3} \xi^{-\frac{1}{3}}(1+\xi)(\psi_1 - \xi^{\frac{4}{3}}\psi_{-1})d\xi, \tag{7.36}$$

where the arguments of ψ_1 and ψ_{-1} become

$$I = 2\xi^{\frac{2}{3}} + \xi^{-\frac{4}{3}}, \quad II = 2\xi^{-\frac{2}{3}} + \xi^{\frac{4}{3}}. \tag{7.37}$$

Then the differential equation for x is

$$\frac{d}{dx}\left\{\left[1 - \left(1 + \frac{\delta}{x^3}\right)^{-\frac{1}{3}}\right] x^3\dot{x}^2\right\} + x^2 g(x, \delta) = \frac{2x^2 P(t)}{\rho R_1^2}, \tag{7.38}$$

where $P(t)$ denotes the difference of pressures on the inner and outer surfaces. While the nature of radial motions seem to be much alike for spheres and cylinders, there is also an important difference: The generalized shear modulus $\hat{\mu}$ does not determine the function g, into whose nature all material properties are condensed as far as radial motions of a sphere are concerned.

As before, we consider the following two cases

I. Free Oscillation.

For free oscillation, $P(t) = 0$. Then the governing equation (7.38) can be integrated at once, yielding the "energy equation"

$$\left[1 - \left(1 + \frac{\delta}{x^3}\right)^{-\frac{1}{3}}\right] x^3\dot{x}^2 + G(x, \delta) = C, \tag{7.39}$$

where

$$G(x, \delta) = \int_1^x \lambda^2 g(\lambda, \delta) d\lambda. \tag{7.40}$$

Again, whether or not the motion is an oscillation depends critically on the form of the "potential energy" G. Indeed, under a given initial condition which determines the value of the "total energy" C, the motion is an oscillation if and only if the equation

$$G(x, \delta) = C \tag{7.41}$$

has exactly two roots a and b such that $a < 1 < b$. These roots are the extremal values of x in the motion.

Now suppose that the response coefficients ψ_1 and ψ_{-1} satisfy the condition (5.14). Then from (7.36) we see that g is positive when $x > 1$

and is negative when $0 < x < 1$. Hence from (7.40) the function G decreases monotonically when $0 < x < 1$, vanishes when $x = 1$, and increases monotonically when $x > 1$. This behavior is the same as that of the function $F(x, \delta)$ in the previous example. Therefore, we can assert that under the condition (5.14) a free oscillation is possible provided that C be sufficiently small.

For an oscillatory motion, the period τ is given by the formula

$$\tau = 2\int_a^b \left[\frac{x^3\left[1-\left(1+\frac{\delta}{x^3}\right)^{-\frac{1}{3}}\right]}{C-G(x,\delta)}\right]^{\frac{1}{2}} dx. \tag{7.42}$$

In the limiting case as $\delta \to 0$, appropriate to a thin shell, the "energy equation" again has the form (7.16) except that the function W is now given by

$$W(x^2) = \frac{4}{\rho R_1^2}\int_1^x \frac{\xi^2-1}{\xi^5}(\psi_1-\xi^2\psi_{-1})d\xi, \tag{7.43}$$

where the arguments of ψ_1 and ψ_{-1} are given by (7.37).

II. Oscillation Induced by an Impulse

To treat the oscillation resulting from an impulse, we assume that the loading and the initial condition are given by (7.22) and (7.23), respectively. Then an integral of the differential equation (7.38) is

$$\left[1-\left(1+\frac{\delta}{x^3}\right)^{-\frac{1}{3}}\right] x^3\dot{x}^2+G(x,\delta) = \tfrac{1}{3}p(x^3-1). \tag{7.44}$$

Thus in this case the amplitude equation is

$$G(x,\delta) = \tfrac{1}{3}p(x^3-1). \tag{7.45}$$

As before, the motion is an oscillation if and only if this equation has a positive root $x = a$ different from the root $x = 1$; moreover, a is greater or less than 1 according as p is positive or negative.

Example 3. Rotation of a Circular Cylinder

While rigid rotations superimposed on some simple quasi-equilibrated motions have not been treated in general, we give here an example[1] of such

[1] The solution described here was first found by Green & Shield (1950), before the general theory of quasi-equilibrated motions had been constructed.

motions. We consider the motion of a circular cylinder rotating and stretching in its axial direction. The components of deformation are

$$r^2 = AR^2,\ \theta = \Theta + G,\ z = FZ, \tag{7.46}$$

where A, F, G are functions of t such that

$$AF = 1 \tag{7.47}$$

for all t. Following the procedure of the preceding section, we determine first the components of the acceleration field:

$$\begin{aligned} a_r &= \left[\frac{\ddot{A}}{2A} - \left(\frac{\dot{A}}{2A}\right)^2 - \dot{G}^2\right] r, \\ a_\theta &= \left[\ddot{G} + \frac{\dot{A}}{A}\dot{G}\right] r^2, \\ a_z &= A\ddot{F}z. \end{aligned} \tag{7.48}$$

Then the restriction due to the condition (6.5) is

$$\ddot{G} + \frac{\dot{A}}{A}\dot{G} = 0. \tag{7.49}$$

The solution of this equation is

$$\begin{aligned} A &= \text{an arbitrary non-vanishing function of } t, \\ F &= \frac{1}{A}, \\ G &= k\int_0^t \frac{1}{A}\,dt + k', \end{aligned} \tag{7.50}$$

where k and k' are constants.

The acceleration-potential of the motion is

$$\zeta = \left[\frac{\ddot{A}}{2A} - \left(\frac{\dot{A}}{2A}\right)^2 - \dot{G}^2\right]\frac{r^2}{2} + A\ddot{F}\frac{z^2}{2} + \Phi(t). \tag{7.51}$$

Then the stress system can be determined in the standard way, giving the result

$$T_{\langle rr\rangle} = T_{\langle\theta\theta\rangle} = \frac{\rho}{2}\left[\frac{\ddot{A}}{2A} - \left(\frac{\dot{A}}{2A}\right)^2 - \dot{G}^2\right] r^2 + \frac{\rho A\ddot{F}}{2} z^2 + \Phi(t),$$

$$T_{\langle zz\rangle} = \frac{\rho}{2}\left[\frac{\ddot{A}}{2A} - \left(\frac{\dot{A}}{2A}\right)^2 - \dot{G}^2\right] r^2 + \frac{\rho A\ddot{F}}{2} z^2 + \left(F^2 - \frac{1}{F}\right)\left(\psi_1 - \frac{1}{F}\psi_{-1}\right) + \Phi(t), \qquad (7.52)$$

$$T_{\langle r\theta\rangle} = T_{\langle rz\rangle} = T_{\langle\theta z\rangle} = 0,$$

where the arguments of ψ_1 and ψ_{-1} are

$$I = F^2 + \frac{2}{F}, \; II = \frac{1}{F^2} + 2F. \qquad (7.53)$$

From $(7.52)_1$ we see that the radial stress is independent of z if and only if $\ddot{F} = 0$.

We shall consider the special case in which

$$F = \frac{1}{A} = \varepsilon = \text{constant}. \qquad (7.54)$$

From $(7.50)_3$, this case is also characterized by

$$\dot{G} = \omega = \text{constant}. \qquad (7.55)$$

As usual we require that the outer surface of the cylinder, say with radius R_0, be free of traction. Then the components of the stress are uniquely determined:

$$\begin{aligned} T_{\langle rr\rangle} &= T_{\langle\theta\theta\rangle} = \frac{1}{2}\frac{\rho\omega^2}{\varepsilon}(R_0^2 - R^2) = \tfrac{1}{2}\rho\omega^2(r_0^2 - r^2), \\ T_{\langle zz\rangle} &= T_{\langle rr\rangle} + \left(\varepsilon^2 - \frac{1}{\varepsilon}\right)\left(\tilde{\psi}_1 - \frac{1}{\varepsilon}\tilde{\psi}_{-1}\right), \end{aligned} \qquad (7.56)$$

where

$$\tilde{\psi}_\Gamma = \psi_\Gamma\left(\varepsilon^2 + \frac{2}{\varepsilon}, \frac{1}{\varepsilon^2} + 2\varepsilon\right), \quad \Gamma = 1, -1. \qquad (7.57)$$

In this case the total axial force on the plane cross-sections vanishes if and only if ε and ω satisfy the condition

$$\frac{1}{4}\frac{\rho\omega^2 R_0^2}{\varepsilon^2} + \left(\varepsilon - \frac{1}{\varepsilon^2}\right)\left(\tilde{\psi}_1 - \frac{1}{\varepsilon}\tilde{\psi}_{-1}\right) = 0. \qquad (7.58)$$

Comparing this result with (5.34), we find the following universal relation:

$$\frac{\pi\rho\omega^2 R_0^6}{8\varepsilon^2 K(\varepsilon)} = \frac{1}{\varepsilon^2} - \varepsilon, \tag{7.59}$$

where $K(\varepsilon)$ denotes the torsional stiffness of the stretched cylinder. If $K(\varepsilon) > 0$, then (7.59) implies that

$$0 < \varepsilon < 1 \tag{7.60}$$

for any angular speed ω other than 0. In the physical interpretation, this fact means that a cylinder whose torsional stiffness is positive for all extensions becomes shorter and thicker when it is spun about its axis at any constant angular speed.

CHAPTER V

Inhomogeneous Elastic Bodies

1. The Material Tangent Bundle of an Elastic Body

We review briefly the concept of a materially uniform smooth body [1], introduced in § IV. 1. A body $\mathscr{B}$ is called *materially uniform* if its body-points are pairwise materially isomorphic. In such a body, then, for any two body-points X and Y there exists a material isomorphism $\boldsymbol{I}(X, Y)$ (*cf.* (IV. 1.1)) such that the identity (IV. 1.2) is satisfied. Equivalently, there exist local reference configurations $\boldsymbol{v}_X$ and $\boldsymbol{v}_Y$ such that the response function of X relative to $\boldsymbol{v}_X$ is the same as that of Y relative to $\boldsymbol{v}_Y$. In this case the composite map

$$\boldsymbol{I}(X, Y) \equiv \boldsymbol{v}_Y^{-1} \circ \boldsymbol{v}_X \tag{1.1}$$

is a material isomorphism of X with Y.

Thus for a materially uniform body $\mathscr{B}$ we can find a field $\boldsymbol{v}$ relative to which the response function is independent of the body-point. In general, such a field is neither smooth nor equal to the gradient of a configuration. However, if $\boldsymbol{v}$ can be chosen to be the gradient of a configuration, then $\mathscr{B}$ is a homogeneous body, for which the theory has been developed in the preceding chapter. In this chapter $\mathscr{B}$ is called a *smooth* body if for every body-point X there exists a subbody $\mathscr{U} \ni X$ and a smooth field of local reference configurations $\boldsymbol{v}$ on $\mathscr{U}$ such that the response function relative to $\boldsymbol{v}$ is independent of the body points in $\mathscr{U}$. A pair $(\mathscr{U}, \boldsymbol{v})$ having the smoothness property just described is called a *material chart* covering the body-point X. Thus $\mathscr{B}$ is a materially uniform smooth body if and only if every body-point of $\mathscr{B}$ can be covered by a material chart. The

[1] The development in this chapter follows the paper of C.–C. WANG, "On the geometric structures of simple bodies, a mathematical foundation for the theory of continuous distributions of dislocations", *Arch. Rational Mech. Anal.* **27**, 33–94 (1967), reprinted in *Continuum Theory of Inhomogeneities in Simple Bodies*, New York, Springer, 1968.

condition of smoothness is local. Of course, some smooth bodies (such as the subbody $\mathscr{U}$) do have global material charts covering all their body-points, but there are generally others which have local material charts only. Naturally we call $\mathscr{B}$ *locally homogeneous* if every body-point of $\mathscr{B}$ can be covered by a material chart $(\mathscr{U}, \boldsymbol{v})$ such that $\boldsymbol{v}$ can be chosen to be the gradient of a configuration of $\mathscr{U}$. Generally, a locally homogeneous body need not be homogeneous.

Two material charts $(\mathscr{U}, \boldsymbol{v})$ and $(\bar{\mathscr{U}}, \bar{\boldsymbol{v}})$ are said to be *compatible* if the response function relative to one is the same as that relative to the other. In § IV. 1 we proved that every materially uniform smooth body $\mathscr{B}$ can be equipped with a maximal collection

$$\mathfrak{A} = \{(\mathscr{U}_\alpha, \boldsymbol{v}_\alpha),\ \alpha \in I\}, \tag{1.2}$$

which is called a *material atlas*, consisting of mutually compatible material charts such that every body-point $X \in \mathscr{B}$ is covered by at least one chart in $\mathfrak{A}$. We proved further that the collection of all material atlases of $\mathscr{B}$ can be put in one-to-one correspondence with the collection of all relative response functions of $\mathscr{B}$. That is to say, two material atlases $\mathfrak{A}$ and $\mathfrak{B}$ are the same if and only if their relative response functions $\boldsymbol{H}_{\mathfrak{A}}$ and $\boldsymbol{H}_{\mathfrak{B}}$ coincide:

$$\mathfrak{A} = \mathfrak{B} \Leftrightarrow \boldsymbol{H}_{\mathfrak{A}}(\boldsymbol{F}) = \boldsymbol{H}_{\mathfrak{B}}(\boldsymbol{F}) \quad \forall \boldsymbol{F}; \tag{1.3}$$

moreover, every relative response function of $\mathscr{B}$ can be realized as a response function relative to a material atlas of $\mathscr{B}$.

The general linear group $\mathfrak{L}$ acts as a transformation group on the material atlases of $\mathscr{B}$ in the sense that if $\boldsymbol{P} \in \mathfrak{L}$, then we define

$$\boldsymbol{P}\mathfrak{A} \equiv \{(\mathscr{U}_\alpha, \boldsymbol{P} \circ \boldsymbol{v}_\alpha),\ \alpha \in I\}. \tag{1.4}$$

In terms of the relative response function,

$$\boldsymbol{H}_{\boldsymbol{P}\mathfrak{A}}(\boldsymbol{F}) = \boldsymbol{H}_{\mathfrak{A}}(\boldsymbol{F}\boldsymbol{P}) \quad \forall \boldsymbol{F} \in \mathfrak{L}. \tag{1.5}$$

Comparing (1.5) with (1.3), we see that $\boldsymbol{P}$ belongs to the symmetry group of $\boldsymbol{H}_{\mathfrak{A}}$ if and only if $\mathfrak{A}$ is invariant under $\boldsymbol{P}$:

$$\boldsymbol{P} \in \mathfrak{G}_{\mathfrak{A}} \Leftrightarrow \mathfrak{A} = \boldsymbol{P}\mathfrak{A}. \tag{1.6}$$

Now since the relative response function is invariant under a change of material chart in $\mathfrak{A}$, the condition (1.6) implies that the transformation

$$\boldsymbol{G}_{\alpha\beta} = \boldsymbol{v}_\alpha \circ \boldsymbol{v}_\beta^{-1} \tag{1.7}$$

defines a smooth field on $\mathscr{U}_\alpha \cap \mathscr{U}_\beta$ having values in $\mathfrak{G}_{\mathfrak{A}}$. Clearly these fields satisfy the conditions

$$\boldsymbol{G}_{\alpha\alpha} = \boldsymbol{1} \text{ on } \mathscr{U}_\alpha, \tag{1.8}$$

$$\boldsymbol{G}_{\alpha\beta} = \boldsymbol{G}_{\beta\alpha}^{-1} \text{ on } \mathscr{U}_\alpha \cap \mathscr{U}_\beta, \tag{1.9}$$

and

$$\boldsymbol{G}_{\alpha\beta}\boldsymbol{G}_{\beta\gamma} = \boldsymbol{G}_{\alpha\gamma} \text{ on } \mathscr{U}_\alpha \cap \mathscr{U}_\beta \cap \mathscr{U}_\gamma. \tag{1.10}$$

In the theory of fibre bundles, these conditions define the collection $\{\boldsymbol{G}_{\alpha\beta}, \alpha, \beta \in I\}$ as the *co-ordinate transformations* of a bundle atlas with the *structure group* $\mathfrak{G}_{\mathfrak{A}}$.

We now explain these concepts in more detail. In differential geometry, the *tangent bundle* $\mathscr{T}(\mathscr{M})$ of a differentiable manifold $\mathscr{M}$ is formed by the disjoint union of the tangent spaces of $\mathscr{M}$ (*cf.* § I. 5):

$$\mathscr{T}(\mathscr{M}) = \bigcup_{\boldsymbol{p}\in\mathscr{M}} \mathscr{M}_{\boldsymbol{p}}. \tag{1.11}$$

For a body manifold $\mathscr{B}$, the tangent bundle $\mathscr{T}(\mathscr{B})$ is then given by

$$\mathscr{T}(\mathscr{B}) = \bigcup_{X\in\mathscr{B}} \mathscr{B}_X = \bigcup_{\substack{X\in\mathscr{B}\\ \boldsymbol{v}\in\mathscr{B}_X}} (X, \boldsymbol{v}). \tag{1.12}$$

The structure of $\mathscr{T}(\mathscr{B})$ is very simple, since $\mathscr{B}$ can be covered by a global co-ordinate system, say

$$\boldsymbol{\kappa} : \mathscr{B} \to \mathscr{R}^3. \tag{1.13}$$

We can regard $\boldsymbol{\kappa}$ as a configuration of $\mathscr{B}$ with respect to the spatial co-ordinate system of the fixed frame of reference. The induced linear map of $\boldsymbol{\kappa}$ then gives rise to a field of isomorphisms of the tangent spaces of $\mathscr{B}$ with $\mathscr{R}^3$:

$$\boldsymbol{\kappa}_* : \mathscr{T}(\mathscr{B}) \to \mathscr{B} \times \mathscr{R}^3 \tag{1.14}$$

such that

$$\boldsymbol{\kappa}_*(X, \boldsymbol{v}) = (X, \boldsymbol{\kappa}_{*X}(\boldsymbol{v})) \tag{1.15}$$

for all $X \in \mathscr{B}$ and all $\boldsymbol{v} \in \mathscr{B}_X$. The inverse of $\boldsymbol{\kappa}_*$ carries the manifold structure of $\mathscr{B} \times \mathscr{R}^3$ onto a manifold structure on $\mathscr{T}(\mathscr{B})$. With respect to this manifold structure, the pair $(\mathscr{B}, \boldsymbol{\kappa}_*)$ becomes a global bundle chart for $\mathscr{T}(\mathscr{B})$.

In general, a *bundle chart* for $\mathscr{T}(\mathscr{B})$ is a pair $(\mathscr{U}, \boldsymbol{\nu})$ consisting in a sub-body $\mathscr{U}$ and a smooth field of linear isomorphisms of the tangent spaces of $\mathscr{U}$ with $\mathscr{R}^3$:

$$\boldsymbol{\nu} : \mathscr{T}(\mathscr{U}) \to \mathscr{U} \times \mathscr{R}^3. \tag{1.16}$$

The collection of all such bundle charts forms the *bundle atlas* of $\mathscr{T}(\mathscr{B})$. The *co-ordinate transformation* from one bundle chart $(\mathscr{U}, \boldsymbol{v})$ to another one $(\bar{\mathscr{U}}, \bar{\boldsymbol{v}})$ is defined in the same way as before:

$$G(\boldsymbol{v}, \bar{\boldsymbol{v}}) \equiv \boldsymbol{v} \circ \bar{\boldsymbol{v}}^{-1}, \tag{1.17}$$

where the domain is the overlap $\mathscr{U} \cap \bar{\mathscr{U}}$. These co-ordinate transformations are fields with values in the general linear group $\mathfrak{L}$, which is the *structure group* of $\mathscr{T}(\mathscr{B})$.

The general structure of $\mathscr{T}(\mathscr{B})$ just described is not of interest to us, since it is based on the body manifold alone and is independent of the distribution of the response functions of the body-points. In order to characterize the material structure of the body, we now introduce the concept of a *material tangent bundle* of $\mathscr{B}$.

Relative to the assigned spatial co-ordinate system on $\mathscr{E}$, we can regard a material chart $(\mathscr{U}_\alpha, \boldsymbol{v}_\alpha)$ as a bundle chart for $\mathscr{T}(\mathscr{B})$. Then a material atlas $\mathfrak{A}$ can be regarded as a subset of the bundle atlas of $\mathscr{T}(\mathscr{B})$. We define the *material tangent bundle* $\mathscr{T}(\mathscr{B}, \mathfrak{A})$ with respect to $\mathfrak{A}$ as the fibre bundle whose bundle space is the same as that of $\mathscr{T}(\mathscr{B})$ but whose bundle atlas is the material atlas $\mathfrak{A}$. Since the co-ordinate transformations $\{\boldsymbol{G}_{\alpha\beta}\}$ for $\mathfrak{A}$ are $\mathfrak{G}_{\mathfrak{A}}$-valued, the structure group of $\mathscr{T}(\mathscr{B}, \mathfrak{A})$ is $\mathfrak{G}_{\mathfrak{A}}$. In the terminology of fibre bundles, we say that $\mathscr{T}(\mathscr{B})$ is *reducible* to $\mathscr{T}(\mathscr{B}, \mathfrak{A})$, since the structure group $\mathfrak{G}_{\mathfrak{A}}$ of $\mathscr{T}(\mathscr{B}, \mathfrak{A})$ is a subgroup of the structure group $\mathfrak{L}$ of $\mathscr{T}(\mathscr{B})$.

The material tangent bundle $\mathscr{T}(\mathscr{B}, \mathfrak{A})$ now characterizes the material structure of $\mathscr{B}$ since its atlas $\mathfrak{A}$ determines completely the distribution of the response functions of the body-points of $\mathscr{B}$. Relative to two material atlases $\mathfrak{A}$ and $\mathfrak{B}$ the bundles $\mathscr{T}(\mathscr{B}, \mathfrak{A})$ and $\mathscr{T}(\mathscr{B}, \mathfrak{B})$ are isomorphic in the sense of fibre bundles. Let $\boldsymbol{P}$ be a tensor such that

$$\mathfrak{B} = \boldsymbol{P}\mathfrak{A}. \tag{1.18}$$

Then the operation (1.4) defines the bundle isomorphism of $\mathscr{T}(\mathscr{B}, \mathfrak{A})$ with $\mathscr{T}(\mathscr{B}, \mathfrak{B})$. Of course, the isomorphism $\boldsymbol{P}$ in (1.18) is not unique, since a bundle may be non-trivially isomorphic to itself. Indeed, from (1.6) we have

$$\boldsymbol{P}\mathfrak{A} = \bar{\boldsymbol{P}}\mathfrak{A} \Leftrightarrow \boldsymbol{P}\bar{\boldsymbol{P}}^{-1} \in \mathfrak{G}_{\mathfrak{A}}. \tag{1.19}$$

A fibre bundle is called a *trivial bundle* if it has a global bundle chart covering the entire base space. For example, the geometric tangent bundle

$\mathscr{T}(\mathscr{B})$ of a body manifold is a trivial bundle. Since all material tangent bundles of $\mathscr{B}$ are bundle-isomorphic, they are either all trivial bundles or all non-trivial bundles. Clearly the material tangent bundles of a homogeneous body are trivial bundles. Conversely, however, a body whose material tangent bundles are trivial bundles need not be a homogeneous body, as is shown by Example 1 below.

We now give some illustrative examples of materially uniform smooth bodies.

Example 1. Continuous Distributions of Dislocations in Solid Crystals.

For definiteness, we consider a triclinic solid crystal body. The material isomorphism $\boldsymbol{I}(X, Y)$ for any two body-points X and Y of this body is unique, since the symmetry group of the triclinic crystal is the trivial group. We choose a reference body-point X and a reference basis $\{\boldsymbol{e}_i(X)\}$ for $\mathscr{B}_X$. Then the material isomorphism $\boldsymbol{I}(X, \cdot)$ maps the basis into a field of bases $\{\boldsymbol{e}_i(\cdot)\}$ on the whole body, called a *crystallographic basis* of the body. Clearly, the material structure of the crystal body is determined by its crystallographic basis.

The crystal is homogeneous if and only if there is a homogeneous configuration in which the crystallographic basis is a constant field. Equivalently, this condition means that the crystallographic basis is the natural basis of a global co-ordinate system. In differential geometry it is shown that this condition is satisfied if and only if the Lie derivative of any basis vector $\boldsymbol{e}_i$ with respect to any other basis vector $\boldsymbol{e}_j$ vanishes identically on the body manifold. In terms of the usual bracket product (*cf.* § I.4),

$$[\boldsymbol{e}_i, \boldsymbol{e}_j] = \boldsymbol{0}, \qquad i, j = 1, 2, 3. \tag{1.20}$$

Now suppose that the crystallographic basis $\{\boldsymbol{e}_i\}$ does not satisfy the condition (1.20). Then the crystal is inhomogeneous. In the terminology of the theory of dislocations in crystals, such a crystal is called a *dislocated crystal*. The *inhomogeneity* of a dislocated crystal is characterized by the third order tensor field $\mathbf{C}$ such that

$$\mathbf{C}(\boldsymbol{e}_i, \boldsymbol{e}_j) = [\boldsymbol{e}_i, \boldsymbol{e}_j]. \tag{1.21}$$

In component form relative to the crystallographic basis $\{\boldsymbol{e}_i\}$, this equation can be written as

$$[\boldsymbol{e}_i, \boldsymbol{e}_j] = C^k_{ij}\, \boldsymbol{e}_k, \tag{1.22}$$

where the components C^k_{ij} are skew-symmetric in i and j for each k. As we shall see in § 6, the tensor $\mathbf{C}$ is the torsion tensor of the unique material connection of this body.

Since the field of material isomorphisms $\boldsymbol{I}(X, \cdot)$ is a global smooth field, the material tangent bundles of a triclinic crystal are trivial, whether or not the crystal be homogeneous. Thus this example shows clearly that a body having a trivial material tangent bundle[1] need not be a homogeneous body.

Example 2. Moebius Crystals

For definiteness, we consider a Moebius cubic crystal. A body-point X is called a *cubic crystal point* if there exists a basis $\boldsymbol{e}(X)$ in $\mathscr{B}_X$, called a *cubic basis* of X, such that a linear transformation

$$\boldsymbol{I} : \mathscr{B}_X \to \mathscr{B}_X \tag{1.23}$$

is a material automorphism of X if and only if it permutes $\boldsymbol{e}(X)$:

$$\boldsymbol{I}(\boldsymbol{e}(X)) = \boldsymbol{e}(X) \tag{1.24}$$

in the sense of set theory. Then in any local configuration $\boldsymbol{v}$ a tensor $\boldsymbol{P}$ is a relative material automorphism if and only if $\boldsymbol{P}$ permutes the basis $\boldsymbol{v} \circ \boldsymbol{e}(X)$. In particular, if $\boldsymbol{v} \circ \boldsymbol{e}(X)$ is an orthonormal basis, then the symmetry group relative to $\boldsymbol{v}$ is a subgroup of the rotational group. Thus a cubic crystal is a solid, and the local configurations which carry a cubic basis into an orthonormal basis are the undistorted states.

Now let $\mathscr{B}$ be a materially uniform smooth cubic crystal body. Then $\mathscr{B}$ can be equipped with a smooth field of cubic bases $\boldsymbol{e}$ such that a linear transformation

$$\boldsymbol{I}(X, Y) : \mathscr{B}_X \to \mathscr{B}_Y \tag{1.25}$$

is a material isomorphism of X with Y if and only if

$$[\boldsymbol{I}(X, Y)]\boldsymbol{e}(X) = \boldsymbol{e}(Y) \tag{1.26}$$

in the sense of set theory. The material isomorphism $\boldsymbol{I}(X, Y)$ is not unique in this case, since the symmetry group is not the trivial group. In

[1] Only this kind of body is contemplated in the theory of inhomogeneities developed by W. Noll in NFTM, Sect. 27, and later in the paper "Materially Uniform Simple Bodies with Inhomogeneities", *Arch. Rational Mech. Anal.* **27**, 1–32 (1967), reprinted in *Continuum Theory of Inhomogeneities in Simple Bodies*, New York, Springer-Verlag, (1968).

general, a field of material isomorphisms $\boldsymbol{I}(X, \cdot)$ from any reference point X need not be a global smooth field. We define a *Moebius crystal* to be a crystal that cannot be equipped with a smooth global field $\boldsymbol{I}(X, \cdot)$. (This condition clearly is independent of the choice of X.)

A typical Moebius cubic crystal can be formed in the following way: Consider a (thick) strip of a homogeneous cubic crystal having its cubic basis $\boldsymbol{e}$ parallel to the edges of the strip. Now twist one edge of the strip through 180°, and then form a Moebius band in the usual way. The resulting crystal body is a Moebius cubic crystal. Clearly, this particular Moebius crystal is a locally homogeneous but globally inhomogeneous body. The material tangent bundles of all Moebius crystals are non-trivial.

Example 3. A Locally Inhomogeneous Transversely Isotropic Laminated Body

A body-point X is *transversely isotropic* if there exist an inner product $g(X)$ on $\mathscr{B}_X$ and a unit vector $\boldsymbol{a}(X)$ such that the symmetry group of X consists in all rotations about $\boldsymbol{a}(X)$. We call the vector $\boldsymbol{a}(X)$ an *axis of symmetry* of X. If a local configuration $\boldsymbol{v}$ carries $g(X)$ into the Euclidean inner product of $\mathscr{R}^3$, then the symmetry group relative to $\boldsymbol{v}$ is the subgroup of the rotational group consisting of all rotations about $\boldsymbol{v}(\boldsymbol{a}(X))$. Hence a transversely isotropic material is a solid, and the local configurations that carry $g(X)$ to the Euclidean inner product on $\mathscr{R}^3$ are the undistorted states.

Let $\mathscr{B}$ be a transversely isotropic body. Then there exist fields g and $\boldsymbol{a}$ on $\mathscr{B}$ such that a transformation $\boldsymbol{I}(X, Y)$ of the form (1.25) is a material isomorphism if and only if it carries $g(X)$ and $\boldsymbol{a}(X)$ to $g(Y)$ and $\boldsymbol{a}(Y)$, respectively. The condition of smoothness on $\mathscr{B}$ implies that g and $\boldsymbol{a}$ are smooth fields.

Now we consider a specific transversely isotropic body $\mathscr{B}$. We assume that $\mathscr{B}$ has a global undistorted configuration $\boldsymbol{\kappa}$ in which it takes the shape of a (thick) spherical shell such that the axis of symmetry $\boldsymbol{a}$ is in the radial direction at all points. We can visualize this body as the union of a family of infinitely thin homogeneous spherical laminae having their axes of symmetry in the normal direction of the laminae. Thus $\mathscr{B}$ deserves to be called a *laminated body*. (In § 7 we shall give a general, formal definition of a laminated body.)

The laminated body $\mathscr{B}$ is a locally inhomogeneous body since it does not have any homogeneous subbody. Indeed, if $\mathscr{U}$ is a homogeneous subbody of $\mathscr{B}$, then there exists a homogeneous undistorted configuration χ which carries the metric $g(X)$ and the vector $\boldsymbol{a}(X)$ to the Euclidean metric and a constant vector for all $X \in \mathscr{U}$. Comparing the configuration χ with the configuration $\boldsymbol{\kappa}$ described above, we see that the deformation from $\boldsymbol{\kappa}$ to $\boldsymbol{\mu}$ must be rigid since it preserves the Euclidean metric. But that deformation must also carry a radial vector field on $\boldsymbol{\kappa}(\mathscr{U})$ into a constant vector field on $\chi(\mathscr{U})$. Thus it is impossible. Consequently the configuration χ does not exist.

Another important condition about the laminated body $\mathscr{B}$ is that its material tangent bundles are non-trivial. To see this fact, suppose that there exist a global material chart $(\mathscr{B}, \boldsymbol{v})$. Without loss of generality we can choose $\boldsymbol{v}$ to be undistorted, so that $\boldsymbol{v}_X$ carries $g(X)$ and $\boldsymbol{a}(X)$ into the Euclidean metric and a constant vector $\boldsymbol{v}$ for all $X \in \mathscr{B}$. We choose an arbitrary non-zero vector $\boldsymbol{u} \in \mathscr{R}^3$ perpendicular to $\boldsymbol{v}$. Then $\boldsymbol{v}^{-1}(\boldsymbol{v})$ is a smooth non-zero vector field on $\mathscr{B}$ orthogonal to the vector field $\boldsymbol{a}$. In the configuration $\boldsymbol{\kappa}$, $\boldsymbol{\kappa}_*(\boldsymbol{v}^{-1}(\boldsymbol{v}))$ is then a smooth non-zero vector field which is tangent to the spherical laminae at every point. But since a sphere is not a parallelizable space, such a vector field does not exist, and hence neither does $\boldsymbol{v}$.

Example 4. A Locally Homogeneous Fluid Crystal Body

We consider a fluid-crystal possessing a preferred plane. Let X be such a fluid-crystal point. Then there exists a plane $\mathscr{P}(X)$ in $\mathscr{B}_X$ such that the symmetry group of X consists in all unimodular transformations of $\mathscr{B}_X$ which leave $\mathscr{P}(X)$ invariant. Let $\mathscr{B}$ be a body made up of such fluid-crystal points. Then $\mathscr{B}$ can be equipped with a smooth volume tensor field $\mathbf{V}$ and a smooth distribution $\mathscr{P}$ such that a transformation $\boldsymbol{I}(X, Y)$ is a material isomorphism of X with Y if and only if it maps $\mathbf{V}(X)$ and $\mathscr{P}(X)$ into $\mathbf{V}(Y)$ and $\mathscr{P}(Y)$, respectively.

Now we assume that in a certain configuration $\boldsymbol{\kappa}$ the body takes the shape of a (thick) spherical shell as in the previous example. We assume also that $\mathscr{P}$ is tangent to the spherical surfaces at every point and that $\mathbf{V}$ coincides with the Euclidean volume tensor of the physical space. In this case, the body is locally homogeneous, because there exist volume-preserving deformations that carry sectors of the spherical laminae in $\boldsymbol{\kappa}(\mathscr{B})$ into parallel plane laminae. For example, if we use a spherical co-ordinate system (R, Θ, Φ) in $\boldsymbol{\kappa}(\mathscr{U})$ and a rectangular Cartesian co-ordinate system (x, y, z) in $\chi(\mathscr{U})$ for a sector $\mathscr{U}$ of $\mathscr{B}$, then we can choose the deformation to be

$$x = \tfrac{1}{3}R^3, \quad y = -\cos\Theta, \quad z = \Phi, \tag{1.27}$$

where Θ and Φ are restricted so as to make the deformation one-to-one. It can be checked easily that (1.27) is volume-preserving and carries a sphere $R = \text{constant}$ into a plane $x = \text{constant}$. Since $\boldsymbol{V}$ and $\mathscr{P}$ are constant in χ, $(\mathscr{U}, \chi_*)$ is a material chart, and $\mathscr{U}$ is homogeneous. This deformation does not imply that the whole body $\mathscr{B}$ is globally homogeneous, however. In fact, there is no way to flatten a whole sphere into a plane without making the points overlap.

By the same argument as in the previous example, we see that the material tangent bundles of this locally homogeneous body are again nontrivial. This fact gives an important counter-example for the next section.

2. The Bundle of Reference Frames

In the preceding section we introduced and explained the concept of the material tangent bundle of a body with respect to a material atlas. In the theory of fibre bundles, every bundle has an *associated principal bundle*. For example, the associated principal bundle of the geometric tangent bundle $\mathscr{T}(\mathscr{B})$ is the *bundle of frames* $\mathscr{E}(\mathscr{B})$. For the material tangent bundle $\mathscr{T}(\mathscr{B}, \mathfrak{A})$ the associated principal bundle is the *bundle of reference frames* $\mathscr{E}(\mathscr{B}, \mathfrak{A})$. We proceed to explain these concepts in detail.

We define first the bundle of frames $\mathscr{E}(\mathscr{B})$. If $X \in \mathscr{B}$, a *frame* at X is an ordered basis

$$\boldsymbol{f}(X) = \{\boldsymbol{f}_1(X), \boldsymbol{f}_2(X), \boldsymbol{f}_3(X)\} \tag{2.1}$$

for the tangent space $\mathscr{B}_X$. We denote the collection of all frames at X by $\mathscr{E}_X$. Then we put

$$\mathscr{E}(\mathscr{B}) = \bigcup_{X \in \mathscr{B}} \mathscr{E}_X. \tag{2.2}$$

The general linear group $\mathfrak{L}$ acts as a transformation group of $\mathscr{E}_X$ by right multiplication in a natural way. If $\boldsymbol{A} \in \mathfrak{L}$, say

$$\boldsymbol{A} = [A^i_j], \tag{2.3}$$

and if $f(X) \in \mathscr{E}_X$ is given by (2.1), we define

$$R_A f(X) \equiv f(X)A = \{A^i_j f_i(X),\ j = 1, 2, 3\}. \tag{2.4}$$

The reader will verify easily that this operation satisfies the axioms of transformation by right multiplication.

Each $e(X) \in \mathscr{E}_X$ gives rise to an isomorphism of $\mathscr{E}_X$ with $\mathfrak{L}$,

$$\eta_X : \mathscr{E}_X \to \mathfrak{L}, \tag{2.5}$$

defined by

$$\eta_X(f(X)) \equiv [f^i_j], \tag{2.6}$$

where f^i_j are the components of $\{f_j(X)\}$ relative to $\{e_j(X)\}$, *viz*

$$f_j(X) = f^i_j e_i(X). \tag{2.7}$$

In terms of the operation R defined by (2.4), (2.7) means

$$R_{\eta_X(f(X))}\, e(X) = f(X) \tag{2.8}$$

for all $f(X) \in \mathscr{E}_X$.

The set $\mathscr{E}_X$ is called the *fibre* of $\mathscr{E}(\mathscr{B})$ at the base point X. Under the isomorphism η_X, the fibre $\mathscr{E}_X$ becomes a copy of $\mathfrak{L}$, which is the *fibre space* of $\mathscr{E}(\mathscr{B})$. As usual, we define a manifold structure on $\mathscr{E}_X$ by requiring the set-isomorphism η_X to be a diffeomorphism. With respect to this induced manifold structure, η_X is then called the *fibre isomorphism* at X induced by the frame $e(X)$.

A smooth field η of fibre isomorphisms defined on a subbody $\mathscr{U} \subset \mathscr{B}$ forms a *bundle chart* $(\mathscr{U}, \eta)$ of $\mathscr{E}(\mathscr{B})$, and the collection of all such bundle charts forms the *bundle atlas* of $\mathbf{E}(\mathscr{B})$. The *co-ordinate transformation* from one bundle chart $(\mathscr{U}, \eta)$ to another bundle chart $(\bar{\mathscr{U}}, \bar{\eta})$ is the smooth field G on the overlap $\mathscr{U} \cap \bar{\mathscr{U}}$ that is given by

$$G_X(\eta, \bar{\eta}) = \eta_X \circ \bar{\eta}_X^{-1}, \quad X \in \mathscr{U} \cap \bar{\mathscr{U}}, \tag{2.9}$$

the values of $G(\eta, \bar{\eta})$ being transformations of $\mathfrak{L}$. From (2.6) and (2.9) we have

$$[G_X(\eta, \bar{\eta})](A) = \eta_X(\bar{e}(X))A \quad \forall A \in \mathfrak{L}, \tag{2.10}$$

where $\bar{e}(X)$ is the frame that induces the fibre isomorphism $\bar{\eta}_X$, and where the operation on the right-hand side of (2.10) is the usual group product for $\mathfrak{L}$. In view of (2.10) we can regard $G_X(\eta, \bar{\eta})$ as an element in $\mathfrak{L}$. In this sense $G(\eta, \bar{\eta})$ is a field:

$$G(\eta, \bar{\eta}) : \mathscr{U} \cap \bar{\mathscr{U}} \to \mathfrak{L}. \tag{2.11}$$

Hence the *structure group* of $\mathscr{E}(\mathscr{B})$ is $\mathfrak{L}$.

Since the fibre space of $\mathscr{E}(\mathscr{B})$ is also $\mathfrak{L}$, it coincides with the structure group. In general, a fibre bundle satisfying this condition is called a *principal bundle*. The principal bundle $\mathscr{E}(\mathscr{B})$ is also the *associated principal bundle* of the tangent bundle $\mathscr{T}(\mathscr{B})$. We proceed to explain this concept in more detail.

Recall that a bundle chart $(\mathscr{U}, \boldsymbol{v})$ for $\mathscr{T}(\mathscr{B})$ is a pair consisting in a subbody $\mathscr{U}$ and a smooth field $\boldsymbol{v}$ of local configurations on $\mathscr{U}$. Such a bundle chart $(\mathscr{U}, \boldsymbol{v})$ can be put in one-to-one correspondence with a bundle chart $(\mathscr{U}, \boldsymbol{\eta})$ for $\mathscr{E}(\mathscr{B})$ in the following way: Given $(\mathscr{U}, \boldsymbol{v})$, we put

$$\boldsymbol{e}_i(X) = \boldsymbol{v}_X^{-1}(\boldsymbol{e}_i), \qquad i = 1, 2, 3, \tag{2.12}$$

for any $X \in \mathscr{U}$, where $\{\boldsymbol{e}_i\}$ is the fixed Cartesian frame of reference for the physical space. Then we define $\boldsymbol{\eta}_X$ as the fibre isomorphism induced by the frame

$$\boldsymbol{e}(X) = \{\boldsymbol{e}_1(X),\ \boldsymbol{e}_2(X),\ \boldsymbol{e}_3(X)\} \tag{2.13}$$

with $\boldsymbol{e}_i(X)$ given by (2.12). Conversely, given $(\mathscr{U}, \boldsymbol{\eta})$, with $\boldsymbol{\eta}_X$ being induced by the frame $\boldsymbol{e}(X)$ for all $X \in \mathscr{U}$, we define

$$\boldsymbol{v}_X(\boldsymbol{e}_i(X)) \equiv \boldsymbol{e}_i \tag{2.14}$$

and extend the mapping $\boldsymbol{v}_X$ to the whole tangent space $\mathscr{B}_X$ by linearity. Thus each $(\mathscr{U}, \boldsymbol{v})$ determines a unique $(\mathscr{U}, \boldsymbol{\eta})$, and conversely, through the relations (2.12) and (2.14).

Next we show that the co-ordinate transformations for the corresponding bundle charts in $\mathscr{T}(\mathscr{B})$ and $\mathscr{E}(\mathscr{B})$ are the same fields. Let $(\mathscr{U}, \boldsymbol{v})$ and $(\bar{\mathscr{U}}, \bar{\boldsymbol{v}})$ be two bundle charts for $\mathscr{T}(\mathscr{B})$, and let $(\mathscr{U}, \boldsymbol{\eta})$ and $(\bar{\mathscr{U}}, \bar{\boldsymbol{\eta}})$ be their corresponding bundle charts for $\mathscr{E}(\mathscr{B})$. As before, we assume that $\boldsymbol{\eta}_X$ and $\bar{\boldsymbol{\eta}}_X$ are induced by the frames $\boldsymbol{e}(X)$ and $\bar{\boldsymbol{e}}(X)$, respectively. Then from (2.10) the co-ordinate transformation from $\boldsymbol{\eta}_X$ to $\bar{\boldsymbol{\eta}}_X$ has the component matrix

$$\boldsymbol{G}_X(\boldsymbol{\eta}, \bar{\boldsymbol{\eta}}) = [\bar{e}^i_j(X)], \tag{2.15}$$

where $\bar{e}^i_j(X)$ are the components of $\{\bar{\boldsymbol{e}}_j(X)\}$ with respect to the frame $\boldsymbol{e}(X)$. But from (1.17) and (2.14), the component matrix of the co-ordinate transformation from $\boldsymbol{v}_X$ to $\bar{\boldsymbol{v}}_X$ is also equal to $[\bar{e}^i_j(X)]$:

$$\boldsymbol{G}_X(\boldsymbol{v}, \bar{\boldsymbol{v}}) = [\bar{e}^i_j(X)]. \tag{2.16}$$

Hence the co-ordinate transformations $\boldsymbol{G}(\boldsymbol{v}, \bar{\boldsymbol{v}})$ and $\boldsymbol{G}(\boldsymbol{\eta}, \bar{\boldsymbol{\eta}})$ are the same field on $\mathscr{U} \cap \bar{\mathscr{U}}$:

$$G(\boldsymbol{v}, \bar{\boldsymbol{v}}) = G(\boldsymbol{\eta}, \bar{\boldsymbol{\eta}}). \tag{2.17}$$

The one-to-one correspondence for the bundle charts $(\mathscr{U}, \boldsymbol{v})$ and $(\mathscr{U}, \boldsymbol{\eta})$ and the equation (2.17) for the co-ordinate transformations establish $\mathscr{E}(\mathscr{B})$ as the *associated principal bundle* of $\mathscr{T}(\mathscr{B})$.

The bundle of frames $\mathscr{E}(\mathscr{B})$, like the tangent bundle $\mathscr{T}(\mathscr{B})$, is a geometric structure of the body manifold and is independent of the response functions of the body-points. As explained in the previous section, the material structure of the body can be characterized by the material tangent bundles, or equivalently, by their associated principal bundles. The associated principal bundles of the material tangent bundles are the *bundles of reference frames*. We proceed to explain this concept.

Let $\mathfrak{A}$ be a particular material atlas, and let $\mathscr{T}(\mathscr{B}, \mathfrak{A})$ be the material tangent bundle with respect to $\mathfrak{A}$. Then $\mathfrak{A}$ corresponds to a particular relative response function $\boldsymbol{H}_{\mathfrak{A}}$. We say that a frame $\boldsymbol{e}(X)$ at X is a *reference frame* relative to $\mathfrak{A}$ if the response function $\boldsymbol{H}_{\boldsymbol{v}_X}$ relative to the local configuration $\boldsymbol{v}_X$ given by (2.14) coincides with $\boldsymbol{H}_{\mathfrak{A}}$:

$$\boldsymbol{H}_{\boldsymbol{v}_X} = \boldsymbol{H}_{\mathfrak{A}}. \tag{2.18}$$

This condition is equivalent to

$$\boldsymbol{e}(X) = \boldsymbol{v}_{\alpha, X}^{-1}(\boldsymbol{e}) \tag{2.19}$$

for some material chart $(\mathscr{U}_\alpha, \boldsymbol{v}_\alpha)$ in $\mathfrak{A}$ covering the body-point X. We denote the collection of all reference frames at X relative to $\mathfrak{A}$ by $\mathscr{E}_X(\mathfrak{A})$. Then we put

$$\mathscr{E}(\mathscr{B}, \mathfrak{A}) = \bigcup_{X \in \mathscr{B}} \mathscr{E}_X(\mathfrak{A}). \tag{2.20}$$

Since

$$\mathscr{E}_X(\mathfrak{A}) \subset \mathscr{E}_X \qquad \forall X \in \mathscr{B}, \tag{2.21}$$

from (2.2) and (2.20) we have

$$\mathscr{E}(\mathscr{B}, \mathfrak{A}) \subset \mathscr{E}(\mathscr{B}). \tag{2.22}$$

We now consider the structure of the bundle $\mathscr{E}(\mathscr{B}, \mathfrak{A})$ in detail. First, we show that the fibre $\mathscr{E}_X(\mathfrak{A})$ is a copy of the symmetry group $\mathfrak{G}_{\mathfrak{A}}$. This fact is more or less obvious, since from (2.18) a frame $\boldsymbol{f}(X)$ at X is a reference frame relative to $\mathfrak{A}$ if and only if the right transformation $\boldsymbol{P}$ from the reference frame $\boldsymbol{e}(X)$ to $\boldsymbol{f}(X)$ belongs to the symmetry group $\mathfrak{G}_{\mathfrak{A}}$. Thus the fibre isomorphism $\boldsymbol{\eta}_X$ induced by $\boldsymbol{e}(X)$, when restricted to the subset $\mathscr{E}_X(\mathfrak{A})$ of $\mathscr{E}_X$, is an isomorphism of $\mathscr{E}_X(\mathfrak{A})$ with $\mathfrak{G}_{\mathfrak{A}}$:

$$\boldsymbol{\eta}_X|_{\mathscr{E}_X(\mathfrak{A})} \equiv \boldsymbol{\eta}_X(\mathfrak{A}) : \mathscr{E}_X(\mathfrak{A}) \to \mathfrak{G}_{\mathfrak{A}}. \tag{2.23}$$

This isomorphism establishes $\mathfrak{G}_{\mathfrak{A}}$ as the fibre space of the bundle $\mathscr{E}(\mathscr{B}, \mathfrak{A})$.

Next we consider the bundle charts for $\mathscr{E}(\mathscr{B}, \mathfrak{A})$. Since $\mathfrak{A}$ is a subset of the bundle atlas of $\mathscr{T}(\mathscr{B})$, it corresponds to a set of bundle charts for $\mathscr{E}(\mathscr{B})$, say

$$\mathfrak{A}' = \{(\mathscr{U}_\alpha, \boldsymbol{\eta}_\alpha),\ \alpha \in I\}. \tag{2.24}$$

We put

$$\mathfrak{A}_{\mathscr{E}} \equiv \{(\mathscr{U}_\alpha, \boldsymbol{\eta}_\alpha(\mathfrak{A})),\ \alpha \in I\}, \tag{2.25}$$

where $\boldsymbol{\eta}_\alpha(\mathfrak{A})$ denotes the restriction of $\boldsymbol{\eta}_\alpha$ to $\mathscr{E}(\mathscr{U}_\alpha, \mathfrak{A})$ as shown in (2.23). Then $\mathfrak{A}_{\mathscr{E}}$ is the bundle atlas for $\mathscr{E}(\mathscr{B}, \mathfrak{A})$. We call $\mathfrak{A}_{\mathscr{E}}$ the *reference atlas* and its members the *reference charts* for $\mathscr{B}$ relative to $\mathfrak{A}$. From (2.17) the co-ordinate transformations for the reference atlas $\mathfrak{A}_{\mathscr{E}}$ are the same as those for the material atlas $\mathfrak{A}$. This fact establishes $\mathfrak{G}_{\mathfrak{A}}$ as the structure group of $\mathscr{E}(\mathscr{B}, \mathfrak{A})$, then $\mathscr{E}(\mathscr{B}, \mathfrak{A})$ as a principal bundle, and finally $\mathscr{E}(\mathscr{B}, \mathfrak{A})$ as the associated principal bundle of $\mathscr{T}(\mathscr{B}, \mathfrak{A})$.

Next we determine the transformation rule for the bundle of reference frames with respect to a change of material atlas. Let $\mathfrak{B}$ be another material atlas of $\mathscr{B}$, and let it be related to $\mathfrak{A}$ by

$$\mathfrak{B} = \boldsymbol{P}\mathfrak{A}, \tag{2.26}$$

where the operation $\boldsymbol{P}\mathfrak{A}$ is given by (1.4). Then the reader will verify easily that $\boldsymbol{f}(X)$ is a reference frame relative to $\mathfrak{B}$ if and only if $\boldsymbol{e}(X)$ as given by

$$\boldsymbol{e}(X) = \boldsymbol{R}_{\boldsymbol{P}}\boldsymbol{f}(X) \tag{2.27}$$

is a reference frame relative to $\mathfrak{A}$. Consequently the bundle spaces of $\mathscr{E}(\mathscr{B}, \mathfrak{A})$ and $\mathscr{E}(\mathscr{B}, \mathfrak{B})$ are related by

$$\mathscr{E}(\mathscr{B}, \mathfrak{B}) = \boldsymbol{R}_{\boldsymbol{P}^{-1}}\mathscr{E}(\mathscr{B}, \mathfrak{A}). \tag{2.28}$$

Further, since every frame is a reference frame with respect to some material atlas, the bundle space of $\mathscr{E}(\mathscr{B})$ is the disjoint union of the bundle spaces of all bundles of reference frames,

$$\mathscr{E}(\mathscr{B}) = \bigcup_{\mathfrak{A}} \mathscr{E}(\mathscr{B}, \mathfrak{A}). \tag{2.29}$$

The transformation rule for the reference atlas can be derived in a similar way. Let $\mathfrak{A}_{\mathscr{E}}$ be given by (2.25). Then $\mathfrak{B}_{\mathscr{E}}$ is given by

$$\mathfrak{B}_{\mathscr{E}} = \{(\mathscr{U}_\alpha, \boldsymbol{C_P} \circ \boldsymbol{\eta}_\alpha(\mathfrak{A}) \circ \boldsymbol{R_P}),\ \alpha \in I\}, \tag{2.30}$$

where $\boldsymbol{C_P}$ denotes the operation of conjugation by $\boldsymbol{P}$ on $\mathfrak{L}$, *i.e.*,

$$\boldsymbol{C_P}(A) = \boldsymbol{P}A\boldsymbol{P}^{-1} \qquad \forall A \in \mathfrak{L}. \tag{2.31}$$

Note. For any body-point $X \in \mathscr{U}_\alpha$, $\boldsymbol{R_P}$ maps $\mathscr{E}_X$ onto $\mathscr{E}_X(\mathfrak{A})$, then $\boldsymbol{\eta}_{\alpha, X}$ maps $\mathscr{E}_X(\mathfrak{A})$ onto $\mathfrak{G}_{\mathfrak{A}}$, and finally[1] $\boldsymbol{C_P}$ maps $\mathfrak{G}_{\mathfrak{A}}$ onto $\mathfrak{G}_{\mathfrak{B}}$. Thus the composition $\boldsymbol{C_P} \circ \boldsymbol{\eta}_\alpha(\mathfrak{A}) \circ \boldsymbol{R_P}$ is a field of fibre isomorphisms on $\mathscr{U}_\alpha$ for the bundle $\mathscr{E}(\mathscr{B}, \mathfrak{B})$.

Since the bundle charts of a fibre bundle and those of the associated principal bundle are in one-to-one correspondence, a fibre bundle is trivial if and only if its associated principal bundle is trivial. Since the bundle charts of the principal bundle $\mathscr{E}(\mathscr{B}, \mathfrak{A})$ are characterized by the condition (2.8), $\mathscr{E}(\mathscr{B}, \mathfrak{A})$ is trivial if and only if $\mathscr{B}$ has a global smooth field $\boldsymbol{e}$ of reference frames relative to $\mathfrak{A}$. Such a global field corresponds to a *cross section* of the bundle $\mathscr{E}(\mathscr{B}, \mathfrak{A})$, *i.e.*, $\boldsymbol{e}$ can be regarded as a smooth mapping

$$\boldsymbol{e} : \mathscr{B} \to \mathscr{E}(\mathscr{B}, \mathfrak{A}) \tag{2.32}$$

satisfying the condition

$$\pi \circ \boldsymbol{e} = \mathrm{id}_{\mathscr{B}}, \tag{2.33}$$

where π is the canonical projection map

$$\pi : \mathscr{E}(\mathscr{B}, \mathfrak{A}) \to \mathscr{B} \tag{2.34}$$

which maps the fibre $\mathscr{E}_X(\mathfrak{A})$ into the base point X for all $X \in \mathscr{B}$.

From the examples given at the end of the preceding section we know that the material tangent bundles, and hence the bundles of reference frames, need not be trivial in general. Thus a global cross section of $\mathscr{E}(\mathscr{B}, \mathfrak{A})$ need not exist. However, for every bundle chart $(\mathscr{U}, \boldsymbol{\eta})$ there is always a canonical cross section $\boldsymbol{e}$ of $\mathscr{E}(\mathscr{U}, \mathfrak{A})$ defined by

$$\boldsymbol{e}(X) \equiv \boldsymbol{\eta}_X^{-1}(\mathbf{1}) \qquad \forall X \in \mathscr{U}. \tag{2.35}$$

Previously we spoke of $\boldsymbol{\eta}_X$ as the fibre isomorphism induced by the frame $\boldsymbol{e}(X)$ (*cf.* (2.8)). In view of the condition (2.35), we call $\boldsymbol{e}$ the *section of*

---

[1] *Cf.* the transformation rule (IV.1.33) for the symmetry group.

the identity for the field of fibre isomorphisms $\boldsymbol{\eta}$. The existence of the local sections of the identity means that $\mathscr{E}(\mathscr{B}, \mathfrak{A})$, and hence $\mathscr{T}(\mathscr{B}, \mathfrak{A})$, are *locally trivial*. (For that matter every fibre bundle is locally trivial.)

The importance of the associated principal bundle for a fibre bundle lies in the fact that the *structural connections* on a fibre bundle can be put in one-to-one correspondence with those on the associated principal bundle. The structural connections on $\mathscr{T}(\mathscr{B}, \mathfrak{A})$ and $\mathscr{E}(\mathscr{B}, \mathfrak{A})$ are the *material connections* of the body, which we shall consider in the next section.

3. The Concept of a Material Connection

We review first the concept of a *connection* on the tangent bundle $\mathscr{T}(\mathscr{B})$. Recall that $\mathscr{T}(\mathscr{B})$ is formed by the disjoint union

$$\mathscr{T}(\mathscr{B}) = \bigcup_{X \in \mathscr{B}} \mathscr{B}_X. \tag{1.12}$$

The *projection map*

$$\pi : \mathscr{T}(\mathscr{B}) \to \mathscr{B} \tag{3.1}$$

is defined by

$$\pi(\mathscr{B}_X) = X \quad \forall X \in \mathscr{B}, \tag{3.2}$$

so that

$$\mathscr{B}_X = \pi^{-1}(X) \quad \forall X \in \mathscr{B}. \tag{3.3}$$

Let $\mathscr{U}$ denote, as usual, a subbody of $\mathscr{B}$. Then a *cross section* of $\mathscr{T}(\mathscr{U})$ is a smooth mapping

$$\boldsymbol{u} : \mathscr{U} \to \mathscr{T}(\mathscr{U}) \tag{3.4}$$

such that

$$\pi \circ \boldsymbol{u} = \mathrm{id}_{\mathscr{U}}. \tag{3.5}$$

Such a cross section is nothing but a smooth vector field on $\mathscr{U}$. A connection is a field of datum spaces on $\mathscr{T}(\mathscr{B})$ relative to which we can determine the slope of a cross section such as $\boldsymbol{u}$.

Specifically, let $(X, \boldsymbol{v})$ be a point of $\mathscr{T}(\mathscr{B})$. We define the *vertical space* $\mathscr{V}_{(X, \boldsymbol{v})}$ at $(X, \boldsymbol{v})$:

$$\mathscr{V}_{(X, \boldsymbol{v})} \equiv \ker \pi_{*(X, \boldsymbol{v})}. \tag{3.6}$$

From (3.3) $\mathscr{V}_{(X, \boldsymbol{v})}$ is the tangent space of the fibre $\mathscr{B}_X$. We define a *datum space* or *horizontal space* at $(X, \boldsymbol{v})$ as a subspace $\mathscr{H}_{(X, \boldsymbol{v})}$ in

$\mathscr{T}(\mathscr{B})_{(X,v)}$ such that $\mathscr{T}(\mathscr{B})_{(X,v)}$ has the following decomposition as a direct sum:

$$\mathscr{T}(\mathscr{B})_{(X,v)} = \mathscr{V}_{(X,v)} \oplus \mathscr{H}_{(X,v)}. \tag{3.7}$$

Unlike the vertical space $\mathscr{V}_{(X,v)}$, the horizontal space $\mathscr{H}_{(X,v)}$ is not unique. A smooth field $\mathscr{H}$ of horizontal spaces is called a *connection* on $\mathscr{T}(\mathscr{B})$.

The function of a connection is to determine the slope of a cross section. Of course, we regard the horizontal space $\mathscr{H}_{(X,v)}$ as the space of slope zero at every point $(X, \boldsymbol{v}) \in \mathscr{T}(\mathscr{B})$. Then the slope of a cross section $\boldsymbol{u}$ with respect to the connection is defined as the slope of the tangent space $\boldsymbol{u}_{*X}(\mathscr{B}_X)$ of $\boldsymbol{u}$ at $(X, \boldsymbol{u}(X))$ with respect to the horizontal space $\mathscr{H}_{(X,\boldsymbol{u}(X))}$. We proceed to explain these concepts in more detail.

We focus our attention on the tangent space $\mathscr{T}(\mathscr{B})_{(X,v)}$ at a particular point $(X, \boldsymbol{v}) \in \mathscr{T}(\mathscr{B})$. Let $\boldsymbol{u}$ be a cross section passing through $(X, \boldsymbol{v})$, *i.e.*,

$$\boldsymbol{u}(X) = \boldsymbol{v}. \tag{3.8}$$

For simplicity, let us denote the tangent space of $\boldsymbol{u}$ at $(X, \boldsymbol{v})$ by $\mathscr{U}_{(X,v)}$:

$$\mathscr{U}_{(X,v)} \equiv \boldsymbol{u}_{*X}(\mathscr{B}_X), \tag{3.9}$$

which is the subspace in $\mathscr{T}(\mathscr{B})_{(X,v)}$ formed by the $\boldsymbol{u}$-lifts $\boldsymbol{u}_{*X}(\boldsymbol{w})$ of base vectors $\boldsymbol{w} \in \mathscr{B}_X$. Clearly, $\mathscr{U}_{(X,v)}$ is also a horizontal space at $(X, \boldsymbol{v})$, but generally it is different from the assigned horizontal space $\mathscr{H}_{(X,v)}$ of the connection $\mathscr{H}$. We can express the relation of $\mathscr{H}_{(X,v)}$, $\mathscr{V}_{(X,v)}$, $\mathscr{U}_{(X,v)}$, and $\mathscr{B}_X$ by the following sketch:

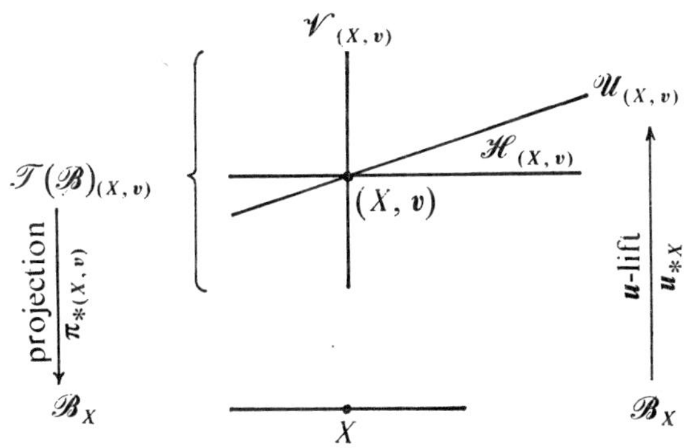

Let

$$\delta_{(X,v)} : \mathscr{T}(\mathscr{B})_{(X,v)} \to \mathscr{V}_{(X,v)} \tag{3.11}$$

denote the projection of $\mathscr{T}(\mathscr{B})_{(X,v)}$ onto $\mathscr{V}_{(X,v)}$ with respect to the decomposition (3.7). Then we define the slope of $\mathscr{U}_{(X,v)}$ with respect to $\mathscr{H}_{(X,v)}$ as the linear map

$$Du(X) : \mathscr{B}_X \to \mathscr{V}_{(X,v)} \tag{3.12}$$

given by

$$Du(X) = \delta_{(X,v)} \circ u_{*X}\,. \tag{3.13}$$

This equation has the usual meaning of the slope: $Du(X)$ assigns to each base vector $w \in \mathscr{B}_X$ a vertical vector $[Du(X)](w)$, which is the vertical component of the u-lift $u_{*X}(w)$ of w. The only difference between the definition (3.13) and the one in calculus is that there the datum space is taken for granted, being the field of tangent spaces of the constant sections. In the present case, the bundle $\mathscr{T}(\mathscr{B})$ has no such canonical "constant sections". In order to determine the slope of a cross section u, a connection $\mathscr{H}$ must be assigned. Of course, the slope Du thus determined depends on the choice of $\mathscr{H}$.

Now since the vertical space $\mathscr{V}_{(X,v)}$ is canonically isomorphic to the fibre $\mathscr{B}_X$, the linear map $Du(X)$ can be regarded as a tensor of type (1, 1) over $\mathscr{B}_X$. In this sense Du is a tensor field on $\mathscr{B}$, called the *covariant derivative* of u with respect to $\mathscr{H}$. The covariant derivative differs from the usual partial derivative in one fundamental way: The field of datum spaces $\mathscr{H}$ on $\mathscr{T}(\mathscr{B})$ generally fails to be the field of tangent spaces of a family of datum cross sections. Those connections which coincide with a field of tangent spaces of a family of cross sections are said to be *integrable*. While we have not yet given a definition for a *material connection*, we remark first that a body $\mathscr{B}$ in general need not have any integrable material connection.

A connection $\mathscr{H}$ gives rise to a parallelism along any curve in $\mathscr{B}$. Let the curve be

$$X : (a, b) \to \mathscr{B}. \tag{3.14}$$

Then we say that a curve

$$\lambda : (a, b) \to \mathscr{T}(\mathscr{B}) \tag{3.15}$$

is a *horizontal lift* of X with respect to $\mathscr{H}$ if the following two conditions are satisfied:

$$\pi \circ \lambda(t) = X(t), \tag{3.16}$$

and

$$\dot{\lambda}(t) \in \mathscr{H}_{\lambda(t)} \tag{3.17}$$

for all $t \in (a, b)$. The first condition says that $\boldsymbol{\lambda}(t)$ belongs to the fibre at $X(t)$ for all t, and the second condition says that $\boldsymbol{\lambda}$ is a horizontal curve with respect to $\mathscr{H}$. Since the curve $X(t)$ is 1-dimensional, and since the connection is smooth, a unique horizontal lift passing through any initial point in a fibre of $X(t)$ exists. By following the horizontal lifts of $X(t)$ passing through initial points in $\mathscr{B}_{X(t_0)}$, we obtain the *parallel transports* $\boldsymbol{\rho}_{t_0,t}$ along $X(t)$ with respect to $\mathscr{H}$:

$$\boldsymbol{\rho}_{t_0,t} : \mathscr{B}_{X(t_0)} \to \mathscr{B}_{X(t)}, \; t \in (a, b) \tag{3.18}$$

such that for each initial point $(X(t_0), \boldsymbol{v}(t_0))$ the curve

$$\boldsymbol{\lambda}(t) \equiv (X(t), \boldsymbol{\rho}_{t_0,t}(\boldsymbol{v}(t_0))), \; t \in (a, b) \tag{3.19}$$

is the horizontal lift passing through $(X(t_0), \boldsymbol{v}(t_0))$. From the uniqueness and the smoothness of $\boldsymbol{\lambda}$ with respect to the initial point, the parallel transport $\boldsymbol{\rho}_{t_0,t}$ is a diffeomorphism of $\mathscr{B}_{X(t_0)}$ with $\mathscr{B}_{X(t)}$ for each $t \in (a, b)$.

The horizontal lifts of $X(t)$ are a family of datum curves covering the cylinder

$$\mathscr{C}_X \equiv \pi^{-1}(X(\cdot)) = \bigcup_{t \in (a,b)} \mathscr{B}_{X(t)}, \tag{3.20}$$

each $\boldsymbol{\lambda}$ being a horizontal cross section of $\mathscr{C}_X$. Let $\hat{\boldsymbol{u}}$ be an arbitrary cross section of $\mathscr{C}_X$. Then $\hat{\boldsymbol{u}}$ is a vector field on the curve $X(t)$. We define the slope of $\hat{\boldsymbol{u}}$ with respect to $\mathscr{H}$ to be the vertical component of $\partial\hat{\boldsymbol{u}}$:

$$\frac{D\hat{\boldsymbol{u}}}{Dt} = \boldsymbol{\delta}_{\hat{\boldsymbol{u}}}(\partial\hat{\boldsymbol{u}}). \tag{3.21}$$

In view of the canonical isomorphism of $\mathscr{V}_{(X,\boldsymbol{v})}$ with $\mathscr{B}_X$, $D\hat{\boldsymbol{u}}/Dt$ is again a vector field on $X(t)$, called the *covariant derivative* of $\hat{\boldsymbol{u}}$ along $X(t)$ with respect to $\mathscr{H}$. In particular, a horizontal lift $\boldsymbol{\lambda}$ is a cross section of $\mathscr{C}_X$ having slope zero, *i.e.*, $\boldsymbol{\lambda}$ is a solution of the *equation of parallel transport* along $X(t)$ with respect to $\mathscr{H}$:

$$\frac{D\boldsymbol{\lambda}}{Dt} = \boldsymbol{0}. \tag{3.22}$$

From (3.19) and (3.22), we can determine the covariant derivative $D\hat{\boldsymbol{u}}/Dt$ by the limit

$$\left.\frac{D\hat{\boldsymbol{u}}}{Dt}\right|_t = \lim_{\Delta t \to 0} \frac{\hat{\boldsymbol{u}}(t) - \boldsymbol{\rho}_{t-\Delta t,t}\,\hat{\boldsymbol{u}}(t-\Delta t)}{\Delta t}, \tag{3.23}$$

where the linear operation under the limit is that of the tangent space $\mathscr{B}_{X(t)}$. For any vector field $\boldsymbol{u}$, we define the *directional covariant derivative* of $\boldsymbol{u}$ along $X(t)$ as the covariant derivative of the restriction $\hat{\boldsymbol{u}}$ of $\boldsymbol{u}$ on $X(t)$. From (3.13) and (3.21), we have

$$\frac{D\hat{\boldsymbol{u}}}{Dt} = [D\boldsymbol{u}](\partial X), \tag{3.24}$$

where $D\boldsymbol{u}$ is the covariant derivative of $\boldsymbol{u}$. This equation shows that $D\hat{\boldsymbol{u}}/Dt$ depends only on the covariant derivative of $\boldsymbol{u}$ and the tangent ∂X of $X(t)$. Conversely, $D\boldsymbol{u}$ is also determined by the directional covariant derivatives $D\hat{\boldsymbol{u}}/Dt$ in all directions.

Next we establish the relation between a connection and a bundle chart. Let $(\mathscr{U}_\alpha, \boldsymbol{v}_\alpha)$ be a bundle chart of $\mathscr{T}(\mathscr{B})$ as before. Then $\boldsymbol{v}_\alpha$ is a diffeomorphism of $\mathscr{T}(\mathscr{U}_\alpha)$ with $\mathscr{U}_\alpha \times \mathscr{R}^3$, so it carries a connection $\mathscr{H}$ on $\mathscr{T}(\mathscr{U}_\alpha)$ onto a connection $\mathscr{H}(\alpha)$ on $\mathscr{U}_\alpha \times \mathscr{R}^3$. Let $X(t)$ be a smooth curve in $\mathscr{U}_\alpha$. Then the parallel transports $\boldsymbol{\rho}_{t_0, t}$ along $X(t)$ with respect to $\mathscr{H}$ and the parallel transports $\boldsymbol{\rho}_{t_0, t}(\alpha)$ along $X(t)$ with respect to $\mathscr{H}(\alpha)$ are related by

$$\boldsymbol{\rho}_{t_0,t}(\alpha) = \boldsymbol{v}_{\alpha, X(t)} \circ \boldsymbol{\rho}_{t_0,t} \circ \boldsymbol{v}^{-1}_{\alpha, X(t_0)} \,. \tag{3.25}$$

We call $\boldsymbol{\rho}_{t_0, t}(\alpha)$ the *relative parallel transports*. Since $\mathscr{U}_\alpha \times \mathscr{R}^3$ is a product manifold, the relative parallel transports $\boldsymbol{\rho}_{t_0, t}(\alpha)$ are automorphisms of the fibre space $\mathscr{R}^3$. From (3.25) and (1.7), the transformation rule for the relative parallel transports with respect to a change of bundle chart is

$$\boldsymbol{\rho}_{t_0,t}(\beta) = \boldsymbol{G}_{\beta\alpha}(X(t))\boldsymbol{\rho}_{t,t_0}(\alpha)\boldsymbol{G}_{\alpha\beta}(X(t_0)), \tag{3.26}$$

where $\boldsymbol{G}_{\alpha\beta}$ is the co-ordinate transformation from $(\mathscr{U}_\alpha, \boldsymbol{v}_\alpha)$ to $(\mathscr{U}_\beta, \boldsymbol{v}_\beta)$.

Now we are ready to define the concept of a *structural connection*. We say that a connection is a *structural* connection on a fibre bundle if the relative parallel transports along any curve belong to the structure group of the bundle. The transformation rule (3.26) shows that this condition is independent of the choice of the bundle chart for the relative parallel transports.

While the condition for a structural connection just stated is meaningful for a fibre bundle in general, for the tangent bundle $\mathscr{T}(\mathscr{B})$ it requires that the relative parallel transports $\boldsymbol{\rho}_{t_0, t}(\alpha)$ along any curve be linear isomorphisms of $\mathscr{R}^3$. This in turn requires that the parallel transports

$\rho_{t_0,t}$ be linear isomorphisms among the fibres of the curve, since the fibre isomorphisms $\nu_{\alpha,X(t)}$ and $\nu_{\alpha,X(t_0)}$ are linear. In view of this condition, a structural connection on the tangent bundle is the same as an *affine connection* in the classical sense.

A structural connection for a material tangent bundle $\mathscr{T}(\mathscr{B}, \mathfrak{A})$ we shall call a *material connection.* Since the structure group of the material tangent bundle $\mathscr{T}(\mathscr{B}, \mathfrak{A})$ is the symmetry group $\mathfrak{G}_{\mathfrak{A}}$, $\mathscr{H}$ is a material connection if and only if the relative parallel transports of $\mathscr{H}$ are material isomorphisms relative to $\mathfrak{A}$.

Now referring to the relation (3.25) of the parallel transports $\rho_{t_0,t}$ with the relative parallel transports $\rho_{t_0,t}(\alpha)$, we see that the $\rho_{t_0,t}(\alpha)$ belong to $\mathfrak{G}_{\mathfrak{A}}$ if and only if $\rho_{t_0,t}$ are material isomorphisms. Consequently a necessary and sufficient condition for a material connection $\mathscr{H}$ is that all parallel transports with respect to $\mathscr{H}$ be material isomorphisms. Since this condition is independent of the choice of a material atlas, a structural connection of any one material tangent bundle is a structural connection of all material tangent bundles.

In the next section we shall prove an existence theorem for material connections in general. Before going into such details we consider first some general properties of the material connections for the bodies given as Examples 1–4 in § 1.

Example 1. Continuous Distributions of Dislocations in Solid Crystals

In this example the material isomorphism between any two body-points is unique – that isomorphism maps the crystallographic basis at the one point onto that at the other. From the necessary and sufficient condition for a material connection given before, that isomorphism must also be the parallel transport along any curve joining the two body-points with respect to a material connection. Thus the material connection is unique in this case; further, it coincides with the field of tangent spaces of the family of *crystallographic vector fields*, which are vector fields whose components are constants with respect to the crystallographic basis. Of course, this material connection is integrable. (The uniqueness and the integrability of the material connection are properties shared by all solid crystal bodies.)

Example 2. Moebius Crystals

In this case the crystallographic basis vector fields are only local smooth fields, which cannot be extended into (single-valued) global smooth fields. The material connection is still unique and integrable, however, since it is locally unique and locally integrable. (A connection is a local structure, and integrability is a local property of a connection.)

As in the previous example, the material connection coincides locally with the field of tangent spaces of the family of local crystallographic vector fields. The only difference between this example and the previous one is that for a Moebius crystal $\mathscr{B}$ the tangent bundle $\mathscr{T}(\mathscr{B})$ has no global horizontal cross section with respect of the material connection.

Example 3. A Locally Inhomogeneous Transversely Isotropic Laminated Body

Since the symmetry group for this body is a continuous group, the material connection is no longer unique. Further, no material connection of this body is integrable. This fact is obvious, since if it were not true, there would be a family of global horizontal sections of $\mathscr{T}(\mathscr{B})$ with respect to a material connection because $\mathscr{B}$ is simply connected. But in view of the fact that the axis of symmetry is in the radial direction at all body-points (*cf.* the description of the body in § 1, Example 3), a horizontal section $\boldsymbol{u}$ of $\mathscr{T}(\mathscr{B})$ whose value $\boldsymbol{u}(X)$ is tangent to the sphere passing through X at any one point X must have the same property at all other points of the body. But this condition clearly contradicts the fact that a sphere is a non-parallelizable space.

Example 4. A Locally Homogeneous Fluid Crystal Body

As in the previous example, the material connection is not unique for this body. Further, by the same argument, no material connection for this body is integrable. This example shows that even a locally homogeneous body need not possess any integrable material connection.

4. The Existence of Material Connections

Having considered at the end of the preceding section the material connections for some specific bodies, we now turn to an existence theorem for material connections in general. In differential geometry the existence of a structural connection on a principal bundle is a standard result. In order to apply this result, we establish first the one-to-one correspondence between the structural connections on a fibre bundle and those on the associated principal bundle. We use the tangent bundle $\mathscr{T}(\mathscr{B})$ and its associated principal bundle $\mathscr{E}(\mathscr{B})$ to illustrate the procedure.

Let $\mathscr{H}$ be a structural connection on $\mathscr{T}(\mathscr{B})$. We show that $\mathscr{H}$ gives rise to a structural connection $\hat{\mathscr{H}}$ on $\mathscr{E}(\mathscr{B})$ in a natural way. Recall that the bundle charts for $\mathscr{T}(\mathscr{B})$ are in one-to-one correspondence with those for $\mathscr{E}(\mathscr{B})$. Let $(\mathscr{U}_\alpha, \boldsymbol{v}_\alpha)$ and $(\mathscr{U}_\alpha, \boldsymbol{\eta}_\alpha)$ be a pair of corresponding bundle charts for $\mathscr{T}(\mathscr{B})$ and $\mathscr{E}(\mathscr{B})$, respectively. Suppose that $X(t)$ be a smooth curve in $\mathscr{U}_\alpha$. Then as explained in the preceding section, $\mathscr{H}$ gives rise to a parallel transport $\boldsymbol{\rho}_{t_0, t}$ along $X(t)$ and a relative parallel transport $\boldsymbol{\rho}_{t_0, t}(\alpha)$ with respect to $(\mathscr{U}_\alpha, \boldsymbol{v}_\alpha)$. By hypothesis, $\mathscr{H}$ is a structural connection, so that $\boldsymbol{\rho}_{t_0, t}(\alpha)$ belongs to the structure group $\mathfrak{L}$ of $\mathscr{T}(\mathscr{B})$.

Since $\mathscr{E}(\mathscr{B})$ is the associated principal bundle of $\mathscr{T}(\mathscr{B})$, its structure group is $\mathfrak{L}$ also. Hence we can regard $\boldsymbol{\rho}_{t_0, t}(\alpha)$ as a family of transformations of the fibre space $\mathfrak{L}$ of $\mathscr{E}(\mathscr{B})$, thus determining a family of relative horizontal lifts of $X(t)$ in $\mathscr{U}_\alpha \times \mathfrak{L}$. The fibre isomorphism $\boldsymbol{\eta}$ then carries these relative horizontal lifts back into a family of horizontal lifts of $X(t)$ in $\mathscr{E}(\mathscr{U}_\alpha)$. Since we can apply this procedure to any smooth curve in $\mathscr{U}_\alpha$, the horizontal lifts in $\mathscr{E}(\mathscr{U}_\alpha)$ determine a unique connection $\hat{\mathscr{H}}$ on $\mathscr{E}(\mathscr{U}_\alpha)$.

Now applying the same procedure to each bundle chart, we can define a connection on each $\mathscr{E}(\mathscr{U}_\alpha)$. In order that these connections correspond to a global connection on $\mathscr{E}(\mathscr{B})$, it is necessary and sufficient that under a change of bundle chart the relative parallel transports satisfy the transformation rule (3.26). But this condition is obviously satisfied, since by construction the relative parallel transports with respect to $\hat{\mathscr{H}}$ are the same as those with respect to $\mathscr{H}$, and likewise, the co-ordinate transformations for $\mathscr{E}(\mathscr{B})$ are the same as those for $\mathscr{T}(\mathscr{B})$. Thus $\hat{\mathscr{H}}$ is well defined on $\mathscr{E}(\mathscr{B})$.

Clearly the procedure leading from the connection $\mathscr{H}$ to the connection $\hat{\mathscr{H}}$ can be worked out in the reversed order, since it depends only on the following three conditions:

i) The structure group $\mathscr{T}(\mathscr{B})$ is the same as that of $\mathscr{E}(\mathscr{B})$.

ii) The bundle charts of $\mathscr{T}(\mathscr{B})$ are in one-to-one correspondence with those of $\mathscr{E}(\mathscr{B})$.

iii) The co-ordinate transformations for $\mathscr{T}(\mathscr{B})$ are the same as those for $\mathscr{E}(\mathscr{B})$.

In each of the three conditions, $\mathscr{T}(\mathscr{B})$ and $\mathscr{E}(\mathscr{B})$ can be interchanged. Thus $\mathscr{H}$ is in one-to-one correspondence with $\hat{\mathscr{H}}$. (For that matter, the structural connections on any fibre bundle are in one-to-one correspondence with those on the associated principal bundle, since the same three conditions are valid in general.)

Applying the result just established to a material connection $\mathscr{H}$, we see that the corresponding $\hat{\mathscr{H}}$ is a structural connection for each bundle of reference frames $\mathscr{E}(\mathscr{B}, \mathfrak{A})$. Thus a material connection $\mathscr{H}$ exists if and only if the bundle of reference frames $\mathscr{E}(\mathscr{B}, \mathfrak{A})$ has a structural connection. We proceed now to prove the existence of a structural connection on a principal bundle. For definiteness, we consider the principal bundle $\mathscr{E}(\mathscr{B})$.

First we derive a necessary and sufficient condition for a structural connection on $\mathscr{E}(\mathscr{B})$. Recall that the structure group $\mathfrak{L}$ acts as a transformation group on $\mathscr{E}(\mathscr{B})$ by right multiplication. For each $A \in \mathfrak{L}$ the operation

$$\boldsymbol{R}_A : \mathscr{E}(\mathscr{B}) \to \mathscr{E}(\mathscr{B}) \tag{4.1}$$

is a fibre-preserving diffeomorphism. We claim that a connection $\hat{\mathscr{H}}$ on $\mathscr{E}(\mathscr{B})$ is a structural connection if and only if it is invariant under $\boldsymbol{R}_A$ for all $\boldsymbol{A} \in \mathfrak{L}$, *i.e.*,

$$\boldsymbol{R}_{A*}(\hat{\mathscr{H}}) = \hat{\mathscr{H}}. \tag{4.2}$$

Equivalently, this condition means that the family of horizontal lifts relative to $\hat{\mathscr{H}}$ of any curve is closed under right multiplication.

To prove the condition (4.2), we recall first that right multiplication commutes with the fibres isomorphism for $\mathscr{E}(\mathscr{B})$, *viz*,

$$\boldsymbol{R}_A \circ \boldsymbol{\eta}_\alpha = \boldsymbol{\eta}_\alpha \circ \boldsymbol{R}_A \tag{4.3}$$

for all $\boldsymbol{A} \in \mathfrak{L}$ and for all bundle charts $(\mathscr{U}_\alpha, \boldsymbol{\eta}_\alpha)$. Suppose that $X(t)$ is a smooth curve in $\mathscr{U}_\alpha$. Let $\boldsymbol{\lambda}(t)$ be a horizontal lift of $X(t)$ relative to $\hat{\mathscr{H}}$. Then a relative horizontal lift of $X(t)$ in $\mathscr{U}_\alpha \times \mathfrak{L}$ is $\boldsymbol{\eta}_\alpha(\boldsymbol{\lambda})$. In terms of the relative parallel transport, the curve $\boldsymbol{\eta}_\alpha(\boldsymbol{\lambda})$ has the representation

$$\boldsymbol{\eta}_{\alpha,\, X(t)}(\boldsymbol{\lambda}(t)) = (X(t), [\boldsymbol{\rho}_{t_0,t}(\alpha)](\boldsymbol{B})), \tag{4.4}$$

where $\boldsymbol{B}$ is an element of $\mathfrak{L}$ such that

$$\boldsymbol{\eta}_{\alpha,\, X(t_0)}(\boldsymbol{\lambda}(t_0)) = (X(t_0), \boldsymbol{B}). \tag{4.5}$$

Suppose that the relative parallel transport $\boldsymbol{\rho}_{t_0,t}(\alpha)$ belongs to $\mathfrak{L}$. Then the form (4.4) is invariant under the right multiplication $\boldsymbol{R}_A$, *viz*

$$\boldsymbol{R}_A(\boldsymbol{\eta}_{\alpha,\, X(t)}(\boldsymbol{\lambda}(t))) = \boldsymbol{\eta}_{\alpha,\, X(t)}(\boldsymbol{R}_A(\boldsymbol{\lambda}(t))) = (X(t), [\boldsymbol{\rho}_{t_0,t}(\alpha)](\boldsymbol{BA})). \tag{4.6}$$

Consequently $\boldsymbol{R}_A \circ \boldsymbol{\lambda}$ is also a horizontal lift of $X(t)$. But since $X(t)$ is arbitrary, the necessity of (4.2) is proved.

Conversely, suppose that the condition (4.2) holds. Using (4.3), we see that (4.6) holds for all $\boldsymbol{A}, \boldsymbol{B} \in \mathfrak{L}$. Thus $\rho_{t_0,t}(\alpha)$ satisfies the condition

$$\boldsymbol{R}_A \circ [\rho_{t_0,t}(\alpha)](\boldsymbol{B}) = [\rho_{t_0,t}(\alpha)](\boldsymbol{BA}), \tag{4.7}$$

or equivalently,

$$[\rho_{t_0,t}(\alpha)](\boldsymbol{B}) = ([\rho_{t_0,t}(\alpha)](\boldsymbol{BA}))(\boldsymbol{A}^{-1}), \qquad \forall \boldsymbol{A}, \boldsymbol{B} \in \mathfrak{L}. \tag{4.8}$$

Setting $\boldsymbol{A} = \boldsymbol{B}^{-1}$ in (4.8), we obtain

$$[\rho_{t_0,t}(\alpha)](\boldsymbol{B}) = ([\rho_{t_0,t}(\alpha)](\boldsymbol{1}))(\boldsymbol{B}) \qquad \forall \boldsymbol{B} \in \mathfrak{L}. \tag{4.9}$$

Since, then, the operator $\rho_{t_0,t}(\alpha)$ is equal to the element $[\rho_{t_0,t}(\alpha)](\boldsymbol{1})$ in $\mathfrak{L}$, the sufficiency of (4.2) is proved.

Next we define the *fundamental fields* on the principal bundle $\mathscr{E}(\mathscr{B})$. Recall first that the structure group $\mathfrak{L}$ of $\mathscr{E}(\mathscr{B})$ is a Lie group[1], so it is equipped with a Lie algebra $\mathfrak{l}$ consisting of left-invariant vector fields on $\mathfrak{L}$. Since $\mathscr{E}(\mathscr{B})$ is a principal bundle, its fibres are copies of $\mathfrak{L}$ under the fibre isomorphisms. We claim that the inverse of these fibre isomorphisms carry $\mathfrak{l}$ onto a family of vector fields $\hat{\mathfrak{l}}$ on $\mathscr{E}(\mathscr{B})$. Specifically, we consider the fibre $\mathscr{E}_X$ and a bundle chart $(\mathscr{U}_\alpha, \boldsymbol{\eta}_\alpha)$ covering X. If $\boldsymbol{v} \in \mathfrak{l}$, we put

$$\hat{\boldsymbol{v}}_X \equiv \boldsymbol{\eta}_{\alpha, X*}^{-1}(\boldsymbol{v}), \tag{4.10}$$

then the vector field $\hat{\boldsymbol{v}}_X$ depends only on $\boldsymbol{v}$ and is independent of the choice of the bundle chart $(\mathscr{U}_\alpha, \boldsymbol{\eta}_\alpha)$.

To see this fact, we consider another bundle chart $(\mathscr{U}_\beta, \boldsymbol{\eta}_\beta)$ also covering X. Then we must show that

$$\boldsymbol{\eta}_{\beta, X*}^{-1}(\boldsymbol{v}) = \boldsymbol{\eta}_{\alpha, X*}^{-1}(\boldsymbol{v}), \tag{4.11}$$

or equivalently,

$$\boldsymbol{v} = \boldsymbol{\eta}_{\alpha, X*} \circ \boldsymbol{\eta}_{\beta, X*}^{-1}(\boldsymbol{v}). \tag{4.12}$$

Since the composite map $\boldsymbol{\eta}_{\alpha,X} \circ \boldsymbol{\eta}_{\beta,X}^{-1}$ is equal to the co-ordinate transformation $\boldsymbol{G}_{\alpha\beta}(X)$, by the chain rule we have

$$\boldsymbol{G}_{\alpha\beta}(X)_* = \boldsymbol{\eta}_{\alpha, X*} \circ \boldsymbol{\eta}_{\beta, X*}^{-1}. \tag{4.13}$$

---

[1] For a brief discussion of the structure of a Lie group, see § 1.6.

But we have also

$$G_{\alpha\beta}(X)_*(\mathfrak{v}) = \mathfrak{v}, \tag{4.14}$$

since $\mathfrak{v}$ is left-invariant. Hence (4.12) holds.

The totality of $\hat{\mathfrak{v}}_X$ for all X in $\mathscr{B}$ now forms a vector field $\hat{\mathfrak{v}}$ on $\mathscr{E}(\mathscr{B})$. For all $\mathfrak{v}$ in the Lie algebra $\mathfrak{l}$ the corresponding $\hat{\mathfrak{v}}$ form the Lie algebra $\hat{\mathfrak{l}}$. From (4.10), the $\wedge$-operation, which sends $\mathfrak{l}$ to $\hat{\mathfrak{l}}$, is an isomorphism of Lie algebras, *i.e.*,

$$\widehat{[\mathfrak{v}, \mathfrak{u}]} = [\hat{\mathfrak{v}}, \hat{\mathfrak{u}}] \qquad \forall \mathfrak{u}, \mathfrak{v} \in \mathfrak{l}. \tag{4.15}$$

We call $\hat{\mathfrak{l}}$ the *Lie algebra* of $\mathscr{E}(\mathscr{B})$ and the elements of $\hat{\mathfrak{l}}$ the *fundamental fields.*

The fundamental fields possess the following two important properties: i) At any point $(X, e) \in \mathscr{E}(\mathscr{B})$ the restriction map $|_{(X,e)}$ is an isomorphism of $\hat{\mathfrak{l}}$ with $\hat{\mathscr{V}}_{(X,e)}$, so every vertical vector $\hat{\mathfrak{v}}_{(X,e)}$ at (X, e) has a unique extension to a fundamental field $\hat{\mathfrak{v}}$ on $\mathscr{E}(\mathscr{B})$, *viz*,

$$\hat{\mathfrak{v}} = |^{-1}_{(X,e)}(\hat{\mathfrak{v}}_{(X,e)}), \tag{4.16}$$

where

$$|_{(X,e)} : \hat{\mathfrak{l}} \to \hat{\mathscr{V}}_{(X,e)}. \tag{4.17}$$

ii) The collection $\hat{\mathfrak{l}}$ is closed with respect to the right multiplications of the elements of $\mathfrak{L}$ on $\mathscr{E}(\mathscr{B})$, *viz*,

$$R_{A*}(\hat{\mathfrak{l}}) = \hat{\mathfrak{l}} \qquad \forall A \in \mathfrak{L}. \tag{4.18}$$

Further, from (4.3) and (4.10), we have

$$R_{A*}(\hat{\mathfrak{v}}) = \widehat{R_{A*}(\mathfrak{v})}. \tag{4.19}$$

For the Lie group $\mathfrak{L}$ the operation R_{A*} on $\mathfrak{l}$ coincides with the adjoint representation of A^{-1} (*cf.* § I.6), *i.e.*,

$$R_{A*}(\mathfrak{v}) = [\mathrm{ad}\,(A^{-1})](\mathfrak{v}), \tag{4.20}$$

so we define $[\mathrm{ad}\,(A^{-1})]$ on $\hat{\mathfrak{l}}$ as R_{A*} also:

$$[\mathrm{ad}\,(A^{-1})](\hat{\mathfrak{v}}) \equiv R_{A*}(\hat{\mathfrak{v}}). \tag{4.21}$$

Since $\hat{\mathscr{V}}_{(X,e)}$ is a copy of $\hat{\mathfrak{l}}$, we can define $[\mathrm{ad}\,(A^{-1})]$ further as an operator on $\hat{\mathscr{V}}_{(X,e)}$:

$$[\mathrm{ad}\,(A^{-1})](\hat{\mathfrak{v}}_{(X,e)}) \equiv (R_{A*}(\hat{\mathfrak{v}}))_{(X,e)} \tag{4.22}$$

for all $\hat{\mathfrak{v}}_{(X,e)} \in \hat{\mathscr{V}}_{(X,e)}$. Then the extension (4.16) has the representation

$$\hat{\mathfrak{v}}_{(X,\boldsymbol{R}_A e)} = \boldsymbol{R}_{A*(X,e)}([\mathrm{ad}\,(A)](\hat{\mathfrak{v}}_{(X,e)})) \tag{4.23}$$

on the fibre $\mathscr{E}_X$.

Next we define the *connection form* of a connection on the principal bundle $\mathscr{E}(\mathscr{B})$. Let $\hat{\mathscr{H}}$ be a connection on $\mathscr{E}(\mathscr{B})$ as usual. Then at any point $(X, \boldsymbol{e}) \in \mathscr{E}(\mathscr{B})$ the tangent space $\mathscr{E}(\mathscr{B})_{(X,e)}$ has the following decomposition as a direct sum:

$$\mathscr{E}(\mathscr{B})_{(X,e)} = \hat{\mathscr{V}}_{(X,e)} \oplus \hat{\mathscr{H}}_{(X,e)}. \tag{4.24}$$

As before, we define a projection map

$$\hat{\boldsymbol{\delta}}_{(X,e)} : \mathscr{E}(\mathscr{B})_{(X,e)} \to \hat{\mathscr{V}}_{(X,e)} \tag{4.25}$$

with respect to the decomposition (4.24). Then the composite map

$$|_{(X,e)}^{-1} \circ \hat{\boldsymbol{\delta}}_{(X,e)} : \mathscr{E}(\mathscr{B})_{(X,e)} \to \hat{\mathfrak{l}} \tag{4.26}$$

is a linear map at each $(X, \boldsymbol{e}) \in \mathscr{E}(\mathscr{B})$. The field $\hat{\omega}$ of such linear maps is a 1-form with value in $\hat{\mathfrak{l}}$, called the *connection form* of $\hat{\mathscr{H}}$:

$$\hat{\omega}_{(X,e)} \equiv |_{(X,e)}^{-1} \circ \hat{\boldsymbol{\delta}}_{(X,e)}. \tag{4.27}$$

The connection form $\hat{\omega}$ characterizes the connection $\hat{\mathscr{H}}$ completely, since we can recover $\hat{\mathscr{H}}$ from $\hat{\omega}$ by

$$\hat{\mathscr{H}}_{(X,e)} = \ker \hat{\omega}_{(X,e)}. \tag{4.28}$$

In terms of $\hat{\omega}$, the condition (4.2) takes the form

$$\boldsymbol{R}_A^*(\hat{\omega}) = [\mathrm{ad}\,(A^{-1})](\hat{\omega}), \tag{4.29}$$

which means that

$$[\hat{\omega}_{(X,\boldsymbol{R}_A e)}](\boldsymbol{R}_{A*}(z)) = [\mathrm{ad}\,(A^{-1})]([\hat{\omega}_{(X,e)}](z)) \tag{4.30}$$

for all $z \in \mathscr{E}(\mathscr{B})_{(X,e)}$. From (4.26) the restriction of $\hat{\omega}_{(X,e)}$ to $\hat{\mathscr{V}}_{(X,e)}$ is the operation of extension:

$$[\hat{\omega}_{(X,e)}](\hat{\mathfrak{v}}_{(X,e)}) = \hat{\mathfrak{v}} \tag{4.31}$$

for all $\hat{\mathfrak{v}} \in \hat{\mathfrak{l}}$.

The conditions (4.31) and (4.29) just proved are not only necessary but also sufficient for a 1-form with value in $\hat{\mathfrak{l}}$ to be the connection form of a structural connection. To prove the sufficiency, since (4.31) and (4.17) imply that the tangent space $\mathscr{E}(\mathscr{B})_{(X,e)}$ has the following decomposition as a direct sum:

$$\mathscr{E}(\mathscr{B})_{(X,e)} = \hat{\mathscr{V}}_{(X,e)} \oplus \ker \hat{\omega}_{(X,e)}, \tag{4.32}$$

the field $\hat{\mathscr{H}}$ defined by (4.28) is a connection on $\mathscr{E}(\mathscr{B})$. From (4.29), that connection satisfies the condition (4.2), so it is a structural connection.

In view of this necessary and sufficient condition for a connection form, we see that the existence of a structural connection $\hat{\mathscr{H}}$ on $\mathscr{E}(\mathscr{B})$ is equivalent to the existence of a 1-form with value in $\hat{\mathfrak{l}}$ satisfying the conditions (4.29) and (4.31). We can prove the latter one by the standard argument based on a *partition of unity* for a manifold (*cf.* § I.4). Specifically, we choose a covering of $\mathscr{E}(\mathscr{B})$ by a family of bundle charts, say $\{(\mathscr{U}_\alpha, \boldsymbol{\eta}_\alpha)\}$. For each bundle chart $(\mathscr{U}_\alpha, \boldsymbol{\eta}_\alpha)$ in this family we define a structural connection $\hat{\mathscr{H}}^\alpha$ on $\mathscr{E}(\mathscr{U}_\alpha)$ by requiring the relative parallel transport of $\hat{\mathscr{H}}^\alpha$ along any curve in $\mathscr{U}_\alpha$ to be the identity map of $\mathfrak{L}$. Clearly such a structural connection exists on each $\mathscr{E}(\mathscr{U}_\alpha)$. Indeed, $\hat{\mathscr{H}}^\alpha$ is the connection whose image on $\mathscr{U}_\alpha \times \mathfrak{L}$ under $\boldsymbol{\eta}_\alpha$ is the field of tangent spaces of the constant sections of the product manifold $\mathscr{U}_\alpha \times \mathfrak{L}$. Let $\hat{\omega}^\alpha$ be the connection form of $\hat{\mathscr{H}}^\alpha$ on $\mathscr{E}(\mathscr{U}_\alpha)$. Then ω^α satisfies the conditions (4.29) and (4.31) for all X belonging to $\mathscr{U}_\alpha$.

Now let $\{f_\alpha\}$ be a partition of unity subordinate to $\{\mathscr{U}_\alpha\}$, so that the following conditions are satisfied:

i) Each f_α is a smooth function whose support is contained in $\mathscr{U}_\alpha$:

$$\operatorname{supp} f_\alpha \subset \mathscr{U}_\alpha. \tag{4.33}$$

ii) The values of f_α range between 0 and 1:

$$0 \leqq f_\alpha \leqq 1. \tag{4.34}$$

iii) For each point $X \in \mathscr{B}$ there exists a neighborhood $\mathscr{U}$ meeting only finitely many $\mathscr{U}_\alpha$ belonging to the covering, so that by (4.33)

$$\mathscr{U} \cap \operatorname{supp} f_\alpha \neq \emptyset \tag{4.35}$$

for finitely many α.

iv) The finite sum $\sum_\alpha f_\alpha(X)$ is equal to unity for all $X \in \mathscr{B}$:

$$\sum_\alpha f_\alpha(X) = 1. \tag{4.36}$$

Using such a partition of unity, we define $\hat{\omega}$ by

$$\hat{\omega}_{(X,e)} \equiv \sum_\alpha f_\alpha(X)\hat{\omega}^\alpha_{(X,e)} \qquad \forall (X,e) \in \mathscr{E}(\mathscr{B}). \tag{4.37}$$

We claim that $\hat{\omega}$ satisfies the conditions (4.30) and (4.31).

Indeed, since each $\hat{\omega}^\alpha$ satisfies (4.30), by linearity we have

$$[f_\alpha(X)\hat{\omega}^\alpha_{(X,\,R_A e)}](R_{A*}(z)) = [\text{ad}\,(A^{-1})]([f_\alpha(X)\hat{\omega}^\alpha_{(X,\,e)}](z)). \tag{4.38}$$

Summing this equation with respect to α and using the condition (4.36) yields (4.30) for $\hat{\omega}$. By the same argument, we have

$$[f_\alpha(X)\hat{\omega}^\alpha_{(X,\,e)}](\hat{\mathfrak{v}}_{(X,\,e)}) = f_\alpha(X)\hat{\mathfrak{v}}. \tag{4.39}$$

Summing with respect to α yields (4.31) for $\hat{\omega}$.

To complete the proof that $\hat{\omega}$ is a connection form, we still have to show that $\hat{\omega}$ is a smooth field on $\mathscr{E}(\mathscr{B})$. But this condition is obvious from the construction of $\hat{\omega}$. Indeed, from (4.35), on the subbundle $\mathscr{E}(\mathscr{U})$ over the neighborhood $\mathscr{U}$ of X the smoothness of $\hat{\omega}$ follows from that of $\hat{\omega}^\alpha$ and f_α. Since smoothness is a local property, $\hat{\omega}$ is smooth on $\mathscr{E}(\mathscr{B})$.

So far we have proved the existence of a structural connection on $\mathscr{E}(\mathscr{B})$ in general. To prove the existence of a material connection, a slight change of the preceding proof suffices: We require that the covering $\{(\mathscr{U}_\alpha, \eta_\alpha)\}$ used in the proof be the bundle atlas $\mathfrak{A}'$ corresponding to the material atlas $\mathfrak{A}$, as shown in (2.24). Because of this choice, the restriction of $\hat{\omega}^\alpha$ to $\mathscr{E}(\mathscr{U}_\alpha, \mathfrak{A})$ has values in the Lie algebra $\hat{\mathfrak{g}}_{\mathfrak{A}}$. The same holds for the connection form $\hat{\omega}$, since $\hat{\omega}$ is a linear combination of the $\hat{\omega}^\alpha$ at each point in $\mathscr{E}(\mathscr{B})$. Thus under the special choice of covering: $\{(\mathscr{U}_\alpha, \eta_\alpha)\} = \mathfrak{A}'$, the general existence theorem implies the existence of a material connection.

Under additional restrictions on the $\hat{\omega}^\alpha$, the proof of existence can be used to establish material connections having special properties. For instance, we shall prove later that every locally homogeneous body can be equipped with a torsion-free material connection. This result is a consequence of the fact that in the construction of ω, if each $\hat{\omega}^\alpha$ is torsion-free, so is $\hat{\omega}$.

In the next section we shall derive a criterion in component form for a structural connection on $\mathscr{E}(\mathscr{B})$ to be a material connection. The condition of integrability for a material connection will be considered in § 6.

5. Material Connection in Component Form

The condition

$$R_{A*}(\hat{\mathscr{H}}) = \hat{\mathscr{H}}, \tag{4.2}$$

or equivalently

$$[\hat{\omega}_{(X, R_A e)}](R_{A*}(z)) = [\text{ad}\,(A^{-1})]([\hat{\omega}_{(X, e)}](z)), \tag{4.30}$$

for a structural connection $\hat{\mathscr{H}}$ on $\mathscr{E}(\mathscr{B})$ implies that the value of $\hat{\omega}$ at any one point in $\mathscr{E}(\mathscr{B})$, say $(X, \boldsymbol{e})$, determines $\hat{\omega}$ completely on the whole fibre $\mathscr{E}_X$ of that point. This fact allows us to express $\hat{\omega}$ in component form, as follows. Let $\{\hat{e}_B^A, A, B = 1, 2, 3\}$ be a basis[1] of $\hat{\mathfrak{l}}$. Then $\hat{\omega}$ has the component form

$$\hat{\omega} = \hat{e}_B^A \otimes \hat{\omega}_A^B, \tag{5.1}$$

where $\hat{\omega}_B^A$, $A, B = 1, 2, 3$, are ordinary 1-forms on $\mathscr{E}(\mathscr{B})$. Since

$$\hat{\mathscr{H}} = \ker \hat{\omega}, \tag{4.28}$$

the 1-forms $\hat{\omega}_B^A$ span the orthogonal complement of $\hat{\mathscr{H}}$ in the cotangent space of $\mathscr{E}(\mathscr{B})$. Clearly the set $\{\hat{\omega}_B^A\}$ is linearly independent at every point $(X, \boldsymbol{e})$, since

$$\langle \hat{\omega}_B^A, \hat{e}_D^C \rangle = \delta_D^A \delta_B^C. \tag{5.2}$$

Now we choose a bundle chart $(\mathscr{U}_\alpha, \boldsymbol{\eta}_\alpha)$ for $\mathscr{E}(\mathscr{B})$ and let $\underset{\alpha}{\boldsymbol{e}}$ be the corresponding section of the identity. Then the induced linear map $\underset{\alpha}{\boldsymbol{e}}^*$ carries the 1-form $\hat{\omega}_B^A$ to a 1-form $\underset{\alpha}{\omega}{}_B^A$:

$$\underset{\alpha}{\omega}{}_B^A = \underset{\alpha}{\boldsymbol{e}}^*(\hat{\omega}_B^A). \tag{5.3}$$

We put

$$\underset{\alpha}{\omega} \equiv \boldsymbol{e}_B^A \otimes \underset{\alpha}{\omega}{}_A^B, \tag{5.4}$$

where $\{\boldsymbol{e}_B^A, A, B = 1, 2, 3\}$ is the basis of $\mathfrak{l}$ corresponding to the basis $\{\hat{e}_B^A, A, B = 1, 2, 3\}$. Then $\underset{\alpha}{\omega}$ is a 1-form on $\mathscr{U}_\alpha$ with values in $\mathfrak{l}$. We claim that $\underset{\alpha}{\omega}$ characterizes $\hat{\omega}$ on $\mathscr{E}(\mathscr{U}_\alpha)$.

Indeed, from (5.3) $\underset{\alpha}{\omega}$ satisfies the condition

$$[\underset{\alpha}{\omega}](\boldsymbol{u}) = [\hat{\omega}](\underset{\alpha}{\boldsymbol{e}}_*(\boldsymbol{u})) \tag{5.5}$$

for all $\boldsymbol{u} \in \mathscr{B}_X$. Since $\underset{\alpha}{\boldsymbol{e}}$ is a cross section, its tangent space $\underset{\alpha}{\boldsymbol{e}}_*(\mathscr{B}_X)$ is a

[1] For a brief review of the structure of the Lie group $\mathfrak{L}$ and its Lie algebra $\mathfrak{l}$, see § I.6.

horizontal space at $(X, \underset{\alpha}{e})$ (although it need not coincide with the connection $\hat{\mathscr{H}}_{(X, \underset{\alpha}{e})}$). Thus together with the condition (4.31), the condition (5.5) determines $\hat{\omega}_{(X, \underset{\alpha}{e})}$ for all $X \subset \mathscr{U}_\alpha$. Then by (4.30), $\hat{\omega}$ is determined on $\mathscr{E}(\mathscr{U}_\alpha)$. Because of this property, $\underset{\alpha}{\omega}$ is called the *representation* of ω relative to the bundle chart $(\mathscr{U}_\alpha, \eta_\alpha)$.

Let $X(t)$ be a smooth curve in $\mathscr{U}_\alpha$. Then we can determine the horizontal lift λ in terms of $\underset{\alpha}{\omega}$ in the following way: First, we express λ by

$$\lambda(t) = (X(t), [\boldsymbol{R}_{\boldsymbol{K}(t)}](\underset{\alpha}{e}(X(t)))), \tag{5.6}$$

where $\boldsymbol{K}(t)$ is a smooth curve in $\mathfrak{L}$. Using the chain rule and the conditions (4.30) and (5.5), we can calculate the vertical component of $\partial\lambda$ by

$$\hat{\delta}(\partial\lambda) = [\text{ad}\,(\boldsymbol{K}^{-1})](\underset{\alpha}{\omega}(\partial X)) + \widehat{\boldsymbol{\theta}(\partial \boldsymbol{K})} \tag{5.7}$$

where $\boldsymbol{\theta}$ is the canonical form of $\mathfrak{L}$ (*cf.* § I.6), *i.e.*,

$$\boldsymbol{\theta}(\partial \boldsymbol{G}) \equiv |_{\boldsymbol{G}}^{-1} : \mathfrak{L}_{\boldsymbol{G}} \to \mathfrak{l} \tag{5.8}$$

for all $\boldsymbol{G} \in \mathfrak{L}$. Since λ is a horizontal curve, (5.7) implies that the equation of parallel transport relative to $(\mathscr{U}_\alpha, \eta_\alpha)$ is

$$[\text{ad}\,(\boldsymbol{K}^{-1})](\underset{\alpha}{\omega}(\partial X)) + \boldsymbol{\theta}(\partial \boldsymbol{K}) = \boldsymbol{0}. \tag{5.9}$$

Any solution $\boldsymbol{K}(t)$ of this equation determines a horizontal lift $\lambda(t)$ of $X(t)$ via the representation (5.6).

We can apply the representation (5.6) to any curve lying above $X(t)$ in $\mathscr{E}(\mathscr{B})$. In particular, for another bundle chart $(\mathscr{U}_\beta, \eta_\beta)$ with the section of the identity $\underset{\beta}{e}$, we have

$$\underset{\beta}{e}(X(t)) = [\boldsymbol{R}_{\boldsymbol{G}_{\alpha\beta}(X(t))}](\underset{\alpha}{e}(X(t))). \tag{5.10}$$

Setting

$$\underset{\alpha}{\lambda}(t) \equiv (X(t), \underset{\alpha}{e}(X(t))),\; \underset{\beta}{\lambda}(t) \equiv (X(t), \underset{\beta}{e}(X(t))), \tag{5.11}$$

we obtain from (5.7)

$$\hat{\delta}(\partial\underset{\beta}{\lambda}) = [\text{ad}\,(\boldsymbol{G}_{\alpha\beta}^{-1})](\underset{\alpha}{\omega}(\partial X)) + \widehat{[\boldsymbol{\theta} \circ \boldsymbol{G}_{\alpha\beta *}](\partial X)}. \tag{5.12}$$

But we can apply (5.7) also to the bundle chart $(\mathscr{U}_\beta, \boldsymbol{\eta}_\beta)$ and obtain

$$\hat{\delta}(\underset{\beta}{\partial\lambda}) = \widehat{\underset{\beta}{\omega}(\partial X)}. \tag{5.13}$$

Since $X(t)$ is arbitrary, the conditions (5.12) and (5.13) imply the following transformation rule for the representations of a connection form:

$$\underset{\beta}{\omega} = [\mathrm{ad}\,(\boldsymbol{G}_{\alpha\beta}^{-1})](\underset{\alpha}{\omega}) + \boldsymbol{\theta} \circ \boldsymbol{G}_{\alpha\beta *}\,. \tag{5.14}$$

This is an equation on the overlap $\mathscr{U}_\alpha \cap \mathscr{U}_\beta$ of the bundle charts.

A convenient bundle chart is a chart induced by a co-ordinate system. Let $(\mathscr{U}_\alpha, \boldsymbol{\eta}_\alpha)$ now be the bundle chart induced by the co-ordinate system (X^A). Then the section of the identity $\underset{\alpha}{\boldsymbol{e}}$ is the field of natural bases:

$$\underset{\alpha}{\boldsymbol{e}} = \{\partial X_A,\ A = 1, 2, 3\}. \tag{5.15}$$

Also, we choose $\{\boldsymbol{e}_B^A\}$ to be the standard basis[1] of $\mathfrak{l}$. Then $\underset{\alpha}{\omega}$ takes the matrix form:

$$\underset{\alpha}{\omega} = \boldsymbol{e}_B^A \otimes \underset{\alpha}{\omega}{}_A^B = [\underset{\alpha}{\omega}{}_A^B]. \tag{5.16}$$

As usual we can express the 1-form $\underset{\alpha}{\omega}{}_B^A$ in component form relative to (X^A):

$$\underset{\alpha}{\omega}{}_B^A = \Gamma_{BC}^A\,\partial X^C. \tag{5.17}$$

The components Γ_{BC}^A defined in this way are called the *connection symbols* of $\mathscr{H}$ relative to (X^A).

In terms of the connection symbols the equation of parallel transport (5.9) takes the form

$$\dot{K}_B^A(t) + \Gamma_{CD}^A(X(t))K_B^C(t)\dot{X}^D(t) = 0, \tag{5.18}$$

where $X^A(t)$, $A = 1, 2, 3$, are the co-ordinates of $X(t)$ in (X^A). A solution $[K_B^A(t)]$ of this equation gives a field of frames on $X(t)$:

$$\boldsymbol{e}(t) = \{K_B^A(t)\partial_{X(t)} X_A,\ B = 1, 2, 3\} \tag{5.19}$$

such that

$$\hat{\rho}_{t_0,t}(\boldsymbol{e}(t_0)) = \boldsymbol{e}(t), \tag{5.20}$$

---

[1] *Cf.* § I.6.

or equivalently,

$$\rho_{t_0,t}(\alpha) = [K^A_B(t)][K^A_B(t_0)]^{-1}. \tag{5.21}$$

Now via the one-to-one correspondence between $\hat{\mathscr{H}}$ and $\mathscr{H}$, the equation (5.21) gives the relative parallel transports for the connection $\mathscr{H}$ also. Then from (5.18) the equation of parallel transport for $\mathscr{H}$ is

$$\dot{v}^A(t) + \Gamma^A_{BC}(X(t))v^B(t)\dot{X}^C(t) = 0. \tag{5.22}$$

A solution of this equation is a vector field on $X(t)$:

$$\boldsymbol{v}(t) = v^A(t)\partial_{X(t)} X_A, \tag{5.23}$$

such that

$$\rho_{t_0,t}(\boldsymbol{v}(t_0)) = \boldsymbol{v}(t). \tag{5.24}$$

We can deduce the transformation rule for the connection symbols from the general transformation rule (5.14). Let $(\mathscr{U}_\beta, \boldsymbol{\eta}_\beta)$ be the bundle chart induced by a co-ordinate system also, say $(\overline{X}^A)$. Then the co-ordinate transformation $\boldsymbol{G}_{\alpha\beta}$ is the Jacobian matrix from $(\overline{X}^A)$ to (X^A):

$$\boldsymbol{G}_{\alpha\beta} = \left[\frac{\partial X^A}{\partial \overline{X}^B}\right]. \tag{5.25}$$

Also, the natural bases $\{\partial X^A\}$ and $\{\partial \overline{X}^A\}$ obey the transformation rule

$$\partial X^A = \frac{\partial X^A}{\partial \overline{X}^B}\partial \overline{X}^B. \tag{5.26}$$

Substituting (5.17), (5.25), and (5.26) in (5.14), we obtain

$$\overline{\Gamma}^A_{BC} = \Gamma^D_{EF}\frac{\partial \overline{X}^A}{\partial X^D}\frac{\partial X^E}{\partial \overline{X}^B}\frac{\partial X^F}{\partial \overline{X}^C} + \frac{\partial \overline{X}^A}{\partial X^D}\frac{\partial^2 X^D}{\partial \overline{X}^B \partial \overline{X}^C}, \tag{5.27}$$

where $\overline{\Gamma}^A_{BC}$ denotes a connection symbol relative to $(\overline{X}^A)$.

A bundle chart $(\mathscr{U}_\alpha, \boldsymbol{\eta}_\alpha)$ induced by the co-ordinate system (X^A) on $\mathscr{U}_\alpha$ gives rise to a *lifted co-ordinate system* (X^A, e^B_C) on $\mathscr{E}(\mathscr{U}_\alpha)$ in an obvious way: The co-ordinates of $(X, \boldsymbol{e})$ are the X^A of X and the components e^B_C of $\boldsymbol{e}(X)$ in (X^A), *viz*

$$\boldsymbol{e} = \{e^B_C \partial X_B,\ C = 1, 2, 3\}. \tag{5.28}$$

In terms of the lifted co-ordinate system the vertical space $\hat{\mathscr{V}}_{(X,\boldsymbol{e})}$ is spanned by the set

$$\{\partial e^C_B,\ B, C = 1, 2, 3\}, \tag{5.29}$$

and the horizontal space $\hat{\mathscr{H}}_{(X,e)}$ is spanned by the set

$$\{\partial_{(X,e)} X_A - \Gamma^B_{CA}(X) e^C_D \partial_{(X,e)} e^D_B,\ A = 1, 2, 3\}. \tag{5.30}$$

The former set is obvious, since (e^B_C) is a co-ordinate system on the fibre. The latter set can be deduced in the following way. From (5.13), the vertical component of the basis vector $\partial_{(X,e)}X_A$ at the section of the identity $\underset{\alpha}{e}$ is given by

$$\hat{\delta}(\partial_{(X,e)} X_A) = \Gamma^B_{CA}(X) \partial_{(X,\underset{\alpha}{e})} e^C_B. \tag{5.31}$$

Hence the horizontal component is

$$\partial_{(X,\underset{\alpha}{e})} X_A - \Gamma^B_{CA}(X) \partial_{(X,\underset{\alpha}{e})} e^C_B. \tag{5.32}$$

From (5.28), the typical point (X, e) is related to the point $(X, \underset{\alpha}{e})$ by

$$e = R_{[e^B{}_C]}(\underset{\alpha}{e}). \tag{5.33}$$

Then from (4.2), the right multiplication $R_{[e^B{}_C]}$ carries the horizontal vector shown in (5.32) into the horizontal vector in (5.30). The condition (5.31) gives a clear geometric meaning to the connection symbols.

Now we can set forth a necessary and sufficient condition for a material connection in terms of the connection symbols. Suppose that $(\mathscr{U}, \eta)$ be a reference chart relative to the material atlas $\mathfrak{A}$, and let f be its section of the identity. Relative to (X^A), f takes the component form

$$f = \{f^A_B \partial X_A,\ B = 1, 2, 3\}. \tag{5.34}$$

Then from (5.30), the horizontal space at f is spanned by the set

$$\{\partial_{(X,f)} X_A - \Gamma^B_{CA}(X) f^C_D(X) \partial_{(X,f)} e^D_B,\ A = 1, 2, 3\}. \tag{5.35}$$

The tangent space of the cross section f is spanned by the set

$$\{f_* \partial X_A,\ A = 1, 2, 3\}. \tag{5.36}$$

Relative to the lifted co-ordinate system (X^A, e^C_D) the component form of this set is

$$\left\{\partial X_A + \left.\frac{\partial f^C_D}{\partial X^A}\right|_X \partial e^D_C,\ A = 1, 2, 3\right\}. \tag{5.37}$$

We can determine the vertical component of this set from the generating set (5.35) for the horizontal space, obtaining

$$\left\{\left(\frac{\partial f_D^C}{\partial X^A}+\Gamma^C_{BA}f_D^B\right)\partial e_C^D,\ A=1,2,3\right\}. \tag{5.38}$$

According to the necessary and sufficient condition for a material connection given at the end of the preceding section, the connection $\hat{\mathscr{H}}$ with the connection symbols Γ^A_{BC} is a material connection if and only if the set (5.38) are the values at $(X, \boldsymbol{f})$ of some fundamental fields of the bundle of reference frames $\mathscr{E}(\mathscr{B}, \mathfrak{A})$. Since the standard representation of the Lie algebra of $\mathscr{E}(\mathscr{B}, \mathfrak{A})$ is the component form of the fundamental field at the section of the identity $\underset{\alpha}{\boldsymbol{e}}$, we must determine the component form of the fundamental field of the set (5.38) at $\underset{\alpha}{\boldsymbol{e}}$. For this purpose we observe first that

$$\underset{\alpha}{\boldsymbol{e}} = \boldsymbol{R}_{[f^A{}_B]^{-1}}(\boldsymbol{f}) \tag{5.39}$$

Expressing the right-multiplication $\boldsymbol{R}_{[f_B{}^A]^{-1}}$ in component form and using the condition (4.21) for the fundamental field, we obtain the criterion that $\mathscr{H}$ is a material connection if and only if the set

$$\left\{\left[f^{-1}{}^E_C(X)\left(\left.\frac{\partial f_D^C}{\partial X^A}\right|_X+\Gamma^C_{BA}(X)f_D^B(X)\right)\right],\ A=1,2,3\right\} \tag{5.40}$$

is contained in the Lie algebra $\mathfrak{g}_{\mathfrak{A}}$ of the symmetry group $\mathfrak{G}_{\mathfrak{A}}$ for all $X \in B$.

While a material atlas $\mathfrak{A}$ and a bundle chart $(\mathscr{U}, \boldsymbol{\eta})$ relative to $\mathfrak{A}$ are used in expressing the criterion for a material connection, the form of that criterion is independent of the choice of $\mathfrak{A}$ and the choice of $(\mathscr{U}, \boldsymbol{\eta})$.

In applications the criterion based on (5.40) is a convenient one, since the classification of the Lie algebras of all types of symmetry groups is known (*cf.* § I.6). In a later section, we shall apply the criterion in the derivation of the field equation for the body $\mathscr{B}$.

For a solid crystal body $\mathscr{B}$, which may be a Moebius crystal, the criterion implies that

$$\frac{\partial f_D^C}{\partial X^A}+\Gamma^C_{BA}f_D{}^B=0, \tag{5.41}$$

since in this case the Lie algebra $\mathfrak{g}_{\mathfrak{A}}$ is the trivial algebra

$$\mathfrak{g}_{\mathfrak{A}} = \{\boldsymbol{0}\}. \tag{5.42}$$

From (5.41), we see that the material connection is unique and that the connection symbols are given by

$$\Gamma^{C}_{BA} = -f^{-1D}_{\ \ B}\frac{\partial f^{C}_{D}}{\partial X^{A}}. \tag{5.43}$$

A formula of this kind is commonly inferred, on one ground or another, in theories of continuous distributions of dislocations.

For a given material atlas $\mathfrak{A}$ the local field $\boldsymbol{f}$ in (5.43) is the local crystallographic basis introduced in Examples 1 and 2 of §§ 1 and 3. For a Moebius crystal, $\boldsymbol{f}$ has no smooth global extension.

The local field $\boldsymbol{f}$ forms a horizontal cross section in $\mathscr{E}(\mathscr{B}, \mathfrak{A})$. Indeed from (5.41) the $\boldsymbol{f}$-lift

$$\boldsymbol{\lambda}(t) \equiv \boldsymbol{f} \circ X(t) \tag{5.44}$$

of any smooth curve $X(t)$ satisfies the equation of parallel transport (5.18), since

$$\frac{df^{C}_{D}}{dt} = \frac{\partial f^{C}_{D}}{\partial X^{A}}\dot{X}^{A} - \Gamma^{C}_{BA}f^{B}_{D}\dot{X}^{A}. \tag{5.45}$$

Hence from (3.24) the covariant derivative of $\boldsymbol{f}$ vanishes:

$$D\boldsymbol{f} = \mathbf{0}. \tag{5.46}$$

This fact is also clear from (5.38), which shows that the tangent space of the cross section $\boldsymbol{f}$ has no vertical component.

For a body having a continuous symmetry group, the Lie algebra $\mathfrak{g}_{\mathfrak{A}}$ is not trivial. In such a body the material connection is not unique, and the criterion limits locally, but does not determine, the connection symbols of a material connection.

6. The Integrability Condition of a Material Connection

A connection $\mathscr{H}$ is called *flat* if every point X can be covered by a local co-ordinate system (X^A) relative to which the connection symbols of $\mathscr{H}$ vanish. If in this case $\mathscr{H}$ is a material connection, then the bundle chart induced by (X^A) is a material chart, and hence the body is locally homogeneous. The existence of a flat material connection is generally sufficient but not necessary for local homogeneity. (See Example 4 of § 3.) For a solid body, however, this condition is necessary also, as we shall see in a later section.

We examine the departure of a connection from a flat connection in two steps. First, we consider the departure of an integrable connection from a flat one. Second, we consider the departure of a non-integrable connection from an integrable one. In each step, we obtain a necessary and sufficicient condition for the departure.

For the first step, we assume that $\mathscr{H}$ is integrable. Then there exists a local horizontal cross section in $\mathscr{E}(\mathscr{B})$ covering any given point X, say $\boldsymbol{f}$. Let (X^A) be a local co-ordinate system near X. Then $\boldsymbol{f}$ has the component form

$$\boldsymbol{f} = \{f_B^A \partial X_A, \ B = 1, 2, 3\}, \tag{5.34}$$

or equivalently

$$\boldsymbol{f} = \boldsymbol{R}_{[f^A{}_B]}(\boldsymbol{e}), \tag{5.39}$$

where $\boldsymbol{e}$ is the natural frame of (X^A). At the end of the preceding section we proved that the connection symbols of $\mathscr{H}$ relative to (X^A) are given by

$$\Gamma^A_{BC} = -f^{-1}{}^D_B \frac{\partial f^A_D}{\partial X^C}. \tag{5.43}$$

From this formula we see that $\mathscr{H}$ is flat if and only if there exists a co-ordinate system in which the components of $\boldsymbol{f}$ are constants; or, equivalently, the frame $\boldsymbol{f}$ coincides with the natural frame of a co-ordinate system.

From the classical Frobenius theorem[1] we know that $\boldsymbol{f}$ is a natural frame if and only if the Lie derivatives of the basis vectors of $\boldsymbol{f}$ with respect to one another vanish identically. That is

$$[\boldsymbol{f}_A, \boldsymbol{f}_B] = \boldsymbol{0} \qquad \forall A, B = 1, 2, 3. \tag{6.1}$$

Relative to the co-ordinate system (X^A) the bracket $[\boldsymbol{f}_A, \boldsymbol{f}_B]$ has the component form

$$[\boldsymbol{f}_A, \boldsymbol{f}_B] = \left(f^D_A \frac{\partial f^C_B}{\partial X^D} - f^D_B \frac{\partial f^C_A}{\partial X^D}\right) \partial X_C. \tag{6.2}$$

---

[1] *Cf.* I.4.

Comparing with (5.43), we see that

$$[\boldsymbol{f}_A, \boldsymbol{f}_B] = (\Gamma^C_{DE} - \Gamma^C_{ED}) f^D_A f^E_B \partial X_c. \tag{6.3}$$

Since $[f^A_B]$ is non-singular, this formula shows that (6.1) holds if and only if the connection symbols Γ^A_{BC} are symmetric with respect to their lower indices.

From the transformation rule (5.27) for the connection symbols, we see that the fields $(\Gamma^A_{BC} - \Gamma^A_{CB})$ satisfy the transformation rule for the components of a tensor field. Hence we can define a tensor field $\mathbf{C}$, called the *torsion tensor* of the connection, by

$$C^A_{BC} \equiv \Gamma^A_{BC} - \Gamma^A_{CB}. \tag{6.4}$$

Then (6.3) becomes the component form of the tensor equation

$$[\boldsymbol{f}_A, \boldsymbol{f}_B] = \mathbf{C}(\boldsymbol{f}_A, \boldsymbol{f}_B), \tag{6.5}$$

which is the previous equation (1.21). In terms of the torsion tensor $\mathbf{C}$, a necessary and sufficient condition for an integrable connection to be flat is

$$\mathbf{C} = \mathbf{0}. \tag{6.6}$$

Since the components of the torsion tensor are linear combinations of the connection symbols, the condition (6.6) is invariant under the linear combination (4.37) by a partition of unity. Thus in the proof of the existence theorem of the previous section, if the local connections $\hat{\mathscr{H}}^\alpha$ are torsion-free, so is the patched-together connection $\hat{\mathscr{H}}$. This remark implies the following

PROPOSITION. *Every locally homogeneous body has a torsion-free material connection.*

In the classical theory of differential geometry, the torsion tensor is interpreted in the following way. Let X be a fixed point, and let $\boldsymbol{u}$ and $\boldsymbol{v}$ be vectors at X. We consider the vectors $\alpha\boldsymbol{u}$ and $\beta\boldsymbol{v}$, where α and β are scalar parameters which approach to zero in a limiting process. Suppose that we take the parallel transport of $\alpha\boldsymbol{u}$ in the direction of $\beta\boldsymbol{v}$ and then the parallel transport of $\beta\boldsymbol{v}$ in the direction of $\alpha\boldsymbol{u}$, as shown in the adjoining sketch:

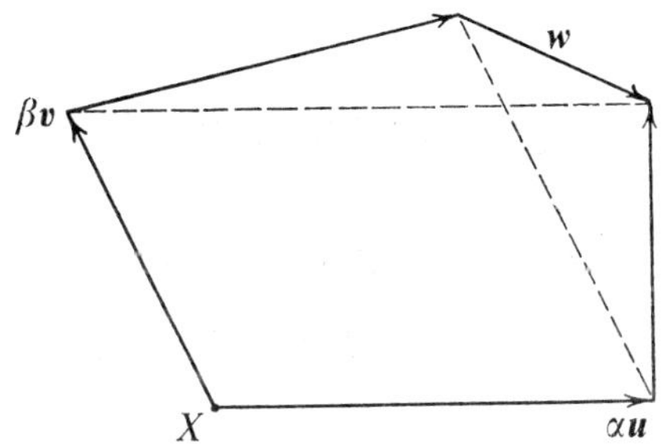

(6.7)

From the equation of parallel transport (5.22), the result of the first parallel transport has the components

$$\alpha u^A - \Gamma^A_{BC}(X)\alpha u^B \beta v^C + o(\alpha) + o(\beta) \tag{6.8}$$

while that of the second parallel transport has the components

$$\beta v^A - \Gamma^A_{BC}(X)\beta v^B \alpha u^C + o(\alpha) + o(\beta), \tag{6.9}$$

for sufficiently small α and β. Hence an estimate for the gap $\boldsymbol{w}$ from a closed infinitesimal parallelogram is

$$w^A = \alpha\beta(\Gamma^A_{BC}(X) - \Gamma^A_{CB}(X))u^A v^B + o(\alpha) + o(\beta), \tag{6.10}$$

or, equivalently [1],

$$\boldsymbol{w} = \alpha\beta[\mathbf{C}(X)](\boldsymbol{u}, \boldsymbol{v}) + \boldsymbol{o}(\alpha) + \boldsymbol{o}(\beta). \tag{6.11}$$

Thus $\mathbf{C}(X)$ characterizes the leading term in an estimate for the gap $\boldsymbol{w}$.

So far we have considered the distinction between an integrable connection and a flat one. For a solid crystal body, the unique material connection is integrable. Thus the torsion tensor of the material connection characterizes the local inhomogeneity of the body completely. For a body having a continuous symmetry group, however, a material connection is neither unique nor integrable in general. We consider the condition of integrability next.

In the preceding section we proved that relative to a lifted co-ordinate system (X^A, e^B_C), $\hat{\mathscr{H}}$ is spanned by the set

$$\{\partial X_A - \Gamma^B_{CA} e^C_D \partial e^D_B,\ A = 1, 2, 3\}, \tag{5.30}$$

so the orthogonal complement of $\hat{\mathscr{H}}$ is spanned by the set

$$\{\partial e^B_C + \Gamma^B_{EF} e^E_C \partial X^F,\ B, C = 1, 2, 3\}. \tag{6.12}$$

[1] It can be shown that the error term in (6.11) is in fact of order $\boldsymbol{o}(\alpha^2 + \beta^2)$.

We put

$$\zeta^B_C \equiv \partial e^B_C + \Gamma^B_{EF}\, e^E_C\, \partial X^F. \tag{6.13}$$

For each B and C, ζ^B_C is a 1-form defined on $\mathscr{E}(\mathscr{U}_\alpha)$, where $\mathscr{U}_\alpha$ is the domain of (X^A). From (6.13) the exterior derivative of ζ^B_C is

$$\partial\zeta^B_C = \tfrac{1}{2} e^A_C R^B_{ADE}\, \partial X^D \wedge \partial X^E + \Gamma^B_{DE}\, \partial e^D_C \wedge \partial X^E, \tag{6.14}$$

where

$$R^A_{BCD} = \frac{\partial \Gamma^A_{BD}}{\partial X^C} - \frac{\partial \Gamma^A_{BC}}{\partial X^D} + \Gamma^A_{EC}\Gamma^E_{BD} - \Gamma^A_{ED}\Gamma^E_{BC}. \tag{6.15}$$

The quantities R^A_{BCD} are called the *curvature symbols* of $\hat{\mathscr{H}}$ relative to (X^A). Because of the transformation rule (5.27) for the connection symbols, the curvature symbols satisfy the transformation rule of the components of a tensor field $\mathbf{R}$, which we call the *curvature tensor* of $\hat{\mathscr{H}}$.

By the classical Frobenius theorem [1], $\hat{\mathscr{H}}$ is integrable if and only if

$$\partial\zeta^B_C \wedge \zeta^1_1 \wedge \zeta^1_2 \ldots \wedge \zeta^3_2 \wedge \zeta^3_3 = \mathbf{0}. \tag{6.16}$$

Substituting (6.14) into this equation yields the following necessary and sufficient condition for integrability:

$$R^A_{BCD} = 0, \tag{6.17}$$

or equivalently,

$$\mathbf{R} = \mathbf{0}. \tag{6.18}$$

The existence of a curvature-free connection is closely related to whether or not the bundle of frames is trivial. If the bundle of reference frames $\mathscr{E}(\mathscr{B}, \mathfrak{A})$ is trivial, then the field of the tangent spaces of a global cross section in $\mathscr{E}(\mathscr{B}, \mathfrak{A})$ gives a curvature-free material connection. Conversely, if the body $\mathscr{B}$ is simply connected and has a curvature-free material connection, then the bundle of reference frames is trivial. Since every bundle is locally trivial, for any point X there exists a material connection which is curvature-free on a neighborhood of X. This condition implies that every retractible body has a curvature-free material connection.

The geometric meaning of the curvature tensor can be interpreted in the following way. We consider first the simple case when the connection is torsion-free, so that the

---

[1] *Cf.* § I.4.

infinitesimal parallelogram (6.7) is closed to within second-order terms in α and β. Now we determine the parallel transport of a vector $\boldsymbol{z}$ first from X to $X+(\alpha\boldsymbol{u}+\beta\boldsymbol{v})$ via $X+\alpha\boldsymbol{u}$, then for the same end points but via $X+\beta\boldsymbol{v}$. The result of the first parallel transport has the components

$$z^A-\Gamma^A_{DC}(X)z^B\alpha u^C+o(\alpha) \tag{6.19}$$

at $X+\alpha\boldsymbol{u}$ and the components

$$z^A-\Gamma^A_{BC}(X)z^B\alpha u^C-\left(\Gamma^A_{BC}(X)+\frac{\partial\Gamma^A_{BC}}{\partial X^D}\alpha u^D\right)(z^B-\Gamma^B_{EF}(X)z^E\alpha u^F). \tag{6.20}$$
$$(\beta v^C-\Gamma^C_{GH}(X)\beta v^G\alpha u^H)+o(\alpha)+o(\beta)$$

at $X+\alpha\boldsymbol{u}+\beta\boldsymbol{v}$. By the same argument, the parallel transport of $\boldsymbol{z}$ via $X+\beta\boldsymbol{v}$ results in a vector with the components

$$z^A-\Gamma^A_{BC}(X)z^B\beta v^C-\left(\Gamma^A_{BC}(X)+\frac{\partial\Gamma^A_{BC}}{\partial X^D}\beta v^D\right)(z^B-\Gamma^B_{EF}(X)z^E\beta v^F). \tag{6.21}$$
$$(\alpha u^C-\Gamma^C_{GH}(X)\alpha u^G\beta v^H)+o(\alpha)+o(\beta)$$

at $X+\alpha\boldsymbol{u}+\beta\boldsymbol{v}$. These two parallel transports can be represented by the following sketch:

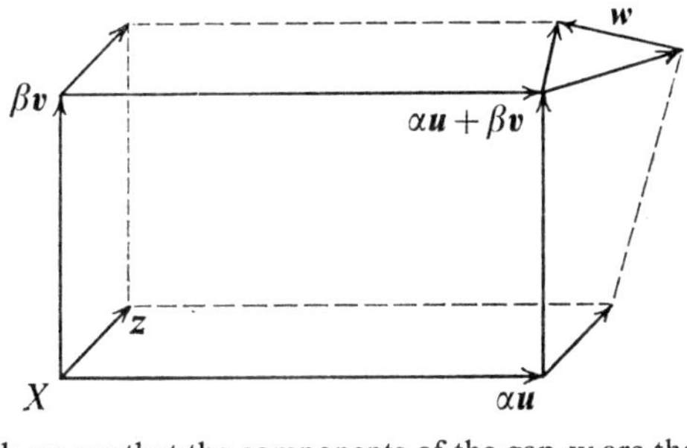

(6.22)

From this sketch we see that the components of the gap $\boldsymbol{w}$ are the differences of those given by (6.21) and those given by (6.20). Specifically, we have

$$w^A = \alpha\beta R^A_{BCD}(X)z^Bu^Cv^D+o(\alpha)+o(\beta), \tag{6.23}$$

or in vector notation[1]

$$\boldsymbol{w} = \alpha\beta[\mathbf{R}(X)](\boldsymbol{z}, \boldsymbol{u}, \boldsymbol{v})+o(\alpha)+o(\beta). \tag{6.24}$$

Thus the curvature tensor characterizes the leading term of an estimate for the gap of the parallel transports of $\boldsymbol{z}$.

Next we consider the general case, in which the connection is not necessarily torsion-

[1] It can be shown that the error term in (6.24) is in fact of order $o(\alpha^2+\beta^2)$.

free. From (6.7), the base parallelogram is not closed but has a gap given by (6.11). We determine one more parallel transport of $\boldsymbol{z}$ along that gap, as shown in the following sketch:

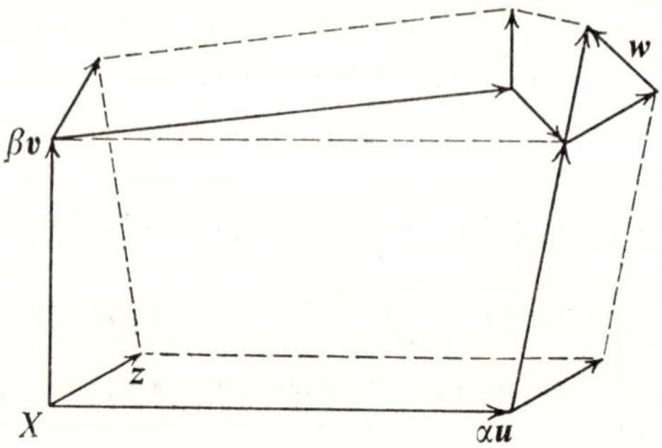

(6.25)

From (6.21), the result of this additional parallel transport has the components

$$z^A - \Gamma^A_{BC}(X) z^B \beta v^C - \left(\Gamma^A_{BC}(X) + \left.\frac{\partial \Gamma^A_{BC}}{\partial X^D}\right|_X \beta v^D \right) (z^B - \Gamma^B_{EF}(X) z^E \beta v^F). \quad (6.26)$$

$$(\alpha u^C - \Gamma^C_{GH}(X) \alpha u^G \beta v^H) - \Gamma^A_{BC}(X) z^B \alpha \beta C^C_{DE}(X) u^D v^E + o(\alpha) + o(\beta).$$

Now the gap $\boldsymbol{w}$ of the parallel transports of $\boldsymbol{z}$ has components equal to the difference of those given by (6.26) and those given by (6.20). The result turns out to be the same as (6.23) found in the previous case. Thus the estimate (6.24) is valid in general.

Summarizing the results for the two steps of departure of a connection in general form a flat connection, we have the following

PROPOSITION. *A connection is flat if and only if its torsion tensor and curvature tensor vanish.*

For a material connection, this proposition gives the following characterization of local homogeneity.

PROPOSTION. *A body is locally homogeneous at a point X if and only if there exists a material connection whose torsion tensor and curvature tensor vanish in a neighborhood of X.*

This proposition means that locally the existence of a flat material connection is equivalent to local homogeneity. In the next section we shall prove that in the case of a solid body, the equivalence is valid globally also. Example 4 of § 3 shows, however, that for fluid crystal bodies in general such global equivalence does not hold.

7. The Concept of a Characteristic Field

In the preceding sections of this chapter we have demonstrated the use of a material connection to characterize the structure of a material body. Knowing a fixed material connection $\mathscr{H}$, we can determine a material isomorphism between any two points X and Y of $\mathscr{B}$ by a parallel transport relative to $\mathscr{H}$ along any curve joining X to Y. If one material isomorphism $\boldsymbol{I}(X, Y)$ of X with Y is known, then the totality of all material isomorphisms of X with Y is given by

$$\{\boldsymbol{I}(X, Y) \circ \boldsymbol{A}_X \quad \forall \boldsymbol{A}_X \in \mathfrak{G}_X\}, \tag{7.1}$$

or by

$$\{\boldsymbol{A}_Y \circ \boldsymbol{I}(X, Y) \quad \forall \boldsymbol{A}_Y \in \mathfrak{G}_Y\}, \tag{7.2}$$

where $\mathfrak{G}_X$ and $\mathfrak{G}_Y$ are the symmetry groups of X and Y, respectively. Thus we can determine the distribution of the response functions on $\mathscr{B}$ from any given material connection. In general, we cannot determine the symmetry group from a material connection, however. We know only that the holonomy group $\mathfrak{H}_X$ of $\mathscr{H}$ at X, consisting in all parallel transports of X with itself along all loops at X, is a subgroup[1] of $\mathfrak{G}_X$.

In this section we introduce a new concept for a material body, called a *characteristic field*, which, as we shall see, determines not only the material structure but also the symmetry group of the body. Before defining this concept in general, we review a few familiar special cases. Recall that a body point X is a fluid point if the symmetry group of X is the unimodular group of $\mathscr{B}_X$. Mathematically this condition means that X has a *characteristic volume tensor* $\mathbf{V}_X$ such that an automorphism of $\mathscr{B}_X$ belongs to $\mathfrak{G}_X$ if and only if it preserves $\mathbf{V}_X$. For a fluid body $\mathscr{B}$ there exists, then, a *characteristic volume tensor field* $\mathbf{V}$ such that an isomorphism $\boldsymbol{I}(X, Y)$ from $\mathscr{B}_X$ to $\mathscr{B}_Y$ is a material isomorphism if and only if it maps $\mathbf{V}_X$ into $\mathbf{V}_Y$. This special case shows clearly that the tensor field $\mathbf{V}$ characterizes both the distribution of the response function and the symmetry group of the body.

---

[1] By choosing $\mathscr{H}$ suitably, we can make the identity component of $\mathfrak{H}_X$ coincide with that of $\mathfrak{G}_X$. Then $\mathfrak{G}_X$ and $\mathfrak{H}_X$ have the same Lie algebra. There need not be any material connection for which $\mathfrak{H}_X$ coincides with $\mathfrak{G}_X$ unless $\mathfrak{G}_X$ is connected. The reader will construct easily a counterexample for this situation in a solid crystal body.

For another example, consider the fluid crystal having one preferred axis. In this case the body point has a characteristic volume tensor $\mathbf{V}_X$ and a characteristic line $\mathscr{L}_X$ such that $\boldsymbol{A}_X \in \mathfrak{G}_X$ if and only if it preserves $\mathbf{V}_X$ and $\mathscr{L}_X$. As in the previous example, the material structure and the symmetry group for a fluid crystal body of this kind are characterized by the fields $\mathbf{V}$ and $\mathscr{L}$.

In general, we say that an object $\mathscr{K}_X$ is *intrinsic* at a point X if

$$\boldsymbol{A}_X \circ \mathscr{K}_X = \mathscr{K}_X \tag{7.3}$$

for all $\boldsymbol{A}_X \in \mathfrak{G}_X$. Here $\mathscr{K}_X$ may be a vector, a covector, a tensor, or a set of vectors or tensors, all belonging to the tangent space or tensor spaces at X. The operation of $\boldsymbol{A}_X$ on these objects are defined in the following way: If $\mathscr{K}_X$ is a vector $\boldsymbol{v} \in \mathscr{B}_X$, then $\boldsymbol{A}_X$ is the usual linear operation:

$$\boldsymbol{A}_X \circ \boldsymbol{v} \equiv \boldsymbol{A}_X \boldsymbol{v}; \tag{7.4}$$

if $\mathscr{K}_X$ is a covector $\boldsymbol{v}^* \in \mathscr{B}_X^*$, then $\boldsymbol{A}_X \circ \boldsymbol{v}^*$ is defined by

$$[\boldsymbol{A}_X \circ \boldsymbol{v}^*](\boldsymbol{v}) \equiv \boldsymbol{v}^*(\boldsymbol{A}_X \boldsymbol{v}) \quad \forall \boldsymbol{v} \in \mathscr{B}_X; \tag{7.5}$$

and, more generally, the operation of $\boldsymbol{A}_X$ on a tensor $♀_X$ of type (r, s) over $\mathscr{B}_X$ is defined by

$$[\boldsymbol{A}_X \circ ♀_X](\boldsymbol{v}_1^*, \ldots, \boldsymbol{v}_r^*, \boldsymbol{v}_1 \ldots, \boldsymbol{v}_s) \equiv ♀_X(\boldsymbol{A}_X \circ \boldsymbol{v}_1^*, \ldots, \boldsymbol{A}_X \circ \boldsymbol{v}_r^*, \boldsymbol{A}_X \boldsymbol{v}_1, \ldots, \boldsymbol{A}_X \boldsymbol{v}_s) \tag{7.6}$$

for all $\boldsymbol{v}_1^*, \ldots, \boldsymbol{v}_r^* \in \mathscr{B}_X^*$ and all $\boldsymbol{v}_1, \ldots, \boldsymbol{v}_s \in \mathscr{B}_X$. Likewise, the operation of $\boldsymbol{A}_X$ on a subset $\mathscr{U}_X \subset \mathscr{B}_X$ is defined by

$$\boldsymbol{A}_X \circ \mathscr{U}_X \equiv \{\boldsymbol{A}_X \boldsymbol{v},\ \boldsymbol{v} \in \mathscr{U}_X\}, \tag{7.7}$$

and so forth.

An intrinsic object $\mathscr{K}_X$ is called a *characteristic object* if among all automorphisms of $\mathscr{B}_X$ an automorphism $\boldsymbol{A}_X$ satisfies the condition (7.3) if and only if $\boldsymbol{A}_X \in \mathfrak{G}_X$. For a characteristic object $\mathscr{K}_X$, we then have

$$\boldsymbol{A}_X \circ \mathscr{K}_X = \mathscr{K}_X \Leftrightarrow \boldsymbol{A}_X \in \mathfrak{G}_X. \tag{7.8}$$

Clearly the volume tensor $\mathbf{V}_X$ is a characteristic tensor for a fluid point X. Likewise, the pair $(\mathbf{V}_X, \mathscr{L}_X)$ is a characteristic object for the kind of fluid crystal point discussed above, though in this case the objects $\mathbf{V}_X$ and $\mathscr{L}_X$ separately are only intrinsic but not characteristic for X.

PROPOSITION. *Every point X has a characteristic object.*

The proof is obvious, since the fibre of reference frames $\mathscr{E}_X(\mathfrak{A})$, which is a particular set of lists of vectors at X, satisfies the condition (7.8).

A field of intrinsic or characteristic objects $\mathscr{K}$ on a body $\mathscr{B}$ is called *intrinsic* or a *characteristic field*, respectively, if any material isomorphism $\boldsymbol{I}(X, Y)$ from X to Y maps $\mathscr{K}_X$ into $\mathscr{K}_Y$. Again, the action of $\boldsymbol{I}(X, Y)$ is understood to be defined suitably in each specific case.

PROPOSITION. *Every intrinsic or characteristic object $\mathscr{K}_X$ at a particular point $X \in \mathscr{B}$ has a unique extension to an intrinsic or characteristic field $\mathscr{K}$ on $\mathscr{B}$ by means of the relation*

$$\mathscr{K}_Y = \boldsymbol{I}(X, Y) \circ \mathscr{K}_X \qquad \forall Y \in \mathscr{B}, \tag{7.9}$$

where $\boldsymbol{I}(X, Y)$ is any material isomorphism of X with Y. In particular, every intrinsic or characteristic field on $\mathscr{B}$ is smooth in the sense that the field $\boldsymbol{I}(X, \cdot)$ can be chosen locally smooth.

The proof is again obvious. First, $\mathscr{K}_Y$ is well defined by (7.9), since from (7.1) and from the fact that $\mathscr{K}_X$ is intrinsic or characteristic, respectively, the result $\boldsymbol{I}(X, Y) \circ \mathscr{K}_X$ is independent of the choice of the material isomorphism $\boldsymbol{I}(X, Y)$. Second, $\mathscr{K}_Y$ is intrinsic or characteristic at Y, since the set (7.1) coincides wih the set (7.2).

Combining the preceding two propositions, we see that *every materially uniform body has a characteristic field.* Further, in the sense explained above, *a materially uniform body is smooth if and only if its characteristic field is smooth.* A characteristic field is not unique, of course. The bundle of reference frames $\mathscr{E}(\mathscr{B}, \mathfrak{A})$ is nothing but a special characteristic field on $\mathscr{B}$. We need not restrict our attention to this special kind of characteristic field. A connection is a material connection if and only if its parallel transports preserve any one, and hence all, characteristic fields.

Besides the characteristic field $\mathbf{V}$ on a fluid body and the characteristic field $(\mathbf{V}, \mathscr{L})$ on a type of fluid crystal body, another important case occurs in an isotropic solid body. We give some preliminary results first.

PROPOSITION. *A body-point X is a solid point if and only if it has an intrinsic inner product. A solid point is isotropic if and only if its intrinsic inner product is a characteristic inner product.*

The proof is again obvious. Indeed, any undistorted state of a solid point X maps the Euclidean inner product of the physical space into an intrinsic inner product on $\mathscr{B}_X$. Conversely, if $\mathscr{B}_X$ has an intrinsic inner product, then any local configuration carrying that inner product onto the Euclidean inner product of the physical space is an undistorted state for X. The proof of the second part is the same.

Now applying the formula (7.9), we obtain the

PROPOSITION. *A body $\mathscr{B}$ is a solid body if and only if it has an intrinsic Riemannian metric. A solid body is isotropic if and only if its intrinsic Riemannian metric is a characteristic field.*

An intrinsic Riemannian metric g on a solid body is not unique, of course. Suppose that $\boldsymbol{v}_X$ is a local configuration of a particular point X such that $\boldsymbol{v}_X$ carries g_X onto the Euclidean inner product of the physical space, *i.e.*,

$$g_X(\boldsymbol{u}, \boldsymbol{v}) = \boldsymbol{v}_X(\boldsymbol{u}) \cdot \boldsymbol{v}_X(\boldsymbol{v}) \tag{7.10}$$

for all $\boldsymbol{u}, \boldsymbol{v} \in \mathscr{B}_X$. Let $\boldsymbol{J}$ be any positive-definite and symmetric tensor over the physical space such that

$$\boldsymbol{QJ} = \boldsymbol{JQ} \tag{7.11}$$

for all $\boldsymbol{Q}$ belonging to the symmetry group $\mathfrak{G}_{\boldsymbol{v}_X}$ relative to $\boldsymbol{v}_X$. Then the inner product $\bar{g}_X$ given by

$$\bar{g}_X(\boldsymbol{u}, \boldsymbol{v}) \equiv \boldsymbol{J}(\boldsymbol{v}_X(\boldsymbol{u}), \boldsymbol{v}_X(\boldsymbol{v})) \quad \forall \boldsymbol{u}, \boldsymbol{v} \in \mathscr{B}_X \tag{7.12}$$

is also intrinsic at X. Finally by (7.9) we can extend $\bar{g}_X$ to an intrinsic Riemannian metric $\bar{g}$ on $\mathscr{B}$. From the commutation theorem (*cf.* § I,2), a symmetric tensor $\boldsymbol{J}$ satisfies (7.11) if and only if its characteristic spaces are invariant under $\boldsymbol{Q}$. In particular,

$$\boldsymbol{J} = c\mathbf{1}, \qquad c > 0, \tag{7.13}$$

is a solution of (7.11). Thus the intrinsic Riemannian metric is not unique for any solid body.

For an isotropic solid or a cubic crystal, the equation (7.13) gives the complete solution for the condition (7.11). Thus for these two types of solids, the intrinsic Riemannian metric is unique to within a positive scalar factor. On the other hand, for a triclinic solid crystal, since the symmetry group is the trivial group, any positive-definite and symmetric tensor $\boldsymbol{J}$ is a solution of (7.11). Thus for this type of solid any inner product on $\mathscr{B}_X$ is intrinsic (for that matter any object is intrinsic at X in this special case) and can be extended to an intrinsic Riemannian metric on $\mathscr{B}$. A characteristic field for a triclinic crystal is the field of crystallographic bases.

In differential geometry it is well known that there exists a unique torsion-free connection, called the *Riemannian connection*, whose parallel transports preserve a given Riemannian metric. Suppose that g is an intrinsic Riemannian metric on $\mathscr{B}$. Then g is invariant under the parallel transports relative to all material connections. In particular, if $\mathscr{B}$ has a torsion-free material connection, then that connection must coincide with the Riemannian connection of g. Further, in this case the Riemannian connections of all intrinsic Riemannian metrics must coincide. We summarize this result in the following

PROPOSITION. *A solid body can have at most one torsion-free material connection. If that material connection exists, it coincides with the Riemannian connection of all intrinsic Riemannian metrics of the body.*

Combining this proposition with the former one, proved in the preceding section, that every locally homogeneous body has a torsion-free material connection, yields immediately the result announced before:

PROPOSITION. *A solid body is locally homogeneous if and only if it has a flat material connection.* (That connection is unique in view of the preceding proposition.)

The proof of sufficiency was given above. For the necessity, it suffices to note that for a locally homogeneous body the representation (7.10) shows an intrinsic Riemannian metric to be locally Euclidean. Consequently the Riemannian connection, which is a material connection on account of the preceding proposition, is flat.

Since an intrinsic Riemannian metric is a characteristic field for an isotropic solid body, we have the following

PROPOSITION. *Every isotropic solid body has a unique torsion-free material connection; further, that material connection is the Riemannian connection of any intrinsic Riemannian metric of the body.*

The connection symbols of the Riemannian connection, of course, are the classical Christoffel symbols (*cf.* § I.3), *viz*,

$$\Gamma^A_{BC} = \begin{Bmatrix} A \\ BC \end{Bmatrix} = \tfrac{1}{2} g^{AD} \left(\frac{\partial g_{BD}}{\partial X^C} + \frac{\partial g_{CD}}{\partial X^B} - \frac{\partial g_{BC}}{\partial X^D} \right), \tag{7.14}$$

where g_{AB} and g^{AB} are the components of the Riemannian metric in the co-ordinate system (X^A). The Christoffel symbols, and hence the Riemannian connection, are invariant under a multiplication of the Riemannian metric by a positive constant.

We have just proved that for an isotropic solid body the Riemannian connection of an intrinsic Riemannian metric is necessarily a material connection. On the other hand, for a solid crystal body the Riemannian connection of an intrinsic metric is a material connection (which is unique and integrable in this case, as we proved before) if and only if the body is locally homogeneous. The only other type[1] of solids are the transversely isotropic solids, for which we have the following

PROPOSITION. *A transversely isotropic body has a torsion-free connection if and only if the Riemannian connections of the intrinsic Riemannian metrics of the body coincide with one another. Then that unique Riemannian connection is the unique torsion-free material connection.*

We note first that uniqueness of the torsion-free connection follows from a previous proposition. Also, necessity was proved above. For sufficiency, we observe that a characteristic field of a transversely isotropic body is a pair $(g, \boldsymbol{a} \otimes \boldsymbol{a})$, where g is an intrinsic Riemannian metric, and where $\boldsymbol{a}$ is the intrinsic field of the axes of symmetry. From (7.9), $\boldsymbol{a}$ is of constant length with respect to g. We denote the plane perpendicular to $\boldsymbol{a}$ with respect to g by $\mathscr{M}$.

Next, from the condition (7.12) we see that another Riemannian metric $\bar{g}$ on $\mathscr{B}$ is intrinsic if and only if the following three conditions are satisfied:

i) The restrictions of $\bar{g}$ and g to $\mathscr{M}$ differ by a positive constant factor only, say

$$\bar{g}(\boldsymbol{u}, \boldsymbol{v}) = \alpha^2 g(\boldsymbol{u}, \boldsymbol{v}), \ \alpha > 0, \tag{7.15}$$

for all $\boldsymbol{u}, \boldsymbol{v} \in \mathscr{M}$.

ii) The lengths of $\boldsymbol{a}$ obtained from $\bar{g}$ and g differ by a positive constant factor, say

$$\bar{g}(\boldsymbol{a}, \boldsymbol{a}) = \beta^2 g(\boldsymbol{a}, \boldsymbol{a}), \ \beta > 0. \qquad (.16)$$

[1] From the classification of the Lie algebras listed in § I.6, we see that there are only three types of skew-symmetric Lie algebras, which correspond to solid crystals, transversely isotropic solids, and isotropic solids, respectively.

iii) $\mathscr{M}$ is orthogonal to $\boldsymbol{a}$:

$$\bar{g}(\boldsymbol{a}, \boldsymbol{u}) = g(\boldsymbol{a}, \boldsymbol{u}) = 0 \tag{7.17}$$

for all $\boldsymbol{u} \in \mathscr{M}$.

Now by hypothesis the Riemannian connection of g coincides with that of $\bar{g}$. Suppose that $\boldsymbol{\rho}_{X,\,Y}$ is a parallel transport relative to that connection. We put

$$\bar{\boldsymbol{a}}_Y \equiv \boldsymbol{\rho}_{X,Y}(\boldsymbol{a}_X). \tag{7.18}$$

We claim that

$$\bar{\boldsymbol{a}}_Y = \boldsymbol{a}_Y. \tag{7.19}$$

To see this we take the usual orthogonal decomposition of $\bar{\boldsymbol{a}}_Y$:

$$\bar{\boldsymbol{a}}_Y = \gamma \boldsymbol{a}_Y + \boldsymbol{u}_Y \tag{7.20}$$

where $\boldsymbol{u}_Y \in \mathscr{M}_Y$. Since $\boldsymbol{\rho}_{X,\,Y}$ is a parallel transport relative to the Riemannian connection of g and $\bar{g}$, we have

$$g_Y(\bar{\boldsymbol{a}}_Y, \bar{\boldsymbol{a}}_Y) = g_X(\boldsymbol{a}_X, \boldsymbol{a}_X) \tag{7.21}$$

and

$$\bar{g}_Y(\boldsymbol{a}_Y, \boldsymbol{a}_Y) = \bar{g}_X(\boldsymbol{a}_X, \boldsymbol{a}_X). \tag{7.22}$$

Now from (7.16) the right-hand side of (7.22) is equal to $\beta^2 g_X(\boldsymbol{a}_X, \boldsymbol{a}_X)$, while from (7.15)–(7.17), (7.20), and (7.21) the left-hand side of (7.22) is equal to $\gamma^2\beta^2 g_X(\boldsymbol{a}_X, \boldsymbol{a}_X) + g_Y(\boldsymbol{u}_Y, \boldsymbol{u}_Y)$, so

$$\gamma^2\beta^2 g_X(\boldsymbol{a}_X, \boldsymbol{a}_X) + g_Y(\boldsymbol{u}_Y, \boldsymbol{u}_Y) = \beta^2 g_X(\boldsymbol{a}_X, \boldsymbol{a}_X). \tag{7.23}$$

But since β is arbitrary, and since $g_X(\boldsymbol{a}_X, \boldsymbol{a}_X)$ is not zero, the equation is satisfied if and only if

$$\gamma^2 = 1, \ \boldsymbol{u}_Y = \boldsymbol{0}. \tag{7.24}$$

Hence from (7.20) $\boldsymbol{\rho}_{X,\,Y}$ preserves the characteristic field $(g, \boldsymbol{a} \otimes \boldsymbol{a})$. This means that the unique Riemannian connection is a material connection for the transversely isotropic body.

The situation with a solid crystal body is quite different. Even if the Riemannian connection of all intrinsic Riemannian metrics is unique, that connection still need not be a material connection. A counterexample to any contrary idea is furnished by an inhomogeneous cubic crystal. From a previous proposition, the intrinsic Riemannian metric for this type of crystal is unique to within a positive constant factor, so the Riemannian connection of all intrinsic Riemannian metrics is indeed unique. But that connection is a material connection if and only if the crystal is locally homogeneous.

From a former proposition a necessary condition for the existence of a torsion-free material connection on a solid body is that the Riemannian connection of any intrinsic Riemannian metric be a material connection. Hence if the Riemannian connection of one intrinsic Riemannian metric is not a material connection, then the solid body does not possess any torsion-free material connection. This remark enables us to construct as a counterexample a solid body whose material connections all have non-zero curvature and torsion.

Example. A Transversely Isotropic Laminated Spherical Shell

This example is the same as Example 3 of § 3. By assumption an intrinsic Riemannian metric coincides with the Euclidean metric in a global configuration $\boldsymbol{\kappa}$ of the body. This condition implies that the Riemannian connection of that intrinsic Riemannian metric is flat. But since we already proved in § 3 that this body had no curvature-free material connection, that Riemannian connection is not a material connection. Now using the remark given just before this example, we see that this body has no torsion-free material connection either.

The Riemannian connection of an intrinsic Riemannian metric need not be integrable in general. Unlike the integrability condition of a material connection, which is a statement of local homogeneity, the curvature tensor $\boldsymbol{R}^*$ of the Riemannian connection is a measure of local inhomogeneity. In other words, for a material connection

$$\boldsymbol{C} = \mathbf{0}, \ \boldsymbol{R} = \mathbf{0} \Rightarrow \text{local homogeneity} \tag{7.25}$$

while for a Riemannian connection of an intrinsic Riemannian metric

$$\boldsymbol{R}^* \neq \mathbf{0} \Rightarrow \text{local inhomogeneity} \tag{7.26}$$

In neither case, in general, can the arrow be reversed, as we have shown already by counterexamples.

For a solid body, however, the arrow in (7.25) may be reversed in the sense that local homogeneity implies the existence of a (unique) flat material connection. Likewise, for an isotropic solid body the arrow in (7.26) may be reversed.

8. The Equations of Motion for an Elastic Body

In this section we derive the equations of motion for an elastic body. For most boundary-value problems it is desirable to have the governing equations in terms of a global co-ordinate system and based on a single response function. We can achieve these conditions by using a material connection.

Let $\mathfrak{A}$ be a given material atlas for $\mathscr{B}$. For simplicity, we suppress the dependence of the relative response function on $\mathfrak{A}$ from the notation and write the constitutive equation as

$$\boldsymbol{T} = \boldsymbol{H}(\boldsymbol{F}). \tag{8.1}$$

Assuming, as usual, that the relative response function $\boldsymbol{H}$ is smooth, we put

$$H^{ij}{}_{kl} \equiv H^{ij}{}_{kl}(\boldsymbol{F}) \equiv \frac{\partial H^{ij}}{\partial F^{kl}}, \tag{8.2}$$

where H^{ij} and F^{kl} are referred to the given frame of reference. The order of the indices is important in this case, since it is understood that the indices may be raised or lowered with respect to the Euclidean metric of the physical space. As before, $\boldsymbol{H}$ satisfies the symmetry condition

$$\boldsymbol{H}(\boldsymbol{FP}) \equiv \boldsymbol{H}(\boldsymbol{F}) \tag{8.3}$$

for all $\boldsymbol{P} \in \mathfrak{G}_{\mathfrak{A}}$. Then the gradient of $\boldsymbol{H}$ satisfies the conditions

$$H^{ij}{}_{kl}(\boldsymbol{F}) F^{km} K_m{}^l = 0 \tag{8.4}$$

for all $\boldsymbol{K}$ belonging to the Lie algebra $\mathfrak{g}_{\mathfrak{A}}$ of $\mathfrak{G}_{\mathfrak{A}}$, and

$$H^{ij}{}_{kl}(\boldsymbol{F}) = H^{ij}{}_{km}(\boldsymbol{FP}) P_l{}^m \tag{8.5}$$

for all $\boldsymbol{P}$ belonging to $\mathfrak{G}_{\mathfrak{A}}$. These two conditions are obtained by differentiating (8.3) with respect to $\boldsymbol{P}$ and $\boldsymbol{F}$, respectively.

Let χ be a configuration of $\mathscr{B}$. Then we can determine the stress tensor in $\chi(\mathscr{B})$ in the following way: We choose an arbitrary material chart $(\mathscr{U}_\alpha, \boldsymbol{v}_\alpha)$ in $\mathfrak{A}$. Then for any point $X \in \mathscr{U}_\alpha$, the deformation gradient from $\boldsymbol{v}_{\alpha, X}$ to χ_{*X} is

$$\boldsymbol{F}(\boldsymbol{x}) = \chi_{*X} \circ \boldsymbol{v}_{\alpha, X}^{-1}, \tag{8.6}$$

where $\boldsymbol{x}$ is the position of X in χ, *viz*

$$\boldsymbol{x} = \chi(X). \tag{8.7}$$

Substituting (8.6) into (8.1) yields the stress tensor at $\boldsymbol{x}$:

$$\boldsymbol{T}(\boldsymbol{x}) = \boldsymbol{H}(\chi_{*X} \circ \boldsymbol{v}_{\alpha, X}^{-1}). \tag{8.8}$$

From (8.6), the components of the deformation gradient are the same as those of the reference frame of $\boldsymbol{v}_{\alpha, X}$ in the co-ordinate system (x^i), *i.e.*,

$$\boldsymbol{v}_{\alpha, X}^{-1}(\boldsymbol{e}) \equiv \boldsymbol{f}(X) = \{F^i{}_j(\boldsymbol{x})\partial_X x_i,\ j = 1, 2, 3\}, \tag{8.9}$$

where $\boldsymbol{e}$ is the given frame of reference. Substituting (8.8) and (8.9) into the general equation of motion (II. 6.21), we obtain

$$H^{ij}{}_{kl} \frac{\partial F^{kl}}{\partial x^j} + \rho b^i = \rho a^i, \tag{8.10}$$

where $\boldsymbol{b}$ is the apparent body force acting on $\chi(\mathscr{B})$, and where $\boldsymbol{a}$ is the acceleration relative to the fixed frame of reference. The argument of the function $H^{ij}{}_{kl}$, of course, is the deformation gradient $\boldsymbol{F}$. The equation (8.10) is not convenient for setting up initial-value or boundary-value problems, however, since the field $\boldsymbol{F}$ is defined on the subbody $\mathscr{U}_\alpha$ only, as shown in (8.9). In order to remove this undesirable feature of (8.10), we introduce a material connection for $\mathscr{B}$.

Let Γ^i_{jk} be the connection symbols of a material connection relative to the co-ordinate system (x^i). From the criterion (5.40) for a material connection we know that the matrices

$$\left[F^{-1i}{}_j \left(\frac{\partial F^j{}_k}{\partial x^l} + \Gamma^j_{ml} F^m{}_k\right)\right],\ l = 1, 2, 3, \tag{8.11}$$

belong to the Lie algebra $\mathfrak{g}_{\mathfrak{A}}$ of $\mathfrak{G}_{\mathfrak{A}}$. Hence from the condition (8.4), we have

$$H^{ij}{}_{kl} \left(\frac{\partial F_j{}^l}{\partial x^m} + \Gamma^k_{nm} F^{nl}\right) = 0. \tag{8.12}$$

Substituting this equation into (8.10) yields

$$-H^{ij}{}_{kl} F^{nl} \Gamma^k_{nj} + \rho b^i = \rho a^i. \tag{8.13}$$

Now this equation has a global form, since the combination

$$\overline{H}^{ij}{}_{k}{}^{n} \equiv H^{ij}{}_{kl}(\boldsymbol{F})F^{nl} \tag{8.14}$$

is independent of the choice of the material chart $(\mathscr{U}_\alpha, \boldsymbol{v}_\alpha)$.

Indeed, let $(\mathscr{U}_\beta, \boldsymbol{v}_\beta)$ be another material chart in $\mathfrak{A}$. Then the co-ordinate transformation

$$\boldsymbol{G}_{\alpha\beta} = \boldsymbol{v}_\alpha \circ \boldsymbol{v}_\beta^{-1} \tag{8.15}$$

belongs to the symmetry group $\mathfrak{G}_{\mathfrak{A}}$. From (8.6), the deformation gradient $\bar{\boldsymbol{F}}$ from $\boldsymbol{v}_\beta$ to $\boldsymbol{\chi}$ is related to $\boldsymbol{F}$ by

$$\bar{\boldsymbol{F}} = \boldsymbol{F}\boldsymbol{G}_{\alpha\beta}. \tag{8.16}$$

Substituting this relation into the condition (8.5), we obtain

$$\overline{H}^{ij}{}_{kl}(\bar{\boldsymbol{F}})\bar{F}^{nl} = H^{ij}{}_{kl}(\boldsymbol{F})F^{nl}, \tag{8.17}$$

which shows that $\overline{H}^{ij}{}_{k}{}^{n}$ is a global field. Substituting (8.14) into (8.13) yields the global equation of motion:

$$-\overline{H}^{ij}{}_{k}{}^{n}\Gamma^{k}_{nj} + \rho b^i = \rho a^i. \tag{8.18}$$

This equation governs the motion of $\mathscr{B}$ in the usual way: At any time t, the leading term of (8.18) depends explicitly on the instantaneous configuration $\boldsymbol{\chi}_t$ through the local formula (8.14) and the formula (5.17) for the connection symbols. Suppose that the body force $\boldsymbol{b}$ be a given field on $\boldsymbol{\chi}_t(\mathscr{B})$. Then (8.18) delivers the instantaneous acceleration field on $\boldsymbol{\chi}_t(\mathscr{B})$.

Now in each new configuration $\boldsymbol{\chi}_t$ of $\mathscr{B}$ the spatial co-ordinate system (x^i) corresponds to a new co-ordinate system on $\mathscr{B}$. Hence the connection symbols and the components of the reference frames also have new values. We can render the changes of Γ^i_{jk} and F^{kl} with respect to t more explicit by using a fixed referential co-ordinate system (X^A) induced by a fixed reference configuration $\boldsymbol{\kappa}$. The components of deformation

$$x^i = x^i(X^A, t) \tag{8.19}$$

then correspond to a 1-parameter family of changes of co-ordinates from the fixed (X^A) to the instantaneous (x^i).

For definiteness we denote the connection symbols of the material connection relative to (X^A) by $\bar{\Gamma}^A_{BC}$. Then from (5.27) $\bar{\Gamma}^A_{BC}$ and Γ^i_{jk} are related by

$$\Gamma^i_{jk} = \bar{\Gamma}^A_{BC} \frac{\partial x^i}{\partial X^A} \frac{\partial X^B}{\partial x^j} \frac{\partial X^C}{\partial x^k} - \frac{\partial^2 x^i}{\partial X^A \partial X^B} \frac{\partial X^A}{\partial x^j} \frac{\partial X^B}{\partial x^k}, \tag{8.20}$$

where we have used the reciprocal relations

$$\frac{\partial x^i}{\partial X^A} \frac{\partial X^A}{\partial x^j} = \delta^i_j, \frac{\partial x^i}{\partial X^A} \frac{\partial X^B}{\partial x^i} = \delta^B_A. \tag{8.21}$$

Also, let $\bar{F}^{Ai}$ be the components of the gradient of the deformation from $\mathfrak{A}$ to χ expressed in the co-ordinate systems (X^A) and (x^i). Then we have

$$F^{ij} = \frac{\partial x^i}{\partial X^A} \bar{F}^{Aj}. \tag{8.22}$$

This equation means that the gradient of the deformation from $\boldsymbol{v}_\alpha$ to χ is equal to the composition of the gradient of the deformation from $\boldsymbol{v}_\alpha$ to (X^A) with that from (X^A) to χ. The fields $\bar{\Gamma}^A_{BC}$ and $\bar{F}^{Aj}$ are those for the fixed reference configuration, so they no longer depend on t.

Substituting (8.20) and (8.22) into the previous field equation (8.13), we obtain a new governing equation of motion:

$$H^{ij}{}_{kl} \bar{F}^{Al} \frac{\partial X^B}{\partial x^j} \left(\frac{\partial^2 x^k}{\partial X^A \partial X^B} - \bar{\Gamma}^C_{AB} \frac{\partial x^k}{\partial X^C} \right) + \rho b^i = \rho a^i. \tag{8.23}$$

By the argument used in connection with (8.13), this equation can be written in the global form

$$\tilde{H}^{ij}{}_k{}^A \frac{\partial X^B}{\partial x^j} \left(\frac{\partial^2 x^k}{\partial X^A \partial X^B} - \bar{\Gamma}^C_{AB} \frac{\partial x^k}{\partial X^C} \right) + \rho b^i = \rho a^i, \tag{8.24}$$

where the global field $\tilde{H}^{ij}{}_k{}^A$ is given by the local formula

$$\tilde{H}^{ij}{}_k{}^A \equiv H^{ij}{}_{kl}(\boldsymbol{F}) \bar{F}^{Al} \tag{8.25}$$

Notice that the field $\tilde{H}^{ij}{}_k{}^A$ depends on X^A (through the dependence of $\bar{F}^{Al}$ on X^A) and $\partial x^k/\partial X^A$ *but not explicitly on t.*

The dependence of $\tilde{H}^{ij}{}_k{}^A$ on (X^A) arises from material inhomogeneity. Indeed, in the special case when $\mathscr{B}$ is a homogeneous body, we choose $\boldsymbol{\kappa}$ to be one of its homogeneous configurations. Using the global material chart $(\mathscr{B}, \boldsymbol{\kappa}_*)$, we have

$$\bar{F}^{Ai} \equiv \delta^{Ai}, \tag{8.26}$$

and using the flat material connection induced by $\boldsymbol{\kappa}$, we have

$$\bar{\Gamma}^{A}_{BC} \equiv 0. \tag{8.27}$$

Then the field equation (8.23) reduces to

$$H^{ij}{}_{k}{}^{A}\frac{\partial X^B}{\partial x^j}\frac{\partial^2 x^k}{\partial X^A \partial X^B} + \rho b^i = \rho a^i, \tag{8.28}$$

where the argument of $H^{ij}{}_{k}{}^{A}$ is the deformation gradient $\boldsymbol{F}$ relative to the homogeneous reference configuration.

Now we return to the general case, in which $\mathscr{B}$ need not be homogeneous. In some problems it is convenient to use the Piola-Kirchhoff stress $\boldsymbol{T}_{\boldsymbol{\kappa}}$ relative to the fixed reference configuration $\boldsymbol{\kappa}$. Recall that $\boldsymbol{T}$ and $\boldsymbol{T}_{\boldsymbol{\kappa}}$ are related by (*cf.* (II. 6.32))

$$T_{\boldsymbol{\kappa}}{}^{kA} = J T^{kl}\frac{\partial X^A}{\partial x^l}, \tag{8.29}$$

where J denotes the determinant of the matrix $[\partial x^i/\partial X^A]$. In terms of $\boldsymbol{T}_{\boldsymbol{\kappa}}$ the equation of motion is (*cf.* (II. 6.26))

$$\frac{\partial T_{\boldsymbol{\kappa}}{}^{kA}}{\partial X^A} + \rho_{\boldsymbol{\kappa}} b^k = \rho_{\boldsymbol{\kappa}} a^k, \tag{8.30}$$

where $\rho_{\boldsymbol{\kappa}}$ denotes the density in the configuration $\boldsymbol{\kappa}$.

In order to express the leading term of (8.30) as a function of the deformation, we introduce a new response function $\boldsymbol{A}$ by

$$\boldsymbol{A} = \boldsymbol{A}(\boldsymbol{F}) \equiv (\det \boldsymbol{F})\boldsymbol{H}(\boldsymbol{F})(\boldsymbol{F}^{-1})^T, \tag{8.31}$$

or in component form

$$A^{ik} \equiv (\det \boldsymbol{F}) H^{ij}(F^{-1})^k{}_j. \tag{8.32}$$

We can solve this equation for $\boldsymbol{H}$, obtaining

$$H^{ij} = \frac{1}{\det \boldsymbol{F}} F^j{}_k A^{ik}. \tag{8.33}$$

Differentiation with respect to $\boldsymbol{F}$ yields

$$\begin{aligned} H^{ij}{}_{kl} = {} & \frac{1}{\det \boldsymbol{F}} F^j{}_m A^{im}{}_{kl} + \frac{1}{\det \boldsymbol{F}} \delta^j_k \delta_{lm} A^{im} \\ & - \frac{1}{\det \boldsymbol{F}} F^j{}_m F^{-1}{}_{lk} A^{im}, \end{aligned} \tag{8.34}$$

where

$$\mathsf{A}^{ij}{}_{kl} \equiv \frac{\partial A^{ij}}{\partial F^{kl}}. \tag{8.35}$$

Substituting (8.34) into (8.13) yields the equation of motion in terms of the response function $\boldsymbol{A}$:

$$-\frac{1}{\det \boldsymbol{F}}[\mathsf{A}^{im}{}_{kl}F^{j}{}_{m}F^{nl}+(\delta^{j}_{k}F^{n}{}_{m}-\delta^{n}_{k}F^{j}{}_{m})A^{im}]\Gamma^{k}_{nj}+\rho b^{i} = \rho a^{i}. \tag{8.36}$$

Again this equation has the global form (8.18).

Now we introduce the reference configuration $\boldsymbol{\kappa}$ and express the fields Γ^{k}_{ij} and $F^{i}{}_{j}$ by (8.20) and (8.22), respectively. Substituting these into (8.36), we obtain

$$\frac{1}{\det \bar{\boldsymbol{F}}}\mathsf{A}^{is}{}_{kl}\bar{F}^{A}{}_{s}\bar{F}^{Bl}\left(\frac{\partial^2 x^k}{\partial X^A \partial X^B}-\bar{\Gamma}^{C}_{BA}\frac{\partial x^k}{\partial X^C}\right)$$
$$-\frac{1}{\det \bar{\boldsymbol{F}}}A^{is}\bar{F}^{A}{}_{s}\bar{C}^{B}_{AB}+\rho_{\boldsymbol{\kappa}}b^{i} = \rho_{\boldsymbol{\kappa}}a^{i}, \tag{8.37}$$

where the $\bar{C}^{B}_{AB}$ are the components of the torsion tensor in (X^A). By exactly the same argument as before, there are smooth global fields given locally by the formulae

$$\bar{\mathsf{A}}^{iA}{}_{k}{}^{B} \equiv \frac{1}{\det \bar{\boldsymbol{F}}}\mathsf{A}^{is}{}_{kl}(\boldsymbol{F})F^{A}{}_{s}F^{Bl} \tag{8.38}$$

and

$$\bar{A}^{iA} \equiv \frac{1}{\det \bar{\boldsymbol{F}}}A^{is}(\boldsymbol{F})\bar{F}^{A}{}_{s} = JH^{ij}(\boldsymbol{F})\frac{\partial X^A}{\partial x^j} = T_{\boldsymbol{\kappa}}{}^{iA}. \tag{8.39}$$

Then the field equation takes the global form

$$\bar{\mathsf{A}}^{iA}{}_{k}{}^{B}\left(\frac{\partial^2 x^k}{\partial X^A \partial X^B}-\bar{\Gamma}^{C}_{BA}\frac{\partial x^k}{\partial X^C}\right)-\bar{A}^{iA}\bar{C}^{B}_{AB}+\rho_{\boldsymbol{\kappa}}b^{i} = \rho_{\boldsymbol{\kappa}}a^{i}. \tag{8.40}$$

This equation was derived first by Noll, using an integrable material connection for an elastic body whose material tangent bundle is a trivial bundle (*cf.* NFTM, § 44).

As before, the inhomogeneity of $\mathscr{B}$ gives rise to the terms $\bar{\Gamma}^{C}_{AB}$, $\bar{C}^{B}_{AB}$, and the dependence of $\bar{\mathsf{A}}^{iA}{}_{k}{}^{B}$ and $\bar{A}^{iA}$ on (X^A) in the equation (8.40). For a homogeneous body, we can use a homogeneous reference configuration. Then (8.40) reduces to

$$A^{iA}{}_{k}{}^{B} \frac{\partial^2 x^k}{\partial X^A \partial X^B} + \rho_{\kappa} b^i = \rho_{\kappa} a^i, \tag{8.41}$$

where the argument of $A^{iA}{}_{k}{}^{B}$ is the deformation gradient relative to the homogeneous reference configuration.

In the next section we shall give some universal solutions for some specific inhomogeneous elastic bodies.

9. Static and Dynamic Universal Solutions for Some Incompressible Laminated Elastic Bodies

In § 7 we introduced the concept of a characteristic field for an elastic body. We proved that for an isotropic solid body an intrinsic Riemannian metric g is a characteristic field, while for a transversely isotropic solid body a pair $(g, \boldsymbol{a} \otimes \boldsymbol{a})$ consisting in an intrinsic Riemannian metric g and a field of tensor products of the axes of symmetry $\boldsymbol{a}$ is a characteristic field. Hence to describe the material structure of an isotropic solid body or of a transversely isotropic solid body, it suffices to specify the characteristic field g or $(g, \boldsymbol{a} \otimes \boldsymbol{a})$, respectively.

In this section we consider a special class of isotropic or transversely isotropic solid bodies called *laminated bodies.*[1] These bodies are made up of a collection of disjoint two-dimensional submanifolds $\mathscr{L}_\xi$, called the *laminae*, say

$$\mathscr{B} = \bigcup_{\xi \in I} \mathscr{L}_\xi, \tag{9.1}$$

where I is an index set. For each lamina $\mathscr{L}_\xi$ there exists locally a configuration $\boldsymbol{\kappa}_\xi$ whose induced local configuration maps the intrinsic metric g on $\mathscr{L}_\xi$ onto the Euclidean metric in the physical space, *i.e.*,

$$g(\boldsymbol{u}, \boldsymbol{v}) = \boldsymbol{\kappa}_{\xi *}(\boldsymbol{u}) \cdot \boldsymbol{\kappa}_{\xi *}(\boldsymbol{v}) \tag{9.2}$$

for all $\boldsymbol{u}, \boldsymbol{v} \in \mathscr{B}_X$, $X \in \mathscr{L}_\xi \cap \mathscr{U}$, where $\mathscr{U}$ is the domain of $\boldsymbol{\kappa}_\xi$, and in the case when $\mathscr{B}$ is transversely isotropic, then in addition to (9.2) $\boldsymbol{\kappa}_{\xi *}$

[1] C.-C. WANG, "Universal solutions for incompressible laminated bodies", *Arch. Rational Mech. Anal.* **29**, 161–192 (1968). Reprinted in *Continuum Theory of Inhomogeneities in Simple Bodies*, New York, Springer-Verlag, 1968.

maps also the intrinsic field $\boldsymbol{a}$ onto a unit normal field of $\boldsymbol{\kappa}_\xi(\mathscr{L}_\xi)$, *i.e.*,

$$\boldsymbol{\kappa}_{\xi*}(\boldsymbol{a}) \cdot \boldsymbol{\kappa}_{\xi*}(\boldsymbol{a}) = 1, \ \boldsymbol{\kappa}_{\xi*}(\boldsymbol{a}) \cdot \boldsymbol{\kappa}_{\xi*}(\boldsymbol{u}) = 0 \tag{9.3}$$

for all $\boldsymbol{u}$ tangent to the submanifold $\mathscr{L}_\xi$. In general, the choice of the configuration $\boldsymbol{\kappa}_\xi$ depends on the point in the lamina $\mathscr{L}_\xi$; there need not be such a configuration $\boldsymbol{\kappa}_\xi$ covering the whole lamina $\mathscr{L}_\xi$. Moreover, a configuration $\boldsymbol{\kappa}_\xi$ satisfying the conditions on a lamina $\mathscr{L}_\xi$ need not satisfy the same conditions on the laminae near to $\mathscr{L}_\xi$.

In the physical interpretation we can visualize a lamina $\mathscr{L}_\xi$ as a locally homogeneous shell, which is infinitely thin, such that the configuration $\boldsymbol{\kappa}_\xi$ is its local homogeneous configuration. For definiteness, we call $\boldsymbol{\kappa}_\xi$ an *initial configuration* for $\mathscr{L}_\xi$, and we call all other configurations the *deformed configurations.* These terms, of course, refer strictly to the particular lamina $\mathscr{L}_\xi$. A laminated body is locally homogeneous if and only if there is an initial configuration $\boldsymbol{\kappa}_\xi$ for $\mathscr{L}_\xi$ which is an initial configuration for all laminae belonging to a subbody near to $\mathscr{L}_\xi$. In general, if $\mathscr{B}$ is not locally homogeneous, then a global configuration $\boldsymbol{\kappa}$ of $\mathscr{B}$ is a deformed configuration for all its laminae. The deformation from an initial configuration $\boldsymbol{\kappa}_\xi$ for $\mathscr{L}_\xi$ to the global configuration $\boldsymbol{\kappa}$ depends on the lamina $\mathscr{L}_\xi$. Thus in some sense we can regard $\mathscr{B}$ as being formed by patching together the laminae at their different deformed states. Such a distribution of incompatible initial states for the various laminae causes the inhomogeneity of $\mathscr{B}$.

A laminated body is said to be *locally rectilinear* if its laminae are initially planes; otherwise, it is said to be *curvilinear.* In this section we consider three classes of laminated bodies, the first two of which are locally rectilinear, while the last is curvilinear.

1. Laminated Plate

In a global reference configuration $\boldsymbol{\kappa}$ with a rectangular Cartesian coordinate system (X, Y, Z), requiring the physical component matrix of the intrinsic Riemannian metric to be

$$[\bar{g}_{\langle AB \rangle}] = \begin{bmatrix} \bar{g}_{\langle 11 \rangle} & 0 & 0 \\ 0 & \bar{g}_{\langle 22 \rangle} & \bar{g}_{\langle 23 \rangle} \\ 0 & \bar{g}_{\langle 32 \rangle} & \bar{g}_{\langle 33 \rangle} \end{bmatrix}, \tag{9.4}$$

where $\bar{g}_{\langle AB \rangle}$ are functions of X only, we obtain a laminated isotropic solid body whose laminae are thin plates normal to the X-axis in $\boldsymbol{\kappa}$. For the transversely isotropic case, we require, in addition to (9.4), that the components of the intrinsic field of axes $\boldsymbol{a}$ be

$$(\bar{a}_{\langle A\rangle}) = (\bar{a}_{\langle 1\rangle}, 0, 0), \tag{9.5}$$

where $\bar{a}_{\langle 1\rangle}$ is a function of X. If $\bar{g}_{\langle AB\rangle} \neq \delta_{\langle AB\rangle}$ and $\bar{a}_{\langle 1\rangle} \neq 1$, then the reference configuration $\boldsymbol{\kappa}$ is not an initial configuration of the laminae. For each fixed X, the departure of $\bar{g}_{\langle AB\rangle}$ at X from δ_{AB} characterizes the deformation of the lamina at X from the initial state. Such a deformation varies from lamina to lamina, resulting in an inhomogeneous body.

If the body is incompressible and if we use an admissible reference configuration $\boldsymbol{\kappa}$, then the component matrix (9.4) satisfies also the condition

$$\det\,[\bar{g}_{\langle AB\rangle}] = 1. \tag{9.6}$$

In this case, the deformations of the laminae from the initial configurations to the reference configuration are isochoric.

2. Laminated Cylindrical Shell

Instead of a Cartesian co-ordinate system (X, Y, Z) we now use a cylindrical co-ordinate system (R, Θ, Z) in the reference configuration $\boldsymbol{\kappa}$. Relative to (R, Θ, Z) we require the physical component matrix of g and the physical components of $\boldsymbol{a}$ to be given by (9.4) and (9.5), respectively, but with $\bar{g}_{\langle AB\rangle}$ and $\bar{a}_{\langle A\rangle}$ functions of R only. In this case, the laminae are thin cylindrical shells parametrized by R. For an incompressible body belonging to this class, the condition (9.6) must be satisfied.

3. Laminated Spherical Shell

We use a spherical co-ordinate system (R, Θ, Φ) and require that the physical component matrix of g be

$$[\bar{g}_{\langle AB\rangle}] = \operatorname{diag}\,[\bar{g}_{\langle 11\rangle}, \bar{g}_{\langle 22\rangle}, \bar{g}_{\langle 33\rangle}], \tag{9.7}$$

where the $\bar{g}_{\langle AB\rangle}$ are functions of R only, and

$$\bar{g}_{\langle 22\rangle} = \bar{g}_{\langle 33\rangle}. \tag{9.8}$$

For transversely isotropic bodies, the physical components of $\boldsymbol{a}$ take the form (9.5) with $\bar{a}_{\langle 1\rangle}$ depending on R. The laminae of this class of laminated bodies are thin spherical shells parametrized by R. Further, for an incompressible body, the condition (9.6) must be satisfied.

We now give some families of static and dynamic universal solutions for the incompressible laminated bodies belonging to the preceding three classes. In § IV.4 we have obtained and described these same families of solutions for incompressible homogeneous isotropic bodies, which are, of course, special cases of laminated bodies.

I. Static Universal Solutions

1. Incompressible Laminated Plate

Family 0. Homogeneous Plane Deformations

Relative to Cartesian co-ordinate systems (X, Y, Z) and (x, y, z) in the reference and the actual configurations, respectively, the components of deformation are

$$x = AX,\ y = BY + CZ,\ z = DY + EZ, \tag{9.9}$$

where $A, \ldots, E$ are constants such that

$$A(BE - CD) = 1. \tag{9.10}$$

We can verify easily that under this deformation the component matrix of the intrinsic metric g in (x, y, z) is

$$[g^{ij}] = \begin{bmatrix} g^{11} & 0 & 0 \\ 0 & g^{22} & g^{23} \\ 0 & g^{32} & g^{33} \end{bmatrix} \tag{9.11}$$

where

$$\begin{aligned} g^{11}(x) &= A^2 \bar{g}^{11}\left(\frac{x}{A}\right), \\ g^{22}(x) &= B^2 \bar{g}^{22}\left(\frac{x}{A}\right) + 2BC\bar{g}^{23}\left(\frac{x}{A}\right) + C^2\bar{g}^{33}\left(\frac{x}{A}\right), \\ g^{23}(x) &= BD\bar{g}^{22}\left(\frac{x}{A}\right) + (CD+BE)\bar{g}^{23}\left(\frac{x}{A}\right) + CE\bar{g}^{33}\left(\frac{x}{A}\right), \\ g^{33}(x) &= D^2\bar{g}^{22}\left(\frac{x}{A}\right) + 2DE\bar{g}^{23}\left(\frac{x}{A}\right) + E^2\bar{g}^{33}\left(\frac{x}{A}\right), \end{aligned} \tag{9.12}$$

and where

$$\bar{g}^{AB} = \bar{g}_{\langle AB \rangle} \tag{9.13}$$

since (X, Y, Z) is a Cartesian co-ordinate system. The dual metric g^* of g then has the component matrix

$$[g^*_{ij}] = \begin{bmatrix} g^*_{11} & 0 & 0 \\ 0 & g^*_{22} & g^*_{23} \\ 0 & g^*_{32} & g^*_{33} \end{bmatrix}, \tag{9.14}$$

where

$$\begin{aligned} g^*_{11} &= \frac{1}{A^2}\bar{g}^*_{11}, \\ g^*_{22} &= A^2[E^2\bar{g}^*_{22} - 2DE\bar{g}^*_{23} + D^2\bar{g}^*_{33}], \\ g^*_{23} &= A^2[-CE\bar{g}^*_{22} + (BE+CD)\bar{g}^*_{23} - BD\bar{g}^*_{33}], \\ g^*_{33} &= A^2[C^2\bar{g}^*_{22} - 2BC\bar{g}^*_{23} + B^2\bar{g}^*_{33}], \end{aligned} \tag{9.15}$$

where the arguments of g^*_{ij} and $\bar{g}^*_{AB}$ are again x and x/A, respectively.

Since we did not give a representation for the constitutive equation of a transversely isotropic solid, for simplicity we assume here that the laminated plate is isotropic. Then from the constitutive equation. (III.2.56) the stress tensor in the actual configuration is

$$\boldsymbol{T} = -p\mathbf{1} + \psi_1 \boldsymbol{g} + \psi_{-1}\boldsymbol{g}^*, \tag{9.16}$$

where ψ_1 and ψ_{-1} are the response coefficients relative to the initial states of the laminae and are functions of the invariants

$$I = \operatorname{tr} \boldsymbol{g}, \; II = \operatorname{tr} \boldsymbol{g}^*. \tag{9.17}$$

Here the trace operation is taken with respect to the Euclidean metric of the physical space in the actual configuration. Substituting (9.11) and (9.14) into (9.16) yields the component matrix of the determinate stress:

$$[S_{\langle ij \rangle}] = \begin{bmatrix} S_{\langle 11 \rangle} & 0 & 0 \\ 0 & S_{\langle 22 \rangle} & S_{\langle 23 \rangle} \\ 0 & S_{\langle 32 \rangle} & S_{\langle 33 \rangle} \end{bmatrix}, \tag{9.18}$$

where $S_{\langle ij \rangle}$ are functions of x only. From the analysis of the planar problems in § IV. 4, we know that the actual configuration is an equilibrium configuration under the pressure field

$$p = p_0 + S_{\langle 11 \rangle}, \tag{9.19}$$

where p_0 is a constant. Thus we have shown that the deformation given by (9.9) is a static universal solution.

Family 1. Relative to the Cartesian co-ordinate system (X, Y, Z) in $\boldsymbol{\kappa}$ and a cylindrical co-ordinate system (r, θ, z) in the actual configuration the components of deformation are

$$r = \sqrt{2AX}, \ \theta = BY + CZ, \ z = DY + EZ, \tag{9.20}$$

where the constants $A, \ldots, E$ satisfy the condition (9.10).

Following the same procedure as for the preceding family, we can prove that relative to (r, θ, z) in the actual configuration the physical component matrix of the determinate stress again takes the form (9.18) with $S_{\langle ij \rangle}$ depending only on r. Then from the analysis of the cylindrical problems in § IV.4, we know that the actual configuration is an equilibrium configuration under the pressure field

$$p = S_{\langle 11 \rangle} + \int \frac{S_{\langle 11 \rangle} - S_{\langle 22 \rangle}}{r} \, dr. \tag{9.21}$$

Thus we have shown that this family is a family of static universal solutions.

2. Incompressible Laminated Cylindrical Shell

Family 2. Relative to the cylinderical co-ordinate system (R, Θ, Z) in $\boldsymbol{\kappa}$ and a Cartesian co-ordinate system (x, y, z) in the actual configuration, the components of deformation are

$$x = \tfrac{1}{2} A R^2, \ y = B\Theta + CZ, \ z = D\Theta + EZ, \tag{9.22}$$

where the constants $A, \ldots, E$ again satisfy the condition (9.10).

The reader will verify easily that the component matrix of the determinate stress relative to the co-ordinate system (x, y, z) in the actual configuration again takes the form (9.18) with $S_{\langle ij \rangle}$ depending only on x. Hence, as before, the actual configuration is in equilibrium under the pressure field p given by (9.19). The same is true for the next family.

Family 3. Relative to cylindrical co-ordinate systems (R, Θ, Z) and (r, θ, z) in the reference and actual configurations, respectively, the components of deformation are

$$r = \sqrt{AR^2 + B}, \ \theta = C\Theta + DZ, \ z = E\Theta + FZ, \tag{9.23}$$

where the constants $A, \ldots, F$ satisfy the condition

$$A(CF-DE) = 1. \tag{9.24}$$

3. Incompressible Laminated Spherical Shell

Family 4. Relative to spherical co-ordinate systems (R, Θ, Φ) and (r, θ, φ) in the reference and actual configurations, respectively, the components of deformation are

$$r = (\pm R^3+A)^{\frac{1}{3}}, \; \theta = \pm\Theta, \; \varphi = \Phi, \tag{9.25}$$

where A is an arbitrary constant.

Using (9.7), (9.8), and (9.25), the reader will verify easily that

$$[S_{\langle ij\rangle}] = \operatorname{diag}[S_{\langle 11\rangle}, S_{\langle 22\rangle}, S_{\langle 33\rangle}] \tag{9.26}$$

with

$$S_{\langle 22\rangle} = S_{\langle 33\rangle} \tag{9.27}$$

in the actual configuration. Hence from the analysis of the spherical problems in § IV.4, the actual configuration is in equilibrium under the pressure field

$$p = S_{\langle 11\rangle}+\int\frac{2S_{\langle 11\rangle}-S_{\langle 22\rangle}-S_{\langle 33\rangle}}{r}\,dr, \tag{9.28}$$

which is a function of r only.

So far, we have considered five families of static universal solutions for homogeneous incompressible isotropic bodies which turn out to be universal solutions for some classes of laminated bodies also. There is, however, one more family of universal solutions for the former (*cf.* (IV. 4.43)):

Family 5. Relative to cylindrical co-ordinate systems (R, Θ, Z) and (r, θ, z), the components of deformation are

$$r = AR, \; \theta = B\log R+C\Theta, \; z = DZ, \tag{9.29}$$

where $A, \ldots, D$ are constants satisfying the condition

$$A^2CD = 1. \tag{9.30}$$

It is not known whether or not this family yields universal solutions for any inhomogeneous laminated body. Even the special case (IV.5.63) of this family, which represents pure bending of a sector of a spherical shell into another one, is not a universal solution for the laminated spherical shell, since the condition (9.27) is not satisfied unless the shell is homogeneous.

II. Dynamic Universal Solutions

Since we showed in the preceding chapter that dynamic universal solutions are quasi-equilibrated, we simply state here the corresponding results for certain laminated bodies.

1. Incompressible Laminated Plate

Family 0. The components of deformation are given by (9.9) with $A, \ldots, E$ now being functions of t. The analysis of this family is the same as for Family 2 of Case 2 in § IV.6.

Family 1. The analysis is the same as for Family 1 of Case 2 in § IV.6.

2. Incompressible Laminated Cylindrical Shell

Family 2 and *Family 3.* The analysis is the same as for the corresponding families of Case 2 in § IV. 6.

3. Incompressible Laminated Spherical Shell

Family 4. The analysis is the same as for Family 4 of Case 2 in § IV. 6.

In each of the preceding families of motions the determinate stress, and hence also the dynamic pressure field, depends on the intrinsic Riemannian metric on the laminated body. Consequently the boundary forces necessary to produce the motion for an inhomogeneous laminated body are different from those for a homogeneous body. In the next section we shall find these boundary forces for some examples.

10. Examples of Static and Dynamic Universal Solutions for Some Incompressible Laminated Bodies

Example 1. Simple Shear of a Laminated Plate

The special case of Family 0 in which the components of deformation are

$$x = X,\ y = Y + DZ,\ z = Z \tag{10.1}$$

is a simple shear of amount D. We focus our attention on the stress tensor at a fixed lamina where the intrinsic metric has the components

$$[\bar{g}_{\langle AB\rangle}] = \text{diag}\,[\bar{g}_{\langle 11\rangle}, \bar{g}_{\langle 22\rangle}, \bar{g}_{\langle 33\rangle}]. \tag{10.2}$$

From the general stress system obtained in the preceding section, we know that the shear stress of this lamina in the actual configuration is

$$\begin{aligned} T_{\langle 23\rangle} &= (\psi_1 \bar{g}_{\langle 33\rangle} - \psi_{-1} \bar{g}^*_{\langle 22\rangle})D, \\ &= \left(\psi_1 \bar{g}_{\langle 33\rangle} - \psi_{-1} \frac{1}{\bar{g}_{\langle 22\rangle}}\right) D, \end{aligned} \tag{10.3}$$

where the arguments of the functions ψ_1 and ψ_{-1} are

$$I = \text{tr}\, \boldsymbol{g} = \bar{g}_{\langle 11\rangle} + \bar{g}_{\langle 22\rangle} + \bar{g}_{\langle 33\rangle} + D^2 \bar{g}_{\langle 33\rangle}, \tag{10.4}$$

and

$$II = \text{tr}\, \boldsymbol{g}^* = \frac{1}{\bar{g}_{\langle 11\rangle}} + \frac{1}{\bar{g}_{\langle 22\rangle}} + \frac{1}{\bar{g}_{\langle 33\rangle}} + \frac{D^2}{\bar{g}_{\langle 22\rangle}}. \tag{10.5}$$

Thus the shear modulus in the reference configuration $\boldsymbol{\kappa}$ is given by

$$\bar{\mu} = \bar{\mu}(D^2) = \psi_1 \bar{g}_{\langle 33\rangle} - \psi_{-1} \frac{1}{\bar{g}_{\langle 22\rangle}}, \tag{10.6}$$

which is different from the shear modulus $\hat{\mu}$ at the initial state as defined by (IV. 3.10):

$$\hat{\mu} = \psi_1 - \psi_{-1}. \tag{10.7}$$

As before, assuming that $\bar{\mu}$ is smooth, we have the estimate of $\bar{\mu}$ up to terms of second order in D:

$$\bar{\mu} = \bar{\mu}_0 + \bar{\mu}_1 D^2 + o(D^2) \tag{10.8}$$

where

$$\bar{\mu}_0 = \bar{\mu}(0) = \lim_{D\to 0} \frac{T_{\langle 23\rangle}}{D}, \tag{10.9}$$

and

$$\bar{\mu}_1 = \bar{\mu}'(0). \tag{10.10}$$

The normal stresses in the simple shear are

$$T_{\langle 11\rangle} = -p + \psi_1 \bar{g}_{\langle 11\rangle} + \psi_{-1} \frac{1}{\bar{g}_{\langle 11\rangle}}, \tag{10.11}$$

$$T_{\langle 22\rangle} = -p + \psi_1(\bar{g}_{\langle 22\rangle} + D^2 \bar{g}_{\langle 33\rangle}) + \psi_{-1} \frac{1}{\bar{g}_{\langle 22\rangle}}, \tag{10.12}$$

and

$$T_{\langle 33\rangle} = -p+\psi_1 \bar{g}_{\langle 33\rangle}+\psi_{-1}\left(\frac{1}{\bar{g}_{\langle 33\rangle}}+\frac{D^2}{\bar{g}_{\langle 22\rangle}}\right), \tag{10.13}$$

where p is the undetermined hydrostatic pressure. These normal stresses, of course, are also different from those of a homogeneous body in the same simple shear (*cf.* § IV. 3).

In order to see the effects of the initial deformation of the lamina on the shear modulus and the normal stresses more explicitly, we consider the case when the material is of the Mooney type, so that ψ_1 and ψ_{-1} are material constants. Then from (10.6 and (10.7), the shear moduli $\hat{\mu}$ and $\bar{\mu}$ are different in general:

$$\bar{\mu} \neq \hat{\mu}. \tag{10.14}$$

In fact, we can adjust the values of $\bar{g}_{\langle 22\rangle}$ and $\bar{g}_{\langle 33\rangle}$ in such a way that $\bar{\mu}$ can take any value we like. The effect of the initial deformation on the normal stresses is similar. Moreover, in general, the normal stresses do not satisfy the universal relation (IV. 3.21) valid for simple shear of a homogeneous plate.

Example 2. Pure Torsion of a Cylindrical Laminated Rod

The components of deformation are

$$r = R,\ \theta = \Theta+DZ,\ z = Z, \tag{10.15}$$

where D is the twist per unit axial length. For simplicity, we assume that the intrinsic Riemannian metric g has the simple component form (10.2) in the co-ordinates (R, Θ, Z). To avoid a singularity at the axis of the cylinder, we assume that

$$\bar{g}_{\langle AB\rangle}|_{R=0} = \delta_{AB}. \tag{10.16}$$

From the preceding example, the shear stress in the actual configuration is

$$T_{\langle 23\rangle} = \left(\psi_1 \bar{g}_{\langle 33\rangle}-\psi_{-1}\frac{1}{\bar{g}_{\langle 22\rangle}}\right) Dr = \bar{\mu} Dr. \tag{10.17}$$

Thus in pure torsion of a laminated cylinder as well as of a homogeneous one, the shear stress is determined by the shear modulus. As before, the twisting moment τ is given by the integral

$$\tau = 2\pi \int_0^a r^2 T_{\langle 23 \rangle}\, dr = 2\pi D \int_0^a r^3 \bar{\mu}(D^2 r^2, r) dr, \tag{10.18}$$

where a is the radius of the cylinder. Here we have indicated the explicit dependence of $\bar{\mu}$ on r, which is the parameter for the lamina. The limiting value of the torsional rigidity τ/D as $D \to 0$ is then given by

$$\lim_{D \to 0} \frac{\tau}{D} = 2\pi \int_0^a r^3 \bar{\mu}_0(r) dr. \tag{10.19}$$

This result shows that, by choosing a suitable lamination so as to adjust the shear modulus $\bar{\mu}_0(r)$, we can vary the torsional rigidity of the laminated cylinder, though the size and the material of the cylinder remain the same. The range of the values of the torsional rigidity, of course, depends on the range of $\bar{\mu}_0$ in different laminations. For the special case of a Mooney material, the range of $\bar{\mu}_0$, and hence that of the torsional rigidity, is infinite.

The axial force of the laminated cylinder, in general, does not vanish either in the reference configuration $\boldsymbol{\kappa}$ or in the deformed configuration χ. Relative to the cylindrical co-ordinate system (R, Θ, Z) in $\boldsymbol{\kappa}$, the physical components of the determinate stress are

$$\bar{S}_{\langle 11 \rangle} = \bar{\psi}_1 \bar{g}_{\langle 11 \rangle} + \bar{\psi}_{-1} \frac{1}{\bar{g}_{\langle 11 \rangle}}, \tag{10.20}$$

$$\bar{S}_{\langle 22 \rangle} = \bar{\psi}_1 \bar{g}_{\langle 22 \rangle} + \bar{\psi}_{-1} \frac{1}{\bar{g}_{\langle 22 \rangle}}, \tag{10.21}$$

$$\bar{S}_{\langle 33 \rangle} = \bar{\psi}_1 \bar{g}_{\langle 33 \rangle} + \bar{\psi}_{-1} \frac{1}{\bar{g}_{\langle 33 \rangle}}, \tag{10.22}$$

where the arguments of $\bar{\psi}_1$ and $\bar{\psi}_{-1}$ are

$$\bar{I} = \overline{\mathrm{tr}}\, g = \bar{g}_{\langle 11 \rangle} + \bar{g}_{\langle 22 \rangle} + \bar{g}_{\langle 33 \rangle}, \tag{10.23}$$

and

$$\overline{II} = \overline{\mathrm{tr}}\, g^* = \frac{1}{\bar{g}_{\langle 11 \rangle}} + \frac{1}{\bar{g}_{\langle 22 \rangle}} + \frac{1}{\bar{g}_{\langle 33 \rangle}}. \tag{10.24}$$

As before, we adopt the boundary condition

$$\bar{T}_{\langle 11 \rangle}|_{R=a} = 0 \tag{10.25}$$

in order to determine the pressure field. Then from (IV. 4.19)$_2$, the radial stress is given by

$$\bar{T}_{\langle 11\rangle} = -\int_R^a \frac{1}{R}\left[\bar{\psi}_1(\bar{g}_{\langle 11\rangle}-\bar{g}_{\langle 22\rangle})+\bar{\psi}_{-1}\left(\frac{1}{\bar{g}_{\langle 11\rangle}}-\frac{1}{\bar{g}_{\langle 22\rangle}}\right)\right] dR. \quad (10.26)$$

Consequently

$$\begin{aligned}\bar{T}_{\langle 33\rangle} &= \bar{\psi}_1(\bar{g}_{\langle 33\rangle}-\bar{g}_{\langle 11\rangle})+\bar{\psi}_{-1}\left(\frac{1}{\bar{g}_{\langle 33\rangle}}-\frac{1}{\bar{g}_{\langle 22\rangle}}\right)\\ &\quad -\int_R^a \left[\bar{\psi}_1(\bar{g}_{\langle 11\rangle}-\bar{g}_{\langle 22\rangle})+\bar{\psi}_{-1}\left(\frac{1}{\bar{g}_{\langle 11\rangle}}-\frac{1}{\bar{g}_{\langle 22\rangle}}\right)\right]\frac{dR}{R}. \end{aligned} \quad (10.27)$$

Finally, the axial force is given by

$$\bar{N} = 2\pi\int_0^a R\bar{T}_{\langle 33\rangle}\, dR. \quad (10.28)$$

Substituting (10.27) into (10.28) and then integrating the result by parts yields

$$\begin{aligned}\bar{N} = \pi\int_0^a R\bigg[&\bar{\psi}_1(2\bar{g}_{\langle 33\rangle}-\bar{g}_{\langle 11\rangle}-\bar{g}_{\langle 22\rangle})\\ &+\bar{\psi}_{-1}\left(\frac{2}{\bar{g}_{\langle 33\rangle}}-\frac{1}{\bar{g}_{\langle 11\rangle}}-\frac{1}{\bar{g}_{\langle 22\rangle}}\right)\bigg] dR. \end{aligned} \quad (10.29)$$

Of course, if we wish, we could assign fields $\bar{g}_{\langle AB\rangle}$ such as to annul $\bar{N}$.

In a similar way, we can compute the axial force N of the cylinder in the actual configuration $\boldsymbol{\chi}$. First, the determinate normal stresses in $\boldsymbol{\chi}$ are

$$S_{\langle 11\rangle} = \psi_1\bar{g}_{\langle 11\rangle}+\psi_{-1}\frac{1}{\bar{g}_{\langle 11\rangle}}, \quad (10.30)$$

$$S_{\langle 22\rangle} = \psi_1(\bar{g}_{\langle 22\rangle}+D^2r^2\bar{g}_{\langle 33\rangle})+\psi_{-1}\frac{1}{\bar{g}_{\langle 22\rangle}}, \quad (10.31)$$

$$S_{\langle 33\rangle} = \psi_1\bar{g}_{\langle 33\rangle}+\psi_{-1}\left(\frac{1}{\bar{g}_{\langle 33\rangle}}+\frac{D^2r^2}{\bar{g}_{\langle 22\rangle}}\right), \quad (10.32)$$

where the arguments of ψ_1 and ψ_{-1} are

$$I = \operatorname{tr}\boldsymbol{g} = \bar{g}_{\langle 11\rangle}+\bar{g}_{\langle 22\rangle}+\bar{g}_{\langle 33\rangle}+D^2r^2\bar{g}_{\langle 33\rangle}, \quad (10.33)$$

and

$$II = \operatorname{tr} \boldsymbol{g}^* = \frac{1}{\bar{g}_{\langle 11\rangle}} + \frac{1}{\bar{g}_{\langle 22\rangle}} + \frac{1}{\bar{g}_{\langle 33\rangle}} + \frac{D^2r^2}{\bar{g}_{\langle 22\rangle}}. \tag{10.34}$$

Imposing the same boundary condition as before:

$$T_{\langle 11\rangle}|_{r=a} = 0, \tag{10.35}$$

we obtain the radial stress

$$T_{\langle 11\rangle} = -\int_r^a \frac{1}{r}\left[\psi_1(\bar{g}_{\langle 11\rangle}-\bar{g}_{\langle 22\rangle})+\psi_{-1}\left(\frac{1}{\bar{g}_{\langle 11\rangle}}-\frac{1}{\bar{g}_{\langle 22\rangle}}\right)\right]dr + \int_r^a \psi_1 D^2 r\bar{g}_{\langle 33\rangle}\,dr. \tag{10.36}$$

Then the axial stress is given by

$$\begin{aligned} T_{\langle 33\rangle} = {} & \psi_1(\bar{g}_{\langle 33\rangle}-\bar{g}_{\langle 11\rangle})+\psi_{-1}\left(\frac{1}{\bar{g}_{\langle 33\rangle}}-\frac{1}{\bar{g}_{\langle 11\rangle}}\right)+\frac{\psi_{-1}D^2r^2}{\bar{g}_{\langle 22\rangle}} \\ & -\int_r^a \frac{1}{r}\left[\psi_1(\bar{g}_{\langle 11\rangle}-\bar{g}_{\langle 22\rangle})+\psi_{-1}\left(\frac{1}{\bar{g}_{\langle 11\rangle}}-\frac{1}{\bar{g}_{\langle 22\rangle}}\right)\right]dr \\ & +\int_r^a \psi_1 D^2 r\bar{g}_{\langle 33\rangle}\,dr. \end{aligned} \tag{10.37}$$

Substituting (10.37) into (10.28) and integrating by parts as before, we obtain

$$\begin{aligned} N = {} & \pi D^2 \int_0^a \left(\frac{2\psi_{-1}}{\bar{g}_{\langle 22\rangle}}-\psi_1\bar{g}_{\langle 33\rangle}\right) r^3 dr \\ & +\pi\int_0^a r\left[\psi_1(2\bar{g}_{\langle 33\rangle}-\bar{g}_{\langle 11\rangle}-\bar{g}_{\langle 22\rangle})+\psi_{-1}\left(\frac{2}{\bar{g}_{\langle 33\rangle}}-\frac{1}{\bar{g}_{\langle 11\rangle}}-\frac{1}{\bar{g}_{\langle 22\rangle}}\right)\right]dr. \end{aligned} \tag{10.38}$$

Of course, N is an even function of D.

Now subtracting $\bar{N}$ from N gives the increment of the axial force due to the pure torsion:

$$\begin{aligned} \Delta N = N-\bar{N} = {} & \pi D^2 \int_0^a \left(\frac{2\psi_{-1}}{\bar{g}_{\langle 22\rangle}}-\psi_1\bar{g}_{\langle 33\rangle}\right) r^3 dr \\ & +\pi\int_0^a r\left[(\psi_1-\bar{\psi}_1)(2\bar{g}_{\langle 33\rangle}-\bar{g}_{\langle 11\rangle}-\bar{g}_{\langle 22\rangle})\right. \\ & \left.+(\psi_{-1}-\bar{\psi}_{-1})\left(\frac{2}{\bar{g}_{\langle 33\rangle}}-\frac{1}{\bar{g}_{\langle 11\rangle}}-\frac{1}{\bar{g}_{\langle 22\rangle}}\right)\right]dr, \end{aligned} \tag{10.39}$$

which is again an even function of D, say

$$\Delta N = f(D^2). \tag{10.40}$$

Since f vanishes at $D = 0$, for sufficiently small D we have the estimate

$$\Delta N = f_1 D^2 + o(D^2), \tag{10.41}$$

where

$$f_1 = f'(0). \tag{10.42}$$

From (10.39), we can compute the value of the constant f_1 by

$$f_1 = \pi \int_0^a \left[\left(\frac{2\bar{\psi}_{-1}}{\bar{g}_{\langle 22 \rangle}} - \bar{\psi}_1 \bar{g}_{\langle 33 \rangle} \right) + \left(\frac{\partial \bar{\psi}_1}{\partial I} \bar{g}_{\langle 33 \rangle} + \frac{\partial \bar{\psi}_1}{\partial II} \frac{1}{\bar{g}_{\langle 22 \rangle}} \right) (2\bar{g}_{\langle 33 \rangle} - \bar{g}_{\langle 11 \rangle} - \bar{g}_{\langle 22 \rangle}) + \left(\frac{\partial \bar{\psi}_{-1}}{\partial I} \bar{g}_{\langle 33 \rangle} + \frac{\partial \bar{\psi}_{-1}}{\partial II} \frac{1}{\bar{g}_{\langle 22 \rangle}} \right) \left(\frac{2}{\bar{g}_{\langle 33 \rangle}} - \frac{1}{\bar{g}_{\langle 11 \rangle}} - \frac{1}{\bar{g}_{\langle 22 \rangle}} \right) \right] \times R^3 dR. \tag{10.43}$$

For sufficiently small departure of $\bar{g}_{\langle AB \rangle}$ from δ_{AB}, the leading term in the integrand of (10.43) dominates the rest. In this case, the usual restrictions

$$\bar{\psi}_1 > 0, \; \bar{\psi}_{-1} \leqq 0 \tag{10.44}$$

imply that f_1, and hence ΔN for sufficiently small D, is negative. On the other hand, if the departure of $\bar{g}_{\langle AB \rangle}$ from δ_{AB} is not small, f_1 depends in a complex way on

$$\bar{\psi}_1, \bar{\psi}_{-1}, \frac{\partial \bar{\psi}_1}{\partial I}, \frac{\partial \bar{\psi}_1}{\partial II}, \frac{\partial \bar{\psi}_{-1}}{\partial I}, \frac{\partial \bar{\psi}_{-1}}{\partial II},$$

as well as on $\bar{g}_{\langle 11 \rangle}$, $\bar{g}_{\langle 22 \rangle}$, and $\bar{g}_{\langle 33 \rangle}$ as functions of R.

Example 3. Torsion, Extension, and Inflation of a Cylindrical Laminated Rod

As explained in Example 2 of § IV. 5, if we choose

$$B = 0, \; C = 1, \; E = 0, \tag{10.45}$$

in (9.23), so that the components of deformation are

$$r = \sqrt{A}R, \; \theta = \Theta + DZ, \; z = FZ, \tag{10.46}$$

where

$$AF = 1, \tag{10.47}$$

the deformation represents an axial stretch F followed by a twist of amount D/F. As in the preceding example, we assume that the intrinsic metric g has the diagonal component form (10.2) in the co-ordinates (R, Θ, Z) and satisfies the condition (10.46). Also, a is the radius of the cylinder in the reference configuration $\boldsymbol{\kappa}$, while $a\sqrt{A}$ is that in the actual configuration $\boldsymbol{\chi}$.

First we determine the torsional rigidity of the cylinder when the axial stretch F is held fixed. We denote that stretched state by $\tilde{\boldsymbol{\chi}}$. Then the components of the intrinsic metric in $\tilde{\boldsymbol{\chi}}$ are given by

$$\tilde{g}_{\langle 11\rangle} = A\bar{g}_{\langle 11\rangle},\ \tilde{g}_{\langle 22\rangle} = A\bar{g}_{\langle 22\rangle},\ \tilde{g}_{\langle 33\rangle} = F^2\bar{g}_{\langle 33\rangle}. \tag{10.48}$$

From (10.6), the shear modulus in $\tilde{\boldsymbol{\chi}}$ is

$$\tilde{\mu} = \psi_1\,\tilde{g}_{\langle 33\rangle} - \psi_{-1}\,\frac{1}{\tilde{g}_{\langle 22\rangle}}, \tag{10.49}$$

where the arguments of ψ_1 and ψ_{-1} are

$$I = \tilde{g}_{\langle 11\rangle} + \tilde{g}_{\langle 22\rangle} + \tilde{g}_{\langle 33\rangle} + \frac{D^2}{F^2}\,r^2\tilde{g}_{\langle 33\rangle}, \tag{10.50}$$

and

$$II = \frac{1}{\tilde{g}_{\langle 11\rangle}} + \frac{1}{\tilde{g}_{\langle 22\rangle}} + \frac{1}{\tilde{g}_{\langle 33\rangle}} + \frac{D^2}{F^2}\,r^2\,\frac{1}{\tilde{g}_{\langle 22\rangle}}. \tag{10.51}$$

Then the shear stress is

$$T_{\langle 23\rangle} = \frac{D}{F}\,r\tilde{\mu}, \tag{10.52}$$

and twisting moment is

$$\tau = 2\pi\,\frac{D}{F}\int_0^{a\sqrt{A}} r^3\tilde{\mu}\left(\frac{D^2}{F^2}\,r^2, r\right)dr. \tag{10.53}$$

From (10.29), the axial force in the configuration $\tilde{\boldsymbol{\chi}}$ is

$$\tilde{N} = \pi\int_0^{a\sqrt{A}} r\left[\tilde{\psi}_1(2\tilde{g}_{\langle 33\rangle} - \tilde{g}_{\langle 11\rangle} - \tilde{g}_{\langle 22\rangle}) + \tilde{\psi}_{-1}\left(\frac{2}{\tilde{g}_{\langle 33\rangle}} - \frac{1}{\tilde{g}_{\langle 11\rangle}} - \frac{1}{\tilde{g}_{\langle 22\rangle}}\right)\right]dr, \tag{10.54}$$

where the arguments of $\tilde{\psi}_1$ and $\tilde{\psi}_{-1}$ are

$$\tilde{I} = \tilde{g}_{\langle 11\rangle} + \tilde{g}_{\langle 22\rangle} + \tilde{g}_{\langle 33\rangle} \tag{10.55}$$

and

$$\tilde{II} = \frac{1}{\tilde{g}_{\langle 11\rangle}} + \frac{1}{\tilde{g}_{\langle 22\rangle}} + \frac{1}{\tilde{g}_{\langle 33\rangle}}. \tag{10.56}$$

Likewise from (10.38), the axial force in the final configuration χ is

$$N = \pi \frac{D^2}{F^2} \int_0^{a\sqrt{A}} \left(\frac{2\psi_{-1}}{\tilde{g}_{\langle 22\rangle}} - \psi_1 \tilde{g}_{\langle 33\rangle} \right) r^3 dr$$
$$+ \pi \int_0^{a\sqrt{A}} r \left[\psi_1 (2\tilde{g}_{\langle 33\rangle} - \tilde{g}_{\langle 11\rangle} - \tilde{g}_{\langle 22\rangle}) \right.$$
$$\left. + \psi_{-1} \left(\frac{2}{\tilde{g}_{\langle 33\rangle}} - \frac{1}{\tilde{g}_{\langle 11\rangle}} - \frac{1}{\tilde{g}_{\langle 22\rangle}} \right) \right] dr. \tag{10.57}$$

We can set up the Poynting problem for the laminated cylinder in the same way as before (*cf.* Example 2 of § IV. 5): We require that in the combined torsion and extension the total increment of axial force be zero, or equivalently

$$N - \tilde{N} = \bar{N} - \tilde{N}. \tag{10.58}$$

This condition determines a relation between the stretch $F-1$ and the twist D/F.

From (10.54) and (10.29), the increment $\tilde{N} - \bar{N}$ in the pure stretch is given by

$$\tilde{N} - \bar{N} = \pi \int_0^a R \left[\tilde{\psi}_1 (2F\bar{g}_{\langle 33\rangle} - A^2 \bar{g}_{\langle 11\rangle} - A^2 \bar{g}_{\langle 22\rangle}) \right.$$
$$\left. + \tilde{\psi}_{-1} \left(\frac{2}{F^3 \bar{g}_{\langle 33\rangle}} - \frac{1}{\bar{g}_{\langle 11\rangle}} - \frac{1}{\bar{g}_{\langle 22\rangle}} \right) \right] dR$$
$$- \pi \int_0^a R \left[\bar{\psi}_1 (2\bar{g}_{\langle 33\rangle} - \bar{g}_{\langle 11\rangle} - \bar{g}_{\langle 22\rangle}) \right.$$
$$\left. + \bar{\psi}_{-1} \left(\frac{2}{\bar{g}_{\langle 33\rangle}} - \frac{1}{\bar{g}_{\langle 11\rangle}} - \frac{1}{\bar{g}_{\langle 22\rangle}} \right) \right] dR$$
$$= \pi (F-1) \int_0^a R \left[\bar{\psi}_1 \left(2\bar{g}_{\langle 33\rangle} + \frac{F+1}{F^2} (\bar{g}_{\langle 11\rangle} + \bar{g}_{\langle 22\rangle}) \right) \right. \tag{10.59}$$
$$\left. - 2\bar{\psi}_{-1} \frac{F^2 + F + 1}{F^3} \bar{g}_{\langle 33\rangle} \right] dR$$
$$+ \pi \int_0^a R \left[(\tilde{\psi}_1 - \bar{\psi}_1)(2F\bar{g}_{\langle 33\rangle} - A^2 \bar{g}_{\langle 11\rangle} - A^2 \bar{g}_{\langle 22\rangle}) \right.$$
$$\left. + (\tilde{\psi}_{-1} - \bar{\psi}_{-1}) \left(\frac{2}{F^3 \bar{g}_{\langle 33\rangle}} - \frac{1}{\bar{g}_{\langle 11\rangle}} - \frac{1}{\bar{g}_{\langle 22\rangle}} \right) \right] dR.$$

For sufficiently small stretch, the differences $\tilde{\psi}_1 - \bar{\psi}_1$ and $\tilde{\psi}_{-1} - \bar{\psi}_{-1}$ have the following approximations:

$$\tilde{\psi}_1 - \bar{\psi}_1 = (F-1)\left[\frac{\partial\bar{\psi}_1}{\partial I}(2\bar{g}_{\langle 33\rangle} - \bar{g}_{\langle 11\rangle} - \bar{g}_{\langle 22\rangle}) - \frac{\partial\bar{\psi}_1}{\partial II}\left(\frac{2}{\bar{g}_{\langle 33\rangle}} - \frac{1}{\bar{g}_{\langle 11\rangle}} - \frac{1}{\bar{g}_{\langle 22\rangle}}\right)\right] + o(F-1), \tag{10.60}$$

and

$$\tilde{\psi}_{-1} - \bar{\psi}_{-1} = (F-1)\left[\frac{\partial\bar{\psi}_{-1}}{\partial I}(2\bar{g}_{\langle 33\rangle} - \bar{g}_{\langle 11\rangle} - \bar{g}_{\langle 22\rangle}) - \frac{\partial\bar{\psi}_{-1}}{\partial II}\left(\frac{2}{\bar{g}_{\langle 33\rangle}} - \frac{1}{\bar{g}_{\langle 11\rangle}} - \frac{1}{\bar{g}_{\langle 22\rangle}}\right)\right] + o(F-1). \tag{10.61}$$

Substituting (10.60) and (10.61) into (10.59) yields the estimate

$$\tilde{N} - \bar{N} = g_1(F-1) + o(F-1), \tag{10.62}$$

where g_1 is a constant given by

$$\begin{aligned} g_1 = \pi\int_0^a \Bigg[& 2\bar{\psi}_1(\bar{g}_{\langle 33\rangle} + \bar{g}_{\langle 22\rangle} + \bar{g}_{\langle 11\rangle}) - 6\bar{\psi}_{-1}\bar{g}_{\langle 33\rangle} \\ & + \frac{\partial\bar{\psi}_1}{\partial I}(2\bar{g}_{\langle 33\rangle} - \bar{g}_{\langle 11\rangle} - \bar{g}_{\langle 22\rangle})^2 \\ & + \left(\frac{\partial\bar{\psi}_{-1}}{\partial I} - \frac{\partial\bar{\psi}_1}{\partial II}\right)(2\bar{g}_{\langle 33\rangle} - \bar{g}_{\langle 11\rangle} - \bar{g}_{\langle 22\rangle})\left(\frac{2}{\bar{g}_{\langle 33\rangle}} - \frac{1}{\bar{g}_{\langle 11\rangle}} - \frac{1}{\bar{g}_{\langle 22\rangle}}\right) \\ & - \frac{\partial\bar{\psi}_{-1}}{\partial II}\left(\frac{2}{\bar{g}_{\langle 33\rangle}} - \frac{1}{\bar{g}_{\langle 11\rangle}} - \frac{1}{\bar{g}_{\langle 22\rangle}}\right)^2\Bigg] R\,dR. \end{aligned} \tag{10.63}$$

As in (10.43), the leading two terms of the integrand of (10.63) dominate the rest when the departure of $\bar{g}_{\langle AB\rangle}$ from δ_{AB} is small. In this case, the restrictions (10.44) imply that g_1, and hence $\tilde{N} - \bar{N}$ for sufficiently small stretch $F-1$, is positive.

From (10.39), the increment $N - \tilde{N}$ in the pure torsion from $\bar{\chi}$ to χ is given by

$$N-\bar{N} = \pi \frac{D^2}{F^2}\int_0^{a\sqrt{A}} \left(\frac{2\psi_{-1}}{\tilde{g}_{\langle 22\rangle}} - \psi_1 \tilde{g}_{\langle 33\rangle}\right) r^3 dr$$

$$+\pi\int_0^{a\sqrt{A}} r\left[(\psi_1-\tilde{\psi}_1)(2\tilde{g}_{\langle 33\rangle}-\tilde{g}_{\langle 11\rangle}-\tilde{g}_{\langle 22\rangle})\right.$$

$$\left.+(\psi_{-1}-\tilde{\psi}_{-1})\left(\frac{2}{\tilde{g}_{\langle 33\rangle}} - \frac{1}{\tilde{g}_{\langle 11\rangle}} - \frac{1}{\tilde{g}_{\langle 22\rangle}}\right)\right] dr. \tag{10.64}$$

As in (10.41), we have the estimate

$$N-\bar{N} = \tilde{f}_1 \frac{D^2}{F^2} + o\left(\frac{D^2}{F^2}\right) \tag{10.65}$$

with F fixed and $D \to 0$. Of course, $\tilde{f}_1$ and $o(D^2/F^2)$ depend on F. From (10.43), $\tilde{f}_1$ is given by

$$\tilde{f}_1 = \pi\int_0^{a\sqrt{A}} \left[\left(\frac{2\tilde{\psi}_{-1}}{\tilde{g}_{\langle 22\rangle}} - \tilde{\psi}_1 \tilde{g}_{\langle 33\rangle}\right)\right.$$

$$+\left(\frac{\partial\tilde{\psi}_1}{\partial I}\tilde{g}_{\langle 33\rangle} + \frac{\partial\tilde{\psi}_1}{\partial II}\frac{1}{\tilde{g}_{\langle 22\rangle}}\right)(2\tilde{g}_{\langle 33\rangle}-\tilde{g}_{\langle 11\rangle}-\tilde{g}_{\langle 22\rangle})$$

$$\left.+\left(\frac{\partial\tilde{\psi}_{-1}}{\partial I}\tilde{g}_{\langle 33\rangle} + \frac{\partial\tilde{\psi}_{-1}}{\partial II}\frac{1}{\tilde{g}_{\langle 22\rangle}}\right)\left(\frac{2}{\tilde{g}_{\langle 33\rangle}} - \frac{1}{\tilde{g}_{\langle 11\rangle}} - \frac{1}{\tilde{g}_{\langle 22\rangle}}\right)\right] r^3 dr. \tag{10.66}$$

Now substituting (10.62) and (10.65) into the condition (10.58), and then taking the limit as $D \to 0$, $F \to 1$, we obtain the Poynting modulus of the laminated rod:

$$\Pi = \lim_{\substack{F\to 1\\ D\to 0}} \frac{F-1}{(D/F)^2} = \lim_{F\to 1}\left(-\frac{\tilde{f}_1(F)}{g_1}\right) = -\frac{f_1}{g_1}, \tag{10.67}$$

where f_1 is given by (10.43) and g_2 is given by (10.63). When $\bar{g}_{\langle AB\rangle} = \delta_{AB}$, this formula reduces to the previously obtained formula (IV. 5.30) for a homogeneous rod.

Example 4. Radial Oscillations of a Cylindrical Laminated Tube

This example generalizes Example 1 of Section IV.7 to laminated tubes. As before, the components of deformation are

$$r^2 = R^2 - R_1{}^2 + r_1{}^2,\ \theta = \Theta,\ z = Z, \tag{10.68}$$

where R_1 and r_1 are the inner radii in the reference and the deformed configuration, respectively (*cf.* (IV. 7.1)). We assume again that the intrinsic metric g has the diagonal component form (10.2) relative to (R, Θ, Z) in $\boldsymbol{\kappa}$.

Following the procedure set out in § IV. 7, we determine first the radial stress in the motion:

$$T_{\langle rr\rangle} = \rho\left[(r_1\ddot{r}_1+\dot{r}_1^2)\log r+\frac{1}{2}\frac{r_1^{\,2}\dot{r}_1^{\,2}}{r^2}\right] + \int\frac{1}{r}\left(\frac{r^2}{R^2}\bar{g}_{\langle 22\rangle}-\frac{R^2}{r^2}\bar{g}_{\langle 11\rangle}\right)(\psi_1-\psi_{-1}\bar{g}_{\langle 33\rangle})dr, \tag{10.69}$$

where the arguments of ψ_1 and ψ_{-1} are

$$I = \frac{R^2}{r^2}\bar{g}_{\langle 11\rangle}+\frac{r^2}{R^2}\bar{g}_{\langle 22\rangle}+\bar{g}_{\langle 33\rangle}, \tag{10.70}$$

and

$$II = \frac{r^2}{R^1}\frac{1}{\bar{g}_{\langle 11\rangle}}+\frac{R^2}{r^2}\frac{1}{\bar{g}_{\langle 22\rangle}}+\frac{1}{\bar{g}_{\langle 33\rangle}}. \tag{10.71}$$

As before, let P be the difference between the pressures on the inner and the outer surfaces:

$$P = T_{\langle rr\rangle}|_{r_2}-T_{\langle rr\rangle}|_{r_1}. \tag{10.72}$$

We set

$$x \equiv \frac{r_1}{R_1}, \tag{10.73}$$

$$\xi \equiv r^2/R^2, \tag{10.74}$$

$$\delta \equiv \frac{R_2^{\,2}}{R_1^{\,2}}-1 > 0, \tag{10.75}$$

and

$$f(x,\delta) = \frac{1}{\rho R_1^{\,2}}\int_{(\delta+x^2)/(\delta+1)}^{x^2}\left(\frac{1}{\xi}\bar{g}_{\langle 11\rangle}-\xi\bar{g}_{\langle 22\rangle}\right)(\psi_1-\psi_{-1}\bar{g}_{\langle 33\rangle})\frac{d\xi}{\xi(1-\xi)}, \tag{10.76}$$

where the arguments of ψ_1 and ψ_{-1} now become

$$I = \frac{1}{\xi}\bar{g}_{\langle 11\rangle} + \xi\bar{g}_{\langle 22\rangle} + \bar{g}_{\langle 33\rangle}, \tag{10.77}$$

and

$$II = \frac{\xi}{\bar{g}_{\langle 11\rangle}} + \frac{1}{\xi\bar{g}_{\langle 22\rangle}} + \bar{g}_{\langle 33\rangle}, \tag{10.78}$$

and the argument of $\bar{g}_{\langle 11\rangle}$, $\bar{g}_{\langle 22\rangle}$, and $\bar{g}_{\langle 33\rangle}$ is

$$R = R_1\sqrt{\frac{x^2-1}{\xi-1}}. \tag{10.79}$$

Then, as before, the governing equation for x is (*cf.* (IV. 7.9))

$$x \log\left(1+\frac{\delta}{x^2}\right)\ddot{x} + \left[\log\left(1+\frac{\delta}{x^2}\right) - \frac{\delta}{\delta+x^2}\right]\dot{x}^2 + f(x,\delta) = \frac{2P}{\rho R_1{}^2}. \tag{10.80}$$

From this point on, the analysis is exactly the same as that of Example 1 in § IV.7.

CHAPTER VI

Wave Propagation in Elastic Bodies

In this chapter we develop the main properties of weak waves in homogeneous or materially uniform elastic bodies. For a more extensive treatment, including the propagation, growth, and decay of shock waves, the reader may consult an article by P. J. CHEN[1].

1. Conditions of Compatibility for Singular Surfaces

In this section we summarize the geometrical and kinematical conditions of compatibility for singular surfaces. For more details about these conditions, we refer the reader to Chapter C of CFT and to the classical treatise by HADAMARD[2].

Let $\mathscr{D}$ be an open set in the physical space $\mathscr{E}$, and let Ψ be a tensor field on $\mathscr{D}$. Then an oriented smooth surface $\mathscr{S}$ in $\mathscr{D}$ is called a *singular surface* for Ψ if Ψ is a smooth field on the complement of $\mathscr{S}$ in $\mathscr{D}$ and has smooth limits Ψ^+ and Ψ^- on $\mathscr{S}$ as it is approached from its side labelled $+$ and its side labelled $-$, respectively. Generally, Ψ^+ is not equal to Ψ^-; otherwise $\mathscr{S}$ would not be called a singular surface. The *jump* of Ψ across $\mathscr{S}$ is then defined by

$$[\Psi] \equiv \Psi^+ - \Psi^- \tag{1.1}$$

and is a smooth field on $\mathscr{S}$. As usual, we denote the positive unit normal vector of $\mathscr{S}$ by $\boldsymbol{n}_{\mathscr{S}}$, or more briefly by $\boldsymbol{n}$ when a particular surface $\mathscr{S}$ has been selected.

The theory of singular surfaces rests upon

---

[1] P. J. CHEN, "Growth and decay of waves in solids", FLÜGGE's *Handbuch der Physik* VIa/3, ed. C. TRUESDELL, Berlin &c., Springer-Verlag, 1973.

[2] J. HADAMARD, *Leçons sur la Propagation des Ondes et les Équations de l'Hydrodynamique*, Paris, Hermann, 1903, variously reprinted.

HADAMARD'S LEMMA. *Let $\mathscr{S}$ be a singular surface for Ψ, and let $\boldsymbol{\lambda} = \boldsymbol{\lambda}(s)$ be a smooth curve on $\mathscr{S}$. Then*

$$\frac{d\Psi^+}{ds} = \left(\frac{\partial \Psi}{\partial x^i}\right)^+ \frac{d\lambda^i}{ds}, \tag{1.2}$$

and

$$\frac{d\Psi^-}{ds} = \left(\frac{\partial \Psi}{\partial x^i}\right)^- \frac{d\lambda^i}{ds}, \tag{1.3}$$

so that

$$\frac{d[\Psi]}{ds} = \left[\frac{\partial \Psi}{\partial x^i}\right] \frac{d\lambda^i}{ds}, \tag{1.4}$$

where (x^i) is the spatial co-ordinate system.

The conditions (1.2) and (1.3) mean that the usual chain rule holds for the limiting values as $\mathscr{S}$ is approached from either side alone. One of HADAMARD's original proofs is reproduced in CFT, § 174.

Now let us introduce a positive local co-ordinate system (y^1, y^2) on the surface $\mathscr{S}$. Then $\mathscr{S}$ can be characterized by certain equations

$$x^i = x^i(y^1, y^2). \tag{1.5}$$

Applying the formula (1.4) to the co-ordinate curves of (y^α), we obtain

$$\frac{\partial[\Psi]}{\partial y^\alpha} = \left[\frac{\partial \Psi}{\partial x^i}\right] \frac{\partial x^i}{\partial y^\alpha}. \tag{1.6}$$

As usual the component form of the surface basis $\{\partial y_1, \partial y_2\}$ in (x^i) is

$$\partial y_\alpha = \frac{\partial x^i}{\partial y^\alpha} \partial x_i = \frac{\partial x^i}{\partial y^\alpha} \boldsymbol{e}_i. \tag{1.7}$$

By the convention of orientation, the ordered basis $\{\partial y_1, \partial y_2, \boldsymbol{n}\}$ has positive orientation in $\mathscr{E}$, and of course

$$\boldsymbol{n} \cdot \partial y_\alpha = 0. \tag{1.8}$$

Using (1.7) and (1.8), we see that the dual basis of $\{\partial y_1, \partial y_2, \boldsymbol{n}\}$ is $\{\partial y^1, \partial y^2, \boldsymbol{n}\}$, where

$$\partial y^\alpha = a^{\alpha\beta} \frac{\partial x^i}{\partial y^\beta} \boldsymbol{e}_i, \tag{1.9}$$

$a^{\alpha\beta}$ being a component of the induced surface metric, *i.e.*,

$$a^{\alpha\beta} = \partial y^{\alpha} \cdot \partial y^{\beta} \tag{1.10}$$

or, equivalently,

$$[a^{\alpha\beta}] = [a_{\alpha\beta}]^{-1}, \tag{1.11}$$

where

$$a_{\alpha\beta} = \partial y_{\alpha} \cdot \partial y_{\beta} = \delta_{ij} \frac{\partial x^i}{\partial y^{\alpha}} \frac{\partial x^j}{\partial y^{\beta}}. \tag{1.12}$$

Using (1.6), (1.7), and (1.9), we have

$$\begin{aligned} \left[\frac{\partial \Psi}{\partial x^i}\right] &= \left(\left[\frac{\partial \Psi}{\partial x^j}\right] n^j \boldsymbol{n} + \left[\frac{\partial \Psi}{\partial x^j}\right] \frac{\partial x^j}{\partial y^{\alpha}} \partial y^{\alpha}\right) \cdot \boldsymbol{e}_i, \\ &= \left[n^j \frac{\partial \Psi}{\partial x^j}\right] n_i + a^{\alpha\beta} \frac{\partial x_i}{\partial y^{\alpha}} \frac{\partial [\Psi]}{\partial y^{\beta}}. \end{aligned} \tag{1.13}$$

This formula, which gives the components of the jump of the gradient of Ψ relative to the basis $\{\partial y_1, \partial y_2, \boldsymbol{n}\}$, is called the *geometrical condition of compatibility* for the singular surface $\mathscr{S}$. An important consequence of (1.13) is

MAXWELL'S THEOREM. *If* $[\Psi] = \mathbf{0}$, *then*

$$\left[\frac{\partial \Psi}{\partial x^i}\right] = \mho n_i, \tag{1.14}$$

where

$$\mho = \left[n^j \frac{\partial \Psi}{\partial x^j}\right]. \tag{1.15}$$

Since the field Ψ in (1.13) is arbitrary, we can use the same formula to determine the jump of the second gradient, *viz*,

$$\left[\frac{\partial^2 \Psi}{\partial x^i \partial x^j}\right] = \left[\frac{\partial}{\partial x^i}\left(\frac{\partial \Psi}{\partial x^j}\right)\right] = \left[n^k \frac{\partial^2 \Psi}{\partial x^k \partial x^j}\right] n_i + a^{\alpha\beta} \frac{\partial x_i}{\partial y^{\alpha}} \frac{\partial}{\partial y^{\beta}} \left[\frac{\partial \Psi}{\partial x^j}\right], \tag{1.16}$$

where $[\partial \Psi / \partial x^j]$ is given by (1.13). In particular, if $[\partial \Psi / \partial x^j] = \mathbf{0}$, then

$$\left[\frac{\partial^2 \Psi}{\partial x^i \partial x^j}\right] = \left[n^k \frac{\partial^2 \Psi}{\partial x^k \partial x^j}\right] n_i. \tag{1.17}$$

Since the left-hand side of this equation is symmetric in i and j, we have also

$$\left[\frac{\partial^2 \Psi}{\partial x^i \partial x^j}\right] = \left[n^k \frac{\partial^2 \Psi}{\partial x^k \partial x^i}\right] n_j. \tag{1.18}$$

Thus

$$\left[n^k \frac{\partial^2 \Psi}{\partial x^k \partial x^i}\right] n_j = \left[n^k \frac{\partial^2 \Psi}{\partial x^k \partial x^j}\right] n_i . \tag{1.19}$$

Multiplying this equation by n^i and summing on i, we see that

$$\left[n^k n^l \frac{\partial^2 \Psi}{\partial x^k \partial x^l}\right] n_j = \left[n^k \frac{\partial^2 \Psi}{\partial x^k x^j}\right]. \tag{1.20}$$

Substituting (1.20) into (1.17), we obtain[1]

$$\left[\frac{\partial^2 \Psi}{\partial x^i \partial x^j}\right] = ♀ n_i n_j , \tag{1.21}$$

where

$$♀ = \left[n^k n^l \frac{\partial^2 \Psi}{\partial x^k \partial x^l}\right]. \tag{1.22}$$

If $[\partial \Psi / \partial x^i] \neq \mathbf{0}$ but $[\Psi] = \mathbf{0}$, then (1.14) holds, and (1.16) reduces to THOMAS's formula:

$$\left[\frac{\partial^2 \Psi}{\partial x^k \partial x^m}\right] = ♀ n_k n_m + n_k x_{m|}{}^{\alpha} ☿_{|\alpha} + n_m x_{k|}{}^{\alpha} ☿_{|\alpha} - x_{k|}{}^{\alpha} x_{m|}{}^{\beta} \Omega_{\alpha\beta} ☿, \tag{1.23}$$

where $\Omega_{\alpha\beta}$ is a component of the second fundamental form of $\mathcal{S}$ relative to (y^γ) and where | denotes the spatial covariant derivative along the surface co-ordinate curves. For the proof of (1.23), we refer the reader to CFT, § 176, or the book by THOMAS.[2]

The same procedure applied to the $(p-1)^{\text{st}}$ gradient of Ψ yields the following recurrence formula:

$$\left[\frac{\partial^p \Psi}{\partial x^{i_1} \ldots \partial x^{i_p}}\right] = \left[n^k \frac{\partial^p \Psi}{\partial x^k \partial x^{i_1} \ldots \partial x^{i_{p-1}}}\right] n_{i_p} + a^{\alpha\beta} \frac{\partial x_{i_p}}{\partial y^\alpha} \frac{\partial}{\partial y^\beta} \left[\frac{\partial^{p-1} \Psi}{\partial x^{i_1} \ldots \partial x^{i_{p-1}}}\right]. \tag{1.24}$$

---

[1] The analysis here improves slightly that of CFT, § 176, where (1.21) is derived under the assumptions

$$[\Psi] = \mathbf{0} \quad \text{and} \quad \left[\frac{\partial \Psi}{\partial x^i}\right] = \mathbf{0};$$

we observe that the first of these assumptions is unnecessary.

[2] T. Y. THOMAS, *Plastic Flow and Fracture in Solids*, New York, Academic Press, 1961.

In particular, if the $(p-1)^{\text{st}}$ gradient is continuous at $\mathscr{S}$, then by the same argument as before we have

$$\left[\frac{\partial^p \Psi}{\partial x^{i_1} \ldots \partial x^{i_p}}\right] = \left[n^{j_1} \ldots n^{j_p} \frac{\partial^p \Psi}{\partial x^{j_1} \ldots \partial x^{j_p}}\right] n_{i_1} \ldots n_{i_p}. \quad (1.25)$$

We call the general formula (1.24) the *iterated geometrical condition of compatibility* for $\mathscr{S}$.

All the conditions derived so far reflect the fact that the discontinuity is smoothly spread out over $\mathscr{S}$, in the sense laid down before the statement of HADAMARD's lemma.

Up to now we have considered the geometric condition of compatibility for a fixed singular surface $\mathscr{S}$ in $\mathscr{E}$. For wave propagation, of course, the singular surface generally moves through $\mathscr{E}$. Then we must consider a 1-parameter family of singular surfaces, say $\mathscr{S} = \mathscr{S}_t$, where t is the time. In addition to the geometrical conditions of compatibility for each $\mathscr{S}_t$, there are *kinematical conditions of compatibility*, conditions which reflect the smooth continuance of the singular surface in an interval of time containing t.

Indeed, let (y^γ) be a surface co-ordinate system on $\mathscr{S}_t$ for each t. Then the moving surface $\mathscr{S} = \mathscr{S}_t$ can be characterized by the equations

$$x^i = x^i(y^1, y^2, t). \quad (1.26)$$

We can regard (y^1, y^2) as the co-ordinates of a fictitious *surface point* $\boldsymbol{y}$ of the moving surface. Then the surface velocity relative to (y^γ) is given by

$$\boldsymbol{c} = \frac{\partial x^i}{\partial t} \boldsymbol{e}_i. \quad (1.27)$$

Since it depends on the choice of the surface co-ordinate system, this surface velocity is not intrinsic to the moving surface.

Suppose that $(\bar{y}^\alpha)$ is another surface co-ordinate system, related to (y^α) as follows:

$$y^\alpha = y^\alpha(\bar{y}^1, \bar{y}^2, t). \quad (1.28)$$

Then the surface velocity $\bar{\boldsymbol{c}}$ relative to $(\bar{y}^\alpha)$ differs from $\boldsymbol{c}$ by a tangential vector:

$$\bar{\boldsymbol{c}} - \boldsymbol{c} = \frac{\partial y^\alpha}{\partial t} \partial y_\alpha = \frac{\partial y^\alpha}{\partial t} \frac{\partial x^i}{\partial y^\alpha} \boldsymbol{e}_i, \quad (1.29)$$

which differerence is the tangential velocity of the $(\bar{y}^\alpha)$-surface point relative to the (y^α)-surface point. In view of the transformation rule (1.29) we see that $\boldsymbol{c}$ and $\bar{\boldsymbol{c}}$ have the same normal component, namely

$$\boldsymbol{u} = u\boldsymbol{n}, \tag{1.30}$$

where

$$u = \boldsymbol{c} \cdot \boldsymbol{n} = \bar{\boldsymbol{c}} \cdot \boldsymbol{n}. \tag{1.31}$$

The quantities $\boldsymbol{u}$ and u, which are thus shown to be intrinsic to the moving surface, are called its *normal velocity* and *normal speed*, respectively.

Resolving the surface velocity $\boldsymbol{c}$ relative to (y^α) into components, we obtain

$$\boldsymbol{c} = \boldsymbol{u} + c^\alpha \partial y_\alpha, \tag{1.32}$$

where the tangential components c^α are given by

$$c^\alpha = \boldsymbol{c} \cdot \partial y^\alpha = \frac{\partial x^i}{\partial t} \frac{\partial x_i}{\partial y^\beta} a^{\alpha\beta}. \tag{1.33}$$

Since

$$c^\alpha \partial y^\alpha = \boldsymbol{c} - \boldsymbol{u}, \tag{1.34}$$

we can regard the tangential vector[1] $c^\alpha \partial y_\alpha$ as the *tangential velocity* of the (y^α)-surface point relative to the normal trajectory of the moving surface.

Next we introduce the concept of the *displacement derivative* $\delta\Psi/\delta t$ of a field $\Psi(x^i, t)$ with respect to the moving surface. We define $\delta\Psi/\delta t$ as the time derivative of Ψ along the normal trajectory, *viz*,

$$\frac{\delta\Psi}{\delta t} \equiv \frac{\partial\Psi}{\partial t} + un^i \frac{\partial\Psi}{\partial x^i}, \tag{1.35}$$

where the argument is the instantaneous position of a point on the moving surface $\mathscr{S}_t$ and the time t. Similarly, if $♃$ is a field defined only on $\mathscr{S}_t$ for all t, say

$$♃ = ♃(y^1, y^2, t), \tag{1.36}$$

then we define the *displacement derivative* $\delta♃/\delta t$ of $♃$ with respect to

[1] In § 177 of CFT the vector $-c^\alpha \partial y_\alpha$ is called the *tangential velocity* of the parametrization (y^α). That vector, of course, is the tangential velocity of the normal trajectory of the surface relative to the (y^α)-surface point.

$\mathscr{S}_t$ by

$$\frac{\delta ♃}{\delta t} = \frac{\partial ♃}{\partial t} + c^\alpha \frac{\partial ♃}{\partial y^\alpha}. \tag{1.37}$$

From the transformation rule (1.29) we see that this definition does not depend on the choice of the co-ordinate system on the surface.

To illustrate the geometric meaning of the displacement derivative, we now derive formulae for $\delta/\delta t(\partial y_\alpha)$ and $\delta/\delta t(\boldsymbol{n})$. We introduce first the concept of a *convected surface co-ordinate system* (z^α) by requiring that the locus of the (z^α)-surface point be a normal trajectory of the surface. Clearly, such a co-ordinate system exists for a sufficiently short interval of time about any given time. As usual, we have the transformation rules

$$\partial z_\alpha = \frac{\partial y^\beta}{\partial z^\alpha} \partial y_\beta, \quad \partial z^\alpha = \frac{\partial z^\alpha}{\partial y_\beta} \partial y^\beta \tag{1.38}$$

and also the component forms

$$\partial z_\alpha = \frac{\partial x^i}{\partial z^\alpha} \boldsymbol{e}_i, \quad \partial z^\alpha = a^{\alpha\beta}(z) \frac{\partial x^i}{\partial z^\beta} \boldsymbol{e}_i, \tag{1.39}$$

where $a^{\alpha\beta}(z)$ is a component of the surface metric relative to (z^γ). Now since the displacement derivative is the time derivative along the normal trajectory, which is the locus of the (z^α)-surface point, we have

$$\frac{\delta}{\delta t}(\partial z_\alpha) = \frac{\partial}{\partial t}\left(\frac{\partial x^i}{\partial z^\alpha}\right) \boldsymbol{e}_i. \tag{1.40}$$

But since

$$\frac{\partial}{\partial t}\left(\frac{\partial x^i}{\partial z^\alpha}\right) = \frac{\partial}{\partial z^\alpha}\left(\frac{\partial x^i}{\partial t}\right) = \frac{\partial}{\partial z^\alpha}(u n^i), \tag{1.41}$$

by use of WEINGARTEN's formula

$$\frac{\partial \boldsymbol{n}}{\partial z^\alpha} = -\Omega_{\alpha\beta}(z) \partial z^\beta, \tag{1.42}$$

in which $\Omega_{\alpha\beta}(z)$ is a component of the second fundamental form of $\mathscr{S}_t$ relative to (z^γ), we obtain

$$\frac{\delta}{\delta t}(\partial z_\alpha) = \frac{\partial u}{\partial z^\alpha} \boldsymbol{n} - u \Omega_{\alpha\beta}(z) \partial z^\beta. \tag{1.43}$$

Now from the conditions

$$\boldsymbol{n} \cdot \partial z_\alpha = 0, \ \boldsymbol{n} \cdot \boldsymbol{n} = 1, \tag{1.44}$$

we obtain

$$\frac{\delta \boldsymbol{n}}{\delta t} \cdot \partial z_\alpha = -\boldsymbol{n} \cdot \frac{\delta}{\delta t}(\partial z_\alpha) \tag{1.45}$$

and

$$\boldsymbol{n} \cdot \frac{\delta \boldsymbol{n}}{\delta t} = 0. \tag{1.46}$$

Substituting (1.43) into (1.45) and using the fact that $\{\partial z_1, \partial z_2, \boldsymbol{n}\}$ is a basis, we obtain

$$\frac{\delta \boldsymbol{n}}{\delta t} = -\frac{\partial u}{\partial z^\alpha} \partial z^\alpha. \tag{1.47}$$

From the transformation rule (1.38), we can rewrite (1.47) in terms of a surface co-ordinate system (y^α) in general, *viz*,

$$\frac{\delta \boldsymbol{n}}{\delta t} = -\frac{\partial u}{\partial y^\alpha} \partial y^\alpha. \tag{1.48}$$

This formula was first stated by HAYES[1] and by THOMAS[2]; proofs of it were given by the latter and in CFT, § 179. The proof here is that of BOWEN & WANG[3].

Now using the conditions

$$\partial z_\alpha \cdot \partial z^\beta = \delta^\beta_\alpha, \ \boldsymbol{n} \cdot \partial z^\beta = 0, \tag{1.49}$$

we have

$$\frac{\delta}{\delta t}(\partial z^\beta) \cdot \partial z_\alpha = -\frac{\delta}{\delta t}(\partial z_\alpha) \cdot \partial z^\beta, \tag{1.50}$$

and

$$\frac{\delta}{\delta t}(\partial z^\beta) \cdot \boldsymbol{n} = -\frac{\delta}{\delta t}(\boldsymbol{n}) \cdot \partial z^\beta. \tag{1.51}$$

---

[1] W. HAYES, "The vorticity jump across a gasdynamic discontinuity", *J. Fluid Mech.* **2**, 595–600 (1957).

[2] T. Y. THOMAS, "Extended compatibility conditions for the study of surfaces of discontinuity in continuum mechanics", *J. Math. Mech.* **6**, 311–322 (1957).

[3] R. M. BOWEN & C.-C. WANG, "On displacement derivatives", *Q. Appl. Math.* **29**, 29–39 (1971).

Substituting (1.43) and (1.47) into (1.50) and (1.51), we obtain

$$\frac{\delta}{\delta t}(\partial z^{\alpha}) = a^{\alpha\beta}(z)\frac{\partial u}{\partial z^{\beta}}\,\boldsymbol{n} + u\Omega^{\alpha\beta}(z)\partial z_{\beta}. \tag{1.52}$$

The formulae (1.43), (1.47), and (1.52) evaluate the displacement derivatives of the bases $\{\partial z_1, \partial z_2, \boldsymbol{n}\}$ and $\{\partial z^1, \partial z^2, \boldsymbol{n}\}$. Finally, using (1.43), (1.52), and the transformation rules (1.38), we obtain for any surface co-ordinate system (y^{γ})

$$\frac{\delta}{\delta t}(\partial y_{\alpha}) = \frac{\partial u}{\partial y^{\alpha}}\,\boldsymbol{n} - u\Omega_{\alpha\beta}(y)\partial y^{\beta} + \frac{\partial c^{\beta}}{\partial y^{\alpha}}\,\partial y_{\beta}, \tag{1.53}$$

and

$$\frac{\delta}{\delta t}(\partial y^{\alpha}) = a^{\alpha\beta}(y)\frac{\partial u}{\partial y^{\beta}}\,\boldsymbol{n} + u\Omega^{\alpha\beta}(y)\partial y_{\beta} - \frac{\partial c^{\alpha}}{\partial y^{\beta}}\,\partial y^{\beta}, \tag{1.54}$$

$\boldsymbol{c}$ being the surface velocity (1.27). The formulae (1.48), (1.53), and (1.54) evaluate the displacement derivatives of the bases $\{\partial y_1, \partial y_2, \boldsymbol{n}\}$ and $\{\partial y^1, \partial y^2, \boldsymbol{n}\}$.

Having determined the displacement derivatives of the bases, we can use the product rule so as to express the displacement derivative of any tensor field Ψ on the moving surface in terms of the displacement derivative of the components of Ψ relative to the bases.

Now suppose that at each time t the surface $\mathscr{S}_t$ is a singular surface for a field $\Psi(\cdot, t)$. Then from HADAMARD's lemma applied to the product space $\mathscr{E} \times \mathscr{R}$, we have

$$\frac{\delta}{\delta t}(\Psi^{+}) = \left(\frac{\partial \Psi}{\partial t}\right)^{+} + un^{i}\left(\frac{\partial \Psi}{\partial x^{i}}\right)^{+}, \tag{1.55}$$

and

$$\frac{\delta}{\delta t}(\Psi^{-}) = \left(\frac{\partial \Psi}{\partial t}\right)^{-} + un^{i}\left(\frac{\partial \Psi}{\partial x^{i}}\right)^{-}, \tag{1.56}$$

so that

$$\frac{\delta}{\delta t}[\Psi] = \left[\frac{\partial \Psi}{\partial t}\right] + un^{i}\left[\frac{\partial \Psi}{\partial x^{i}}\right]. \tag{1.57}$$

This relation is called the *kinematical condition of compatibility* for the jumps of Ψ on the moving surface. In particular, if $[\Psi] = \boldsymbol{0}$, then (1.57) reduces to

$$\left[\frac{\partial \Psi}{\partial t}\right] = -un^i\left[\frac{\partial \Psi}{\partial x^i}\right] = -u☿, \tag{1.58}$$

where ☿ is given by (1.15).

We can iterate the kinematical condition of compatibility so as to determine the conditions on the jumps of higher derivatives of Ψ. For example, the jumps of the second derivatives satisfy the conditions

$$\begin{aligned}\left[\frac{\partial^2 \Psi}{\partial t\,\partial x^k}\right] &= -u\left[n^i\frac{\partial^2 \Psi}{\partial x^i \partial x^k}\right] + \frac{\delta}{\delta t}\left[\frac{\partial \Psi}{\partial x^k}\right] \\ &= \left[n^j\frac{\partial^2 \Psi}{\partial x^j \partial t}\right]n_k + a^{\alpha\beta}\frac{\partial x_k}{\partial y^\alpha}\frac{\partial}{\partial y^\beta}\left[\frac{\partial \Psi}{\partial t}\right]\end{aligned} \tag{1.59}$$

and

$$\left[\frac{\partial^2 \Psi}{\partial t^2}\right] = -u\left[n^i\frac{\partial^2 \Psi}{\partial x^i \partial t}\right] + \frac{\delta}{\delta t}\left[\frac{\partial \Psi}{\partial t}\right]. \tag{1.60}$$

In particular if the first derivatives of Ψ are continuous at $\mathscr{S}_t$ for all t, then by the same argument as before we have

$$\left[\frac{\partial^2 \Psi}{\partial x^k \partial t}\right] = -u♀n_k, \tag{1.61}$$

and

$$\left[\frac{\partial^2 \Psi}{\partial t^2}\right] = u^2♀, \tag{1.62}$$

where ♀ is given by (1.22).

If Ψ is continuous but its first derivatives suffer jumps across $\mathscr{S}_t$, then (1.59) and (1.60) reduce to THOMAS's formulae:

$$\left[\frac{\partial^2 \Psi}{\partial x^k \partial t}\right] = \left(-u♀+\frac{\delta ☿}{\delta t}\right)n_k - a^{\alpha\beta}x_{k|\alpha}(u☿)_{|\beta}, \tag{1.63}$$

and

$$\left[\frac{\partial^2 \Psi}{\partial t^2}\right] = u^2♀ - 2u\frac{\delta ☿}{\delta t} - ☿\frac{\delta u}{\delta t}. \tag{1.64}$$

For the proof of these conditions, see CFT, § 181, or the book by THOMAS[1].

[1] T. Y. THOMAS, p. 48 of the work cited on p. 420.

For jumps of higher derivatives of Ψ, we use the following recurrence formulae:

$$\left[\frac{\partial^{p+q}\Psi}{\partial x^{i_1}\dots\partial x^{i_p}\partial t^q}\right] = -u\left[n^k\frac{\partial^{p+q}\Psi}{\partial x^k\partial x^{i_1}\dots\partial x^{i_p}\partial t^{q-1}}\right] + \frac{\delta}{\delta t}\left[\frac{\partial^{p+q-1}\Psi}{\partial x^{i_1}\dots\partial x^{i_p}\partial t^{q-1}}\right], \tag{1.65}$$

and

$$\left[\frac{\partial^{p+q}\Psi}{\partial x^{i_1}\dots\partial x^{i_p}\partial t^q}\right] = \left[n^k\frac{\partial^{p+q}\Psi}{\partial x^k\partial x^{i_1}\dots\partial x^{i_{p-1}}\partial t^{q}}\right]n_{i_p} + a^{\alpha\beta}\frac{\partial x_{i_p}}{\partial y^\alpha}\frac{\partial}{\partial y^\beta}\left[\frac{\partial^{p+q-1}\Psi}{\partial x^{i_1}\dots\partial x^{i_{p-1}}\partial t^{q}}\right]. \tag{1.66}$$

In particular, if all $(p+q-1)^{\text{st}}$-order derivatives of Ψ are continuous at $\mathscr{S}_t$ for all t, then

$$\left[\frac{\partial^{p+q}\Psi}{\partial x^{i_1}\dots\partial x^{i_p}\partial t^q}\right] = (-u)^q\left[n^{k_1}\dots n^{k_{p+q}}\frac{\partial^{p+q}\Psi}{\partial x^{k_1}\dots\partial x^{k_{p+q}}}\right]n_{i_1}\dots n_{i_p}. \tag{1.67}$$

The conditions (1.65) and (1.66) are called the *iterated kinematical conditions of compatibility* for the jumps of Ψ across the moving singular surface.

2. Waves in Deformable Bodies[1]

In the preceding section we have obtained the conditions of compatibility for a moving surface singular with respect to some field Ψ. Now we shall apply these results to particular singular surfaces defined in terms of the motion of a deformable body $\mathscr{B}$. As before, we choose a frame of reference $\boldsymbol{\varphi}$. Then the motion of $\mathscr{B}$ can be represented by the components of motion

$$x^i = \chi^i(X, t), \qquad X \in \mathscr{B}, \tag{2.1}$$

[1] For more detail, and for corresponding results for shock waves, see §§ 187–191 of CFT.

where (x^i) denotes the spatial co-ordinate system of $\boldsymbol{\varphi}$. Let $\boldsymbol{\kappa}$ be a fixed reference configuration for $\mathscr{B}$ with referential co-ordinates (X^A). Then (2.1) is equivalent to the component formulae

$$x^i = \chi^i(X^A, t) \tag{2.2}$$

with $(X^A) \in \boldsymbol{\kappa}(\mathscr{B})$.

In the theory of elastic bodies developed so far, we have required that the components of displacement χ^i be of class C^∞ and that the transformation from $\boldsymbol{\kappa}(\mathscr{B})$ to $\boldsymbol{\chi}_t(\mathscr{B})$ be nonsingular, so that the motion corresponds to a 1-parameter family of diffeomorphisms from $\boldsymbol{\kappa}(\mathscr{B})$ into the physical space $\mathscr{E}$. For the theory of wave propagation we relax the condition of smoothness in order to consider jump discontinuities for various kinematical quantities. Specifically, we now require that while $\boldsymbol{\chi}_t$ is a diffeomorphism of class C^∞ in general, on a certain singular surface it need be only of class C^1. This relaxed condition turns out to be appropriate to the theory of *acceleration waves*, which furnishes the major subject of this chapter.

Under this weaker assumption, we still can transform an arbitrary C^1-function

$$f = f(x^i, t) \tag{2.3}$$

on $\boldsymbol{\chi}_t(\mathscr{B})$ into a C^1-function

$$f = \bar{f}(X^A, t) \equiv f(\chi^i(X^A, t), t) \tag{2.4}$$

on $\boldsymbol{\kappa}(\mathscr{B})$. In particular, if $\mathscr{S}_t$ is a C^1 surface in $\boldsymbol{\chi}_t(\mathscr{B})$ characterized by the condition

$$f(x^i, t) = 0, \tag{2.5}$$

then the image of $\mathscr{S}_t$ in $\boldsymbol{\kappa}(\mathscr{B})$ is a C^1 surface $\bar{\mathscr{S}}_t$ characterized by the condition

$$\bar{f}(X^A, t) = 0. \tag{2.6}$$

We denote the unit normals of $\mathscr{S}_t$ and $\bar{\mathscr{S}}_t$ by $\boldsymbol{n}$ and $\bar{\boldsymbol{n}}$, respectively. If $\operatorname{grad} f \neq \mathbf{0}$ on $\mathscr{S}_t$, then [1]

$$\boldsymbol{n} = \pm \frac{\operatorname{grad} f}{\|\operatorname{grad} f\|}, \quad \bar{\boldsymbol{n}} = \pm \frac{\operatorname{Grad} \bar{f}}{\|\operatorname{Grad} \bar{f}\|}, \tag{2.7}$$

[1] Strictly we do not need two different notations for the gradient operator, since the notation distinguishes the function $\bar{f}$ from f. Thus the symbols "Grad" and "grad" serve only to remind the reader which variables are being used.

where the sign is + if $f > 0$ on the positive side of $\mathscr{S}_t$ and $f < 0$ on the negative side of $\mathscr{S}_t$; otherwise, the sign is negative.

In general $\mathscr{S}_t$ and $\bar{\mathscr{S}}_t$ depend on t. We can introduce the normal speeds u and $\bar{u}$ and the normal velocities $\boldsymbol{u}$ and $\bar{\boldsymbol{u}}$ for $\mathscr{S}_t$ and $\bar{\mathscr{S}}_t$, respectively, as before. In view of the representations (2.5) and (2.6),

$$u = \mp \frac{\dfrac{\partial f}{\partial t}}{||\text{grad } f||}, \quad \bar{u} = \mp \frac{\dfrac{\partial \bar{f}}{\partial t}}{||\text{Grad } \bar{f}||}, \tag{2.8}$$

where the signs are to be taken as the opposites of those in (2.7).

Since f and $\bar{f}$ are related by $(2.4)_2$, by the chain rule

$$\frac{\partial \bar{f}}{\partial X^A} = \frac{\partial f}{\partial x^i}\frac{\partial x^i}{\partial X^A}. \tag{2.9}$$

Then from (2.7) it follows that

$$\bar{n}_A = n_i \frac{\partial x^i}{\partial X^A}\frac{||\text{grad } f||}{||\text{Grad } \bar{f}||}, \tag{2.10}$$

or equivalently,

$$n_i = \bar{n}_A \frac{\partial X^A}{\partial x^i}\frac{||\text{Grad } \bar{f}||}{||\text{grad } f\,||}. \tag{2.11}$$

Similarly, by the chain rule with respect to t,

$$\frac{\partial \bar{f}}{\partial t} = \frac{\partial f}{\partial t} + \frac{\partial f}{\partial x^i} v^i, \tag{2.12}$$

where v^i is the ith component of the velocity field $\boldsymbol{v}$:

$$v^i = \dot{x}^i = \frac{\partial \chi^i}{\partial t}. \tag{2.13}$$

By (2.8), then,

$$\bar{u} = (u - v^i n_i)\frac{||\text{grad } f||}{||\text{Grad } \bar{f}||}. \tag{2.14}$$

As before (*cf.* (II. 6.11)), we set

$$U \equiv u - v^i n_i. \tag{2.15}$$

U is called the *intrinsic speed* of $\mathscr{S}_t$.

Now we assume that the moving surface $\bar{\mathscr{S}}_t$ is singular with respect to the second derivative of the components of deformation χ^i, this assumption being consistent with the relaxed condition of smoothness on the motion. Since the first derivative of χ^i is continuous at the surface at all t, by (1.21), (1.61), and (1.62) there exists a vector field $\bar{\mathbf{s}}$ defined on $\bar{\mathscr{S}}_t$ such that

$$\left[\frac{\partial^2 x^k}{\partial X^A \partial X^B}\right] = \bar{s}^k \bar{n}_A \bar{n}_B, \tag{2.16}$$

$$\left[\frac{\partial^2 x^k}{\partial X^A \partial t}\right] = -\bar{u}\bar{s}^k \bar{n}_A, \tag{2.17}$$

$$[a^k] = \left[\frac{\partial^2 x^k}{\partial t^2}\right] = \bar{u}^2 \bar{s}^k, \tag{2.18}$$

where a^k is a component of the acceleration field. These jump conditions, called *Hugoniot's conditions* (1885), are expressed in terms of the referential co-ordinates (X^A) for the moving singular surface $\bar{\mathscr{S}}_t$ in $\boldsymbol{\kappa}(\mathscr{B})$.

If $\bar{u} \neq 0$, the singular surface propagates in $\boldsymbol{\kappa}(\mathscr{B})$. Propagating singular surfaces are called *waves*. From (2.16)–(2.18) it follows that if the acceleration field is continuous at a wave $\bar{\mathscr{S}}_t$, then $\bar{\mathscr{S}}_t$ is not singular with respect to any second derivative of any χ^i. Thus if $\bar{\mathscr{S}}_t$ is a wave, it carries a non-zero jump in the acceleration. For this reason the waves under discussion here are called *acceleration waves.* (It remains possible that $\bar{u} = 0$ but $\bar{\mathbf{s}} \neq \mathbf{0}$, in which case the singular surface, which is at rest in $\boldsymbol{\kappa}(\mathscr{B})$, can carry non-zero jumps in the second spatial derivatives of χ^k at most.)

We can express the jump conditions in terms of the spatial co-ordinates (x^i) for the singular surface $\mathscr{S}_t$ in $\chi_t(\mathscr{B})$ also. First, (2.17) is equivalent to

$$\left[\frac{\partial v^k}{\partial x^i}\right] = -\bar{u}\bar{s}^k \bar{n}_A \frac{\partial X^A}{\partial x^i}\,. \tag{2.19}$$

By use of (2.11), (2.14), and (2.15) we have

$$\left[\frac{\partial v^k}{\partial x^i}\right] = -U s^k n_i, \tag{2.20}$$

where U is the intrinsic speed of $\mathscr{S}_t$ and where

$$s^k = \bar{s}^k \frac{\|\text{grad } f\|^2}{\|\text{Grad } \bar{f}\|^2}. \tag{2.21}$$

The independent variables in the jump condition (2.20) are now x^i and t. Further, the vector field $\boldsymbol{s}$ with components s^k on $\mathscr{S}_t$ is intrinsic to the acceleration, since the velocity field is independent of the choice of the reference configuration. We call $\boldsymbol{s}$ the *amplitude vector* of the acceleration wave in the configuration $\chi_t(\mathscr{B})$.

Next, from (2.18), the jump of the acceleration field across $\mathscr{S}_t$ can be expressed by

$$[a^i] = U^2 s^i. \tag{2.22}$$

This equation gives a clear meaning to the vector $\boldsymbol{s}$. Finally, from (2.16), we have

$$\left[\frac{\partial F^k_B}{\partial x^i}\right] = F^j_B s^k n_i n_j, \tag{2.23}$$

where F^k_B is a component of the deformation gradient $\boldsymbol{F}$ as usual:

$$F^k_B = \frac{\partial x^k}{\partial X^B}. \tag{2.24}$$

Of course, (2.20) is equivalent to

$$[\dot{F}^k_B] = -U F^j_B s^k n_j. \tag{2.25}$$

The equations (2.20), (2.22), (2.23), and (2.25) express HUGONIOT's basic jump conditions at an acceleration wave in terms of the spatial coordinates x^i for the singular surface $\mathscr{S}_t$ in $\chi_t(\mathscr{B})$.

In view of the condition (2.22) we call $\mathscr{S}_t$ a *longitudinal wave* if $\boldsymbol{s}$ is parallel to $\boldsymbol{n}$, a *transverse wave* if $\boldsymbol{s}$ is orthogonal to $\boldsymbol{n}$. From (2.20) we have

$$[\text{div } \boldsymbol{v}] = -U\boldsymbol{s}\cdot\boldsymbol{n}, \tag{2.26}$$

$$[\text{curl } \boldsymbol{v}] = -U\boldsymbol{s}\times\boldsymbol{n}. \tag{2.27}$$

Thus across a longitudinal wave the vorticity is continuous, while across a transverse wave the expansion is continuous. The elegant formulae (2.26) and (2.27) were discovered by WEINGARTEN (1901). The result (2.26) shows that in an incompressible body there can be no longitudinal acceleration waves.

From the jump condition for the balance of mass (*cf.* (II. 6.38)) we see that the density ρ is continuous at an acceleration wave:

$$[\rho] = 0. \tag{2.28}$$

Hence the field equation (II. 6.20) and the condition (2.26) show that

$$[\dot{\rho}] = \rho U \boldsymbol{s} \cdot \boldsymbol{n}. \tag{2.29}$$

In view of this result, we call an acceleration wave *expansive* if $\boldsymbol{s} \cdot \boldsymbol{n} > 0$, *compressive* if $\boldsymbol{s} \cdot \boldsymbol{n} < 0$. A transverse wave, of course, is neither expansive nor compressive.

We close this section with a sketch of the properties of higher-order waves. If for some integer p greater than or equal to 2 the displacement functions are of class C^{p-1} at a certain surface but at least one of their p^{th} derivatives suffers a jump discontinuity there, the surface is called a singular surface of pth order. The transformation (2.2) preserves smoothness classes of order equal to or less than $p-1$. As before, we denote the spatial and the referential positions of a moving singular surface for the pth-order derivatives of the displacement functions by $\mathscr{S}_t$ and $\overline{\mathscr{S}}_t$, respectively. When $u \neq 0$, such a singular surface is called a *p^{th}-order wave* in the deformable body $\mathscr{B}$.

By an analysis parallel to that given above for the case when $p = 2$, we obtain the basic conditions of compatibility for a pth-order singular surface first in referential form:

$$\begin{aligned}
\left[\frac{\partial^p x^k}{\partial X^{A_1} \ldots \partial X^{A_p}}\right] &= \bar{s}^k \bar{n}_{A_1} \ldots \bar{n}_{A_p}, \\
\left[\frac{\partial^p x^k}{\partial X^{A_1} \ldots \partial X^{A_{p-1}} \partial t}\right] &= (-\bar{u}) \bar{s}^k \bar{n}_{A_1} \ldots \bar{n}_{A_{p-1}}, \\
&\vdots \\
\left[\frac{\partial^p x^k}{\partial X^{A_1} \ldots \partial X^{A_{p-q}} \partial t^q}\right] &= (-\bar{u})^q \bar{s}^k \bar{n}_{A_1} \ldots \bar{n}_{A_{p-q}}, \\
&\vdots \\
\left[\frac{\partial^p x^k}{\partial t^p}\right] &= (-\bar{u})^p \bar{s}^k.
\end{aligned} \tag{2.30}$$

Then we set

$$s^k = \bar{s}^k \frac{\|\operatorname{grad} f\|^p}{\|\operatorname{Grad} \bar{f}\|^p} \tag{2.31}$$

and obtain the conditions of compatibility in spatial form:

$$\left[\frac{\partial^p x^k}{\partial X^{A_1}\dots\partial X^{A_p}}\right] = s^k n_{i_1}\dots n_{i_p} F^{i_1}_{A_1}\dots F^{i_p}_{A_p},$$

$$\left[\frac{\partial^p x^k}{\partial X^{A_1}\dots\partial X^{A_{p-1}}\partial t}\right] = (-U) s^k n_{i_1}\dots n_{i_{p-1}} F^{i_1}_{A_1}\dots F^{i_{p-1}}_{A_{p-1}}, \tag{2.32}$$

$$\vdots$$

$$\left[\frac{\partial^p x^k}{\partial X^{A_1}\dots\partial X^{A_{p-q}}\partial t^q}\right] = (-U)^q s^k n_{i_1}\dots n_{i_{p-q}} F^{i_1}_{A_1}\dots F^{i_{p-q}}_{A_{p-q}},$$

$$\vdots$$

$$\left[\frac{\partial^p x^k}{\partial t^p}\right] = (-U)^p s^k.$$

In view of the last condition we call the vector field **s** the *amplitude* of the pth-order wave. Of course, the amplitude is intrinsic to the wave and is independent of the choice of the reference configuration. We can rewrite (2.32) also in the following spatial form:

$$\left[\frac{\partial^{p-1}(F^k_A)}{\partial x^{i_1}\dots\partial x^{i_{p-1}}}\right] = s^k n_{i_1}\dots n_{i_{p-1}} n_i F^i_A,$$

$$\left[\frac{\partial^{p-2}(\dot{F}^k_A)}{\partial x^{i_1}\dots\partial x^{i_{p-2}}}\right] = (-U) s^k n_{i_1}\dots n_{i_{p-2}} n_i F^i_A,$$

$$\vdots$$

$$\left[\frac{\partial^{p-q}(\overset{(q-1)}{F^k_A})}{\partial x^{i_1}\dots\partial x^{i_{p-q}}}\right] = (-U)^{q-1} s^k n_{i_1}\dots n_{i_{p-q}} n_i F^i_A, \tag{2.33}$$

$$\vdots$$

$$[\overset{(p-1)}{F^k_A}] = (-U)^{p-1} s^k n_i F^i_A.$$

Further,

$$\left[\frac{\partial \overset{(p-1)}{x^k}}{\partial x^i}\right] = (-U)^{p-1} s^k n_i, \tag{2.34}$$

where $\overset{(p-1)}{x^k}$ denotes the $(p-1)^{\text{st}}$-order time derivative of the kth component of deformation:

$$\overset{(p-1)}{x^k} \equiv \frac{\partial^{p-1}\chi^k}{\partial t^{p-1}}. \tag{2.35}$$

These results are due to HADAMARD (1903).

3. The Propagation Condition

The theory of wave propagation developed in the preceding two sections makes no use of the constitutive relations of the deformable body. In this section we find the restrictions that the constitutive relations of elasticity add to that theory. As before, we assume that $\mathscr{B}$ is an elastic body and that $\boldsymbol{\kappa}$ is a reference configuration for it. Then the constitutive relation for $\mathscr{B}$ is

$$\boldsymbol{T}_{\kappa} = \boldsymbol{A}(\boldsymbol{F}, X^A), \tag{3.1}$$

where $\boldsymbol{T}_{\kappa}$ denotes the Piola-Kirchhoff stress tensor relative to $\boldsymbol{\kappa}$ (*cf.* (II. 6.32)). In Chapters IV and V the elastic body was assumed to be both smooth and materially uniform. For the general theory of wave propagation smoothness suffices. In other words, now we require only that the response function $\boldsymbol{A}$ be smooth with respect to $\boldsymbol{F}$ and X^A. Clearly this condition does not depend on the choice of the reference configuration, and the results we shall now obtain apply in particular to the smooth and materially uniform bodies considered hitherto in this treatise.

Now let $\bar{\mathscr{S}}_t$ be an acceleration wave in $\boldsymbol{\kappa}(\mathscr{B})$. From (3.1) we see that

$$[\boldsymbol{T}_{\kappa}] = \boldsymbol{0}, \tag{3.2}$$

since $\boldsymbol{F}$ is continuous at the wave. In particular, the jump condition (II. 6.39) for the balance of linear momentum is satisfied. Substituting (3.1) into the field equation (II. 6.26), we obtain

$$\mathsf{A}_{k\ m}^{\ K\ M} \frac{\partial^2 x^m}{\partial X^K \partial X^M} + A_k + \rho_{\kappa} b_k = \rho_{\kappa} a_k, \tag{3.3}$$

where

$$\mathsf{A}_{k\ m}^{\ K\ M} \equiv \frac{\partial A_k^{\ K}}{\partial F^m_{\ M}}, \tag{3.4}$$

and

$$A_k \equiv \frac{\partial A_k^{\ K}}{\partial X^K}. \tag{3.5}$$

By the condition of smoothness for $\mathscr{B}$, both A_k and $\mathsf{A}_{k\ m}^{\ K\ M}$ are continuous at the wave. For simplicity, we assume that the body force is continuous[1].

---

[1] The effect of discontinuities in $\boldsymbol{b}$ has been analysed by M. E. GURTIN & E. K. WALSH, "Extrinsically induced acceleration waves in elastic bodies", *J. Acoust. Soc. Am.* **41**, 1320–1324 (1967).

Then (3.3) implies that

$$\mathsf{A}_{k\ m}^{\ K\ M}\left[\frac{\partial^2 x^m}{\partial X^K \partial X^M}\right] = \rho_\kappa [a_k]. \tag{3.6}$$

Now from the conditions of compatibility (2.16) and (2.18), we can rewrite (3.6) as

$$\mathsf{A}_{k\ m}^{\ K\ M}\bar{s}^m \bar{n}_K \bar{n}_M = \rho_\kappa \bar{u}^2 \bar{s}_k. \tag{3.7}$$

If we put

$$\bar{Q}_{km} \equiv \mathsf{A}_{k\ m}^{\ K\ M}\bar{n}_K \bar{n}_M, \tag{3.8}$$

(3.7) becomes the *propagation condition*:

$$\bar{Q}_{km}\bar{s}^m = \rho_\kappa \bar{u}^2 \bar{s}_k. \tag{3.9}$$

Thus $\bar{\boldsymbol{s}}$ is a proper vector of the tensor $\bar{\boldsymbol{Q}}$, and the corresponding proper number is $\rho_\kappa \bar{u}^2$. The dependence of $\bar{\boldsymbol{Q}}$ on $\boldsymbol{F}$, X^L, and $\bar{\boldsymbol{n}}$ is given explicitly by (3.8). In particular, $\bar{\boldsymbol{Q}}$ is symmetric for all $\bar{\boldsymbol{n}}$ if

$$\mathsf{A}_{k\ m}^{\ K\ M} = \mathsf{A}_{m\ k}^{\ M\ K}. \tag{3.10}$$

We can express the condition (3.6) in terms of the field co-ordinates x^i also. Indeed, from (2.21) and (2.22) we obtain

$$\mathsf{A}_{k\ m}^{\ K\ M} s^m F^i_K F^j_M n_i n_j = \rho_\kappa U^2 s_k, \tag{3.11}$$

where $\boldsymbol{n}$ is the wave-normal. If we put

$$Q_{km} \equiv \frac{\rho}{\rho_\kappa} \mathsf{A}_{k\ m}^{\ K\ M} F^i_K F^j_M n_i n_j, \tag{3.12}$$

(3.11) becomes

$$Q_{km} s^m = \rho U^2 s_k, \tag{3.13}$$

also called the *propagation condition*. Thus $\boldsymbol{s}$ is a proper vector of $\boldsymbol{Q}$, and the corresponding proper number is ρU^2. Comparing (3.8) with (3.12), we see that

$$\bar{\boldsymbol{Q}} = \frac{\rho_\kappa \bar{u}^2}{\rho U^2} \boldsymbol{Q}. \tag{3.14}$$

In particular, $\boldsymbol{Q}$ is symmetric if and only if $\bar{\boldsymbol{Q}}$ is. We call $\boldsymbol{Q}$ the *acoustic tensor* for the acceleration wave.

We summarize the preceding analysis in the following

FRESNEL-HADAMARD THEOREM. *The amplitude* $\mathbf{s}$ *of an acceleration wave travelling in the direction of* $\mathbf{n}$ *must be a proper vector of the acoustic tensor* $\boldsymbol{Q}$; *the corresponding proper number is* ρU^2, *where* U *is the intrinsic speed of the wave. The acoustic tensor* $\boldsymbol{Q}$ *is determined through (3.12) and (3.4) by the constitutive equation of the elastic point* X, *the deformation gradient* $\boldsymbol{F}$, *and the wave-normal* $\mathbf{n}$.

From (3.13), we can determine the intrinsic speed U through the characteristic equation

$$\det(\boldsymbol{Q}-\rho U^2\mathbf{1}) = 0. \tag{3.15}$$

While, in general, this equation need not have any positive root, in the physical interpretation only real speeds and hence only positive roots of (3.15) can correspond to real acceleration waves. If $\boldsymbol{s}$ is a proper vector of $\boldsymbol{Q}$ corresponding to a positive proper number, the local speed of the acceleration wave with amplitude $\boldsymbol{s}$ in the direction of $\boldsymbol{n}$ is given by

$$U = \sqrt{\frac{\boldsymbol{s}\cdot\boldsymbol{Q}\boldsymbol{s}}{\rho\boldsymbol{s}\cdot\boldsymbol{s}}}. \tag{3.16}$$

As we shall see later, the amplitudes and intrinsic speeds of higher-order waves also satisfy (3.13) and hence (3.16).

From the basic formula (3.12) we can prove the following

PROPOSITION (HADAMARD, B. BERNSTEIN). *The acoustic tensor* $\boldsymbol{Q}$ *is symmetric for all* $\mathbf{F}$ *and* $\mathbf{n}$ *if and only if the elastic material is hyperelastic.*

Proof of necessity. Since the constitutive relation for a hyperelastic material can be written in the form (*cf.* (III. 4.4))

$$\boldsymbol{T}_{\boldsymbol{\kappa}} = \rho_{\boldsymbol{\kappa}}\left(\frac{\partial\hat{\sigma}}{\partial\boldsymbol{F}}\right), \tag{3.17}$$

where $\hat{\sigma}$ denotes the stored-energy function, we can express the function $\mathsf{A}_{k\ m}^{\ K\ M}$ by

$$\mathsf{A}_{k\ m}^{\ K\ M} = \rho_{\boldsymbol{\kappa}}\frac{\partial^2\hat{\sigma}}{\partial F^k_K\,\partial F^m_M}. \tag{3.18}$$

Thus the condition (3.10) is satisfied, and hence $\boldsymbol{Q}$ is symmetric.

Sufficiency. By use of (II. 6.32) and (III. 1.27) we can express the response function $\boldsymbol{A}$ in terms of a function $\boldsymbol{t}$ that maps symmetric tensors onto symmetric tensors:

$$\boldsymbol{A}(\boldsymbol{F}, X^A) = \boldsymbol{F}\boldsymbol{t}(\boldsymbol{C}, X^A), \tag{3.19}$$

where $\boldsymbol{C}$ is the right Cauchy-Green tensor of $\boldsymbol{F}$. Differentiating (3.19) with respect to $\boldsymbol{F}$ yields

$$\mathsf{A}_{k\ m}^{\ K\ M} = \delta_{km}\, t^{KM} + 2F_{kP}\, F_{mQ}\, \frac{\partial t^{KP}}{\partial C_{MQ}}, \tag{3.20}$$

or, equivalently,

$$\mathsf{B}^{kpmq} = \delta^{km} T^{pq} + \mathsf{C}^{kpmq}, \tag{3.21}$$

where T^{pq} is a component of the Cauchy stress tensor and where

$$\mathsf{B}_{k\ m}^{\ p\ q} = \frac{\rho}{\rho_{\kappa}}\, \mathsf{A}_{k\ m}^{\ K\ M} F_K^p F_M^q, \tag{3.22}$$

and

$$\mathsf{C}^{kpmq} = 2\, \frac{\rho}{\rho_{\kappa}}\, F_K^k F_M^p F_P^m F_Q^q\, \frac{\partial t^{KM}}{\partial C_{PQ}}. \tag{3.23}$$

Since $\boldsymbol{t}$ and $\boldsymbol{C}$ are symmetric, we have

$$\mathsf{C}^{kpmq} = \mathsf{C}^{pkmq} = \mathsf{C}^{kpqm}. \tag{3.24}$$

Substituting these symmetry conditions into (3.21), we obtain

$$\mathsf{B}^{kpmq} - \mathsf{B}^{pkqm} = \delta^{km} T^{pq} - \delta^{pq} T^{km}. \tag{3.25}$$

Now in terms of B^{kpmq} the acoustic tensor $\boldsymbol{Q}$ is given by

$$Q^{km} = \mathsf{B}^{kpmq} n_p n_q. \tag{3.26}$$

Thus $\boldsymbol{Q}$ is symmetric if and only if

$$\mathsf{B}^{kpmq} n_p n_q = \mathsf{B}^{mpkq} n_p n_q. \tag{3.27}$$

Since the direction of $\boldsymbol{n}$ is arbitrary, (3.27) implies that

$$\mathsf{B}^{kpmq} - \mathsf{B}^{mpkq} + \mathsf{B}^{kqmp} - \mathsf{B}^{mqkp} = 0, \tag{3.28}$$

or, equivalently,

$$\mathsf{B}^{pkqm} - \mathsf{B}^{qkpm} + \mathsf{B}^{pmqk} - \mathsf{B}^{qmpk} = 0. \tag{3.29}$$

But from (3.25) we have also

$$\mathsf{B}^{kpmq} - \mathsf{B}^{pkqm} + \mathsf{B}^{qmpk} - \mathsf{B}^{mqkp} = 0, \tag{3.30}$$

or, equivalently,

$$\mathsf{B}^{mpkq} - \mathsf{B}^{pmqk} + \mathsf{B}^{qkpm} - \mathsf{B}^{kqmp} = 0. \tag{3.31}$$

Forming the sum of the last four equations yields

$$2(B^{kpmq} - B^{mqkp}) = 0. \tag{3.32}$$

In view of (3.22), this result shows that the condition (3.10) is satisfied. Thus (3.10) is not only sufficient but also necessary in order for the acoustic tensor to be symmetric for all $\boldsymbol{n}$.

Now since the condition (3.10) is the integrability condition for (3.17), the domain of $\hat{\sigma}$ being simply connected, the proof is complete.

It remains possible that in an elastic body which is not hyperelastic, $\boldsymbol{Q}$ may be symmetric for certain wave-normals $\boldsymbol{n}$ but not for others. We shall see an example of this kind in § 5.

Suppose that the acoustic tensor $\boldsymbol{Q}$ be symmetric. Then it has an orthonormal principal basis, and the corresponding proper numbers are positive if and only if $\boldsymbol{Q}$ is positive-definite, that is,

$$Q^{km} v_k v_m > 0 \tag{3.33}$$

for all non-zero vectors $\boldsymbol{v}$. In terms of the tensor B^{kpmq}, this condition becomes

$$B^{kpmq} v_k n_p v_m n_q > 0. \tag{3.34}$$

The inequality (3.34) is called the *strong ellipticity* condition for an elastic material. Whether or not $\boldsymbol{Q}$ be symmetric, this condition suffices for all real proper numbers of $\boldsymbol{Q}$ to be positive. Since the characteristic equation (3.15) for $\boldsymbol{Q}$ is a cubic equation, the strong ellipticity condition alone ensures that the propagation condition has a solution appropriate to an acceleration wave propagating in any desired direction $\boldsymbol{n}$.

The normal direction $\boldsymbol{n}$ of an acceleration wave, in general, need not be a proper vector of its corresponding acoustic tensor. Thus the waves considered here are generally not longitudinal waves. A sufficient condition for a longitudinal amplitude is furnished by the following

PROPOSITION (STIPPES, TRUESDELL[1]). *Suppose that for a given deformation gradient* **F** *and for each wave-normal* **n**, *the acoustic tensor* **Q** *satisfy*

---

[1] C. TRUESDELL, "Existence of longitudinal waves", *J. Acoust. Soc. Am.* **40**, 729–730 (1966), generalizing analysis by STIPPES for the classical infinitesimal theory. The result of STIPPES had been obtained a little earlier by FEDOROV, using a different method. For the theory of waves according to the infinitesimal theory see §§ 70–74 of the article by GURTIN, cited above, p. vi.

the condition

$$\boldsymbol{n} \cdot \boldsymbol{Q}\boldsymbol{n} > 0. \tag{3.35}$$

Then at least one wave-normal $\boldsymbol{n}$ *allows a longitudinal amplitude.* In other words, for at least one unit vector $\boldsymbol{n}_l$ the acoustic tensor $\boldsymbol{Q}(\boldsymbol{n}_l)$ has $\boldsymbol{n}_l$ itself as a proper vector corresponding to a positive proper number.

Proof. The assumption (3.35) allows us, for any $\boldsymbol{n}$, to set

$$\boldsymbol{m}(\boldsymbol{n}) \equiv \frac{\boldsymbol{Q}(\boldsymbol{n})\boldsymbol{n}}{||\boldsymbol{Q}(\boldsymbol{n})\boldsymbol{n}||} \tag{3.36}$$

and to conclude that the unit vector $\boldsymbol{m}(\boldsymbol{n})$ subtends an acute angle from $\boldsymbol{n}$. Therefore, $\boldsymbol{m}(\boldsymbol{n})$ maps the surface of the unit sphere into itself and maps no point into its antipode. Since $\boldsymbol{Q}(\boldsymbol{n})\boldsymbol{n}$ is a non-vanishing polynomial in $\boldsymbol{n}$, the map defined by (3.36) is continuous. According to a fixed-point theorem[1], any such map has a fixed point. Thus there exists a unit vector $\boldsymbol{n}_l$ such that $\boldsymbol{m}(\boldsymbol{n}_l) = \boldsymbol{n}_l$. By (3.36), $\boldsymbol{n}_l$ is in fact a longitudinal amplitude: $\boldsymbol{Q}(\boldsymbol{n}_l)\boldsymbol{n}_l \parallel \boldsymbol{n}_l$.

The strong ellipticity condition (3.34) obviously implies (3.35). Hence follows the

COROLLARY (TRUESDELL). *In a hyperelastic material satisfying the strong ellipticity condition, there is at least one normal direction which allows not only a longitudinal amplitude but also two mutually orthogonal transverse amplitudes.*

A much stronger hypothesis yields a somewhat stronger result of the same kind, which we merely state here, for proof referring the reader to its source.

PROPOSITION (KOLODNER[2]). *Suppose that the elastic material be hyperelastic and satisfy the condition* (3.35). *Then at any unstressed point in*

---

[1] Theorem 8.7 on p. 132 of D. G. BOURGIN, *Modern Algebraic Topology*, New York, MacMillan, 1963.

[2] I. I. KOLODNER, "Existence of longitudinal waves in anisotropic media", *J. Acoust. Soc. Amer.* **40**, 730–731 (1966). The work of KOLODNER, valid only for the classical infinitesimal theory, was shown to admit the above interpretation in the general theory by C. TRUESDELL, "Comment on longitudinal waves", *J. Acoust. Soc. Amer.* **43**, 170 (1968). For the infinitesimal theory the existence of at least two distinct wave normals which allow longitudinal amplitudes was observed by FEDOROV, *Theory of Elastic Waves in Crystals*, § 18, English translation, Plenum Press, New York, 1968.

the body there are at least three distinct wave normals that allow longitudinal amplitudes.

COROLLARY. *At an unstressed point in a hyperelastic body satisfying the condition of strong ellipticity, there are at least three distinct wave normals that allow a longitudinal amplitude and two mutually orthogonal transverse amplitudes.*

The above propositions are strictly local results. In general, for a given deformation gradient field the longitudinal amplitudes $\boldsymbol{n}_l$ delivered need not form a normal field of any surface. Thus even when there is a field of possible longitudinal amplitudes, it need not possess a normal congruence and thus correspond to possible longitudinal waves. Of course, for a homogeneous body in a homogeneous configuration the solution $\boldsymbol{n}_l$ is a homogeneous field, so that TRUESDELL's result then guarantees the existence of at least one family of parallel planes which may be the seats of longitudinal waves. KOLODNER's result, since it refers to an *unstressed* point only, is particularly appropriate to the classical infinitesimal theory, which presumes the elastic body to be in a natural state. In this theory, for a homogeneous body of arbitrary symmetry KOLODNER's theorem delivers at least three directions in which longitudinal plane waves may propagate.

So far, we have considered only the propagation condition for an acceleration wave. Our analyses and results, however, can be generalized easily for higher-order waves. Indeed, for a pth-order wave we simply differentiate the field equation (3.3) with respect to X^A $p-2$ times, obtaining

$$\mathsf{A}_k{}^K{}_m{}^M \frac{\partial^p x^m}{\partial X^{Q_1} \dots \partial X^{Q_{p-2}} \partial X^K \partial X^M} + \dots = \rho_\kappa \frac{\partial^p x^m}{\partial X^{Q_1} \dots \partial X^{Q_{p-2}} \partial t^2} + \dots \tag{3.37}$$

where the dots denote terms continuous across the wave. Taking the jump of this equation, we then get

$$\mathsf{A}_k{}^K{}_m{}^M \left[\frac{\partial^p x^m}{\partial X^{Q_1} \dots \partial X^{Q_{p-2}} \partial X^K \partial X^M}\right] = \rho_\kappa \left[\frac{\partial^p x^m}{\partial X^{Q_1} \dots \partial X^{Q_{p-2}} \partial t^2}\right]. \tag{3.38}$$

Now from the conditions of compatibility (2.32) for the pth-order wave, we can rewrite the equation (3.38) as

$$\begin{aligned}\mathsf{A}_k{}^K{}_m{}^M s^m F^i_K F^j_M n_i n_j F^{i_1}_{Q_1} \dots F^{i_{p-2}}_{Q_{p-2}} n_{i_1} \dots n_{i_{p-2}} \\ = \rho_\kappa U^2 s_k F^{i_1}_{Q_1} \dots F^{i_{p-2}}_{Q_{p-2}} n_{i_1} \dots n_{i_{p-2}}.\end{aligned} \tag{3.39}$$

Since F is non-singular, this condition is equivalent to (3.11). Again defining the acoustic tensor by (3.12), we may reduce (3.39) to (3.13). The preceding analysis proves the following

EQUIVALENCE THEOREM (TRUESDELL). *If p and q are any two integers greater than* 1, *the propagation conditions for waves of orders p and q are the same.*

The propagation condition (3.13) relates s and U at any given time but does not restrict the way either may change as the wave does in fact propagate through the body. To determine the growth or decay of the amplitudes of waves of specified forms requires a further analysis, which we shall present in § 6. In particular, we shall see there that despite the equivalence theorem, the time-dependence of the amplitude of a pth-order wave is generally different from that of a qth-order wave in the same direction unless $q = p$.

All considerations in this section refer to unconstrained elastic bodies. For incompressible bodies results of the same kind may be obtained; they are somewhat simpler because all waves in such bodies are necessarily transverse. For details and references, *cf.* § 78 of NFTM, where the basic work of ERICKSEN on this problem is presented and extended.

4. Acceleration Waves in Isotropic Elastic Bodies, I. Gradients of the Response Function

In the preceding section we have derived and exhibited the properties of the acoustic tensor for an acceleration wave in an elastic body; it depends on the direction of the wave and the first gradient of the response function of the material. In § 6 we shall see that the displacement derivative of the amplitude vector of an acceleration wave depends on the second gradient of the response function also. Since we shall apply the theory to acceleration waves in a materially uniform smooth isotropic elastic body as defined in Chapter V, the response function of the body relative to an undistorted material atlas is given by

$$T = H(B) = \varphi_0 \mathbf{1} + \varphi_1 B + \varphi_2 B^2, \tag{4.1}$$

where B denotes the left Cauchy-Green tensor as usual, and the response coefficients φ_0, φ_1, and φ_2 are functions of the principal invariants I,

II, and *III* of $\boldsymbol{B}$. We assume that φ_0, φ_1, and φ_2 are smooth functions, so that they are uniquely determined by $\boldsymbol{H}(\boldsymbol{B})$.

Taking the gradient of $\boldsymbol{H}$ with respect to $\boldsymbol{B}$, we obtain the component formula[1]

$$\begin{aligned} H^{km}{}_{pq} \equiv \frac{\partial T^{km}}{\partial B^{pq}} &= \tfrac{1}{2}\varphi_1(\delta^k_p\delta^m_q+\delta^m_p\delta^k_q)+\tfrac{1}{2}\varphi_2(\delta^k_p B^m_q+\delta^k_q B^m_p+\delta^m_p B^k_q+\delta^m_q B^k_p) \\ &+\delta^{km}\left\{\frac{\partial\varphi_0}{\partial I}\delta_{pq}+\frac{\partial\varphi_0}{\partial II}(I\delta_{pq}-B_{pq})+III\frac{\partial\varphi_0}{\partial III}B^{-1}_{pq}\right\} \\ &+B^{km}\left\{\frac{\partial\varphi_1}{\partial I}\delta_{pq}+\frac{\partial\varphi_1}{\partial II}(I\delta_{pq}-B_{pq})+III\frac{\partial\varphi_1}{\partial III}B^{-1}_{pq}\right\} \\ &+B^k_r B^{rm}\left\{\frac{\partial\varphi_2}{\partial I}\delta_{pq}+\frac{\partial\varphi_2}{\partial II}(I\delta_{pq}-B_{pq})+III\frac{\partial\varphi_2}{\partial III}B^{-1}_{pq}\right\}. \end{aligned} \tag{4.2}$$

If we take the partial derivative of this formula with respect to a component of $\boldsymbol{B}$ one more time, we obtain the component formula for the second gradient of $\boldsymbol{H}$. As we can see from (4.2), the result will be rather complex.

There is a simple interpretation of (4.2) in terms of the principal components of $\boldsymbol{T}$ and $\boldsymbol{B}$, however. From (4.1), the principal stress t_i is the value of a function of the principal stretches b_1, b_2, and b_3, *viz*,

$$\begin{aligned} t_i &= t_i(b_1, b_2, b_3) \\ &= \varphi_0+\varphi_1 b_i+\varphi_2 b_i^2, \end{aligned} \tag{4.3}$$

where the invariants *I*, *II*, and *III* are related to the b_i as follows:

$$I = b_1+b_2+b_3, \quad II = b_1 b_2+b_2 b_3+b_3 b_1, \quad III = b_1 b_2 b_3. \tag{4.4}$$

Now differentiating $(4.3)_2$ with respect to b_j, we obtain when $i = j = 1$

$$\begin{aligned} \frac{\partial t_1}{\partial b_1} &= \varphi_1+2b_1\varphi_2+\left\{\frac{\partial\varphi_0}{\partial I}+(b_2+b_3)\frac{\partial\varphi_0}{\partial II}+b_2 b_3\frac{\partial\varphi_0}{\partial III}\right\} \\ &+b_1\left\{\frac{\partial\varphi_1}{\partial I}+(b_2+b_3)\frac{\partial\varphi_1}{\partial II}+b_2 b_3\frac{\partial\varphi_1}{\partial III}\right\} \\ &+b_1^2\left\{\frac{\partial\varphi_2}{\partial I}+(b_2+b_3)\frac{\partial\varphi_2}{\partial II}+b_2 b_3\frac{\partial\varphi_2}{\partial III}\right\} \end{aligned} \tag{4.5}$$

---

[1] This formula was derived originally by TRUESDELL; *cf.* NFTM, Eq. (48.13).

and when $i = 1, j = 2$,

$$\frac{\partial t_1}{\partial b_2} = \left\{\frac{\partial \varphi_0}{\partial I} + (b_1+b_3)\frac{\partial \varphi_0}{\partial II} + b_1 b_3 \frac{\partial \varphi_0}{\partial III}\right\}$$

$$+ b_1 \left\{\frac{\partial \varphi_1}{\partial I} + (b_1+b_3)\frac{\partial \varphi_1}{\partial II} + b_1 b_3 \frac{\partial \varphi_1}{\partial III}\right\}$$

$$+ b_1^2 \left\{\frac{\partial \varphi_2}{\partial I} + (b_1+b_3)\frac{\partial \varphi_2}{\partial II} + b_1 b_3 \frac{\partial \varphi_2}{\partial III}\right\}. \tag{4.6}$$

Also, if $b_1 \neq b_2$, then we obtain directly from $(4.3)_2$

$$\frac{t_1 - t_2}{b_1 - b_2} = \varphi_1 + (b_1+b_2)\varphi_2, \tag{4.7}$$

while if $b_1 = b_2$, then we obtain from (4.5) and (4.6)

$$\frac{\partial t_1}{\partial b_1} - \frac{\partial t_1}{\partial b_2} = \varphi_1 + 2b_1 \varphi_2. \tag{4.8}$$

The result

$$\lim_{b_1 \to b_2} \frac{t_1 - t_2}{b_1 - b_2} = \frac{\partial t_1}{\partial b_1} - \frac{\partial t_1}{\partial b_2} \tag{4.9}$$

is also a consequence of a simple rule of calculus.

Now comparing the formulae (4.5)–(4.8) with (4.2), we see that the components $H_{\langle ijkl \rangle}$ of the first gradient of $\boldsymbol{H}$ relative to a principal basis of $\boldsymbol{B}$ are given by[1]

$$H_{\langle iijj \rangle} = \frac{\partial t_i}{\partial b_j}, \tag{4.10}$$

where i and j are free indices, and

$$H_{\langle ijji \rangle} = H_{\langle ijij \rangle} = \begin{cases} \frac{1}{2}\left(\frac{t_i - t_j}{b_i - b_j}\right) & \text{if } b_i \neq b_j, \\ \frac{1}{2}\left(\frac{\partial t_i}{\partial b_i} - \frac{\partial t_i}{\partial b_j}\right) & \text{if } b_i = b_j, \end{cases} \tag{4.11}$$

where i and j are distinct free indices. Further, all components $H_{\langle ijkl \rangle}$

[1] The formulae (4.10) and $(4.11)_1$ were derived originally by TRUESDELL; *cf.* NFTM, Eq. (48.16).

not given by (4.10) and (4.11) vanish identically. Since the components $H_{\langle ijkl \rangle}$ and $H^{ij}{}_{kl}$ are related by the usual transformation rule, we can recover the formula (4.2) from the formulae (4.10) and (4.11) provided that a principal basis of $\boldsymbol{B}$ be given. We call $H_{\langle ijkl \rangle}$ the *principal components* of the first gradient of $\boldsymbol{H}$.

Since (4.10) and (4.11) do not depend explicitly on the representation $(4.1)_2$, they might be derived without use of the formula (4.2) in general co-ordinates. Indeed, a derivation of this kind has been given by BOWEN & WANG[1]; we summarize their analysis here.

From $(4.1)_1$, for any smooth curve given by

$$\boldsymbol{B} = \boldsymbol{B}(s), \tag{4.12}$$

the image curve

$$\boldsymbol{T} = \boldsymbol{T}(s) = \boldsymbol{H}(\boldsymbol{B}(s)) \tag{4.13}$$

is a smooth curve provided that $\boldsymbol{H}$ be a smooth function. Further, the tangent vectors $\partial \boldsymbol{T}$ and $\partial \boldsymbol{B}$ with respect to the parameter s are related by

$$\partial \boldsymbol{T} = [\nabla_{\boldsymbol{B}} \boldsymbol{H}](\partial \boldsymbol{B}). \tag{4.14}$$

In component form relative to a principal basis of $\boldsymbol{B}$, this condition has the form

$$\partial T_{\langle ij \rangle} = H_{\langle ijkl \rangle} \partial B_{\langle kl \rangle}, \tag{4.15}$$

where k and l are summed from 1 to 3. We now compute the components $\partial T_{\langle ij \rangle}$ and $\partial B_{\langle ij \rangle}$.

Let $\{\boldsymbol{e}_i(s)\}$ be a principal basis of $\boldsymbol{B}(s)$; we choose $\{\boldsymbol{e}_i(s)\}$ in such a way that each $\boldsymbol{e}_i(s)$ is smooth with respect to s. Then we apply the product rule to the spectral form

$$\boldsymbol{B}(s) = b_i(s) \boldsymbol{e}_i(s) \otimes \boldsymbol{e}_i(s), \tag{4.16}$$

yielding the result

$$\partial \boldsymbol{B} = \{\dot{b}_i \delta_{ij} + (b_i - b_j)\omega_{ij}\} \boldsymbol{e}_i \otimes \boldsymbol{e}_j, \tag{4.17}$$

where

$$\omega_{ij} = \partial \boldsymbol{e}_i \cdot \boldsymbol{e}_j, \tag{4.18}$$

so that

$$\partial \boldsymbol{e}_i = \omega_{ij} \boldsymbol{e}_j, \; \omega_{ij} = -\omega_{ji}. \tag{4.19}$$

---

[1] RAY M. BOWEN & C.-C. WANG, "Acceleration waves in inhomogeneous isotropic elastic bodies", *Arch. Rational Mech.Anal.* **38**, 13–45 (1970).

Now since $\{e_i(s)\}$ is also a principal basis for $T(s)$, the same calculation yields

$$\partial T = \{\dot{t}_i \delta_{ij} + (t_i - t_j)\omega_{ij}\} e_i \otimes e_j. \tag{4.20}$$

But from $(4.3)_1$, $\dot{t}_i$ is given by

$$\dot{t}_i = \frac{\partial t_i}{\partial b_k} \dot{b}_k. \tag{4.21}$$

Substituting (4.21) into (4.20) and then comparing the result with (4.15) and (4.17), we obtain

$$\frac{\partial t_i}{\partial b_k} \dot{b}_k = H_{\langle iikk \rangle} \dot{b}_k + H_{\langle iikl \rangle}(b_k - b_l)\omega_{kl}, \tag{4.22}$$

where i is a free index, and

$$(t_i - t_j)\omega_{ij} = H_{\langle ijkk \rangle} \dot{b}_k + H_{\langle ijkl \rangle}(b_k - b_l)\omega_{kl}, \tag{4.23}$$

where i and j are distinct free indices. Since $\dot{b}_k$ and ω_{ij} are arbitrary, (4.22) and (4.23) imply (4.10) and (4.11); further, all principal components of $\nabla_B H$ not given by (4.10) and (4.11) vanish. Thus we have derived the special formulae (4.10) and (4.11) without using the representation $(4.1)_2$ or the general formula (4.2).

The advantage of this approach is that from the start it makes use of the simplifying speciality of a principal basis of B. We can use the same technique to calculate the principal components of the second gradient of H with respect to B. Taking the derivative of (4.14) with respect to s, we obtain

$$\partial^2 T = [[\nabla_B^2 H](\partial B)](\partial B) + [\nabla_B H](\partial^2 B), \tag{4.24}$$

which has the component form

$$\partial^2 T_{\langle ij \rangle} = H_{\langle ijklmn \rangle} \partial B_{\langle kl \rangle} \partial B_{\langle mn \rangle} + H_{\langle ijkl \rangle} \partial^2 B_{\langle kl \rangle} \tag{4.25}$$

relative to a principal basis of B. We can compute the components $\partial^2 T_{\langle ij \rangle}$ and $\partial^2 B_{\langle ij \rangle}$ by differentiating (4.17) and (4.20) with respect to s as before, and thus we obtain

$$\partial^2 B = \{\ddot{b}_i \delta_{ij} + 2(\dot{b}_i - \dot{b}_j)\omega_{ij} + (b_i - b_j)\dot{\omega}_{ij} + (2b_k - b_i - b_j)\omega_{ik}\omega_{jk}\} e_i \otimes e_j, \tag{4.26}$$

and

$$\partial^2 \boldsymbol{T} = \{\ddot{t}_i \delta_{ij} + 2(\dot{t}_i - \dot{t}_j)\omega_{ij} + (t_i - t_j)\dot{\omega}_{ij} + (2t_k - t_i - t_j)\omega_{ik}\,\omega_{jk}\}\boldsymbol{e}_i \otimes \boldsymbol{e}_j, \tag{4.27}$$

where $\ddot{t}_i$, of course, is given by

$$\ddot{t}_i = \frac{\partial^2 t_i}{\partial b_r \partial b_s}\,\dot{b}_r \dot{b}_s + \frac{\partial t_i}{\partial b_r}\,\ddot{b}_r. \tag{4.28}$$

Substituting (4.28) into (4.27) and then comparing the result with (4.25) and (4.26), we obtain

$$\begin{aligned}\left(\frac{\partial^2 t_i}{\partial b_s \partial b_r}\,\dot{b}_r \dot{b}_s + \frac{\partial t_i}{\partial b_r}\,\ddot{b}_r\right)\delta_{ij} &+ 2\left(\frac{\partial t_i}{\partial b_r} - \frac{\partial t_j}{\partial b_r}\right)\dot{b}_r\,\omega_{ij} \\ &+ (t_i - t_j)\dot{\omega}_{ij} + (2t_r - t_i - t_j)\omega_{ir}\,\omega_{jr} \\ &= H_{\langle ijklmn\rangle}\{\dot{b}_k\,\delta_{kl} + (b_k - b_l)\omega_{kl}\}\{\dot{b}_m\,\delta_{mn} + (b_m - b_n)\omega_{mn}\} \\ &+ H_{\langle ijkl\rangle}\{\ddot{b}_k\,\delta_{kl} + 2(\dot{b}_k - \dot{b}_l)\omega_{kl} + (b_k - b_l)\dot{\omega}_{kl} \\ &+ (2b_r - b_k - b_l)\omega_{kr}\,\omega_{lr}\},\end{aligned} \tag{4.29}$$

where i and j are free indices, and where r, s, k, l, m, and n are summed from 1 to 3. As before, $\dot{b}_i$, $\ddot{b}_i$, ω_{rs}, and $\dot{\omega}_{rs}$ are arbitrary, so we can choose appropriate values for them and determine from (4.29) the components $H_{\langle ijklmn\rangle}$ and $H_{\langle ijkl\rangle}$. Of course, the latter must square with the previous formulae (4.10) and (4.11).

Indeed, if we choose $\omega_{rs} = \dot{\omega}_{rs} = \dot{b}_i = 0$, then (4.29) reduces to (4.10), and if we choose $\omega_{rs} = \dot{b}_i = \ddot{b}_i = 0$, then (4.29) reduces to (4.11). Having checked (4.10) and (4.11), we can now delete the terms in (4.29) depending on $\ddot{b}_r$ or $\dot{\omega}_{rs}$, since they cancel out. Then (4.29) reduces to

$$\begin{aligned}\frac{\partial^2 t_i}{\partial b_r \partial b_s}\,\dot{b}_r \dot{b}_s \delta_{ij} &+ 2\left(\frac{\partial t_i}{\partial b_r} - \frac{\partial t_j}{\partial b_r}\right)\dot{b}_r\,\omega_{ij} + (2t_r - t_i - t_j)\omega_{ir}\,\omega_{jr} \\ &= H_{\langle ijklmn\rangle}\{\dot{b}_k\,\delta_{kl} + (b_k - b_l)\omega_{kl}\}\{\dot{b}_m\,\delta_{mn} + (b_m - b_n)\omega_{mn}\} \\ &+ H_{\langle ijkl\rangle}\{2(\dot{b}_k - \dot{b}_l)\omega_{kl} + (2b_r - b_k - b_l)\omega_{kr}\,\omega_{lr}\}.\end{aligned} \tag{4.30}$$

Now we choose appropriate values for i, j, $\dot{b}_k$, and ω_{rs} so as to show that (4.30) implies the formulae

$$H_{\langle iijjkk\rangle} = \frac{\partial^2 t_i}{\partial b_j \partial b_k}, \tag{4.31}$$

where i, j, k are free indices;

$$H_{\langle iiijij\rangle}=\begin{cases}\dfrac{1}{2}\dfrac{1}{(b_i-b_j)}\left(\dfrac{\partial t_i}{\partial b_i}-\dfrac{\partial t_i}{\partial b_j}-\dfrac{t_i-t_j}{b_i-b_j}\right) & \text{if } b_i\neq b_j,\\ \dfrac{1}{4}\left(\dfrac{\partial^2 t_i}{\partial b_i^2}+\dfrac{\partial^2 t_j}{\partial b_i^2}\right)-\dfrac{1}{2}\dfrac{\partial^2 t_i}{\partial b_i\partial b_j} & \text{if } b_i=b_j,\end{cases}\tag{4.32}$$

where i, j are free indices but $i\neq j$;

$$H_{\langle iijkjk\rangle}=\begin{cases}\dfrac{1}{2}\dfrac{1}{(b_j-b_k)}\left(\dfrac{\partial t_i}{\partial b_j}-\dfrac{\partial t_i}{\partial b_k}\right) & \text{if } b_j\neq b_k,\\ \dfrac{1}{2}\left(\dfrac{\partial^2 t_i}{\partial b_j^2}-\dfrac{\partial^2 t_i}{\partial b_j\partial b_k}\right) & \text{if } b_j=b_k,\end{cases}\tag{4.33}$$

where i, j, k are distinct free indices;

$$H_{\langle ijijii\rangle}=\begin{cases}\dfrac{1}{2}\dfrac{1}{(b_i-b_j)}\left(\dfrac{\partial t_i}{\partial b_i}-\dfrac{\partial t_j}{\partial b_i}-\dfrac{t_i-t_j}{b_i-b_j}\right) & \text{if } b_i\neq b_j,\\ \dfrac{1}{4}\left(\dfrac{\partial^2 t_i}{\partial b_i^2}-\dfrac{\partial^2 t_j}{\partial b_i^2}\right) & \text{if } b_i=b_j,\end{cases}\tag{4.34}$$

where i, j are distinct free indices;

$$H_{\langle ijijkk\rangle}=\begin{cases}\dfrac{1}{2}\dfrac{1}{(b_i-b_j)}\left(\dfrac{\partial t_i}{\partial b_k}-\dfrac{\partial t_j}{\partial b_k}\right) & \text{if } b_i\neq b_j,\\ \dfrac{1}{2}\left(\dfrac{\partial^2 t_i}{\partial b_i\partial b_k}-\dfrac{\partial^2 t_j}{\partial b_i\partial b_k}\right) & \text{if } b_i=b_j,\end{cases}\tag{4.35}$$

where i, j, k are distinct free indices, and[1]

$$H_{\langle ijjkki\rangle}=\begin{cases}\dfrac{1}{4}\left[\dfrac{t_i}{(b_i-b_j)(b_j-b_k)}+\dfrac{t_j}{(b_j-b_i)(b_j-b_k)}+\dfrac{t_k}{(b_k-b_i)(b_k-b_j)}\right]\\ \qquad\qquad\text{if } b_i, b_j, b_k \text{ are distinct},\\ \dfrac{1}{4}\dfrac{1}{(b_i-b_k)}\left[\dfrac{\partial t_i}{\partial b_i}-\dfrac{\partial t_j}{\partial b_i}-\dfrac{t_i-t_k}{b_i-b_k}\right] & \text{if } b_i=b_j\neq b_k,\\ \dfrac{1}{4}\left(\dfrac{\partial^2 t_i}{\partial b_i\partial b_k}-\dfrac{\partial^2 t_j}{\partial b_i\partial b_k}\right)+\dfrac{1}{8}\left(\dfrac{\partial^2 t_i}{\partial b_k^2}-\dfrac{\partial^2 t_k}{\partial b_k^2}\right) & \text{if } b_i=b_j=b_k,\end{cases}\tag{4.36}$$

---

[1] This formula was obtained by P. CHADWICK & R. W. OGDEN; *cf.* the Corrigendum of BOWEN & WANG, *Arch. Rational Mech. Anal.* **40**, 403 (1971).

where i, j, k are distinct free indices. Of course, $H_{\langle ijklmn\rangle}$ satisfies the usual symmetry conditions:

$$H_{\langle ijklmn\rangle} = H_{\langle jiklmn\rangle} = H_{\langle ijlkmn\rangle} = H_{\langle ijklnm\rangle} = H_{\langle ijmnkl\rangle}. \qquad (4.37)$$

The components of the second gradient not obtainable from the formulae (4.31)–(4.36) by the conditions (4.37) vanish identically.

We remark that the principal stress function $t_i(b_1, b_2, b_3)$ obeys the following symmetry conditions:

$$t_1(b_1, b_2, b_3) = t_2(b_3, b_1, b_2) = t_3(b_2, b_3, b_1) = t_1(b_1, b_3, b_2). \quad (4.38)$$

In particular, if $b_1 = b_2 = b$, then

$$t_1(b, b, b_3) = t_2(b, b, b_3) = t_3(b_3, b, b), \qquad (4.39)$$

and the derivatives of t_i satisfy such symmetry conditions as

$$\begin{aligned}
\left.\frac{\partial t_1}{\partial b_1}\right|_{(b,b,b_3)} &= \left.\frac{\partial t_2}{\partial b_2}\right|_{(b,b,b_3)}, \\
\left.\frac{\partial t_1}{\partial b_2}\right|_{(b,b,b_3)} &= \left.\frac{\partial t_2}{\partial b_1}\right|_{(b,b,b_3)}, \\
\left.\frac{\partial t_3}{\partial b_1}\right|_{(b,b,b_3)} &= \left.\frac{\partial t_3}{\partial b_2}\right|_{(b,b,b_3)}, \\
\left.\frac{\partial^2 t_1}{\partial b_1^2}\right|_{(b,b,b_3)} &= \left.\frac{\partial^2 t_2}{\partial b_2^2}\right|_{(b,b,b_3)},
\end{aligned} \qquad (4.40)$$

etc. These conditions confirm the symmetry conditions (4.37) and $(4.11)_1$ in the limiting cases of (4.32)–(4.36) and $(4.11)_2$.

Unlike the principal components of the first gradient, the principal components of the second gradient of the response function cannot be assigned arbitrarily but must obey some identities[1]. For example, from (4.34), (4.10), and (4.11) we have

$$2(b_j - b_i)H_{\langle ijijii\rangle} = H_{\langle jjii\rangle} - H_{\langle iiii\rangle} + 2H_{\langle ijij\rangle}, \; i \neq j; \qquad (4.41)$$

---

[1] RAY M. BOWEN & C.-C. WANG "Acceleration waves in orthotropic elastic materials", *Arch. Rational Mech. Anal.* **47**, 149–170 (1972).

from (4.32), (4.10), and (4.11) we have

$$2(b_j - b_i)H_{\langle iiijij\rangle} = H_{\langle iijj\rangle} - H_{\langle iiii\rangle} + 2H_{\langle ijij\rangle} \text{ if } i \neq j; \tag{4.42}$$

from (4.36), (4.10), and (4.11) we have

$$2(b_k - b_i)H_{\langle ijjkki\rangle} = H_{\langle jkjk\rangle} - H_{\langle ijij\rangle}, \text{ if } i, j, k \text{ are distinct}; \tag{4.43}$$

and from (4.31), (4.35), (4.10), and (4.11) we have

$$H_{\langle ijijkk\rangle} = \frac{\partial H_{\langle ijij\rangle}}{\partial b_k}. \tag{4.44}$$

In (4.41)–(4.44) the indices i, j, and k are free. These identities are useful in the derivation of the amplitude equation for acceleration waves, as we shall see in § 6.

5. Acceleration Waves in Isotropic Elastic Bodies, II. The Propagation Condition

We now apply the component formulae derived in the preceding section to the propagation condition for acceleration waves in a materially uniform smooth isotropic elastic body $\mathscr{B}$. As in Chapter V, we let

$$\mathfrak{A} = \{(\mathscr{U}_\alpha, \boldsymbol{\nu}_\alpha),\ \alpha \in I\} \tag{5.1}$$

be an undistorted material atlas for $\mathscr{B}$. Then $\mathfrak{A}$ corresponds to a characteristic Riemannian metric $g_{\mathfrak{A}}$ in such a way that for any body-point X and any vectors $\boldsymbol{u}$, $\boldsymbol{v}$ in $\mathscr{B}_X$

$$g_{\mathfrak{A}}(\boldsymbol{u}, \boldsymbol{v}) = \boldsymbol{\nu}_{\alpha X}(\boldsymbol{u}) \cdot \boldsymbol{\nu}_{\alpha X}(\boldsymbol{v}) \tag{5.2}$$

for any material chart $(\mathscr{U}_\alpha, \boldsymbol{\nu}_\alpha)$ in $\mathfrak{A}$ covering X, and, conversely, any pair $(\mathscr{U}, \boldsymbol{\nu})$ consisting in a subbody $\mathscr{U}$ and a smooth field of local configurations $\boldsymbol{\nu}$ on $\mathscr{U}$ is a material chart belonging to $\mathfrak{A}$ if $\boldsymbol{\nu}$ satisfies the condition (5.2) for all $\boldsymbol{u}, \boldsymbol{v} \in \mathscr{B}_X$ and for all $X \in \mathscr{U}$.

Now let $\boldsymbol{\kappa}$ be a reference configuration of $\mathscr{B}$ corresponding to the referential co-ordinate system (X^A). Then from (5.2), the components of $g_{\mathfrak{A}}$ relative to (X^A) are given by

$$g^{AB} = \delta^{ij} K_{\alpha i}^{\ A} K_{\alpha j}^{\ B}, \tag{5.3}$$

where $K_{\alpha i}^{A}$ denotes a component of the deformation gradient from $\boldsymbol{v}_{\alpha}$ to $\boldsymbol{\kappa}$, *viz*

$$\boldsymbol{K}_{\alpha} = \boldsymbol{\kappa}_{*} \circ \boldsymbol{v}_{\alpha}^{-1}. \tag{5.4}$$

The right-hand side of (5.3) does not depend on the choice of the material chart $(\mathscr{U}_{\alpha}, \boldsymbol{v}_{\alpha})$ in $\mathfrak{A}$, of course. Indeed, if $(\mathscr{U}_{\beta}, \boldsymbol{v}_{\beta})$ is any material chart in $\mathfrak{A}$, then we have

$$\boldsymbol{K}_{\beta} = \boldsymbol{\kappa}_{*} \circ \boldsymbol{v}_{\beta}^{-1} = \boldsymbol{\kappa}_{*} \circ \boldsymbol{v}_{\alpha}^{-1} \circ \boldsymbol{v}_{\alpha} \circ \boldsymbol{v}_{\beta}^{-1} = \boldsymbol{K}_{\alpha} \boldsymbol{G}_{\alpha\beta}, \tag{5.5}$$

where $\boldsymbol{G}_{\alpha\beta}$ is the co-ordinate transformation from $(\mathscr{U}_{\alpha}, \boldsymbol{v}_{\alpha})$ to $(\mathscr{U}_{\beta}, \boldsymbol{v}_{\beta})$, so that it is an orthogonal tensor field on $\mathscr{U}_{\alpha} \cap \mathscr{U}_{\beta}$. (Since $\mathfrak{A}$ is undis orted, the symmetry group $\mathfrak{G}_{\mathfrak{A}}$ is the rotational group.) From (5.5) we obtain

$$\delta^{ij} K_{\beta i}^{A} K_{\beta j}^{B} = \delta^{ij} K_{\alpha k}^{A} (G_{\alpha\beta})_{i}^{k} K_{\alpha l}^{B} (G_{\alpha\beta})_{j}^{l}, = \delta^{kl} K_{\alpha k}^{A} K_{\alpha l}^{B}, \tag{5.6}$$

from which we see that g^{AB} is well defined by (5.3).

Now let $\boldsymbol{\chi}$ be an arbitrary deformed configuration of $\mathscr{B}$ with deformation gradient $\boldsymbol{F}$ relative to the reference configuration $\boldsymbol{\kappa}$, *viz*

$$\boldsymbol{F} = \boldsymbol{\chi}_{*} \circ \boldsymbol{\kappa}_{*}^{-1}. \tag{5.7}$$

Then the deformation gradient of $\boldsymbol{\chi}$ relative to any material chart $(\mathscr{U}_{\alpha}, \boldsymbol{v}_{\alpha})$ in $\mathfrak{A}$ is given by

$$\boldsymbol{F}_{\alpha} = \boldsymbol{\chi}_{\alpha} \circ \boldsymbol{v}_{\alpha}^{-1} = \boldsymbol{\chi}_{*} \circ \boldsymbol{\kappa}_{*}^{-1} \circ \boldsymbol{\kappa}_{*} \circ \boldsymbol{v}_{\alpha} = \boldsymbol{F}\boldsymbol{K}_{\alpha}. \tag{5.8}$$

Hence the left Cauchy-Green tensor of $\boldsymbol{F}_{\alpha}$ is given by

$$\boldsymbol{B} = \boldsymbol{F}\boldsymbol{K}_{\alpha}\boldsymbol{K}_{\alpha}^{T}\boldsymbol{F}^{T}, \tag{5.9}$$

which has the component form

$$B^{ij} = g^{AB} \frac{\partial x^{i}}{\partial X^{A}} \frac{\partial x^{j}}{\partial X^{B}}, \tag{5.10}$$

showing that $\boldsymbol{B}$ is independent of the choice of $(\mathscr{U}_{\alpha}, \boldsymbol{v}_{\alpha})$ in $\mathfrak{A}$. Thus $\boldsymbol{B}$ is called the *Cauchy-Green tensor of* $\boldsymbol{\chi}$ *relative to* $\mathfrak{A}$.

Since $\mathfrak{A}$ is undistorted, the response function of $\mathscr{B}$ relative to $\mathfrak{A}$ is given by (4.1), where the argument of $\boldsymbol{H}(\boldsymbol{B})$ is given by (5.9). Hence in any motion $\boldsymbol{\chi}_{t}$ the acceleration field $\boldsymbol{a}$ satisfies the field equation

$$H^{km}{}_{pq} \frac{\partial B^{pq}}{\partial x^{m}} + \rho f^{k} = \rho a^{k}, \tag{5.11}$$

where f^k denotes a component of the body force. (To avoid confusion, we here relinquish the previous notation b_i for components of body force, since we are using that symbol now to denote a squared principal stretch.) From (5.10) we have

$$\frac{\partial B^{pq}}{\partial x^m} = \frac{\partial X^A}{\partial x^m} g^{BC} \left(\frac{\partial x^p}{\partial X^C} \frac{\partial^2 x^q}{\partial X^A \partial X^B} + \frac{\partial x^q}{\partial X^C} \frac{\partial^2 x^p}{\partial X^A \partial X^B} \right) + \frac{\partial X^A}{\partial x^m} \frac{\partial g^{BC}}{\partial X^A} \frac{\partial x^p}{\partial X^B} \frac{\partial x^q}{\partial X^C}. \tag{5.12}$$

Taking the jump of (5.11) and (5.12) at a singular surface and using the condition of compatibility for an acceleration wave as before, we obtain the propagation condition

$$Q^k{}_m s^m = \rho U^2 s^k, \tag{5.13}$$

where the acoustic tensor for the direction $\boldsymbol{n}$ is given by[1]

$$Q^k{}_m = 2H^{kp}{}_{mq} B^{qs} n_p n_s. \tag{5.14}$$

An acceleration wave travelling in the direction of a proper vector of $\boldsymbol{B}$ is called a *principal wave*. Let $\{\boldsymbol{e}_i\}$ be a principal basis of $\boldsymbol{B}$ as before, and let $\boldsymbol{n}$ coincide with $\boldsymbol{e}_1$. Then from (5.14), (4.10) and (4.11), the components of $\boldsymbol{Q}(\boldsymbol{e}_1)$ relative to $\{\boldsymbol{e}_i\}$ form a diagonal matrix:

$$[Q_{\langle km \rangle}] = 2b_1 \operatorname{diag} [H_{\langle 1111 \rangle}, H_{\langle 1212 \rangle}, H_{\langle 1313 \rangle}]. \tag{5.15}$$

This means that $\{\boldsymbol{e}_i\}$ is also a principal basis for $\boldsymbol{Q}(\boldsymbol{e}_1)$. Thus, apart from a trivial exception[2], *every principal wave is either longitudinal or transverse.* Further, the squared speed of the longitudinal wave in the direction of $\boldsymbol{e}_1$ is given by[3]

$$U_{11}^2 = \frac{2b_1}{\rho} H_{\langle 1111 \rangle} = \frac{2b_1}{\rho} \frac{\partial t_1}{\partial b_1}, \tag{5.16}$$

---

[1] This formula was derived originally by TRUESDELL, generalizing results by many predecessors; *cf.* NFTM, Eq. (74.1).

[2] The trivial exception occurs when $H_{\langle 1111 \rangle} = H_{\langle 1212 \rangle}$ or $H_{\langle 1313 \rangle}$, since then any vector in the plane spanned by $\boldsymbol{e}_1$ and $\boldsymbol{e}_2$ or $\boldsymbol{e}_1$ and $\boldsymbol{e}_3$, respectively, is a proper vector of $\boldsymbol{Q}(\boldsymbol{e}_2)$.

[3] The formulae (5.16) and $(5.18)_1$ were attributed to ERICKSEN by TRUESDELL; *cf.* NFTM, Eq. (74.7).

while the squared speed of the transverse wave with

$$s = s e_2 \tag{5.17}$$

is given by[1]

$$U_{12}^2 = \frac{2b_1}{\rho} H_{\langle 1212 \rangle} = \begin{cases} \dfrac{b_1}{\rho}\left(\dfrac{t_1 - t_2}{b_1 - b_2}\right) \text{ if } b_1 \neq b_2, \\ \dfrac{b_1}{\rho}\left(\dfrac{\partial t_1}{\partial b_1} - \dfrac{\partial t_1}{\partial b_2}\right) \text{ if } b_1 = b_2. \end{cases} \tag{5.18}$$

Having determined the speeds of the principal waves, we now determine the principal components of the acoustic tensor $\boldsymbol{Q}(\boldsymbol{n})$ for an arbitrary direction $\boldsymbol{n}$. Let $\cos \theta_i$ be the direction cosine of $\boldsymbol{n}$ relative to $\boldsymbol{e}_i$, *viz*

$$\boldsymbol{n} = n_{\langle i \rangle} \boldsymbol{e}_i = \cos \theta_i \boldsymbol{e}_i. \tag{5.19}$$

Then we obtain from (5.14)

$$Q_{\langle km \rangle} = 2b_r H_{\langle kpmr \rangle} \cos \theta_p \cos \theta_r, \tag{5.20}$$

which means that[1]

$$Q_{\langle kk \rangle} = \rho U_{pk}^2 \cos \theta_p^2, \tag{5.21}$$

where k is a free index and p is summed from 1 to 3, and

$$Q_{\langle km \rangle} = (2b_m H_{\langle kkmm \rangle} + \rho U_{km}^2) \cos \theta_k \cos \theta_m \tag{5.22}$$

k and m being distinct free indices. Unlike the acoustic tensor for a principal wave, the acoustic tensor $\boldsymbol{Q}(\boldsymbol{n})$ for general $\boldsymbol{n}$ need not be symmetric.

If $b_1 = b_2 = b_3 = b$, then the actual configuration is undistorted, and hence every wave is a principal wave. In this circumstance, the acoustic tensor is given by

$$\boldsymbol{Q}(\boldsymbol{n}) = b\left(\frac{\partial t_1}{\partial b_1} - \frac{\partial t_1}{\partial b_2}\right) \mathbf{1} + b\left(\frac{\partial t_1}{\partial b_1} + \frac{\partial t_1}{\partial b_2}\right) \boldsymbol{n} \otimes \boldsymbol{n}, \tag{5.23}$$

where the argument of $\partial t_i / \partial b_j$ is (b, b, b). From (5.23), we see that the squared speed of the longitudinal wave is given by[2]

$$U_L^2 = \frac{2b}{\rho} \left.\frac{\partial t_1}{\partial b_1}\right|_{(b,b,b)}, \tag{5.24}$$

---

[1] NFTM, Eq. (74.11).

[2] The formulae (5.24) and (5.25) were originally derived by TRUESDELL; *cf.* NFTM, § 75.

and the squared speed of the transverse wave by

$$U_T^2 = \frac{b}{\rho}\left(\frac{\partial t_1}{\partial b_1} - \frac{\partial t_1}{\partial b_2}\right)\bigg|_{(b,\,b,\,b)}, \tag{5.25}$$

In the undistorted state with $b_1 = b_2 = b_3$, the stress tensor is a hydrostatic stress $-p(b)\mathbf{1}$, of course, and $p(b)$ is given by

$$p(b) = -t_1(b, b, b). \tag{5.26}$$

We can replace the argument of p by the density ρ, since

$$b = \left(\frac{\rho_{\mathfrak{A}}}{\rho}\right)^{\frac{2}{3}}, \tag{5.27}$$

where $\rho_{\mathfrak{A}}$ denotes the density in any material chart of $\mathfrak{A}$. Differentiating (5.26) with respect to b and using the change of variable given by (5.27), we obtain

$$\frac{dp}{d\rho} = \frac{2b}{3\rho}\left(\frac{\partial t_1}{\partial b_1} + \frac{\partial t_1}{\partial b_2} + \frac{\partial t_1}{\partial b_3}\right)\bigg|_{(b,\,b,\,b)}. \tag{5.28}$$

Now using the symmetry condition (4.40) for the undistorted state and observing the formulae (5.24) and (5.25), we see that[1]

$$U_L^2 = \tfrac{4}{3}U_T^2 + \frac{dp}{d\rho}. \tag{5.29}$$

On the assumption, which seems to be justified by the applications, that $dp/d\rho > 0$, (5.29) implies that

$$U_L^2 > \tfrac{4}{3}U_T^2, \tag{5.30}$$

and

$$U_L^2 > \frac{dp}{d\rho}, \tag{5.31}$$

provided that the transverse wave speed be real.

Before closing this section, we remark here that the formulae (5.16) and (5.18) can be regarded as differential equations for $t_i(b_1, b_2, b_3)$

[1] The universal relation (5.29) is due to TRUESDELL; *cf.* NFTM, Eq. (75.4). Special cases based upon various approximations and formal series had appeared earlier in the literature of seismology. For the infinitesimal theory, of course, the result is obvious and trivial.

provided that the squared speeds be given. The solution of the differential equations, then, determines t_i to within some constants of integration. Since there are nine squared speeds and only three unknown functions, certain compatibility conditions relating the squared speeds must be satisfied in order that the data be consistent with the theory of isotropic elastic bodies. For more details, we refere the reader to NFTM, § 76, and to the original memoir[1] of TRUESDELL.

6. Acceleration Waves in Isotropic Elastic Bodies, III. The Displacement Derivative of the Amplitude Vector[2]

In this section we compute the displacement derivative of the amplitude vector of an acceleration wave propagating into a region at rest. Since the velocity field is continuous at an acceleration wave, we obtain from (2.15)

$$U = u. \tag{6.1}$$

Thus the intrinsic speed is the normal speed of the wave in this case.

Since we are interested in the growth of the jump of $\boldsymbol{a}$, we put

$$[a^i] = S^i, \tag{6.2}$$

and we regard S^i as a function of t and the co-ordinates (x^1, x^2, x^3) of the wave in the deformed configuration. From (2.22), we see that S^i and s^i are related by

$$S^i = U^2 s^i. \tag{6.3}$$

Since the acceleration field is given by

$$a^i = \frac{\partial v^i}{\partial t} + \frac{\partial v^i}{\partial x^j} v^j, \tag{6.4}$$

[1] C. TRUESDELL, "General and exact theory of waves in finite elastic strain", *Arch. Rational Mech. Anal.* **8**, 263–296 (1961). Reprinted in *Continuum Mechanics IV, Problems of Non-linear Elasticity*, Intern. Sci. Rev. Ser., Vol. VIII, ed. C. TRUESDELL, New York, Gordon and Breach, 1965. Not all the contents of this paper is reproduced in NFTM.

[2] The development in this section follows R. M. BOWEN & C.-C. WANG, "Acceleration waves in inhomogeneous isotropic elastic bodies", *Arch. Rational Mech. Anal.* **38**, 13–45 (1970).

by taking the jump of this equation and using the fact that v^j vanishes at the wave we obtain

$$S^i = \left[\frac{\partial v^i}{\partial t}\right]. \tag{6.5}$$

Now from THOMAS's formula (1.64), the displacement derivative of S^i is given by

$$2\frac{\delta S^i}{\delta t} = \frac{S^i}{U}\frac{\delta U}{\delta t} + \left[\frac{\partial^2 v^i}{\partial t^2}\right] - U^2\left[\frac{\partial^2 v^i}{\partial x^j \partial x^k}\right] n^j n^k, \tag{6.6}$$

where we have used the condition (6.1). The jumps of the second derivatives of the velocity field are related through the equation of motion (5.11), as we now show.

First, from (5.10), the material derivative of $\boldsymbol{B}$ is given by

$$\dot{B}^{ij} = B^{ik}\frac{\partial v^j}{\partial x^k} + B^{kj}\frac{\partial v^i}{\partial x^k}. \tag{6.7}$$

Taking the jump of this equation, we obtain

$$[\dot{B}^{ij}] = B^{ik}\left[\frac{\partial v^j}{\partial x^k}\right] + B^{kj}\left[\frac{\partial v^i}{\partial x^k}\right]. \tag{6.8}$$

From the kinematical condition of compatibility (2.20) we obtain

$$[\dot{B}^{ij}] = -\frac{1}{U}(B^{ik}S^j n_k + B^{kj}S^i n_k), \tag{6.9}$$

where we have used (6.3). A similar calculation gives a formula for the jump of the gradient of $\boldsymbol{B}$:

$$\left[\frac{\partial B^{ij}}{\partial x^k}\right] = \frac{1}{U^2}(B^{lj}S^i n_l n_k + B^{li}S^j n_l n_k). \tag{6.10}$$

Next we compute the jump of the gradient of $\dot{\boldsymbol{B}}$. Calculating the gradient of $\dot{\boldsymbol{B}}$ from (6.7), we obtain

$$\frac{\partial \dot{B}^{ij}}{\partial x^l} = \frac{\partial B^{ik}}{\partial x^l}\frac{\partial v^j}{\partial x^k} + B^{ik}\frac{\partial^2 v^j}{\partial x^k \partial x^l} + \frac{\partial B^{kj}}{\partial x^l}\frac{\partial v^i}{\partial x^k} + B^{kj}\frac{\partial^2 v^i}{\partial x^k \partial x^l}, \tag{6.11}$$

and taking the jump of this equation yields

$$\left[\frac{\partial \dot{B}^{ij}}{\partial x^l}\right] = B^{ik}\left[\frac{\partial^2 v^j}{\partial x^k \partial x^l}\right] + B^{kj}\left[\frac{\partial^2 v^i}{\partial x^k \partial x^l}\right] + \left[\frac{\partial B^{ik}}{\partial x^l}\frac{\partial v^j}{\partial x^k}\right] + \left[\frac{\partial B^{kj}}{\partial x^l}\frac{\partial v^i}{\partial x^k}\right]. \tag{6.12}$$

Now from (1.1) the jump of a product fg can be calculated by the rule

$$[fg] = f^+[g] + g^+[f] - [f][g]. \tag{6.13}$$

Applying this rule to the last two terms of (6.12) and using (2.20) and (6.10), we find that

$$\left[\frac{\partial \dot{B}^{ij}}{\partial x^l}\right] = B^{ik}\left[\frac{\partial^2 v^j}{\partial x^k \partial x^l}\right] + B^{kj}\left[\frac{\partial^2 v^i}{\partial x^k \partial x^l}\right] - \frac{1}{U}\mathsf{D}^{ik}{}_l S^j n_k - \frac{1}{U}\mathsf{D}^{kj}{}_l S^i n_k + \frac{1}{U^3}(B^{mk}S^i + B^{mi}S^k)S^j n_k n_l n_m + \frac{1}{U^3}(B^{mk}S^j + B^{mj}S^k)S^i n_k n_l n_m, \tag{6.14}$$

where

$$\mathsf{D}^{ij}{}_k \equiv \left(\frac{\partial B^{ij}}{\partial x^k}\right)^+. \tag{6.15}$$

Clearly the tensor field $\mathbf{D}$ with components $\mathsf{D}^{ij}{}_k$ characterizes the inhomogeneity of the region at rest in front of the wave.

We notice now that the gradient of $\dot{\boldsymbol{B}}$ is related to the material derivative of the gradient of $\boldsymbol{B}$ by the relation

$$\frac{d}{dt}\left(\frac{\partial B^{ij}}{\partial x^l}\right) = \frac{\partial \dot{B}^{ij}}{\partial x^l} - \frac{\partial B^{ij}}{\partial x^k}\frac{\partial v^k}{\partial x^l}. \tag{6.16}$$

Taking the jump of this equation and using (6.13) and (6.14), we obtain

$$\left[\frac{d}{dt}\left(\frac{\partial B^{ij}}{\partial x^l}\right)\right] = B^{ik}\left[\frac{\partial^2 v^j}{\partial x^k \partial x^l}\right] + B^{kj}\left[\frac{\partial^2 v^i}{\partial x^k \partial x^l}\right] - \frac{1}{U}\mathsf{D}^{ik}{}_l S^j n_k - \frac{1}{U}\mathsf{D}^{kj}{}_l S^i n_k + \frac{1}{U}\mathsf{D}^{ij}{}_k S^k n_l + \frac{2}{U^3}B^{mk}S^i S^j n_k n_l n_m. \quad (6.17$$

Having derived the formula (6.17) for the jump of the material derivative of the gradient of $\boldsymbol{B}$, we now take the material derivative of the equation of motion (5.11), obtaining

$$H^{km}{}_{pqij}\dot{B}^{ij}\frac{\partial B^{pq}}{\partial x^m}+H^{km}{}_{pq}\frac{d}{dt}\left(\frac{\partial B^{pq}}{\partial x^m}\right)+\dot{\rho}f^k+\rho\dot{f}^k=\dot{\rho}a^k+\rho\dot{a}^k. \tag{6.18}$$

Taking the jump of this equation, we obtain

$$H^{km}{}_{pqij}\left[\dot{B}^{ij}\frac{\partial B^{pq}}{\partial x^m}\right]+H^{km}{}_{pq}\left[\frac{d}{dt}\left(\frac{\partial B^{pq}}{\partial x^m}\right)\right]+[\dot{\rho}]f^k=[\dot{\rho}a^k]+\rho[\dot{a}^k], \tag{6.19}$$

where we have assumed that $\dot{\boldsymbol{f}}$ is continuous at the wave. Now by (2.29) and (6.3),

$$[\dot{\rho}]=\frac{\rho}{U}S^jn_j. \tag{6.20}$$

By (5.11), then,

$$[\dot{\rho}]f^k=-\frac{1}{U}H^{km}{}_{pq}D^{pq}{}_mS^jn_j. \tag{6.21}$$

Taking the jump of the material derivative of (6.4) yields

$$[\dot{a}^k]=\left[\frac{\partial^2v^k}{\partial t^2}\right]-\frac{1}{U}S^kS^jn_j. \tag{6.22}$$

From (6.20) and the product rule (6.13), we get

$$[\dot{\rho}a^k]=-\frac{\rho}{U}S^kS^jn_j. \tag{6.23}$$

Substituting (6.8), (6.10), (6.17), (6.21), and (6.23) into (6.19), we finally obtain

$$\begin{aligned}\rho\left[\frac{\partial^2v^k}{\partial t^2}\right]=H^{km}{}_{pq}\Big\{&2B^{pr}\left[\frac{\partial^2v^q}{\partial x^r\partial x^m}\right]-\frac{2}{U}D^{pr}{}_mS^qn_r\\&+\frac{1}{U}D^{pq}{}_rS^rn_m-\frac{1}{U}D^{pq}{}_mS^2n_r+\frac{2}{U^3}B^{rs}S^pS^qn_mn_rn_s\Big\}\\&+H^{km}{}_{pqij}\left\{\frac{4}{U^3}B^{ir}B^{sp}S^jS^qn_mn_rn_s-\frac{2}{U}D^{pq}{}_mB^{ir}S^jn_r\right\}.\end{aligned} \tag{6.24}$$

This equation gives the desired relation between the jumps of the second derivatives of the velocity field.

Substituting (6.24) into (6.6), we obtain

$$2\rho \frac{\delta S^i}{\delta t} = \frac{\rho S^i}{U}\frac{\delta U}{\delta t} + 2H^{im}{}_{pq} B^{pr}\left[\frac{\partial^2 v^q}{\partial x^r \partial x^m}\right] - \rho U^2 n^j n^m \left[\frac{\partial^2 v^i}{\partial x^j \partial x^m}\right]$$
$$- H^{im}{}_{pq}\left\{\frac{2}{U} D^{pr}{}_m S^q n_r - \frac{1}{U} D^{pq}{}_r S^r n_m + \frac{1}{U} D^{pq}{}_m S^r n_r\right.$$
$$\left. - \frac{2}{U^3} B^{rs} S^p S^q n_m n_r n_s\right\} - H^{im}{}_{pqlj}\left\{\frac{2}{U} D^{pq}{}_m B^{lr} S^j n_r\right.$$
$$\left. - \frac{4}{U^3} B^{lr} B^{sp} S^j S^q n_m n_r n_s\right\}. \tag{6.25}$$

The jump of the second gradient of $\boldsymbol{v}$ can be determined to within its normal component

$$C^i \equiv \left[\frac{\partial^2 v^i}{\partial x^r \partial x^m} n^r n^m\right] \tag{6.26}$$

by THOMAS's formula (1.23). Multiplying (6.25) by S_i and summing on i, we see that the left-hand side of the equation reduces to

$$\rho \frac{\delta S^2}{\delta t}, \tag{6.27}$$

while the coefficient of the undetermined normal component C^q on the right-hand side of the equation is

$$2H^{im}{}_{pq} B^{pr} n_r n_m S_i - \rho U^2 S_q , \tag{6.28}$$

which vanishes if the acoustic tensor is symmetric[1] (*cf.* (5.13) and (5.14)). In particular, the quantity (6.28) vanishes if the material is hyperelastic or if the wave is a principal one.

Now we apply the formula (6.25) to a longitudinal principal wave, for which

$$\boldsymbol{n} = \boldsymbol{e}_1, \tag{6.29}$$

and

$$\boldsymbol{S} = S\boldsymbol{e}_1, \tag{6.30}$$

[1] If the acoustic tensor is not symmetric, the coefficient (6.28) vanishes provided that the S_i be components of a proper vector of the transpose of the acoustic tensor.

where $\{\boldsymbol{e}_i\}$ denotes a principal basis of $\boldsymbol{B}$ as before. By (6.29), $\{\boldsymbol{e}_2, \boldsymbol{e}_3\}$ is a basis on the wave surface. Hence from (6.15) and (4.17) we have

$$\mathsf{D}_{\langle 122\rangle} = (b_2 - b_1)\Omega_{\langle 22\rangle}, \quad \mathsf{D}_{\langle 133\rangle} = (b_3 - b_1)\Omega_{\langle 33\rangle}, \tag{6.31}$$

and

$$\mathsf{D}_{\langle ii1\rangle} = \frac{1}{U_{11}}\,\frac{\delta b_i}{\delta t}, \quad i = 1, 2, 3. \tag{6.32}$$

In deriving these formulae, we have used the technique leading to (4.17) for a curve on the wave surface tangent to $\boldsymbol{e}_2$ or $\boldsymbol{e}_3$. As before, $\boldsymbol{\Omega}$ is the second fundamental form of the wave surface. Now substituting (6.29)–(6.32) to (6.25), using (1.23), (5.16), and the identities (4.41)–(4.44), we obtain[1]

$$\begin{aligned}2\rho\frac{\delta S}{\delta t} = &-\frac{S}{U_{11}}\,\frac{\delta(\rho U_{11})}{\delta t} + \frac{2S}{U_{11}}\{(b_1 \mathsf{H}_{\langle 1111\rangle} + b_2 \mathsf{H}_{\langle 1122\rangle} - b_1 \mathsf{H}_{\langle 2211\rangle}\\ &+(b_1 - b_2)\mathsf{H}_{\langle 1212\rangle})\Omega_{\langle 22\rangle} + (b_1 \mathsf{H}_{\langle 1111\rangle} + b_3 \mathsf{H}_{\langle 1133\rangle} - b_1 \mathsf{H}_{\langle 3311\rangle}\\ &+(b_1 - b_3)\mathsf{H}_{\langle 1313\rangle})\Omega_{\langle 33\rangle}\} + \frac{2S^2 b_1}{U_{11}^3}\{\mathsf{H}_{\langle 1111\rangle} + 2b_1 \mathsf{H}_{\langle 111111\rangle}\}.\end{aligned} \tag{6.33}$$

This equation governs the growth or decay of the amplitude of a longitudinal principal wave. In deriving (6.33), we have used the results of § 4, where we observed that many principal components of the gradient of $\boldsymbol{H}$ vanish identically. In the next section we shall apply (6.33) to waves in certain laminated bodies.

Next, we apply (6.25) to a transverse principal wave, for which $\boldsymbol{n}$ is given by (6.29) and $\boldsymbol{S}$ is given by

$$\boldsymbol{S} = S\boldsymbol{e}_2. \tag{6.34}$$

Then, following the same procedure as before, we obtain[1]

$$2\rho\frac{\delta S}{\delta t} = -\frac{S}{U_{12}}\,\frac{\delta(\rho U_{12})}{\delta t} + \rho U_{12} S K, \tag{6.35}$$

where K is the mean curvature of the wave surface, *i.e.*,

$$K = \Omega_{\langle 22\rangle} + \Omega_{\langle 33\rangle} = \operatorname{tr}\boldsymbol{\Omega}. \tag{6.36}$$

---

[1] Ray M. Bowen & C.–C. Wang, "Acceleration waves in orthotropic elastic materials", *Arch. Rational Mech. Anal.* **47**, 149–170 (1972).

The equation (6.35) governs the growth or decay of the amplitude of a transverse principal wave. In the following section we shall apply (6.35) to waves in certain laminated bodies.

7. Acceleration Waves in Homogeneous or Laminated Bodies[1]

I. Principal Waves in a Laminated Plate

Let χ be the configuration in which the component matrix of the characteristic Riemannian metric g relative to the spatial co-ordinate system (x, y, z) takes the form

$$[g^{ij}] = \begin{bmatrix} g^{11} & 0 & 0 \\ 0 & g^{22} & g^{23} \\ 0 & g^{32} & g^{33} \end{bmatrix}, \tag{7.1}$$

where g^{ij} is a function of x only. Thus, in the spectral decomposition of $\boldsymbol{B}$

$$\boldsymbol{B} = b_1(x)\boldsymbol{e}_1 \otimes \boldsymbol{e}_1 + b_2(x)\boldsymbol{e}_2(x) \otimes \boldsymbol{e}_2(x) + b_3(x)\boldsymbol{e}_3(x) \otimes \boldsymbol{e}_3(x), \tag{7.2}$$

$\boldsymbol{e}_1$ may be taken to point in the x-direction. In particular, if $b_1(x)$, $b_2(x)$, $b_3(x)$, $\boldsymbol{e}_2(x)$, and $\boldsymbol{e}_3(x)$ are independent of x, then the body is homogeneous, and χ is a homogeneous configuration. If, further, $b_1 = b_2 = b_3$, then χ is an undistorted configuration.

As we pointed out in the preceding section, the inhomogeneity can be characterized by the tensor field D defined by (6.15), since χ is a homogeneous configuration if and only if D vanishes. From (7.2), the principal components of D are given by

$$\mathsf{D}_{\langle ii1 \rangle} = \frac{db_i}{dx}, \tag{7.3}$$

and

$$\mathsf{D}_{\langle 231 \rangle} = \mathsf{D}_{\langle 321 \rangle} = (b_2 - b_3)\frac{d\boldsymbol{e}_2}{dx} \cdot \boldsymbol{e}_3, \tag{7.4}$$

and all other principal components of D vanish. Now checking the governing equations (6.33) and (6.35), we see that the effect of inhomogen-

[1] *Cf.* Ray M. Bowen & C.-C. Wang *op. cit. ante.*, p. 454.

eity is characterized by the term $\delta(\rho U)/\delta t$, which depends on the gradient of the principal stretches but not on the gradient of the principal directions. Hence from (7.3) and (7.4) the amplitude is affected by the principal components $D_{\langle ii1\rangle}$, $i = 1, 2, 3$, but not affected by the principal components $D_{\langle 231\rangle}$ and $D_{\langle 321\rangle}$.

Now we consider the amplitude of a longitudinal principal wave in the $\boldsymbol{e}_1$-direction. From (6.33), using the fact that the curvature of a plane vanishes, we obtain

$$\frac{\partial S}{\partial x} = PS + QS^2, \tag{7.5}$$

where

$$P = -\frac{1}{2\rho U_{11}} \frac{d(\rho U_{11})}{dx}, \tag{7.6}$$

and

$$Q = \frac{1}{2U_{11}{}^2}\left(1 + \frac{2H_{\langle 111111\rangle}b_1}{H_{\langle 1111\rangle}}\right). \tag{7.7}$$

Here we have used the conditions

$$\frac{\delta S}{\delta t} = U_{11}\frac{\partial S}{\partial x}, \tag{7.8}$$

and

$$\frac{\delta(\rho U_{11})}{\delta t} = U_{11}\frac{d(\rho U_{11})}{dx}, \tag{7.9}$$

since the normal trajectory of the wave is parallel to the x-axis.

In view of the formulae (7.5)–(7.7), we see that the effect of inhomogeneity on the growth of S can be characterized by the function P. If $P > 0$, or, equivalently, $d(\rho U_{11})/dx < 0$, then PS and S have the same sign, so that $PS > 0$, and the term PS contributes to the growth of S. On the other hand, if $P < 0$, or, equivalently, $d(\rho U_{11})/dx > 0$, then PS and S have opposite signs, so that $PS < 0$, and the term PS contributes to the decay of S.

If the body is homogeneous, then, among other things, P vanishes identically, and Q reduces to a constant. In this case the governing equation (7.5) takes the simple form[1]

$$\frac{\partial S}{\partial x} = QS^2, \tag{7.10}$$

which has the general solution

$$\frac{1}{S} = -Qx+C, \tag{7.11}$$

where C is an arbitrary function of y and z. We can evaluate C if an initial condition such as

$$S = S_0 \text{ at } x = 0 \tag{7.12}$$

is given. Then the solution (7.13) becomes

$$\frac{S}{S_0} = \frac{1}{1-QS_0 x}, \tag{7.13}$$

which shows that the amplitude will grow to infinity at

$$x = \frac{1}{QS_0} \tag{7.14}$$

if QS_0 is positive, while the amplitude will decay to zero as x tends to infinity if QS_0 is negative. It is customary to interpret the former case as expressing the reinforcement of an acceleration wave so as to grow into a shock wave when (7.14) holds, although up to the present time the details of shock formation have not been made clear mathematically.

For a laminated plate in general, P and Q are functions of x. Then we divide (7.5) by S^2, obtaining

$$\frac{\partial}{\partial x}\left(\frac{1}{S}\right) + P\left(\frac{1}{S}\right) + Q = 0, \tag{7.15}$$

which is a linear first-order differential equation for $1/S$, the solution of which can be gotten by quadratures in terms of P and Q.

---

[1] For waves in a homogeneous body, the governing equation (7.10) and its general solution (7.13) were obtained by W. A. Green, "The growth of plane discontinuities propagating into a homogeneously deformed elastic material", *Arch. Rational Mech. Anal.* **16**, 79–88, (1964). Addendum, *ibid.* **19**, 20–23, (1965).

Next we consider transverse waves in the x-direction with S given by (6.34). From (6.35) the governing equation for the amplitude is

$$\frac{\partial S}{\partial x} = RS \tag{7.16}$$

where

$$R = -\frac{1}{2\rho U_{12}} \frac{d(\rho U_{12})}{dx}. \tag{7.17}$$

As before, we have used the conditions

$$\frac{\delta S}{\delta t} = U_{12} \frac{\partial S}{\partial x} \tag{7.18}$$

and

$$\frac{\delta(\rho U_{12})}{\delta t} = U_{12} \frac{d(\rho U_{12})}{dt}. \tag{7.19}$$

From (7.17) the solution of (7.16) is given by

$$\rho U_{12} S^2 = C, \tag{7.20}$$

where C is an arbitrary function of y and z.

For a homogeneous body, ρ and U_{12} are constant. Then (7.20) shows that S, too, remains constant[1]. Thus transverse plane waves propagate at constant strength in a homogeneous body.

II. Principal Waves in a Laminated Cylindrical Shell

We assume that χ is the configuration in which the characteristic Riemannian metric g has the component matrix

$$[g_{\langle ij \rangle}] = \operatorname{diag}(g_{\langle 11 \rangle}, g_{\langle 22 \rangle}, g_{\langle 33 \rangle}) \tag{7.21}$$

relative to a cylindrical co-ordinate system (r, θ, z) with $g_{\langle ii \rangle}$ a function of r only. Then $\boldsymbol{B}$ has a principal basis $\{\boldsymbol{e}_i\}$, where $\boldsymbol{e}_1$, $\boldsymbol{e}_2$, and $\boldsymbol{e}_3$ are unit vectors in the r, θ, and z directions, respectively. Further, the principal values b_i of $\boldsymbol{B}$ are functions of r only. For such a field $\boldsymbol{B}$, we obtain from (6.15)

$$\mathsf{D}_{\langle ii1 \rangle} = \frac{db_i}{dr}, \tag{7.22}$$

[1] *Cf.* W. A. Green, *op. cit.* p. 462.

and

$$\mathsf{D}_{\langle 212\rangle} = \mathsf{D}_{\langle 122\rangle} = \frac{b_1 - b_2}{r}, \tag{7.23}$$

and all other principal components of $\mathbf{D}$ vanish.

Now we consider a longitudinal wave in the radial direction. From the general amplitude equation (6.33), by using the conditions

$$\frac{\delta S}{\delta t} = U_{11}\frac{\partial S}{\partial r}, \quad \frac{\delta(\rho U_{11})}{\delta t} = U_{11}\frac{d(\rho U_{11})}{dr}, \tag{7.24}$$

and

$$\Omega_{\langle 22\rangle} = -\frac{1}{r}, \ \Omega_{\langle 33\rangle} = 0, \tag{7.25}$$

we obtain the governing equation

$$\frac{\partial S}{\partial r} = \frac{MS}{r} + PS + QS^2, \tag{7.26}$$

where Q is given by (7.7), P is given by

$$P = -\frac{1}{2\rho U_{11}}\frac{d(\rho U_{11})}{dr}, \tag{7.27}$$

and M is given by

$$M = -\frac{1}{\rho U_{11}{}^2}\{(b_1 - b_2)\mathsf{H}_{\langle 1212\rangle} + b_2\mathsf{H}_{\langle 1122\rangle} + b_1(\mathsf{H}_{\langle 1111\rangle} - \mathsf{H}_{\langle 2211\rangle})\}. \tag{7.28}$$

If the body is homogeneous and if χ is an undistorted configuration, then P vanishes identically, and Q and M reduce to constants, with

$$M = -\tfrac{1}{2}. \tag{7.29}$$

In this case, the governing equation for S takes the simple form[1]

$$\frac{\partial S}{\partial r} = -\frac{S}{2r} + QS^2, \tag{7.30}$$

---

[1] For cylindrical waves in an undistorted state of a homogeneous isotropic elastic body, the governing equation (7.30) and its solution (7.31) were given by P. J. Chen "The growth of acceleration waves of arbitrary form in homogeneously deformed elastic materials", *Arch. Rational Mech. Anal.* **30**, 81–89 (1968).

which has the general solution

$$\frac{1}{S} = -2Qr + Cr^{\frac{1}{2}}, \tag{7.31}$$

where C is an arbitrary function of θ and z. If an initial condition such as

$$S = S_0 \quad \text{at} \quad r = r_0 \tag{7.32}$$

is given, then the function C can be determined:

$$C = \frac{1}{r_0^{\frac{1}{2}}}\left(\frac{1}{S_0} + 2Qr_0\right), \tag{7.33}$$

and the solution has the explicit form

$$\frac{Sr^{\frac{1}{2}}}{S_0\, r_0^{\frac{1}{2}}} = \frac{1}{1 - 2QS_0\, r_0^{\frac{1}{2}}(r^{\frac{1}{2}} - r_0^{\frac{1}{2}})}, \tag{7.34}$$

which shows that the amplitude will grow to infinity at

$$r = r_0\left(1 + \frac{1}{2QS_0\, r_0}\right)^2 \tag{7.35}$$

if QS_0 is positive, and that the amplitude will decay to zero as r tends to infinity if QS_0 is zero or negative. Here we have assumed that $r \geqq r_0$ in (7.34).

If the body is homogeneous, but if χ is a state of uniaxial stretch in the z-direction, then $b_1 = b_2$, but generally $b_1 \neq b_3$. From (7.28), $(4.40)_2$, and (5.16), we see that the condition (7.29) still holds. Further, P vanishes, and Q reduces to a constant. Thus the governing equation again takes the form (7.30).

Next, if the body is inhomogeneous, but if b_1, b_2, and b_3 are constants independent of r, then P vanishes as before, and Q and M reduce to constants. In this case, the general solution is

$$\frac{1}{S} = \begin{cases} -\dfrac{Q}{1+M}\, r + Cr^{-M} & \text{if } M \neq -1, \\ -Qr \log r + Cr & \text{if } M = -1, \end{cases} \tag{7.36}$$

where C is an arbitrary function of θ and z. Substituting the initial condition (7.32) into (7.30), we see that the solution can be written explicitly as

$$\frac{S}{S_0} = \begin{cases} \dfrac{1}{1-\dfrac{QS_0}{(1+M)r_0^M}(r^{M+1}-r_0^{M+1})}\left(\dfrac{r}{r_0}\right)^M & \text{if } M \neq -1, \\ \dfrac{1}{1-QS_0\, r_0(\log r-\log r_0)}\left(\dfrac{r_0}{r}\right) & \text{if } M = -1, \end{cases} \tag{7.37}$$

which shows that the amplitude will grow to infinity at

$$r = \begin{cases} r_0\left(1+\dfrac{1+M}{QS_0\, r_0}\right)^{1/(1+M)} & \text{if } M \neq -1, \\ r_0 \exp\left(\dfrac{1}{QS_0\, r_0}\right) & \text{if } M = -1, \end{cases} \tag{7.38}$$

if QS_0 is positive, and that the amplitude will decay to zero as r tends to infinity if QS_0 is negative. Here, again, we have assumed that $r \geqq r_0$.

In the general case with b_1, b_2, and b_3 depending on r, we divide the governing equation (7.26) by S^2, obtaining

$$\frac{\partial}{\partial r}\left(\frac{1}{S}\right) + \left(\frac{M}{r}+P\right)\frac{1}{S} + Q = 0, \tag{7.39}$$

which can then be integrated by quadrature as before.

Next, we consider transverse waves with $\boldsymbol{S}$ in the θ-direction. Substituting (7.25) and (6.36) into (6.35) yields the governing equation

$$\frac{\partial S}{\partial r} = -\frac{S}{2r} + RS, \tag{7.40}$$

where, as in the derivation of (7.16), R is given by

$$R = -\frac{1}{2\rho U_{12}}\frac{d}{dr}(\rho U_{12}). \tag{7.41}$$

Substituting (7.41) into (7.40), we see that the general solution of (7.43) is

$$\rho U_{12} r^{\frac{1}{2}} S^2 = C = \text{arbitrary function of } \theta \text{ and } z. \tag{7.42}$$

If the body is homogeneous and if χ is an undistorted state or a state of uniaxial stretch in the z-direction, then R vanishes, and the governing equation for S takes the simple form[1]

$$\frac{\partial S}{\partial r} = -\frac{S}{2r}, \tag{7.43}$$

and its solution is

$$S = Cr^{-\frac{1}{2}}, \tag{7.44}$$

which shows that S will decay monotonically to zero as r tends to infinity. This solution remains valid if the body is inhomogeneous but b_1, b_2, and b_3 are constants, since in this case R still vanishes.

Of course, there are also transverse waves with $\boldsymbol{S}$ pointing in the z-direction. For such waves the governing equation for the amplitude S again takes the form (7.40), but R is now given by

$$R = -\frac{1}{2\rho U_{13}} \frac{d}{dr}(\rho U_{13}), \tag{7.45}$$

so in this case the general solution is

$$\rho U_{13} r S^2 = C = \text{arbitrary function of } \theta \text{ and } z. \tag{7.46}$$

In particular, if the body is homogeneous and if χ is an undistorted state, or if the body is inhomogeneous but b_1, b_2, and b_3 are constants, then (7.43) and (7.44) hold as before.

III. Principal Waves in a Laminated Spherical Shell

Now let χ be the configuration in which the component matrix of the characteristic Riemannian metric g is given by (7.21) relative to a spherical co-ordinate system (r, θ, φ), with $g_{\langle ii \rangle}$ a function of r only; further,

$$g_{\langle 22 \rangle} = g_{\langle 33 \rangle}. \tag{7.47}$$

Then $\boldsymbol{B}$ has the spectral form

$$\boldsymbol{B} = (b_1(r) - b(r))\boldsymbol{e}_1 \otimes \boldsymbol{e}_1 + b(r)\mathbf{1}, \tag{7.48}$$

where $\boldsymbol{e}_1$ is a unit vector in the r-direction. From (7.48) and (6.15), we have

$$\mathsf{D}_{\langle 111 \rangle} = \frac{db_1}{dr}, \quad \mathsf{D}_{\langle 221 \rangle} = \mathsf{D}_{\langle 331 \rangle} = \frac{db}{dr}, \tag{7.49}$$

---

[1] For waves in an undistorted state of a homogeneous body, the governing equation (7.43) and its solution (7.44) were given by P. J. CHEN, *op. cit.* p. 464.

and

$$\mathsf{D}_{\langle 122\rangle} = \mathsf{D}_{\langle 133\rangle} = \frac{b_1 - b}{r}, \tag{7.50}$$

and all other principal components of **D** vanish.

We consider first longitudinal principal waves in the radial direction. From the general amplitude equation (6.33), and using the fact that the sectional curvature of a sphere is equal to $-1/r$, we obtain

$$\frac{\partial S}{\partial r} = \frac{LS}{r} + PS + QS^2, \tag{7.51}$$

where Q and P are given as before by (7.7) and (7.27), and where L is given by

$$L = -\frac{2}{\rho U_{11}{}^2}\{(b_1 - b)H_{\langle 1212\rangle} + bH_{\langle 1122\rangle} + b_1(H_{\langle 1111\rangle} - H_{\langle 2211\rangle})\}. \tag{7.52}$$

In particular, if the body is homogeneous and if χ is an undistorted state, then P vanishes, Q reduces to a constant, and

$$L = -1. \tag{7.53}$$

In this case, the governing equation for S takes the simple form[1]

$$\frac{\partial S}{\partial r} = -\frac{S}{r} + QS^2, \tag{7.54}$$

and its solution is given by (7.36) with $M = -1$.

If the body is inhomogeneous but b_1 and b are constants, then P vanishes as before, and Q and L reduce to constants. In this case the solution is given by (7.36) with M replaced by L.

Next, we consider spherical transverse waves. From (6.35) the amplitude equation is

$$\frac{\partial S}{\partial r} = -\frac{S}{r} + RS, \tag{7.55}$$

where R is given by (7.41). The solution of (7.55) is

$$\rho U_{12} r S^2 = C = \text{arbitrary function of } \theta \text{ and } \varphi.$$

---

[1] *Cf.* P. J. Chen, *op. cit.* p. 464.

If the body is homogeneous and if χ is an undistorted configuration, or if the body is inhomogeneous but b and b_1 are constants, then R vanishes. In these cases, the governing equation for S becomes[1]

$$\frac{\partial S}{\partial r} = -\frac{S}{r}, \tag{7.56}$$

and its solution is

$$S = Cr^{-1}, \tag{7.57}$$

which shows that the amplitude will decay to zero as r tends to infinity.

Since a sphere is not a parallelizable surface, the amplitude S of a spherical transverse wave must vanish at some point on the sphere.

---

[1] *Cf.* P. J. Chen, *op. cit.* p. 464.

CHAPTER VII

Theorems of Existence, Uniqueness and Stability in Elasticity

1. Formulation of Boundary-value Problems

Let $\mathscr{B}$ be an elastic body, and let $\boldsymbol{\kappa}$ be a fixed reference configuration for $\mathscr{B}$. Then for each $X \in \mathscr{B}$, the position of which is $\boldsymbol{X} = \boldsymbol{\kappa}(X)$ in $\boldsymbol{\kappa}(\mathscr{B})$, we have a constitutive equation of the form (III. 8A.2), *i.e.*,

$$\boldsymbol{T}_{\boldsymbol{\kappa}} = \boldsymbol{A}(\boldsymbol{F}, \boldsymbol{X}), \tag{1.1}$$

where $\boldsymbol{T}_{\boldsymbol{\kappa}}$ and $\boldsymbol{F}$ denote, respectively, the Piola-Kirchhoff stress tensor and the deformation gradient relative to $\boldsymbol{\kappa}$. In any motion $\boldsymbol{x} = \chi(\boldsymbol{X}, t)$ of $\mathscr{B}$ the following *differential equation of motion* is satisfied:

$$\operatorname{Div} \boldsymbol{A}(\boldsymbol{F}, \boldsymbol{X}) + \rho_{\boldsymbol{\kappa}} \boldsymbol{b} = \rho_{\boldsymbol{\kappa}} \boldsymbol{a}, \tag{1.2}$$

$\boldsymbol{a}$ being the acceleration. When the body force $\boldsymbol{b}$ is a given function of $\boldsymbol{x}$ and t, (1.2) is the field equation to be solved in order to determine the motion corresponding to prescribed initial and boundary conditions.

In Chapters IV and V we have seen that relative to a homogeneous reference configuration of a homogeneous elastic body $\mathscr{B}$, (1.2) takes the co-ordinate form (V. 8.41):

$$\mathsf{A}^{iA}{}_{k}{}^{B} \frac{\partial^2 x^k}{\partial X^A \partial X^B} + \rho_{\boldsymbol{\kappa}} b^i = \rho_{\boldsymbol{\kappa}} a^i, \tag{1.3}$$

where $\mathsf{A}^{iA}{}_{k}{}^{B}$ is a component of the gradient of $\boldsymbol{A}(\boldsymbol{F})$:

$$\mathsf{A}_i{}^A{}_k{}^B = \mathsf{A}_i{}^A{}_k{}^B(\boldsymbol{F}) \equiv \frac{\partial A_i^A}{\partial F^k{}_B}, \tag{1.4}$$

a quantity which is independent of X^A because we have assumed the reference configuration to be homogeneous. Likewise, relative to a material atlas $\mathfrak{A}$ of a materially uniform inhomogeneous body $\mathscr{B}$, (1.2) takes the component form (V. 8.40):

$$\bar{\mathsf{A}}^{iA}{}_{k}{}^{B} \left(\frac{\partial^2 x^k}{\partial X^A \partial X^B} - \bar{\Gamma}^C_{BA} \frac{\partial x^k}{\partial X^C} \right) - \bar{A}^{iA} \bar{C}^B_{AB} + \rho_{\boldsymbol{\kappa}} b^i = \rho_{\boldsymbol{\kappa}} a^i, \tag{1.5}$$

where $\bar{\mathsf{A}}^{iA}{}_{k}{}^{B}$ is a component of a fourth-order tensor function of $\boldsymbol{F}$ given by (V. 8.38), $\bar{\Gamma}^{C}_{BA}$ and $\bar{\mathsf{C}}^{C}_{BA}$ denote a connection symbol and a component of the torsion tensor of a fixed material connection relative to (X^A), respectively, and, finally, $\bar{A}_i^{\,A}$ is a component of the Piola-Kirchhoff stress. Generally $\bar{\mathsf{A}}^{iA}{}_{k}{}^{B}$, $\bar{\Gamma}^{C}_{BA}$, $\bar{\mathsf{C}}^{C}_{BA}$, and $\bar{A}_i^{\,A}$ all depend on X^A, *but they do not depend explicitly on t.*

If, still more generally, $\mathscr{B}$ is not even materially uniform, we have the following co-ordinate form for (1.2):

$$\mathsf{A}^{kA}{}_{m}{}^{B}\frac{\partial^2 x^m}{\partial X^A \partial X^B}+A^k+\rho_{\kappa} b^k = \rho_{\kappa} a^k, \tag{1.6}$$

where

$$\mathsf{A}_k{}^{A}{}_{m}{}^{B} = \mathsf{A}_k{}^{A}{}_{m}{}^{B}(\boldsymbol{F}, \boldsymbol{X}) \equiv \frac{\partial A_k{}^A}{\partial F^m{}_B}, \tag{1.7}$$

and

$$A_k = A_k(\boldsymbol{F}, \boldsymbol{X}) \equiv \frac{\partial A_k{}^A}{\partial X^A}. \tag{1.8}$$

We have assumed here, as we shall assume henceforth in this chapter, that the response function and the components of deformation are smooth.

A boundary-value problem in elasticity is a problem of finding solutions of the differential equation (1.2), or explicitly (1.3), (1.5), or (1.6), that satisfy certain boundary conditions. The following four kinds of boundary conditions are often considered.

First, a *boundary condition of traction* is defined by the surface traction $\boldsymbol{t}_{\kappa}(\boldsymbol{X}, t)$, $\boldsymbol{X} \in \partial\boldsymbol{\kappa}(\mathscr{B})$, measured per unit surface area in the reference configuration $\boldsymbol{\kappa}$. The solution is required to satisfy the condition

$$\boldsymbol{A}(\boldsymbol{F}(\boldsymbol{X}, t), \boldsymbol{X})\boldsymbol{n}_{\kappa}(\boldsymbol{X}) = \boldsymbol{t}_{\kappa}(\boldsymbol{X}, t) \quad \forall \boldsymbol{X} \in \partial\boldsymbol{\kappa}(\mathscr{B}) \tag{1.9}$$

and for times t in some interval, perhaps $[0, \infty)$. Here $\boldsymbol{A}(\boldsymbol{F}, \boldsymbol{X})$ and $\boldsymbol{F}(\boldsymbol{X}, t)$ for any $\boldsymbol{X} \in \partial\boldsymbol{\kappa}(\mathscr{B})$ are defined as the limits of $\boldsymbol{A}(\boldsymbol{F}, \tilde{\boldsymbol{X}})$ and $\boldsymbol{F}(\tilde{\boldsymbol{X}}, t)$, respectively, as $\tilde{\boldsymbol{X}}$ tends to $\boldsymbol{X}$ from the interior of $\boldsymbol{\kappa}(\mathscr{B})$, these limits being assumed to exist.

Second, a *boundary condition of pressure* is defined by a function $p(t)$. The traction vector $\boldsymbol{t}$ on the boundary $\partial\boldsymbol{\chi}_t(\mathscr{B})$ of the actual configuration is now required to be normal to the surface and have the magnitude $p(t)$ for all time t:

$$\boldsymbol{H}(\boldsymbol{F}(\boldsymbol{x}, t), \boldsymbol{x})\boldsymbol{n}(\boldsymbol{x}) = -p(t)\boldsymbol{n}(\boldsymbol{x}) \quad \forall \boldsymbol{x} \in \partial\boldsymbol{\chi}_t(\mathscr{B}). \tag{1.10}$$

Here $\boldsymbol{H}(\boldsymbol{F}, \boldsymbol{x})$ is the response function giving the Cauchy stress in the actual configuration. (*Cf.* (III. 1.9). $\boldsymbol{H}$ and $\boldsymbol{A}$ are related by (III. 8A.5).) We may require the function $p(t)$ to be determined by the volume of $\mathscr{B}$ in the actual configuration χ_t.

Third, a *boundary condition of place* corresponds to constraining the boundary $\partial\chi_t(\mathscr{B})$ to assume a given shape. This shape may be described by a function $\bar{\chi}(X, t)$, $X \in \partial\kappa(\mathscr{B})$, and the condition may be written

$$\chi(X, t) = \bar{\chi}(X, t) \quad \forall X \in \partial\kappa(\mathscr{B}) \tag{1.11}$$

and for all times t of the motion.

Finally, a *mixed boundary condition* corresponds to conditions of the preceding three types imposed separately on different portions of the boundary.

For an elastic body subject to given body forces, the left-hand side of (1.2) is a known function of $\boldsymbol{x}$ and t in any given configuration of the body. Through (1.2) the acceleration field is then determined also in that configuration. Such is not the case for the velocity field, however. Indeed, on any given configuration we can choose the velocity field arbitrarily without violating the field equation (1.2). Hence in order to determine the motion the boundary conditions must be supplemented by some initial conditions. We may define an initial condition by a configuration $\chi_0(\mathscr{B})$ and a vector field $\boldsymbol{v}_0$ on $\chi_0(\mathscr{B})$. The solution is then required to satisfy also the conditions

$$\chi(X, t_0) = \chi_0(X), \ \boldsymbol{v}(\boldsymbol{x}, t_0) = \boldsymbol{v}_0(\boldsymbol{x}) \tag{1.12}$$

at a particular initial instant t_0. We may choose χ_0 to be the reference configuration κ, but such a choice is not necessary.

The concept of boundary conditions in $(\boldsymbol{x}, t)$-space subsumes formally both initial conditions and what we regard here as boundary conditions, but the different roles played by space and time make separate treatments of the corresponding conditions more natural.

We next consider problems of equilibrium, for which $\boldsymbol{a} = \boldsymbol{0}$, so (1.2) reduces to the *differential equation of equilibrium*:

$$\operatorname{Div} \boldsymbol{A}(\boldsymbol{F}, X) + \rho_\kappa \boldsymbol{b} = \boldsymbol{0}, \tag{1.13}$$

which has the co-ordinate forms

$$\mathsf{A}^{iA}{}_{k}{}^{B} \frac{\partial^2 x^k}{\partial X^A \partial X^B} + \rho_\kappa b^i = 0, \tag{1.14}$$

$$\bar{\mathsf{A}}^{iA}{}_{k}{}^{B} \left(\frac{\partial^2 x^k}{\partial X^A \partial X^B} - \bar{\Gamma}^C_{BA} \frac{\partial x^k}{\partial X^C} \right) - \bar{A}^{iA} \bar{C}^B_{AB} + \rho_\kappa b^i = 0, \tag{1.15}$$

or

$$\mathsf{A}^{kA}{}_{m}{}^{B} \frac{\partial^2 x^k}{\partial X^A \partial X^B} + A^k + \rho_\kappa b^k = 0, \tag{1.16}$$

according as the body is homogeneous, materially uniform but inhomogeneous, or materially non-uniform, respectively.

For problems of equilibrium the boundary conditions must be time-independent, of course. Further, from (II. 3.16) and (II. 3.17) the external loads in the actual configuration must be equilibrated. That is, we must have

$$\int_{\partial\chi(\mathscr{B})} \boldsymbol{t}\, dA + \int_{\chi(\mathscr{B})} \rho \boldsymbol{b}\, dV = \boldsymbol{0}, \tag{1.17}$$

and

$$\int_{\partial\chi(\mathscr{B})} \boldsymbol{r} \wedge \boldsymbol{t}\, dA + \int_{\chi(\mathscr{B})} \rho \boldsymbol{r} \wedge \boldsymbol{b}\, dV = \boldsymbol{0}, \tag{1.18}$$

where $\boldsymbol{r}$ denotes the position vector of a typical place $\boldsymbol{x}$ in $\chi(\mathscr{B})$ relative to a fixed origin. The conditions (1.17) and (1.18) are expressed in terms of the actual configuration, which is unknown. The former condition, expressing equilibrium of forces, is equivalent to

$$\int_{\partial\kappa(\mathscr{B})} \boldsymbol{t}_\kappa\, dA_\kappa + \int_{\kappa(\mathscr{B})} \rho_\kappa \boldsymbol{b}\, dV_\kappa = \boldsymbol{0}, \tag{1.19}$$

which is independent of the deformation and hence can be checked at once for the boundary conditions (1.9), but when we transform to the reference configuration the latter condition, which expresses equilibrium of moments, we obtain

$$\int_{\partial\kappa(\mathscr{B})} \boldsymbol{r} \wedge \boldsymbol{t}_\kappa\, dA_\kappa + \int_{\kappa(\mathscr{B})} \rho_\kappa \boldsymbol{r} \wedge \boldsymbol{b}\, dV_\kappa = \boldsymbol{0}, \tag{1.20}$$

which still depends on the actual and unknown deformation through the field of position vectors $\boldsymbol{r}$. We cannot see at the outset whether or not the given surface traction $\boldsymbol{t}_\kappa$ satisfies this condition, yet if it is not satisfied, there can be no solution to the boundary-value problem of traction corresponding to $\boldsymbol{t}_\kappa$. As was remarked by SIGNORINI, we should regard (1.20) as a *condition of compatibility* to be satisfied by any solution to a traction boundary-value problem.

Unfortunately, the presently known theorems of existence and uniqueness of the solutions of partial differential equations are not strong enough to handle the various general types of boundary-value problems in elasticity we have just defined. In the next few sections we shall summarize some particular, special theorems of existence and uniqueness and develop some systematic methods of approximation for traction boundary-value problems involving small but finite deformations.

If the theory of elasticity is to model well the deformations of real bodies, it is clear that unqualified uniqueness of solution to the boundary-value problems is not to be desired. Indeed, as ERICKSEN has remarked, since the boundary condition of traction is defined by the surface traction $\boldsymbol{t}_\kappa$ on $\partial\boldsymbol{\kappa}(\mathscr{B})$ in the reference configuration $\boldsymbol{\kappa}$, a rod subjected to equal but opposite axial tractions on its ends clearly should be represented by at least two solutions: one with the rod under axial tension and another with the rod rotated through a straight angle and under axial compression of equal amount. For these two possible solutions the surface tractions $\boldsymbol{t}_\kappa$ on $\partial\boldsymbol{\kappa}(\mathscr{B})$ are the same, as shown in the adjoining sketch:

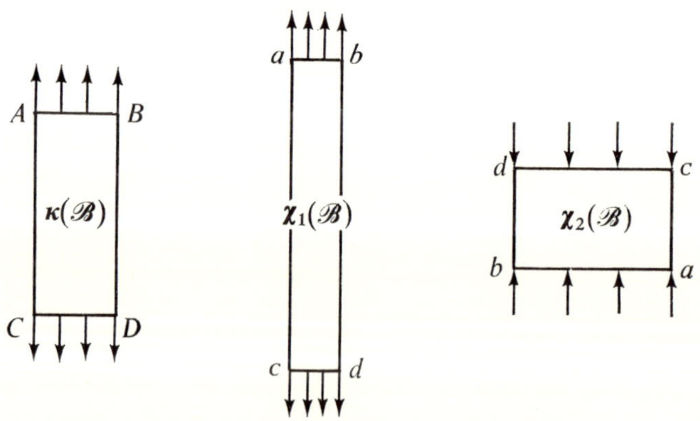

Likewise, a spherical shell or a cylindrical tube may be turned inside out into an everted equilibrium configuration whose bounding surface is altogether free of traction, though there will be non-vanishing stresses in the interior. Hence in this case the traction boundary problem corresponding to vanishing surface traction ought to have at least two solutions. In §§ IV. 4 and IV.5 we have even exhibited solutions corresponding to eversion of isotropic incompressible cylinders and spheres in general, and such solutions are known also for unconstrained isotropic bodies having certain particular stored-energy functions.

Moreover, we know that both the natural and the everted configurations of a thin hemispherical shell are stable in the sense that if we forcibly deform the shell slightly from either, it will snap back, but a suitably directed greater force will cause it to snap from the everted to the natural configuration or *vice versa.* Thus there must exist a third configuration which serves as a boundary of stability between the other two. This configuration, the shape of which might well be corrugated rather than hemispherical or nearly so, should also be a configuration of equilibrium, but an unstable one. A small deformation from this third configuration should cause the shell to snap back to one or the other of the two stable configurations. With a little care one can sometimes cause half a tennis ball to take on and retain, for a second or two, such a corrugated shape even after all force on its surface has been released. Also, not all long thin tubes of rubber take on when everted the cylindrical shape the theory of § IV.4 suggests is possible. Some tubes buckle when everted, and for them the cylindrical everted shape, while a possible one, is unstable. Thus experience suggests that if eversion is possible at all, there might be three solutions rather than merely two for the corresponding boundary-value problem of traction.

Unqualified uniqueness cannot be expected for arbitrary boundary-value problems of place, either. Indeed, there are many deformations that leave the entire boundary of certain body unmoved but nevertheless deform the interior. For example, as has been remarked by JOHN, after we have rotated the inner boundary or the outer boundary of an elastic spherical shell by an angle that is an arbitrary integral multiple of 2π about an arbitrary axis passing through the center of the sphere, the points upon the boundary return to their original positions, but of course the interior points will have moved and the interior of the body will have been severely deformed. Thus we expect that the theory should allow infinitely many equilibrium configurations for an elastic shell the points on whose inner and outer boundaries occupy fixed postions upon two spheres.

2. Signorini's Expansion and Theorems of Uniqueness and Compatibility[1]

A systematic method of approximate solution of boundary-value problems of traction was developed by SIGNORINI (1930). As we explained in the preceding section, the differential equation and the boundary condition are given by (1.2) and (1.9), respectively. For simplicity we shall use the initial configuration as the reference configuration $\boldsymbol{\kappa}$; further, we assume that $\boldsymbol{\kappa}$ is a natural state for every point of the elastic body $\mathscr{B}$.

As usual, we denote the displacement relative to $\boldsymbol{\kappa}$ by $\boldsymbol{u}$ and the displacement gradient by $\boldsymbol{H}$. Then

$$\boldsymbol{F} = \mathbf{1}+\boldsymbol{H}. \tag{2.1}$$

We assume that the response function $\boldsymbol{A}$ in (1.1) is an analytic function of $\boldsymbol{F}$, so that

$$\boldsymbol{A}(\boldsymbol{F};\boldsymbol{X}) = \sum_{s=1}^{\infty} \boldsymbol{A}_s(\boldsymbol{H};\boldsymbol{X}) \qquad \forall \boldsymbol{X} \in \boldsymbol{\kappa}(\mathscr{B}), \tag{2.2}$$

where $\boldsymbol{A}_s = \boldsymbol{A}_s(\boldsymbol{H}, \boldsymbol{X})$ is a homogeneous polynomial of degree s in $\boldsymbol{H}$ with coefficients depending on $\boldsymbol{X}$. We assume also that $\boldsymbol{A}$ is smooth with respect to $\boldsymbol{X}$, so we require that the series (2.2) and its derivative with respect to $\boldsymbol{X}$ converge uniformly on $\boldsymbol{\kappa}(\mathscr{B})$.

Since the principle of material frame-indifference must be satisfied, the response function $\boldsymbol{A}$, and hence the polynomials $\boldsymbol{A}_s$ in (2.2), cannot be arbitrary. To find the restriction imposed by that principle, we use the reduced form (III. 1.27), assuming also that the function $\boldsymbol{J}(\boldsymbol{C};\ \boldsymbol{X})$ there has the expansion

$$\boldsymbol{J}(\boldsymbol{C};\boldsymbol{X}) = \sum_{s=1}^{\infty} \boldsymbol{J}_s(\boldsymbol{G};\boldsymbol{X}), \tag{2.3}$$

where $\boldsymbol{G}$ is the Green-St. Venant strain tensor, which is defined by (III. 8C.43):

$$\boldsymbol{G} \equiv \tfrac{1}{2}(\boldsymbol{C}-\mathbf{1}) = \tfrac{1}{2}(\boldsymbol{H}+\boldsymbol{H}^T+\boldsymbol{H}^T\boldsymbol{H}), \tag{2.4}$$

and where $\boldsymbol{J}_s$ is a homogeneous polynomial of degree s in $\boldsymbol{G}$ with coef-

---

[1] We follow NFTM §§ 63 and 64, where the work of SIGNORINI is generalized as well as condensed. SIGNORINI employed only the first terms in the series (2.8).

ficients depending on $\boldsymbol{X}$. In view of (III. 1.31) and (III. 1.34) we see that

$$\boldsymbol{J}_1(\boldsymbol{G}; \boldsymbol{X}) = [\mathbf{L}(\boldsymbol{X})](\boldsymbol{G}), \tag{2.5}$$

where $\mathbf{L}(\boldsymbol{X})$ is the classical fourth-order tensor of elasticity at the point $\boldsymbol{X}$. From (2.4), (1.5), (III. 1.29), and (III. 8A.5), we then have

$$\boldsymbol{A}_1(\boldsymbol{H}; \boldsymbol{X}) = [\mathbf{L}(\boldsymbol{X})](\boldsymbol{E}), \tag{2.6}$$

where $\boldsymbol{E}$ is the classical strain tensor defined by (III. 1.32):

$$\boldsymbol{E} = \tfrac{1}{2}(\boldsymbol{H}+\boldsymbol{H}^T). \tag{2.7}$$

Thus the leading term in the expansion (2.2) is the stress tensor given by the classical linearized theory. The remaining terms in the expansion (2.2) can be expressed in terms of $\boldsymbol{J}_s$ in (2.3) also, but their forms become increasingly complex.

In the systematic method of approximation to be developed, we assume that the body force $\boldsymbol{b}$ and the surface traction $\boldsymbol{t}_\kappa$ depend analytically on a parameter ε and that they vanish when $\varepsilon = 0$. Thus we set up a one-parameter family of boundary-value problems corresponding to a one-parameter family of body forces and surface tractions given by the series expansions

$$\boldsymbol{b} = \sum_{n=1}^{\infty} \varepsilon^n \boldsymbol{b}_n, \qquad \boldsymbol{t}_\kappa = \sum_{n=1}^{\infty} \varepsilon^n \boldsymbol{t}_{\kappa n}. \tag{2.8}$$

Corresponding to this one-parameter family of boundary-value problems we seek a one-parameter family of solutions $\boldsymbol{u}$ which depend analytically on the parameter ε and vanish when $\varepsilon = 0$:

$$\boldsymbol{u} = \sum_{n=1}^{\infty} \varepsilon^n \boldsymbol{u}_n, \tag{2.9}$$

where $\boldsymbol{u}$ and $\boldsymbol{u}_n$ are smooth fields on $\boldsymbol{\kappa}(\mathscr{B})$ in such a way that the series (2.9) and its gradient converge uniformly on $\boldsymbol{\kappa}(\mathscr{B})$. SIGNORINI's systematic and formal method delimits the values possible for $\boldsymbol{u}_n$ in order that such a solution exist. The problem of existence of analytic solutions will be considered later. In SIGNORINI's method ε is left unspecified. In a particular problem, however, ε may arise naturally as a measure of strain which increases from zero, such as an extension or a twist.

From (2.9) we see that the displacement gradient $\boldsymbol{H}$ has a series expansion

$$\boldsymbol{H} = \sum_{n=1}^{\infty} \varepsilon^n \boldsymbol{H}_n, \tag{2.10}$$

where $\boldsymbol{H}_n$ is the gradient of $\boldsymbol{u}_n$. Substituting (2.10) into (2.2) yields the corresponding series expansion of the Piola-Kirchhoff stress $\boldsymbol{T}_\kappa$:

$$\boldsymbol{T}_\kappa = \sum_{n=1}^{\infty} \varepsilon^n \boldsymbol{T}_{\kappa n}, \tag{2.11}$$

where

$$\begin{aligned} \boldsymbol{T}_{\kappa n}(\boldsymbol{X}) &= \boldsymbol{A}_1(\boldsymbol{H}_n; \boldsymbol{X}) + \boldsymbol{B}_n(\boldsymbol{H}_1, \ldots, \boldsymbol{H}_{n-1}; \boldsymbol{X}) \\ &= [\mathbf{L}(\boldsymbol{X})](\boldsymbol{E}_n) + \boldsymbol{B}_n(\boldsymbol{H}_1, \ldots, \boldsymbol{H}_{n-1}; \boldsymbol{X}). \end{aligned} \tag{2.12}$$

Here $\boldsymbol{B}_n$ is a polynomial of degree n in $\boldsymbol{H}_1, \ldots, \boldsymbol{H}_{n-1}$ with coefficients depending on $\boldsymbol{X}$. The polynomial $\boldsymbol{B}_n$, which generally is not homogeneous, is determined uniquely by $\boldsymbol{A}_1, \ldots, \boldsymbol{A}_n$.

Substituting (2.8), (2.9), (2.11), and $(2.12)_1$ into (1.2) and (1.9) and then equating like powers of ε yields the following successive systems of differential equations and associated boundary conditions:

$$\left.\begin{aligned} \operatorname{Div} \boldsymbol{T}_{\kappa n} + \rho_\kappa \boldsymbol{b}_n &= \rho_\kappa \ddot{\boldsymbol{u}}_n, \\ \boldsymbol{T}_{\kappa n} \boldsymbol{n}_\kappa &= \boldsymbol{t}_{\kappa n}, \end{aligned}\right\} \quad n = 1, 2, 3, \ldots \tag{2.13}$$

When $(2.12)_2$ is used, these systems become

$$\left.\begin{aligned} \operatorname{Div} \mathbf{L}(\boldsymbol{E}_n) + \rho_\kappa \boldsymbol{b}_n^* &= \rho_\kappa \ddot{\boldsymbol{u}}_n, \\ \mathbf{L}(\boldsymbol{E}_n)\boldsymbol{n}_\kappa &= \boldsymbol{t}_{\kappa n}^*, \end{aligned}\right\} \quad n = 1, 2, 3, \ldots \tag{2.14}$$

where

$$\left.\begin{aligned} \rho_\kappa \boldsymbol{b}_n^* &\equiv \rho_\kappa \boldsymbol{b}_n + \operatorname{Div} \boldsymbol{B}_n(\boldsymbol{H}_1, \ldots, \boldsymbol{H}_{n-1}; \boldsymbol{X}), \\ \boldsymbol{t}_{\kappa n}^* &\equiv \boldsymbol{t}_{\kappa n} - \boldsymbol{B}_n(\boldsymbol{H}_1, \ldots, \boldsymbol{H}_{n-1}; \boldsymbol{X})\boldsymbol{n}_\kappa, \end{aligned}\right\} \quad n = 1, 2, 3, \ldots \tag{2.15}$$

Since $\boldsymbol{B}_1 = \boldsymbol{0}$, when $n = 1$ we have

$$\boldsymbol{b}_1^* = \boldsymbol{b}_1, \quad \boldsymbol{t}_{\kappa 1}^* = \boldsymbol{t}_{\kappa 1}. \tag{2.16}$$

Hence $(2.14)_1$ are precisely the equations that define the initial-value and boundary-value problem of traction in the classical infinitesimal theory of elasticity, where $\boldsymbol{b}_1$ and $\boldsymbol{t}_{\kappa 1}$ are the assigned body force and surface traction, respectively, for that theory. More generally, as soon as displacements $\boldsymbol{u}_1, \ldots, \boldsymbol{u}_{n-1}$ satisfying $(2.14)_1$–$(2.14)_{n-1}$ have been found, the displacement $\boldsymbol{u}_n$ may be obtained by solving $(2.14)_n$, with $\boldsymbol{b}_n^*$ and $\boldsymbol{t}_{\kappa n}^*$ given by $(2.15)_n$. Since $\mathbf{L}$ and ρ_κ do not depend on n, (2.14) defines an infinite sequence of traction boundary-value and initial-value problems in linear elasticity for *the same material and the same boundary*, though subject to body forces and surface tractions generally depending on n. From (2.14), the body forces $\boldsymbol{b}_n^*$ and the surface tractions $\boldsymbol{t}_{\kappa n}^*$ to be used in $(2.14)_n$ are determined explicitly by $\boldsymbol{b}_n$, $\boldsymbol{t}_{\kappa n}$, and $\boldsymbol{u}_1, \ldots, \boldsymbol{u}_{n-1}$. Thus *the calculation of n terms in the expansion* (2.9) *is reduced, formally, to the solution of n initial-value and boundary-value problems of traction in the classical infinitesimal theory of elasticity for the same body.*

SIGNORINI's systematic method of approximation for solving initial-value and boundary-value problems of traction, as summarized above, applies to homogeneous or inhomogeneous elastic bodies of isotropic or anisotropic material alike. Now we restrict it to statics. We may then encounter some difficulty, since, as we remarked in the preceding section, in order that the system $(2.14)_n$ with $\ddot{\boldsymbol{u}}_n = \mathbf{0}$ can have any solution at all, the $\boldsymbol{b}_n^*$ and the $\boldsymbol{t}_{\kappa n}^*$ must form an equilibrated system on $\boldsymbol{\kappa}(\mathscr{B})$, the the reference configuration common to all the systems (2.14). As we shall see, when the assigned $\boldsymbol{t}_\kappa$ and $\boldsymbol{b}_\kappa$ form an equilibrated system on $\boldsymbol{\kappa}(\mathscr{B})$, the $\boldsymbol{t}_{\kappa n}^*$ and the $\boldsymbol{b}_n^*$ given by (2.15) have no resultant force but may have a non-zero resultant couple on $\boldsymbol{\kappa}(\mathscr{B})$. The requirement that the resultant couple on $\boldsymbol{\kappa}(\mathscr{B})$ vanish is SIGNORINI's *condition of compatibility.*

To see that the $\boldsymbol{t}_{\kappa n}^*$ and the $\boldsymbol{b}_n^*$ defined by (2.15) have no resultant force, we substitute (2.8) into (1.17) and then equate like powers of ε, obtaining

$$\int_{\partial\boldsymbol{\kappa}(\mathscr{B})} \boldsymbol{t}_{\kappa n}\, dA_\kappa + \int_{\boldsymbol{\kappa}(\mathscr{B})} \rho_\kappa\, \boldsymbol{b}_n\, dV_\kappa = \mathbf{0}. \tag{2.17}$$

By the divergence theorem

$$\int_{\partial\boldsymbol{\kappa}(\mathscr{B})} -\boldsymbol{B}_n \boldsymbol{n}_\kappa\, dA_\kappa + \int_{\boldsymbol{\kappa}(\mathscr{B})} \operatorname{Div} \boldsymbol{B}_n\, dV_\kappa = \mathbf{0}. \tag{2.18}$$

Adding (2.17) and (2.18) yields the desired result:

$$\int_{\partial\boldsymbol{\kappa}(\mathscr{B})} \boldsymbol{t}_{\kappa n}^*\, dA_\kappa + \int_{\boldsymbol{\kappa}(\mathscr{B})} \rho_\kappa\, \boldsymbol{b}_n^*\, dV_\kappa = \mathbf{0}. \tag{2.19}$$

The corresponding analysis for the resultant couple of the $\boldsymbol{t}_{\kappa n}^*$ and the $\boldsymbol{b}_n^*$ is not so simple. Following SIGNORINI, we assume first that the assigned t_κ and b satisfy the condition

$$\int_{\partial\kappa(\mathscr{B})} r_\kappa \wedge t_\kappa \, dA_\kappa + \int_{\kappa(\mathscr{B})} \rho_\kappa r_\kappa \wedge b \, dV_\kappa = 0, \tag{2.20}$$

and hence for each n

$$\int_{\partial\kappa(\mathscr{B})} r_\kappa \wedge t_{\kappa n} \, dA_\kappa + \int_{\kappa(\mathscr{B})} \rho_\kappa r_\kappa \wedge b_n \, dV_\kappa = 0. \tag{2.21}$$

Here r_κ denotes the position vector in the reference configuration κ. The conditions (1.19) and (2.21) simply mean that $t_{\kappa n}$ and b_n form an equilibrated system on $\kappa(\mathscr{B})$. This assumption is inessential, since, as was shown long ago by DA SILVA, even if it does not hold in $\kappa(\mathscr{B})$, nevertheless the assumed and necessary condition (1.19) allows us to find a rotation that carries $\kappa(\mathscr{B})$ into another reference configuration in which the moments of t_κ and b, which are held fixed in the rotation, are equilibrated. Any important property a reference configuration might have, such as being a natural state or an undistorted state, is invariant under rotation. To determine an appropriate rotation, we define the *astatic load*[1]:

$$A \equiv \int_{\partial\kappa(\mathscr{B})} r_\kappa \otimes t_\kappa \, dA_\kappa + \int_{\kappa(\mathscr{B})} \rho_\kappa r_\kappa \otimes b \, dV_\kappa. \tag{2.22}$$

Then (2.20) is equivalent to

$$A - A^T = 0. \tag{2.23}$$

From (2.22), if $\kappa(\mathscr{B})$ is rotated by Q but t_κ and b are kept fixed, then the astatic load on $Q \circ \kappa(\mathscr{B})$ becomes QA. Hence Q is an appropriate rotation if

$$QA - A^T Q^T = 0. \tag{2.24}$$

By using a polar decomposition $A = PS$, where P is a rotation and S is symmetric, we see that a solution[2] of (2.24) is given by $Q = P^T$.

Now assuming that the condition (2.20) holds, we can express the condition (1.20) in terms of the displacement u:

$$\int_{\partial\kappa(\mathscr{B})} u \wedge t_\kappa \, dA_\kappa + \int_{\kappa(\mathscr{B})} \rho_\kappa u \wedge b \, dV_\kappa = 0. \tag{2.25}$$

---

[1] *Cf.* CFT § 220.

[2] In general there are further solutions.

Substituting (2.8) and (2.9) into (2.25) and expanding, we obtain

$$\sum_{m=1}^{n-1}\left[\int_{\partial\kappa(\mathscr{B})} \boldsymbol{u}_{n-m}\wedge \boldsymbol{t}_{\kappa m}\,dA_\kappa+\int_{\kappa(\mathscr{B})}\rho_\kappa \boldsymbol{u}_{n-m}\wedge \boldsymbol{b}_m\,dV_\kappa\right]=\mathbf{0},\ n=2,3,\ldots \tag{2.26}$$

which is SIGNORINI's *condition of compatibility* for the $\boldsymbol{u}_m$. In order that the system (2.14) have a static solution $\boldsymbol{u}_n$, it is *necessary* that the solutions $\boldsymbol{u}_1, \boldsymbol{u}_2, \ldots, \boldsymbol{u}_{n-1}$ at the preceding stages of the process satisfy (2.26).

We shall now show that (2.26) is equivalent to the requirement that the resultant couple of the $\boldsymbol{t}^*_{\kappa n}$ and the $\boldsymbol{b}^*_n$ vanishes on $\kappa(\mathscr{B})$, *i.e.*,

$$\int_{\partial\kappa(\mathscr{B})} \boldsymbol{r}_\kappa\wedge \boldsymbol{t}^*_{\kappa n}\,dA_\kappa+\int_{\kappa(\mathscr{B})}\rho_\kappa \boldsymbol{r}_\kappa\wedge \boldsymbol{b}^*_n\,dV_\kappa=\mathbf{0},\ n=1,2,3,\ldots \tag{2.27}$$

By (2.21) we see that (2.27) is equivalent to

$$\int_{\partial\kappa(\mathscr{B})} \boldsymbol{r}_\kappa\wedge (\boldsymbol{t}^*_{\kappa n}-\boldsymbol{t}_{\kappa n})\,dA_\kappa+\int_{\kappa(\mathscr{B})}\rho_\kappa \boldsymbol{r}_\kappa\wedge (\boldsymbol{b}^*_n-\boldsymbol{b}_n)\,dV_\kappa=\mathbf{0},\ n=1,2,3,\ldots \tag{2.28}$$

Then (2.15) shows that (2.28) is equivalent to

$$-\int_{\partial\kappa(\mathscr{B})} \boldsymbol{r}_\kappa\wedge \boldsymbol{B}_n\boldsymbol{n}_\kappa\,dA_\kappa+\int_{\kappa(\mathscr{B})} \boldsymbol{r}_\kappa\wedge \operatorname{Div}\boldsymbol{B}_n\,dV_\kappa=\mathbf{0},\ n=1,2,3,\ldots \tag{2.29}$$

To prove that (2.29) is equivalent to (2.26), we shall make repeated use of the identity

$$\int_{\kappa(\mathscr{B})}(\operatorname{Grad}\boldsymbol{s})\boldsymbol{S}^T dV_\kappa=\int_{\partial\kappa(\mathscr{B})}\boldsymbol{s}\otimes\boldsymbol{S}\boldsymbol{n}_\kappa\,dA_\kappa-\int_{\kappa(\mathscr{B})}\boldsymbol{s}\otimes\operatorname{Div}\boldsymbol{S}\,dV_\kappa \tag{2.30}$$

for any smooth vector field $\boldsymbol{s}$ and tensor field $\boldsymbol{S}$ on $\kappa(\mathscr{B})$. Applying (2.30) with the choices $\boldsymbol{s}=\boldsymbol{r}_\kappa$ and $\boldsymbol{S}=\boldsymbol{B}_n$, we see that (2.29) is equivalent to

$$\int_{\kappa(\mathscr{B})}\boldsymbol{B}_n\,dV_\kappa=\text{a symmetric tensor,} \tag{2.31}$$

and by (2.21) this condition is equivalent to

$$\int_{\kappa(\mathscr{B})}\boldsymbol{T}_{\kappa n}\,dV_\kappa=\text{a symmetric tensor,} \tag{2.32}$$

since the leading term on the right-hand side of (2.12) is a symmetric tensor. By (II. 6.27)

$$T_\kappa F^T = T_\kappa(1+H^T) = \text{a symmetric tensor.} \tag{2.33}$$

Substitution of (2.11) and (2.10) into (2.33) and collecting terms multiplied by like powers of ε shows that

$$T_{\kappa n} + \sum_{m=1}^{n-1} T_{\kappa m} H^T_{n-m} = \text{a symmetric tensor.} \tag{2.34}$$

Hence (2.32) is equivalent to

$$\sum_{m=1}^{n-1} \int_{\kappa(\mathscr{B})} H_{n-m} T^T_{\kappa m} \, dV_\kappa = \text{a symmetric tensor.} \tag{2.35}$$

Now applying the identity (2.30) with the choice $s = u_{n-m}$ and $S = T_{\kappa m}$, we see that (2.35) can be rewritten as

$$\sum_{m=1}^{n-1} \left[\int_{\partial\kappa(\mathscr{B})} u_{n-m} \wedge T_{\kappa m} n_\kappa \, dA_\kappa - \int_{\kappa(\mathscr{B})} u_{n-m} \wedge \operatorname{Div} T_{\kappa m} \, dV_\kappa \right] = 0, \tag{2.36}$$

which is the same as (2.26) since u_m is a static solution corresponding to the body force b_m and the boundary traction $t_{\kappa m}$.

In order to use SIGNORINI's condition (2.26) to select the u_n inductively, we observe first that if u_n is any equilibrium solution of the linear boundary-value problem of traction defined by $(2.14)_n$, so is $u_n + W_n r_\kappa$ for any infinitesimal constant rotation W_n. This is a standard result in the classical infinitesimal theory of elasticity. Now taking as the hypothesis of induction that $u_1, \ldots, u_{p-1}$ are equilibrium solutions satisfying SIGNORINI's condition (2.26) if $n = 2, 3, \ldots, p-1$ and that $\bar{u}_p$ is an arbitrary equilibrium solution of $(2.14)_p$, we shall determine an infinitesimal rotation W_p such that the equilibrium solution u_p given by

$$u_p = \bar{u}_p + W_p r_\kappa, \tag{2.37}$$

together with the equilibrium solutions $u_1, \ldots, u_{p-1}$, shall satisfy (2.26) when $n = p$. To determine W_p, we put

$$R_p \equiv \sum_{m=1}^{p} \left[\int_{\partial\kappa(\mathscr{B})} \bar{u}_{p+1-m} \wedge t_{\kappa m} \, dA_\kappa + \int_{\kappa(\mathscr{B})} \rho_\kappa \bar{u}_{p+1-m} \wedge b_m \, dV_\kappa \right], \tag{2.38}$$

where $\bar{u}_m \equiv u_m$ for $m = 1, \ldots, p-1$. Then the u_p given by (2.37) satisfies the condition

$$\mathbf{0} = \sum_{m=1}^{p} \left[\int_{\partial\kappa(\mathscr{B})} \boldsymbol{u}_{p+1-m} \wedge \boldsymbol{t}_{\kappa m}\, dA_\kappa + \int_{\kappa(\mathscr{B})} \rho_\kappa \boldsymbol{u}_{p+1-m} \wedge \boldsymbol{b}_m\, dV_\kappa \right] \tag{2.39}$$

if and only if $\boldsymbol{W}_p$ is a solution of the condition

$$-\boldsymbol{R}_p = \int_{\partial\kappa(\mathscr{B})} (\boldsymbol{W}_p \boldsymbol{r}_\kappa) \wedge \boldsymbol{t}_{\kappa 1}\, dA_\kappa + \int_{\kappa(\mathscr{B})} \rho_\kappa (\boldsymbol{W}_p \boldsymbol{r}_\kappa) \wedge \boldsymbol{b}_1\, dV_\kappa = \tfrac{1}{2}(\boldsymbol{W}_p \boldsymbol{A}_1 + \boldsymbol{A}_1 \boldsymbol{W}_p), \tag{2.40}$$

where $\boldsymbol{A}_1$ is the astatic load (2.22) corresponding to $\boldsymbol{t}_{\kappa 1}$ and $\boldsymbol{b}_1$. Here we have used the fact that an infinitesimal rotation $\boldsymbol{W}_p$ is skew-symmetric and that the astatic load $\boldsymbol{A}_1$ is symmetric, as required by (2.21). In terms of the axial vectors $\boldsymbol{r}_p$ and $\boldsymbol{w}_p$ corresponding to the skew-symmetric tensors $\boldsymbol{R}_p$ and $\boldsymbol{W}_p$, the condition (2.40) takes the form

$$(\boldsymbol{A}_1 - (\operatorname{tr} \boldsymbol{A}_1)\mathbf{1})\boldsymbol{w}_p ,= -2\boldsymbol{r}_p . \tag{2.41}$$

Hence in order that (2.41) possess a unique solution $\boldsymbol{w}_p$, or, equivalently, that (2.40) possess a unique solution $\boldsymbol{W}_p$, it is necessary and sufficient that

$$\det(\boldsymbol{A}_1 - (\operatorname{tr} \boldsymbol{A}_1)\mathbf{1}) \neq 0. \tag{2.42}$$

We summarize the preceding analysis of the systematic method of approximate solution to equilibrium boundary-value problems of traction by the following

THEOREM OF COMPATIBILITY AND UNIQUENESS (SIGNORINI). *Let it be supposed that the load* $(\boldsymbol{t}_{\kappa 1}, \boldsymbol{b}_1)$ *satisfy the condition* (2.42). *Then*

i) *The equilibrium system of* (2.13) *is compatible*; i.e., *if equilibrium solutions exist for the traction boundary-value problem in the infinitesimal theory, then there are equilibrium solutions* $\boldsymbol{u}_1, \boldsymbol{u}_2, \ldots$ *of* (2.13).

ii) *If the classical condition*[1] *of uniqueness for the traction boundary-*

[1] For hyperelastic materials in the classical infinitesimal theory a solution of the traction boundary-value problem is unique to within a uniform infinitesimal rigid displacement if the stored-energy function is definite, positive or negative. This condition is not necessary for such uniqueness, however. For isotropic materials a necessary and sufficient condition for such uniqueness is the requirement that the Poisson modulus $\xi = \frac{1}{2}\lambda/(\lambda+\mu)$ satisfy the condition $\xi < 1$. *Cf.* NFTM, § 51. For a modern exposition of the analytical theory of solutions in the classical infinitesimal theory, see M. GURTIN, "The linear theory of elasticity", and G. FICHERA, "Existence theorems in elasticity", both in FLÜGGE's *Encyclopedia of Physics*, Vol. VIa/2, ed. C. TRUESDELL, Berlin *etc.*, Springer-Verlag, 1972.

value problem in the infinitesimal theory holds, then the equilibrium solution $\boldsymbol{u}_1, \boldsymbol{u}_2, \ldots$ *of the system* (2.13) *is unique to within a uniform translation.*

The theorem just stated does not imply that the equilibrium system of (2.13) necessarily fails to be compatible when the condition (2.42) does not hold, *i.e.*, when

$$\det(\boldsymbol{A}_1 - (\operatorname{tr} \boldsymbol{A}_1)\mathbf{1}) = 0. \tag{2.43}$$

The analysis merely shows that in this case (2.40), or, equivalently, the equation (2.41), does not determine a unique infinitesimal rotation. Then there may be no solution or more than one solution. The system (2.13) is incompatible only if for some n the equation (2.40) possesses no solution at all.

In the case excluded by the theorem, we see from (2.43) that $\operatorname{tr} \boldsymbol{A}_1$ is a proper number of $\boldsymbol{A}_1$. This means that a spectral form of the symmetric tensor $\boldsymbol{A}_1$ is

$$\boldsymbol{A}_1 = \alpha \boldsymbol{e}_1 \otimes \boldsymbol{e}_1 + \beta(\boldsymbol{e}_2 \otimes \boldsymbol{e}_2 - \boldsymbol{e}_3 \otimes \boldsymbol{e}_3), \tag{2.44}$$

where $\alpha = \operatorname{tr} \boldsymbol{A}_1$. It is well known that a necessary and sufficient condition for (2.44) is that

$$\boldsymbol{Q}\boldsymbol{A}_1 - \boldsymbol{A}_1 \boldsymbol{Q}^T = \mathbf{0} \tag{2.45}$$

for all rotations $\boldsymbol{Q}$ about $\boldsymbol{e}_1$. Comparing (2.45) with (2.24), we see that (2.43) asserts the existence of an axis about which any rotation of $\boldsymbol{\kappa}(\mathscr{B})$ while $\boldsymbol{t}_{\kappa 1}$ and $\boldsymbol{b}_1$ are held fixed in space maintains the equilibrium of the load system $(\boldsymbol{t}_{\kappa 1}, \boldsymbol{b}_1)$. A load system of this kind is said to possess an *axis of equilibrium*.

The analysis of SIGNORINI's systematic expansion in the case when the load system $(\boldsymbol{t}_{\kappa 1}, \boldsymbol{b}_1)$ possesses an axis of equilibrium is complex. For some discussion of it and references the reader may consult NFTM, § 64.

An adjustment[1] of SIGNORINI's expansion was effected by RIVLIN & TOPAKOGLU (1954). They first take partial sums of the expansions (2.2), (2.8), (2.9), (2.10), and (2.11) and call them n^{th}*-order approximations.*

[1] We follow NFTM, § 65, in which RIVLIN & TOPAKOGLU's method, which was proposed independently of SIGNORINI's but long afterward, was first shown to be a variant of it.

We denote these by $\boldsymbol{A}^{(n)}$, $\boldsymbol{b}^{(n)}$, $\boldsymbol{t}_{\kappa}{}^{(n)}$, $\boldsymbol{u}^{(n)}$, $\boldsymbol{H}^{(n)}$, and $\boldsymbol{T}_{\kappa}{}^{(n)}$, respectively. For example,

$$\boldsymbol{A}^{(n)} = \boldsymbol{A}^{(n)}(\boldsymbol{H}, \boldsymbol{X}) \equiv \sum_{s=1}^{n} \boldsymbol{A}_s(\boldsymbol{H}; \boldsymbol{X}). \tag{2.46}$$

Multiplying (2.13) by ε^n and adding with respect to n yields

$$\left.\begin{aligned} \operatorname{Div} \boldsymbol{T}_{\kappa}{}^{(n)} + \rho_{\kappa} \boldsymbol{b}^{(n)} &= \rho_{\kappa} \ddot{\boldsymbol{u}}^{(n)}, \\ \boldsymbol{T}_{\kappa}{}^{(n)} \boldsymbol{n}_{\kappa} &= \boldsymbol{t}_{\kappa}{}^{(n)}, \end{aligned}\right\} \quad n = 1, 2, 3, \ldots, \tag{2.47}$$

which may be called the nth*-order system*. By (2.2)

$$\boldsymbol{A}^{(n)}(\boldsymbol{H}^{(n)}; \boldsymbol{X}) = \boldsymbol{T}_{\kappa}{}^{(n)} + \boldsymbol{O}(\varepsilon^{n+1}). \tag{2.48}$$

Hence (2.47) can be written also as

$$\left.\begin{aligned} \operatorname{Div} \boldsymbol{A}^{(n)} + \rho_{\kappa} \boldsymbol{b}^{(n)} &= \rho_{\kappa} \ddot{\boldsymbol{u}}^{(n)} + \boldsymbol{O}(\varepsilon^{n+1}), \\ \boldsymbol{A}^{(n)} \boldsymbol{n}_{\kappa} &= \boldsymbol{t}_{\kappa}{}^{(n)} + \boldsymbol{O}(\varepsilon^{n+1}), \end{aligned}\right\} \quad n = 1, 2, 3, \ldots \tag{2.49}$$

RIVLIN & TOPAKOGLU define a pth*-order solution* to be an expansion $\boldsymbol{u} = \sum \varepsilon^m \boldsymbol{u}_m$, which may be finite or infinite, such that its partial sum $\boldsymbol{u}^{(n)}$ satisfies (2.49) for all $n = 1, 2, \ldots, p$. Of course, the coefficients of order higher than p in the expansion of a pth-order solution are arbitrary. A judicious choice of those coefficients may lighten the calculation by an inductive procedure, as we shall see. Clearly, the SIGNORINI expansion (2.9), when it satisfies (2.13), or equivalently (2.14), is a pth-order solution for all p. Conversely, any infinite expansion $\boldsymbol{u} = \sum \varepsilon^m \boldsymbol{u}_m$ which is a pth-order solution for all p is a SIGNORINI expansion satisfying (2.13).

The inductive procedure worked out by RIVLIN & TOPAKOGLU for the determination of a pth-order solution may be summarized as follows:

i) Assume that a $(p-1)^{\text{st}}$-order solution

$$\boldsymbol{u} = \boldsymbol{u}^{(p-1)} + \varepsilon^p \bar{\boldsymbol{u}}_p + \boldsymbol{O}(\varepsilon^{p+1}) \tag{2.50}$$

has been found. Substituting the solution into the pth-order system $(2.29)_p$, find the loads $(-\varepsilon^p \tilde{\boldsymbol{t}}_{\kappa p}, -\varepsilon^p \tilde{\boldsymbol{b}}_p)$ required to maintain the acceleration $\ddot{\boldsymbol{u}}^{(p-1)}$ unchanged. The loads $(\tilde{\boldsymbol{t}}_{\kappa p}, \tilde{\boldsymbol{b}}_p)$ generally depend on the pth-order coefficient $\bar{\boldsymbol{u}}_p$ of the $(p-1)^{\text{st}}$-order solution. If $\bar{\boldsymbol{u}}_p = \boldsymbol{0}$[1], then the loads $(\tilde{\boldsymbol{t}}_{\kappa p}, \tilde{\boldsymbol{b}}_p)$ are equal to the $(\boldsymbol{t}_{\kappa p}^*, \boldsymbol{b}_p^*)$ defined by $(2.15)_p$.

[1] This is the case considered in NFTM, § 65. The more general case with a non-vanishing $\bar{\boldsymbol{u}}_p$ is used in the two examples in NFTM, § 66, however.

ii) Find the solution $\varepsilon^p \tilde{\boldsymbol{u}}_p$ of the infinitesimal theory corresponding to the reversed loads $(\varepsilon^p \tilde{\boldsymbol{t}}_{\kappa p}, \varepsilon^p \tilde{\boldsymbol{b}}_p)$.

iii) Then a pth-order solution is given by

$$\boldsymbol{u} = \boldsymbol{u}^{(p-1)} + \varepsilon^p(\tilde{\boldsymbol{u}}_p + \bar{\boldsymbol{u}}_p) + \boldsymbol{O}(\varepsilon^{p+1}). \tag{2.51}$$

The advantage of this procedure over SIGNORINI's lies in step ii), where we have the freedom to choose a non-vanishing p^{th}-order coefficient $\bar{\boldsymbol{u}}_p$ in the $(p-1)^{\text{st}}$-order solution. This means that instead of solving the linear elastic system for the displacement $\boldsymbol{u}_p$ corresponding to the load system $(\boldsymbol{t}^*_{\kappa p}, \boldsymbol{b}^*_p)$, we may solve the linear system for the displacement $\tilde{\boldsymbol{u}}_p$ corresponding to the loads $(\tilde{\boldsymbol{t}}_{\kappa p}, \tilde{\boldsymbol{b}}_p)$. In the next section we shall work out an example to show that solution of the linear problem may be made easier in this way. Of course, if we choose $\bar{\boldsymbol{u}}_p = \boldsymbol{0}$, then RIVLIN & TOPAKOGLU's procedure becomes exactly the same as SIGNORINI's.

To prove that (2.51) is a pth-order solution, from (2.12) and (2.48) we observe that

$$\begin{aligned} \boldsymbol{A}^{(p)}(\boldsymbol{H}^{(p)}; X) &= \boldsymbol{T}_\kappa^{(p-1)} + \varepsilon^p \boldsymbol{T}_{\kappa p} + \boldsymbol{O}(\varepsilon^{p+1}) \\ &= \boldsymbol{T}_\kappa^{(p-1)} + \varepsilon^p[\mathsf{L}(X)](\boldsymbol{E}_p) + \varepsilon^p \boldsymbol{B}_p(\boldsymbol{H}_1, \ldots, \boldsymbol{H}_{p-1}; X) + \boldsymbol{O}(\varepsilon^{p+1}). \end{aligned} \tag{2.52}$$

Substituting the gradient of the $(p-1)^{\text{st}}$-order solution (2.50) into (2.52), we obtain

$$\begin{aligned} \boldsymbol{A}^{(p)}(\boldsymbol{H}^{(p-1)} + \varepsilon^p \bar{\boldsymbol{H}}_p; X) = \boldsymbol{T}_\kappa^{(p-1)} &+ \varepsilon^p[\mathsf{L}(X)](\bar{\boldsymbol{E}}_p) \\ &+ \varepsilon^p \boldsymbol{B}_p(\boldsymbol{H}_1, \ldots, \boldsymbol{H}_{p-1}; X) + \boldsymbol{O}(\varepsilon^{p+1}). \end{aligned} \tag{2.53}$$

Now taking the divergence of (2.53) and using the $(p-1)^{\text{st}}$-order system $(2.47)_{p-1}$ to eliminate $\boldsymbol{T}_\kappa^{(p-1)}$, we find that

$$\begin{aligned} \rho_\kappa \tilde{\boldsymbol{b}}_p &= \rho_\kappa \boldsymbol{b}_p + \operatorname{Div} \boldsymbol{B}_p(\boldsymbol{H}_1, \ldots, \boldsymbol{H}_{p-1}; X) + \operatorname{Div} \mathsf{L}(\bar{\boldsymbol{E}}_p) \\ &= \rho_\kappa \boldsymbol{b}^*_p + \operatorname{Div} \mathsf{L}(\bar{\boldsymbol{E}}_p) \\ \tilde{\boldsymbol{t}}_{\kappa p} &= \boldsymbol{t}_{\kappa p} - \boldsymbol{B}_p(\boldsymbol{H}_1, \ldots, \boldsymbol{H}_{p-1}; X)\boldsymbol{n}_\kappa - [\mathsf{L}(\bar{\boldsymbol{E}}_p)]\boldsymbol{n}_\kappa \\ &= \boldsymbol{t}^*_{\kappa p} - [\mathsf{L}(\bar{\boldsymbol{E}}_p)]\boldsymbol{n}_\kappa \end{aligned} \tag{2.54}$$

Hence $\tilde{\boldsymbol{u}}_p$ is a solution of the linear elastic system

$$\begin{aligned} \operatorname{Div} \mathsf{L}(\tilde{\boldsymbol{E}}_p) + \rho_\kappa \tilde{\boldsymbol{b}}_p &= \rho_\kappa \ddot{\tilde{\boldsymbol{u}}}_p, \\ \mathsf{L}(\tilde{\boldsymbol{E}}_p)\boldsymbol{n}_\kappa &= \tilde{\boldsymbol{t}}_{\kappa p}, \end{aligned} \tag{2.55}$$

if and only if $\boldsymbol{u}_p$ as defined by

$$\boldsymbol{u}_p \equiv \bar{\boldsymbol{u}}_p + \tilde{\boldsymbol{u}}_p \tag{2.56}$$

is a solution of the linear elastic system $(2.14)_p$. Consequently the expansion (2.51) coincides with SIGNORINI's expansion to within terms of order $p+1$ in ε, and hence (2.51) is a pth-order solution.

For a static problem, RIVLIN & TOPAKOGLU's procedure can be carried out only if the loads $(\tilde{\boldsymbol{t}}_{\kappa p}, \tilde{\boldsymbol{b}}_p)$ form an equilibrated system on $\boldsymbol{\kappa}(\mathscr{B})$ at each step of the induction. In view of (2.54), this condition is equivalent to $(2.27)_p$, since the loads $([\mathbf{L}(\overline{\boldsymbol{E}}_p)]\boldsymbol{n}_\kappa, -\operatorname{Div}\mathbf{L}(\overline{\boldsymbol{E}}_p))$, being the surface traction and the body force required to maintain the displacement $\overline{\boldsymbol{u}}_p$ in equilibrium, automatically form an equilibrated system on $\boldsymbol{\kappa}(\mathscr{B})$. Consequently the leading p coefficients $\boldsymbol{u}_1, \ldots, \boldsymbol{u}_p$ of a pth-order equilibrium solution must satisfy the SIGNORINI condition (2.26) for all $n = 2, \ldots, p+1$.

Before closing this section we remark that the expansions here developed are not directly applicable to incompressible bodies, since a deformation which is isochoric to within terms of order n is generally not isochoric to a higher order and thus not possible subject to finite forces. Systematic methods of iteration may be set up directly for incompressible materials, however, as is indicated in NFTM, § 67.

3. A Central Example: Second-order Effects in Torsion and Extension of an Isotropic Circular-Cylindrical Tube

Several examples of application of the systematic procedure we have outlined at the end of the last section were given by RIVLIN (1953). Here we shall summarize one central example[1]. In order to do so, we first derive the explicit form of the expansion (2.2) up to terms of second order for isotropic elastic materials not subject to internal constraints. First, to within error of the third order in $\boldsymbol{H}$

$$\begin{aligned} \boldsymbol{B} &= \mathbf{1}+2\boldsymbol{E}+\boldsymbol{H}\boldsymbol{H}^T, \\ \boldsymbol{B}^{-1} &= \mathbf{1}-2\boldsymbol{E}-\boldsymbol{H}\boldsymbol{H}^T+4\boldsymbol{E}^2+\ldots, \\ \mathrm{I}_{\boldsymbol{G}} &= \mathrm{I}_{\boldsymbol{E}}+\tfrac{1}{2}\mathrm{I}_{\boldsymbol{H}\boldsymbol{H}^T}, \\ \mathrm{II}_{\boldsymbol{G}} &= \mathrm{II}_{\boldsymbol{E}}+\ldots, \\ \mathrm{III}_{\boldsymbol{G}} &= 0+\ldots, \end{aligned} \tag{3.1}$$

[1] We follow the presentation of NFTM, § 66, which slightly generalizes RIVLIN's work.

where $\boldsymbol{G}$ is the Green-St. Venant strain tensor, defined by (2.4); $\boldsymbol{E}$ is the classical strain tensor, defined by (2.7), and $\boldsymbol{B}$ is the left Cauchy-Green tensor. As usual, for any symmetric tensor $\boldsymbol{S}$ we denote the three principal invariants of $\boldsymbol{S}$ by $\mathrm{I}_{\boldsymbol{S}}$, $\mathrm{II}_{\boldsymbol{S}}$, and $\mathrm{III}_{\boldsymbol{S}}$.

For an isotropic elastic material the response function that delivers the Cauchy stress tensor $\boldsymbol{T}$ has the representation (III. 2.50):

$$\boldsymbol{T} = \psi_0 \mathbf{1} + \psi_1 \boldsymbol{B} + \psi_{-1} \boldsymbol{B}^{-1}, \tag{3.2}$$

where ψ_Γ is a function of the principal invariants of $\boldsymbol{B}$, or, equivalently, a function of the principal invariants of $\boldsymbol{G}$. Under the usual assumption of smoothness we have the following expansion for ψ_Γ:

$$\psi_\Gamma = \mu(\alpha_{\Gamma 0} + \alpha_{\Gamma 1} \mathrm{I}_{\boldsymbol{G}} + \alpha_{\Gamma 2} \mathrm{I}_{\boldsymbol{G}}^2 + \alpha_{\Gamma 3} \mathrm{II}_{\boldsymbol{G}} + \ldots). \tag{3.3}$$

Substituting (3.1) and (3.3) into (3.2) shows that, to within terms of the third order in $\boldsymbol{H}$, the stress relation for isotropic elastic materials is

$$\begin{aligned} \boldsymbol{T}/\mu = {} & \alpha_1 \mathrm{I}_{\boldsymbol{E}} \mathbf{1} + 2\alpha_2 \boldsymbol{E} + (\tfrac{1}{2}\alpha_1 \mathrm{I}_{\boldsymbol{H}\boldsymbol{H}^T} + \alpha_3 \mathrm{I}_{\boldsymbol{E}}^2 + \alpha_4 \mathrm{II}_{\boldsymbol{E}})\mathbf{1} \\ & + \alpha_5 \mathrm{I}_{\boldsymbol{E}} \boldsymbol{E} + \alpha_2 \boldsymbol{H}\boldsymbol{H}^T + \alpha_6 \boldsymbol{E}^2 + \ldots, \end{aligned} \tag{3.4}$$

where we have assumed that the reference configuration is a natural state as before, so that

$$\alpha_{00} + \alpha_{10} + \alpha_{-10} = 0, \tag{3.5}$$

and where

$$\begin{aligned} \alpha_1 &= \alpha_{01} + \alpha_{11} + \alpha_{-11}, & \alpha_2 &= \alpha_{10} - \alpha_{-10}, \\ \alpha_3 &= \alpha_{02} + \alpha_{12} + \alpha_{-12}, & \alpha_4 &= \alpha_{03} + \alpha_{13} + \alpha_{-13}, \\ \alpha_5 &= 2(\alpha_{11} - \alpha_{-11}), & \alpha_6 &= 4\alpha_{-10}. \end{aligned} \tag{3.6}$$

From (2.6), the first-order terms in the expansion are equal to $\mathbf{L}(\boldsymbol{E})/\mu$. Hence if we interpret μ as the classical shear modulus, as we shall do, then

$$\lambda = \alpha_1 \mu, \qquad \alpha_2 = 1. \tag{3.7}$$

The remaining four independent coefficients in (3.4), namely α_3, α_4, α_5, α_6, are called the *second-order elasticities* of the isotropic elastic material.

From (III. 8A.5) and (3.4), the expansion for the Piola-Kirchhoff stress tensor $\boldsymbol{T}_\kappa$ to within terms of the third order in $\boldsymbol{H}$ is given by

$$\begin{aligned} \boldsymbol{T}_\kappa/\mu &= (1+\operatorname{tr}\boldsymbol{H}+\ldots)(\boldsymbol{T}/\mu)(\mathbf{1}-\boldsymbol{H}^T+\ldots) \\ &= \alpha_1 \mathrm{I}_{\boldsymbol{E}}\mathbf{1}+2\boldsymbol{E}+\left[\frac{\alpha_1}{2}(\mathrm{I}_{\boldsymbol{H}\boldsymbol{H}^T}+2\mathrm{I}_{\boldsymbol{E}}^2)+\alpha_3\mathrm{I}_{\boldsymbol{E}}^2+\alpha_4\mathrm{II}_{\boldsymbol{E}}\right]\mathbf{1} \\ &\quad +(\alpha_5+2)\mathrm{I}_{\boldsymbol{E}}\boldsymbol{E}-\alpha_1\mathrm{I}_{\boldsymbol{E}}\boldsymbol{H}^T-(\boldsymbol{H}^T)^2+\alpha_6\boldsymbol{E}^2+\ldots, \end{aligned} \tag{3.8}$$

which shows clearly that while $\boldsymbol{T}$ and $\boldsymbol{T}_\kappa$ coincide in the linearized theory they are distinct in the second-order theory.

Turning now to the problem of torsion and extension of a cylindrical tube, we consider the deformation of a cylinder obtained by superimposing a twist of amount ε upon an extension of amount δ accompanied by a lateral contraction of amount $\xi\delta$. So as to obtain a family of deformations depending upon a single parameter, we assume that δ is given by a power series in ε:

$$\delta = \delta_1\varepsilon+\delta_2\varepsilon^2+\ldots. \tag{3.9}$$

We assume also that the body force vanishes and that the mantle of the cylinder is free of surface traction.

It is well known that the classical linearized theory requires the contraction ratio ξ be equal to the Poisson modulus:

$$\xi = \frac{\lambda}{2(\lambda+\mu)} = \frac{\alpha_1}{2(\alpha_1+1)}. \tag{3.10}$$

Then the displacement $\boldsymbol{u}$ given by the following component form is a first-order solution:

$$\begin{aligned} u \equiv u_1 &= (1-\xi\delta)\{X\cos[(1+\delta)\varepsilon Z]-Y\sin[(1+\delta)\varepsilon Z]\}-X \\ &= -(\xi\delta X+\varepsilon ZY)-[\tfrac{1}{2}\varepsilon^2 XZ^2+(1-\xi)\delta\varepsilon YZ]+\ldots \\ v \equiv u_2 &= (1-\xi\delta)\{X\sin[(1+\delta)\varepsilon Z]+Y\cos[(1+\delta)\varepsilon Z]\}-Y \\ &= -(\xi\delta Y-\varepsilon ZX)-[\tfrac{1}{2}\varepsilon^2 YZ^2-(1-\xi)\delta\varepsilon XZ]+\ldots \\ w \equiv u_3 &= \delta Z. \end{aligned} \tag{3.11}$$

Here X, Y, Z are the Cartesian co-ordinates of $\boldsymbol{X}$ as usual. Since we have included a non-vanishing $\bar{\boldsymbol{u}}_2$ in the first-order solution, (3.11) is the component form of a particular case of (2.50) with $p = 2$. As we remarked before, $\bar{\boldsymbol{u}}_2$ is not determined by the linearized theory, but by including it in the first-order solution we can lighten the problem of solving the linear system in step ii) of the iterative procedure developed in the last section, as we now see.

For simplicity, from this point on in this section the sign "=" is to be interpreted as equality to within an error of order ε^3.

We compute first the component matrix of the gradient $\boldsymbol{H}$ of the first-order solution (3.11):

$$[\boldsymbol{H}] = -\begin{bmatrix} \xi\delta & \varepsilon Z & \varepsilon Y \\ -\varepsilon Z & \xi\delta & -\varepsilon X \\ 0 & 0 & 0 \end{bmatrix}$$

$$-\begin{bmatrix} \tfrac{1}{2}\varepsilon^2 Z^2 & (1-\xi)\delta\varepsilon Z & \varepsilon^2 XZ+(1-\xi)\delta\varepsilon Y \\ -(1-\xi)\delta\varepsilon Z & \tfrac{1}{2}\varepsilon^2 Z^2 & \varepsilon^2 YZ-(1-\xi)\delta\varepsilon X \\ 0 & 0 & 0 \end{bmatrix}. \qquad (3.12)$$

Substituting (3.12) into (3.8) gives the following expansion for the Piola-Kirchhoff stress $\boldsymbol{T}_\kappa$:

$$\boldsymbol{T}_\kappa/\mu = 2(\xi\delta\mathbf{1}+\boldsymbol{E}_1)+\zeta\mathbf{1}+\delta(1-2\xi)(\alpha_5+2)\boldsymbol{E}_1+\boldsymbol{K}+\alpha_6\boldsymbol{E}_1^2, \qquad (3.13)$$

where

$$\zeta = \delta^2\left[\frac{\xi(1+2\xi^2)}{1-2\xi}+2\xi(1-2\xi)+\alpha_3(1-2\xi)^2-\alpha_4\xi(2-\xi)\right] + \left(\frac{\xi}{1-2\xi}-\frac{\alpha_4}{4}\right)\varepsilon^2(X^2+Y^2), \qquad (3.14)$$

and where the matrices of components of $\boldsymbol{E}_1$ and $\boldsymbol{K}$ are given by

$$[\boldsymbol{E}_1] = \begin{bmatrix} -\xi\delta & 0 & -\tfrac{1}{2}\varepsilon Y \\ 0 & -\xi\delta & \tfrac{1}{2}\varepsilon X \\ -\tfrac{1}{2}\varepsilon Y & \tfrac{1}{2}\varepsilon X & \delta \end{bmatrix},$$

$$[\boldsymbol{K}] = \begin{bmatrix} \xi^2\delta^2 & 0 & -\varepsilon^2 XZ-(1-\xi)\delta\varepsilon Y \\ 0 & \xi^2\delta^2 & -\varepsilon^2 YZ-(1-\xi)\delta\varepsilon X \\ 2\xi\delta\varepsilon Y & -2\xi\delta\varepsilon X & -(2\xi+1)\delta^2 \end{bmatrix}. \qquad (3.15)$$

From (3.13) we see that the load $(-\varepsilon^2 \boldsymbol{t}_{\kappa 2}, -\varepsilon^2 \tilde{\boldsymbol{b}}_2)$ required to maintain the displacement (3.11) in equilibrium is given by

$$
\begin{aligned}
\varepsilon^2\tilde{\boldsymbol{b}}_2 &= \mu[\operatorname{Grad}\zeta+\delta(1-2\xi)(\alpha_5+2)\operatorname{Div}\boldsymbol{E}_1+\operatorname{Div}\boldsymbol{K}+\alpha_6\operatorname{Div}(\boldsymbol{E}_1{}^2)],\\
&= -\varepsilon^2\mu\left(\frac{1-4\xi}{1-2\xi}+\frac{\alpha_4}{2}+\frac{\alpha_6}{4}\right)R\boldsymbol{e}_R, \qquad (3.16)\\
\varepsilon^2\tilde{\boldsymbol{t}}_{\mathrm{I}} &= -\mu\left[\left(\frac{\xi}{1-2\xi}-\frac{\alpha_4}{2}\right)\varepsilon^2R_{\mathrm{I}}^2+\gamma\delta^2\right]\boldsymbol{e}_R,\\
\varepsilon^2\tilde{\boldsymbol{t}}_{\mathrm{II}} &= -\mu\left[\left(\frac{\xi}{1-2\xi}-\frac{\alpha_4}{2}\right)\varepsilon^2R_{\mathrm{II}}^2+\gamma\delta^2\right]\boldsymbol{e}_R,
\end{aligned}
$$

where $R \equiv \sqrt{X^2+Y^2}$; where R_{I} and R_{II} are the inner and the outer radii of the cylindrical tube in the reference configuration $\boldsymbol{\kappa}$; where $\boldsymbol{e}_{\mathrm{R}}$ is the unit vector in the radial direction; where $\tilde{\boldsymbol{t}}_{\mathrm{I}}$ and $\tilde{\boldsymbol{t}}_{\mathrm{II}}$ are the values of $\tilde{\boldsymbol{t}}_{\boldsymbol{\kappa}2}$ on the inner and the outer mantle $R = R_{\mathrm{I}}$ and $R = R_{\mathrm{II}}$; and where, finally,

$$
\gamma = \frac{\xi(1+\xi)}{1-2\xi}+\alpha_3(1-2\xi)^2-\alpha_4\,\xi(2-\xi)-\alpha_5\,\xi(1-2\xi)+\alpha_6\,\xi^2, \qquad (3.17)
$$

a constant.

Having completed step i) of the iterative procedure, we must now determine the solution $\varepsilon^2\tilde{\boldsymbol{u}}_2$ by solving the equations of equilibrium of the infinitesimal theory corresponding to the loads (3.16). At this point the advantage of having a non-vanishing second-order coefficient $\bar{\boldsymbol{u}}_2$ in the first-order solution (3.11) is apparent: The loads $\tilde{\boldsymbol{b}}_2, \tilde{\boldsymbol{t}}_{\mathrm{I}}, \tilde{\boldsymbol{t}}_{\mathrm{II}}$ induced by that first-order solution are radial. Hence, first, the loads form an equilibrated system, and, second, the displacement $\tilde{\boldsymbol{u}}_2$ may be assumed to be radial, *i.e.*,

$$
\tilde{\boldsymbol{u}}_2 = U(R)\boldsymbol{e}_R. \qquad (3.18)
$$

Had we used a first-order solution with no second-order coefficient, from (2.56) we should have had to solve a linear system whose solution would be the total second-order displacement $\boldsymbol{u}_2$, which is the sum of the radial displacement $\tilde{\boldsymbol{u}}_2$ and the tangential displacement $\bar{\boldsymbol{u}}_2$ included in (3.11).

After routine calculation, the differential equations of the infinitesimal theory for a displacement of the form (3.18) and body forces $(3.16)_2$ lead to the condition

$$
\begin{aligned}
U''+\frac{U'}{R}-\frac{U}{R^2} &= 8\eta R,\\
\eta &= \frac{1}{16(1-\xi)}\left[1-4\xi+\tfrac{1}{2}(\alpha_4+\tfrac{1}{2}\alpha_6)(1-2\xi)\right].
\end{aligned} \qquad (3.19)
$$

The general solution of (3.19) is

$$U(R) = \eta R^3 + aR + \frac{b}{R}, \tag{3.20}$$

where a and b are constants of integration. The particular solution satisfying the boundary conditions corresponding to $(3.16)_{3,4}$ is

$$U(R) = \eta R^3 - \beta\left[(R_{\mathrm{I}}^2 + R_{\mathrm{II}}^2)R + \frac{R_{\mathrm{I}}^2 R_{\mathrm{II}}^2}{(1-2\xi)R}\right] - \delta_1^2\left(\frac{1-2\xi}{2}\right)\gamma R, \tag{3.21}$$

where δ_1 is the leading coefficient in the expansion (3.9), and where β is a constant given by

$$\beta = \frac{1-2\xi}{16(1-\xi)}\left[3 - \tfrac{1}{2}(1-2\xi)\alpha_4 + \frac{3-2\xi}{4}\alpha_6\right]. \tag{3.22}$$

If the cylindrical tube is a solid rod, we set $R_{\mathrm{I}} = 0$; then the solution (3.21) becomes

$$U(R) = \eta R^3 - \left[\beta R_{\mathrm{II}}^2 + \delta_1^2\left(\frac{1-2\xi}{2}\right)\gamma\right] R. \tag{3.23}$$

Now by (2.56) the complete second-order solution has the components

$$\begin{aligned}
u^{(2)} &= -[\xi\delta_1 X + ZY]\varepsilon - \left[\tfrac{1}{2}XZ^2 + (1-\xi)\delta_1 YZ - \frac{X}{R}U(R)\right]\varepsilon^2,\\
v^{(2)} &= -[\xi\delta_1 Y - ZX]\varepsilon - \left[\tfrac{1}{2}YZ^2 - (1-\xi)\delta_1 XZ - \frac{Y}{R}U(R)\right]\varepsilon^2,\\
w^{(2)} &= \delta_1 Z\varepsilon + \delta_2 Z\varepsilon^2,
\end{aligned} \tag{3.24}$$

where $U(R)$ is given by (3.21) or (3.23). Since we shall stop the iterative procedure at the second order, we give here only the components of the partial sum $\boldsymbol{u}^{(2)}$ of the second-order solution.

Having found the second-order solution (3.24), after routine calculation we can determine the surface tractions that must be applied at the plane ends of the tube or the rod in order to maintain the deformation according to the second-order theory. These tractions consist in an azimuthal shear stress of amount

$$t_\Theta = \mu\varepsilon R\{1 + \delta_1[(2-3\xi) + \tfrac{1}{2}\alpha_5(1-2\xi) = \tfrac{1}{2}\alpha_6(1-\xi)]\}, \tag{3.25}$$

and a normal traction of amount

$$t_Z = \mu\left\{2(1+\xi)\delta_1\,\varepsilon + \left[\beta' R^2 - \frac{4\xi\beta}{1-2\xi}(R_{\rm I}^2+R_{\rm II}^2)+\delta_1^2\gamma' + 2(1+\xi)\delta_2\right]\varepsilon^2\right\}, \tag{3.26}$$

where

$$\begin{aligned}\beta' &= \frac{1}{\delta(1-\xi)}[12\xi+2(2\xi-1)\alpha_4+(2-\xi)\alpha_6],\\ \gamma' &= (\xi+1)(1-4\xi)+\alpha_3(1-2\xi)^3+\alpha_4\,\xi(2-\xi)(2\xi-1)\\ &\qquad +\alpha_5(1+2\xi^2)(1-2\xi)+\alpha_6(1-2\xi^3).\end{aligned} \tag{3.27}$$

These tractions are taken per unit area in the reference configuration $\boldsymbol{\kappa}$.

The total torque τ necessary to produce the twist is obtained from (3.25):

$$\tau = \frac{\pi}{2}\mu\varepsilon(R_{\rm II}^4-R_{\rm I}^4)\{1+\delta_1[(2-3\xi)+\tfrac{1}{2}\alpha_5(1-2\xi)+\tfrac{1}{2}\alpha_6(1-\xi)]\}. \tag{3.28}$$

Likewise, the total longitudinal tension N needed to maintain the extension can be calculated from (3.26):

$$N = \pi\mu(R_{\rm II}^2-R_{\rm I}^2)\{2\delta_1(1+\xi)\varepsilon+[\beta''(R_{\rm I}^2+R_{\rm II}^2)+\delta_1^2\gamma'+2\delta_2(1+\xi)]\varepsilon^2\}, \tag{3.29}$$

where

$$\beta'' = \tfrac{1}{8}[-(1-2\xi)\alpha_4+(1-\xi)\alpha_6]. \tag{3.30}$$

If no longitudinal tension is supplied, $N = 0$, so that by (3.29)

$$\begin{aligned}\delta_1 &= 0,\\ \delta_2 &= -\frac{\beta''}{2(1+\xi)}(R_{\rm I}^2+R_{\rm II}^2).\end{aligned} \tag{3.31}$$

If (3.31) holds, then (3.28), to within terms of order ε^3, reduces to

$$\tau = \frac{\pi}{2}\mu\varepsilon(R_{\rm II}^4-R_{\rm I}^4), \tag{3.32}$$

which is identical with the formula for the torque given by the infinitesimal theory. Hence to find deviation from this classical formula, a theory taking into account terms of order ε^3 is needed. An extension of amount

$$\delta = \delta_2\varepsilon^2, \tag{3.33}$$

which is neglected by the infinitesimal theory, is predicted by the second-order theory. From (3.30), the tube or the rod when twisted will lengthen, shorten, or neither according as $\beta'' < 0$, $\beta'' > 0$, or $\beta'' = 0$. Further, the constant β'' depends only on the second-order elasticities α_4 and α_6 and the Poisson modulus ξ.

Although POYNTING, who was the first to report experimental measurement of this effect (1905–1913), concluded that wires of steel, copper, brass, and rubber lengthened when twisted, recent experiments do not lead to any definite conclusion of this kind except for rubber, which does indeed lengthen.

From (3.33) and (3.24) we can calculate the change of volume, per unit original volume, due to a twist:

$$\frac{\Delta V}{V} = \pi\beta'''(R_{\mathrm{II}}^2 + R_{\mathrm{I}}^2)\varepsilon^2,$$

$$\beta''' = -\tfrac{1}{4} + \frac{1}{16(1+\xi)}\left[(3-4\xi-4\xi^2)\alpha_4 + (2\xi^2+2\xi-2)\alpha_6\right]. \tag{3.34}$$

Thus the volume increases, decreases, or neither according as $\beta''' > 0$, $\beta''' < 0$, or $\beta''' = 0$. The constant β''' depends again on α_4, α_6, and ξ.

Also the problem of simultaneous extension and torsion of a prism of arbitrary cross section was studied by RIVLIN, using his iterative procedure. For more detail, other examples, and discussion of the historical development of the second-order theory of elasticity we refer the reader to NFTM, §§ 66, 67, 77, and 93.

4. Stoppelli's Theorems of the Existence, Uniqueness, and Analyticity of the Solution of a Class of Traction Boundary-value Problems[1]

A theorem of existence and uniqueness for certain boundary-value problems of equilibrium was proved by STOPPELLI (1954). He considered a family of problems depending upon a parameter ε; the loads are assumed to be of the form

$$\boldsymbol{b} = \varepsilon \boldsymbol{b}_1, \quad \boldsymbol{t}_\kappa = \varepsilon \boldsymbol{t}_{\kappa 1}, \tag{4.1}$$

where $\boldsymbol{b}_1$ and $\boldsymbol{t}_{\kappa 1}$ are prescribed functions of $\boldsymbol{X}$ for $\boldsymbol{X} \in \kappa(\mathscr{B})$ and $\boldsymbol{X} \in \partial\kappa(\mathscr{B})$,

[1] *Cf.* NFTM, § 46.

respectively, relative to a given reference configuration $\boldsymbol{\kappa}$. Clearly (4.1) is a special case of (2.8). As we remarked in § 1, the condition (1.19) is necessary for the existence of a solution. Following SIGNORINI, we assume also the condition (2.20). When (4.1) holds, these two conditions are equivalent to

$$\begin{aligned} \int_{\partial\boldsymbol{\kappa}(\mathscr{B})} \boldsymbol{t}_{\kappa 1}\, dA_\kappa + \int_{\boldsymbol{\kappa}(\mathscr{B})} \rho_\kappa \boldsymbol{b}_1\, dV_\kappa &= \mathbf{0}, \\ \int_{\partial\boldsymbol{\kappa}(\mathscr{B})} \boldsymbol{r}_\kappa \wedge \boldsymbol{t}_{\kappa 1}\, dA_\kappa + \int_{\boldsymbol{\kappa}(\mathscr{B})} \rho_\kappa \boldsymbol{r}_\kappa \wedge \boldsymbol{b}_1\, dV_\kappa &= \mathbf{0}. \end{aligned} \tag{4.2}$$

The idea is to reduce the problem of existence and uniqueness in the general theory, provided that $|\varepsilon|$ be sufficiently small, to that for the theory obtained by linearization. This kind of reduction is commonly used to treat non-linear systems; we shall now summarize it in general.

An infinite-dimensional topological vector space $\mathscr{S}$ is called a *potential Banach space* if its topology can be defined by a norm[1] and if the metric[1] induced by the norm is complete. If a particular norm, denoted by $\|\ \|$, has been selected for $\mathscr{S}$, then $\mathscr{S}$ is called a *Banach space.*

Let $\mathscr{S}$ and $\mathscr{S}'$ be Banach spaces. Then $\mathscr{L}(\mathscr{S};\mathscr{S}')$ denotes the potential Banach space of continuous[2] linear maps from $\mathscr{S}$ to $\mathscr{S}'$. We may use the norm

$$\|\boldsymbol{A}\| \equiv \inf_{\mathbf{0}\neq \boldsymbol{u}\in\mathscr{S}} \left(\frac{\|\boldsymbol{A}\boldsymbol{u}\|}{\|\boldsymbol{u}\|}\right) \tag{4.3}$$

for any $\boldsymbol{A}\in\mathscr{L}(\mathscr{S};\mathscr{S}')$. Let $\mathscr{D}$ be an open set of $\mathscr{S}$, and let $\boldsymbol{\Theta}:\mathscr{D}\to\mathscr{S}'$. Then $\boldsymbol{\Theta}$ is said to be *differentiable* at a point $\boldsymbol{u}\in\mathscr{D}$ if there exists a continuous linear map $\nabla_{\boldsymbol{u}}\boldsymbol{\Theta}\in\mathscr{L}(\mathscr{S};\mathscr{S}')$ such that

$$\boldsymbol{\Theta}(\boldsymbol{u}+\boldsymbol{v}) = \boldsymbol{\Theta}(\boldsymbol{u}) + [\nabla_{\boldsymbol{u}}\boldsymbol{\Theta}](\boldsymbol{v}) + \boldsymbol{o}(\|\boldsymbol{v}\|) \tag{4.4}$$

for all $\boldsymbol{v}\to\mathbf{0}$ in $\mathscr{S}$. Clearly, if $\nabla_{\boldsymbol{u}}\boldsymbol{\Theta}$ exists then it is unique; further, the operation ∇ here generalizes the gradient operation, which was denoted by the same symbol, from finite-dimensional spaces to Banach spaces. The linear map $\nabla_{\boldsymbol{u}}\boldsymbol{\Theta}$ is often called the *Fréchet derivative* of $\boldsymbol{\Theta}$ at $\boldsymbol{u}$.

---

[1] For the definition of these terms, *cf.* §§ I.1–3.

[2] For mappings of infinite-dimensional spaces, linearity does not imply continuity.

If $\boldsymbol{\Theta}$ is differentiable at each $\boldsymbol{u} \in \mathscr{D}$, and if the mapping

$$\nabla\boldsymbol{\Theta} : \mathscr{D} \to \mathscr{L}(\mathscr{S}; \mathscr{S}') \tag{4.5}$$

defined by

$$(\nabla\boldsymbol{\Theta})(\boldsymbol{u}) \equiv \nabla_{\boldsymbol{u}}\boldsymbol{\Theta} \quad \forall \boldsymbol{u} \in \mathscr{D} \tag{4.6}$$

is continuous, then $\boldsymbol{\Theta}$ is said to be of *class* C^1; in symbols: $\boldsymbol{\Theta} \in \mathscr{C}^1(\mathscr{D};\mathscr{S}')$. More generally, if $\nabla^k\boldsymbol{\Theta}$, $\mathscr{L}^k(\mathscr{S}; \mathscr{S}')$, and $\mathscr{C}^k(\mathscr{D}; \mathscr{S}')$ have been defined for some $k \geqq 1(\nabla^1\boldsymbol{\Theta} \equiv \nabla\boldsymbol{\Theta}, \mathscr{L}^1(\mathscr{S}; \mathscr{S}') \equiv \mathscr{L}(\mathscr{S}; \mathscr{S}')$, of course), then we define

$$\begin{aligned} \nabla^{k+1}\boldsymbol{\Theta} &\equiv \nabla(\nabla^k\boldsymbol{\Theta}), \\ \mathscr{L}^{k+1}(\mathscr{S}; \mathscr{S}') &\equiv \mathscr{L}(\mathscr{S}; \mathscr{L}^k(\mathscr{S}; \mathscr{S}')), \end{aligned} \tag{4.7}$$

and we say that $\boldsymbol{\Theta} \in \mathscr{C}^{k+1}(\mathscr{D}; \mathscr{S}')$ if the mapping

$$\nabla^{k+1}\boldsymbol{\Theta} : \mathscr{D} \to \mathscr{L}^{k+1}(\mathscr{S}; \mathscr{S}') \tag{4.8}$$

is continuous. We denote the set of continuous mappings from $\mathscr{D}$ to $\mathscr{S}'$ by $\mathscr{C}^0(\mathscr{D}; \mathscr{S}')$ and the set of mappings $\boldsymbol{\Theta}$ such that $\boldsymbol{\Theta} \in \mathscr{C}^k(\mathscr{D}; \mathscr{S}')$ for all $k = 1, 2, \ldots$, by $\mathscr{C}^\infty(\mathscr{D}; \mathscr{S}')$.

Most of the rules of differential calculus can be generalized from Euclidean space to Banach space. For example, if $\boldsymbol{\Theta} \in \mathscr{C}^k(\mathscr{D}; \mathscr{S}')$, then the kth-order Fréchet derivative $\nabla^k_{\boldsymbol{u}}\boldsymbol{\Theta}$ is necessarily a symmetric k-linear map. There are also chain rules, product rules, the Taylor formula, the mean value theorem, *etc.*, for mappings of Banach spaces[1].

The central result needed for the reduction of a non-linear system to a linear one is the following INVERSE MAPPING THEOREM[2]. *Let $\mathscr{S}$ and $\mathscr{S}'$ be Banach spaces, let $\mathscr{D}$ be an open subset of $\mathscr{S}$, and let $\boldsymbol{\Theta} : \mathscr{D} \to \mathscr{S}'$ be a C^k-differentiable map with $k \geqq 1$. Assume that for some point $\boldsymbol{u}_0 \in \mathscr{D}$ the Fréchet derivative $\nabla_{\boldsymbol{u}_0}\boldsymbol{\Theta}$ is an isomorphism of $\mathscr{S}$ with $\mathscr{S}'$. Then $\boldsymbol{\Theta}$ is a local C^k-diffeomorphism near $\boldsymbol{u}_0$.*

---

[1] For details the reader may consult, *e.g.*, J. DIEUDONNÉ, *Foundations of Modern Analysis*, New York, Academic Press 1960; S. LANG *Introduction to Differentiable Manifolds*, New York, Interscience, 1962; R. ABRAHAM & J. ROBBIN, *Transversal Mappings and Flows*, New York and Amsterdam, Benjamin 1967; M. M. VAINBERG, *Variational Methods for the Study of Nonlinear Operators*, San Francisco, London, and Amsterdam, Holden Day (1964).

[2] *Cf.* § 5 of S. LANG, *op. cit.*, and § VIII. 2 of DIEUDONNÉ, *op. cit.*

A *local C^k-diffeomorphism* near $\boldsymbol{u}_0$ is a map $\boldsymbol{\Theta}$ whose restriction to some open neighborhood $\mathcal{N}$ of $\boldsymbol{u}_0$ in $\mathcal{D}$ establishes a C^k-diffeomorphism between $\mathcal{N}$ and an open neighborhood of $\boldsymbol{\Theta}(\boldsymbol{u}_0)$ in $\mathcal{S}'$.

In order to apply the preceding theorem to a non-linear system of differential equations and boundary conditions, we set up the system first as a non-linear mapping $\boldsymbol{\Theta} : \mathcal{D} \to \mathcal{S}'$ relative to a "space of trial solutions" $\mathcal{S} \supset \mathcal{D}$ and a "space of induced loads" $\mathcal{S}'$; that is, we interpret the non-linear system as

$$\boldsymbol{\Theta}(\boldsymbol{u}) = \boldsymbol{f}, \quad \boldsymbol{u} \in \mathcal{D},\ \boldsymbol{f} \in \mathcal{S}'. \tag{4.9}$$

We assume that a particular solution $\boldsymbol{u}_0$ corresponding to a particular load $\boldsymbol{f}_0$ has been found:

$$\boldsymbol{\Theta}(\boldsymbol{u}_0) = \boldsymbol{f}_0. \tag{4.10}$$

Then we show that the linearization of (4.9) about the particular solution $\boldsymbol{u}_0$ is given by the linear map

$$\nabla_{\boldsymbol{u}_0} \boldsymbol{\Theta} : \mathcal{S} \to \mathcal{S}', \tag{4.11}$$

or, equivalently, by the linear system

$$[\nabla_{\boldsymbol{u}_0} \boldsymbol{\Theta}](\delta \boldsymbol{u}) = \delta \boldsymbol{f}, \quad \delta \boldsymbol{u} \in \mathcal{S},\ \delta \boldsymbol{f} \in \mathcal{S}'. \tag{4.12}$$

Now suppose that the linear system (4.12) be known to have a unique solution. This means that the Fréchet derivative $\nabla_{\boldsymbol{u}_0} \boldsymbol{\Theta}$ is an isomorphism. By applying the inverse mapping theorem, we may assert that the non-linear map $\boldsymbol{\Theta}$ is a local diffeomorphism near $\boldsymbol{u}_0$; equivalently, that the non-linear system (4.9) has a unique solution $\boldsymbol{u}$ near $\boldsymbol{u}_0$ provided the load $\boldsymbol{f}$ be close enough to the load $\boldsymbol{f}_0$. A theorem of existence and uniqueness of the solution of the non-linear system (4.9) is then established for small but finite departure from the particular solution $\boldsymbol{u}_0$.

Before we may apply the preceding method to traction boundary-value problems in non-linear elasticity, we must meet an objection: In the classical infinitesimal theory, the solution of the linear elastic traction boundary-value problem is unique only to within an arbitrary infinitesimal rigid displacement. In the non-linear theory, however, there is no reason why an arbitrary rotation should remain in the solution. Following STOPPELLI, we shall now resolve the difference by a technique that is essentially the same as that of SIGNORINI, presented earlier in § 2.

Let the desired solution of the traction boundary-value problem corresponding to the load (4.1) be denoted by $\boldsymbol{u}^*$. Since we expect the solution to be unique to within an arbitrary rigid translation only, by requiring that

$$\boldsymbol{u}^*(\boldsymbol{0}; \varepsilon) = \boldsymbol{0} \tag{4.13}$$

relative to a particular point whose position is chosen as the origin in the reference configuration $\boldsymbol{\kappa}(\mathscr{B})$, we expect to find a unique one-parameter family of solutions $\boldsymbol{u}^* = \boldsymbol{u}^*(X; \varepsilon)$ satisfying the system

$$\begin{aligned} \operatorname{Div} \boldsymbol{A}(\boldsymbol{F}^*) + \rho_\kappa \varepsilon \boldsymbol{b}_1 &= \boldsymbol{0}, \\ \boldsymbol{A}(\boldsymbol{F}^*)\boldsymbol{n}_\kappa &= \varepsilon \boldsymbol{t}_{\kappa 1}. \end{aligned} \tag{4.14}$$

If we superimpose a rigid rotation $\boldsymbol{Q}$ upon $\boldsymbol{u}^*$, we obtain a new displacement $\boldsymbol{u}$ satisfying the system

$$\begin{aligned} \operatorname{Div} \boldsymbol{A}(\boldsymbol{F}) + \rho_\kappa \varepsilon \boldsymbol{Q} \boldsymbol{b}_1 &= \boldsymbol{0}, \\ \boldsymbol{A}(\boldsymbol{F})\boldsymbol{n}_\kappa &= \varepsilon \boldsymbol{Q} \boldsymbol{t}_{\kappa 1}. \end{aligned} \tag{4.15}$$

Conversely, if $\boldsymbol{u}$ is a solution of (4.15), then the $\boldsymbol{u}^*$ obtained from $\boldsymbol{u}$ by the rigid rotation $\boldsymbol{Q}^T$ is a solution of (4.14). So as to make $\boldsymbol{Q}$, and hence $\boldsymbol{u}$, unique, we require that the gradient $\boldsymbol{H}(\boldsymbol{0}, \varepsilon)$ of $\boldsymbol{u}$ at $\boldsymbol{0}$ be a pure stretch. This means that $\boldsymbol{Q}^T$ is the local rotation of $\boldsymbol{F}^*(\boldsymbol{0}, \varepsilon)$.

Now so as to solve (4.15), we choose an arbitrary $\boldsymbol{u}$ and introduce the additional load which would be required to equilibrate it:

$$\begin{aligned} \boldsymbol{h} &\equiv \operatorname{Div} \boldsymbol{A}(\boldsymbol{F}) + \rho_\kappa \varepsilon \boldsymbol{Q} \boldsymbol{b}_1, \\ \boldsymbol{g} &\equiv -\boldsymbol{A}(\boldsymbol{F})\boldsymbol{n}_\kappa + \varepsilon \boldsymbol{Q} \boldsymbol{t}_{\kappa 1}. \end{aligned} \tag{4.16}$$

Then the system (4.15) would be solved by $\boldsymbol{u}$ if the load $(\boldsymbol{g}, \boldsymbol{h})$ turned out to be $(\boldsymbol{0}, \boldsymbol{0})$. Of course the load $(\boldsymbol{0}, \boldsymbol{0})$ forms an equilibrated system. Hence, in order that $\boldsymbol{u}$ might possibly be a solution of (4.15), it is necessary that $(\boldsymbol{g}, \boldsymbol{h})$ be equilibrated on $\boldsymbol{\kappa}(\mathscr{B})$ in the first place. We shall now use this condition to select a rotation $\boldsymbol{Q}$ for each trial displacement field $\boldsymbol{u}$.

Since the condition $(4.2)_1$ is invariant under any rotation $\boldsymbol{Q}$, from (4.16) we have

$$\int_{\partial\kappa(\mathscr{B})} \boldsymbol{g}\, dA_\kappa + \int_{\kappa(\mathscr{B})} \boldsymbol{h}\, dV_\kappa = \boldsymbol{0}. \tag{4.17}$$

The condition $(4.2)_2$, however, is generally not invariant under a rotation

$\boldsymbol{Q}$. A straightforward calculation using (4.16) and $(4.2)_2$ shows that the condition

$$\int_{\partial\kappa(\mathscr{B})} \boldsymbol{r}_\kappa \wedge \boldsymbol{g}\, dA_\kappa + \int_{\kappa(\mathscr{B})} \boldsymbol{r}_\kappa \wedge \boldsymbol{h}\, dV_\kappa = \boldsymbol{0} \tag{4.18}$$

is equivalent to the requirement that

$$\varepsilon(\boldsymbol{A}_1 \boldsymbol{Q}^T - \boldsymbol{Q}\boldsymbol{A}_1) = \boldsymbol{M} \equiv \int_{\kappa(\mathscr{B})} [A(F)^T - A(F)]\, dV_\kappa, \tag{4.19}$$

where $\boldsymbol{A}_1$ is the astatic load (*cf.* (2.22)) corresponding to $(\boldsymbol{t}_{\kappa 1}, \boldsymbol{b}_1)$. The condition $(4.2)_2$ ensures that $\boldsymbol{A}_1$ is a symmetric tensor, and the condition $(4.19)_2$ shows that $\boldsymbol{M}$ is a skew tensor.

Now for each trial solution $\boldsymbol{u}$, $(4.19)_2$ determines $\boldsymbol{M}$, and then $(4.19)_1$ becomes an equation to be solved for $\boldsymbol{Q}$. This equation has the form

$$\boldsymbol{A}_1 \boldsymbol{Q}^T - \boldsymbol{Q}\boldsymbol{A}_1 \equiv \boldsymbol{\Phi}(\boldsymbol{Q}) = \frac{1}{\varepsilon} \boldsymbol{M}, \tag{4.20}$$

in which $\boldsymbol{\Phi}$ is a mapping from the rotational group $\mathfrak{O}(3)$ to the space of skew tensors $\mathscr{W}$. Clearly, $\boldsymbol{\Phi}(\boldsymbol{1}) = \boldsymbol{0}$. By the classical inverse mapping theorem (*cf.* § I.3) $\boldsymbol{\Phi}$ is a local diffeomorphism near $\boldsymbol{1}$ if and only if the gradient of $\boldsymbol{\Phi}$ at $\boldsymbol{1}$ is an automorphism of $\mathscr{W}$. Here we have used the fact that the tangent space of $\mathfrak{O}(3)$ at $\boldsymbol{1}$ may be identified with $\mathscr{W}$ (*cf.* (I. 6.80)). Now from $(4.20)_1$ the gradient of $\boldsymbol{\Phi}$ at $\boldsymbol{1}$ is given by

$$[\nabla_{\boldsymbol{1}} \boldsymbol{\Phi}](\boldsymbol{W}) = -(\boldsymbol{A}_1 \boldsymbol{W} + \boldsymbol{W}\boldsymbol{A}_1) \quad \forall \boldsymbol{W} \in \mathscr{W}. \tag{4.21}$$

In view of SIGNORINI's analysis of $(2.40)_2$, we see that $\nabla_{\boldsymbol{1}} \boldsymbol{\Phi}$ is an isomorphism, or, equivalently, $\boldsymbol{\Phi}$ is a local diffeomorphism near $\boldsymbol{1}$, if and only if the condition (2.42) holds, *i.e.*, if and only if the load $(\boldsymbol{t}_{\kappa 1}, \boldsymbol{b}_1)$ does not possess an axis of equilibrium. Accordingly, so as to continue the procedure, we shall assume henceforth that the load $(\boldsymbol{t}_{\kappa 1}, \boldsymbol{b}_1)$ does not possess an axis of equilibrium.

Under this assumption we can solve (4.20) uniquely for $\boldsymbol{Q}$ provided that $\boldsymbol{M}/\varepsilon$ be sufficiently close to $\boldsymbol{1}$. This means that there exist positive numbers α and β, independent of ε, such that if

$$||\boldsymbol{M}|| < \alpha|\varepsilon|, \tag{4.22}$$

the equation (4.20) has a unique solution $\boldsymbol{Q}$ with the property

$$||\boldsymbol{Q} - \boldsymbol{1}|| < \beta. \tag{4.23}$$

Having found a way to choose a unique $\boldsymbol{Q}$ for each $\boldsymbol{u}$ that satisfies (4.22), we return to (4.16), which now defines a transformation from $\boldsymbol{u}$ to $(\boldsymbol{g}, \boldsymbol{h})$. As remarked before, we shall set up the transformation as a non-linear system of the form (4.9). To ensure the necessary degree of smoothness in order to apply the inverse mapping theorem, we assume the following conditions of regularity:

i) The body $\mathscr{B}$ is homogeneous, and the reference configuration $\boldsymbol{\kappa}$ is a homogeneous natural state of $\mathscr{B}$.

ii) The elasticities of $\mathscr{B}$ in $\boldsymbol{\kappa}$ are sufficient that a solution of the traction boundary-value problem in the classical infinitesimal theory[1] exist and be unique to within an infinitesimal rotation.

iii) The region $\boldsymbol{\kappa}(\mathscr{B})$ occupied by $\mathscr{B}$ in the reference configuration is bounded.

iv) The boundary $\partial\boldsymbol{\kappa}(\mathscr{B})$ has Hölder-continuous[2] principal curvatures with exponent λ, and it is possible to cover $\partial\boldsymbol{\kappa}(\mathscr{B})$ by a finite number of regions $\mathscr{R}^{(p)}$, each of which can be represented parametrically on a disc $\mathscr{C}^{(p)}$, and every point of $\partial\boldsymbol{\kappa}(\mathscr{B})$ is interior to at least one $\mathscr{R}^{(p)}$.

v) $\boldsymbol{b}_1$ is Hölder-continuous in the closure $\overline{\boldsymbol{\kappa}(\mathscr{B})}$ of $\boldsymbol{\kappa}(\mathscr{B})$ with exponent λ; $\boldsymbol{t}_{\kappa 1}$ and its derivatives with respect to the parameters of some parametric representation of $\partial\boldsymbol{\kappa}(\mathscr{B})$(as in iv) above) are Hölder-continuous with exponent λ.

vi) The response function $\boldsymbol{A}(\boldsymbol{F})$ and its derivatives up to the third order are continuous in the region of the 9-dimensional space of non-singular second-order tensors $\boldsymbol{F}$ satisfying the inequality $\|\boldsymbol{F}-\mathbf{1}\| \leqq \gamma$, where γ is a given positive number.

Now let $\mathscr{S}$ be the class of all vector-valued functions $\boldsymbol{u}$ on $\overline{\boldsymbol{\kappa}(\mathscr{B})}$ that have Hölder-continuous second derivatives, with exponent λ, on $\overline{\boldsymbol{\kappa}(\mathscr{B})}$, and that satisfy the conditions

$$\boldsymbol{u}(\mathbf{0}) = \mathbf{0}, \quad \boldsymbol{H}(\mathbf{0}) = \boldsymbol{H}(\mathbf{0})^T. \tag{4.24}$$

Using a single Cartesian co-ordinate system, we define the norm of $\boldsymbol{u}$ thus:

$$\|\boldsymbol{u}\| \equiv \sum_{k=1}^{3} \underset{X\in\overline{\boldsymbol{\kappa}(\mathscr{B})}}{\operatorname{Max}} |u^k(X)| + \sum_{k,A=1}^{3} \underset{X\in\overline{\boldsymbol{\kappa}(\mathscr{B})}}{\operatorname{Max}} |u^k,_A(X)| + \sum_{k,A,B=1}^{3} [\underset{X\in\overline{\boldsymbol{\kappa}(\mathscr{B})}}{\operatorname{Max}} |u^k,_{AB}(X)| + B^k_{AB}], \tag{4.25}$$

[1] *Cf.* footnote 1 on p. 483.

[2] In a space with metric d, the function F is Hölder-continuous with exponent λ in a domain $\mathscr{D}$ if the quotient $|F(x)-F(x')|/|d(x, x')|^\lambda$ is bounded as x and x' vary over $\mathscr{D}$. The least upper bound of the quotient is the *Hölder coefficient* of F in the domain $\mathscr{D}$.

where the comma denotes partial differentiation with respect to (X^1, X^2, X^3), and where B^k_{AB} is the Hölder coefficient of $u^k,_{AB}$ in $\overline{\kappa(\mathscr{B})}$. With (4.25) as a norm it can be shown that the class $\mathscr{S}$ becomes a Banach space. We shall regard $\mathscr{S}$ as the "space of trial solutions", denoted earlier by the same symbol, for the system (4.9).

Next, let $\mathscr{S}'$ be the class of all pairs $(\boldsymbol{g}, \boldsymbol{h})$ of vector-valued functions with the following properties:

a) $\boldsymbol{h}$ is defined on $\overline{\kappa(\mathscr{B})}$ and is Hölder-continuous with exponent λ in $\overline{\kappa(\mathscr{B})}$.

b) $\boldsymbol{g}$ is defined on the boundary surface $\partial\kappa(\mathscr{B})$; $\boldsymbol{g}$ and its derivatives $g^k,_\Gamma$ with respect to the parameters y^Γ, $\Gamma = 1, 2$, of some parametric representation of $\partial\kappa(\mathscr{B})$(as in iv) above) are Hölder-continuous with exponent λ.

c) Equations (4.17) and (4.18) are satisfied.

Using the same Cartesian co-ordinates as before, we define the norm of $(\boldsymbol{g}, \boldsymbol{h})$ thus:

$$||(\boldsymbol{g}, \boldsymbol{h})|| \equiv \sum_{k=1}^{3} [\underset{\boldsymbol{X}\in\overline{\kappa(\mathscr{B})}}{\text{Max}} |h^k(\boldsymbol{X})| + \underset{\boldsymbol{X}\in\partial\kappa(\mathscr{B})}{\text{Max}} |g^k(\boldsymbol{X})| + C^k + D^k], \tag{4.26}$$

where C^k is the Hölder coefficient of h^k, and where D^k is the greatest of the Hölder coefficients of the derivatives $g^k,_\Gamma$ in the various discs $\mathscr{C}^{(p)}$ (as in iv) above). It can be verified that $\mathscr{S}'$, with (4.26) as a norm, is also a Banach space. We shall regard $\mathscr{S}'$ as the "space of induced loads", denoted earlier by the same symbol, for the system (4.9).

Now we set up the non-linear map $\boldsymbol{\Theta}$ corresponding to (4.16). First, each member $\boldsymbol{u} \in \mathscr{S}$ can be regarded as a possible displacement of $\mathscr{B}$ relative to $\boldsymbol{\kappa}$. By means of $(4.19)_2$, $\boldsymbol{u}$ gives rise to a skew tensor $\boldsymbol{M}$. From the conditions of regularity iii) and vi) STOPPELLI shows that there exists a positive number η, independent of $\boldsymbol{u}$, such that

$$||\boldsymbol{M}|| < \eta ||\boldsymbol{u}||^2 \tag{4.27}$$

provided that $||\boldsymbol{u}|| < \gamma$ (as in vi) above). Hence (4.22) is satisfied if

$$||\boldsymbol{u}|| < \sqrt{\frac{\alpha}{\eta}}\,|\varepsilon|, \qquad |\varepsilon| < \frac{\eta}{\alpha}\gamma^2. \tag{4.28}$$

It follows that if (4.28) holds, there is a unique rotation $\boldsymbol{Q}$ that has the property (4.23) and solves the equation (4.20). With this choice of $\boldsymbol{Q}$, the load $(\boldsymbol{g}, \boldsymbol{h})$ determined by (4.16) is equilibrated; in view of the conditions of regularity, it defines a point in the Banach space $\mathscr{S}'$. In this manner (4.16) establishes a mapping $\boldsymbol{\Theta}$ of a neighborhood $\mathscr{D}$ of $\boldsymbol{0} \in \mathscr{S}$ into the space $\mathscr{S}'$:

$$\boldsymbol{\Theta}(\boldsymbol{u}) \equiv (\boldsymbol{g}, \boldsymbol{h}), \ \boldsymbol{u} \in \mathscr{D}, \ (\boldsymbol{g}, \boldsymbol{h}) \in \mathscr{S}'. \tag{4.29}$$

The open set $\mathscr{D}$ is defined by $(4.28)_1$. The condition of regularity i) and $(4.19)_2$ show that $\boldsymbol{M} = \boldsymbol{0}$ if $\boldsymbol{u} = \boldsymbol{0}$. The corresponding solution $\boldsymbol{Q}$ of (4.20) is $\boldsymbol{Q} = \boldsymbol{1}$. Hence (4.16) yields

$$\boldsymbol{\Theta}(\boldsymbol{0}) = \varepsilon(\boldsymbol{t}_{\kappa 1}, \rho_\kappa \boldsymbol{b}_1) = (\boldsymbol{t}_\kappa, \rho_\kappa \boldsymbol{b}). \tag{4.30}$$

This means that $\boldsymbol{u}_0 = \boldsymbol{0}$ is a pre-image of $\boldsymbol{f}_0 = (\boldsymbol{t}_\kappa, \rho_\kappa \boldsymbol{b})$ under the mapping $\boldsymbol{\Theta}$ as in (4.10).

Through a long chain of estimates, STOPPELLI proves that the transformation $\boldsymbol{\Theta}$ is of class $\mathscr{C}^1(\mathscr{D}; \mathscr{S}')$; further, the linear system (4.12) associated with the non-linear system (4.29) has the explicit form

$$\begin{aligned} \operatorname{Div} \mathbf{L}(\delta \boldsymbol{E}) &= \delta \boldsymbol{h}, \\ -\mathbf{L}(\delta \boldsymbol{E}) \boldsymbol{n}_\kappa &= \delta \boldsymbol{g}, \end{aligned} \tag{4.31}$$

where $\mathbf{L}$ is the fourth-order tensor of linear elasticity at the natural state κ.

The system (4.31) is the very system that defines the traction boundary-value problem of classical infinitesimal elasticity for the load $(-\delta \boldsymbol{g}, -\delta \boldsymbol{h})$. By property c) of $\mathscr{S}'$, the load is equilibrated. Hence by the condition of regularity ii) the linear system (4.31) possesses a unique solution $\delta \boldsymbol{u}$ satisfying the condition (4.24), with $\boldsymbol{u}$ replaced by $\delta \boldsymbol{u}$. This means that the Fréchet derivative $\nabla_{\boldsymbol{0}} \boldsymbol{\Theta}$ is an isomorphism. The inverse mapping theorem ensures that $\boldsymbol{\Theta}$ is a C^1-diffeomorphism near $\boldsymbol{u} = \boldsymbol{0}$, say over the neighborhood $\mathscr{N}$ of $\boldsymbol{0}$ in $\mathscr{D}$.

STOPPELLI then shows that the zero element $(\boldsymbol{0}, \boldsymbol{0})$ of $\mathscr{S}'$ belongs to the image $\boldsymbol{\Theta}(\mathscr{N})$, provided $|\varepsilon|$ be sufficiently small. Accordingly, for small $|\varepsilon|$ the equation

$$\boldsymbol{\Theta}(\boldsymbol{u}) = (\boldsymbol{0}, \boldsymbol{0}), \tag{4.32}$$

or, equivalently, the system (4.15), has a unique solution $\boldsymbol{u}$ belonging to $\mathscr{N}$. Since this solution belongs to $\mathscr{S}$, it satisfies the condition (4.24); in view of $(4.28)_1$ and (4.25), its gradient $\boldsymbol{H}$ is uniformly close to $\boldsymbol{0}$ when

$|\varepsilon|$ is sufficiently small. Hence, for small enough $|\varepsilon|$, $\boldsymbol{F} = \boldsymbol{H}+\mathbf{1}$ is nonsingular at all $\boldsymbol{X} \in \boldsymbol{\kappa}(\mathscr{B})$. From the remark following (4.15), the inverse of the rotation $\boldsymbol{Q}$ corresponding to the solution $\boldsymbol{u}$ sends $\boldsymbol{u}$ into a solution $\boldsymbol{u}^*$ of the original system (4.14).

In summary we have the following

THEOREM OF EXISTENCE AND UNIQUENESS (STOPPELLI). *Let the conditions of regularity i) – vi) be satisfied, and suppose that the loads* $(\boldsymbol{t}_{\kappa 1}, \boldsymbol{b}_1)$ *have no axis of equilibrium; then it is possible to find positive numbers* ξ, ζ *such that for all* ε *for which* $|\varepsilon| < \xi$, *the system* (4.14) *has a unique solution* $\boldsymbol{u}^*$ *satisfying the conditions*

$$||\boldsymbol{u}^*|| < \zeta, \quad \boldsymbol{u}^*(\mathbf{0}, \varepsilon) = \mathbf{0}. \tag{4.33}$$

Here $||\boldsymbol{u}^*||$ is defined by (4.25) with $\boldsymbol{u}$ replaced by $\boldsymbol{u}^*$. It is easily seen that $\boldsymbol{u}^*$ tends to zero in norm as $\varepsilon \to 0$, since the local inverse of $\boldsymbol{\Theta}$ is continuous.

STOPPELLI next considers the case when the condition of regularity vi) above is replaced by the stronger condition

via) The response function $\boldsymbol{A}(\boldsymbol{F})$ is analytic in the region defined by the inequality $||\boldsymbol{F}-\mathbf{1}|| < \gamma$.

Under this assumption STOPPELLI shows that the solution $\boldsymbol{u}^*$ whose existence and uniqueness was established above depends analytically on ε when $|\varepsilon| < \xi$. The method for constructing the coefficients of a power series expansion of $\boldsymbol{u}^*$ with respect to ε is that given by SIGNORINI, which we have summarized in § 2. In this sense, SIGNORINI's formal calculation is justified by STOPPELLI's theorem.

In a sequence of further papers STOPPELLI[1] has investigated the nature of solutions in the case when the loads have an axis of equilibrium. The results are too complicated to be summarized here. In some cases a unique solution exists and is an analytic function of ε. In others there are two or three distinct solutions, all or some of which are analytic functions of ε, or of $\sqrt{\varepsilon}$ when $\varepsilon > 0$, or of $\sqrt{-\varepsilon}$ when $\varepsilon < 0$. In some cases a solution may exist for given forces but fail to exist for the negatives of those forces. For details the reader should consult the memoirs of STOPPELLI.

[1] F. STOPPELLI, "Su un sistema di equazioni integrodifferenziali interessante l'elastostatica", *Ricerche Mat.* **6**, 11–26 (1957); "Sull'esistenza di soluzioni delle equazioni dell'elastostatica isoterma nel caso di sollecitazioni dotate di assi di equilibrio, I, II e III", *ibid.* **6**, 241–287 (1957); **7**, 71–101, 138–152 (1958).

5. Van Buren's Analysis of Existence and Uniqueness of Solutions of Boundary-value Problems in Elasticity [1]

The general technique developed in the preceding section for reducing the proof of existence and uniqueness of the solution of a non-linear system to that of an associated linear system has been applied by W. VAN BUREN to boundary-value problems of place and mixed boundary-value problems. We summarize his results for boundary-value problems of place[2] in this section; these results are similar to those of STOPPELLI but somewhat stronger.

In order to compare the analysis and the results with STOPPELLI's, we consider here also a one-parameter family of problems defined by

$$\begin{aligned} \operatorname{Div} \boldsymbol{A}(\boldsymbol{F}^*)+\rho_{\kappa}\varepsilon \boldsymbol{b}_1 &= \boldsymbol{0} \text{ in } \kappa(\mathscr{B}), \\ \boldsymbol{u}^* &= \varepsilon \boldsymbol{u}_1 \text{ on } \partial\kappa(\mathscr{B}), \end{aligned} \tag{5.1}$$

where $\boldsymbol{b}_1$ is a prescribed vector field in $\kappa(\mathscr{B})$ and $\boldsymbol{u}_1$ is a prescribed vector field on $\partial\kappa(\mathscr{B})$. We seek a one-parameter family of solutions $\boldsymbol{u}^* = \boldsymbol{u}^*(X, \varepsilon)$ of the system (5.1).

As before, we choose an arbitrary trial solution $\boldsymbol{u}$ and compute the pair $(\boldsymbol{h}, \boldsymbol{g})$ as follows:

$$\begin{aligned} \boldsymbol{h} &\equiv \operatorname{Div} \boldsymbol{A}(\boldsymbol{F})+\rho_{\kappa}\varepsilon \boldsymbol{b}_1, \\ \boldsymbol{g} &\equiv -\boldsymbol{u}+\varepsilon \boldsymbol{u}_1. \end{aligned} \tag{5.2}$$

The system (5.1) would be solved by $\boldsymbol{u}$ if the pair $(\boldsymbol{h}, \boldsymbol{g})$ should turn out to be $(\boldsymbol{0}, \boldsymbol{0})$.

---

[1] W. VAN BUREN, *On the Existence and Uniqueness of Solutions to Boundary Value Problems in Finite Elasticity*, Thesis, Department of Mathematics, Carnegie-Mellon University, 1968. Research Report 68-ID7-MEKMA-RI, Westinghouse Research Laboratories, Pittsburgh, Pa. 1968.

[2] VAN BUREN's analysis, like STOPPELLI's, is based on an assumption of existence and uniqueness of smooth solutions to boundary-value problems in the linear theory. Unfortunately, for the mixed boundary-value problem, VAN BUREN's assumption of smoothness seems to be too strong to be satisfied in general by solutions of the linear problems. As remarked by FICHERA, "Existence Theorems in Elasticity", FLÜGGE's Handbuch der Physik, Vol. VIa/2, ed. C. TRUESDELL, Berlin *etc.*, Springer-Verlag, 1972, solutions of the mixed boundary-value problems in the linear theory are generally smooth in the interior but on the boundary are only continuous. Consequently, a solution need not have the finite $C^{2+\lambda}$ norm required by VAN BUREN.

Next, we must set up the system (5.2) as a non-linear map of the form (4.9) which transforms $\boldsymbol{u}$ into the pair $(\boldsymbol{h}, \boldsymbol{g})$. To ensure the necessary degree of smoothness in order to apply the inverse mapping theorem, we assume the following conditions of regularity:

i)[1] The body $\mathscr{B}$ is homogeneous, and the reference configuration $\boldsymbol{\kappa}$ is a homogeneous natural state of $\mathscr{B}$.

ii) The region $\boldsymbol{\kappa}(\mathscr{B})$ occupied by $\mathscr{B}$ in the reference configuration $\boldsymbol{\kappa}$ is bounded.

iii) The boundary $\partial\boldsymbol{\kappa}(\mathscr{B})$ is a smooth surface of class $C^{2+\lambda}$; *i.e.*, it can be covered by an atlas of local co-ordinate systems $\{(y^1, y^2, z)\}$ such that the co-ordinate transformations $X^A = X^A(y^1, y^2, z)$ are of class C^2 and have Hölder-continuous second derivatives with exponent λ; further, a point $(y^1, y^2, z) \in \boldsymbol{\kappa}(\mathscr{B})$ if and only if $z > 0$, while a point $(y^1, y^2, z) \in \partial\boldsymbol{\kappa}(\mathscr{B})$ if and only if $z = 0$, and in this case (y^Γ) becomes a local surface co-ordinate system on $\partial\boldsymbol{\kappa}(\mathscr{B})$. In view of condition ii) above, $\partial\boldsymbol{\kappa}(\mathscr{B})$ is compact, so that we may assume that the atlas $\{(y^1, y^2, z)\}$ is finite. This condition is essentially the same as condition iv) in STOPPELLI's analysis.

iv) The response function $\boldsymbol{A}(\boldsymbol{F})$ is of class C^3 in a region of the 9-dimensional space of deformation gradients $\boldsymbol{F}$ satisfying the inequality $\|\boldsymbol{F}-\boldsymbol{1}\| \leqq \gamma$, where γ is a given positive number.

v) $\boldsymbol{b}_1$ is of class $C^{0+\lambda}$ in $\boldsymbol{\kappa}(\mathscr{B})$, and $\boldsymbol{u}_1$ is of class $C^{2+\lambda}$ in $\partial\boldsymbol{\kappa}(\mathscr{B})$.

vi) The elasticities of $\mathscr{B}$ in $\boldsymbol{\kappa}$ are such that a solution of class $C^{2+\lambda}$ of the boundary-value problem of place in the classical infinitesimal theory exists and is unique[2].

When these conditions are satisfied, we define the Banach space $\mathscr{S}$ to be the space of all vector fields of class $C^{2+\lambda}$ on $\boldsymbol{\kappa}(\mathscr{B})$ with respect to the uniform norm $\|\boldsymbol{u}\|$ defined by (4.25). Likewise, we define $\mathscr{S}'$ to be the Banach space of all pairs $(\boldsymbol{h}, \boldsymbol{g})$ where $\boldsymbol{h}$ is of class $C^{0+\lambda}$ in $\overline{\boldsymbol{\kappa}(\mathscr{B})}$ and $\boldsymbol{g}$ is of class $C^{2+\lambda}$ in $\partial\boldsymbol{\kappa}(\mathscr{B})$. We define the norm of $(\boldsymbol{h}, \boldsymbol{g})$ thus:

$$\|(\boldsymbol{h}, \boldsymbol{g})\| \equiv \|\boldsymbol{h}\|_{\mathscr{C}^{0+\lambda}(\overline{\boldsymbol{\kappa}(\mathscr{B})})} + \|\boldsymbol{g}\|_{\mathscr{C}^{2+\lambda}(\partial\boldsymbol{\kappa}(\mathscr{B}))}, \tag{5.3}$$

---

[1] This condition may be dropped or modified if suitable results for linear problems of infinitesimal deformations superimposed upon arbitrary finite deformations are known.

[2] VAN BUREN does not refer to any specific theorem of existence and uniqueness in the classical infinitesimal theory or in the more general theory of infinitesimal deformations superimposed on a finite deformation. As he points out, if condition vi) cannot be satisfied, then the results of his analysis are vacuous.

where the norms on the right-hand side are the usual ones, *e.g.*, $||\boldsymbol{u}||_{\mathscr{C}^{2+\lambda}(\overline{\kappa(\mathscr{B})})}$ is given by (4.25). As before, we define the open set $\mathscr{D}$ in $\mathscr{S}$ by the requirement that $\boldsymbol{u} \in \mathscr{D}$ if and only if $||\boldsymbol{u}|| < \gamma$, where γ is the given positive number in condition iv) above.

Now we set up the non-linear map $\boldsymbol{\Theta}$ corresponding to the system (5.2). First we choose any trial solution $\boldsymbol{u}$ belonging to $\mathscr{D}$. Substituting $\boldsymbol{u}$ into (5.2) yields a pair $(\boldsymbol{h}, \boldsymbol{g})$ belonging to $\mathscr{S}'$; the required conditions of continuity and differentiability of $\mathscr{S}'$ are obviously satisfied by $(\boldsymbol{h}, \boldsymbol{g})$. Estimates of the Hölder coefficients for $(h^i, g^i{}_{,\Gamma\Delta})$ are given by VAN BUREN for the case when $\kappa(\mathscr{B})$ is a convex region. From the condition of regularity i) and (5.2), we see that

$$\boldsymbol{\Theta}(\boldsymbol{0}) = (\rho_\kappa \varepsilon \boldsymbol{b}_1, \varepsilon \boldsymbol{u}_1). \tag{5.4}$$

This means that $\boldsymbol{u}_0 = \boldsymbol{0}$ is a pre-image of $\boldsymbol{f}_0 = (\rho_\kappa \varepsilon \boldsymbol{b}_1, \varepsilon \boldsymbol{u}_1)$ under $\boldsymbol{\Theta}$ as in (4.10). In order to solve (5.1), we must show that $\boldsymbol{\Theta}$ is a local diffeomorphism near $\boldsymbol{u} = \boldsymbol{0}$, say over an open neighborhood $\mathscr{N}$ of $\boldsymbol{0}$ in $\mathscr{D}$; further, the image $\boldsymbol{\Theta}(\mathscr{N})$ must contain the zero element $(\boldsymbol{0}, \boldsymbol{0})$ of $\mathscr{S}'$.

To prove that $\boldsymbol{\Theta}$ is a local diffeomorphism near $\boldsymbol{0}$, we proceed as before by showing first that $\boldsymbol{\Theta}$ is of class $\mathscr{C}^1(\mathscr{D}; \mathscr{S}')$. We select $\boldsymbol{u}$ and $\Delta \boldsymbol{u}$ such that $\boldsymbol{u} \in \mathscr{D}$, $\Delta \boldsymbol{u} \in \mathscr{S}$, $\boldsymbol{u}+\Delta \boldsymbol{u} \in \mathscr{D}$. We put

$$\boldsymbol{\Theta}(\boldsymbol{u}+\Delta \boldsymbol{u}) = (\boldsymbol{h}+\Delta \boldsymbol{h}, \boldsymbol{g}+\Delta \boldsymbol{g}). \tag{5.5}$$

Then from (5.2) we have the following co-ordinate forms for $(\Delta \boldsymbol{h}, \Delta \boldsymbol{g})$:

$$\mathsf{A}^{iA}{}_k{}^B \frac{\partial^2 \Delta u^k}{\partial X^A \partial X^B} + \mathsf{A}^{iA}{}_k{}^B{}_j{}^C \frac{\partial^2 u^k}{\partial X^A \partial X^B} \frac{\partial \Delta u^j}{\partial X^C} + O(||\Delta u||) = \Delta h^i, \; -\Delta u^k = \Delta g^k, \tag{5.6}$$

where $\mathsf{A}^{iA}{}_k{}^B$ is a component of the first gradient of $\boldsymbol{A}(\boldsymbol{F})$ defined as before, and where $\mathsf{A}^{iA}{}_k{}^B{}_j{}^C$ is a component of the second gradient of $\boldsymbol{A}(\boldsymbol{F})$:

$$\mathsf{A}^{iA}{}_k{}^B{}_j{}^C = \mathsf{A}^{iA}{}_k{}^B{}_j{}^C(\boldsymbol{F}) = \frac{\partial^2 A^{iA}}{\partial F^k_B \partial F^j_C}. \tag{5.7}$$

Equation (5.6) is a local result, valid for each point $\boldsymbol{X} \in \kappa(\mathscr{B})$. However, in view of the conditions of regularity iii) and iv) the error term on the left-hand side of (5.6) is of order $\boldsymbol{O}(||\Delta \boldsymbol{u}||)$ in the norm of the image space $\mathscr{S}'$ also. Hence the non-linear map $\boldsymbol{\Theta}$ in (5.5) can be approximated at $\boldsymbol{u}$ by the linear map $\nabla_{\boldsymbol{k}} \boldsymbol{\Theta}$:

$$[\nabla_{\boldsymbol{u}} \boldsymbol{\Theta}](\delta \boldsymbol{u}) = (\delta \boldsymbol{h}, \delta \boldsymbol{g}), \qquad \delta \boldsymbol{u} \in \mathscr{S}, \tag{5.8}$$

where $\delta\boldsymbol{h}$ and $\delta\boldsymbol{g}$ have the co-ordinate forms

$$\mathsf{A}^{iA}{}_{k}{}^{B}\frac{\partial^2\delta u^k}{\partial X^A\partial X^B}+\mathsf{A}^{iA}{}_{k}{}^{B}{}_{j}{}^{C}\frac{\partial^2 u^k}{\partial X^A\partial X^B}\frac{\partial\delta u^j}{\partial X^C}=\delta h^i,\qquad -\delta u^k=\delta g^k. \tag{5.9}$$

In view of (5.9) and the norms (4.25) and (5.3), we see that $(\delta\boldsymbol{h}, \delta\boldsymbol{g}) \to (\boldsymbol{0}, \boldsymbol{0})$ as $\delta\boldsymbol{u} \to \boldsymbol{0}$ in $\mathscr{S}$. Hence the linear map $\nabla_{\boldsymbol{u}}\boldsymbol{\Theta}$ is continuous.

Now let $\bar{\boldsymbol{u}}$ be another point in $\mathscr{D}$. We put

$$[\nabla_{\bar{\boldsymbol{u}}}\boldsymbol{\Theta}](\delta\boldsymbol{u}) = (\overline{\delta\boldsymbol{h}}, \overline{\delta\boldsymbol{g}}). \tag{5.10}$$

From (5.9) we see that $\nabla_{\bar{\boldsymbol{u}}}\boldsymbol{\Theta}$ has the co-ordinate form

$$\bar{\mathsf{A}}^{iA}{}_{k}{}^{B}\frac{\partial^2\delta u^k}{\partial X^A\partial X^B}+\bar{\mathsf{A}}^{iA}{}_{k}{}^{B}{}_{j}{}^{C}\frac{\partial^2 \bar{u}^k}{\partial X^A\partial X^B}\frac{\partial\delta u^j}{\partial X^C}=\overline{\delta h}^i,\ -\delta u^k=\overline{\delta g}^k, \tag{5.11}$$

where $\bar{\mathsf{A}}^{iA}{}_{k}{}^{B}$, $\bar{\mathsf{A}}^{iA}{}_{k}{}^{B}{}_{j}{}^{C}$ denote the values of $\mathsf{A}^{iA}{}_{k}{}^{B}$, $\mathsf{A}^{iA}{}_{k}{}^{B}{}_{j}{}^{C}$ at $\bar{\boldsymbol{F}}$, the deformation gradient of $\bar{\boldsymbol{u}}$. Subtracting (5.8) from (5.10), we obtain

$$[\nabla_{\bar{\boldsymbol{u}}}\boldsymbol{\Theta}-\nabla_{\boldsymbol{u}}\boldsymbol{\Theta}](\delta\boldsymbol{u}) = (\overline{\delta\boldsymbol{h}}-\delta\boldsymbol{h}, \overline{\delta\boldsymbol{g}}-\delta\boldsymbol{g}), \tag{5.12}$$

which has a co-ordinate form given by the difference of (5.11) and (5.9), *i.e.*,

$$(\bar{\mathsf{A}}^{iA}{}_{k}{}^{B}-\mathsf{A}^{iA}{}_{k}{}^{B})\frac{\partial^2\delta u^k}{\partial X^A\partial X^B}+\left(\bar{\mathsf{A}}^{iA}{}_{k}{}^{B}{}_{j}{}^{C}\frac{\partial^2 \bar{u}^k}{\partial X^A\partial X^B}-\mathsf{A}^{iA}{}_{k}{}^{B}{}_{j}{}^{C}\frac{\partial^2 u^k}{\partial X^A\partial X^B}\right)\frac{\partial\delta u^j}{\partial X^C}$$
$$=\overline{\delta h}^i-\delta h^i,\quad 0=\overline{\delta g}^i-\delta g^i. \tag{5.13}$$

In view of the norms (4.25) and (5.3) we see that

$$\|\nabla_{\bar{\boldsymbol{u}}}\boldsymbol{\Theta}-\nabla_{\boldsymbol{u}}\boldsymbol{\Theta}\| \to 0 \quad\text{as}\quad \bar{\boldsymbol{u}}\to\boldsymbol{u}\quad\text{in}\quad \mathscr{D}.$$

Hence the Fréchet derivative

$$\nabla\boldsymbol{\Theta} : \mathscr{D} \to \mathscr{L}(\mathscr{S}; \mathscr{S}') \tag{5.14}$$

is continuous. This means that $\boldsymbol{\Theta} \in \mathscr{C}^1(\mathscr{D}; \mathscr{S}')$.

Now from (5.9) we see that the linear system (5.8) at the origin $\boldsymbol{0} \in \mathscr{S}$ has the explicit form

$$\begin{aligned}\text{Div}\,\mathsf{L}(\delta E) &= \delta\boldsymbol{h},\\ -\delta\boldsymbol{u} &= \delta\boldsymbol{g},\end{aligned} \tag{5.15}$$

which is, of course, nothing but the system that defines the boundary-value problem of place in the classical infinitesimal theory. By condition vi), the linear map $\nabla_0 \boldsymbol{\Theta}$ is invertible. In virtue of the inverse mapping theorem stated in the preceding section $\boldsymbol{\Theta}$ is a local diffeomorphism near $\mathbf{0}$. This means that the system (5.2) has a unique solution in $\mathscr{N}$ for each $(\boldsymbol{h}, \boldsymbol{g})$ sufficiently near to $(\rho_\kappa \varepsilon \boldsymbol{b}_1, \varepsilon \boldsymbol{u}_1)$. Therefore, if $|\varepsilon|$ is sufficiently small, the system (5.2) has a unique solution for $(\boldsymbol{h}, \boldsymbol{g}) = (\mathbf{0}, \mathbf{0})$. Thus follows the THEOREM OF EXISTENCE AND UNIQUENESS (VAN BUREN). *Let conditions i)–vi) be satisfied. Then it is possible to find positive numbers ξ and ζ such that for all ε for which $|\varepsilon| < \xi$ the system* (5.1) *has a unique solution $\boldsymbol{u}^* \in \mathscr{S}$ satisfying the condition*

$$||\boldsymbol{u}^*|| < \zeta. \tag{5.16}$$

It is clear that the solution $\boldsymbol{u}^*$ will converge to $\mathbf{0}$ in $\mathscr{S}$ as ε tends to 0, since the local inverse of $\boldsymbol{\Theta}$ is continuous. Although we have worded the theorem so as to be parallel to STOPPELLI's, in fact a somewhat stronger result is true. In STOPPELLI's theorem the small positive numbers ξ and ζ depend on the load $(\boldsymbol{b}_1, \boldsymbol{t}_{\kappa 1})$. If we change $(\boldsymbol{b}_1, \boldsymbol{t}_{\kappa 1})$ to $\alpha(\boldsymbol{b}_1, \boldsymbol{t}_{\kappa 1})$ for any non-zero constant α, then ζ remains unchanged, but ξ must be replaced by $\xi/|\alpha|$. Thus STOPPELLI's theorem establishes existence and uniqueness of solution only on an interval about $(\mathbf{0}, \mathbf{0})$ on some straight line passing through $(\mathbf{0}, \mathbf{0})$ in $\mathscr{S}'$. In general, the length of the interval of existence and uniqueness of solution depends in a complicated way on the direction of the straight line because of the transformation $\boldsymbol{\Phi}$ defined by (4.20). Further, the length of the interval might shrink to zero in a straight line corresponding to loads that possess an axis of equilibrium, as explained in the preceding section. Hence for boundary-value problems of traction there is not always a neighborhood of loads near $(\mathbf{0}, \mathbf{0})$, all of which give rise to a unique solution. It is a different matter with boundary-value problems of place, since the existence and uniqueness is ensured in an entire neighborhood of loads near $(\mathbf{0}, \mathbf{0})$ in $\mathscr{S}'$. This fact is obvious from the proof of the preceding theorem. Indeed, if $\mathscr{N}$ is the neighborhood of $\mathbf{0}$ in $\mathscr{S}$ over which $\boldsymbol{\Theta}$ is a diffeomorphism, then $\boldsymbol{\Theta}(\mathscr{N})$ contains a ball of radius ξ about $\boldsymbol{\Theta}(\mathbf{0})$. Thus the system (5.1) has a unique solution if $||(\rho_\kappa \varepsilon \boldsymbol{b}_1, \varepsilon \boldsymbol{u}_1)|| < \xi$. Therefore, the previous theorem can be reworded as follows:

THEOREM OF EXISTENCE AND UNIQUENESS (VAN BUREN). *Let conditions i) – vi) be satisfied. Then there exist positive numbers ξ and ζ such that for all $(\boldsymbol{b}, \boldsymbol{u})$ for which*

$$||(\rho_\kappa \boldsymbol{b}, \boldsymbol{u})|| < \xi \tag{5.17}$$

the system

$$\text{Div}\ \boldsymbol{A}(\boldsymbol{F}^*)+\rho_\kappa \boldsymbol{b} = \boldsymbol{0} \ \text{in}\ \kappa(\mathscr{B}),$$

$$\boldsymbol{u}^* - \boldsymbol{u} \ \text{on}\ \partial\kappa(\mathscr{B}),$$

has a unique solution $\boldsymbol{u}^*$ *in* $\mathscr{S}$ *satisfying the condition*

$$||\boldsymbol{u}^*|| < \zeta.$$

The positive numbers ξ *and* ζ *depend only on the shape of the body and the response function of the material.*

The theorems of STOPPELLI and VAN BUREN do not solve the basic problems of the general theory of elasticity, since they do not allow general prescribed loads and prescribed boundary displacements. Only by limitation to *sufficiently small* data do they avoid having to face the basic uncertainties set forth in § VII.1 and § III.7: Neither a proper statement of the kind of uniqueness fit for the general theory nor *a priori* inequalities sufficient to guarantee such uniqueness are yet known. Rather, VAN BUREN's theorem provides a *justification for the infinitesimal theory in terms of the general theory* by showing that when the infinitesimal theory delivers unique and sufficiently smooth solutions of all of its own boundary-value problems of place, each of those solutions is in a specified sense an approximation to the solution of a corresponding boundary-value problem in the general theory, provided the prescribed loads and prescribed displacements be sufficiently small. STOPPELLI's theorem for the boundary-value problem of traction does not go so far, since the difficulties associated with an axis of equilibrium make the relation between the infinitesimal theory and the general theory more delicate, as we have said above. SIGNORINI and his disciples[1] devoted much effort to elucidating this "incompatibility".

---

[1] The literature is cited in § 64 of NFTM, where an estimate of the problem is given. SIGNORINI's work on incompatibility goes back to 1930. His final exposition is given in Cap. I, § 3, of "Trasformazioni termoelastiche finite, Memoria 2a", *Annali di Mat. Pur. Appl.* (4) **30**, 1–72 (1949).

6. Beju's Analysis of Existence, Uniqueness, and Stability of Solutions of Boundary-value Problems of Place for Hyperelastic Bodies[1]

Unlike the procedures of STOPPELLI and VAN BUREN, BEJU's analysis of boundary-value problems of place in the non-linear theory is not based on any theorem in the linear theory. Beyond the usual conditions of smoothness his basic assumption is a condition of stability, which he shows to be sufficient for the existence and uniqueness of a generalized solution. The condition of stability he assumes refers to the material and the shape of the body together and is somewhat stronger than the usual conditions of infinitesimal stability or finite stability. As we shall explain by a specific example, BEJU's results cannot be applied to some elastic bodies which may be perfectly reasonable physically, such as elastic fluid bodies.

BEJU's analysis rests upon four basic theorems[2] of LANGENBACH in a theory of minimization of functionals. Before stating these theorems we review some basic concepts. An inner product on a finite-dimensional vector space was introduced in § I.2. For an infinite-dimensional vector space $\mathscr{S}$, the definition of an inner product is similar. Letting $\boldsymbol{u}$ and $\boldsymbol{v}$ be elements of $\mathscr{S}$, we denote their inner product by $(\boldsymbol{u}, \boldsymbol{v})$ and require it to obey the standard rules:

i) $(\boldsymbol{u}, \boldsymbol{v})$ is bilinear in $\boldsymbol{u}$ and $\boldsymbol{v}$.

ii) $(\boldsymbol{u}, \boldsymbol{v})$ is symmetric with respect to interchange of $\boldsymbol{u}$ and $\boldsymbol{v}$.

iii) $(\boldsymbol{u}, \boldsymbol{u})$ is positive-definite, *i.e.*, $(\boldsymbol{u}, \boldsymbol{u}) > 0$ unless $\boldsymbol{u} = \boldsymbol{0}$. Of course an inner product induces a norm:

$$||\boldsymbol{u}|| = (\boldsymbol{u}, \boldsymbol{u})^{\frac{1}{2}}. \tag{6.1}$$

A potential Banach space was defined in § 4. Such a space is called a *Hilbert space* when it is endowed with a norm induced by an assigned inner product.

Now let $\mathscr{D}$ be a bounded domain with smooth boundary in a Euclidean space $\mathscr{E}$, say $\mathscr{E} = \mathscr{R}^3$. Then the space of all square-integrable functions

---

[1] I. BEJU, "Theorems on existence, uniqueness, and stability of the solution of the place boundary-value problem in statics for hyperelastic materials", *Arch. Rational Mech. Anal.* **42**, 1–23 (1971).

[2] Proofs of these theorems are given by M. M. VAINBERG, *op. cit.* p. 496. The original paper by A. LANGENBACH is in Russian.

on $\mathscr{D}$ is a Hilbert space $\mathscr{L}^2(\mathscr{D})$ with respect to the inner product defined by

$$(f, g) = \int_{\mathscr{D}} f(x)g(x)dV \quad \forall f, g \in \mathscr{L}^2(\mathscr{D}). \tag{6.2}$$

Likewise, the space of all square-integrable vector fields on $\mathscr{D}$ also forms a Hilbert space $\mathscr{S}$, and the inner product is defined by

$$(\boldsymbol{u}, \boldsymbol{v}) = \int_{\mathscr{D}} \boldsymbol{u}(x) \cdot \boldsymbol{v}(x)dV \quad \forall \boldsymbol{u}, \boldsymbol{v} \in \mathscr{S}. \tag{6.3}$$

Here the dot in the integrand denotes the inner product on $\mathscr{V}$, the translation space of $\mathscr{E}$. The Hilbert space $\mathscr{S}$ just defined is the underlying function space in BEJU's analysis, the bounded domain $\mathscr{D}$ being the domain $\boldsymbol{\kappa}(\mathscr{B})$ occupied by the body $\mathscr{B}$ in the reference configuration $\boldsymbol{\kappa}$.

BEJU considers a non-linear system $\boldsymbol{\Theta}$ together with some linear homogeneous boundary conditions:

$$\left.\begin{aligned} \boldsymbol{\Theta}(\boldsymbol{u}) &= \boldsymbol{f}, \\ \boldsymbol{L}_i(\boldsymbol{u}) &= \boldsymbol{0}, \quad i = 1, \ldots, p, \end{aligned}\right\} \tag{6.4}$$

where the domain $\mathscr{D}(\boldsymbol{\Theta})$ and the range $\mathscr{R}(\boldsymbol{\Theta})$ of the non-linear map are contained in $\mathscr{S}$. We put

$$\mathscr{D}_0(\boldsymbol{\Theta}) \equiv \{\boldsymbol{u} \in \mathscr{D}(\boldsymbol{\Theta}) : \boldsymbol{L}_i(\boldsymbol{u}) = \boldsymbol{0}, \quad i = 1, \ldots, p\}. \tag{6.5}$$

Given any data $\boldsymbol{f} \in \mathscr{S}$, we seek a solution $\boldsymbol{u}_0 \in \mathscr{D}_0(\boldsymbol{\Theta})$ for the system (6.4). As formulated by BEJU, the boundary-value problem of place can be regarded as a special case of the general non-linear problem (6.4).

In § 4 we defined the concept of Fréchet derivative for a non-linear map. Here we need a somewhat more general concept. We assume first that $\mathscr{D}(\boldsymbol{\Theta})$ is a linear manifold in $\mathscr{S}$. If $\boldsymbol{u} \in \mathscr{D}(\boldsymbol{\Theta})$, we say that $\boldsymbol{\Theta}$ has a *linear Gâteaux differential* $D_{\boldsymbol{u}}\boldsymbol{\Theta}$ at $\boldsymbol{u}$ if $D_{\boldsymbol{u}}\boldsymbol{\Theta}$ is a linear map from $\mathscr{D}(\boldsymbol{\Theta})$ into $\mathscr{S}$ such that

$$\lim_{t \to 0} \frac{\boldsymbol{\Theta}(\boldsymbol{u}+t\boldsymbol{v}) - \boldsymbol{\Theta}(\boldsymbol{u})}{t} = D_{\boldsymbol{u}}\boldsymbol{\Theta}(\boldsymbol{v}) \quad \forall \boldsymbol{v} \in \mathscr{D}(\boldsymbol{\Theta}), \tag{6.6}$$

the limit on the left-hand side being that of the difference quotient of $\boldsymbol{\Theta}$, at $\boldsymbol{u}$, for any $\boldsymbol{v} \in \mathscr{D}(\boldsymbol{\Theta})$, relative to the topology of $\mathscr{S}$. We do not require a linear Gâteaux differential $D_{\boldsymbol{u}}\boldsymbol{\Theta}$ to be a bounded linear map. If $D_{\boldsymbol{u}}\boldsymbol{\Theta}$ is continuous, then it is called a *Gâteaux derivative*. In this case, its domain can be extended to the closure of $\mathscr{D}(\boldsymbol{\Theta})$ in $\mathscr{S}$.

Naturally, $\boldsymbol{\Theta}$ is said to have a linear Gâteaux differential or a Gâteaux derivative in $\mathscr{D}(\boldsymbol{\Theta})$ if it has a linear Gâteaux differential or a Gâteaux derivative at every point in $\mathscr{D}(\boldsymbol{\Theta})$. Then we denote the field of linear Gâteaux differentials or the field of Gâteaux derivatives by $D\boldsymbol{\Theta}$. If $\boldsymbol{\Theta}$ has a Fréchet derivative $\nabla\boldsymbol{\Theta}$, then it clearly has a Gâteaux derivative, and the two are equal:

$$D\boldsymbol{\Theta} = \nabla\boldsymbol{\Theta}. \tag{6.7}$$

Conversely, however, a mapping that has a linear Gâteaux differential or a Gâteaux derivative need not have a Fréchet derivative.

Having introduced the necessary basic concepts, we can now state the theorems of LANGENBACH:

THEOREM 1. *Let the following conditions be satisfied by the non-linear system* (6.4):

i) $\mathscr{D}_0(\boldsymbol{\Theta})$ *and* $\mathscr{D}(\boldsymbol{\Theta})$ *are linear manifolds dense in* $\mathscr{S}$.

ii) $\boldsymbol{\Theta}$ *has a linear Gâteaux differential* $D\boldsymbol{\Theta}$ *such that for each* $\boldsymbol{v} \in \mathscr{D}(\boldsymbol{\Theta})$ *the field* $D_{\boldsymbol{u}}\boldsymbol{\Theta}(\boldsymbol{v})$ *is continuous on each 2-dimensional hyperplane passing through* $\boldsymbol{u}$ *in* $\mathscr{D}(\boldsymbol{\Theta})$.

iii) $\boldsymbol{\Theta}(\mathbf{0}) = \mathbf{0}$. (6.8)

iv) $D\boldsymbol{\Theta}$ *is symmetric with respect to* $\mathscr{D}_0(\boldsymbol{\Theta})$:

$$(D\boldsymbol{\Theta}(\boldsymbol{v}), \boldsymbol{w}) = (\boldsymbol{v}, D\boldsymbol{\Theta}(\boldsymbol{w})) \quad \forall \boldsymbol{v},\ \boldsymbol{w} \in \mathscr{D}_0(\boldsymbol{\Theta}). \tag{6.9}$$

v) $D\boldsymbol{\Theta}$ *is positive-definite with respect to* $\mathscr{D}_0(\boldsymbol{\Theta})$:

$$(D\boldsymbol{\Theta}(\boldsymbol{v}), \boldsymbol{v}) > 0 \tag{6.10}$$

for all non-zero $\boldsymbol{v} \in \mathscr{D}_0(\boldsymbol{\Theta})$.

Under these conditions the following assertions are true:

a) *If a solution* $\boldsymbol{u}_0$ *of the system* (6.4) *exists, then it is unique; further, it establishes on* $\mathscr{D}_0(\boldsymbol{\Theta})$ *an absolute minimum of the functional* $F(\boldsymbol{u})$ *defined by*

$$F(\boldsymbol{u}) = J(\boldsymbol{u}) - (\boldsymbol{f}, \boldsymbol{u}), \tag{6.11}$$

where

$$J(\boldsymbol{u}) \equiv \int_0^1 (\boldsymbol{\Theta}(t\boldsymbol{u}), \boldsymbol{u})\, dt. \tag{6.12}$$

b) *Conversely, if on* $\mathscr{D}_0(\boldsymbol{\Theta})$ $F(\boldsymbol{u})$ *attains an absolute minimum at* $\boldsymbol{u}_0$, *then* $\boldsymbol{u}_0$ *is a solution of* (6.4).

We note that the preceding is a theorem of uniqueness, not a theorem of existence. In general, the functional $F(\boldsymbol{u})$ need not attain an absolute minimum on $\mathscr{D}_0(\boldsymbol{\Theta})$.

THEOREM 2. *If condition v) of Theorem* 1 *is replaced by*

v)′ $D\boldsymbol{\Theta}$ *is strictly positive with respect to* $\mathscr{D}_0(\boldsymbol{\Theta})$; i.e., *there exists a positive constant c such that*

$$(D\boldsymbol{\Theta}(\boldsymbol{v}), \boldsymbol{v}) \geqq c\|\boldsymbol{v}\|^2 \quad \forall \boldsymbol{v} \in \mathscr{D}_0(\boldsymbol{\Theta}), \tag{6.13}$$

then the following additional assertions are true:

c) *The functional* $F(\boldsymbol{u})$ *is bounded from below on* $\mathscr{D}_0(\boldsymbol{\Theta})$.

d) *The functional* $F(\boldsymbol{u})$ *is strictly convex on* $\mathscr{D}_0(\boldsymbol{\Theta})$, i.e., *for each unequal* $\boldsymbol{u}, \boldsymbol{v} \in \mathscr{D}_0(\boldsymbol{\Theta})$ *and each* $t \in (0, 1)$

$$F(t\boldsymbol{u}+(1-t)\boldsymbol{v}) < tF(\boldsymbol{u})+(1-t)F(\boldsymbol{v}). \tag{6.14}$$

e) *Any minimizing sequence for* $F(\boldsymbol{u})$ *on* $\mathscr{D}_0(\boldsymbol{\Theta})$ *is convergent in* $\mathscr{S}$.

By assertion e) above, a minimizing sequence for $F(\boldsymbol{u})$ has a limit in $\mathscr{S}$. If that limit belongs to $\mathscr{D}_0(\boldsymbol{\Theta})$, then by Theorem 1 it is the solution of the system (6.4). On the other hand, if that limit does not belong to $\mathscr{D}_0(\boldsymbol{\Theta})$, then the system (6.4) has no solution, but we call the limit of a minimizing sequence a *generalized solution.*

THEOREM 3. *Under the assumptions of Theorem* 2

f) *The generalized solution of* (6.4) *is unique.*

THEOREM 4. *If condition v)′ of Theorem* 2 *is replaced by*

v)″ *There exist an element* $\bar{\boldsymbol{u}} \in \mathscr{D}_0(\boldsymbol{\Theta})$ *and positive numbers* c_1, c_2, *such that*

$$(D\boldsymbol{\Theta}(\boldsymbol{v}), \boldsymbol{v}) \geqq c_1(D_{\bar{\boldsymbol{u}}}\boldsymbol{\Theta}(\boldsymbol{v}), \boldsymbol{v}) \geqq c_2\|\boldsymbol{v}\|^2 \quad \forall \boldsymbol{v} \in \mathscr{D}_0(\boldsymbol{\Theta}), \tag{6.15}$$

then

g) *The generalized solution of* (6.4) *belongs to the completion of* $\mathscr{D}_0(\boldsymbol{\Theta})$ *with respect to the inner product* $[\boldsymbol{u}, \boldsymbol{v}] \equiv (D_{\bar{\boldsymbol{u}}}\boldsymbol{\Theta}(\boldsymbol{u}), \boldsymbol{v})$ *induced by* $D_{\bar{\boldsymbol{u}}}\boldsymbol{\Theta}$.

Having stated the theorems of LANGENBACH, we now summarize BEJU's formulation of the boundary-value problems of place as a special case of (6.4). As we have explained in §1, the boundary-value problem of place is given by the system

$$\begin{aligned} \operatorname{Div} \boldsymbol{A}(\boldsymbol{F}^*)+\rho_\kappa \boldsymbol{b} &= \boldsymbol{0} \text{ in } \kappa(\mathscr{B}), \\ \boldsymbol{u}^* &= \boldsymbol{u} \text{ on } \partial\kappa(\mathscr{B}), \end{aligned} \tag{6.16}$$

where $\boldsymbol{b}$ is a prescribed vector field in $\boldsymbol{\kappa}(\mathscr{B})$ and $\boldsymbol{u}$ is a prescribed vector field on $\partial\boldsymbol{\kappa}(\mathscr{B})$. For simplicity, we assume that the body $\mathscr{B}$ is homogeneous and that the reference configuration $\boldsymbol{\kappa}$ is a homogeneous configuration of $\mathscr{B}$. To reduce (6.16) to a special case of (6.4), we choose first a kinematical solution $\boldsymbol{w}$ which is of class C^2 in $\boldsymbol{\kappa}(\mathscr{B})$ and which attains the boundary-value $\boldsymbol{u}$ on $\partial\boldsymbol{\kappa}(\mathscr{B})$. We put

$$\boldsymbol{w}^* \equiv \boldsymbol{u}^* - \boldsymbol{w}. \tag{6.17}$$

Then the boundary condition in terms of $\boldsymbol{w}^*$ becomes the linear homogeneous condition

$$\boldsymbol{w}^* = \boldsymbol{0} \quad \text{on} \quad \partial\boldsymbol{\kappa}(\mathscr{B}). \tag{6.18}$$

Now let $\mathscr{S}$ denote the Hilbert space of square-integrable vector fields on $\boldsymbol{\kappa}(\mathscr{B})$ defined as before. We define $\mathscr{D}(\boldsymbol{\Theta})$ as the linear manifold of $\mathscr{S}$ consisting in all vector fields of class C^2 on $\overline{\boldsymbol{\kappa}(\mathscr{B})}$, and we define $\mathscr{D}_0(\boldsymbol{\Theta})$ as the submanifold of $\mathscr{D}(\boldsymbol{\Theta})$ subject to the linear homogeneous boundary condition (6.18). Then $\mathscr{D}(\boldsymbol{\Theta})$ and $\mathscr{D}_0(\boldsymbol{\Theta})$ are dense in $\mathscr{S}$. We define the non-linear map[1]

$$\boldsymbol{\Theta} : \mathscr{D}(\boldsymbol{\Theta}) \to \mathscr{S} \tag{6.19}$$

by

$$\boldsymbol{\Theta}(\boldsymbol{w}^*) \equiv \frac{1}{\rho_\kappa} \operatorname{Div} \boldsymbol{A}(\boldsymbol{F}(\boldsymbol{w})) + \boldsymbol{b}(\boldsymbol{w}) - \frac{1}{\rho_\kappa} \operatorname{Div} \boldsymbol{A}(\boldsymbol{F}(\boldsymbol{u}^*)) - \boldsymbol{b}(\boldsymbol{u}^*), \tag{6.20}$$

where $\boldsymbol{u}^*$ is related to $\boldsymbol{w}^*$ by (6.17), $\boldsymbol{w}$ being the particular kinematical solution chosen to reduce the boundary condition to the homogeneous form (6.18). From (6.20), it is clear that the mapping $\boldsymbol{\Theta}$ satisfies the condition (6.8). In view of (6.20), we define the data $\boldsymbol{f}$ by

$$\boldsymbol{f} \equiv \frac{1}{\rho_\kappa} \operatorname{Div} \boldsymbol{A}(\boldsymbol{F}(\boldsymbol{w})) + \boldsymbol{b}(\boldsymbol{w}). \tag{6.21}$$

As usual, we assume that $\boldsymbol{b}$ is continuous, so that $\boldsymbol{f}$ belongs to $\mathscr{S}$.

---

[1] Since $\boldsymbol{A}(\boldsymbol{F})$ is defined for non-singular tensors $\boldsymbol{F}$ only, the non-linear map $\boldsymbol{\Theta}$ is defined by (6.20) for those $\boldsymbol{w}^* \in \mathscr{D}(\boldsymbol{\Theta})$ such that the displacement field $\boldsymbol{u}^*$ given by (6.17) is a kinematical solution. Then it is assumed that $\boldsymbol{\Theta}$ can be extended smoothly to the entire domain $\mathscr{D}(\boldsymbol{\Theta})$.

In terms of the mapping $\boldsymbol{\Theta}$ and the data $\boldsymbol{f}$ defined by (6.20) and (6.21), the system (6.16) now takes the form

$$\begin{aligned} \boldsymbol{\Theta}(\boldsymbol{w}^*) &= \boldsymbol{f} \quad \text{in} \quad \boldsymbol{\kappa}(\mathscr{B}), \\ \boldsymbol{w}^* &= \mathbf{0} \quad \text{on} \quad \partial\boldsymbol{\kappa}(\mathscr{B}), \end{aligned} \tag{6.22}$$

which clearly is a special case of (6.4). As before, we seek a solution of (6.22) in the linear manifold $\mathscr{D}_0(\boldsymbol{\Theta})$.

Since the domain $\mathscr{D}(\boldsymbol{\Theta})$ consists of smooth fields only, by using the same argument as before we obtain from (6.20) the co-ordinate form for the linear Gâteaux differential of $\boldsymbol{\Theta}$ at any point $\boldsymbol{v} \in \mathscr{D}(\boldsymbol{\Theta})$:

$$D_{\boldsymbol{v}}\boldsymbol{\Theta}(\delta\boldsymbol{v})^i = -\frac{1}{\rho_{\boldsymbol{\kappa}}}\left[\mathsf{A}^{iA\ B}_{\ \ k}\frac{\partial^2\delta v^k}{\partial X^A\partial X^B} + \mathsf{A}^{iA\ B\ C}_{\ \ k\ j}\frac{\partial^2 v^k}{\partial X^A\partial X^B}\frac{\partial\delta v^j}{\partial X^C}\right] - \frac{\partial b^i}{\partial x^k}\delta v^k. \tag{6.23}$$

From (6.23) it is clear that for each $\delta\boldsymbol{v}$ the differential $D_{\boldsymbol{v}}\boldsymbol{\Theta}(\delta\boldsymbol{v})$ is continuous on each 2-dimensional hyperplane passing through $\boldsymbol{v}$.

So far we have verified that the mapping $\boldsymbol{\Theta}$ defined by (6.20) satisfies Conditions i)–iii) of Theorem 1. Next we consider Conditions iv) and v). To prove these it suffices to assume that the material is hyperelastic and that the body force is conservative, so that

$$\mathsf{A}_{i\ j}^{\ A\ B} = \mathsf{A}_{j\ i}^{\ B\ A}, \quad \frac{\partial b^i}{\partial x^j} = \frac{\partial b^j}{\partial x^i} \equiv -\frac{\partial^2 \pi}{\partial x^i \partial x^j}, \tag{6.24}$$

where π denotes a potential function for $\boldsymbol{b}$. Substituting (6.24) into (6.23) and using the divergence theorem, we obtain

$$\begin{aligned} (D\boldsymbol{\Theta}(\boldsymbol{v}), \boldsymbol{w}) &= \int_{\boldsymbol{\kappa}(\mathscr{B})}\left[\frac{1}{\rho_{\boldsymbol{\kappa}}}\mathsf{A}_{i\ k}^{\ A\ B}\frac{\partial v^i}{\partial X^A}\frac{\partial w^k}{\partial X^B} + \frac{\partial^2\pi}{\partial x^i\partial x^k}v^i w^k\right]dV_{\boldsymbol{\kappa}}, \\ &= (D\boldsymbol{\Theta}(\boldsymbol{w}), \boldsymbol{v}) \quad \forall \boldsymbol{v}, \boldsymbol{w} \in \mathscr{D}_0(\boldsymbol{\Theta}). \end{aligned} \tag{6.25}$$

Thus $\boldsymbol{\Theta}$ satisfies Condition iv).

Finally, from (6.25), Condition v) is satisfied if and only if

$$\int_{\boldsymbol{\kappa}(\mathscr{B})}\left[\frac{1}{\rho_{\boldsymbol{\kappa}}}\mathsf{A}_{i\ k}^{\ A\ B}\frac{\partial v^i}{\partial X^A}\frac{\partial v^k}{\partial X^B} + \frac{\partial^2\pi}{\partial x^i\partial x^k}v^i v^k\right]dV_{\boldsymbol{\kappa}} > 0 \tag{6.26}$$

for all non-zero $\boldsymbol{v} \in \mathscr{D}_0(\boldsymbol{\Theta})$. Similarly, condition v)′ is satisfied if and only if

$$\int_{\kappa(\mathscr{B})} \left[\frac{1}{\rho_\kappa} \mathsf{A}_i{}^A{}_k{}^B \frac{\partial v^i}{\partial X^A} \frac{\partial v^k}{\partial X^B} + \frac{\partial^2 \pi}{\partial x^i \partial x^k} v^i v^k \right] dV_\kappa \geqq c \int_{\kappa(\mathscr{B})} v^i v_i \, dV_\kappa, \tag{6.27}$$

and condition v)″ is satisfied if and only if

$$\begin{aligned} &\int_{\kappa(\mathscr{B})} \left[\frac{1}{\rho_\kappa} \mathsf{A}_i{}^A{}_k{}^B \frac{\partial v^i}{\partial X^A} \frac{\partial v^k}{\partial X^B} + \frac{\partial^2 \pi}{\partial x^i \partial x^k} v^i v^k \right] dV_\kappa \\ &\qquad \geqq c_1 \int_{\kappa(\mathscr{B})} \left[\frac{1}{\rho_\kappa} \mathsf{A}_i{}^A{}_k{}^B(\boldsymbol{F}(\overline{\boldsymbol{w}})) \frac{\partial v^i}{\partial X^A} \frac{\partial v^k}{\partial X^B} + \frac{\partial^2 \pi(\overline{\boldsymbol{w}})}{\partial x^i \partial x^k} v^i v^k \right] dV_\kappa, \\ &\qquad \geqq c_2 \int_{\kappa(\mathscr{B})} v^i v_i \, dV_\kappa, \end{aligned} \tag{6.28}$$

where $\overline{\boldsymbol{w}}$ is a particular element of $\mathscr{D}_0(\boldsymbol{\Theta})$.

Having set up assumptions sufficient to satisfy the premisses of LANGENBACH's theorems for the system (6.22), we can now read off the following results:

THEOREM 5 (BEJU). *Let $\mathscr{B}$ be a homogeneous hyperelastic body subject to conservative body force. If the response function and the body force together satisfy the inequality* (6.26), *then assertions a) and b) of Theorem* 1 *hold for the system* (6.22).

THEOREM 6 (BEJU). *If the inequality* (6.26) *in Theorem* 5 *is replaced by the inequality* (6.27), *then also assertions c) – f) of Theorems* 2 *and* 3 *hold for the system* (6.22).

THEOREM 7 (BEJU). *If the inequality* (6.27) *in Theorem* 6 *is replaced by the inequality* (6.28), *then also assertion g) of Theorem* 4 *holds for the system* (6.22).

The inequality (6.26) is the strict inequality corresponding to the condition of infinitesimal stability introduced in § III.8B, except that here the "ground state" is arbitrary. Of course, we assume here also that the body force $\boldsymbol{b}$ is a given function of $\boldsymbol{x}$, the place in the actual configuration, while in § III.8B we considered only dead loads. For dead loads (6.26) reduces to

$$\int_{\kappa(\mathscr{B})} \mathsf{A}_i{}^A{}_k{}^B \frac{\partial v^i}{\partial X^A} \frac{\partial v^k}{\partial X^A} \, dV_\kappa > 0, \tag{6.29}$$

which is the strict inequality corresponding to (III. 8B.7).

A configuration in which the inequality (6.29) holds is called *infinitesimally superstable*[1] for the boundary condition being considered. In the theory of infinitesimal deformation superimposed on a finite deformation, a theorem of ERICKSEN & TOUPIN[2] shows that infinitesimal superstability implies uniqueness for the solution of the boundary-value problem of place. This theorem has been generalized to the mixed boundary-value problem by HILL.[2]

As we have stated in § III. 8B, HADAMARD's theorem shows that if a configuration $\boldsymbol{\kappa}$ is infinitesimally stable for boundary condition of place, then the Hadamard condition (III. 8B.3) must hold at every point in $\boldsymbol{\kappa}(\mathscr{B})$. It follows by exactly the same argument that if the configuration is infinitesimally superstable, then the S-E condition (III. 8B.1) must hold at every point in $\boldsymbol{\kappa}(\mathscr{B})$. As we have shown in § III.9B, the S-E condition implies the B-E inequality (III. 9B.13), which is impossible for elastic fluids. Consequently, (6.29) cannot be satisfied, so BEJU's theorems do not yield the assertions a) – g) for fluid bodies. In fact, they cannot generally apply to such bodies. So as to establish this last claim, we consider an example. Let $\mathscr{B}$ be an elastic fluid body, and suppose that $\boldsymbol{\kappa}$ is a homogeneous reference configuration of $\mathscr{B}$. We shall now construct a one-parameter family of classical solutions of the boundary-value problem of place for the homogeneous boundary condition

$$\boldsymbol{u} = \mathbf{0} \text{ on } \partial\boldsymbol{\kappa}(\mathscr{B}). \tag{6.30}$$

We choose a point $\boldsymbol{X}_0 \in \boldsymbol{\kappa}(\mathscr{B})$ and assume that the ball $\mathscr{B}(\boldsymbol{X}_0, \varepsilon)$ centered at $\boldsymbol{X}_0$ with radius ε is contained in $\boldsymbol{\kappa}(\mathscr{B})$. Using a spherical co-ordinate system centered at $\boldsymbol{X}_0$, we define a solution $\boldsymbol{u}$ in $\mathscr{B}(\boldsymbol{X}_0, \varepsilon)$ by the following components of deformation:

$$r = R, \; \theta = \Theta, \; \varphi = \Phi + 2n\pi f(R), \; n = 1, 2, 3, \ldots, \tag{6.31}$$

where $f(R)$ is any C^∞ function such that $0 \leqq f(R) \leqq 1$ and

$$\begin{aligned} f(R) &\equiv 1 \text{ if } R \in [0, \varepsilon/3], \\ f(R) &\equiv 0 \text{ if } R \in [2\varepsilon/3, \varepsilon]. \end{aligned} \tag{6.32}$$

[1] NFTM, § 68 bis. BEJU's assumption (6.26), in effect, means that every configuration of $\mathscr{B}$ compatible with the boundary condition is infinitesimally superstable under the given body force.

[2] NFTM, § 68bis.

For X outside $\mathscr{B}(X_0, \varepsilon)$, we simply put

$$\boldsymbol{u}(X) \equiv \mathbf{0}. \tag{6.33}$$

Then $\boldsymbol{u}$ is a non-trivial isochoric C^∞ deformation of $\boldsymbol{\kappa}(\mathscr{B})$, and it satisfies the boundary condition (6.30). Consequently $\boldsymbol{u}$ is a static universal solution corresponding to the particular conservative body force

$$\boldsymbol{b} = \mathbf{0}. \tag{6.34}$$

This example shows clearly that the assertions of Theorem 5 cannot be satisfied by any elastic fluid body.

While BEJU's theorems are important in that they provide the first positive results on existence and uniqueness of generalized solutions of boundary-value problems in finite elasticity, they rest upon hypotheses so strong as to exclude not only some important particular elastic materials but also some of the main problems elasticity theory is designed to represent: instability and failure of uniqueness in the classical sense.

7. The Concept of Dynamic Stability in Elasticity

In § III.8B we have introduced the concept of infinitesimal stability for an equilibrium configuration of an elastic body, and at the end of the preceding section the somewhat more restrictive concept of infinitesimal superstability. These concepts of stability are all based upon the idea that an equilibrium configuration is stable if the increment of the elastic stored-energy of the body equals or exceeds the work done by the loading in any virtual displacement compatible with some given boundary conditions. Clearly, such a concept of stability is strictly a static one, depending only on the stored-energy and the loading in the virtual displacement. Hence we may call this type of stability *static stability*.

A more intuitive idea of stability, however, is dynamic: An equilibrium configuration or a motion is stable if small disturbances applied to the configuration or the motion produce only small departure from the configuration or the motion. On the other hand if some disturbances tend to grow, then the equilibrium configuration or the motion is unstable. To make this idea precise, measures for disturbances and departures must be chosen; further, dynamic stability or instability depend not only on the equilibrium configuration or the motion but also on the choice of the measures. It is possible for an equilibrium configuration or a motion to be stable with respect to certain measures but become unstable with respect to others.

Not only that, a static concept of equilibrium makes use of the constitutive relation of a material only in statics. Since static elasticity is synonymous with the statics of simple materials, the criteria of elastostatic stability can be applied to *all simple materials* and do not require us even to know the nature of the response of such materials to deformations which change as time proceeds. A given equilibrium configuration which is statically stable may be, with exactly the same measure of dynamic stability, dynamically stable for one material and dynamically unstable for another whose equilibrium equations are just the same. To every static problem of stability correspond infinitely many different problems of dynamic stability. We sometimes think we should not have to know the constitutive equation of a body in dynamics in order to determine whether or not a certain configuration of equilibrium be stable. In this sense a static criterion of stability may be a more natural one.

These possibilities are easily illustrated by taking the simple pendulum as a prototype. Static stability requires that the potential energy of the bob be a minimum. Thus the lowest possible configuration of the bob is the only statically stable one. If we choose as the criterion of dynamic stability that height of the bob above the stable configuration shall never exceed twice the length of the rod, then, trivially, the statically stable configuration is dynamically stable, no matter what be the dynamical equations. If the criterion of dynamic stability be that the position of the bob shall approach the lowest possible one as $t \to \infty$, then the statically stable configuration is dynamically stable for a pendulum with positive linear damping but dynamically unstable for the ideal, undamped pendulum. Of course, it is possible to choose other measures in terms of which the statically stable configuration of the ideal pendulum is also dynamically stable.

Intuitively we are likely to think that the lowest possible configuration of the bob is stable *no matter what be the constitutive relation* of the pendulum in dynamics. We know this cannot be true without qualification, since negative damping contradicts it. We tend to think that negative damping is unreal and unnatural, yet we do not have any way of specifying precisely the infinitely many possible or reasonable "laws of friction" for the manifold kinds and circumstances of natural bodies we should like to represent as a mass-point constrained to move on a circle.

An extensive article on the various ideas of stability, whether static or dynamic, has been written recently by KNOPS & WILKES[1]. Since so far we

[1] R. J. KNOPS & E. W. WILKES, "Elastic Stability", FLÜGGE's *Handbuch der Physik* VIa/3, ed. C. TRUESDELL, Berlin etc., Springer-Verlag, 1973.

have not considered dynamic stability, we shall explore this idea first in this section.

According to KNOPS & WILKES a dynamical system $\mathscr{D}$ is a class of mappings

$$\mathscr{D} \equiv \{\xi : \mathscr{I} \to \mathscr{S}\} \tag{7.1}$$

from a given interval of time $\mathscr{I}$ into a function space $\mathscr{S}$. Within the context of elasticity, $\mathscr{S}$ may be the space of displacements, velocities, or some other kinematical quantities relative to a reference configuration of an elastic body $\mathscr{B}$. $\mathscr{I}$ may be a finite interval or a semi-infinite interval of time, say $[0, \infty)$. An element $\xi \in \mathscr{D}$ is called a *motion*, and its value $\xi(t)$ is called the *configuration* at time $t \in \mathscr{I}$. Of course, these terms are used here in senses more general than their usual ones, since $\xi(t)$, being an element in $\mathscr{S}$, may contain the velocity field or other fields in addition to the displacement field at the time t. Motions in $\mathscr{D}$ are generally required to correspond to solutions of some dynamical initial-value and boundary-value problems of the elastic body as we have set them up in § 1. If there exists a constant motion $\xi \in \mathscr{D}$ such that

$$\xi(t) = \xi_0 \quad \forall t \in \mathscr{I}, \tag{7.2}$$

the constant configuration ξ_0 is called an *equilibrium configuration.* Since ξ always contains the displacement field, an equilibrium configuration defined by (7.2) corresponds to an equilibrium configuration in the usual sense. We call a reference time $t_0 \in \mathscr{I}$ an *intitial time* since normally we choose t_0 to be the beginning of $\mathscr{I}$. The configuration $\xi(t_0)$ of $\xi \in \mathscr{D}$ is then called the *initial configuration.* Note again that $\xi(t_0)$ may contain the initial velocity field or other fields in addition to the initial displacement field.

Next we must define some measures for the initial disturbance and the departure of one motion from another one during the time interval $\mathscr{I}$. For this purpose we introduce the concept of a *positive-definite function* on $\mathscr{S}$. We say that a function

$$\mu : \mathscr{S} \times \mathscr{S} \to \mathscr{R} \tag{7.3}$$

is *positive-definite* on $\mathscr{S}$ if

$$\mu(\xi_1, \xi_2) > 0 \quad \forall \xi_1, \xi_2 \in \mathscr{S} \quad \xi_1 \neq \xi_2 \tag{7.4}$$

and

$$\mu(\xi, \xi) = 0 \quad \forall \xi \in \mathscr{S}.$$

Clearly, a positive-definite function is more general than a distance function, since it need not satisfy the condition of symmetry and the triangle inequality.

Given any positive-definite function on $\mathscr{S}$, we define an extended positive definite function $\tilde{\mu}$ on $\mathscr{D}$,

$$\tilde{\mu} : \mathscr{D} \times \mathscr{D} \to \mathscr{R} \cup \{\infty\}, \tag{7.5}$$

by

$$\tilde{\mu}(\xi, \boldsymbol{\eta}) \equiv \sup_{t \in \mathscr{I}} \mu(\xi(t), \boldsymbol{\eta}(t)). \tag{7.6}$$

From (7.4) and (7.6) it is clear that $\tilde{\mu}(\xi, \boldsymbol{\eta})$ is positive if $\xi \neq \boldsymbol{\eta}$, and $\tilde{\mu}(\xi, \xi) = 0$, for all $\xi, \boldsymbol{\eta} \in \mathscr{D}$. However, the supremum on the right-hand side of (7.6) may be infinite, in which case, of course, we put

$$\tilde{\mu}(\xi, \boldsymbol{\eta}) = \infty. \tag{7.7}$$

When $\mathscr{S}$ is equipped with a positive-definite function μ, we define convergence in $\mathscr{S}$ in the usual way: A sequence $\{\xi_n\}$ converges to ξ in $\mathscr{S}$ if

$$\mu(\xi, \xi_n) \to 0 \text{ as } n \to \infty. \tag{7.8}$$

We use $\mu(\xi(t), \boldsymbol{\eta}(t))$ to measure the instantaneous departure of $\boldsymbol{\eta}$ from ξ. Convergence in $\mathscr{D}$ is defined in a similar way: A sequence of motions $\{\xi_n\}$ converges to a motion ξ in $\mathscr{D}$ if

$$\tilde{\mu}(\xi, \xi_n) \to 0 \text{ as } n \to \infty. \tag{7.9}$$

We use $\tilde{\mu}(\xi, \boldsymbol{\eta})$ to measure the overall departure of the motion $\boldsymbol{\eta}$ from the motion ξ, over the entire time interval $\mathscr{I}$. These measures depend on the choice of the positive-definite function μ on $\mathscr{S}$.

In stability analysis sometimes another positive-definite function μ_0 on $\mathscr{S}$ is chosen, independent of μ, to measure the initial disturbance. That is, we choose to interpret an initial disturbance $\boldsymbol{\eta}(t_0)$ from $\xi(t_0)$ as being small if $\mu_0(\xi(t_0), \boldsymbol{\eta}(t_0))$ is small, t_0 being the initial time. So as to avoid instability at time t_0, we generally assume that μ is continuous with respect to μ_0; *i.e.*, if $\mu_0(\xi(t_0), \boldsymbol{\eta}(t_0))$ is small, so is $\mu(\xi(t_0), \boldsymbol{\eta}(t_0))$. The converse need not hold, of course; otherwise, the measures μ and μ_0 would be topologically equivalent, and hence there would be no need to introduce them separately. We assume also that the motions in $\mathscr{D}$ are continuous with respect to μ; *i.e.*, for each $\xi \in \mathscr{D}$ we have

$$\mu(\xi(t), \xi(t_n)) \to 0 \text{ as } n \to \infty \tag{7.10}$$

for any sequence $\{t_n\}$ in $\mathscr{I}$ convergent to $t \in \mathscr{I}$. Under such a continuity assumption, the extended positive-definite function $\tilde{\mu}$ on $\mathscr{D}$ is finite if the time interval $\mathscr{I}$ is compact.

Having introduced the functions μ, $\tilde{\mu}$, and μ_0 to measure the departures of motions and the initial disturbances, we can now define the concept of (*dynamic*) *stability*. We say that[1] a motion ξ (possibly a constant) in $\mathscr{D}$ is *stable* with respect to the measures μ, $\tilde{\mu}$, and μ_0 if for each positive number ε there exists a positive number δ such that

$$\mu_0(\xi(t_0), \boldsymbol{\eta}(t_0)) < \delta \tag{7.11}$$

for any $\boldsymbol{\eta} \in \mathscr{D}$ implies

$$\tilde{\mu}(\xi, \boldsymbol{\eta}) < \varepsilon. \tag{7.12}$$

In other words, ξ is stable if a small initial disturbance, which here is characterized by the condition (7.11), implies a small overall departure from ξ, characterized here by the condition (7.12). Naturally, we say that ξ is *unstable* if it is not stable, and we say that the entire dynamical system $\mathscr{D}$ is *stable* or *unstable* if every motion in $\mathscr{D}$ is stable or unstable, respectively.

Since the above criterion for stability is nothing but a condition of continuity of the motions in $\mathscr{D}$ relative to the initial configurations, that criterion clearly depends on the measures $\tilde{\mu}$ and μ_0, which define topologies for $\mathscr{D}$ and $\mathscr{S}$, respectively. As we have remarked before, it is possible for a motion to be stable with respect to certain measures μ and μ_0 but unstable with respect to other measures. Indeed, such is the case for an equilibrium configuration of an isotropic homogeneous elastic body in a natural state. As has been observed by SHIELD & GREEN[2], even if the stored-energy function of the body is positive-definite at the natural state, it is possible to produce displacements and velocities that are not small for all time and at all places by focussing an initially small spherical wave. Thus if μ and μ_0 are chosen to be the uniform norms of the displacements and the velocities, then the equilibrium configuration in a natural state must be regarded as unstable. On the other hand, as shown by KOITER[3], if

---

[1] This definition is due to GILBERT & KNOPS, who allow the initial time to spread over a time interval also; *cf.* Chapter II of KNOPS & WILKES, *op. cit.* p. 519.

[2] R. T. SHIELD & A. E. GREEN, "On certain methods in the stability theory of continuous systems", *Arch. Rational Mech. Anal.* **12**, 354–360 (1963).

[3] W. T. KOITER, "On the concept of stability of equilibrium for continuous bodies", *Proc. Kon. Ned. Ak. Wet.* **B66**, 174–177 (1963). "The energy criterion of stability for continuous elastic bodies. I", *ibid.* **B68**, 178–202 (1965).

μ and μ_0 are chosen to be the L_2-norms of the displacements and the velocities, then that configuration is stable. We shall prove KOITER's stability theorem later. The example shows clearly the importance of the choice of the measures μ and μ_0 in the stability analysis.

The definition of dynamic stability given above is just one of several possible ones that are consistent with the intuitive idea sketched at the beginning of this section. We shall state here only one refinement of the preceding definitions; for others we refer the reader to Chapter II of the article by KNOPS & WILKES already mentioned. Let ξ be stable with respect to μ and μ_0, and suppose that $\mathscr{I}$ is the semi-infinite interval $[t_0, \infty)$. Then we say that ξ is *asymptotically stable* if there exists a positive number δ such that

$$\mu(\xi(t), \boldsymbol{\eta}(t)) \to 0 \text{ as } t \to \infty \tag{7.13}$$

for each $\boldsymbol{\eta} \in \mathscr{D}$ satisfying the condition

$$\mu_0(\xi(t_0), \boldsymbol{\eta}(t_0)) < \delta. \tag{7.14}$$

An obvious consequence of dynamic stability is uniqueness. In general, a motion $\xi \in \mathscr{D}$ is said to be *unique* with respect to initial configuration if ξ is the only motion in $\mathscr{D}$ having the initial configuration $\xi(t_0)$, *i.e.*,

$$\boldsymbol{\eta}(t_0) = \xi(t_0) \tag{7.15}$$

for any $\boldsymbol{\eta} \in \mathscr{D}$ implies

$$\boldsymbol{\eta}(t) = \xi(t) \quad \forall t \in \mathscr{I}. \tag{7.16}$$

Clearly we have the following

THEOREM (MOVCHAN, GILBERT & KNOPS). *If $\xi \in \mathscr{D}$ is stable, then it is unique.*

The proof is obvious, since if ξ is not unique, then there exists in $\mathscr{D}$ a motion $\boldsymbol{\eta} \neq \xi$ such that

$$\xi(t_0) = \boldsymbol{\eta}(t_0), \tag{7.17}$$

so that

$$\mu_0(\xi(t_0), \boldsymbol{\eta}(t_0)) = 0, \tag{7.18}$$

but

$$\tilde{\mu}(\xi, \boldsymbol{\eta}) > 0. \tag{7.19}$$

Obviously, the conditions (7.18) and (7.19) contradict the hypothesis that ξ is stable.

COROLLARY. *If a motion $\xi \in \mathscr{D}$ is not unique, then it is unstable with respect to every μ and μ_0.*

Next we state some stability theorems[1].

MAXIMAL PRINCIPLE. *A motion $\xi \in \mathscr{D}$ is stable if there exists a bounded positive function*

$$M \equiv M_\xi : \mathscr{I} \to \mathscr{R} \tag{7.20}$$

such that

$$\mu(\xi(t), \boldsymbol{\eta}(t)) \leqq M(t)\mu_0(\xi(t_0), \boldsymbol{\eta}(t_0)) \tag{7.21}$$

for all $t \in \mathscr{I}$ and all $\boldsymbol{\eta} \in \mathscr{D}$. Further, if $\mathscr{I} = [t_0, \infty)$, and if

$$M(t) \to 0 \text{ as } t \to \infty, \tag{7.22}$$

then ξ is asymptotically stable.

This principle is a direct consequence of the definition.

The next stability theorem is usually referred to as LIAPOUNOV's *second method* for stability:

THEOREM (MOVCHAN, GILBERT & KNOPS). *A motion $\xi \in \mathscr{D}$ is stable if and only if there exist positive-definite functions v_t, $t \in \mathscr{I}$, called* LIAPOUNOV *functions, such that the following two conditions are satisfied:*

i) *Given any positive number ε, there exists a positive number $\delta = \delta(\varepsilon)$ such that*

$$\mu_0(\xi(t_0), \eta(t_0)) < \delta \tag{7.23}$$

implies

$$\tilde{v}(\xi, \boldsymbol{\eta}) \equiv \sup_{t \in \mathscr{I}} v_t(\xi(t), \boldsymbol{\eta}(t)) < \varepsilon. \tag{7.24}$$

ii) *Given any positive number $\bar{\varepsilon}$, there exists a positive number $\bar{\delta} = \bar{\delta}(\bar{\varepsilon})$ such that*

$$\tilde{v}(\xi, \boldsymbol{\eta}) < \bar{\delta} \tag{7.25}$$

implies

$$\tilde{\mu}(\xi, \boldsymbol{\eta}) < \bar{\varepsilon}. \tag{7.26}$$

The proof is again obvious. If the LIAPOUNOV functions exist, (7.23) then implies (7.26), proving that ξ is stable. Conversely, if ξ is stable, then the functions $v_t \equiv \mu \; \forall \in \mathscr{I}$ can be regarded as the LIAPOUNOV functions. Of course, such a choice annuls the whole advantage of the second meth-

[1] *Cf.* Chapter III of KNOPS & WILKES, *op. cit.* p. 519.

oa, since Condition ii) is then automatically satisfied while Condition i) is simply the original definition of stability.

Since the LIAPOUNOV functions are nothing but some auxiliary functions employed to establish stability, we may require that they satisfy additional conditions so as to aid in the proof of Conditions i) or ii). For example, the following two conditions[1] are sufficient for Condition i):

A) *Given any positive number* ε, *there exists a positive number* $\delta = \delta(\varepsilon)$ *such that*

$$\mu_0(\xi(t_0), \boldsymbol{\eta}(t_0)) < \delta \tag{7.27}$$

implies

$$v_0(\xi(t_0), \boldsymbol{\eta}(t_0)) \equiv v_{t_0}(\xi(t_0), \boldsymbol{\eta}(t_0)) < \varepsilon. \tag{7.28}$$

B) $v_t(\xi(t), \boldsymbol{\eta}(t))$ *is non-increasing with respect to* $t \in \mathscr{I}$ *for all* $\boldsymbol{\eta} \in \mathscr{D}$.

Likewise, the following condition[1] is sufficient for Condition ii):

C) *There exists a strictly increasing continuous function* $\lambda : \mathscr{R} \to \mathscr{R}$ *satisfying the conditions* $\lambda(0) = 0$ *and* $\lambda(x) \to \infty$ *as* $x \to \infty$ *such that*

$$\tilde{v}(\xi, \boldsymbol{\eta}) \geqq \lambda(\tilde{\mu}(\xi, \boldsymbol{\eta})) \tag{7.29}$$

for all $\boldsymbol{\eta} \in \mathscr{D}$.

By exactly the same argument as before, we have the following theorem for instability:

THEOREM (GILBERT & KNOPS) *A motion* $\xi \in \mathscr{D}$ *is unstable if and only if there exist positive-definite functions* v_t, $t \in \mathscr{I}$ *such that the following two conditions are satisfied*:

ia) *There exists a positive number* ε *such that for each positive number* δ *the inequalities*

$$\mu_0(\xi(t_0), \boldsymbol{\eta}(t_0)) < \delta \tag{7.30}$$

and

$$\tilde{v}(\xi, \boldsymbol{\eta}) \geqq \varepsilon \tag{7.31}$$

hold for at least one motion $\boldsymbol{\eta} \in \mathscr{D}$.

iia) *Given any positive number* $\hat{\varepsilon}$, *there exists a positive number* $\hat{\delta} = \hat{\delta}(\hat{\varepsilon})$ *such that*

$$\tilde{v}(\xi, \boldsymbol{\eta}) < \hat{\varepsilon} \tag{7.32}$$

for all $\boldsymbol{\eta} \in \mathscr{D}$ *satisfying*

$$\tilde{\mu}(\xi, \boldsymbol{\eta}) < \hat{\delta}. \tag{7.33}$$

---

[1] *Cf.* KNOPS & WILKES, *op. cit.*, who attribute Conditions *A* and *B* to ZUBOV and MOVCHAN and Condition *C* to BHATIA & SZEGÖ.

As before, we may require v_t to satisfy additional conditions in order to establish Conditions ia) or iia).

Having set up the general framework of dynamic stability, we now apply the analysis to elasticity. We consider hyperelastic bodies only.

Let $\mathscr{B}$ be a hyperelastic body, and let $\boldsymbol{\kappa}$ be a reference configuration. Then a motion of $\mathscr{B}$ can be described by a one-parameter family of deformations χ_t relative to $\boldsymbol{\kappa}$. We can represent χ_t by the components of deformation

$$x^i = x^i(X^1, X^2, X^3, t). \tag{7.34}$$

Under the usual assumptions of smoothness the motion can be regarded as a solution of the dynamical equation

$$\text{Div}\, \boldsymbol{T}_\kappa + \rho_\kappa \boldsymbol{b}_\kappa = \rho_\kappa \ddot{\boldsymbol{x}} \tag{7.35}$$

corresponding to some body forces

$$\boldsymbol{b}_\kappa = \boldsymbol{b}_\kappa(\boldsymbol{X}, t) \tag{7.36}$$

and some boundary tractions

$$\boldsymbol{t}_\kappa = \boldsymbol{t}_\kappa(\boldsymbol{X}, t) = \boldsymbol{T}_\kappa(\boldsymbol{X}, t)\boldsymbol{n}_\kappa(\boldsymbol{X}), \quad \boldsymbol{X} \in \partial\kappa(\mathscr{B}), \tag{7.37}$$

where $\boldsymbol{n}_\kappa$ denotes the unit outward normal on the boundary surface $\partial\kappa(\mathscr{B})$, which is assumed to be smooth. The Piola-Kirchhoff stress tensor $\boldsymbol{T}_\kappa$ is given by the motion through the constitutive equation:

$$\boldsymbol{T}_\kappa = \boldsymbol{T}_\kappa(\boldsymbol{X}, t) = \rho_\kappa \left(\frac{\partial \sigma_\kappa}{\partial \boldsymbol{F}}\right), \tag{7.38}$$

where σ_κ denotes the stored-energy function relative to the reference configuration $\boldsymbol{\kappa}$, *viz*

$$\sigma_\kappa = \sigma_\kappa(\boldsymbol{F}, \boldsymbol{X}). \tag{7.39}$$

The argument of σ_κ in (7.38) is the deformation gradient $\boldsymbol{F}$ of χ_t relative to $\boldsymbol{\kappa}$, at the time t. For definiteness, we shall refer to such a motion as the motion ξ, and another such motion with components of deformation

$$y^i = y^i(X^1, X^2, X^3, t) \tag{7.40}$$

we shall refer to as the motion $\boldsymbol{\eta}$. We denote the values of the Piola-Kirchhoff stress tensor, the body force, the surface traction, and the stored-energy function corresponding to $\boldsymbol{\eta}$ by $\boldsymbol{T}^*_\kappa$, $\boldsymbol{b}^*_\kappa$, $\boldsymbol{t}^*_\kappa$, and σ^*_κ, respectively, and we denote the displacement vector from $\boldsymbol{x}$ to $\boldsymbol{y}$ by $\boldsymbol{u}$, *viz*,

$$\boldsymbol{u} = \boldsymbol{u}(X, t) = \boldsymbol{y}(X, t) - \boldsymbol{x}(X, t). \tag{7.41}$$

We choose $t_0 = 0$ to be the initial time.

Now to measure the initial disturbance we may choose μ_0 to be the difference of the initial energies[1], *viz*,

$$\mu_0(\xi(0), \boldsymbol{\eta}(0)) \equiv \left\{ \int_{\boldsymbol{\kappa}(\mathscr{B})} \rho_{\kappa}(\sigma_{\kappa}^* - \sigma_{\kappa})\, dV_{\kappa} + \frac{1}{2} \int_{\boldsymbol{\kappa}(\mathscr{B})} \rho_{\kappa}(\dot{y}^2 - \dot{x}^2)\, dV_{\kappa} \right\}_{t=0}. \tag{7.42}$$

From the very form of (7.42), we see that we must choose $\mathscr{S}$ to be the function space consisting in smooth displacement fields and velocity fields on $\boldsymbol{\kappa}(\mathscr{R})$; further, μ_0 is positive if the following two conditions are satisfied:

1) *ξ is an equilibrium solution.*
2) *σ_{κ} is a minimum in the configuration of ξ.*

Thus the measure μ_0 is appropriate for the stability analysis of the natural state.

From the requirement that both ξ and $\boldsymbol{\eta}$ satisfy the principle of balance of energy:

$$\frac{d}{dt} \int_{\boldsymbol{\kappa}(\mathscr{B})} \rho_{\kappa}(\tfrac{1}{2}\dot{x}^2 + \sigma_{\kappa})\, dV_{\kappa} = \int_{\boldsymbol{\kappa}(\mathscr{B})} \rho_{\kappa}\, \boldsymbol{b}_{\kappa} \cdot \dot{\boldsymbol{x}}\, dV_{\kappa} + \int_{\partial\boldsymbol{\kappa}(\mathscr{B})} \boldsymbol{t}_{\kappa} \cdot \dot{\boldsymbol{x}}\, dA_{\kappa} \tag{7.43}$$

and

$$\frac{d}{dt} \int_{\boldsymbol{\kappa}(\mathscr{B})} \rho_{\kappa}(\tfrac{1}{2}\dot{y}^2 + \sigma_{\kappa}^*)\, dV_{\kappa} = \int_{\boldsymbol{\kappa}(\mathscr{B})} \rho_{\kappa}\, \boldsymbol{b}_{\kappa}^* \cdot \dot{\boldsymbol{y}}\, dV_{\kappa} + \int_{\partial\boldsymbol{\kappa}(\mathscr{B})} \boldsymbol{t}_{\kappa}^* \cdot \dot{\boldsymbol{y}}\, dA_{\kappa}, \tag{7.44}$$

we obtain the following representation of μ_0 for all positive time t:

$$\begin{aligned} \mu_0(\xi(0), \boldsymbol{\eta}(0)) = & \int_{\boldsymbol{\kappa}(\mathscr{B})} \rho_{\kappa}(\sigma_{\kappa}^* - \sigma_{\kappa})\, dV_{\kappa} + \frac{1}{2} \int_{\boldsymbol{\kappa}(\mathscr{B})} \rho_{\kappa}(\dot{y}^2 - \dot{x}^2)\, dV_{\kappa} \\ & - \int_0^t \int_{\boldsymbol{\kappa}(\mathscr{B})} \rho_{\kappa}(\boldsymbol{b}_{\kappa}^* \cdot \dot{\boldsymbol{y}} - \boldsymbol{b}_{\kappa} \cdot \dot{\boldsymbol{x}})\, dV_{\kappa}\, dt - \int_0^t \int_{\boldsymbol{\kappa}(\mathscr{B})} (\boldsymbol{t}_{\kappa}^* \cdot \dot{\boldsymbol{y}} - \boldsymbol{t}_{\kappa} \cdot \dot{\boldsymbol{x}})\, dA_{\kappa}\, dt \end{aligned} \tag{7.45}$$

[1] This choice is consistent with some measures given by ERICKSEN in thermo-elastic stability analysis; *cf.* "Thermoelastic stability", *Proc. Fifth U.S. Nat. Congr. Appl. Mech.*, 187–193 (1966) and "A thermokinetic view of elastic stability theory", *Int. J. Solids Structures* **2**, 575–580 (1966). In the thermo-elastic case ERICKSEN's measures correspond to some LIAPOUNOV functions satisfying the ZUBOV-MOVCHAN Conditions A) and B) given above.

or, equivalently, when ξ is an equilibrium solution we have

$$\mu_0(\xi(0), \boldsymbol{\eta}(0)) = \int_{\kappa(\mathscr{B})} \rho_\kappa(\sigma_\kappa^* - \sigma_\kappa)\, dV_\kappa + \frac{1}{2}\int_{\kappa(\mathscr{B})} \rho_\kappa \dot{\boldsymbol{u}}^2 dV_\kappa$$
$$- \int_0^t \int_{\kappa(\mathscr{B})} \rho_\kappa \boldsymbol{b}_\kappa^* \cdot \dot{\boldsymbol{u}}\, dV_\kappa\, dt - \int_0^t \int_{\partial\kappa(\mathscr{B})} \boldsymbol{t}_\kappa^* \cdot \dot{\boldsymbol{u}}\, dA_\kappa\, dt. \tag{7.46}$$

So far we have not imposed any restrictions on the body forces $\boldsymbol{b}_\kappa^*$ and the surface tractions $\boldsymbol{t}_\kappa^*$ corresponding to the perturbed motions $\boldsymbol{\eta}$, so the dynamical system $\mathscr{D}$ is only loosely connected. A possible restriction is the requirement that the loads possess a potential in the following sense:

$$\int_{\kappa(\mathscr{B})} \rho_\kappa \boldsymbol{b}_\kappa^* \cdot \dot{\boldsymbol{u}}\, dV_\kappa + \int_{\partial\kappa(\mathscr{B})} \boldsymbol{t}_\kappa^* \cdot \dot{\boldsymbol{u}}\, dA_\kappa = -\frac{d\zeta}{dt}, \tag{7.47}$$

where ζ is path-independent and is given by

$$\zeta = \zeta(\boldsymbol{u}) \equiv \int_{\kappa(\mathscr{B})} \rho_\kappa \zeta_1\, dV_\kappa + \int_{\partial\kappa(\mathscr{B})} \zeta_2\, dA_\kappa \tag{7.48}$$

ζ_1 and ζ_2 being smooth functions which depend on the displacement $\boldsymbol{u}$, and possibly also on the position $\boldsymbol{X} \in \kappa(\mathscr{B})$, but not explicitly on the time t. A dynamical system satisfying the restriction (7.47) may be called a (*weakly*) *conservative system*. For such a system, the body forces and the boundary tractions are determined by

$$\boldsymbol{b}_\kappa^* = -\frac{\partial\zeta_1}{\partial\boldsymbol{u}}, \quad \boldsymbol{t}_\kappa^* = -\frac{\partial\zeta_2}{\partial\boldsymbol{u}}, \tag{7.49}$$

and the representation (7.46) reduces to

$$\mu_0(\xi(0), \boldsymbol{\eta}(0)) = \int_{\kappa(\mathscr{B})} \rho_\kappa(\sigma_\kappa^* - \sigma_\kappa)\, dV_\kappa + \frac{1}{2}\int_{\kappa(\mathscr{B})} \rho_\kappa \dot{\boldsymbol{u}}^2 dV_\kappa + \zeta, \tag{7.50}$$

where we have assumed that ζ is equal to zero when $\boldsymbol{u}$ vanishes. Since the leading term on the right-hand side of (7.50) is also a functional of $\boldsymbol{u}$, we can combine it with ζ and define a *total potential* λ by

$$\lambda = \lambda(\boldsymbol{u}) \equiv \int_{\kappa(\mathscr{B})} \rho_\kappa(\sigma_\kappa^* - \sigma_\kappa)\, dV_\kappa + \zeta. \tag{7.51}$$

Then (7.50) becomes

$$\mu_0(\xi(0), \eta(0)) = \lambda + \frac{1}{2}\int_{\kappa(\mathscr{B})} \rho_\kappa \dot{u}^2 dV_\kappa . \tag{7.52}$$

Now we introduce the usual L_2-norm of the displacement field $\boldsymbol{u}$:

$$||\boldsymbol{u}||^2 \equiv \int_{\kappa(\mathscr{B})} \rho_\kappa \boldsymbol{u}^2 dV_\kappa , \tag{7.53}$$

and we regard λ as a functional on the L_2-space completed by the elements $\boldsymbol{u}$. Then we say that λ is *positive-definite* if the following two conditions are satisfied:

1) *On each sphere* $\{\boldsymbol{u} : ||\boldsymbol{u}|| = c > 0\}$ *there exists a positive upper bound* $d(c)$ *for* λ.

2) $d(c)$ *is a non-decreasing continuous function of* c *on some interval* $[0, a]$, $a > 0$.

THEOREM (KOITER). *An equilibrium configuration of a conservative system whose total potential is positive-definite stable with respect to the measure*[1]

$$\bar{\mu}_0(\xi(0), \eta(0)) \equiv \mu_0(\xi(0), \eta(0)) + ||\boldsymbol{u}(0)|| \tag{7.54}$$

for the initial disturbance and the measure

$$\mu(\xi(t), \eta(t)) \equiv ||\boldsymbol{u}|| + ||\dot{\boldsymbol{u}}|| \tag{7.55}$$

for the subsequent departure of $\eta(t)$ *from* $\xi(t)$.

The proof of this theorem is exactly the same as the classical LAGRANGE-DIRICHLET argument for the stability of a dynamical system in analytical mechanics, where the function space $\mathscr{S}$ is the finite-dimensional $(\boldsymbol{q}, \dot{\boldsymbol{q}})$-space. Indeed, since the initial disturbance is small as measured by $\bar{\mu}_0$, by the representation (7.52) the kinetic energy and the total potential energy are small for all subsequent time t. Now the initial L_2-norm of $\boldsymbol{u}$ is small because of the definition (7.54); since λ is positive-definite, $||\boldsymbol{u}||$ must remain small for all subsequent time t. Thus $\tilde{\mu}(\xi, \eta)$ is small, and the proof is complete.

---

[1] By adding the term $||\boldsymbol{u}(0)||$ in (7.54), we make the measure μ continuous with respect to $\bar{\mu}_0$, this being a necessary condition for stability, as we have explained before. If μ is continuous with respect to μ_0, that extra term may be dropped.

The problem of verifying that the functional λ is positive-definite is clearly a problem in the calculus of variations. For smooth functionals we have the following general theorems:

THEOREM[1]. *Let λ be a functional on a Banach space $\mathscr{X}$, and suppose that λ is of class C^2 on a neighborhood of the origin $\mathbf{0} \in \mathscr{X}$. Then in order that λ be positive-definite at the origin the following two conditions*

$$\begin{aligned} &1)\ [\nabla_0 \lambda](\boldsymbol{u}) = \mathbf{0} \qquad \forall \boldsymbol{u} \in \mathscr{X}, \\ &2)\ [\nabla_0^2 \lambda](\boldsymbol{u}, \boldsymbol{u}) \geqq 0 \quad \forall \boldsymbol{u} \in \mathscr{X}, \end{aligned} \tag{7.56}$$

are necessary conditions, and Condition 1) *and the following condition*:

$$3)\ \exists \alpha > 0 : [\nabla_0^2 \lambda](\boldsymbol{u}, \boldsymbol{u}) \geqq \alpha \|\boldsymbol{u}\|^2 \quad \forall \boldsymbol{u} \in \mathscr{X} \tag{7.57}$$

are sufficient conditions.

Combining this theorem with KOITER's stability theorem, we can obtain some necessary conditions and some sufficient conditions for stability in terms of the Fréchet derivatives of the total potential function λ. Since Condition 2) is considerably weaker than Condition 3), the positive-definiteness of λ, and hence the stability of the equilibrium solution, is not certain when

$$[\nabla_0^2 \lambda](\boldsymbol{u}, \boldsymbol{u}) = 0 \quad \forall \boldsymbol{u} \in \mathscr{X}. \tag{7.58}$$

In this circumstance we have to examine higher-order Fréchet derivatives of λ. This matter has been studied extensively by KOITER[2], especially in post-buckling behavior.

For a more detailed treatment of dynamic stability we refer the reader to the article by KNOPS & WILKES.

8. Static Stability and the Energy Criterion.

KOITER's theorem at the end of the last section shows that a sufficient condition for dynamic stability of an equilibrium configuration

[1] *Cf.* § III.9 of VAINBERG, *op. cit.* p. 510.

[2] *Cf.* W. T. KOITER, *Over de stabiliteit van het elastisch evenwicht*, Proefschrift Delft. H. J. Paris, Amsterdam, 1945, and "Elastic stability and post-buckling behavior", *Proc. Symp. Nonlinear Problems*, University of Wisconsin Press, Madison (1963), 257–272.

with respect to certain measures μ and $\bar{\mu}_0$ is that the total potential be positive-definite at the equilibrium configuration. In the static theory, a condition similar to this but somewhat weaker is the definition of the static stability. The major difference between the two types of stability, however, is that dynamic stability is defined relative to some measures, and hence is not an intrinsic property of the configuration and the potential, while static stability is a property of the configuration and the potential, independent of any measure. Thus, as we remarked in the preceding section, even for the same elastic material dynamic stability of a given static configuration depends on the choice of measures.

This difference does not appear for the classical dynamical systems of analytical mechanics, since, as KNOPS & WILKES have remarked, the measures used in analytical mechanics are usually norms of some finite dimensional spaces, and all norms on a finite-dimensional space are topologically equivalent. Hence there is but one equivalence class of measures that can be used in dynamic stability criteria for analytical mechanics. Further, by the argument of LAGRANGE and DIRICHLET, dynamic stability with respect to that class of measures is essentially the same as static stability. This rule is not valid for a dynamical system in continuum mechanics, however, since the measures now are defined on some infinite-dimensional function spaces, and norms on such a space are not topologically equivalent, an obvious counterexample being the uniform norm and the L_2-norm of a vector field. In view of this basic difference, we must regard static stability and dynamic stability as independent concepts in elasticity.

We shall not summarize the historical development of the concept of static stability here. The reader will find some remarks in NFTM §§ 68 bis and 89 and a more complete discussion in § 8.1 of the article by KNOPS & WILKES. The general idea for static stability is the statement that *an equilibrium configuration of a hyperelastic body is stable if and only if the total potential energy does not decrease in any virtual displacement compatible with given boundary conditions on the equilibrium configuration.* Of course, as it stands the statement has no content whatsoever unless the total potential energy, and in particular the potential energy of the loads, is defined for every virtual configuration. HADAMARD overcame this difficulty by considering virtual displacements satisfying the boundary condition of place and by requiring the body force be conservative. Then the condition $(7.49)_1$ is satisfied, and the load has a potential ζ

in the sense that the equation

$$\int_{\kappa(\mathscr{B})} \rho_\kappa \boldsymbol{b}_\kappa^*(s) \cdot \frac{d\boldsymbol{u}}{ds}\, dV_\kappa = -\frac{d\zeta}{ds} \tag{8.1}$$

holds for any one-parameter family of virtual displacements $\boldsymbol{u} = \boldsymbol{u}(s)$ satisfying the boundary condition

$$\boldsymbol{u} = \boldsymbol{0} \text{ on } \partial\kappa(\mathscr{B}). \tag{8.2}$$

The equation (8.1) is integrable for the functional $\zeta(\boldsymbol{u})$ because by (7.49), the left-hand side of (8.1) is a total differential with respect to s; *i.e.*, we have

$$\int_{\kappa(\mathscr{B})} \rho_\kappa \boldsymbol{b}_\kappa^*(s) \cdot \frac{d\boldsymbol{u}}{ds}\, dV_\kappa = \frac{d}{ds}\int_{\kappa(\mathscr{B})} \zeta_1^*\, dV_\kappa. \tag{8.3}$$

Then we can define the total potential λ simply by

$$\lambda = \lambda(\boldsymbol{u}) \equiv \int_{\kappa(\mathscr{B})} \rho_\kappa(\sigma_\kappa^* - \sigma_\kappa)\, dV_\kappa + \zeta(\boldsymbol{u}). \tag{8.4}$$

Accordingly the condition for static stability now is

$$\lambda(\boldsymbol{u}) \geqq 0 \tag{8.5}$$

for all $\boldsymbol{u}$ satisfying the boundary condition (8.2). In particular, when the body force vanishes, (8.5) reduces to

$$\int_{\kappa(\mathscr{B})} \rho_\kappa \sigma_\kappa^*\, dV_\kappa \geqq \int_{\kappa(\mathscr{B})} \rho_\kappa \sigma_\kappa\, dV_\kappa, \tag{8.6}$$

which says that the stored-energy of the body has a weak minimum in a stable equilibrium configuration. TRUESDELL & NOLL call stability defined by the condition (8.6) or more generally the condition (8.5) *finite stability* since the virtual displacements $\boldsymbol{u}$ are no longer required to be infinitesimal.

Of course, finite stability implies infinitesimal stability, as it should. Indeed, from the theorem at the end of the last section, necessary conditions for finite stability are the vanishing of the first Fréchet derivative and the non-negativeness of the second Fréchet derivative of the total potential functional with respect to all infinitesimal virtual displacements $\boldsymbol{u}$ satisfying the boundary condition. Clearly, the former is the condition of equilibrium, since from (8.3) and (8.4) the first Fréchet derivative of λ is given by

$$[\nabla_0 \lambda](\boldsymbol{u}) = \int_{\boldsymbol{\kappa}(\mathscr{B})} (\operatorname{Div} \boldsymbol{T}_\kappa + \rho_\kappa \boldsymbol{b}_\kappa)\, dV_\kappa, \tag{8.7}$$

while the latter is the condition of infinitesimal stability, since the second Fréchet derivative of λ is given by

$$[\nabla_0^2 \lambda](\boldsymbol{u}, \boldsymbol{u}) = \int_{\boldsymbol{\kappa}(\mathscr{B})} \operatorname{tr} \{\mathsf{A}_0(\boldsymbol{H})\boldsymbol{H}^T\}\, dV_\kappa, \tag{8.8}$$

where $\boldsymbol{H}$ is the gradient of $\boldsymbol{u}$.

As we remarked in § III.8B, HADAMARD found the following

THEOREM [1] *In order that an equilibrium configuration $\boldsymbol{\kappa}$ of an elastic body $\mathscr{B}$ be infinitesimally stable for the boundary condition of place, or of traction, or of the mixed type, it is necessary that the Hadamard condition*

$$\mathring{\mathsf{A}}_k{}^A{}_m{}^B u^k u^m v_A v_B \geqq 0 \quad \forall \boldsymbol{u}, \boldsymbol{v} \tag{8.9}$$

holds at each point of $\boldsymbol{\kappa}(\mathscr{B})$.

To prove HADAMARD's theorem, assume that $\boldsymbol{\kappa}$ is an infinitesimally stable configuration of $\mathscr{B}$. We consider infinitesimal displacements of the form

$$\boldsymbol{u} = \varphi \boldsymbol{u}_0 \tag{8.10}$$

where φ is a scalar field with compact support in $\boldsymbol{\kappa}(\mathscr{B})$ such that

$$\varphi(\boldsymbol{X}_0) = 1 \tag{8.11}$$

at an arbitrary reference point $\boldsymbol{X}_0 \in \boldsymbol{\kappa}(\mathscr{B})$, and where $\boldsymbol{u}_0 = \boldsymbol{u}(\boldsymbol{X}_0)$ is an arbitrary non-zero constant vector. Since the displacement $\boldsymbol{u}$ of the form (8.10) vanishes identically near $\partial\boldsymbol{\kappa}(\mathscr{B})$, its gradient

$$\boldsymbol{H} = (\operatorname{Grad} \varphi)\boldsymbol{u}_0 \tag{8.12}$$

is also zero on $\partial\boldsymbol{\kappa}(\mathscr{B})$ so that $\boldsymbol{u}$ satisfies zero boundary condition of place, or of traction, or of mixed type. Substituting (8.12) into (8.8), we obtain

$$\int_{\boldsymbol{\kappa}(\mathscr{B})} \operatorname{Grad} \varphi \cdot \boldsymbol{S} \operatorname{Grad} \varphi\, dV_\kappa \geqq 0, \tag{8.13}$$

[1] *Cf.* the footnote to the statement of this theorem in § III.8B.

where $\boldsymbol{S}$ is a symmetric tensor field defined by

$$S^{AB} \equiv \tfrac{1}{2}(\mathsf{A}_{k\ m}^{\ A\ B} + \mathsf{A}_{k\ m}^{\ B\ A})\, u_0^k\, u_0^m. \tag{8.14}$$

We set

$$\boldsymbol{S}_0 \equiv \boldsymbol{S}(\boldsymbol{X}_0). \tag{8.15}$$

Since $\mathsf{A}_{k\ m}^{\ A\ B}$, and hence $\boldsymbol{S}$, is assumed to be continuous, for every $\varepsilon > 0$ there exists a neighborhood $\mathcal{N}$ of $\boldsymbol{X}_0$ such that[1]

$$\boldsymbol{S}(\boldsymbol{X}) \leqq \boldsymbol{S}_0 + \varepsilon \mathbf{1} \tag{8.16}$$

for all $\boldsymbol{X} \in \mathcal{N}$. Substituting (8.15) into (8.13), we obtain

$$\begin{aligned} 0 &\leqq \int_{\kappa(\mathscr{B})} \operatorname{Grad} \varphi \cdot \boldsymbol{S} \operatorname{Grad} \varphi \, dV_{\kappa} \\ &\leqq \int_{\kappa(\mathscr{B})} \{\operatorname{Grad} \varphi \cdot \boldsymbol{S}_0 \operatorname{Grad} \varphi + \varepsilon(\operatorname{Grad} \varphi)^2\} \, dV_{\kappa} \end{aligned} \tag{8.17}$$

provided that the support of φ be contained in the neighborhood $\mathcal{N}$. We employ a Cartesian co-ordinate system X^A whose origin is $\boldsymbol{X}_0$ and whose axes point in the directions of three orthogonal proper vectors of $\boldsymbol{S}_0$. Then

$$\begin{aligned} 0 &\leqq \int_{\kappa(\mathscr{B})} \{\operatorname{Grad} \varphi \cdot \boldsymbol{S}_0 \operatorname{Grad} \varphi + \varepsilon(\operatorname{Grad} \varphi)^2\} \, dV_{\kappa} \\ &= \sum_{A=1}^{3} (s_A + \varepsilon) \int_{\kappa(\mathscr{B})} \left(\frac{\partial \varphi}{\partial X^A}\right)^2 dX^1 \, dX^2 \, dX^3, \end{aligned} \tag{8.18}$$

where s_1, s_2, s_2 are the proper numbers of $\boldsymbol{S}_0$.

We now show that (8.18) implies

$$s_A \geqq 0, \quad A = 1, 2, 3. \tag{8.19}$$

To see this, let $f(x)$ be an arbitrary C^∞ function whose support is contained in the interval $(-1, 1)$ and whose value at $x = 0$ is equal to 1. We put

$$\alpha \equiv \int_{-1}^{1} f(x)^2 dx > 0, \quad \beta \equiv \int_{-1}^{1} f'(x)^2 dx > 0. \tag{8.20}$$

---

[1] We use the usual partial ordering for symmetric tensors, *viz*, $\boldsymbol{A} \leqq \boldsymbol{B}$ means $\boldsymbol{B} - \boldsymbol{A}$ is positive semi-definite.

Now for any $\delta > 0$ we define a function $f_\delta(x)$ by

$$f_\delta(x) \equiv f\left(\frac{x}{\delta}\right). \tag{8.21}$$

Then the value of f_δ at $x = 0$ is still equal to 1, but the support for f_δ is contained in the interval $(-\delta, \delta)$; further, we have

$$f'_\delta(x) = \frac{1}{\delta} f'\left(\frac{x}{\delta}\right), \tag{8.22}$$

so that

$$\int_{\delta}^{\delta} f_\delta(x)^2 dx = \delta \int_{-1}^{1} f\left(\frac{x}{\delta}\right)^2 d\left(\frac{x}{\delta}\right) = \alpha\delta, \tag{8.23}$$

and

$$\int_{-\delta}^{\delta} f'_\delta(x)^2 dx = \frac{1}{\delta}\int_{-1}^{1} f'\left(\frac{x}{\delta}\right)^2 d\left(\frac{x}{\delta}\right) = \frac{\beta}{\delta}. \tag{8.24}$$

We choose the function φ in (8.10) as follows:

$$\varphi(X^1, X^2, X^3) \equiv f_{\delta_1}(X^1) f_{\delta_2}(X^2) f_{\delta_3}(X^3), \tag{8.25}$$

where δ_1, δ_2, and δ_3 are sufficiently small that the rectangular parallelepiped whose center is at $\boldsymbol{X}_0$ and whose edges are parallel to the coordinate axes and have lengths $2\delta_1$, $2\delta_2$, and $2\delta_3$, respectively, be contained in the neighborhood $\mathcal{N}$. Clearly, φ is of class C^∞ and has value 1 at the origin $\boldsymbol{X}_0$. With this choice of φ, (8.23) and (8.24) give

$$\int_{\boldsymbol{\kappa}(\mathscr{B})} \left(\frac{\partial \varphi}{\partial X^1}\right)^2 dX^1 dX^2 dX^3 = \alpha^2 \beta \frac{\delta_2 \delta_3}{\delta_1}, \tag{8.26}$$

and similar expressions when $A = 2, 3$. Substituting (8.26) into (8.18), we get

$$0 \leqq \alpha^2 \beta \left\{(s_1 + \varepsilon)\frac{\delta_2 \delta_3}{\delta_1} + (s_2 + \varepsilon)\frac{\delta_1 \delta_3}{\delta_2} + (s_3 + \varepsilon)\frac{\delta_1 \delta_2}{\delta_3}\right\}. \tag{8.27}$$

Since this inequality must hold for all sufficiently small positive δ_1, δ_2, δ_3, and ε, it is clear that s_1, s_2, s_3 must be non-negative. Therefore, the tensor $\boldsymbol{S}_0$ defined by (8.14) and (8.15) must be positive semi-definite. Since $\boldsymbol{X}_0$ and $\boldsymbol{u}_0$ were chosen arbitrarily, it follows that the H-condition must hold at each point of $\boldsymbol{\kappa}(\mathscr{B})$. Thus the proof is complete.

Now suppose that the reference configuration is homogeneous[1], so that the stored-energy function is independent of the position $\boldsymbol{X} \in \kappa(\mathscr{B})$. Then by a theorem of VAN HOVE in the calculus of variations or more generally by GÅRDING's inequality, we know that the condition of strong ellipticity:

$$\mathsf{A}_i{}^A{}_j{}^B w^i w^j v_A v_B \geqq c w_i w^i v_A v^A \quad c > 0, \tag{8.28}$$

for all vectors $\boldsymbol{w}$ and $\boldsymbol{v}$, implies the inequality

$$\int_{\kappa(\mathscr{B})} \mathsf{A}_i{}^A{}_j{}^B H^i_A H^j_B \, dV_\kappa \geqq c_0 \int_{\kappa(\mathscr{B})} \rho_\kappa H^i_A H_i{}^A \, dV_\kappa, \quad c_0 > 0, \tag{8.29}$$

for all $\boldsymbol{H} = \operatorname{Grad} \boldsymbol{u}$, where $\boldsymbol{u}$ satisfies the boundary condition (8.2). Now the inequality (8.29) means simply that the second Fréchet derivative of the stored-energy functional is positive-definite with respect to the L_2-norm of Grad $\boldsymbol{u}$. By POINCARÉ's lemma[2], there exists a positive number c_1 such that:

$$\int_{\kappa(\mathscr{B})} \rho_\kappa H^i_A H_i{}^A \, dV_\kappa \geqq c_1 \int_{\kappa(\mathscr{B})} \rho_\kappa u^i u_i \, dV_\kappa \tag{8.30}$$

since $\boldsymbol{u}$ obeys (8.2). Combining (8.29) and (8.30), we see that the second Fréchet derivative of the stored-energy functional is positive-definite with respect to the L_2-norm of $\boldsymbol{u}$ also. Then by the theorem given at the end of the last section (8.28) implies that the homogeneous equilibrium configuration is not only *infinitesimally superstable* as defined by the condition

$$\int_{\kappa(\mathscr{B})} \mathsf{A}_i{}^A{}_j{}^B H^i{}_A H^j{}_B \, dV_\kappa > 0, \tag{8.31}$$

for all $\boldsymbol{H} = \operatorname{Grad} \boldsymbol{u}$ with $\boldsymbol{u} \neq 0$, but also *finitely superstable* as defined by the condition

$$\int_{\kappa(\mathscr{B})} \rho_\kappa \sigma^*_\kappa \, dV_\kappa > \int_{\kappa(\mathscr{B})} \rho_\kappa \sigma_\kappa \, dV_\kappa \quad \forall \boldsymbol{u} \neq \boldsymbol{0}. \tag{8.32}$$

---

[1] The condition of homogeneity assumed in VAN HOVE's theorem may be relaxed to a condition of small inhomogeneities to the extent that the simple form of GÅRDING's inequality holds; *cf.* the paper of G. FICHERA cited on p. 221.

[2] *Cf.* G. FICHERA, *Linear Elliptic Differential Systems and Eigenvalue Problems*, Lecture Notes in Mathematics, No. 8, Berlin etc., Springer-Verlag, 1965.

Because of the boundary condition (8.2), surface tractions on the boundary of the deformed configurations corresponding to the virtual displacements need not be specified since they do no work in a zero displacement. If more general boundary conditions are used, such as

$$\boldsymbol{u} = \boldsymbol{0} \text{ on } \partial_1 \boldsymbol{\kappa}(\mathscr{B}), \tag{8.33}$$

then on the complementary boundary $\partial_2 \boldsymbol{\kappa}(\mathscr{B})$ where $\boldsymbol{u}$ need not vanish the surface tractions $\boldsymbol{t}_\kappa^*$ do effect work and hence must be specified. Of course, one possible solution is to require that the surface traction be conservative in the sense of (7.49) also. More specifically, we require that the work done by the loads be given by

$$\int_{\boldsymbol{\kappa}(\mathscr{B})} \rho_\kappa \boldsymbol{b}_\kappa^*(s) \cdot \frac{d\boldsymbol{u}}{ds} \, dV_\kappa + \int_{\partial_2 \boldsymbol{\kappa}(\mathscr{B})} \boldsymbol{t}_\kappa^*(s) \cdot \frac{d\boldsymbol{u}}{ds} \, dA_\kappa = -\frac{d\zeta}{ds} \tag{8.34}$$

in any one-parameter family of virtual displacement $\boldsymbol{u} = \boldsymbol{u}(., s)$ satisfying the boundary condition (8.33). Then the total potential functional λ is still given by (8.4), and the statement of static stability is still the inequality (8.5).

At this point we remark the role of rigid virtual displacement in the stability analysis. Suppose that $\partial_1 \boldsymbol{\kappa}(\mathscr{B})$ is an empty set, a segment of a straight line[1], or just one point. Then there exist certain non-trivial rigid virtual displacements $\boldsymbol{u}_0$ compatible with (8.33). From the principle of material frame-indifference, we know that the stored-energy functional does not change value in such rigid displacements. Consequently, a necessary condition for stability is that the equilibrium configuration be stable with respect to the potential ζ within the class of admissible rigid virtual displacements $\boldsymbol{u}_0$. This means (8.5) holds only if

$$\zeta(\boldsymbol{u}_0) \geqq 0 \tag{8.35}$$

for all admissible rigid virtual displacements $\boldsymbol{u}_0$ in the first place. If (8.35) fails, then the equilibrium configuration must be regarded as unstable. So as to exclude such rigid instabilities, we may impose some additional conditions on the class of virtual displacements to be considered in the stability analysis.

---

[1] Static solutions corresponding to these boundary-conditions are generally not smooth near $\partial_1 \boldsymbol{\kappa}(\mathscr{B})$, so that weak solutions must be used. The theory of weak solutions is beyond the scope of this book, however.

We give some possible additional conditions for the three cases:

I) $\partial_1 \boldsymbol{\kappa}(\mathscr{B}) = \emptyset$. For this case we may require the virtual displacements $\boldsymbol{u}$ to be such that the loads in the deformed configurations are equilibrated, *i.e.*, in stability analysis we consider only those virtual displacements $\boldsymbol{u}$ for which the conditions

$$\int_{\boldsymbol{\kappa}(\mathscr{B})} \rho_{\kappa} \boldsymbol{b}^*_{\kappa} dV_{\kappa} + \int_{\partial\boldsymbol{\kappa}(\mathscr{B})} \boldsymbol{t}^*_{\kappa} dA_{\kappa} = \mathbf{0} \tag{8.36}$$

and

$$\int_{\boldsymbol{\kappa}(\mathscr{B})} \rho_{\kappa} \boldsymbol{b}^*_{\kappa} \wedge (\boldsymbol{u}+\boldsymbol{X}) dV_{\kappa} + \int_{\partial\boldsymbol{\kappa}(\mathscr{B})} \boldsymbol{t}^*_{\kappa} \wedge (\boldsymbol{u}+\boldsymbol{X}) dA_{\kappa} = \mathbf{0} \tag{8.37}$$

are satisfied. In particular, if $\boldsymbol{u}_0(s)$ is a one-parameter family of rigid virtual displacements satisfying the above conditions, then by (8.34) we have

$$\frac{d\zeta}{ds} = 0. \tag{8.38}$$

Consequently within the class of admissible virtual displacements the undesirable rigid instabilities, for which

$$\zeta(\boldsymbol{u}_0) < 0, \tag{8.39}$$

are excluded.

II) $\partial_1 \boldsymbol{\kappa}(\mathscr{B})$ is a segment of a straight line in the direction of unit vector $\boldsymbol{e}$. For this case we may require the virtual displacements $\boldsymbol{u}$ to be such that the loads in the deformed configurations have no resultant moment in the direction of $\boldsymbol{e}$, *i.e.*, they satisfy the condition

$$\left[\int_{\boldsymbol{\kappa}(\mathscr{B})} \rho_{\kappa} \boldsymbol{b}^*_{\kappa} \wedge (\boldsymbol{u}+\boldsymbol{X}) dV_{\kappa} + \int_{\partial\boldsymbol{\kappa}(\mathscr{B})} \boldsymbol{t}^*_{\kappa} \wedge (\boldsymbol{u}+\boldsymbol{X}) dA_{\kappa}\right] \wedge \boldsymbol{e} = \mathbf{0}. \tag{8.40}$$

By the same argument as before, if $\boldsymbol{u}_0(s)$ is a one-parameter family of rigid rotations about the unit vector $\boldsymbol{e}$ such that the condition (8.40) is satisfied, then (8.38) holds.

III) $\partial_1 \boldsymbol{\kappa}(\mathscr{B})$ is a single point $\boldsymbol{X}_0$. For this case we may require the virtual displacements $\boldsymbol{u}$ to be such that the loads have no resultant moment in the deformed configurations, *i.e.*, (8.37) holds. Then the same argument as before still can be applied in this case.

The assumption that the loads be conservative in any virtual displacement is just one possible way to specify the total potential in the actual configuration of the virtual displacement. Another possible way is to require that the loads to be dead in all virtual displacements. Under this assumption we replace the body force $\boldsymbol{b}_\kappa^*(s)$ and the surface traction $\boldsymbol{t}_\kappa^*(s)$ in (8.34) by the body force $\boldsymbol{b}_\kappa$ and the surface traction $\boldsymbol{t}_\kappa$ in the equilibrium configuration. Then (8.34) can be integrated with respect to s, yielding

$$\int_{\kappa(\mathscr{B})} \rho_\kappa \boldsymbol{b}_\kappa \cdot \boldsymbol{u}\, dV_\kappa + \int_{\partial_2\kappa(\mathscr{B})} \boldsymbol{t}_\kappa \cdot \boldsymbol{u}\, dA_\kappa = -\zeta(\boldsymbol{u}). \tag{8.41}$$

Again we define the total potential functional λ by (8.4) with ζ given by (8.41), and we define static stability with respect to dead loads by (8.5) using this λ.

By the same remark as before, if $\partial_1\boldsymbol{\kappa}(\mathscr{B})$ is not large enough to rule out any non-trivial rigid virtual displacements $\boldsymbol{u}_0$, then in order to exclude rigid instabilities under dead loads, we may impose some additional conditions on the virtual displacements. Again we consider three cases.

Id) $\partial_1\boldsymbol{\kappa}(\mathscr{B}) = \emptyset$ and IIId) $\partial_1\boldsymbol{\kappa}(\mathscr{B}) = \boldsymbol{X}_0$. For these two cases we may require the virtual displacement to satisfy the zero moment condition[1]:

$$\int_{\kappa(\mathscr{B})} \rho_\kappa \boldsymbol{b}_\kappa \wedge (\boldsymbol{u}+\boldsymbol{X})\, dV_\kappa + \int_{\partial\kappa(\mathscr{B})} \boldsymbol{t}_\kappa \wedge (\boldsymbol{u}+\boldsymbol{X})\, dA_\kappa = \boldsymbol{0}. \tag{8.42}$$

Since the dead loads are equilibrated in $\boldsymbol{\kappa}(\mathscr{B})$, this condition is equivalent to

$$\int_{\kappa(\mathscr{B})} \rho_\kappa \boldsymbol{b}_\kappa \wedge \boldsymbol{u}\, dV_\kappa + \int_{\partial\kappa(\mathscr{B})} \boldsymbol{t}_\kappa \wedge \boldsymbol{u}\, dA_\kappa = \boldsymbol{0}. \tag{8.43}$$

IId) $\partial_1\boldsymbol{\kappa}(\mathscr{B})$ is a segment of a straight line in the direction of a unit vector $\boldsymbol{e}$. For this case, we may require that the dead loads have no resultant moment in the direction of $\boldsymbol{e}$ in any virtual displacement:

$$\left[\int_{\kappa(\mathscr{B})} \rho_\kappa \boldsymbol{b}_\kappa \wedge (\boldsymbol{u}+\boldsymbol{X})\, dV_\kappa + \int_{\partial\kappa(\mathscr{B})} \boldsymbol{t}_\kappa \wedge (\boldsymbol{u}+\boldsymbol{X})\, dA_\kappa\right] \wedge \boldsymbol{e} = \boldsymbol{0}, \tag{8.44}$$

---

[1] Since the reference configuration is the equilibrium configuration, the dead loads have no resultant force. The zero moment condition for dead loads was proposed by BEATTY, "Some static and dynamic implications of the general theory of elastic stability", *Arch. Rational Mech. Anal.* **19**, 167–188 (1965).

or, equivalently,

$$\left[\int_{\kappa(\mathscr{B})} \rho_\kappa \boldsymbol{b}_\kappa \wedge \boldsymbol{u}\, dV_\kappa + \int_{\partial\kappa(\mathscr{B})} \boldsymbol{t}_\kappa \wedge \boldsymbol{u}\, dA_\kappa\right] \wedge \boldsymbol{e} = \boldsymbol{0}. \tag{8.45}$$

The assumption of dead loads and the assumption of conservative loads may be combined in various ways. For example[1], we may require that the body forces be conservative and the surface tractions be dead in all virtual displacements. In this case, the potential functional for the body force is given by (8.3), and the potential functional for the surface traction is given by the second term on the left-hand side of (8.41). Adding these two functionals gives the potential functional ζ for the loads. Then (8.4) with this ζ defines the total potential functional λ, and finally (8.5) defines the stability. Also if $\partial_1 \kappa(\mathscr{B})$ degenerates into one of the three cases, then we may impose some additional conditions as before so as to exclude rigid instabilities.

So far we have considered static stability of an equilibrium configuration with respect to some total potential functional, which is defined to be the sum of the increment of the stored-energy functional and the potential energy functional of the loads in any virtual displacement. Now if the loads are not conservative or dead, then the latter functional does not exist, since the work done by the loads in any virtual displacement is path-dependent. If the loads are specified in each path $\boldsymbol{u} = \boldsymbol{u}(s)$, $s \in [0, s_0]$, then we can define a path-dependent potential ζ by

$$-\zeta(\boldsymbol{u}(s), s \in [0, s_0]) \equiv \int_0^{s_0} \int_{\kappa(\mathscr{B})} \rho_\kappa \boldsymbol{b}_\kappa^*(s) \cdot \frac{d\boldsymbol{u}}{ds}\, dV_\kappa\, ds + \int_0^{s_0} \int_{\partial_2\kappa(\mathscr{B})} \boldsymbol{t}_\kappa^*(s) \cdot \frac{d\boldsymbol{u}}{ds}\, dA_\kappa\, ds. \tag{8.46}$$

Likewise, if the stored-energy function does not exist, we can define a path-dependent stress potential σ by

$$\sigma(\boldsymbol{u}(s), s \in [0, s_0]) \equiv \int_0^{s_0} \int_{\kappa(\mathscr{B})} \operatorname{tr}\left[(\boldsymbol{T}_\kappa^*(s) - \boldsymbol{T}_\kappa)\operatorname{Grad} \boldsymbol{u}\right] dV_\kappa\, ds, \tag{8.47}$$

where $\boldsymbol{T}_\kappa^*(s)$ is the Piola-Kirchhoff stress tensor in the actual con-

[1] This kind of static stability is attributed to TRUESDELL & NOLL by KNOPS & WILKES, *op. cit.* p. 519.

figuration of the virtual displacement $\boldsymbol{u}(s)$. Combining (8.46) and (8.47), we then define a path-dependent total potential λ by

$$\lambda(\boldsymbol{u}(s),\, s \in [0, s_0]) \equiv \sigma(\boldsymbol{u}(s),\, s \in [0, s_0]) + \zeta(\boldsymbol{u}(s),\, s \in [0, s_0]). \tag{8.48}$$

Naturally, we say that the equilibrium configuration is *stable with respect to the path* $\boldsymbol{u} = \boldsymbol{u}(s)$, $s \in [0, s_0]$, if the path-dependent potential λ satisfies the inequality

$$\lambda(\boldsymbol{u}(s),\, s \in [0, s_0]) \geqq 0. \tag{8.49}$$

If on a class of paths of virtual displacements the loads are specified, then the equilibrium configuration is said to be *stable*[1] *with respect to the class* if (8.49) is satisfied on each path $\boldsymbol{u}(s)$, $s \in [0, s_0]$, belonging to the class.

The various types of static stabilities considered so far are characterized by the condition that the total potential functional $\lambda(\boldsymbol{u})$ be a weak minimum at $\boldsymbol{u} = \boldsymbol{0}$ with respect to virtual displacements $\boldsymbol{u}$ belonging to certain class $\mathscr{X}$. Now if $\mathscr{X}$ can be completed to a Banach space and if λ can be extended to a smooth functional, then we say that the equilibrium configuration is *infinitesimally superstable* if the following two conditions are satisfied:

$$[\nabla_{\boldsymbol{0}}\lambda](\boldsymbol{u}) = \boldsymbol{0} \quad \forall \boldsymbol{u} \in \mathscr{X}, \tag{8.50}$$

and

$$[\nabla_{\boldsymbol{0}}^2\lambda](\boldsymbol{u}, \boldsymbol{u}) > 0 \quad \forall \boldsymbol{u} \in \mathscr{X},\ \boldsymbol{u} \not\equiv \boldsymbol{0}. \tag{8.51}$$

These conditions mean that the total potential functional has a strict minimum at $\boldsymbol{u} = \boldsymbol{0}$ with respect to all infinitesimal virtual displacements. More generally, the equilibrium configuration is said to be *finitely superstable* if λ satisfies the strict inequality

$$\lambda(\boldsymbol{u}) > 0 \quad \forall \boldsymbol{u} \neq \boldsymbol{0}. \tag{8.52}$$

By the theorem at the end of the preceding section, if λ is a smooth functional on $\mathscr{X}$, then the condition (8.50) and the condition

$$\exists \alpha > 0 : [\nabla_{\boldsymbol{0}}^2\lambda](\boldsymbol{u}, \boldsymbol{u}) \geqq \alpha \|\boldsymbol{u}\|^2 \quad \forall \boldsymbol{u} \in \mathscr{X} \tag{8.53}$$

are sufficient for finite superstability, and hence for infinitesimal superstability.

---

[1] This general concept of static stability is given essentially by BEATTY, *op. cit.* p. 539.

We make two remarks about infinitesimal superstability with respect to dead loads. First, since in virtue (8.41) ζ is a linear functional of $\boldsymbol{u}$, the second Fréchet derivative $\nabla_0^2\zeta$ vanishes. Hence $\nabla_0^2\lambda$ is equal to the second Fréchet derivative of the stored-energy functional. This means that infinitesimal superstability with respect to dead loads is characterized by the condition (8.31) relative to any boundary condition. Second, so as to exclude rigid instabilities, if $\partial_1\boldsymbol{\kappa}(\mathscr{B})$ degenerates into one of the three cases considered previously, we impose additional conditions on the virtual displacements. For dead loads, the additional condition for Cases Id) and IIId) is (8.42) and for Case IId) it is (8.45). Notice that these conditions are linear in $\boldsymbol{u}$. This means that in a degenerate case, the condition (8.31) for infinitesimal superstability is required to hold only for all non-zero $\boldsymbol{u}$ on some linear subspaces characterized by the additional conditions (8.42) or (8.45), since, as explained before, the condition (8.52) for superstability, and hence the condition (8.31) for infinitesimal superstability, cannot possibly be satisfied if $\boldsymbol{u}$ is a rigid virtual displacement or an infinitesimal rigid virtual displacement.

With these remarks in mind we can now state some uniqueness theorems:

THEOREM (ERICKSEN & TOUPIN, HILL, BEATTY). *If an equilibrium configuration is infinitesimally superstable, then the mixed boundary-value problem of equilibrium in the theory of infinitesimal deformation superimposed on the equilibrium configuration has at most one solution, to within an arbitrary rotation compatible with the boundary conditions.*

THEOREM (HILL, BEATTY). *Corresponding to prescribed initial velocity and infinitesimal displacement from an infinitesimally stable equilibrium configuration, there exists at most one solution.*

Note that only stability, not superstability, is assumed in the second theorem.

Proofs of these uniqueness theorems can be found in NFTM § 68 bis and in BEATTY's paper[1].

---

[1] BEATTY, Part 1, § 5, and Part 2, § 1, of *op. cit.* p. 539. While the statement in NFTM does not mention the existence of a stored-energy function, the proof given there in fact requires it, and the theorem seems not to be true for elastic materials that are not hyperelastic.

Note: Many of the best recent contributions to the theory of elastic stability[1] rest upon thermodynamics. Pertaining essentially to thermo-elasticity, they do not have counterparts or special cases in elasticity. A good account of researches on thermo-elastic stability through 1971 may be found in the article by KNOPS & WILKES cited on p. 519.

[1] In particular, we mention the paper of ERICKSEN cited on p. 527 and the paper of M. E. GURTIN, "Thermodynamics and the potential energy of an elastic body", *Journal of Elasticity*, in press.

Index of Persons Named or Cited

Index of Matters Treated